Bureau of Flora and Fauna, Canberra

ZOOLOGICAL CATALOGUE OF AUSTRALIA

Volume 1

AMPHIBIA AND REPTILIA

Australian Government Publishing Service
Canberra 1983

ISBN for series: 0 644 02840 8
ISBN for Volume 1: 0 644 02839 4

Zoological Catalogue of Australia

The compilation of the *Zoological Catalogue of Australia* is conducted under the auspices of the Bureau of Flora and Fauna, B. J. Richardson, Assistant Director (Fauna).

D. W. Walton, Executive Editor

Vol. 1

AMPHIBIA AND REPTILIA

by

Harold G. Cogger
Elizabeth E. Cameron and Heather M. Cogger
Department of Reptiles and Amphibians
The Australian Museum
Sydney, N.S.W.

Printed at Griffin Press Limited, Netley, South Australia.

CONTENTS

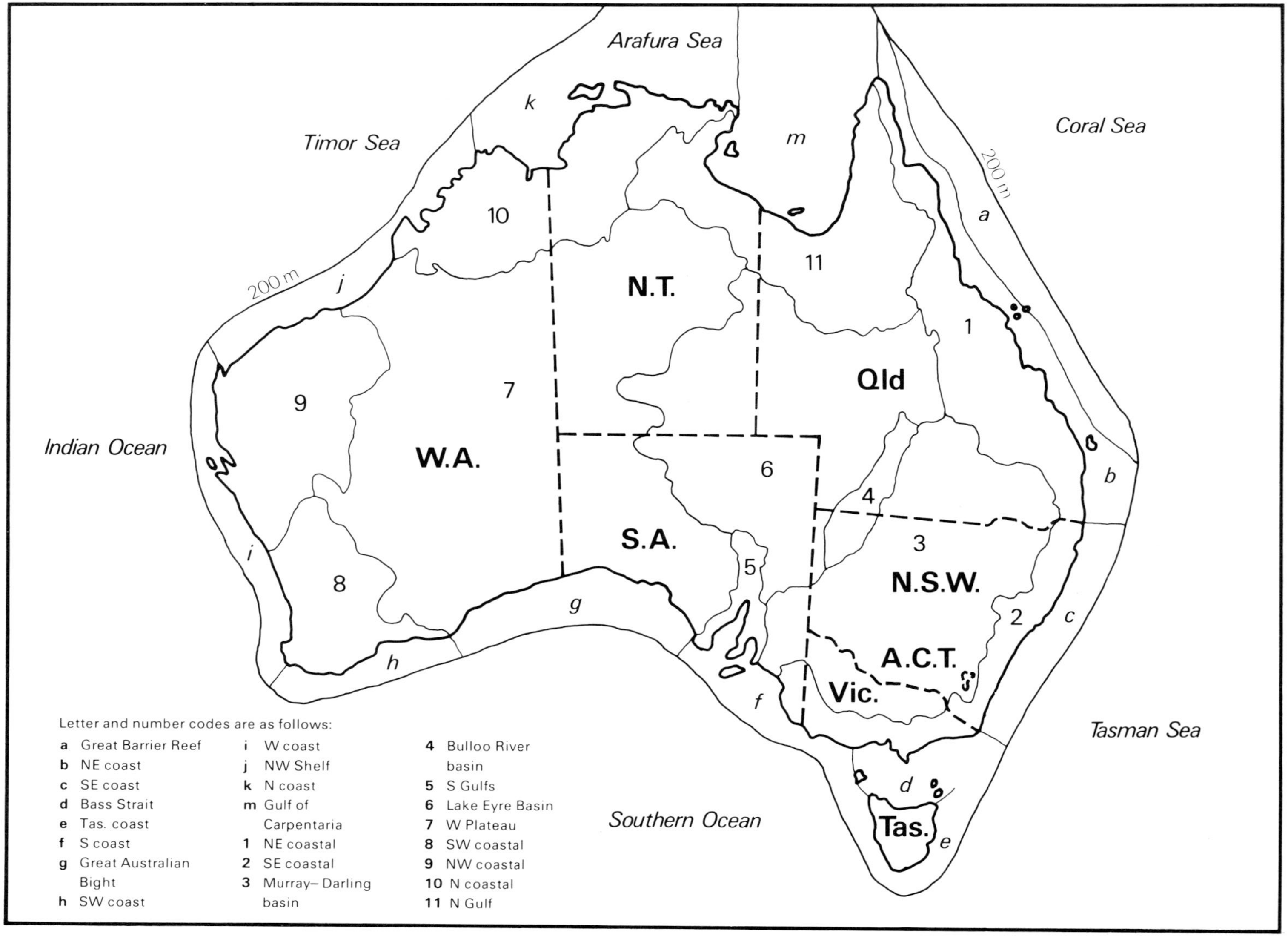

Map 1. States, catchment areas and coastal zones of Australia

EDITORIAL PREFACE

INTRODUCTION

An objective of the Australian Biological Resources Study is to stimulate research and publications on the taxonomy and distribution of Australian fauna and flora. Consistent with this aim, the *Zoological Catalogue of Australia* was conceived as a concise, computer-based data bank consisting of current taxonomic and biological knowledge of the Australian fauna, accessible to all interested in such information. As the project developed, the advantages of publication of this information were recognised.

Data for inclusion in the *Catalogue* are assembled in four separate files: a genus taxonomic arrangement file, a species taxonomic arrangement file, a genus available name file and a species available name file. The contents of appropriate files are then integrated by computer. This methodology yields a standard format which will be maintained throughout the volumes of the *Catalogue* and provides consistency in the data. The format and style of presentation are, therefore, the responsibility of the Bureau of Flora and Fauna. The authors are responsible only for the information content.

Each volume of the *Catalogue*, treating specific taxa, will cite by name and original reference all species known from Australia. The species are arranged taxonomically by family and genus. Information for each species includes synonymy, literature citation, location/status of the type material and type locality for each available name in synonymy, a brief summary of geographical distribution and ecological attributes, and important references on various aspects of the biology. It is designed to serve primarily as a bibliographic directory to the most comprehensive and recent information available on each species.

This data base is intended to provide a substantial assessment of current knowledge and to stimulate and provide a starting point for future investigations. It is estimated that the Australian fauna exceeds 150 000 species of which about half have yet to be recognised and described. As knowledge of the Australian fauna advances, the data base will be updated and expanded.

TAXONOMIC INFORMATION

Nomenclature in the *Catalogue* adheres to the provisions established in the International Code of Zoological Nomenclature. The author and date of all names appearing in the *Catalogue* are presented so that the user may understand the nature and relationships of the names and all names appear in their legitimate form, not as they appeared in their original presentation. The valid genus and species group names and their allocation to families are determined by the contributors. No new genus or species group names are introduced in the *Catalogue* although new combinations may be established. Synonymies do not include new combinations. Treatment of family group names is not included.

ECOLOGY AND DISTRIBUTION

Information on ecology and distribution is given with each valid species. The ecological descriptors are general terms derived from a list prepared by the Bureau of Flora and Fauna. These descriptors act as computer search terms for use with the data base.

Distribution data are based on a standardised list of computer search terms established by the Bureau of Flora and Fauna. Both political and geographical region descriptors are included (see map). Political areas include the adjacent waters. Terrestrial geographical terms are based on the drainage systems of continental Australia, while marine terms are self explanatory except as follows: the boundary between the coastal and the oceanic zones is the 200 m contour; the Arafura Sea extends from Cape York to 124°E longitude; and the boundary between the Tasman and Coral Seas is considered to be the latitude of Fraser Island, also regarded as the southern terminus of the Great Barrier Reef. Ecological or distributional terms in parentheses imply that the information is unconfirmed but, in the opinion of the contributor, likely to be correct. Terms for terrestrial habitat or vegetation type follow Specht, R.L. (1970). Vegetation. pp 44–67 *in* Leeper, G.W. (ed.) *The Australian Environment*. 4th edn. Melbourne : CSIRO-Melbourne Univ. Press.

BIBLIOGRAPHIC INFORMATION

Where possible, selected references are provided as an introduction to the biology of a species. Literature citations throughout the *Catalogue* are given in full. Older works, with extended subtitles, in some cases have been shortened but only if their identity is preserved. Serial titles are abbreviated in a manner designed to facilitate library research. The number and variety of sources for serial abbreviations employed by workers of different nationalities and among the various taxa precluded use of a standard guide. References or titles originally issued in a script other than Roman and lacking a Romanised translation are transliterated with the original language shown in brackets. Common abbreviations are listed in Appendix I. Acronyms of museums or collections are defined in Appendix II.

ACKNOWLEDGEMENTS

The *Zoological Catalogue of Australia* developed through the ideas and efforts of a broad segment of the zoological community of Australia. David Ride, with the assistance of Murray McDonald, did much to organise and develop these ideas and efforts into a workable framework. The highly complex integration program for treating the data base was the creation of John Busby. The format and cover design was established with the co-operation of the Australian Government Publishing Service. Within the Bureau of Flora and Fauna, Richard Longmore and Barry Richardson assisted in the editing of the volume. David Berman managed the data base and Wendy Riley and Cindy Warhurst the entry and revisions of the data.

To all those involved, grateful acknowledgement is extended.

D. W. Walton

AMPHIBIA AND REPTILIA

Harold G. Cogger

Elizabeth E. Cameron and Heather M. Cogger

INTRODUCTION

A century has passed since G.A. Boulenger began to publish his land-mark catalogues of the reptiles and amphibians in the British Museum (Natural History). These catalogues provided the first comprehensive taxonomic treatment of the known Australian herpetofauna, albeit integrated into a world fauna. Boulenger recognised some 340 species of Australian reptiles and frogs.

Since Boulenger's time, a vast amount of taxonomic research has been carried out on the Australian herpetofauna. The number of species recognised has climbed dramatically, to a total of 807 in this *Catalogue*, while a better understanding of phyletic relationships led to many changes in the higher classification of this fauna.

Despite these changes, the only checklists dealing specifically with major elements of the Australian herpetofauna published since Boulenger's catalogues are an abbreviated checklist of Australian lizards (Zietz, 1920) and a more extensive list of Australian frogs (Moore, 1961). These have been supplemented by world-wide checklists which, varying widely in their nomenclatural thoroughness, cover about 75% of Australian families.

To the 807 species of reptiles and amphibians currently recognised from Australia and its territories more than 2,000 separate trivial names have been applied since 1758 - the year which formally marks the establishment of the Linnaean binomial system of classification. Thus Australian herpetology has been hampered by the lack of a comprehensive checklist. With this goal in mind, I began preparing a primary checklist of Australian reptiles and amphibians as a spare-time task in the late 1960's. This project had reached a final draft stage in 1980 when Dr. Barry Richardson of the Bureau of Flora and Fauna in Canberra suggested that it might be appropriate to publish it as part of the *Zoological Catalogue of Australia* — a long-term project initiated by the Bureau for a computer-based checklist of the Australian fauna.

The organisation of the *Catalogue* which follows is a major departure from my original plan. A large volume of distributional and basic ecological information, together with extensive references, have been added to meet the format specifications of the *Catalogue*.

Considerable effort went into locating type material, checking original descriptions, and, in the case of older taxa or obscure recent taxa, examining type specimens to determine their taxonomic status. The aim of this catalogue is to provide a working document for those concerned, either directly or indirectly, with the taxonomy and nomenclature of Australian reptiles and frogs. The approach is parochial as it is geared to the needs of herpetologists and students concerned with the regional Australian herpetofauna rather than to the needs of specialists in particular groups.

In attempting to cover the entire herpetofauna of Australia, the rigour of specialist reviews is lacking. This catalogue, therefore, should be used in conjunction with the available world-wide or regional monographs and/or checklists of elapid and hydrophiid snakes (Klemmer, 1963), pygopodid lizards (Wermuth, 1965; Kluge, 1974), gekkonid lizards (Wermuth, 1965), agamid lizards (Wermuth, 1967), varanid lizards (Mertens, 1963), chelonians and crocodilians (Wermuth & Mertens, 1977), leptodactylid frogs (Lynch, 1971; Gorham, 1966), hylid frogs (Duellman, 1977), typhlopid snakes (Hahn, 1980) and boid snakes (Stimson, 1969).

The original intention was to produce concurrently a checklist and complementary identification manual. The vicissitudes of publication schedules, however, resulted in the checklist lagging several years behind the manual (Cogger, 1975). Unfortunately, a number of nomenclatural changes, including the resurrection of old names and some new combinations, were first presented in the identification manual without explanation; it was anticipated that this checklist would appear before the manual. Some of those nomenclatural changes, therefore, are justified here for the first time. All nomenclatural changes were based on the application of the International Code of Zoological Nomenclature (second edition, 1964) and its subsequent amendments. All of the *taxonomic* decisions made in the identification manual (other than the reallocation of synonyms) were based on studies by other workers and, though not original in the manual, they clearly represented one author's (HGC) view of any disputed opinions or interpretations. Explanations for the more contentious of these actions are provided below or in the body of the text.

Most of the collections of reptiles and frogs made in the early days of European settlement found their way to the universities and museums of Europe, and modern taxonomic studies within Australia often have been hampered seriously by the inaccessability of critical type material. A more serious problem has been the loss or misplacement of many types from both Australian and overseas collections. As many early descriptions were brief to the point of obscurity, the identity of the described animals may never be resolved.

The compiler of any checklist is constantly faced with decisions as to the relative value of further effort, *e.g.*, at what stage does the continued search for type material become futile or even counter-productive? When two names published in the same year compete for priority, how can one ensure that all avenues have been explored for establishing precise dates of publication?

In order to accommodate publication schedules, the termination point for entry of recently published information is 30 June 1982.

Whereas the numerous tasks involved in compilation, location and examination of type specimens, and bibliographic research, were shared by all three authors, the taxonomic content is the sole responsibility of the senior author.

Our hope is that this catalogue, while leaving many questions unanswered, will provide a nomenclatural background to facilitate the application of correct and appropriate names by future taxonomists and thus contribute to greater stability in Australian herpetological nomenclature.

COMPILATION GUIDELINES

In addition to the compilation standards established by the Bureau of Flora and Fauna, the guidelines below were used to determine the scope and content of the catalogue and to establish criteria on which to base taxonomic, nomenclatural and organisational decisions.

Fossil-based Taxa

Those taxa based solely on fossil material are excluded from the synonymies. In practice this decision results in the exclusion of no more than a handful of names mostly in the generic synonymies of crocodiles and turtles. Fossil taxa in these groups are cited in Romer (1956).

Priority in Nomenclature

One of the major problems in compiling this, as in other checklists, has been to decide which cases warrant the application of the normal laws of priority and, conversely, which junior synonyms should be retained on the basis of usage in so far as such decisions are consistent with Article 23 (and its amendments) of the International Code of Zoological Nomenclature. During the period in which this checklist was compiled, several major changes have been made to Article 23, thus complicating the problem.

Fortunately, replacement of Articles 23(a) and 23(b) of the 1964 Code by a new Article 23(a-b) was adopted by the XVII International Congress of Zoology at Monaco in 1972. This new section states that "The Law of Priority is to be used to promote stability and is not intended to be used to upset a long-established name in its accustomed meaning through the introduction of an unused name which is its senior synonym. A zoologist who considers that the application of the Law of Priority would in his judgement disturb stability or universality or cause confusion is to maintain existing usage and must refer the case to the Commission for a decision under the plenary powers."

Because it is often difficult to define "existing usage" in the terms of Article 23, specialists frequently disagree on the preservation of particular junior synonyms. It is my firm belief that the strict application of the Law of Priority, despite some initial inconvenience, will lead most effectively to long-term stability in nomenclature. This is not to deny, however, that there are important exceptions and many of the nomenclatural decisions made in the following pages will be subject to dispute.

Homonymy

Until 1960 the Law of Homonymy of the International Code of Zoological Nomenclature, as expressed through Article 59(c) of the 1964 Code, precluded the resurrection of any name once it had been rejected as a junior secondary homonym. In practice, this rule made many names unavailable, often through some trivial or inept taxonomic decision. Since 1961, it has been permissible under the Code to resurrect junior secondary homonyms when the species to which they apply have been transferred to different genera.

The criteria laid down in Article 59(c) have been applied throughout this checklist, often against my better judgement. In scincid lizards, especially, the expansion of the genus *Lygosoma* by Boulenger (1887) to encompass the majority of the lygosomine skinks created a host of homonyms which have been variously rejected or reinstated, with a complete lack of consistency, by later workers.

First Citation of Current Combination

It is customary in many modern checklists to cite the first reference in which the current genus/species combination appeared, no matter how trivial such a reference might be or whether the new combination was explained or justified. This practice I have found to be of dubious value and scarcely worth the effort involved in compilation. In a region such as Australia, where systematic herpetology is in such a state of flux, combinations change rapidly. For example, nearly all first citations of current combinations for frogs, which were until recently placed in the genus *Hyla* Laurenti, 1768, would have had to be replaced when most Australian hylids were transferred to the genus *Litoria* Tschudi, 1838. Such citations, therefore, have little intrinsic interest for the nomenclaturist or taxonomist.

Partim* References and *Nomina Auctorum

Throughout the taxonomic and general biological literature there have been numerous cases in which an author, believing his material to refer to a single species, has subsequently been shown to be referring to two or more species under the one name. Such *partim* or "in part" references are so numerous that to document them not only requires enormous bibliographic effort, but more importantly, it also requires specialist knowledge of the taxonomy, biology and literature of the particular assemblages of species concerned.

For this reason the only *partim* references which are cited are those in which there is a description of a new taxon which has been shown, or can be shown, to have been based on type specimens belonging to more than that one taxon.

Similarly, authors frequently associate their observations with particular taxa, but it is subsequently discovered that the observations were made on a species different from the one named. Apart from actual misidentifications of taxa (many of which may be perpetuated by later authors) the problem usually arises when a name long applied to a particular taxon, often from a particular geographic region, is shown to apply either to another or several different taxa. As a consequence, a great body of literature may be built up concerning a particular species but under the wrong binomen.

Such references, no matter how extensive, generally have been excluded from the catalogue except where they can be covered by a general statement. Like *partim* references, the identification of this literature often requires specialist knowledge of a particular group and goes beyond the scope of this work. Where *nomina auctorum*, however, have been fully documented by a specialist taxonomist, reference is made to such paper(s).

ORGANISATION OF THE CATALOGUE ENTRIES

In preparing the individual entries to the format specifications of the Bureau of Flora and Fauna, all information was entered into a series of predefined data fields prior to its transfer to a number of computer-based files. These files were eventually integrated, by family, to produce the final catalogue.

The following notes are intended to acquaint the reader with the criteria adopted when compiling the information to be entered in each of the major data fields.

Family Introductions

The brief family introductions are derived in part from Cogger (1979). Each describes briefly the size and global distribution of the family, the numbers of Australian genera and species and provides a morphological diagnosis to distinguish its Australian members from those of other families. The family introduction also may include some explanation of major taxonomic or nomenclatural problems encountered in compiling the relevant entries and the manner in which these problems have been resolved. In some cases it has been possible merely to identify the problems.

Each introduction concludes with one or more general references, each of which has been selected on the basis that, whatever its subject, it covers or is relevant to all of the Australian members of the family.

Synonymies

The generic and specific synonymies are as complete as I have been able to achieve, within the constraints imposed by our decision to exclude those *partim* references not based on primary type material and most *nomina auctorum* (see *Partim* References and *Nomina Auctorum*). The synonymies presented are therefore lists of primary or "absolute" synonyms. In the case of genera, this means that only those names whose type species are regarded as belonging to the genus in question, together with *nomina nuda*, are included.

In the case of species and subspecies, only those whose original or subsequent primary types (holotype, syntypes, lectotype or neotype) are considered to belong to the species in question are included in synonymy. The species synonymies include numerous errors and misspellings which have no formal standing in nomenclature (*i.e.* are invalid, unavailable names), but which are likely to confuse any worker who discovers them in the literature for the first time. *Partim* references are included only where the taxon under discussion is represented in the original type series.

Whenever possible the individual taxa making up a synonymy have been carefully checked and, in the case of species, the available types examined and their identities confirmed. In such cases the synonymy is attributed to one of us (HGC), but this does not imply that the synonymy is in any way original or unique. Where we have been unable to confirm the allocation of every entry in a synonymy, then the authors responsible for all or part of the synonymy are also cited.

Citation of Taxon Names

The editors of the *Catalogue* have opted to cite each name in its currently legitimate form (*i.e.* without the diacritic marks, and without the capitalization or hyphenation of the specific name as it appeared in the original reference). While we believe that this can result in loss of information, and is contrary to convention, we have complied with editorial policy.

Author's Name(s)

These are normally cited in the form in which they appear in the references being quoted. Exceptions are explained where they occur.

Date(s) of Publication

Each reference is preceded by the year of publication. Where the year cited on the published paper, or on the title page of the volume in which it appears, is known to differ from the actual year of publication, this is indicated in the text. Every effort has been made to ascertain the precise publication dates (cited where appropriate) when these are necessary to determine priority in nomenclature. Nevertheless, some inconsistencies remain. For example, in 1863 Lütken, in *Vidensk. Meddr. Dansk Naturh. Foren.* for 1862 described two new taxa: the pygopodid lizard *Aprasia pulchella* var. *striolata* and the scincid lizard *Cyclodus fasciatus*. In the same year (1863) Peters, in *Mber. K. Preuss. Akad. Wiss. Berl.*, also described two new taxa: the pygopodid lizard *Aprasia octolineata* and the scincid lizard *Cyclodus occipitalis*. Each of Peters' taxa is synonymous with the relevant taxon described by Lütken in the same year, yet whereas Lütken's description of *Aprasia pulchella striolata* long has been given priority over Peters' description of *Aprasia octolineata* (Kluge, 1974), Peters' description of *Cyclodus occipitalis* has been given precedence equally long over Lütken's *Cyclodus fasciatus*. Such inconsistencies are difficult to resolve and I have hedged the question by accepting "existing usage" in each case. All references to volume numbers and figure or plate numbers which were originally printed as roman numerals have been converted to arabic numerals, unless to do so would cause ambiguity or confusion.

Type Species

In the generic synonymies every attempt has been made to ascertain the source and basis of type species designation. This, however, has not been possible always, especially for older, extra-limital synonyms.

Type Specimens

Wherever practicable, the location, catalogue number(s) and status of the primary type specimens are cited. Less effort was made to locate or examine the types of taxa originally described from extra-limital (*i.e.* non-Australian) material. In the case of many taxa described in the eighteenth and nineteenth centuries, type specimens were not clearly indicated by their describers, while specimens labelled "types" of a particular taxon may be found in two or more widely dispersed collections (*e.g. Chondropython azureus* Meyer). Wherever possible, attempts were made to verify the status of labelled type specimens, or to locate unlabelled types by reconciling original descriptions with specimens available to the original author. Type status often can be resolved only by specialists in a particular group; the doubtful or probable status of types is indicated. Nevertheless, even this status has not been accepted without substantial supporting evidence.

In a few cases only have lectotypes been designated from syntypic series. This has been done only in groups with which I am familiar, and only when I felt strongly that important nomenclatural stability was at stake. In a few cases I have indicated for information, Mr P.R. Rawlinson's intention of designating certain specimens as lectotypes; although such designations have no status until formally published, they will alert others to work currently in progress.

Every attempt has been made to cite a catalogue number for each type specimen, and each number is prefixed by one of the institutional acronyms listed in Appendix II.

While no special effort has been made to examine types of more recently described taxa in Australian institutions, every effort was made during two visits to Europe and two to North America to locate and examine all type specimens of Australian species described during the 18th, 19th and early 20th centuries. Many of these proved to be synonyms of species other than those to which they had long been assigned. Much of the information garnered about these earlier types on the first European visit (by HGC and HMC) in 1970 has been passed on to colleagues and has since been incorporated into a number of publications. Photographs were taken of many of these early types and are on file in the Department of Herpetology at the Australian Museum.

Type Localities

Type localities have normally been anglicised and/or modernised to conform with current geographic place names within Australia. For example, "New Holland" has been converted to Australia, "Van Diemen's Land" to Tasmania and "Neu Sud Wales" to New South Wales. In rare cases, where the modern equivalent of the type locality is uncertain, or where modernisation or anglicisation would cause confusion, the type locality is quoted in its original form. Although restricted type localities (and their authorities) often are cited, it should be remembered that such restrictions have no status under the International Code of Zoological Nomenclature, despite Recommendation 72E, except where associated with a neotype or lectotype designation.

Distribution

Under this heading are included terms selected from the standardised list provided by the Bureau of Flora and Fauna. Any additional and/or explanatory notes follow a semicolon.

Ecological and Biological Descriptors

The ecological and biological descriptor terms were selected from two lists; the first was provided by the Bureau of Flora and Fauna as a series of computer search terms, and we have selected these on the basis of our understanding of their most likely meaning to ecologists and biologists. All terms selected from this first list appear before the semicolon. The second list, which follows the semicolon, we compiled as additional broad terms generally applicable to reptiles and amphibians.

We have avoided well-known but specific terms such as heliotherm or thigmotherm, because our knowledge of most species is too fragmentary to permit such terms to be applied with confidence to all but a few species.

The absence of any ecological descriptive term for a particular taxon is to be interpreted in only one way: that we have no knowledge of the ecology of that taxon and are unaware of any published data on its ecology. If ecological information about a species is not known for certain, however, it appears in parenthesis.

Biological References
Under this heading those literature sources appear which, in my judgement, will best serve as points of entry into the wider body of taxonomic and biological literature of the particular taxon. Articles of broad scope, or which survey the literature, have been selected in preference to those of a more specialised nature.

No claim to completeness is made, nor is it implied that those references cited are more useful or comprehensive than those not cited.

CLASSIFICATION

The classification used in the checklist is primarily that adopted by Cogger (1979). Currently recognised genera are listed alphabetically within their respective families and currently recognised species are listed alphabetically within genera. Synonyms are listed under each genus or species in chronological order.

Family Level
There is a great need in Australian herpetology for broader evolutionary studies leading to an understanding of the phyletic relationships of higher taxa. The last decade has seen major changes proposed in the family level classification of several groups of Australian reptiles and amphibians. These changes overturn long standing classifications and have tended to emphasise regional endemism at the family level. In general, few of these proposals have become accepted and used. The family classification used here is a fairly conservative one, but the major contentious issues are:

1. Savage (1973) proposed that Australian frogs previously assigned to the southern hemisphere family Leptodactylidae warranted recognition as a distinct regional family, the Myobatrachidae. This action has received wide support and has been adopted in this catalogue.

Long included in the family Myobatrachidae (as Leptodactylidae) has been the genus *Cyclorana* (Steindachner, 1867). Following the transfer of particular species of *Cyclorana* to the family Hylidae (Straughan, 1969; Tyler, 1974; Tyler & Martin, 1977), the genus with the remaining species was transferred (Tyler, 1978) to the Hylidae.

2. Savage (1973) further proposed that those Australian frogs previously included in the widespread family Hylidae also warranted recognition as a discrete regional family and resurrected the name Pelodryadidae Günther, 1858. This proposal received a mixed response from Australian herpetologists and has not been adopted in this work.

3. The family classification of colubroid snakes has been the subject of much conjecture and remains unresolved. I have adopted the most conservative approach by regarding all caenophidian snakes which lack fangs at the front of the upper jaw, except the aquatic file snakes of the genus *Acrochordus*, as members of the family Colubridae.

4. The family classification of Australia's venomous front-fanged land snakes remains obscure. Until recently considered to be members of the family Elapidae (traditionally a group including the African mambas, the Afro-Asian cobras, the Asian kraits and the New World coral snakes), Smith *et al.* (1977) have proposed that the Australian front-fanged land snakes are not only confamilial with the true sea-snakes but that these two groups of snakes together constitute the separate family Hydrophiidae.

5. Smith (1926), in his classic monograph on sea snakes, placed all of the paddle-tailed marine snakes in a single family - the Hydrophiidae. McDowell (1969) showed that the oviparous sea-kraits of the genus *Laticauda* were not true sea-snakes, and placed them in a separate family level group, the Laticaudinae. He also reduced the elapid and hydrophiid snakes to the same subfamilial level - the Elapinae and Hydrophiinae - within the family Elapidae. Subsequently, Smith *et al.* (1977), brought together the Hydrophiinae and the Australian terrestrial "elapids" in an expanded family, Hydrophiidae (see 4 above). On the other hand, Burger and Natsuno (1974) elevated McDowell's sub–families to full family status - the Elapidae, Hydrophiidae and Laticaudidae. Pending critical review of the work of Smith *et al.* (1977), I have adopted Burger and Natsuno's more traditional classification, but the familial relationships of these venomous snakes remain obscure.

Generic Level

As in all classifications which purport or attempt to be phyletic, recognition of particular taxa will depend on which point on a hypothetical dendrogram or cladogram a worker chooses to select as the level distinguishing a higher from a lower taxonomic grouping. Nowhere is this process of selection more subjective than in the recognition of genera, and most of the differences between the genera recognised in this checklist and those recognised by other workers fall into an area as much philosophic as scientific.

For example, I have retained the monotypic genus *Trachydosaurus* for the shingle-back lizard. This does not deny the very close affinity of this lizard with the blue-tongued lizards of the genus *Tiliqua* (Mitchell, 1950), and there is no doubt that both lines are derived from a common ancestor. I believe, however, that the available morphological, biological and geographic evidence suggests that the shingle-back/blue-tongue divergence was earlier than, rather than approximately contemporaneous with, the radiation of the blue-tongued lizards in Australia. It represents, therefore, an earlier offshoot from the blue-tongued lineage than any of the present species of blue-tongues, and I believe that this relationship is more accurately expressed by recognition of the genus *Trachydosaurus*. Conversely, recognition of the genus *Omolepida* by Storr (1976) is more difficult to sustain only because there is a continuum of character states linking the extreme expression of *Tiliqua* via *Hemisphaeriodon* with that of *Omolepida* (= *Cyclodomorphus*).

Species Level

Species, provided that they have been adequately diagnosed in the literature, generally have been accepted at face value and listed in the catalogue. Equivocal cases such as those regarded variously as either subspecies or full species, have been treated either as the majority of workers have regarded them or, if rarely cited, assessed on the basis of the available data. Similarly, inferred species (by which I mean the brief, undiagnosed, and often tentative resurrection of names from synonymy) usually have not been recognised.

Subspecies

For the reasons cited in Cogger (1975), subspecies have not been treated separately and all subspecies are included within the general synonymy of the species within which they fall. A list of the subspecies currently recognised within each species is provided in Cogger (1979).

Despite the preceding comments, a checklist is concerned with names rather than classifications, and the checklist component of the *Catalogue* is intended to be no more than a tool for taxonomists and nomenclaturists. Its value should be diminished only marginally by changes in classification.

ACKNOWLEDGEMENTS

Many long-standing problems in Australian systematic herpetology were resolved during visits to major museum collections in North America in June-July, 1967, in Europe and North America during the period October, 1970 to April, 1971 and in Europe in July-August, 1979. These visits were essential to the completion of the present catalogue, and were supported in part by grants to one of us (HGC) by the Science and Industry Endowment Fund of CSIRO, the Explorer's Club (New York), the British Council, the Australian Museum Trust and the New South Wales Government, to each of which we express our gratitude.

Other taxonomic studies which have been utilised, at least in part, for this catalogue have been supported by grants from the following: the Australian Museum, the Australian Research Grants Scheme, the Australian Biological Resources Study, the Japan Society for the Promotion of Science, and the National Science Foundation under grants NSF GA35835 to the Pennsylvania State University and grants NSF GB8400, GA34948 and GD34462 to the Scripps Institution of Oceanography for operation of the *Alpha Helix* Research Program.

We are also grateful to the following individuals and institutions for providing us with information, specimens or facilities: Dr. E.N. Arnold (BMNH), Dr. W. Böhme (ZFMK), Dr. F.W. Braestrup (UZM), Prof. L.D. Brongersma (RMNH), Dr. W.C. Brown (Menlo College, California, U.S.A.), Dr. L. Capocaccia (MCG), Ms. J. Covacevich (QM), Mr J. Coventry (NMV), Dr. J. Eiselt (NHMW), Miss A.G.C. Grandison (BMNH), Prof. J. Guibé (MNHP), Prof. H. Heatwole (University of New England, Armidale, N.S.W.), Dr. D. Hillenius (ZMA), Dr. R. Inger (FMNH), Mr. G. Ingram (QM), Dr. K. Klemmer (SMF), Prof. A.G. Kluge (University of Michigan, Ann Arbor, U.S.A.), Dr. P. Kuenzer (ZMG), Dr. A.E. Leviton (CAS), Mr. C. McCarthy (BMNH), Dr. S.B. McDowell (AMNH), Mr. E. Malnate (ANSP), Dr. H. Marx (FMNH), the late Prof. R. Mertens (SMF), Dr. P. Pethon (UZMO), Dr. G. Peters (ZMB), the late Dr. J.A. Peters (USNM), Dr. R. Roux-Estève (MNHP), Dr. L. Saldanha (MB), Dr. R. Shine (University of Sydney, N.S.W.), Mr. A.F. Stimson (BMNH), Dr. E.E. Williams (MCZ), Mr J. Wombey (ANWC), and Dr. R.G. Zweifel (AMNH).

Many of the problems which we were unable to resolve during overseas visits were taken up by Mr Peter Rawlinson of La Trobe University on a subsequent visit to many of the same institutions; we are indebted to him for generously making available the results of his studies. All errors, however, are our responsibility.

The late Mr. Gilbert P. Whitley gave invaluable bibliographic and other advice until his untimely death in 1975. Opinions and/or advice on nomenclatural matters were kindly offered by Dr. J.H. Calaby, Dr. D.K. McAlpine and Dr. W.D.L. Ride. We are also grateful to Dr. Dan Walton and Mr. Richard Longmore of the Bureau of Flora and Fauna whose editorial attention to detail materially improved this catalogue.

We are grateful to past and present library staff at the Australian Museum, especially Ms. N. Bain, Ms G. Baker, Miss M. Davies, Mrs. A.-M. McConochie, Mrs. J. Watson and to Mrs. R. Brewer and Miss S. Rogers for undertaking much of the routine library work, and to Mrs. Brewer, Mrs. A. Thomas, Mrs. C. Sinclair, Ms. P. Clark, Ms. K. Brady and Mr. T. May for processing various drafts of the manuscript.

Finally, we are especially indebted to past and present colleagues in the Department of Herpetology at the Australian Museum - Dr. A.E. Greer, the late Mr. P.R. Rankin, Mr. R. Sadlier, Mr. G. Shea, Mr. P. Webber - and to Dr. F.H. Talbot and Dr. D.J.G. Griffin, respectively immediate past and present Directors of the Australian Museum.

H.G.C.

References

Boulenger, G.A. (1887). *Catalogue of the Lizards in the British Museum (Natural History).* 3. London : British Museum xii 575 pp. pls 40

Burger, W.L. & Natsuno, T. (1974). A new genus for the Arafura smooth sea snake and redefinitions of other sea snake genera. *The Snake* **6**: 61–75

Cogger, H.G. (1975). *The Reptiles and Amphibians of Australia.* Sydney : A.H. & A.W. Reed 584 pp.

Cogger, H.G. (1979). *Reptiles and Amphibians of Australia.* rev. edn. Sydney : A.H. & A.W. Reed 608 pp.

Duellman, W.E. (1977). Liste der rezenten Amphibien und Reptilien; Hylidae, Centrolenidae, Pseudidae. *Das Tierreich* **95**: i–xix 1–225

Gorham, S.W. (1966). Liste der rezenten Amphibien und Reptilien; Ascaphidae, Leiopelmatidae, Pipidae, Discoglossidae, Pelobatidae, Leptodactylidae, Rhinophrynidae. *Das Tierreich* **85**: i–xvi 1–222

Hahn, D.E. (1980). List der rezenten Amphibien und Reptilien; Anomalepididae, Leptotyphlopidae, Typhlopidae. *Das Tierreich* **101**: i–xii 1–93

The International Commission on Zoological Nomenclature. (1964). *International Code of Zoological Nomenclature, adopted by the XV International Congress of Zoology.* 2nd edn. London : Internat. Trust for Zool. Nomenclature 176 pp.

Klemmer, K. (1963). Liste der rezenten Giftschlangen; Elapidae, Hydrophiidae, Viperidae und Crotalidae. Marburg : N.G. Elwert Universitäts- U. Verlagsbuchhand pp. 253–464

Kluge, A.G. (1974). A taxonomic revision of the lizard family Pygopodidae. *Misc. Publ. Mus. Zool. Univ. Mich.* 147: 1–221

Lütken, C. (1863). Nogle nye krybdyr og padder. *Vidensk. Meddr. Dansk Naturh. Foren.* **1862**: 292–331

Lynch, J.D. (1971). Evolutionary relationships, osteology, and zoogeography of leptodactyloid frogs. *Misc. Publs. Univ. Kansas Mus. Nat. Hist.* **53**: 1–238

McDowell, S.B. (1969). Notes on the Australian sea-snake *Ephalophis greyi* M. Smith (Serpentes : Elapidae : Hydrophiinae) and the origin and classification of sea-snakes. *J. Linn. Soc. Zool.* **48**: 333–349

Mertens, R. (1963). Liste der rezenten Amphibien und Reptilien; Helodermatidae, Varanidae, Lanthanotidae. *Das Tierreich* **79**: i–x 1–26

Mitchell, F.J. (1950). The scincid genera *Egernia* and *Tiliqua* (Lacertilia). *Rec. S. Aust. Mus.* **9**: 275–308

Moore, J.A. (1961). The frogs of eastern New South Wales. *Bull. Am. Mus. Nat. Hist.* **121**: 149–386

Neave, A.S. (1939). *Nomenclator Zoologicus.* London : Zoolological Society of London Vol. 2 1025 pp.

Peters, W. (1863). Übersicht der von Hrn. Richard Schomburgk an das zoologische Museum eingesandten Amphibien, aus Buchsfelde bei Adelaide in Südaustralien. *Mber. K. Preuss. Akad. Wiss. Berl.* **1863**: 228–236

Romer, A.S. (1956). *Osteology of the Reptiles.* Chicago : Univ. Chicago Press xxii 772 pp.

Savage, J.M. (1973). The geographic distribution of frogs: patterns and predictions. pp. 351–455 *in* Vial, J.L. (ed.) *Evolutionary Biology of the Anurans, Contemporary Research on Major Problems.* Columbia : Univ. Missouri Press

Sherborn, C.D. (1925). *Index Animalium.* London : British Museum 2: 1773–2880

Smith, H.M., Smith, R.B. & Sawin, H.L. (1977). A summary of snake classification (Reptilia, Serpentes). *J. Herpet.* **11**: 115–121

Smith, M.A. (1926). *Monograph on the Sea Snakes. (Hydrophiidae).* London : British Museum xvii 130 pp.

Stimson, A.F. (1969). Liste der rezenten Amphibien und Reptilien; Boidae. *Das Tierreich* **89**: 1–49

Storr, G.M. (1976). The genus *Omolepida* (Lacertilia, Scincidae) in Western Australia. *Rec. West. Aust. Mus.* **4**: 163–170

Straughan, I.R. (1969). *Hyla inermis* (Peters), a species hitherto erroneously referred to the leptodactylid genus *Cyclorana* (Anura : Hylidae/Leptodactylidae). *Zool. Meded.* **43**: 207–212

Tyler, M.J. (1974). The systematic position and geographic distribution of the Australian frog *Chiroleptes alboguttatus* Günther. *Proc. R. Soc. Qd.* **85**: 27–32

Tyler, M.J. (1978). *Amphibians of South Australia.* Adelaide : South Australian Government 84 pp.

Tyler, M.J. & Martin, A.A. (1977). Taxonomic studies of some Australian leptodactylid frogs of the genus *Cyclorana* Steindachner. *Rec. S. Aust. Mus.* **17**: 261–276

Wermuth, H. (1965). Liste der rezenten Amphibien und Reptilien; Gekkonidae, Pygopodidae, Xantusiidae. *Das Tierreich* **80**: i–xxii 1–246

Wermuth, H. (1967). Liste der rezenten Amphibien und Reptilien; Agamidae. *Das Tierreich* **86**: 1–127

Wermuth, H. & Mertens, R. (1961). *Schildkröten, Krokodile, Brückenechsen.* Jena : Gustav Fischer 422 pp.

Wermuth, H. & Mertens, R. (1977). Liste der rezenten Amphibien und Reptilien; Testudines, Crocodylia, Rhynchocephalia. *Das Tierreich* **100**: i–xxvii 1–174

Zietz, F.R. (1920). Catalogue of Australian lizards. *Rec. S. Aust. Mus.* **1**: 181–228

MYOBATRACHIDAE

INTRODUCTION

Members of this family are generally moderate to small terrestrial frogs confined to the Australian region. Until recently they were regarded as a subfamily within the family Leptodactylidae (Southern Frogs).

Currently recognized are 21 genera and approximately 100 species; of these, all are confined to Australia except 3 genera and 4 species which either also occur in New Guinea or are confined to New Guinea.

Characterised in Australia by: arciferal pectoral girdle; intercalary cartilages absent; finger and toe discs small or absent, without lateral grooves; sacral diapophyses more or less dilated; eight presacral vertebrae, the first two fused in some genera; maxillary teeth present or, if absent, the toes never more than about half-webbed.

References

(as Leptodactylidae)

Barker, J. & Grigg, G. (1977). *A Field Guide to Australian Frogs.* Adelaide : Rigby Ltd. 229 pp.

Brattstrom, B.H. (1970). Thermal acclimation in Australian amphibians. *Comp. Biochem. Physiol.* **35**: 69–103

Cogger, H.G. (1979). *Reptiles and Amphibians of Australia.* Sydney : A.H. & A.W. Reed 608 pp.

Gorham, S.W. (1974). *Checklist of World Amphibians up to January 1, 1970.* Saint John : The New Brunswick Museum 172 pp.

Lynch, J.D. (1971). Evolutionary relationships, osteology, and zoogeography of leptodactylid frogs. *Univ. Kansas Mus. Nat. Hist. Misc. Publ.* **53**: 1–238

Parker, H.W. (1940). The Australasian frogs of the family Leptodactylidae. *Novit. Zool.* **42**: 1–106

Tyler, M.J., Watson, F.G. & Martin, A.A. (1981). The Amphibia: diversity and distribution. pp. 1275–1302 *in* Keast, A., (ed.) *Ecological Biogeography of Australia.* The Hague : Junk

Watson, G.F. & Martin, A.A. (1973). Life history, larval morphology and relationships of Australian leptodactylid frogs. *Trans. R. Soc. S. Aust.* **97**: 33–45

Adelotus Ogilby, 1907

Cryptotis Günther, A. (1863). On new species of batrachians from Australia. *Ann. Mag. Nat. Hist. (3)* **11**: 26–28 [27] [*non Cryptotis* Pomel, 1848]. Type species *Cryptotis brevis* Günther, 1863 by monotypy.

Adelotus Ogilby, J.D. (1907). A new tree frog from Brisbane. *Proc. R. Soc. Qd.* **20**: 31–32 [32] [*nom. nov. pro Cryptotis* Günther, 1863].

Synonymy that of Cogger, H.G., this work.

Adelotus brevis (Günther, 1863)

Cryptotis brevis Günther, A. (1863). On new species of batrachians from Australia. *Ann. Mag. Nat. Hist. (3)* **11**: 26–28 [28 pl 4 fig B]. Type data: holotype, BMNH 1947.2.20.3, from Clarence River, N.S.W.

Distribution: SE coastal, NE coastal; N.S.W., Qld. Ecology: amphibious, noctidiurnal, closed forest, tall forest, (open forest), lentic freshwater, lotic freshwater, predator; winter dormant, seasonal breeder, oviparous, sexually dimorphic, free-living tadpole, arthropod-feeder, cryptozoic, riparian habitats. Biological references: Tyler, M.J. & Davies, M. (1979). Foam nest construction by Australian leptodactylid frogs (Amphibia, Anura, Leptodactylidae). *J. Herpet.* **13**: 509–510 (reproduction); Moore, J.A. (1961). The frogs of eastern New South Wales. *Bull. Am. Mus. Nat. Hist.* **121**: 149–386 (taxonomy, natural history); Watson, G.F. & Martin, A.A. (1973). Life history, larval morphology and relationships of Australian leptodactylid frogs. *Trans. R. Soc. S. Aust.* **97**: 33–45.

Arenophryne Tyler, 1976

Arenophryne Tyler, M.J. (1976). A new genus and two new species of leptodactylid frogs from Western Australia. *Rec. West. Aust. Mus.* **4**: 45–52 [45]. Type species *Arenophryne rotunda* Tyler, 1976 by original designation.

Arenophryne rotunda Tyler, 1976

Arenophryne rotunda Tyler, M.J. (1976). A new genus and two new species of leptodactylid frogs from Western Australia. *Rec. West. Aust. Mus.* **4**: 45–52 [46 fig 1]. Type data: holotype, WAM R39120, from approx. 100 m from False Entrance Well Tank, Carrarang Station, 320 km NNW of Geraldton, W.A.

Arenophryne robusta Tyler, M.J. (1976). A new genus and two new species of leptodactylid frogs from Western Australia. *Rec. West. Aust. Mus.* **4**: 45–52 [49] [invalid emend., *lapsus pro Arenophryne rotunda* Tyler, 1976].

Synonymy that of Cogger, H.G., this work.

Distribution: NW coastal, W.A. Ecology: terrestrial, burrowing, nocturnal, soil, predator; walking, oviparous, arthropod-feeder, cryptozoic. Biological references: Tyler, M.J., Roberts, J.D. & Davies, M. (1980). Field observations on *Arenophryne rotunda* Tyler, a leptodactylid frog of coastal sandhills. *Aust. Wildl. Res.* **7**: 295–304 (natural history).

Assa Tyler, 1972

Assa Tyler, M.J. (1972). A new genus for the Australian leptodactylid frog *Crinia darlingtoni*. *Zool. Meded.* **47**: 193–201 [200]. Type species *Crinia darlingtoni* Loveridge, 1933 by monotypy.

Assa darlingtoni (Loveridge, 1933)

Crinia darlingtoni Loveridge, A. (1933). Four new crinine frogs from Australia. *Occ. Pap. Bost. Soc. Nat. Hist.* **8**: 55–60 [57]. Type data: holotype, MCZ 18390, from Queensland National Park, McPherson Range, Qld.

Distribution: SE coastal, NE coastal, N.S.W., Qld. Ecology: terrestrial, nocturnal, litter, closed forest, (tall forest), predator; winter dormant, saltatorial, seasonal breeder, arthropod-feeder, cryptozoic, tadpoles develop in pair of inguinal brood pouches of male. Biological references: Straughan, I.R. & Main, A.R. (1966). Speciation and polymorphism in the genus *Crinia* Tschudi (Anura : Leptodactylidae) in Queensland. *Proc. R. Soc. Qd.* **78**: 11–28; Leptodactylidae. *Novit. Zool.* **42**: 1–106 (taxonomy); Parker, H.W. (1940). The Australasian frogs of the family Leptodactylidae. *Novit. Zool.* **42**: 1–106 (taxonomy); Ingram, G.J., Anstis, M. & Corben, C.J. (1975). Observations on the Australian leptodactylid frog, *Assa darlingtoni*. *Herpetologica* **31**: 425–429 (natural history); Watson, G.F. & Martin, A.A. (1973). Life history, larval morphology and relationships of Australian leptodactylid frogs. *Trans. R. Soc. S. Aust.* **97**: 33–45.

Crinia Tschudi, 1838

Crinia Tschudi, J.J. (1838). *Classification der Batrachier, mit Berücksichtigung der fossilen Tiere dieser Abteilung der Reptilien.* Neuchâtel : Petitpierre 102 pp. [38, 78]. Type species *Crinia georgiana* Tschudi, 1838 by monotypy.

Crinia georgiana Tschudi, 1838

Crinia georgiana Tschudi, J.J. (1838). *Classification der Batrachier, mit Berücksichtigung der fossilen Tiere dieser Abteilung der Reptilien.* Neuchâtel : Petitpierre 102 pp. [38,78]. Type data: holotype, MNHP 759, from King George's Sound, W.A.

Pterophrynus affinis Günther, A. (1864). Third contribution to our knowledge of batrachians from Australia. *Proc. Zool. Soc. Lond.* **1864**: 46–49 [47 pl 7 fig 2]. Type data: holotype, BMNH 1947.2.20.2, from Western Australia.

Crinia ignata Cope, E.D. (1866). On the structure and distribution of the genera of arciferous Anura. *J. Acad. Nat. Sci. Philad. (2)* **6**: 67–112 [95]. Type data: holotype, ANSP 10058, from West Australia.

Crinia stolata Cope, E.D. (1867). On the families of raniform Anura. *J. Acad. Nat. Sci. Philad. (2)* **6**: 189–206

[201]. Type data: holotype, ANSP 10061, from West Australia.

Synonymy that of Cogger, H.G., this work.

Distribution: SW coastal, W.A. Ecology: terrestrial, nocturnal, open forest, woodland, (lentic freshwater), swamp, predator; seasonal breeder, oviparous, free-living tadpole, arthropod-feeder, cryptozoic. Biological references: Main, A.R. (1965). Frogs of southern Western Australia. *West. Aust. Nat. Club Handbook* 8: 1–73 (natural history).

Geocrinia Blake, 1973

Geocrinia Blake, A.J.D. (1973). Taxonomy and relationships of myobatrachine frogs (Leptodactylidae): a numerical approach. *Aust. J. Zool.* **21**: 119–149 [142]. Type species *Pterophrynus laevis* Günther, 1864 by original designation.

Geocrinia laevis (Günther, 1864)

Pterophrynus laevis Günther, A. (1864). Third contribution to our knowledge of batrachians from Australia. *Proc. Zool. Soc. Lond.* **1864**: 46–49 pl 7 [48 pl 7 fig 4]. Type data: holotype, BMNH 1947.2.19.65, from Tas.

Distribution: SE coastal, S.A., Tas., Vic. Ecology: terrestrial, nocturnal, litter, open forest, (woodland), (alpine), predator; winter dormant, seasonal breeder, oviparous, free-living tadpole, arthropod-feeder, cryptozoic. Biological references: Littlejohn, M.J. & Martin, A.A. (1964). The *Crinia laevis* complex (Anura : Leptodactylidae) in south-eastern Australia. *Aust. J. Zool.* **12**: 70–83 (taxonomy); Littlejohn, M.J., Watson, G.F. & Loftus-Hills, J.J. (1971). Contact hybridization in the *Crinia laevis* complex (Anura : Leptodactylidae). *Aust. J. Zool.* **19**: 85–100; Littlejohn, M.J. & Watson, G.F. (1974). Mating call discrimination and phonotaxis by females of the *Crinia laevis* complex (Anura : Leptodactylidae). *Copeia* **1974**: 71–175; Littlejohn, M.J. & Martin, A.A. (1974). The Amphibia of Tasmania. pp. 251–289 *in* Williams, W.D. (ed.) *Biogeography and Ecology in Tasmania.* The Hague : Junk; Gartside, D.F., Littlejohn, M.J. & Watson, G.F. (1979). Structure and dynamics of a narrow hybrid zone between *Geocrinia laevis* and *G. victoriana* (Anura : Leptodactylidae) in south-eastern Australia. *Heredity* **43**: 165–177; Tyler, M.J. (1978). *Amphibians of South Australia.* Adelaide : South Australian Government 84 pp. (natural history).

Geocrinia leai (Fletcher, 1898)

Crinia leai Fletcher, J.J. (1898). Contributions to a more exact knowledge of the geographical distribution of Australian Batrachia. 5. A. Batrachia of Tasmania. *Proc. Linn. Soc. N.S.W. (2)* **12**: 660–684 [677]. Type data: syntypes, AM R8337, AM R10324, AM R49819–24, from Bridgetown and Pipe Clay Creek (near Jarrahdale), W.A.

Crinia michaelseni Werner, F. (1914). Amphibia. *in* Michaelsen, W. & Hartmeyer, R. (eds.) *Die Fauna Südwest-Australiens.* Jena : Gustav Fischer 4: 403–426 [416]. Type data: holotype (probable), AMNH 23451, from Donnybrook, W.A.

Synonymy that of Cogger, H.G., this work.

Distribution: SW coastal, W.A. Ecology: terrestrial, nocturnal, litter, open forest, woodland, swamp, predator; seasonal breeder, oviparous, free-living tadpole, arthropod-feeder, cryptozoic. Biological references: Main, A.R. (1965). Frogs of southern Western Australia. *West. Aust. Nat. Club Handbook* 8: 1–73 (natural history).

Geocrinia lutea (Main, 1963)

Crinia lutea Main, A.R. (1963). A new species of *Crinia* (Anura : Leptodactylidae) from National Park, Nornalup. *West. Aust. Nat.* **8**: 143–144 [143 fig 1]. Type data: holotype, WAM R17616, from National Park, Nonalup, W.A.

Distribution: SW coastal, W.A. Ecology: terrestrial, tall forest, (open forest), predator; seasonal breeder, oviparous, direct development, arthropod-feeder, cryptozoic. Biological references: Main, A.R. (1965). Frogs of southern Western Australia. *West. Aust. Nat. Club Handbook* 8: 1–73 (natural history).

Geocrinia rosea (Harrison, 1927)

Crinia rosea Harrison, L. (1927). Notes on some western Australian frogs, with descriptions of new species. *Rec. Aust. Mus.* **15**: 277–287 [279]. Type data: syntypes, MMUS (not found), from Pemberton, W.A.

Distribution: SW coastal, W.A. Ecology: terrestrial, tall forest, (open forest), predator; seasonal breeder, oviparous, direct development, arthropod-feeder, cryptozoic. Biological references: Main, A.R. (1965). Frogs of southern Western Australia. *West. Aust. Nat. Club Handbook* 8: 1–73 (natural history).

Geocrinia victoriana (Boulenger, 1888)

Crinia victoriana Boulenger, G.A. (1888). Descriptions of two new Australian frogs. *Ann. Mag. Nat. Hist. (6)* **2**: 142–143 [142]. Type data: holotype, BMNH 1947.2.19.66, from Warragul, Vic.

Crinia froggatti Fletcher, J.J. (1891). Description of a supposed new cystignathoid frog. *Proc. Linn. Soc. N.S.W. (2)* **6**: 275–276 [275]. Type data: syntypes, BMNH 1947.2.19.67 (provenance of specimen not specified), AM R10326, AM R49776–83 (provenance of individual specimens not specified), one syntype not found, provenance unknown, all from Buninyong and Gong Gong, near Ballarat, Vic.

Synonymy that of Cogger, H.G., this work.

Distribution: SE coastal, Murray-Darling basin, Vic. Ecology: terrestrial, nocturnal, litter, open forest, alpine, (woodland), predator; winter dormant, seasonal breeder, oviparous, free-living tadpole, arthropod-feeder, cryptozoic. Biological references: Littlejohn, M.J. & Martin, A.A. (1964). The *Crinia laevis* complex (Anura : Leptodactylidae) in south-eastern Australia. *Aust. J. Zool.* **12**: 70–83 (taxonomy); Littlejohn, M.J., Watson, G.F. & Loftus-Hills, J.J. (1971). Contact hybridization in the *Crinia laevis* complex (Anura : Leptodactylidae). *Aust. J. Zool.* **19**: 85–100; Gartside, D.F., Littlejohn, M.J. & Watson, G.F. (1979). Structure and dynamics of a narrow hybrid zone between *Geocrinia laevis* and *G. victoriana* (Anura : Leptodactylidae) in south-eastern Australia. *Heredity* **43**: 165–177; Littlejohn, M.J. & Harrison, P.A. (1981). Accoustic communication in *Geocrinia victoriana* (Anura : Leptodactylidae). p. 6 *in* Banks, C.B. & Martin, A.A. (eds.) *Proceedings of the Melbourne Herpetological Symposium, 1980.* Melbourne : Zoological Board of Victoria.

Heleioporus Gray, 1841

Heleioporus Gray, J.E. (1841). Description of some new species and four new genera of reptiles from Western Australia, discovered by John Gould, Esq. *Ann. Mag. Nat. Hist. (7)* **42**: 86–91 [91]. Type species *Heleioporus albopunctatus* Gray, 1841 by monotypy.

Helioporus Gray, J.E. (1841). A catalogue of the species of reptiles and Amphibia hitherto described as inhabiting Australia, with a description of some new species from Western Australia. Appendix E *in* Grey, G. *Journals of Two Expeditions of Discovery in North-west and Western Australia, during the years 1837, 38, and 39, under the authority of Her Majesty's Government.* London : T. & W. Boone 2: 422–449 [447] [invalid emend. *pro Heleioporus* Gray, 1841].

Perialia Gray, J.E. (1845). Description of some new Australian animals. Appendix *in* Eyre, E.J. *Journals of Expeditions of Discovery into Central Australia and Overland from Adelaide to King George's Sound in the Years 1840–1.* London : T. & W. Boone 1: 405–411 [406]. Type species *Perialia eyrei* Gray, 1845 by subsequent designation, Parker, H.W. (1940). The Australasian frogs of the family Leptodactylidae. *Novit. Zool.* **42**: 1–106.

Heleioforus Kre[f]ft, G. (1865). The frogs of Australia. *Pap. Proc. R. Soc. Tas.* **1865**: 16–20 [17] [invalid emend., *errore pro Heleioporus* Gray, 1841].

Philocryphus Fletcher, J.J. (1894). Description of a new cystignathoid frog from New South Wales. *Proc. Linn. Soc. N.S.W. (2)* **8**: 229–236 [233]. Type species *Philocryphus flavoguttatus* Fletcher, 1894 by monotypy.

Synonymy that of Cogger, H.G., this work.

Heleioporus albopunctatus Gray, 1841

Heleioporus albopunctatus Gray, J.E. (1841). Description of some new species and four new genera of reptiles from Western Australia, discovered by John Gould, Esq. *Ann. Mag. Nat. Hist. (7)* **42**: 86–91 [91]. Type data: holotype, BMNH 1947.2.19.19, from W.A.

Heleioforus alleopunctatus Kre[f]ft, G. (1865). The frogs of Australia. *Pap. Proc. R. Soc. Tas.* **1865**: 16–20 [17] [invalid emend. *errore pro Heleioporous albopunctatus* Gray, 1841].

Synonymy that of Cogger, H.G., this work.

Distribution: (NW coastal), SW coastal, W.A. Ecology: terrestrial, burrowing, nocturnal, woodland, open forest, swamp, (open heath), predator; seasonal breeder, oviparous, free-living tadpole, arthropod-feeder. Biological references: Main, A.R. (1965). Frogs of southern Western Australia. *West. Aust. Nat. Club Handbook* 8: 1–73 (natural history); Lee, A.K. (1967). Studies in Australian Amphibia. II. Taxonomy, ecology, and evolution of the genus *Heleioporus* Gray (Anura : Leptodactylidae). *Aust. J. Zool.* **15**: 367–439.

Heleioporus australiacus (Shaw and Nodder, 1795)

Rana australiaca Shaw, G. & Nodder, F.P. (1795). Rana australiaca. *Naturalist's Miscellany* London 6: pl 200. Type data: description based on a drawing *vide* Shaw, G. (1802). *General Zoology*, or Systematic Natural History. Vol. 3 Pt. 2 Amphibia. London : Kearsley vii pp. 313–615, sent from Australia.

Rana spinipes Schneider, J.G. (1799). *Historiae Amphibiorum naturalis et literariae.* fasc. prim. Jena : Frommani xvi 266 pp. [129,139]. Type data: description based on the illustration of *Rana australiaca* in Shaw, G. & Nodder, F.P. (1795). Rana australiaca. *Naturalist's Miscellany* London 6: pl 200, from Australia.

Philocryphus flavoguttatus Fletcher, J.J. (1894). Description of a new cystignathoid frog from New South Wales. *Proc. Linn. Soc. N.S.W. (2)* **8**: 229–236 [233]. Type data: syntypes, presumed lost (4 specimens), from Mt. Victoria, Thornleigh, Dural and Manly, N.S.W.

Synonymy that of Cogger, H.G., this work.

Distribution: SE coastal, Vic., N.S.W. Ecology: terrestrial, burrowing, nocturnal, tall forest, open forest, predator; seasonal breeder, oviparous, free-living tadpole, arthropod-feeder. Biological references: Moore, J.A. (1961). The frogs of eastern New South Wales. *Bull. Am. Mus. Nat. Hist.* **121**: 149–386 (taxonomy, natural history); Lee, A.K. (1967). Studies in Australian Amphibia. II. Taxonomy, ecology, and evolution of the genus *Heleioporus* Gray (Anura : Leptodactylidae). *Aust. J. Zool.* **15**: 367–439; Littlejohn, M.J. & Martin, A.A. (1967). The rediscovery of *Heleioporus australiacus* (Shaw) (Anura : Leptodactylidae) in eastern Victoria. *Proc. R. Soc. Vict.* **80**: 31–35 (distribution).

Heleioporus barycragus Lee, 1967

Heleioporus barycragus Lee, A.K. (1967). Studies in Australian Amphibia. II. Taxonomy, ecology, and evolution of the genus *Heleioporus* Gray (Anura : Leptodactylidae). *Aust. J. Zool.* **15**: 367–439 [390]. Type data: holotype, WAM R21869, from Boya, W.A.

Distribution: SW coastal, W.A. Ecology: terrestrial, burrowing, nocturnal, open forest, predator; seasonal breeder, oviparous, free-living tadpole, arthropod-feeder.

Heleioporus eyrei (Gray, 1845)

Perialia eyrei Gray, J.E. (1845). Description of some new Australian animals. Appendix *in* Eyre, E.J. *Journals of Expeditions of Discovery into Central Australia and Overland from Adelaide to King George's Sound in the Years 1840–1.* London : T. & W. Boone 1: 405–411 [407 pl 2 fig 3]. Type data: syntypes, BMNH 1947.2.19.80–83, from Murray River, W.A., see Glauert, L. (1947). Some unfortunate errors in collecting localities. *West. Aust. Nat.* **1**: 48.

Heleioporus insularis Loveridge, A. (1933). A new genus and three new species of crinine frogs from Australia. *Occ. Pap. Bost. Soc. Nat. Hist.* **8**: 89–94 [92]. Type data: holotype, WAM R4303, from Rottnest Is., W.A.

Synonymy that of Cogger, H.G., this work.

Distribution: SW coastal, W.A. Ecology: terrestrial, burrowing, nocturnal, open forest, (woodland), swamp, predator; seasonal breeder in swamps, oviparous, free-living tadpole, arthropod-feeder. Biological references: Main, A.R. (1965). Frogs of southern Western Australia. *West. Aust. Nat. Club Handbook* 8: 1–73 (natural history); Lee, A.K. (1967). Studies in Australian Amphibia. II. Taxonomy, ecology, and evolution of the genus *Heleioporus* Gray (Anura : Leptodactylidae). *Aust. J. Zool.* **15**: 367–439; Lee, A.K. (1968). Water economy of the burrowing frog, *Heleioporus eyrei* (Gray). *Copeia* **1963**: 741–745 (physiology); King, M., Tyler, M.J., Davies, M. & King, D. (1979). Karyotypic studies on *Cyclorana* and associated genera of Australian frogs. *Aust. J. Zool.* **27**: 699–708.

Heleioporus inornatus (Lee and Main, 1954)

Helioporus inornatus Lee, A.K. & Main, A.R. (1954). Two new species of burrowing frogs of the genus *Helioporus* Gray from south western Australia. *West. Aust. Nat.* **4**: 156–158 [157]. Type data: holotype, WAM R11428, from Beechina, W.A.

Distribution: SW coastal, W.A. Ecology: terrestrial, burrowing, nocturnal, open forest, (woodland), swamp, predator; seasonal breeder in swamps, oviparous, free-living tadpole, arthropod-feeder. Biological references: Main, A.R. (1965). Frogs of southern Western Australia. *West. Aust. Nat. Club Handbook* 8: 1–73 (natural history); Lee, A.K. (1967). Studies in Australian Amphibia. II. Taxonomy, ecology, and evolution of the genus *Heleioporus* Gray (Anura : Leptodactylidae). *Aust. J. Zool.* **15**: 367–439.

Heleioporus psammophilus (Lee and Main, 1954)

Helioporus psammophilus Lee, A.K. & Main, A.R. (1954). Two new species of burrowing frogs of the genus *Helioporus* Gray from south western Australia. *West. Aust. Nat.* **4**: 156–158 [157]. Type data: holotype, WAM R11427, from Beechina, W.A.

Distribution: SW coastal, W.A. Ecology: terrestrial, burrowing, nocturnal, open forest, (woodland), swamp, predator; seasonal breeder in swamps, oviparous, free-living tadpole, arthropod-feeder. Biological references: Main, A.R. (1965). Frogs of southern Western Australia. *West. Aust. Nat. Club Handbook* 8: 1–73 (natural history); Lee, A.K. (1967). Studies in Australian Amphibia. II. Taxonomy, ecology, and evolution of the genus *Heleioporus* Gray (Anura : Leptodactylidae). *Aust. J. Zool.* **15**: 367–439.

Lechriodus Boulenger, 1882

Batrachopsis Boulenger, G.A. (1882). *Catalogue of the Batrachia Gradientia s. Caudata in the collection of the British Museum.* 2nd edn. London : British Museum xvi 503 pp. [439] [*non Batrachopsis* Fitzinger, 1843]. Type species *Asterophrys melanopyga* Doria, 1874 by monotypy, under Opinion 1063 of the International Commission on Zoological Nomenclature (1976).

Lechriodus Boulenger, G.A. (1882). *Catalogue of the Batrachia Gradientia s. Caudata in the collection of the British Museum.* 2nd edn. London : British Museum viii 127 pp. [116] [*nom. nov. pro Batrachopsis* Boulenger, 1882].

Phanerotis Boulenger, G.A. (1890). Description of a new genus of cystignathoid frogs from New South Wales. *Proc. Linn. Soc. N.S.W. (2)* **5**: 593–594 [593]. Type species *Phanerotis fletcheri* Boulenger, 1890 by monotypy.

Synonymy that of Cogger, H.G., this work.

This group is also found in New Guinea.

Lechriodus fletcheri (Boulenger, 1890)

Phanerotis fletcheri Boulenger, G.A. (1890). Description of a new genus of cystignathoid frogs from New South Wales. *Proc. Linn. Soc. N.S.W. (2)* **5**: 593–594 [594]. Type data: holotype, BMNH 1947.2.18.67, from Dunoon, Richmond River, N.S.W.

Distribution: SE coastal, NE coastal, N.S.W., Qld. Ecology: terrestrial, nocturnal, litter, closed forest, tall forest, (open forest), lentic freshwater, predator; saltatorial, seasonal breeder in lentic fresh water, oviparous, free-living tadpole, arthropod-feeder. Biological references: Moore, J.A. (1961). The frogs of eastern New South Wales. *Bull. Am. Mus. Nat. Hist.* **121**: 149–386 (taxonomy, natural history); Zweifel, R.G. (1972).

A review of the frog genus *Lechriodus* (Leptodactylidae) of New Guinea and Australia. *Am. Mus. Novit.* 2507: 1–41 (taxonomy).

Limnodynastes Fitzinger, 1843

Limnodynastes Fitzinger, L.J. (1843). *Systema Reptilium.* Vienna : Braümüller u. Seidel vi 106 pp. [31]. Type species *Cystign[athus] peronii* Duméril and Bibron, 1841 by original designation.

Wagleria Girard, C. (1853). Descriptions of new species of reptiles, collected by the U.S. Exploring Expedition, under the command of Capt. Charles Wilkes, U.S.N. Second Part. Including the species of batrachians, exotic to North America. *Proc. Acad. Nat. Sci. Philad.* **6**: 420–424 [421] [1854 on title page]. Type species *Cystign[athus] peronii* Duméril and Bibron, 1841 by original designation.

Platyplectrum Günther, A. (1863). On new species of batrachians from Australia. *Ann. Mag. Nat. Hist. (3)* **11**: 26–28 [27]. Type species *Platyplectrum marmoratum* Günther, 1863 by monotypy.

Platyplectron Peters, W. (1863). Übersicht der von Hrn. Richard Schomburgk an das zoologische Museum eingesandten Amphibien, aus Buchsfelde bei Adelaide in Südaustralien. *Mber. K. Preuss. Akad. Wiss. Berl.* **1863**: 228–236 [235] [described with subgeneric rank; emend. *pro Platyplectrum* Günther, 1863 (*vide* Neave, S.A. (ed.) (1940). *Nomenclator Zoologicus.* London : Zoological Society London Vol. 3 1065 pp.) without supporting evidence; here regarded as *subgen. nov.*]. Type species *Limnodynastes (Platyplectron) dumerilii* Peters, 1863 by monotypy.

Lymnodynastes Cope, E. D. (1865). Sketch of the primary groups of Batrachia Salientia. *Nat. Hist. Rev.* **5**: 97–120 [113] [invalid emend. *pro Limnodynastes* Fitzinger, 1843].

Heliorana Steindachner, F. (1867). Amphibien. *in, Reise der österreichischen Fregatte* Novara *um die Erde in den Jahren 1857, 1858, 1859.* Zoologie 1(4): 1–70 Vienna : State Printer [32] [1869 on title page]. Type species *Heliorana grayi* Steindachner, 1867 by monotypy.

Opisthodon Steindachner, F. (1867). Amphibien. *in, Reise der österreichischen Fregatte* Novara *um die Erde in den Jahren 1857, 1858, 1859.* Zoologie 1(4): 1–70 Vienna : State Printer [9] [1869 on title page]. Type species *Opisthodon frauenfeldi* Steindachner, 1867 by monotypy.

Ranaster Macleay, W. (1877). The batrachians of the *Chevert* Expedition. *Proc. Linn. Soc. N.S.W.* **2**: 135–138 [135]. Type species *Ranaster convexiusculus* Macleay, 1877 by monotypy.

Synonymy that of Cogger, H.G., this work.

This group is also found in southern New Guinea.

Limnodynastes convexiusculus (MacLeay, 1877)

Ranaster convexiusculus Macleay, W. (1877). The batrachians of the *Chevert* Expedition. *Proc. Linn. Soc. N.S.W.* **2**: 135 138 [136]. Type data: holotype, presumed lost, from Mawatta, Binaturi River, near Daru, Papua New Guinea (as Katow).

Limnodynastes olivaceus De Vis, C.W. (1884). On some new batrachians from Queensland. *Proc. Linn. Soc. N.S.W.* **9**: 65–68 [66] [1885 on title page]. Type data: holotype, BMNH 1947.2.18.78, from Mackay, Qld.

Phanerotis novaeguineae Kampen, P.N. van (1909). Die Amphibienfauna von Neu-Guinea, nach der Ausbeute der neiderlandischen Süd-Neu-Guinea-Expeditionen von 1904–1905 und 1907. *Nova Guinea* **9**: 31–49 [36 pl 2 fig 4]. Type data: syntypes, ZMA 5874, AM R6693, three syntypes not found, from Merauke, Irian Jaya, Indonesia.

Synonymy that of Cogger, H.G., this work.

Distribution: NE coastal, N Gulf, N coastal, Qld., W.A., N.T. Ecology: terrestrial, nocturnal, litter, open forest, woodland, swamp, predator; perennially active, seasonal breeder, oviparous, free-living tadpole, arthropod-feeder. Biological references: Cogger, H.G. & Lindner, D.A. (1974). Frogs and reptiles. *in* Fauna survey of the Port Essington District, Cobourg Peninsula, Northern Territory of Australia. Frith, H.J. & Calaby, J.H. (eds.) *CSIRO Div. Wildl. Res. Tech. Pap.* **28**: 63–107 (natural history).

Limnodynastes depressus Tyler, 1976

Limnodynastes depressus Tyler, M.J. (1976). A new genus and two new species of leptodactylid frogs from Western Australia. *Rec. West. Aust. Mus.* **4**: 45–52 [50 fig 3]. Type data: holotype, WAM R43896, from near Argyle Downs homestead (now submerged in Lake Argyle), Ord River, W.A.

Distribution: N coastal, W.A.

Limnodynastes dorsalis (Gray, 1841)

Cystignathus dorsalis Gray, J.E. (1841). Description of some new species and four new genera of reptiles from Western Australia, discovered by John Gould, Esq. *Ann. Mag. Nat. Hist.* **7**: 86–91 [91]. Type data: holotype, BMNH 1947.2.19.7, from W.A.

Limnodynastes dorsalis typica Fry, D.B. (1913). On a *Varanus* and a frog from Burnett River, Queensland, and a revision of the variations in *Limnodynastes dorsalis* Gray. *Rec. Aust. Mus.* **10**: 17–34 [24]. Type data: that of *Limnodynastes dorsalis* Gray, 1841.

Synonymy that of Cogger, H.G., this work.

Distribution: SW coastal, W plateau, W.A. Ecology: terrestrial, burrowing, nocturnal, open forest, woodland, lentic freshwater, lotic freshwater, swamp, predator; seasonal breeder in lentic and lotic freshwater, oviparous, free-living tadpole, arthropod-feeder. Biological references: Main, A.R. (1965). Frogs of southern Western Australia. *West. Aust. Nat. Club Handbook* 8: 1–73 (natural history); Martin, A.A. (1972). Studies in Australian Amphibia. III. The *Limnodynastes dorsalis* complex (Anura : Leptodactylidae). *Aust. J. Zool.* **20**: 165–211 (taxonomy).

Limnodynastes dumerilii Peters, 1863

Limnodynastes (Platyplectron) dumerilii Peters, W. (1863). Übersicht der von Hrn. Richard Schomburgk an das zoologische Museum eingesandten Amphibien, aus Buchsfelde bei Adelaide in Südaustralien. *Mber. K. Preuss. Akad. Wiss. Berl.* **1863**: 228–236 [235] [1864 on title page]. Type data: syntypes ZMB 4727 (3 specimens, not found), ZMB 4728 (3 specimens), AMNH A23590, from Loos, 4.5 km W Gawler (as Buchsfelde), S.A., see, Praite, R. & Tolley, J.C. (1970). *Place Names of South Australia.* Adelaide : Rigby 203 pp.

Limnodynastes bibronii Kre[f]ft, G. (1865). The frogs of Australia. *Pap. Proc. R. Soc. Tas.* **1865**: 16–20 [16] [*lapsus pro Limnodynastes dumerilii* Peters, 1863].

Platyplectrum superciliare Keferstein, W. (1867). Ueber einige neue oder seltene Batrachier aus Australien und dem tropischen Amerika. *Nachr. Ges. Wiss. Göttingen* **1867**: 341–361 [346]. Type data: holotype, ZFMK 28331, from Australia.

Heliorana grayi Steindachner, F. (1867). Amphibien. *in, Reise der österreichischen Fregatte* Novara *um die Erde in den Jahren 1857, 1858, 1859.* Zoologie 1(4): 1–70 Vienna : State Printer [32 pl 2 figs 11–14] [1869 on title page]. Type data: holotype, NHMW 4695, from N.S.W.

Limnodynastes dorsalis insularis Parker, H.W. (1940). The Australasian frogs of the family Leptodactylidae. *Novit. Zool.* **42**: 1–106 [47]. Type data: syntypes, BMNH 1947.2.19.8 M immature (Launceston), MCZ 19238 F adult (Eagle Hawk Neck), MCZ 19330 M adult (Tasmania), from Eagle Hawk Neck, near Launceston, Tas.

Limnodynastes dumerili variegatus Martin, A.A. (1972). Studies in Australian Amphibia. III. The *Limnodynastes dorsalis* complex (Anura : Leptodactylidae). *Aust. J. Zool.* **20**: 165–211 [181 fig 7B]. Type data: holotype, AM R32742, from 6.4 km N of Cape Otway, Vic.

Limnodynastes dumerili fryi Martin, A.A. (1972). Studies in Australian Amphibia. III. The *Limnodynastes dorsalis* complex (Anura : Leptodactylidae). *Aust. J. Zool.* **20**: 165–211 [182]. Type data: holotype, AM R32747, from 9.6 km NE of Thredbo Village, Snowy Mts., N.S.W.

Synonymy that of Cogger, H.G., this work.

Distribution: SE coastal, S Gulfs, Murray-Darling basin, S.A., Vic., N.S.W., Qld., A.C.T. Ecology: terrestrial, burrowing, nocturnal, (tall forest), open forest, woodland, (open heath), (alpine), lentic freshwater, lotic freshwater, swamp, predator; breeds in lentic and lotic freshwater, seasonal breeder, oviparous, free-living tadpole, arthropod-feeder. Biological references: Martin, A.A. (1972). Studies in Australian Amphibia. III. The *Limnodynastes dorsalis* complex (Anura : Leptodactylidae). *Aust. J. Zool.* **20**: 165–211 (taxonomy); Littlejohn, M.J. & Martin, A.A. (1974). The Amphibia of Tasmania. pp. 251–289 *in* Williams, W.D. (ed.) *Biogeography and Ecology in Tasmania.* The Hague : Junk; Crook, G.A. & Tyler, M.J. (1981). Structure and function of the tibial gland of the Australian frog *Limnodynastes dumerilii* Peters. *Trans. R. Soc. S. Aust.* **105**: 49–52; Tyler, M.J. (1978). *Amphibians of South Australia.* Adelaide : South Australian Government 84 pp. (natural history); Moore, J.A. (1961). The frogs of eastern New South Wales. *Bull. Am. Mus. Nat. Hist.* **121**: 149–386 (taxonomy, natural history, as *L. dorsalis*).

Limnodynastes fletcheri Boulenger, 1888

Limnodynastes fletcheri Boulenger, G.A. (1888). Descriptions of two new Australian frogs. *Ann. Mag. Nat. Hist. (6)* **2**: 142–143 [142]. Type data: syntypes, BMNH 1947.2.19.9–10, from Guntawang, near Mudgee, N.S.W.

Limnodynastes marmoratus Lamb, J. (1911). Description of three new batrachians from southern Queensland. *Ann. Qd. Mus.* **10**: 26–28 [28]. Type data: holotype, QM J12597, from Goondiwindi, Qld.

Synonymy that of Cogger, H.G., this work.

Distribution: Murray-Darling basin, Bulloo River basin, Lake Eyre basin, S.A., Vic., N.S.W., Qld. Ecology: terrestrial, burrowing, nocturnal, woodland, low woodland, low open woodland, open scrub, swamp, predator; seasonal breeder in swamps, oviparous, free-living tadpole, arthropod-feeder. Biological references: Moore, J.A. (1961). The frogs of eastern New South Wales. *Bull. Am. Mus. Nat. Hist.* **121**: 149–386 (natural history); Tyler, M.J. (1978). *Amphibians of South Australia.* Adelaide : South Australian Government 84 pp. (natural history).

Limnodynastes interioris Fry, 1913

Limnodynastes dorsalis interioris Fry, D.B. (1913). On a *Varanus* and a frog from Burnett River, Queensland, and a revision of the variations in *Limnodynastes dorsalis* Gray. *Rec. Aust. Mus.* **10**: 17–34 [33 pl 3 fig 1]. Type data: holotype, AM R5869, from Merool Creek, Riverina, N.S.W.; [incorrectly cited as AM R9615 by Martin, A.A. (1972). Studies in Australian Amphibia. III. The *Limnodynastes dorsalis* complex (Anura : Leptodactylidae). *Aust. J. Zool.* **20**: 165–211].

Distribution: Murray-Darling basin, Vic., N.S.W. Ecology: terrestrial, burrowing, nocturnal, woodland, low woodland, low open woodland, open scrub, low shrubland, swamp, predator; seasonal breeder in swamps, oviparous, free-living tadpole, arthropod-feeder. Biological references: Moore, J.A. (1961). The frogs of eastern New South Wales. *Bull. Am. Mus. Nat. Hist.* **121**: 149–386 (natural history); Martin, A.A. (1972). Studies in Australian Amphibia. III. The *Limnodynastes dorsalis* complex (Anura : Leptodactylidae). *Aust. J. Zool.* **20**: 165–211 (taxonomy).

Limnodynastes ornatus (Gray, 1842)

Discoglossus ornatus Gray, J.E. (1842). Description of some hitherto unrecorded species of Australian reptiles and batrachians. *in* Gray, J.E. (ed.) *Zoological Miscellany*. London : Treuttel, Würtz & Co. pp. 51–57 [56]. Type data: holotype, BMNH 1947.2.19.11, from Port Essington, N.T.

Platyplectrum marmoratum Günther, A. (1863). On new species of batrachians from Australia. *Ann. Mag. Nat. Hist. (3)* **11**: 26–28 [27 pl 4 fig A]. Type data: holotype, BMNH 1947.2.19.12, from Clarence River, N.S.W.

P[latyplectrum] occidentale Cope, E.D. (1866). On the structure and distribution of the genera of arciferous Anura. *J. Acad. Nat. Sci. Philad. (2)* **6**: 67–112 [90]. Type data: syntypes (probable), not located, from W.A.

Opisthodon frauenfeldi Steindachner, F. (1867). Amphibien. *in, Reise der österreichischen Fregatte* Novara *um die Erde in den Jahren 1857, 1858, 1859.* Zoologie 1(4): 1–70 Vienna : State Printer [9 pl 1 figs 1–4] [1869 on title page]. Type data: syntypes, NHMW 4699 (2 specimens), from Rockhampton, Qld. (as N.S.W.).

Synonymy that of Cogger, H.G., this work.

Distribution: SE coastal, NE coastal, N Gulf, N coastal, Murray-Darling basin, Lake Eyre basin, N.S.W., Qld., W.A., N.T. Ecology: terrestrial, burrowing, nocturnal, open forest, woodland, low woodland, low open woodland, lentic freshwater, swamp, predator; seasonal breeder in lentic freshwater and swamps, oviparous, free-living tadpole, arthropod-feeder. Biological references: Moore, J.A. (1961). The frogs of eastern New South Wales. *Bull. Am. Mus. Nat. Hist.* **121**: 149–386 (taxonomy, natural history).

Limnodynastes peronii (Duméril and Bibron, 1841)

Cystignathus peronii Duméril, A.M.C. & Bibron, G. (1841). *Erpétologie Générale* ou Histoire Naturelle Complète des Reptiles. Paris : Roret 8: ii 792 pp. [409]. Type data: syntypes, MNHP 4499 (2 specimens), MNHP 4500, all from Australia?, MNHP 4500 erroneously cited as holotype by Guibé, J. (1950). *Catalogue des types d'Amphibiens du Muséum National d'Histoire Naturelle.* Paris : Imprimerie Nationale 71 pp.

Limnodynastes krefftii Günther, A. (1863). On new species of batrachians from Australia. *Ann. Mag. Nat. Hist. (3)* **11**: 26–28 [26]. Type data: syntypes, BMNH 1947.2.18.71 M adult, BMNH 1947.2.18.72–74 young, BMNH 1947.2.18.75 F adult, from Sydney, N.S.W.

Limnodynastes kreftii Steindachner, F. (1867). Amphibien. *in, Reise der österreichischen Fregatte* Novara *um die Erde in den Jahren 1857, 1858, 1859.* Zoologie 1(4): 1–70 Vienna : State Printer [26] [1869 on title page; invalid emend. *errore pro Limnodynastes krefftii* Günther, 1863].

Limnodynastes lineatus De Vis, C.W. (1884). On some new batrachians from Queensland. *Proc. Linn. Soc. N.S.W.* **9**: 65–68 [65] [1885 on title page]. Type data: syntypes, BMNH 1947.2.18.76–77, from Mackay, Qld.

Synonymy that of Cogger, H.G., this work.

Distribution: SE coastal, NE coastal, Murray-Darling basin, Tas., Vic., N.S.W., Qld., A.C.T. Ecology: terrestrial, nocturnal, tall forest, open forest, woodland, lentic freshwater, swamp, predator; seasonal breeder in swamps and lentic freshwater, oviparous, free-living tadpole, arthropod-feeder. Biological references: Moore, J.A. (1961). The frogs of eastern New South Wales. *Bull. Am. Mus. Nat. Hist.* **121**: 149–386 (taxonomy, natural history); Littlejohn, M.J. & Martin, A.A. (1974). The Amphibia of Tasmania. pp. 251–289 *in* Williams, W.D. (ed.) *Biogeography and Ecology in Tasmania.* The Hague : Junk; Tyler, M.J. (1978). *Amphibians of South Australia.* Adelaide : South Australian Government 84 pp. (natural history).

Limnodynastes salmini Steindachner, 1867

Limnodynastes salmini Steindachner, F. (1867). Amphibien. *in, Reise der österreichischen Fregatte* Novara *um die Erde in den Jahren 1857, 1858, 1859.* Zoologie 1(4): 1–70 Vienna : State Printer [27 pl 4 figs 12–15] [1869 on title page]. Type data: syntypes, NHMW 14849 (2 specimens), from Cape York, Qld. (as Australia), see Haupl, M. & Tiedemann, F. (1978). Typenkatalog der Herpetologischen Sammlung. Amphibia. *Kat. Wiss. Samml. Naturhist. Mus. Wien* **2** (Vertebrata 1) :1–35.

Distribution: SE coastal, NE coastal, Murray-Darling basin, N.S.W., Qld. Ecology: terrestrial, nocturnal, open forest, woodland, low woodland, (low open woodland), swamp, predator; seasonal breeder in swamps, oviparous, free-living tadpole, arthropod-feeder. Biological references: Moore, J.A. (1961). The frogs of eastern New South Wales. *Bull. Am. Mus. Nat. Hist.* **121**: 149–386 (taxonomy, natural history).

Limnodynastes spenceri Parker, 1940

Limnodynastes spenceri Parker, H.W. (1940). The Australasian frogs of the family Leptodactylidae. *Novit. Zool.* **42**: 1–106 [50]. Type data: holotype, BMNH 1947.2.18.79, from Alice Springs, N.T.

Distribution: NW coastal, W plateau, Lake Eyre basin, S.A., W.A., N.T., (Qld.). Ecology: terrestrial, burrowing, nocturnal, desert, tall shrubland, predator; opportunistic breeder, oviparous, free-living tadpole, arthropod-feeder. Biological references: Moore, J.A. (1961). The frogs of eastern New South Wales. *Bull. Am. Mus. Nat. Hist.* **121**: 149–386 (taxonomy, natural history); Main, A.R. (1965). Frogs of southern Western Australia. *West. Aust. Nat. Club Handbook* 8: 1–73 (natural history); Main, A.R. & Storr, G.M. (1966). Range extension and notes on the biology of frogs from the Pilbara Region, Western Australia. *West. Aust. Nat.* **10**: 53–61 (distribution); Tyler, M.J. (1978). *Amphibians of South Australia.* Adelaide : South Australian Government 84 pp. (natural history).

Limnodynastes tasmaniensis Günther, 1858

Limnodynastes tasmaniensis Günther, A. (1858). *Catalogue of the Batrachia Salientia in the collection of the British Museum.* London : British Museum xvi 160 pp. [33]. Type data: syntypes, BMNH 1947.2.18.96–99, BMNH 1947.2.19.1–3 (7 specimens, but original description cites only 6 specimens), from Tas. and Australia.

Limnodynastes affinis Günther, A. (1863). On new species of batrachians from Australia. *Ann. Mag. Nat. Hist. (3)* **11**: 26–28 [27]. Type data: holotype, BMNH 1947.2.19.6, from Clarence River, N.S.W.

Limnodynastes platycephalus Günther, A. (1867). Additions to the knowledge of Australian reptiles and fishes. *Ann. Mag. Nat. Hist. (3)* **20**: 45–68 [54]. Type data: syntypes, BMNH 1947.2.19.4–5, from Adelaide, S.A.

Limnodynastes peronii rugulosus Keferstein, W. (1867). Ueber einige neue oder seltene Batrachier aus Australien und dem tropischen Amerika. *Nachr. Ges. Wiss. Göttingen* **1867**: 341–361 [344]. Type data: syntypes, ZFMK 28332–28337, from Sydney, N.S.W.

Limnodynastes tasmanicus McCoy, F. (1867). On the recent zoology and palaeontology of Victoria. *Ann. Mag. Nat. Hist. (3)* **20**: 175–202 [182] [invalid emend. *lapsus pro Limnodynastes tasmaniensis* Günther, 1858].

Synonymy that of Cogger, H.G., this work.

Distribution: SE coastal, NE coastal, N coastal, W plateau, N Gulf, S Gulfs, Murray-Darling basin, Bulloo River basin, Lake Eyre basin, S.A., Tas., Vic., N.S.W., Qld., W.A., A.C.T. Ecology: terrestrial, nocturnal, tall forest, open forest, woodland, low woodland, low open woodland, (open scrub), swamp, predator; seasonal breeder, oviparous, free-living tadpole, arthropod-feeder. Biological references: Moore, J.A. (1961). The frogs of eastern New South Wales. *Bull. Am. Mus. Nat. Hist.* **121**: 149–386 (taxonomy, natural history); Tyler, M.J. (1971). Voluntary control of the shape of the inflated vocal sac by the Australian leptodactylid frog *Limnodynastes tasmaniensis. Trans. R. Soc. S. Aust.* **95**: 49–52; Littlejohn, M.J. & Martin, A.A. (1974). The Amphibia of Tasmania. pp. 251–289 *in* Williams, W.D. (ed.) *Biogeography and Ecology in Tasmania.* The Hague : Junk; Martin, A.A. & Tyler, M.J. (1978). The introduction into Western Australia of the frog *Limnodynastes tasmaniensis* Günther. *Aust. Zool.* **19**: 321–325 (distribution); Tyler, M.J. (1978). *Amphibians of South Australia.* Adelaide : South Australian Government 84 pp. (natural history).

Limnodynastes terraereginae Fry, 1915

Limnodynastes dorsalis terraereginae Fry, D.B. (1915). Herpetological notes. *Proc. R. Soc. Qd.* **27**: 60–95 [67 fig 2a]. Type data: holotype, AM R4525, from Somerset, Cape York, Qld.

Distribution: SE coastal, NE coastal, N Gulf, Murray-Darling basin, N.S.W., Qld. Ecology: terrestrial, burrowing, nocturnal, closed forest, tall forest, open forest, woodland, lentic freshwater, lotic freshwater, predator; seasonal breeder in lentic and lotic freshwater, oviparous, free-living tadpole, arthropod-feeder. Biological references: Martin, A.A. (1972). Studies in Australian Amphibia. III. The *Limnodynastes dorsalis* complex (Anura : Leptodactylidae). *Aust. J. Zool.* **20**: 165–211 (taxonomy).

Megistolotis Tyler, Martin and Davies, 1979

Megistolotis Tyler, M.J., Martin, A.A. & Davies, M. (1979). Biology and systematics of a new limnodynastine genus (Anura : Leptodactylidae) from north-western Australia. *Aust. J. Zool.* **27**: 135–150 [136]. Type species *Megistolotis lignarius* Tyler, Martin and Davies, 1979 by original designation.

Megistolotis lignarius Tyler, Martin and Davies, 1979

Megistolotis lignarius Tyler, M.J., Martin, A.A. & Davies, M. (1979). Biology and systematics of a new limnodynastine genus (Anura : Leptodactylidae) from north-western Australia. *Aust. J. Zool.* **27**: 135–150 [137 figs 1–2,5–6,8]. Type data: holotype, WAM R58299, from Lake Argyle - Kununurra road, 6.5 km N of Lake Argyle Tourist Village, Lake Argyle, Kimberley Division, W.A.

Distribution: N coastal, W.A., N.T. Ecology: terrestrial, nocturnal, (open forest), woodland, lotic freshwater, predator; seasonal breeder in lotic freshwater, oviparous, free-living tadpole, arthropod-feeder. Biological references: Cogger, H.G. (1979). *Reptiles and Amphibians of Australia.* Sydney : A.H. & A.W. Reed 608 pp. (as *Heleioporus* sp.).

Metacrinia Parker, 1940

Metacrinia Parker, H.W. (1940). The Australasian frogs of the family Leptodactylidae. *Novit. Zool.* **42**: 1–106 [93]. Type species *Pseudophryne nichollsi* Harrison, 1927 by original designation.

Metracrinia Parker, H.W. (1940). The Australasian frogs of the family Leptodactylidae. *Novit. Zool.* **42**: 1–106 [93] [incorrect original spelling].

Synonymy that of Cogger, H.G., this work.

Metacrinia nichollsi (Harrison, 1927)

Pseudophryne nichollsi Harrison, L. (1927). Notes on some western Australian frogs, with descriptions of new species. *Rec. Aust. Mus.* **15**: 277–287 [284]. Type data: syntypes, MMUS (not found), from Pemberton, W.A.

Distribution: SW coastal, W.A. Ecology: terrestrial, nocturnal, tall forest, (open forest), predator; seasonal breeder, oviparous, (direct development), arthropod-feeder, cryptozoic. Biological references: Main, A.R. (1965). Frogs of southern Western Australia. *West. Aust. Nat. Club Handbook* 8: 1–73 (natural history); Blake, A.J.D. (1973). Taxonomy and relationships of myobatrachine frogs (Leptodactylidae): a numerical approach. *Aust. J. Zool.* **21**: 119–149 (taxonomy).

Mixophyes Günther, 1864

Mixophyes Günther, A. (1864). Third contribution to our knowledge of batrachians from Australia. *Proc. Zool. Soc. Lond.* **1864**: 46–49 pl 7 [46]. Type species *Mixophyes fasciolatus* Günther, 1864 by monotypy.

Myxophyes Cope, E.D. (1864). On the limits and relations of the Raniformes. *Proc. Acad. Nat. Sci. Philad.* **1864**: 181–183 [182] [invalid emend. *pro Mixophyes* Günther, 1864].

Synonymy that of Cogger, H.G., this work.

Mixophyes balbus Straughan, 1968

Mixophyes balbus Straughan, I.R. (1968). A taxonomic review of the genus *Mixophyes* (Anura : Leptodactylidae). *Proc. Linn. Soc. N.S.W.* **93**: 52–59 [55]. Type data: holotype, AM R25922, from Point Lookout, New England National Park, N.S.W.

Distribution: SE coastal, NE coastal, Vic., N.S.W., Qld. Ecology: terrestrial, nocturnal, closed forest, tall forest, open forest, lotic freshwater, predator; saltatorial, seasonal breeder in lotic freshwater, oviparous, free-living tadpole, arthropod-feeder.

Mixophyes fasciolatus Günther, 1864

Mixophyes fasciolatus Günther, A. (1864). Third contribution to our knowledge of batrachians from Australia. *Proc. Zool. Soc. Lond.* **1864**: 46–49 pl 7 [46 pl 7 fig 1]. Type data: syntypes, BMNH 1947.2.19.89–90, from Clarence River, N.S.W.

Hyla fenestrata De Vis, C.W. (1884). On a new species of *Hyla. Proc. R. Soc. Qd.* **1**: 128–130 [128]. Type data: syntypes (probable), QM (presumed lost), from Tweed River, N.S.W.

Synonymy that of Cogger, H.G., this work.

Distribution: SE coastal, NE coastal, N.S.W., Qld. Ecology: terrestrial, nocturnal, closed forest, tall forest, open forest, lotic freshwater, predator; saltatorial, seasonal breeder in lotic freshwater, oviparous, free-living tadpole, arthropod-feeder. Biological references: Moore, J.A. (1961). The frogs of eastern New South Wales. *Bull. Am. Mus. Nat. Hist.* **121**: 149–386 (taxonomy, natural history); Straughan, I.R. (1968). A taxonomic review of the genus *Mixophyes* (Anura : Leptodactylidae). *Proc. Linn. Soc. N.S.W.* **93**: 52–59 (taxonomy); Wotherspoon, D. (1981). The great barred frog - mouth breeder? *Herpetofauna* **12**: 30 (reproduction).

Mixophyes iteratus Straughan, 1968

Mixophyes iteratus Straughan, I.R. (1968). A taxonomic review of the genus *Mixophyes* (Anura : Leptodactylidae). *Proc. Linn. Soc. N.S.W.* **93**: 52–59 [54]. Type data: holotype, AM R25929, from Tweed River, Mt. Warning, N.S.W.

Distribution: SE coastal, NE coastal, (Murray-Darling basin), N.S.W., Qld. Ecology: terrestrial, nocturnal, closed forest, tall forest, open forest, lotic freshwater, predator; saltatorial, seasonal breeder in lotic freshwater, oviparous, free-living tadpole, arthropod-feeder. Biological references: King, M., Tyler, M.J., Davies, M. & King, D. (1979). Karyotypic studies on *Cyclorana* and associated genera of Australian frogs. *Aust. J. Zool.* **27**: 699–708.

Mixophyes schevilli Loveridge, 1933

Mixophyes fasciolatus schevilli Loveridge, A. (1933). Four new crinine frogs from Australia. *Occ. Pap. Bost. Soc. Nat. Hist.* **8**: 55–60 [55]. Type data: holotype, MCZ 18150, from Millaa Millaa, Qld.

Distribution: NE coastal, Qld. Ecology: terrestrial, nocturnal, closed forest, tall forest, lotic freshwater, predator; saltatorial, seasonal breeder in lotic freshwater, oviparous, free-living tadpole, arthropod-feeder. Biological references: Straughan, I.R. (1968). A taxonomic review of the genus *Mixophyes* (Anura : Leptodactylidae). *Proc. Linn. Soc. N.S.W.* **93**: 52–59 (taxonomy); King, M., Tyler, M.J., Davies, M. & King, D. (1979). Karyotypic studies on *Cyclorana* and associated genera of Australian frogs. *Aust. J. Zool.* **27**: 699–708.

Myobatrachus Schlegel, 1850

Myobatrachus Schlegel, H. (1850). Description of a new genus of batrachians from Swan River. *Proc. Zool. Soc. Lond.* **1850**: 9–10 [9]. Type species *Myobatrachus paradoxus* Schlegel, 1850 by monotypy.

Myiobatrachus Schlegel, H. (1858). *Handleiding tot de beoefening der Dierkunde.* Vol. 2 Breda : Koninklijke Militaire Akademie xii 532 pp. [59] [invalid emend. *pro Myobatrachus* Schlegel, 1850].

Chelydobatrachus Günther, A. (1858). *Catalogue of the Batrachia Salientia in the collection of the British Museum.* London : British Museum xvi 160 pp. [53]. Type species *Breviceps gouldii* Gray, 1841 by monotypy.

Synonymy that of Cogger, H.G., this work.

Myobatrachus gouldii (Gray, 1841)

Breviceps gouldii Gray, J.E. (1841). Description of some new species and four new genera of reptiles from Western Australia, discovered by John Gould, Esq. *Ann. Mag. Nat. Hist.* **7**: 86–91 [89]. Type data: holotype, BMNH 1947.2.20.5, from Western Australia.

Breviceps heliogabali Gray, J.E. (1841). List of Illustrations. Reptiles and Amphibia. *in* Grey, G. *Journals of Two Expeditions of Discovery in North-west and Western Australia, during the years 1837, 38, and 39, under the authority of Her Majesty's Government.* London : T. & W. Boone 1: xiv [*nom. nud.*].

Myobatrachus paradoxus Schlegel, H. (1850). Description of a new genus of batrachians from Swan River. *Proc. Zool. Soc. Lond.* **1850**: 9–10 [10]. Type data: syntypes (probable), RMNH (not found), from Swan River, W.A.

Synonymy that of Cogger, H.G., this work.

Distribution: (NW coastal), SW coastal, (W plateau), W.A. Ecology: terrestrial, burrowing, nocturnal, open forest, predator; walking, oviparous, seasonal breeder, direct development, feeds primarily on ant and termite larvae and pupae. Biological references: Calaby, J.H. (1956). The food habits of the frog, *Myobatrachus gouldii* (Gray). *West. Aust. Nat.* **5**: 93–96; Main, A.R. (1965). Frogs of southern Western Australia. *West. Aust. Nat. Club Handbook* 8: 1–73 (natural history); Roberts, J.D. (1981). Terrestrial breeding in the Australian leptodactylid frog *Myobatrachus gouldii* (Gray). *Aust. Wildl. Res.* **8**: 451–462.

Neobatrachus Peters, 1863

Neobatrachus Peters, W. (1863). Übersicht der von Hrn. Richard Schomburgk an das zoologische Museum eingesandten Amphibien, aus Buchsfelde bei Adelaide in Südaustralien. *Mber. K. Preuss. Akad. Wiss. Berl.* **1863**: 228–236 [234] [1864 on title page]. Type species *Neobatrachus pictus* Peters, 1863 by monotypy.

Neobatrachus aquilonius Tyler, Davies and Martin, 1981

Neobatrachus aquilonius Tyler, M.J., Davies, M. & Martin, A.A. (1981). New and rediscovered species of frogs from the Derby - Broome area of Western Australia. *Rec. West. Aust. Mus.* **9**: 147–172 [155 figs 6–7]. Type data: holotype, WAM R71005, from between 22 and 41 km S of Derby, Kimberley Division, W.A.

Distribution: N coastal, W.A. Ecology: terrestrial, burrowing, nocturnal, predator; oviparous, free-living tadpole, arthropod-feeder.

Neobatrachus centralis (Parker, 1940)

Heleioporus centralis Parker, H.W. (1940). The Australasian frogs of the family Leptodactylidae. *Novit. Zool.* **42**: 1–106 [35]. Type data: holotype, BMNH 1947.2.19.13, from 100 mi E of Lake Eyre, S.A.

Distribution: NW coastal, SW coastal, W plateau, (S Gulfs), Murray-Darling basin, Bulloo River basin, Lake Eyre basin, S.A., Vic., N.S.W., Qld., W.A., N.T. Ecology: terrestrial, burrowing, nocturnal, low open woodland, (open scrub), tall shrubland, desert, swamp, predator; opportunistic breeder, oviparous, free-living tadpole, arthropod-feeder. Biological references: Main, A.R. (1965). Frogs of southern Western Australia. *West. Aust. Nat. Club Handbook* 8: 1–73 (natural history); Tyler, M.J. (1978). *Amphibians of South Australia.* Adelaide : South Australian Government 84 pp. (natural history); Mahoney, M.J. & Robinson, E.S. (1980). Polyploidy in the Australian leptodactylid frog genus *Neobatrachus. Chromosoma (Berl.)* **81**: 199–212 (karyology); Roberts, J.D. (1978). Redefinition of the Australian leptodactylid frog *Neobatrachus pictus* Peters. *Trans. R. Soc. S. Aust.* **102**: 97–105 (taxonomy); King, M., Tyler, M.J., Davies, M. & King, D. (1979). Karyotypic studies on *Cyclorana* and associated genera of Australian frogs. *Aust. J. Zool.* **27**: 699–708; Warburg, M.R. (1965). Studies on the water economy of some Australian frogs. *Aust. J. Zool.* **13**: 317–330.

Neobatrachus pelobatoides (Werner, 1914)

Heleioporus albopunctatus pelobatoides Werner, F. (1914). Amphibia. *in* Michaelsen, W. & Hartmeyer, R. (eds.) *Die Fauna Südwest-Australiens.* Jena : Gustav Fischer 4: 403–426 [418]. Type data: syntypes (probable), not found, from Beverley and Broome Hill, W.A.

Distribution: NW coastal, SW coastal, W plateau, W.A. Ecology: terrestrial, burrowing, nocturnal, open forest, woodland, (low woodland), swamp, predator; seasonal breeder in swamps, oviparous, free-living tadpole, arthropod-feeder. Biological references: Main, A.R. (1965). Frogs of southern Western Australia. *West. Aust. Nat. Club Handbook* 8: 1–73 (natural history).

Neobatrachus pictus Peters, 1863

Neobatrachus pictus Peters, W. (1863). Übersicht der von Hrn. Richard Schomburgk an das zoologische Museum eingesandten Amphibien, aus Buchsfelde bei Adelaide in Südaustralien. *Mber. K. Preuss. Akad. Wiss. Berl.* **1863**: 228–236 [235] [1864 on title page; Roberts, J.D. ((1978). Redefinition of the Australian leptodactylid frog *Neobatrachus pictus* Peters. *Trans. R. Soc. S. Aust.* **102**: 97–105 includes ZMB 9705 (one *Neobatrachus pictus* and one *Notaden melanoscaphus* as part of syntype series, but ZMB records are equivocal on this point]. Type data: syntypes, ZMB 4725–6, from Loos, 4.5 km W Gawler, S.A. (as Buchsfelde), see Praite, R. & Tolley, J.C. (1970). *Place Names of South Australia.* Adelaide : Rigby 203 pp.

Neobatrachus fictus Kre[f]ft, G. (1865). The frogs of Australia. *Pap. Proc. R. Soc. Tas.* **1865**: 16–20 [17] [invalid emend. *errore pro Neobatrachus pictus* Peters, 1863].

Synonymy that of Cogger, H.G., this work.

Distribution: W plateau, S Gulfs, Murray-Darling basin, Lake Eyre basin, S.A., Vic., (N.S.W.). Ecology: terrestrial, burrowing, nocturnal, open forest, woodland, low woodland, low open woodland, open scrub, swamp, predator; seasonal breeder in swamps, oviparous, free-living tadpole, arthropod-feeder. Biological references: Moore, J.A. (1961). The frogs of eastern New South Wales. *Bull. Am. Mus. Nat. Hist.* **121**: 149–386 (taxonomy, natural history); Tyler, M.J. (1978). *Amphibians of South Australia.* Adelaide : South Australian Government 84 pp. (natural history); Mahoney, M.J. & Robinson, E.S. (1980). Polyploidy in the Australian leptodactylid frog genus *Neobatrachus. Chromosoma (Berl.)* **81**: 199–212 (karyology); Warburg, M.R. (1965). Studies on the water economy of some Australian frogs. *Aust. J. Zool.* **13**: 317–330.

Neobatrachus sudelli (Lamb, 1911)

Heleioporus sudelli Lamb, J. (1911). Description of three new batrachians from southern Queensland. *Ann. Qd. Mus.* **10**: 26–28 [26]. Type data: holotype, QM J78, from Warwick, Qld.

Distribution: Murray-Darling basin, Lake Eyre basin, Bulloo River basin, (Vic.), N.S.W., Qld., (A.C.T.). Ecology: terrestrial, burrowing, nocturnal, open forest, woodland, low woodland, low open woodland, swamp, predator; seasonal breeder in swamps, oviparous, free-living tadpole, arthropod-feeder. Biological references: Roberts, J.D. (1978). Redefinition of the Australian leptodactylid frog *Neobatrachus pictus* Peters. *Trans. R. Soc. S. Aust.* **102**: 97–105 (taxonomy); Hosmer, W. (1958). A note on the identity of *Heleioporus sudelli* Lamb. *N. Qd. Nat.* **121** : 1–2.

Neobatrachus sutor Main, 1957

Neobatrachus sutor Main, A.R. (1957). A new burrowing frog from Western Australia. *West. Aust. Nat.* **6**: 23–24 [24]. Type data: holotype, WAM R32892, from Gnoolowa Hill, 18 mi N of Mingenew, W.A.

Distribution: NW coastal, SW coastal, W plateau, W.A. Ecology: terrestrial, burrowing, nocturnal, (woodland), low open woodland, (open scrub), tall shrubland, low shrubland, desert, low open shrubland, swamp, predator; breeds in swamps, oviparous, free-living tadpole, arthropod-feeder. Biological references: Main, A.R. (1965). Frogs of southern Western Australia. *West. Aust. Nat. Club Handbook* 8: 1–73 (natural history).

Neobatrachus wilsmorei (Parker, 1940)

Heleioporus wilsmorei Parker, H.W. (1940). The Australasian frogs of the family Leptodactylidae. *Novit. Zool.* **42**: 1–106 [36]. Type data: holotype, BMNH 1947.2.18.68, from Wiraga, N of Yalgoo, Murchison, W.A.

Distribution: NW coastal, W plateau, W.A. Ecology: terrestrial, burrowing, nocturnal, woodland, low woodland, tall shrubland, (low shrubland), open scrub, swamp, predator; breeds in swamps, oviparous, free-living tadpole, arthropod-feeder. Biological references: Main, A.R. (1965). Frogs of southern Western Australia. *West. Aust. Nat. Club Handbook* 8: 1–73 (natural history).

Notaden Günther, 1873

Notaden Günther, A. (1873). Description of two new species of frogs from Australia. *Ann. Mag. Nat. Hist. (4)* **11**: 349–350 [349]. Type species *Notaden bennettii* Günther, 1873 by monotypy.

Notaden bennettii Günther, 1873

Notaden bennettii Günther, A. (1873). Description of two new species of frogs from Australia. *Ann. Mag. Nat. Hist. (4)* **11**: 349–350 [350]. Type data: lectotype, BMNH 1947.2.18.56, from Castlereagh River, N.S.W., designation by Cogger, H.G. this work.

Distribution: Murray-Darling basin, Bulloo River basin, (Lake Eyre basin), (S.A.), N.S.W., Qld. Ecology: terrestrial, burrowing, nocturnal, woodland, low woodland, tussock grassland, low open woodland, low open shrubland, open scrub, swamp, predator; seasonal breeder in swamps, oviparous, arthropod-feeder. Biological references: Moore, J.A. (1961). The frogs of eastern New South Wales. *Bull. Am. Mus. Nat. Hist.* **121**: 149–386 (taxonomy, natural history); King, M., Tyler, M.J., Davies, M. & King, D. (1979). Karyotypic studies on *Cyclorana* and associated genera of Australian frogs. *Aust. J. Zool.* **27**: 699–708 (karyology); Heatwole, H., Cameron, E. & Webb, G.J.W. (1971). Studies on anuran water balance. II. Evaporative loss, vital limit and behavioral response to desiccation in *Notaden bennetti. Herpetologica* **27**: 365–378.

Notaden melanoscaphus Hosmer, 1962

Notaden melanoscaphus Hosmer, W. (1962). A new leptodactylid frog of the genus *Notaden* from northern Australia. *Am. Mus. Novit.* 2077: 1–8 [2 figs 1–6]. Type data: holotype, NMV D10144 (*ex* AMNH 67161), from Borroloola, N.T.

Distribution: N Gulf, N coastal, Qld., W.A., N.T. Ecology: terrestrial, burrowing, nocturnal, woodland, swamp, predator; seasonal breeder in swamps, oviparous, free-living tadpole, arthropod-feeder. Biological references: Tyler, M.J.

(1967). *Notaden melanoscaphus*, a species of frog new to the fauna of Queensland. *N. Qd. Nat.* **34**(144): 7 (distribution); Parker, F. & Tanner, C. (1971). Some frogs from southern Cape York Peninsula. *N. Qd. Nat.* **39**(155): 4–8 (distribution); Day, K. (1980). Northern spade-foot toad (*Notaden melanoscaphus*). *Vict. Herp. Soc. Newsletter* **17**: 11–12 (natural history).

Notaden nichollsi Parker, 1940

Notaden nichollsi Parker, H.W. (1940). The Australasian frogs of the family Leptodactylidae. *Novit. Zool.* **42**: 1–106 [63]. Type data: holotype, BMNH 1947.2.18.57, from Roebuck Bay, W.A.

Distribution: NW coastal, W plateau, Lake Eyre basin, Qld., W.A., N.T. Ecology: terrestrial, burrowing, nocturnal, desert, tall shrubland, low shrubland, swamp, predator; seasonal breeder in swamps, oviparous, free-living tadpole, arthropod-feeder. Biological references: Main, A.R. (1965). Frogs of southern Western Australia. *West. Aust. Nat. Club Handbook* 8: 1–73 (natural history); Main, A.R. & Storr, G.M. (1966). Range extensions and notes on the biology of frogs from the Pilbara Region, Western Australia. *West. Aust. Nat.* **10**: 53–61 (distribution); Slater, P. & Main, A.R. (1963). Notes on the biology of *Notaden nichollsi* Parker (Anura : Leptodactylidae). *West. Aust. Nat.* **8**: 163–166 (natural history).

Paracrinia Heyer and Liem, 1976

Paracrinia Heyer, W.R. & Liem, D.S. (1976). Analysis of the intergeneric relationships of the Australian frog family Myobatrachidae. *Smithson. Contrib. Zool.* 233: 1–29 [12]. Type species *Crinia haswelli* Fletcher, 1894 by original designation.

Paracrinia haswelli (Fletcher, 1894)

Crinia haswelli Fletcher, J.J. (1894). Description of a new cystignathoid frog from New South Wales. *Proc. Linn. Soc. N.S.W. (2)* **8**: 522–523 [522]. Type data: syntypes, AM R10355, AM R49816–49818 (one additional syntype presumed lost), from Jervis Bay, N.S.W.

Distribution: SE coastal, N.S.W., Vic. Ecology: terrestrial, nocturnal, tall forest, open forest, swamp, predator; saltatorial, seasonal breeder, oviparous, free-living tadpole, arthropod-feeder. Biological references: Moore, J.A. (1961). The frogs of eastern New South Wales. *Bull. Am. Mus. Nat. Hist.* **121**: 149–386 (natural history, taxonomy).

Philoria Spencer, 1901

Philoria Spencer, B. (1901). Two new species of frogs from Victoria. *Proc. R. Soc. Vict. (ns)* **13**: 175–178 [176]. Type species *Philoria frosti* Spencer, 1901 by monotypy.

Kyarranus Moore, J. A. (1958). A new genus and species of leptodactylid frog from Australia. *Am. Mus. Novit.* 1919: 1–7 [3]. Type species *Kyarranus sphagnicolus* Moore, 1958 by original designation.

Synonymy that of Cogger, H.G., this work.

Philoria frosti Spencer, 1901

Philoria frosti Spencer, B. (1901). Two new species of frogs from Victoria. *Proc. R. Soc. Vict. (ns)* **13**: 175–178 [176]. Type data: lectotype, NMV D8497, from Mt. Baw Baw, Vic., designation by Coventry, A.J. (1970). Reptile and amphibian type specimens housed in the National Museum of Victoria. *Mem. Natl. Mus. Vict.* **31**: 115–124.

Distribution: SE coastal, Vic. Ecology: terrestrial, nocturnal, litter, alpine, predator; hibernating, seasonal breeder, oviparous, direct development, arthropod-feeder, cryptozoic. Biological references: Littlejohn, M.J. (1963). The breeding biology of the Baw Baw frog *Philoria frosti* Spencer. *Proc. Linn. Soc. N.S.W.* **88**: 273–276.

Philoria kundagungan (Ingram and Corben, 1975)

Kyarranus kundagungan Ingram, G.J. & Corben, C.J. (1975). A new species of *Kyarranus* (Anura : Leptodactylidae) from Queensland, Australia. *Mem. Qd. Mus.* **17**: 335–339 [335 pl 42]. Type data: holotype, QM J23944, from Mistake Mt., in 27°53'S, 152°21'E, 83 km SW of Brisbane, Qld.

Distribution: NE coastal, Qld. Ecology: terrestrial, nocturnal, closed forest, (tall forest), predator; seasonal breeder, oviparous, direct development, arthropod-feeder, cryptozoic.

Philoria loveridgei Parker, 1940

Philoria loveridgei Parker, H.W. (1940). The Australasian frogs of the family Leptodactylidae. *Novit. Zool.* **42**: 1–106 [60]. Type data: holotype, BMNH 1947.2.19.94, from McPherson Range, 3–4000 ft, S Qld.

Distribution: SE coastal, NE coastal, N.S.W., Qld. Ecology: terrestrial, nocturnal, closed forest, tall forest, predator; seasonal breeder, oviparous, direct development, arthropod-feeder, cryptozoic. Biological references: Moore, J. A. (1958). A new genus and species of leptodactylid frog from Australia. *Am. Mus. Novit.* 1919: 1–7 (taxonomy); Moore, J.A. (1961). The frogs of eastern New South Wales. *Bull. Am. Mus. Nat. Hist.* **121**: 149–386 (taxonomy, natural history).

Philoria sphagnicolus (Moore, 1958)

Kyarranus sphagnicolus Moore, J. A. (1958). A new genus and species of leptodactylid frog from Australia. *Am. Mus. Novit.* 1919: 1–7 [4 fig 2]. Type data: holotype, AM R16005, from Point Lookout, near Ebor, N.S.W.

Distribution: SE coastal, N.S.W. Ecology: terrestrial, nocturnal, closed forest, tall forest, predator; seasonal breeder, oviparous, direct development, arthropod-feeder, cryptozoic.

Biological references: Moore, J.A. (1961). The frogs of eastern New South Wales. *Bull. Am. Mus. Nat. Hist.* **121**: 149–386 (taxonomy, natural history); Wotherspoon, D. (1981). Note on the range of the sphagnum frog, *Philoria sphagnicolus*. *Herpetofauna* **12**: 36; Anstis, M. (1981). Breeding biology and range extension for the New South Wales frog *Kyarranus sphagnicolus* (Anura : Leptodactylidae). *Aust. J. Herp.* **1**: 1–9.

Pseudophryne Fitzinger, 1843

Pseudophryne Fitzinger, L.J. (1843). *Systema Reptilium.* Vienna : Braümüller u. Seidel vi 106 pp. [32]. Type species *Phrynisc[us] australis* (*non* Gray) Duméril and Bibron, 1841 (=*Pseudophryne semimarmorata* Lucas, 1892 *vide* Parker, H.W. (1940). The Australasian frogs of the family Leptodactylidae. *Novit. Zool.* **42**: 1–106) by original designation.

Bufonella Girard, C. (1853). Descriptions of new species of reptiles, collected by the U.S. Exploring Expedition, under the command of Capt. Charles Wilkes, U.S.N. Second Part. Including the species of batrachians, exotic to North America. *Proc. Acad. Nat. Sci. Philad.* **6**: 420–424 [424] [1854 on title page]. Type species *Bufonella crucifera* Girard, 1853 by monotypy.

Kankanophryne Heyer, W.R. & Liem, D.S. (1976). Analysis of the intergeneric relationships of the Australian frog family Myobatrachidae. *Smithson. Contrib. Zool.* 233: 1–29 [12]. Type species *Pseudophryne occidentalis* Parker, 1940 by original designation.

Synonymy that of Cogger, H.G., this work.

Pseudophryne australis (Gray, 1835)

Bombinator australis Gray, J.E. (1835). Mr. Gray exhibited a specimen of a toad. *Proc. Zool. Soc. Lond.* **1835**: 57 [57]. Type data: holotype, BMNH 1947.2.20.17, from Swan River, W.A. (in error, as known only from N.S.W.).

Bufonella crucifera Girard, C. (1853). Descriptions of new species of reptiles, collected by the U.S. Exploring Expedition, under the command of Capt. Charles Wilkes, U.S.N. Second Part. Including the species of batrachians, exotic to North America. *Proc. Acad. Nat. Sci. Philad.* **6**: 420–424 [424] [1854 on title page]. Type data: syntypes (probable), not found, from Australia.

Phryniscus albifrons Duméril, A.M.C., Bibron, G. & Duméril, A. (1854). *Erpétologie Générale* ou Histoire Naturelle Complète des Reptiles. Paris : Roret 9: xx 440 pp. [413]. Type data: syntypes (probable), not found, from unknown locality.

Bufonella crucigera Keferstein, W. (1868). Über die Batrachier Australiens. *Arch. Naturgesch.* **34**: 253–290 [272] [invalid emend. *errore pro Bufonella crucifera* Girard, 1853].

Synonymy that of Cogger, H.G., this work.

Distribution: SE coastal, N.S.W. Ecology: terrestrial, noctidiurnal, litter, (tall forest), open forest, lotic freshwater, predator; opportunistic breeder in lotic freshwater, oviparous, free-living tadpole, arthropod-feeder, cryptozoic. Biological references: Moore, J.A. (1961). The frogs of eastern New South Wales. *Bull. Am. Mus. Nat. Hist.* **121**: 149–386 (taxonomy, natural history); Harrison, L. (1922). On the breeding habits of some Australian frogs. *Aust. Zool.* **3**: 17–34; Woodruff, D.S. (1978). Hybridization between two species of *Pseudophryne* (Anura : Leptodactylidae) in the Sydney Basin, Australia. *Proc. Linn. Soc. N.S.W.* **102**: 131–147.

Pseudophryne bibronii Günther, 1858

***Pseudophryne bibronii* (part.)** Günther, A. (1858). *Catalogue of the Batrachia Salientia in the collection of the British Museum.* London : British Museum xvi 160 pp. [46]. [original description refers to 4 syntypes, but BMNH records indicate that BMNH 1947.2.20.20 (which is also a syntype of *Pseudophryne guentheri* Boulenger, 1882) is the only extant syntype; consequently the name *bibronii* should apply strictly to the species currently known as *guentheri*, but as this would overturn existing usage it has not been adopted here on the grounds that the status of the remaining syntype is uncertain]. Type data: syntypes, BMNH, from Australia and Tas.

***Pseudophryne coriacea* (part.)** Keferstein, W. (1868). Über die Batrachier Australiens. *Arch. Naturgesch.* **34**: 253–290 [272 pl 6 fig 15]. Type data: syntype, ZFMK 28204, from Clarence River, N.S.W. (for remaining syntypes see *Pseudophryne coriacea* Keferstein, 1868).

Synonymy that of Cogger, H.G., this work.

Distribution: SE coastal, NE coastal, S Gulfs, Murray-Darling basin, S.A., Tas., Vic., N.S.W., Qld., A.C.T. Ecology: terrestrial, nocturnal, litter, (tall forest), open forest, woodland, low woodland, low open woodland, tussock grassland, predator; seasonal breeder, oviparous, free-living tadpole, arthropod-feeder, cryptozoic. Biological references: Moore, J.A. (1961). The frogs of eastern New South Wales. *Bull. Am. Mus. Nat. Hist.* **121**: 149–386 (taxonomy, natural history); Pengilley, R.K. (1971). Calling and associated behaviour of some species of *Pseudophryne* (Anura : Leptodactylidae). *J. Zool.* **163**: 73–92; Pengilley, R.K. (1971). The food of some Australian anurans (Amphibia). *J. Zool.* **163**: 93–103; Pengilley, R.K. (1973). Breeding biology of some species of *Pseudophryne* (Anura : Leptodactylidae) of the southern highlands, New South Wales. *Aust. Zool.* **18**: 15–30; McDonnell, L.J., Gartside, D.F. & Littlejohn, M.J. (1978). Analysis of a narrow hybrid zone between two species of *Pseudophryne* (Anura : Leptodactylidae) in south-eastern Australia. *Evolution* **32**: 602–612; Littlejohn, M.J. & Martin, A.A. (1974). The Amphibia of Tasmania. pp. 251–289 *in* Williams, W.D. (ed.) *Biogeography and Ecology in Tasmania.* The Hague : Junk (natural history); Tyler, M.J. (1978). *Amphibians of South Australia.* Adelaide : South

Australian Government 84 pp. (natural history); Woodruff, D.S. (1978). Hybridization between two species of *Pseudophryne* (Anura : Leptodactylidae) in the Sydney Basin, Australia. *Proc. Linn. Soc. N.S.W.* **102**: 131–147; Wiltshire, D.J. & Bull, C.M. (1977). Potential competitive interactions between larvae of *Pseudophryne bibroni* and *P. semimarmorata* (Anura : Leptodactylidae). *Aust. J. Zool.* **25**: 449–454; Harrison, L. (1922). On the breeding habits of some Australian frogs. *Aust. Zool.* **3**: 17–34; Warburg, M.R. (1965). Studies on the water economy of some Australian frogs. *Aust. J. Zool.* **13**: 317–330.

Pseudophryne coriacea Keferstein, 1868

***Pseudophryne coriacea* (part.)** Keferstein, W. (1868). Über die Batrachier Australiens. *Arch. Naturgesch.* **34**: 253–290 [272 pl 6 fig 15]. Type data: syntypes (6 specimens), ZFMK 28203, ZFMK 28205–28207, from Clarence River, N.S.W., (remaining syntype, ZFMK 28204 is *Pseudophryne bibronii* Günther, 1858).

Distribution: SE coastal, NE coastal, N.S.W., Qld. Ecology: terrestrial, nocturnal, tall forest, open forest, woodland, swamp, predator; seasonal breeder in swamps and inundated grassland, oviparous, free-living tadpole, arthropod-feeder, cryptozoic. Biological references: Moore, J.A. (1961). The frogs of eastern New South Wales. *Bull. Am. Mus. Nat. Hist.* **121**: 149–386 (natural history); Frauca, H. (1965). Notes on the breeding behaviour of a frog (*Pseudophryne*) in central Queensland. *Qd. Nat.* **17**: 94–98.

Pseudophryne corroboree Moore, 1953

Pseudophryne corroboree Moore, J.A. (1953). A new species of *Pseudophryne* from Victoria. *Proc. Linn. Soc. N.S.W.* **78**: 179–180 [179]. Type data: holotype, AM R13103, from Round Mt., N.S.W. (as Towong Hill Station, Corryong, Vic.), see Colefax, A.N. (1956). New information on the corroboree frog (*Pseudophryne corroboree* Moore). *Proc. Linn. Soc. N.S.W.* **80**: 258–266.

Distribution: SE coastal, Murray-Darling basin, N.S.W., A.C.T. Ecology: terrestrial, noctidiurnal, litter, alpine, (open forest), predator; seasonal breeder, oviparous, free-living tadpole, arthropod-feeder, cryptozoic. Biological references: Colefax, A.N. (1956). New information on the corroboree frog (*Pseudophryne corroboree* Moore). *Proc. Linn. Soc. N.S.W.* **80**: 258–266 (natural history); Moore, J.A. (1961). The frogs of eastern New South Wales. *Bull. Am. Mus. Nat. Hist.* **121**: 149–386 (taxonomy, natural history); Pengilley, R.K. (1971). Calling and associated behaviour of some species of *Pseudophryne* (Anura : Leptodactylidae). *J. Zool.* **163**: 73–92; Pengilley, R.K. (1971). The food of some Australian anurans (Amphibia). *J. Zool.* **163**: 93–103; Pengilley, R.K. (1973). Breeding biology of some species of *Pseudophryne* (Anura : Leptodactylidae) of the southern highlands, New South Wales. *Aust. Zool.* **18**: 15–30.

Pseudophryne dendyi Lucas, 1892

Pseudophryne dendyi Lucas, A.H.S (1892). Notes on the distribution of Victorian batrachians with descriptions of two new species. *Proc. R. Soc. Vict. (ns)* **4**: 59–64 [62]. Type data: holotype, presumed lost, from Wellington River, N Gippsland, Vic.

Distribution: SE coastal, Murray-Darling basin, Vic., N.S.W. Ecology: terrestrial, nocturnal, tall forest, open forest, predator; seasonal breeder, oviparous, free-living tadpole, arthropod-feeder, cryptozoic. Biological references: Moore, J.A. (1961). The frogs of eastern New South Wales. *Bull. Am. Mus. Nat. Hist.* **121**: 149–386 (taxonomy, natural history).

Pseudophryne douglasi Main, 1964

Pseudophryne douglasi Main, A.R. (1964). A new species of *Pseudophryne* (Anura : Leptodactylidae) from north-western Australia. *West. Aust. Nat.* **9**: 66–72 [66]. Type data: holotype, WAM R21255, from Kookhabinna Gorge (23°10′S, 115°58′E), W.A.

Distribution: NW coastal, W.A. Ecology: terrestrial, nocturnal, low open shrubland, tall shrubland, (open heath), predator; seasonal breeder, oviparous, free-living tadpole, arthropod-feeder, cryptozoic. Biological references: Main, A.R. (1965). Frogs of southern Western Australia. *West. Aust. Nat. Club Handbook* 8: 1–73 (natural history).

Pseudophryne guentheri Boulenger, 1882

***Pseudophryne bibronii* (part.)** Günther, A. (1858). *Catalogue of the Batrachia Salientia in the collection of the British Museum.* London : British Museum xvi 160 pp. [46] [see comments under *Pseudophryne bibronii* Günther, 1858]. Type data: syntype, BMNH 1947.2.20.20, from Australia.

Pseudophryne guentheri Boulenger, G.A. (1882). *Catalogue of the Batrachia Salientia s. Ecaudata in the collection of the British Museum.* 2nd edn. London : British Museum xvi 503 pp. [279 pl 18 fig 2]. Type data: syntypes, BMNH 1947.2.20.18–21, from Swan River, W.A., NW Australia and Australia.

Pseudophryne brooksi Loveridge, A. (1933). Four new crinine frogs from Australia. *Occ. Pap. Bost. Soc. Nat. Hist.* **8**: 55–60 [59]. Type data: holotype, MCZ 13025, from Manjimup, near Pemberton, W.A.

Synonymy that of Cogger, H.G., this work.

Distribution: NW coastal, SW coastal, (W plateau), W.A. Ecology: terrestrial, nocturnal, (tall forest), open forest, woodland, predator; seasonal breeder, oviparous, free-living tadpole, arthropod-feeder, cryptozoic. Biological references: Main, A.R.

(1965). Frogs of southern Western Australia. *West. Aust. Nat. Club Handbook* 8: 1–73 (natural history).

Pseudophryne major Parker, 1940

Pseudophryne major Parker, H.W. (1940). The Australasian frogs of the family Leptodactylidae. *Novit. Zool.* **42**: 1–106 [98]. Type data: holotype, BMNH 1947.2.20.24, from Gayndah, SE Qld.

Distribution: NE coastal, Qld. Ecology: terrestrial, nocturnal, open forest, woodland, predator; oviparous, free-living tadpole, arthropod-feeder, cryptozoic. Biological references: Woodruff, D.S. (1976). North Queensland toadlets of the genus *Pseudophryne*. *Qd. Nat.* **21**: 142–143 (taxonomy).

Pseudophryne occidentalis Parker, 1940

Pseudophryne occidentalis Parker, H.W. (1940). The Australasian frogs of the family Leptodactylidae. *Novit. Zool.* **42**: 1–106 [97]. Type data: holotype, BMNH 1947.2.20.27, from Bruce Rock, W.A.

Distribution: SW coastal, W plateau, S.A., W.A. Ecology: terrestrial, burrowing, nocturnal, desert, tall shrubland, open scrub, predator; oviparous, free-living tadpole, arthropod-feeder, cryptozoic. Biological references: Main, A.R. (1965). Frogs of southern Western Australia. *West. Aust. Nat. Club Handbook* 8: 1–73 (natural history); Tyler, M.J. (1971). Discovery in the Everard Ranges of a species of leptodactylid frog new to the fauna of South Australia. *Trans. R. Soc. S. Aust.* **95**: 215–217 (distribution); Tyler, M.J. (1978). *Amphibians of South Australia.* Adelaide : South Australian Government 84 pp. (natural history); Tyler, M.J. & Davies, M. (1980). Systematic status of *Kankanophryne* Heyer and Liem (Anura : Leptodactylidae). *Trans. R. Soc. S. Aust.* **104**: 17–20 (taxonomy).

Pseudophryne semimarmorata Lucas, 1892

Pseudophryne semimarmorata Lucas, A.H.S (1892). Notes on the distribution of Victorian batrachians with descriptions of two new species. *Proc. R. Soc. Vict. (ns)* **4**: 59–64 [63]. Type data: lectotype, NMV D7196, from Grampian Ranges, Vic. (as "South-eastern parts of the colony, south of the Divide"), designation by Coventry, A.J. (1970). Reptile and amphibian type specimens housed in the National Museum of Victoria. *Mem. Natl. Mus. Vict.* **31**: 115–124.

Pseudophryne blanchardi Loveridge, A. (1933). A new genus and three new species of crinine frogs from Australia. *Occ. Pap. Bost. Soc. Nat. Hist.* **8**: 89–94 [91]. Type data: holotype, MCZ 19259, from Millgrove, Vic.

Synonymy that of Cogger, H.G., this work.

Distribution: SE coastal, S.A., Tas., Vic. Ecology: terrestrial, nocturnal, open forest, (woodland), predator; seasonal breeder, oviparous, free-living tadpole, arthropod-feeder, cryptozoic. Biological references: Littlejohn, M.J. & Martin, A.A. (1974). The Amphibia of Tasmania. *in* Williams, W.D. (ed.) *Biogeography and Ecology in Tasmania.* The Hague : Junk pp. 251–289 (natural history); Tyler, M.J. (1978). *Amphibians of South Australia.* Adelaide : South Australian Government 84 pp. (natural history); McDonnell, L.J., Gartside, D.F. & Littlejohn, M.J. (1978). Analysis of a narrow hybrid zone between two species of *Pseudophryne* (Anura : Leptodactylidae) in south-eastern Australia. *Evolution* **32**: 602–612; Wiltshire, D.J. & Bull, C.M. (1977). Potential competitive interactions between larvae of *Pseudophryne bibroni* and *P. semimarmorata* (Anura : Leptodactylidae). *Aust. J. Zool.* **25**: 449–454.

Ranidella Girard, 1853

Ranidella Girard, C. (1853). Descriptions of new species of reptiles, collected by the U.S. Exploring Expedition, under the command of Capt. Charles Wilkes, U.S.N. Second Part. Including the species of batrachians, exotic to North America. *Proc. Acad. Nat. Sci. Philad.* **6**: 420–424 [421] [as subgenus of *Crinia*; 1854 on title page]. Type species *Crinia (Ranidella) signifera* Girard, 1853 by original designation.

Camariolius Peters, W. (1863). Übersicht der von Hrn. Richard Schomburgk an das zoologische Museum eingesandten Amphibien, aus Buchsfelde bei Adelaide in Südaustralien. *Mber. K. Preuss. Akad. Wiss. Berl.* **1863**: 228–236 [236] [1864 on title page]. Type species *Camariolius varius* Peters, 1863 by monotypy.

Pterophrynus Lütken, C. (1863). Nogle nye krybdyr og padder. *Vidensk. Meddr. Dansk Naturh. Foren.* **1862**: 292–331 [302]. Type species *Pterophrynus verrucosus* Lütken, 1863 by monotypy.

Pterophryne Cope, E. D. (1865). Sketch of the primary groups of Batrachia Salientia. *Nat. Hist. Rev.* **5**: 97–120 [114] [*non Pterophryne* Gill 1864; invalid emend. *pro Pterophyrnus* Lütken, 1863].

Australocrinia Heyer, W.R. & Liem, D.S. (1976). Analysis of the intergeneric relationships of the Australian frog family Myobatrachidae. *Smithson. Contrib. Zool.* 233: 1–29 [12]. Type species *Pterophrynus tasmaniensis* Günther, 1864 by original designation.

Synonymy that of Cogger, H.G., this work.

This group is also found in southern New Guinea.

Ranidella bilingua Martin, Tyler and Davies, 1980

Ranidella bilingua Martin, A.A., Tyler, M.J. & Davies, M. (1980). A new species of *Ranidella* (Anura : Leptodactylidae) from northwestern Australia. *Copeia* **1980**: 93–99 [94]. Type data: holotype, WAM R59775 M, from swamp on E side of Spillway Creek at Spillway Bridge (16°01'47"S, 128°47'08"E), 11.5 km by road N of the Lake Argyle Tourist Village, Kimberley Division, W.A.

Distribution: N coastal, (N Gulf), W.A., N.T., (Qld.). Ecology: terrestrial, nocturnal, low woodland, woodland, predator; seasonal breeder, oviparous, free-living tadpole, arthropod-feeder.

Ranidella deserticola Liem and Ingram, 1977

Ranidella deserticola Liem, D.S. & Ingram, G.J. (1977). Two new species of frogs (Anura : Myobatrachidae, Pelodryadidae) from Queensland and New South Wales. *Vict. Nat.* **94**: 255–262 [255 figs 1–2]. Type data: holotype, QM J22654, from Charleville, SW Qld.

Distribution: W plateau, Murray-Darling basin, Bulloo River basin, Lake Eyre basin, S.A., N.S.W., Qld., N.T. Ecology: terrestrial, nocturnal, tall shrubland, low open woodland, (low open shrubland), (low shrubland), predator; oviparous, free-living tadpole, arthropod-feeder, cryptozoic. Biological references: Tyler, M.J., Davies, M. & Martin, A.A. (1981). Frog fauna of the Northern Territory: new distributional records and the description of a new species. *Trans. R. Soc. S. Aust.* **105**: 149–154.

Ranidella glauerti (Loveridge, 1933)

Crinia glauerti Loveridge, A. (1933). Four new crinine frogs from Australia. *Occ. Pap. Bost. Soc. Nat. Hist.* **8**: 55–60 [57]. Type data: holotype, MCZ 18420, from Mundaring Weir, *ca.* 30 mi NE Perth, W.A.

Distribution: SW coastal, W.A. Ecology: terrestrial, nocturnal, (open heath), woodland, open forest, predator; seasonal breeder, oviparous, free-living tadpole, arthropod-feeder, cryptozoic. Biological references: Main, A.R. (1957). Studies in Australian Amphibia. 1. The genus *Crinia* Tschudi in south western Australia and some species from south eastern Australia. *Aust. J. Zool.* **5**: 30–55; Main, A.R. (1965). Frogs of southern Western Australia. *West. Aust. Nat. Club Handbook* **8**: 1–73.

Ranidella insignifera (Moore, 1954)

Crinia insignifera Moore, J.A. (1954). Geographic and genetic isolation in Australian Amphibia. *Am. Nat.* **88**(839): 65–74 [71]. Type data: holotype, AM R16007, from Armadale (as Attadale on type label), W.A.

Distribution: SW coastal, W.A. Ecology: terrestrial, nocturnal, open forest, woodland, (open heath), swamp, predator; seasonal breeder in swamps, oviparous, free-living tadpole, arthropod-feeder, cryptozoic. Biological references: Main, A.R. (1957). Studies in Australian Amphibia. 1. The genus *Crinia* Tschudi in south western Australia and some species from south eastern Australia. *Aust. J. Zool.* **5**: 30–55 (taxonomy); Main, A.R. (1961). *Crinia insignifera* Moore (Anura : Leptodactylidae) on Rottnest Island. *J. R. Soc. West. Aust.* **44**: 10–13 (distribution); Main, A.R. (1965). Frogs of southern Western Australia. *West Aust. Nat. Club Handbook* **8**: 1–73 (natural history); Main, A.R. (1965). The inheritance of dorsal pattern in *Crinia* species (Anura : Leptodactylidae). *J. R. Soc. West. Aust.* **48**: 60–64 (genetics); Littlejohn, M.J. (1959). Call differentiation in a complex of seven species of *Crinia* (Anura : Leptodactylidae). *Evolution* **13**: 452–468; Littlejohn, M.J. (1961). Age and origin of some southwestern Australian species of *Crinia* (Anura : Leptodactylidae). pp. 514–536 *in* Blair, W.F. (ed.) *Vertebrate Speciation.* Austin : Univ. of Texas Press (biogeography); Bull, C.M. (1979). A narrow hybrid zone between two Western Australian frog species. *Ranidella insignifera* and *R. pseudinsignifera*: the fitness of hybrids. *Heredity* **42**: 381–389; Blackwell, J.M. & Bull, C.M. (1978). A narrow hybrid zone between two Western Australian frog species *Ranidella insignifera* and *R. pseudinsignifera*: the extent of introgression. *Heredity* **40**: 13–25; Blackwell, J.M. (1978). Intra-specific divergence in the Western Australian frog species *Ranidella insignifera*. 1. The evidence from gene frequencies and genetic distance. *Heredity* **40**: 339–348.

Ranidella parinsignifera (Main, 1957)

Crinia parinsignifera Main, A.R. (1957). Studies in Australian Amphibia. 1. The genus *Crinia* Tschudi in south western Australia and some species from south eastern Australia. *Aust. J. Zool.* **5**: 30–55 [53]. Type data: holotype, WAM R38735, from Kingston-on-Murray, S.A.

Distribution: SE coastal, NE coastal, Murray-Darling basin, Bulloo River basin, Lake Eyre basin, S.A., Vic., N.S.W., Qld., (A.C.T.). Ecology: terrestrial, nocturnal, open forest, woodland, low woodland, low open woodland, tall shrubland, (open scrub), predator; seasonal breeder, oviparous, free-living tadpole, arthropod-feeder, cryptozoic. Biological references: Main, A.R. (1965). The inheritance of dorsal pattern in *Crinia* species (Anura : Leptodactylidae). *J. R. Soc. West. Aust.* **48**: 60–64 (genetics); Littlejohn, M.J. (1959). Call differentiation in a complex of seven species of *Crinia* (Anura : Leptodactylidae). *Evolution* **13**: 452–468; Littlejohn, M.J. (1961). Age and origin of some southwestern Australian species of *Crinia* (Anura : Leptodactylidae). pp. 514–536 *in* Blair, W.F. (ed.) *Vertebrate Speciation.* Austin : Univ. of Texas Press (biogeography); MacNally, R.C. (1979). Social organization and interspecific interactions in two sympatric species of *Ranidella* (Anura). *Oecologia* **42**: 293–306; Tyler, M.J. (1978). *Amphibians of South Australia.* Adelaide : South Australian Government 84 pp. (natural history).

Ranidella pseudinsignifera (Main, 1957)

Crinia pseudinsignifera Main, A.R. (1957). Studies in Australian Amphibia. 1. The genus *Crinia* Tschudi in south western Australia and some species from south eastern Australia. *Aust. J. Zool.* **5**: 30–55 [52]. Type data: holotype, WAM R36153, from Yorkrakine Rock, 15 mi N of Tammin, W.A.

Distribution: SW coastal, W.A. Ecology: terrestrial, nocturnal, woodland, swamp, predator; breeds in swamps, oviparous, free-living tadpole, arthropod-feeder, cryptozoic,. Biological references: Main, A.R. (1965). The inheritance of dorsal pattern in *Crinia* species (Anura : Leptodactylidae). *J. R. Soc. West. Aust.* **48**: 60–64 (genetics); Littlejohn, M.J. (1959). Call differentiation in a complex of seven species of *Crinia* (Anura : Leptodactylidae). *Evolution* **13**: 452–468; Littlejohn, M.J. (1961). Age and origin of some southwestern Australian species of *Crinia* (Anura : Leptodactylidae). pp. 514–536 *in* Blair, W.F. (ed.) *Vertebrate Speciation.* Austin : Univ. of Texas Press (biogeography); Main, A.R. (1965). Frogs of southern Western Australia. *West. Aust. Nat. Club Handbook* 8: 1–73 (natural history); Bull, C.M. (1979). A narrow hybrid zone between two Western Australian frog species. *Ranidella insignifera* and *R. pseudinsignifera*: the fitness of hybrids. *Heredity* **42**: 381–389; Blackwell, J.M. & Bull, I.M. (1978). A narrow hybrid zone between two Western Australia frog species *Ranidella insignifera* and *R. pseudinsignifera*: the extent of introgression. *Heredity* **40**: 13–25.

Ranidella remota Tyler and Parker, 1974

Ranidella remota Tyler, M.J. & Parker, F. (1974). New species of hylid and leptodactylid frogs from southern New Guinea. *Trans. R. Soc. S. Aust.* **98**: 71–78 [74]. Type data: holotype, SAMA R13524, from Morehead, Papua New Guinea.

Distribution: (NE coastal), (N Gulf), (Qld.). Ecology: terrestrial, nocturnal, woodland, predator; oviparous, free-living tadpole, arthropod-feeder, cryptozoic.

Ranidella riparia (Littlejohn and Martin, 1965)

Crinia riparia Littlejohn, M.J. & Martin, A.A. (1965). A new species of *Crinia* (Anura : Leptodactylidae) from South Australia. *Copeia* **1965**: 319–324 [319 figs 1–5]. Type data: holotype, AM R26161, from Alligator Gorge, 8.5 mi SSW of Wilmington, lower Flinders Ranges, S.A.

Distribution: S Gulfs, (Lake Eyre basin), S.A.; Ecology: low open woodland, lotic freshwater, tall shrubland, terrestrial, nocturnal; breeds in lotic fresh water, seasonal breeder, oviparous, free-living tadpole, arthropod-feeder, cryptozoic. Biological references: MacNally, R.C. (1979). Social organization and interspecific interactions in two sympatric species of *Ranidella* (Anura). *Oecologia* **42**: 293–306; Tyler, M.J. (1978). *Amphibians of South Australia.* Adelaide : South Australian Government 84 pp. (natural history); Odendaal, F.J. & Bull, C.M. (1980). Influence of water speed on tadpoles of *Ranidella signifera* and *R. riparia* (Anura : Leptodactylidae). *Aust. J. Zool.* **28**: 79–82 (larvae); Thompson, M.B. (1981). The systematic status of the genus *Australocrinia* Heyer and Liem (Anura : Leptodactylidae). *Aust. J. Zool.* **29**: 93–102 (taxonomy).

Ranidella signifera (Girard, 1853)

Crinia (Ranidella) signifera Girard, C. (1853). Descriptions of new species of reptiles, collected by the U.S. Exploring Expedition, under the command of Capt. Charles Wilkes, U.S.N. Second Part. Including the species of batrachians, exotic to North America. *Proc. Acad. Nat. Sci. Philad.* **6**: 420–424 [422] [1854 on title page]. Type data: holotype, presumed lost, from Australia.

Camariolus varius Peters, W. (1863). Übersicht der von Hrn. Richard Schomburgk an das zoologische Museum eingesandten Amphibien, aus Buchsfelde bei Adelaide in Südaustralien. *Mber. K. Preuss. Akad. Wiss. Berl.* **1863**: 228–236 [236] [1864 on title page]. Type data: holotype, ZMB 4708 (not found), from Loos, 4.5 km W Gawler (as Buchsfelde), S.A., see Praite, R. & Tolley, J.C. (1970). *Place Names of South Australia.* Adelaide : Rigby 203 pp.

Pterophrynus verrucosus Lütken, C. (1863). Nogle nye krybdyr og padder. *Vidensk. Meddr. Dansk Naturh. Foren.* **1862**: 292–331 [302]. Type data: holotype, UZM R1181, from Australia.

Crinia georgiana laevipes Keferstein, W. (1867). Ueber einige neue oder seltene Batrachier aus Australien und dem tropischen Amerika. *Nachr. Ges. Wiss. Göttingen* **1867**: 341–361 [348]. Type data: syntypes, ZFMK 28246–28247, from Sydney, N.S.W.

Cystignathus sydneyensis Keferstein, W. (1867). Ueber einige neue oder seltene Batrachier aus Australien und dem tropischen Amerika. *Nachr. Ges. Wiss. Göttingen* **1867**: 341–361 [348] [*nom. nud.*; Krefft ms name introduced in synonymy].

Camariolus pictus Cope, E.D. (1867). On the families of raniform Anura. *J. Acad. Nat. Sci. Philad. (2)* **6**: 189–206 [202] [invalid emend. *lapsus pro Camariolius varius* Peters, 1863].

Crinia stictiventris Cope, E.D. (1867). On the families of raniform Anura. *J. Acad. Nat. Sci. Philad. (2)* **6**: 189–206 [202]. Type data: holotype, not found; cited as MCZ 672 in error by Barbour, T. & Loveridge, A. (1929). Typical reptiles and amphibians in the Museum of Comparative Zoology. *Bull. Mus. Comp. Zool.* **69**: 205–360, from Australia.

Pterophrynus fasciatus Steindachner, F. (1867). Amphibien. *in, Reise der österreichischen Fregatte* Novara *um die Erde in den Jahren 1857, 1858, 1859.* Zoologie 1(4): 1–70 Vienna : State Printer [31 pl 5 figs 3–4] [1869 on title page]. Type data: syntypes, ?NHMW 4707 (13 specimens; the original description cites only 4 specimens in NHMW and these may be included in this series), from N.S.W.

Crinia signifera englishi Parker, H.W. (1940). The Australasian frogs of the family Leptodactylidae. *Novit. Zool.* **42**: 1–106 [89]. Type data: syntypes, BMNH 1947.2.19.46–64, BMNH 1947.2.19.76–79, AM R6041–5, AM R7601 (2 specimens), AM R10353, AM R10356–7, AM R49825–30, from near Launceston, near Devonport, Launceston, Eagle Hawk Neck, National Park, Ulverstone and Georges Bay, Tas.

Crinia signifera montana Parker, H.W. (1940). The Australasian frogs of the family Leptodactylidae. *Novit. Zool.* **42**: 1–106 [90]. Type data: syntypes, MCZ 26761–2, AM R579, AM R4647, AM R5044, AM R5046–48, AM R5050–53, AM R7439, AM R9741, AM R9742 (2 specimens), AM R9743, AM R54676–92, from Mt. Kosciusko region, N.S.W.

Crinia affinis halmaturina Condon, H.T. (1941). Further records of lizards and frogs from Kangaroo Island. *Rec. S. Aust. Mus.* **7**: 111–116 [114]. Type data: holotype, SAMA R2165, (from Flinders Chase, Kangaroo Is., S.A. presumed lost *vide* Tyler, M.J. (1976). Vertebrate type-specimens in the South Australian Museum. II. Amphibians. *Rec. S. Aust. Mus.* **17**: 177–180).

Synonymy that of Cogger, H.G., this work.

Distribution: SE coastal, NE coastal, S Gulfs, Murray-Darling basin, S.A., Tas., Vic., N.S.W., Qld., A.C.T. Ecology: terrestrial, nocturnal, tall forest, open forest, woodland, swamp, predator; seasonal breeder, oviparous, free-living tadpole, arthropod-feeder, cryptozoic. Biological references: Moore, J.A. (1954). Geographic and genetic isolation in Australian Amphibia. *Am. Nat.* **88**(839): 65–74 (genetics); Moore, J.A. (1961). The frogs of eastern New South Wales. *Bull. Am. Mus. Nat. Hist.* **121**: 149–386 (taxonomy, natural history); Main, A.R. (1957). Studies in Australian Amphibia. 1. The genus *Crinia* Tschudi in south western Australia and some species from south eastern Australia. *Aust. J. Zool.* **5**: 30–55 (taxonomy, biology); Littlejohn, M.J. (1959). Call differentiation in a complex of seven species of *Crinia* (Anura : Leptodactylidae). *Evolution* **13**: 452–468 (taxonomy); Littlejohn, M.J. (1964). Geographic isolation and mating call differentiation in *Crinia signifera*. *Evolution* **18**: 262–266; Odendaal, F.J. & Bull, C.M. (1980). Influence of water speed on tadpoles of *Ranidella signifera* and *R. riparia* (Anura : Leptodactylidae). *Aust. J. Zool.* **28**: 79–82 (larvae); Tyler, M.J. (1978). *Amphibians of South Australia.* Adelaide : South Australian Government 84 pp. (natural history); Littlejohn, M.J. & Martin, A.A. (1974). The Amphibia of Tasmania. pp. 251–289 *in* Williams, W.D. (ed.) *Biogeography and Ecology in Tasmania.* The Hague : Junk (natural history).

Ranidella sloanei (Littlejohn, 1958)

Crinia sloanei Littlejohn, M.J. (1958). A new species of frog of the genus *Crinia* Tschudi from south-eastern Australia. *Proc. Linn. Soc. N.S.W.* **83**: 222–226 [225]. Type data: holotype, AM R19610, from Murray River, near Tocumwal, N.S.W.

Distribution: Murray-Darling basin, Vic., N.S.W. Ecology: terrestrial, nocturnal, woodland, swamp, low open woodland, predator; seasonal breeder in swamps, oviparous, free-living tadpole, arthropod-feeder, cryptozoic. Biological references: Littlejohn, M.J. (1959). Call differentiation in a complex of seven species of *Crinia* (Anura : Leptodactylidae). *Evolution* **13**: 452–468 (taxonomy).

Ranidella subinsignifera (Littlejohn, 1957)

Crinia subinsignifera Littlejohn, M.J. (1957). A new species of frog of the genus *Crinia*. *West. Aust. Nat.* **6**: 18–23 [22]. Type data: holotype, WAM R12786, from Wilgarup, 7 mi N of Manjimup, W.A.

Distribution: SW coastal, W.A. Ecology: terrestrial, nocturnal, open forest, (woodland), predator; oviparous, free-living tadpole, arthropod-feeder, cryptozoic. Biological references: Littlejohn, M.J. (1959). Call differentiation in a complex of seven species of *Crinia* (Anura : Leptodactylidae). *Evolution* **13**: 452–468 (taxonomy); Main, A.R. (1965). The inheritance of dorsal pattern in *Crinia* species (Anura : Leptodactylidae). *J. R. Soc. West. Aust.* **48**: 60–64 (genetics); Main, A.R. (1965). Frogs of southern Western Australia. *West. Aust. Nat. Club Handbook* 8: 1–73 (natural history).

Ranidella tasmaniensis (Günther, 1864)

Pterophrynus tasmaniensis Günther, A. (1864). Third contribution to our knowledge of batrachians from Australia. *Proc. Zool. Soc. Lond.* **1864**: 46–49 [48 pl 7 fig 3]. Type data: syntypes, BMNH 1947.2.19.68–75, from Tas.

Distribution: Tas.; Ecology: alpine, open forest, swamp, aquatic, terrestrial, nocturnal; breeds in swamps, seasonal breeder, oviparous, free-living tadpole, arthropod-feeder, cryptozoic. Biological references: Littlejohn, M.J. (1970). *Crinia tasmaniensis* (Anura : Leptodactylidae): geographic distribution, mating call structure, and relationships. *Proc. Linn. Soc. N.S.W.* **94**: 119–127; Littlejohn, M.J. & Martin, A.A. (1974). The Amphibia of Tasmania. *in* Williams, W.D. (ed.) *Biogeography and Ecology in Tasmania.* The Hague : Junk pp. 251–289 (natural history); Thompson, M.B. (1981). The systematic status of the genus *Australocrinia* Heyer and Liem (Anura : Leptodactylidae). *Aust. J. Zool.* **29**: 93–102 (taxonomy).

Ranidella tinnula (Straughan and Main, 1966)

Crinia tinnula Straughan, I.R. & Main, A.R. (1966). Speciation and polymorphism in the genus *Crinia* Tschudi (Anura : Leptodactylidae) in Queensland. *Proc.*

R. Soc. Qd. **78**: 11–28 [19 pl 1 fig 1, pl 3 fig 1, pl 6B]. Type data: holotype, QM J13546, from Rose Creek, Beerburrum, SE Qld.

Distribution: SE coastal, NE coastal, N.S.W., Qld. Ecology: terrestrial, nocturnal, woodland, swamp, predator; breeds in swamps, oviparous, free-living tadpole, arthropod-feeder, cryptozoic. Biological references: Ingram, G.J. & Corben, C.J. (1975). The frog fauna of North Stradbroke Island with comments on the acid frogs of the wallum. *Proc. R. Soc. Qd.* **86**: 49–54 (natural history).

Rheobatrachus Liem, 1973

Rheobatrachus Liem, D.S. (1973). A new genus of frog of the family Leptodactylidae from SE. Queensland, Australia. *Mem. Qd. Mus.* **16**: 459–470 [463]. Type species *Rheobatrachus silus* Liem, 1973 by original designation.

Rheobatrachus silus Liem, 1973

Rheobatrachus silus Liem, D.S. (1973). A new genus of frog of the family Leptodactylidae from SE. Queensland, Australia. *Mem. Qd. Mus.* **16**: 459–470 [467]. Type data: holotype, QM J22489, from Kondalilla, 3 km SW Montville, Qld.

Distribution: NE coastal, Qld. Ecology: aquatic, nocturnal, (closed forest), tall forest, lotic freshwater, predator; seasonal breeder, oviparous, torrent dweller, gastric brooder. Biological references: Corben, C.J., Ingram, G.J. & Tyler, M.J. (1974). Gastric brooding: unique form of parental care in an Australian frog. *Science* **186**: 946–947.

Taudactylus Straughan and Lee, 1966

Taudactylus Straughan, I.R. & Lee, A.K. (1966). A new genus and species of leptodactylid frog from Queensland. *Proc. R. Soc. Qd.* **77**: 63–66 [63]. Type species *Taudactylus diurnus* Straughan and Lee, 1966 by original designation.

Taudactylus acutirostris (Andersson, 1916)

Crinia acutirostris Andersson, L.G. (1916). Results of Dr. E. Mjöberg's Swedish Scientific Expeditions to Australia 1910–1913. 9. Batrachians from Queensland. *K. Svenska Vetensk.-Akad. Handl. (ns)* **52**: 1–20 [8 fig 2]. Type data: syntypes, NHRM 1624 (2 specimens), from Malanda, Qld.

Distribution: NE coastal, Qld. Ecology: amphibious, noctidiurnal, closed forest, (tall forest), lotic freshwater, predator; oviparous, free-living tadpole, arthropod-feeder, torrent dweller. Biological references: Straughan, I.R. & Lee, A.K. (1966). A new genus and species of leptodactylid frog from Queensland. *Proc. R. Soc. Qd.* **77**: 63–66 (taxonomy); Liem, D.S. & Hosmer, W. (1973). Frogs of the genus *Taudactylus* with descriptions of two new species (Anura : Leptodactylidae). *Mem. Qd. Mus.* **16**: 435–457 (taxonomy).

Taudactylus diurnus Straughan and Lee, 1966

Taudactylus diurnus Straughan, I.R. & Lee, A.K. (1966). A new genus and species of leptodactylid frog from Queensland. *Proc. R. Soc. Qd.* **77**: 63–66 [63]. Type data: holotype, QM J13398, from Green's Falls, Maiali National Park, Mt. Glorious, SE Qld.

Distribution: NE coastal, Qld. Ecology: amphibious, noctidiurnal, closed forest, (tall forest), lotic freshwater, predator; oviparous, free-living tadpole, arthropod-feeder, torrent dweller. Biological references: Johnson, C.R. (1971). Thermal relations and water balance in the day frog, *Taudactylus diurnus*, from an Australian rainforest. *Aust. J. Zool.* **19**: 35–39; Liem, D.S. & Hosmer, W. (1973). Frogs of the genus *Taudactylus* with descriptions of two new species (Anura : Leptodactylidae). *Mem. Qd. Mus.* **16**: 435–457; Ingram, G. (1980). A new frog of the genus *Taudactylus* (Myobatrachidae) from mid-eastern Queensland with notes on the other species of the genus. *Mem. Qd Mus.* **20**: 111–119.

Taudactylus eungellensis Liem and Hosmer, 1973

Taudactylus eungellensis Liem, D.S. & Hosmer, W. (1973). Frogs of the genus *Taudactylus* with descriptions of two new species (Anura : Leptodactylidae). *Mem. Qd. Mus.* **16**: 435–457 [455 figs 1C 2D 3D 8]. Type data: holotype, QM J22433, from Eungella, 75 km W Mackay, Qld.

Distribution: NE coastal, Qld. Ecology: amphibious, noctidiurnal, closed forest, (tall forest), lotic freshwater, predator; oviparous, free-living tadpole, arthropod-feeder, torrent dweller. Biological references: Ingram, G. (1980). A new frog of the genus *Taudactylus* (Myobatrachidae) from mid-eastern Queensland with notes on the other species of the genus. *Mem. Qd. Mus.* **29**: 111–119.

Taudactylus liemi Ingram, 1980

Taudactylus liemi Ingram, G. (1980). A new frog of the genus *Taudactylus* (Myobatrachidae) from mid-eastern Queensland with notes on the other species of the genus. *Mem. Qd Mus.* **20**: 111–119 [111 fig 1]. Type data: holotype, QM J32625, from Crediton (21°12′S, 148°33′E), Qld.

Distribution: NE coastal, Qld. Ecology: amphibious, nocturnal, closed forest, (tall forest), predator, lotic freshwater; oviparous, free-living tadpole, arthropod-feeder, torrent-dweller.

Taudactylus rheophilus Liem and Hosmer, 1973

Taudactylus rheophilus Liem, D.S. & Hosmer, W. (1973). Frogs of the genus *Taudactylus* with descriptions of two new species (Anura : Leptodactylidae). *Mem. Qd. Mus.* **16**: 435–457 [450 figs 1A 2A 2C 2E 3A 3E 4C 5, pl 29]. Type data: holotype, QM J22418, from Mt. Lewis, 1200 m altitude, Qld.

Distribution: NE coastal, Qld. Ecology: amphibious, nocturnal, closed forest, (tall forest), lotic freshwater, predator; oviparous, free-living tadpole, arthropod-feeder, torrent dweller. Biological references: Ingram, G. (1980). A new frog of the genus *Taudactylus* (Myobatrachidae) from mid-eastern Queensland with notes on the other species of the genus. *Mem. Qd. Mus.* **20**:111–119.

Uperoleia Gray, 1841

Uperoleia Gray, J.E. (1841). Description of some new species and four new genera of reptiles from Western Australia, discovered by John Gould, Esq. *Ann. Mag. Nat. Hist.* **7**: 86–91 [90]. Type species *Uperoleia marmorata* Gray, 1841 by monotypy.

Uperoleja Gray, J.E. (1841). A catalogue of the species of reptiles and Amphibia hitherto described as inhabiting Australia, with a description of some new species from Western Australia. Appendix E *in* Grey, G. *Journals of Two Expeditions of Discovery in North-west and Western Australia, during the years 1837, 38, and 39, under the authority of Her Majesty's Government.* London : T. & W. Boone 2: 422–449 [436] [invalid emend. *lapsus pro Uperoleia* Gray, 1841].

Hyperolia Cope, E. D. (1865). Sketch of the primary groups of Batrachia Salientia. *Nat. Hist. Rev.* **5**: 97–120 [108] [*non Hyperolius* Rapp, 1842; invalid emend. *pro Uperoleia* Gray, 1841].

Glauertia Loveridge, A. (1933). A new genus and three new species of crinine frogs from Australia. *Occ. Pap. Bost. Soc. Nat. Hist.* **8**: 89–94 [89]. Type species *Glauertia russelli* Loveridge, 1933 by monotypy.

Synonymy that of Tyler, M.J., Davies, M. & Martin, A.A. (1981). Australian frogs of the leptodactylid genus *Uperoleia* Gray. *Aust. J. Zool. Suppl. Ser.* **79**: 1–64; Cogger, H.G., this work.

This group is also found in southern New Guinea.

Uperoleia arenicola Tyler, Davies and Martin, 1981

Uperoleia arenicola Tyler, M.J., Davies, M. & Martin, A.A. (1981). Australian frogs of the leptodactylid genus *Uperoleia* Gray. *Aust. J. Zool. Suppl. Ser.* **79**: 1–64 [26 fig 9A]. Type data: holotype, SAMA R16991 M, from Birndu, East Alligator River Region, in 12°28'S, 132°56'E, N.T.

Distribution: N coastal, N.T. Ecology: burrowing, terrestrial, nocturnal, predator, woodland, swamps; oviparous, arthropod-feeder, cryptozoic, breeds in swamps.

Uperoleia aspera Tyler, Davies and Martin, 1981

Uperoleia aspera Tyler, M.J., Davies, M. & Martin, A.A. (1981). New and rediscovered species of frogs from the Derby - Broome area of Western Australia. *Rec. West. Aust. Mus.* **9**: 147–172 [159 figs 8–12]. Type data: holotype, WAM R69648 M, from 28 km S of Derby, in 123°43'S[*sic*], 17°30'E[*sic*], W.A.

Distribution: N coastal, (W plateau), W.A. Ecology: terrestrial, burrowing, nocturnal, predator, woodland, swamp; oviparous, arthropod-feeder, cryptozoic, breeds in swamps.

Uperoleia borealis Tyler, Davies and Martin, 1981

Uperoleia borealis Tyler, M.J., Davies, M. & Martin, A.A. (1981). Australian frogs of the leptodactylid genus *Uperoleia* Gray. *Aust. J. Zool. Suppl. Ser.* **79**: 1–64 [30 fig 9B]. Type data: holotype, WAM R62474 M, from Dead Horse Spring, 3.7 km NE of Lake Argyle Tourist Village, in 16°06'S, 128°45'E, W.A.

Distribution: N coastal, W.A. Ecology: terrestrial, burrowing, nocturnal, predator, woodland, swamp; oviparous, arthropod-feeder, cryptozoic, breeds in swamps.

Uperoleia crassa Tyler, Davies and Martin, 1981

Uperoleia crassa Tyler, M.J., Davies, M. & Martin, A.A. (1981). Australian frogs of the leptodactylid genus *Uperoleia* Gray. *Aust. J. Zool. Suppl. Ser.* **79**:1–64 [34 fig 12A]. Type data: holotype, WAM R59951 M, from Amax Campsite, Mitchell Plateau, Kimberley Division, in 14°15'S, 125°50'E, W.A.

Distribution: N coastal, W.A. Ecology: terrestrial, burrowing, nocturnal, woodland, predator, swamp; oviparous, arthropod-feeder, cryptozoic, breeds in swamps.

Uperoleia fimbrianus (Parker, 1926)

Pseudophryne fimbrianus Parker, H.W. (1926). New reptiles and a new frog from Queensland. *Ann. Mag. Nat. Hist. (9)* **17**: 665–670 [669]. Type data: holotype, BMNH 1947.2.18.70, from St. George district, Qld.

Distribution: Murray-Darling basin, Qld. Biological references: Tyler, M.J., Davies, M. & Martin, A.A. (1981). Australian frogs of the leptodactylid genus *Uperoleia* Gray. *Aust. J. Zool. Suppl. Ser.* **79**: 1–64.

Uperoleia inundata Tyler, Davies and Martin, 1981

Uperoleia inundata Tyler, M.J., Davies, M. and Martin, A.A. (1981). Australian frogs of the leptodactylid genus *Uperoleia* Gray. *Aust. J. Zool. Suppl. Ser.* **79**: 1–64 [39 fig 9C]. Type data: holotype, SAMA R17182 M, from Jabiru, East Alligator River Region, in 12°28'S, 132°56'E, N.T.

Distribution: N coastal, N.T. Ecology: terrestrial, burrowing, nocturnal, woodland, predator, swamp; oviparous, arthropod-feeder, cryptozoic, breeds in swamps.

Uperoleia laevigata Keferstein, 1867

Uperoleia marmorata laevigata Keferstein, W. (1867). Ueber einige neue oder seltene Batrachier aus Australien und dem tropischen Amerika. *Nachr. Ges. Wiss. Göttingen* **1867**: 341–361 [349] [in not applying any of the available specific names in *Uperoleia* from eastern Australia to other than their type specimens, Tyler, M.J., Davies, M. & Martin, A.A. (1981). Australian frogs of the leptodactylid genus *Uperoleia* Gray. *Aust. J. Zool. Suppl. Ser.* **79**: 1–64 have effectively left all eastern populations of this abundant and widespread genus without names. Until this situation has been rectified we believe that it is appropriate to maintain existing usage by retaining the names *Uperoleia rugosa* and *U. laevigata*]. Type data: syntypes, ZFMK 26304–26312, from Randwick (as Raudewick), N.S.W.

Distribution: SE coastal, NE coastal, Vic., N.S.W., Qld. Ecology: terrestrial, burrowing, nocturnal, predator, tall forest, open forest, swamp; seasonal breeder in swamps, oviparous, free-living tadpole, arthropod-feeder, cryptozoic.

Uperoleia lithomoda Tyler, Davies and Martin, 1981

Uperoleia lithomoda Tyler, M.J., Davies, M. & Martin, A.A. (1981). Australian frogs of the leptodactylid genus *Uperoleia* Gray. *Aust. J. Zool. Suppl. Ser.* **79**: 1–64 [43 fig 9D]. Type data: holotype, WAM R61620 M, from Spillway Bridge, 11.5 km NE of Lake Argyle Tourist Village, Kimberley Division, in 16°02'S, 128°47'E, W.A.

Distribution: N coastal, W.A., N.T. Ecology: terrestrial, burrowing, nocturnal, predator, woodland, swamp; seasonal breeder in swamps, oviparous, free-living tadpole, arthropod-feeder, cryptozoic. Biological references: Tyler, M.J., Davies, M. & Martin, A.A. (1981). Frog fauna of the Northern Territory: new distributional records and the description of a new species. *Trans. R. Soc. S. Aust.* **105**: 149–154.

Uperoleia marmorata Gray, 1841

Uperoleia marmorata Gray, J.E. (1841). Description of some new species and four new genera of reptiles from Western Australia, discovered by John Gould, Esq. *Ann. Mag. Nat. Hist.* **7**: 86–91 [90]. Type data: holotype, BMNH 1947.2.19.88, from W.A.

Distribution: NW coastal, W.A. Ecology: terrestrial, burrowing, nocturnal, predator; oviparous, arthropod-feeder, cryptozoic. Biological references: Tyler, M.J., Davies, M. & Martin, A.A. (1981). Australian frogs of the leptodactylid genus *Uperoleia* Gray. *Aust. J. Zool. Suppl. Ser.* **79**: 1–64 (taxonomy).

Uperoleia micromeles Tyler, Davies and Martin, 1981

Uperoleia micromeles Tyler, M.J., Davies, M. & Martin, A.A. (1981). Australian frogs of the leptodactylid genus *Uperoleia* Gray. *Aust. J. Zool. Suppl. Ser.* **79**: 1–64 [46 figs 5F 16A-B]. Type data: holotype, SAMA R17175 F, from Tanami Desert, in 28°38'S, 130°25'E, N.T.

Distribution: W plateau, N.T. Ecology: terrestrial, burrowing, nocturnal, predator, desert, swamp; oviparous, arthropod-feeder, cryptozoic, breeds in swamps.

Uperoleia minima Tyler, Davies and Martin, 1981

Uperoleia minima Tyler, M.J., Davies, M. & Martin, A.A. (1981). Australian frogs of the leptodactylid genus *Uperoleia* Gray. *Aust. J. Zool. Suppl. Ser.* **79**: 1–64 [49 fig 12B]. Type data: holotype, WAM R62482 M, from Amax Crusher Site, 5 km SW of the Mining Camp, Mitchell Plateau Kimberley Division, in 14°50'S, 125°50'E, W.A.

Distribution: N coastal, W.A. Ecology: terrestrial, burrowing, nocturnal, predator, woodland, swamp; oviparous, arthropod-feeder, cryptozoic, breeds in swamps.

Uperoleia mjobergii (Andersson, 1913)

Pseudophryne mjobergii Andersson, L.G. (1913). Results of Dr. E. Mjöberg's Swedish Scientific Expeditions to Australia 1910–1913. 4. Batrachians. *K. Svenska Vetensk.-Akad. Handl. (ns)* **52**: 1–26 [19]. Type data: syntypes, NHRM 1567 (two specimens), from Noonkambah, Kimberley Division, north-western Australia, W.A.

Distribution: N coastal, W.A. Ecology: terrestrial, burrowing, nocturnal, swamp, predator; breeds in swamps, oviparous, arthropod-feeder, cryptozoic. Biological references: Main, A.R. (1965). Frogs of southern Western Australia. *West. Aust. Nat. Club Handbook* 8: 1–73 (natural history); Tyler, M.J., Davies, M. & Martin, A.A. (1981). New and rediscovered species of frogs from the Derby - Broome area of Western Australia. *Rec. West. Aust. Mus.* **9**: 147–172; Tyler, M.J., Davies, M. & Martin, A.A. (1981). Australian frogs of the leptodactylid genus *Uperoleia* Gray. *Aust. J. Zool. Suppl. Ser.* **79**: 1–64 (taxonomy).

Uperoleia orientalis (Parker, 1940)

Glauertia orientalis Parker, H.W. (1940). The Australasian frogs of the family Leptodactylidae. *Novit. Zool.* **42**: 1–106 [67]. Type data: holotype, BMNH 1947.2.19.84, from Alexandria Station, N.T.

Distribution: N coastal, N Gulf, N.T. Ecology: terrestrial, nocturnal, woodland, swamp, burrowing, predator; seasonal breeder in swamps, oviparous, free-living tadpole, arthropod-feeder, cryptozoic. Biological references: Tyler, M.J., Davies, M. & Martin, A.A. (1981). Australian frogs of the leptodactylid genus *Uperoleia* Gray. *Aust. J. Zool. Suppl. Ser.* **79**: 1–64 (taxonomy).

Uperoleia rugosa (Andersson, 1916)

Pseudophryne rugosa Andersson, L.G. (1916). Results of Dr. E. Mjöberg's Swedish Scientific Expeditions to Australia 1910–1913. 9. Batrachians from Queensland. *K. Svenska Vetensk.-Akad. Handl. (ns)* **52**: 1–20 [13 fig 4] [see comment under *Uperoleia laevigata* Keferstein, 1867]. Type data: holotype, NHRM 1630, from Colosseum, S Qld.

Distribution: NE coastal, SE coastal, Murray-Darling basin, Bulloo River basin, Qld., N.S.W., (A.C.T.). Ecology: tall open forest, open forest, woodland, low woodland, low open woodland, open scrub, tall shrubland, burrowing, terrestrial, nocturnal, predator; oviparous, opportunistic breeder, free-living tadpole, cryptozoic, arthropod-feeder. Biological references: Tyler, M.J., Davies, M. & Martin, A.A. (1981). Australian frogs of the leptodactylid genus *Uperoleia* Gray. *Aust. J. Zool. Suppl. Ser.* **79**: 1–64 (taxonomy).

Uperoleia russelli (Loveridge, 1933)

Glauertia russelli Loveridge, A. (1933). A new genus and three new species of crinine frogs from Australia. *Occ. Pap. Bost. Soc. Nat. Hist.* **8**: 89–94 [89]. Type data: holotype, WAM R2608, from Gascoyne River near Landor Station, W.A.

Distribution: NW coastal, W.A. Ecology: terrestrial, nocturnal, tall shrubland, (low shrubland), swamp, burrowing, predator; seasonal breeder in swamps, oviparous, free-living tadpole, arthropod-feeder, cryptozoic. Biological references: Main, A.R. (1965). Frogs of southern Western Australia. *West. Aust. Nat. Club Handbook* 8: 1–73 (natural history); Tyler, M.J., Davies, M. & Martin, A.A. (1981). Australian frogs of the leptodactylid genus *Uperoleia* Gray. *Aust. J. Zool. Suppl. Ser.* **79**: 1–64 (taxonomy).

Uperoleia talpa Tyler, Davies and Martin, 1981

Uperoleia talpa Tyler, M.J., Davies, M. & Martin, A.A. (1981). Australian frogs of the leptodactylid genus *Uperoleia* Gray. *Aust. J. Zool. Suppl. Ser.* **79**: 1–64 [52 figs 5B 17A-B]. Type data: holotype, WAM R62472 M, from 24km S of Derby, in 17°37′S, 123°36′E, W.A.

Distribution: N coastal, W.A. Ecology: terrestrial, burrowing, nocturnal, swamp, predator; oviparous, arthropod-feeder, cryptozoic, breeds in swamps.

Uperoleia trachyderma Tyler, Davies and Martin, 1981

Uperoleia trachyderma Tyler, M.J., Davies, M. & Martin, A.A. (1981). Frog fauna of the Northern Territory: new distributional records and the description of a new species. *Trans. R. Soc. S. Aust.* **105**: 149–154 [149 figs 1–4]. Type data: holotype, SAMA R20374 M, from Newcastle Creek floodplain at the George Redman Causeway, in 17°14′S, 133°28′E, 37 km N Elliot, N.T.

Distribution: W plateau, N.T. Ecology: terrestrial, burrowing, predator, nocturnal, swamp; oviparous, arthropod-feeder, cryptozoic, breeds in swamps.

Uperoleia variegata Tyler, Davies and Martin, 1981

Uperoleia variegata Tyler, M.J., Davies, M. & Martin, A.A. (1981). Australian frogs of the leptodactylid genus *Uperoleia* Gray. *Aust. J. Zool. Suppl. Ser.* **79**: 1–64 [55 figs 51 17C–D]. Type data: holotype, WAM R62461 M, from Gibb River Station, Kimberley Division, in 16°26′S, 126°26′E, W.A.

Distribution: N coastal, W.A. Ecology: terrestrial, burrowing, nocturnal, predator, swamp; oviparous, arthropod-feeder, cryptozoic, breeds in swamps.

HYLIDAE

INTRODUCTION

A large family of frogs found throughout most of the tropical and temperate regions of the world. Usually termed tree frogs because many species possess sticky discs on the tips of their fingers and toes which make them agile climbers. The family contains about 640 species in 40 genera of which 68 species and 3 genera occur in Australia. Proposals have been made to recognise a regional family, the Pelodryadidae, to accommodate the Australasian forms, but these have not gained widespread acceptance.

Characterised in Australia by: arciferal pectoral girdle; intercalary cartilages usually present; finger and toe discs usually present, though sometimes very poorly developed; sacral diapophyses dilated; maxillary teeth present.

References

Barker, J. & Grigg, G. (1977). *A Field Guide to Australian Frogs.* Adelaide : Rigby Ltd. 229 pp.

Brattstrom, B.H. (1970). Thermal acclimation in Australian amphibians. *Comp. Biochem. Physiol.* **35**: 69–103

Cogger, H.G. (1979). *Reptiles and Amphibians of Australia.* Sydney : A.H. & A.W. Reed 608 pp.

Copland, S.J. (1957). Presidential address. Australian tree frogs of the genus *Hyla. Proc. Linn. Soc. N.S.W.* **82**: 9–108

Duellman, W.E. (1977). Liste der rezenten Amphibien und Reptilien; Hylidae, Centrolenidae, Pseudidae. *Das Tierreich* **95**: i–xix 1–225

Tyler, M.J. (1970). Patterns of distribution and the origins of the Papuan hylid frog fauna. *Search* **1**: 246–247

Tyler, M.J. (1982). *Frogs.* 2nd edn. Sydney : William Collins 256 pp.

Cyclorana Steindachner, 1867

Chiroleptes Günther, A. (1858). *Catalogue of the Batrachia Salientia in the collection of the British Museum.* London : British Museum xvi 160 pp. [34] [*non Chiroleptes* Kirby, 1837]. Type species *Alytes australis* Gray, 1842 by monotypy.

Cyclorana Steindachner, F. (1867). Amphibien. *in, Reise der österreichischen Fregatte* Novara *um die Erde in den Jahren 1857, 1858, 1859.* Zoologie 1(4): 1–70 Vienna : State Printer [29] [1869 on title page]. Type species *Cyclorana novaehollandiae* Steindachner, 1867 by monotypy.

Phractops Peters, W. (1867). Herpetologische Notizen. *Mber. K. Preuss. Akad. Wiss. Berl.* **1867**: 13–37 [30] [1868 on title page]. Type species *Phractops alutaceus* Peters, 1867 by monotypy.

Cheiroleptes Spencer, B. (1901). Two new species of frogs from Victoria. *Proc. R. Soc. Vict. (ns)* **13**: 175–178 [176] [invalid emend. *lapsus pro Chiroleptes* Günther, 1858].

Synonymy that of Cogger, H.G., this work.

Cyclorana australis (Gray, 1842)

Alytes australis Gray, J.E. (1842). Description of some hitherto unrecorded species of Australian reptiles and batrachians. *in* Gray, J.E. (ed.) *Zoological Miscellany.* London : Treuttel, Würtz & Co. pp. 51–57 [56]. Type data: syntypes, BMNH 1947.2.18.43–44, from Port

Essington, N.T. (as N coast of Australia), see Boulenger, G.A. (1882). *Catalogue of the Batrachia Salientia s. Ecaudata in the collection of the British Museum.* 2nd edn. London : British Museum xvi 503 pp.

Distribution: N Gulf, N coastal, W plateau, Lake Eyre basin, Qld., N.T., W.A. Ecology: burrowing, predator, terrestrial, nocturnal, open forest, woodland, low woodland, low open woodland, low open shrubland, swamp; arthropod-feeder, seasonal breeder in swamps, oviparous, free-living tadpole, anuriphagous. Biological references: Parker, H.W. (1940). The Australasian frogs of the family Leptodactylidae. *Novit. Zool.* **42**: 1–106; Main, A.R. (1965). Frogs of southern Western Australia. *West. Aust. Nat. Club Handbook* 8: 1–73; Tyler, M.J. & Martin, A.A. (1975). Australian leptodactylid frogs of the *Cyclorana australis* complex. *Trans. R. Soc. S. Aust.* **99**: 93–99; King, M., Tyler, M.J., Davies, M. & King, D. (1979). Karyotypic studies on *Cyclorana* and associated genera of Australian frogs. *Aust. J. Zool.* **27**: 699–708.

Cyclorana brevipes (Peters, 1871)

Chiroleptes brevipes Peters, W. (1871). Über einige Arten der herpetologischen Sammlung des Berliner Zoologischen Museums. *Mber. K. Preuss. Akad. Wiss. Berl.* **1871**: 644–652 [648] [1872 on title page]. Type data: holotype, ZMB 7063 (not found), from Port Clinton, Qld. (as Port Bowen).

Chiroleptes brevipalmatus Günther, A. (1876). Descriptions of new species of reptiles from Australia. *J. Mus. Godeffroy* **5**: 45–47 [47]. Type data: syntypes, BMNH 1947.2.18.52 (Port Denison), BMNH 1947.2.18.53 (Peak Downs), BMNH 1947.2.18.54 (Gayndah), from Bowen (as Port Denison), Peak Downs and Gayndah, Qld.

Synonymy that of Cogger, H.G., this work.

Distribution: NE coastal, N gulf, Qld. Ecology: burrowing, predator, terrestrial, burrowing, nocturnal, tall forest, open forest, woodland; arthropod-feeder, seasonal breeder, oviparous, free-living tadpole. Biological references: Parker, H.W. (1940). The Australasian frogs of the family Leptodactylidae. *Novit. Zool.* **42**: 1–106; King, M., Tyler, M.J., Davies, M. & King, D. (1979). Karyotypic studies on *Cyclorana* and associated genera of Australian frogs. *Aust. J. Zool.* **27**: 699–708.

Cyclorana cryptotis Tyler and Martin, 1977

Cyclorana cryptotis Tyler, M.J. & Martin, A.A. (1977). Taxonomic studies of some Australian leptodactylid frogs of the genus *Cyclorana* Steindachner. *Rec. S. Aust. Mus.* **17**: 261–276 [271 fig 7a–c]. Type data: holotype, SAMA R14716, from Daly Waters, N.T.

Distribution: N Gulf, (N coastal), N.T. Ecology: terrestrial, burrowing, nocturnal; oviparous, arthropod-feeder. Biological references: Tyler, M.J., Davies, M. & Martin, A.A. (1981). Frog fauna of the Northern Territory: new distributional records and the description of a new species. *Trans. R. Soc. S. Aust.* **105**: 149–154; King, M., Tyler, M.J., Davies, M. & King, D. (1979). Karyotypic studies on *Cyclorana* and associated genera of Australian frogs. *Aust. J. Zool.* **27**: 699–708.

Cyclorana cultripes Parker, 1940

Cyclorana cultripes Parker, H.W. (1940). The Australasian frogs of the family Leptodactylidae. *Novit. Zool.* **42**: 1–106 [22]. Type data: holotype, BMNH 1947.2.18.45, from Alexandria, N.T.

Distribution: W plateau, Lake Eyre basin, Bulloo River basin, Qld., N.T., S.A. Ecology: burrowing, terrestrial, nocturnal, tall shrubland, tussock grassland, low open shrubland, desert, predator; opportunistic breeder, arthropod-feeder, oviparous, free-living tadpole. Biological references: Main, A.R. (1965). Frogs of southern Western Australia. *West. Aust. Nat. Club Handbook* 8: 1–73; Tyler, M.J. & Martin, A.A. (1977). Taxonomic studies of some Australian leptodactylid frogs of the genus *Cyclorana* Steindachner. *Rec. S. Aust. Mus.* **17**: 261–276.

Cyclorana longipes Tyler and Martin, 1977

Cyclorana longipes Tyler, M.J. & Martin, A.A. (1977). Taxonomic studies of some Australian leptodactylid frogs of the genus *Cyclorana* Steindachner. *Rec. S. Aust. Mus.* **17**: 261–276 [271 figs 1a–b,d,7b,d]. Type data: holotype, WAM R43258, from Mitchell Plateau "(140°52'S, 125°50'W)", Kimberley Division, W.A.

Distribution: N coastal, W.A., N.T. Ecology: terrestrial, burrowing, nocturnal; oviparous, arthropod-feeder. Biological references: King, M., Tyler, M.J., Davies, M. & King, D. (1979). Karyotypic studies on *Cyclorana* and associated genera of Australian frogs. *Aust. J. Zool.* **27**: 699–708.

Cyclorana maculosa Tyler and Martin, 1977

Cyclorana maculosus Tyler, M.J. & Martin, A.A. (1977). Taxonomic studies of some Australian leptodactylid frogs of the genus *Cyclorana* Steindachner. *Rec. S. Aust. Mus.* **17**: 261–276 [269 fig 6a,d]. Type data: holotype, SAMA R14719, from Daly Waters, N.T.

Distribution: N Gulf, (W plateau), Qld., N.T. Ecology: terrestrial, burrowing, nocturnal, low woodland, woodland; oviparous, arthropod-feeder.

Cyclorana maini Tyler and Martin, 1977

Cyclorana maini Tyler, M.J. & Martin, A.A. (1977). Taxonomic studies of some Australian leptodactylid frogs of the genus *Cyclorana* Steindachner. *Rec. S. Aust. Mus.* **17**: 261–276 [273 fig 8]. Type data: holotype, SAMA R15191, from Barrow Creek, N.T.

Distribution: NW coastal, W plateau, Lake Eyre basin, W.A., N.T. Ecology: terrestrial, burrowing, nocturnal, desert, low open shrubland, tall shrubland; oviparous, opportunistic breeder, free-living tadpole, arthropod-feeder. Biological references: Main, A.R. & Storr, G.M. (1966). Range extension and notes on the biology of frogs from the Pilbara Region, Western Australia. *West. Aust. Nat.* **10**: 53–61 (as *Cyclorana cultripes*); King, M., Tyler, M.J., Davies, M. & King, D. (1979). Karyotypic studies on *Cyclorana* and associated genera of Australian frogs. *Aust. J. Zool.* **27**: 699–708.

Cyclorana manya van Beurden and McDonald, 1980

Cyclorana manya van Beurden, E. & McDonald, K.R. (1980). A new species of *Cyclorana* (Anura: Hylidae) from north Queensland. *Trans. R. Soc. S. Aust.* **104**: 193–195 [193]. Type data: holotype, QM J34886, from between Coen airport and Deep Creek crossing, 25 km N of Coen township, Qld. in 13°52'S, 143°12'E.

Distribution: (NE coastal), N Gulf, Qld. Ecology: woodland, terrestrial, burrowing, nocturnal; oviparous, arthropod-feeder.

Cyclorana novaehollandiae Steindachner, 1867

Cyclorana novaehollandiae Steindachner, F. (1867). Amphibien. *in, Reise der österreichischen Fregatte* Novara *um die Erde in den Jahren 1857, 1858, 1859.* Zoologie 1(4): 1–70 Vienna : State Printer [29 pl 2 figs 7–10] [1869 on title page]. Type data: syntypes (probable), NHMW 4716–7, from Rockhampton, Qld.

Phractops alutaceus Peters, W. (1867). Herpetologische Notizen. *Mber. K. Preuss. Akad. Wiss. Berl.* **1867**: 13–37 [31] [1868 on title page; there is evidence that this name was published prior to *Cyclorana novaehollandiae* Steindachner, 1867]. Type data: holotype (probable), ZMB 5653, syntypes (possible) RMNH 3867, AMNH 23581, from Rockhampton, Qld.

Synonymy that of Cogger, H.G., this work.

Distribution: NE coastal, N Gulf, Murray-Darling basin, Bulloo River basin, Lake Eyre basin, N.S.W., Qld. Ecology: terrestrial, burrowing, nocturnal, closed forest, open forest, woodland, low woodland, open scrub, low open shrubland, tall shrubland, tussock grassland; seasonal breeder, oviparous, free-living tadpole, anuriphagous, arthropod-feeder. Biological references: Parker, H.W. (1940). The Australasian frogs of the family Leptodactylidae. *Novit. Zool.* **42**: 1–106; Tyler, M.J. & Martin, A.A. (1975). Australian leptodactylid frogs of the *Cyclorana australis* complex. *Trans. R. Soc. S. Aust.* **99**: 93–99.

Cyclorana platycephala (Günther, 1873)

Chiroleptes platycephalus Günther, A. (1873). Description of two new species of frogs from Australia. *Ann. Mag. Nat. Hist. (4)* **11**: 349–350 [350]. Type data: holotype, BMNH 1947.2.18.42, from Fort Bourke, N.S.W.

Cyclorana slevini Loveridge, A. (1950). New frogs of the genera *Cyclorana* and *Hyla* from south-eastern Australia. *Proc. Biol. Soc. Wash.* **63**: 131–138 [131]. Type data: holotype, CAS 82052, from Noondoo, SE Qld.

Synonymy that of Cogger, H.G., this work.

Distribution: NE coastal, NW coastal, W plateau, Murray-Darling basin, Bulloo River basin, Lake Eyre basin, S.A., N.S.W., Qld., W.A., N.T. Ecology: burrowing, predator, terrestrial, nocturnal, desert, low shrubland, tall shrubland, low open shrubland, hummock grass-land, low woodland; opportunistic aquatic breeder, oviparous, arthropod-feeder. Biological references: Parker, H.W. (1940). The Australasian frogs of the family Leptodactylidae. *Novit. Zool.* **42**: 1–106; Main, A.R. (1965). Frogs of southern Western Australia. *West. Aust. Nat. Club Handbook* 8: 1–73; Dawson, A.B. (1951). Functional and degenerate or rudimentary glomeruli in the kidney of two species of Australian frog, *Cyclorana* (*Chiroleptes*) *platycephalus* and *alboguttatus* (Günther). *Anat. Rec.* **109**: 417–430; van Beurden, E.K. (1980). Energy metabolism of dormant Australian water-holding frogs (*Cyclorana platycephalus*). *Copeia* **1980**: 787–799; King, M., Tyler, M.J., Davies, M. & King, D. (1979). Karyotypic studies on *Cyclorana* and associated genera of Australian frogs. *Aust. J. Zool.* **27**: 699–708.

Cyclorana vagita Tyler, Davies and Martin, 1981

Cyclorana vagitus Tyler, M.J., Davies, M. & Martin, A.A. (1981). New and rediscovered species of frogs from the Derby - Broome area of Western Australia. *Rec. West. Aust. Mus.* **9**: 147–172 [148 figs 1–4]. Type data: holotype, WAM R71037, from junction of the Great Northern Highway and the road to Derby, 41 km S of Derby, Kimberley District, W.A., in 124°38'E, 17°44'S.

Distribution: N coastal, W.A., (N.T.). Ecology: hummock grassland, tussock grassland, terrestrial, burrowing, nocturnal; "flooded grassland ... and on a flooded spinifex plain", oviparous, arthropod-feeder.

Cyclorana verrucosa Tyler and Martin, 1977

Cyclorana verrucosus Tyler, M.J. & Martin, A.A. (1977). Taxonomic studies of some Australian leptodactylid frogs of the genus *Cyclorana* Steindachner. *Rec. S. Aust. Mus.* **17**: 261–276 [267 fig 2c–d 5]. Type data: holotype, QM J18105, from 18 km W of Dalby, Qld.

Distribution: Murray-Darling basin, Bulloo River basin, Lake Eyre basin, N.S.W., Qld. Ecology: terrestrial, burrowing, nocturnal, woodland, low open woodland, low open shrubland, tall shrubland, low woodland; oviparous, free-living tadpole, arthropod-feeder.

Litoria Tschudi, 1838

Litoria Tschudi, J.J. (1838). *Classification der Batrachier, mit Berücksichtigung der fossilen Tiere dieser Abteilung der Reptilien.* Neuchâtel : Petitpierre 102 pp. [36,77]. Type species *Litoria freycineti* Tschudi, 1838 by monotypy.

Ranoidea Tschudi, J.J. (1838). *Classification der Batrachier, mit Berücksichtigung der fossilen Tiere dieser Abteilung der Reptilien.* Neuchâtel : Petitpierre 102 pp. [35,76]. Type species *Ranoidea jacksoniensis* Tschudi, 1838 by monotypy.

Dryopsophus Fitzinger, L.J. (1843). *Systema Reptilium.* Vienna : Braümüller u. Seidel vi 106 pp. [30]. Type species "*Hyla citropa* Péron" (=*Hyla citropa* Duméril and Bibron, 1841) by original designation.

Euscelis Fitzinger, L.J. (1843). *Systema Reptilium.* Vienna : Braümüller u. Seidel vi 106 pp. [31]. Type species *Hyla lesueurii* Duméril and Bibron, 1841 by original designation.

Pelobius Fitzinger, L.J. (1843). *Systema Reptilium.* Vienna : Braümüller u. Seidel vi 106 pp. [31]. Type species "*Litor[ia] freycineti* Duméril and Bibron" (=*Litoria freycineti* Tschudi, 1838) by original designation.

Pelodryas Günther, A. (1858). *Catalogue of the Batrachia Salientia in the collection of the British Museum.* London : British Museum xvi 160 pp. [119]. Type species *Rana caerulea* White, 1790 by monotypy.

Chirodryas Keferstein, W. (1867). Ueber einige neue oder seltene Batrachier aus Australien und dem tropischen Amerika. *Nachr. Ges. Wiss. Göttingen* **1867**: 341–361 [358]. Type species *Chirodryas raniformis* Keferstein, 1867 by monotypy.

Hylomantis Peters, W. (1880). Mitteilung über neue oder weniger bekannte Amphibien des Berliner Zoologischen Museums. *Mber. K. Preuss. Akad. Wiss. Berl.* **1880**: 217–224 [224] [*non Hylomantis* Peters 1872; 1881 on title page]. Type species *Hylomantis fallax* Peters, 1880 by monotypy.

Drymomantis Peters, W. (1882). [Name change of genus from *Hylomantis* to *Drymomantis*]. *Sber. Ges. Naturf. Freunde Berl.* **1882**: 8 [8] [*nom. nov. pro Hylomantis* Peters, 1880].

Fanchonia Werner, F. (1893). Herpetologische nova. *Zool. Anz.* **16**: 81–84 [82]. Type species *Fanchonia elegans* Werner, 1893 by monotypy.

Mitrolysis Cope, E.D. (1889). The Batrachia of North America. *Bull. U.S. Natl. Mus.* **34**: 1–515 [312]. Type species *Chiroleptes alboguttatus* Günther, 1867 by original designation.

Synonymy that of Cogger, H.G., this work.

This group is also found in eastern Indonesia and the New Guinea region, see Tyler, M.J. (1968). Papuan hylid frogs of the genus *Hyla. Zool. Verh.* **96**: 1–203; van Kampen, P.N. (1923). *The Amphibia of the Indo-Australian Archipelago.* Leiden : E.J. Brill xii 304 pp.

Litoria adelaidensis (Gray, 1841)

Hyla adelaidensis Gray, J.E. (1841). Description of some new species and four new genera of reptiles from Western Australia, discovered by John Gould, Esq. *Ann. Mag. Nat. Hist.* **7**: 86–91 [90]. Type data: holotype, BMNH 1947.2.22.80, from Western Australia.

Hyla bioculata Gray, J.E. (1841). Description of some new species and four new genera of reptiles from Western Australia, discovered by John Gould, Esq. *Ann. Mag. Nat. Hist.* **7**: 86–91 [90]. Type data: syntypes, BMNH 1947.2.23.68–74, from Western Australia.

Hyla binoculata Gray, J.E. (1841). A catalogue of the species of reptiles and Amphibia hitherto described as inhabiting Australia, with a description of some new species from Western Australia. Appendix E *in* Grey, G. *Journals of Two Expeditions of Discovery in North-west and Western Australia, during the years 1837, 38, and 39, under the authority of Her Majesty's Government.* London : T. & W. Boone 2: 422–449 [447] [invalid emend.; ?*lapsus pro Hyla bioculata* Gray, 1841].

Hyla schuetteii Keferstein, W. (1868). Über die Batrachier Australiens. *Arch. Naturgesch.* **34**: 253–290 [279 pl 8 figs 27–28]. Type data: holotype, ZFMK (*ex* ZMG 108b), from Sydney, N.S.W.

Synonymy that of Cogger, H.G., this work.

Distribution: SW coastal, W.A. Ecology: tall forest, open forest, woodland, swamp, nocturnal, predator; seasonal breeder, oviparous, free-living tadpole, arthropod-feeder, breeds in swamps. Biological references: Copland, S.J. (1957). Presidential address. Australian tree frogs of the genus *Hyla. Proc. Linn. Soc. N.S.W.* **82**: 9–108 (taxonomy); Main, A.R. (1965). Frogs of southern Western Australia. *West. Aust. Nat. Club Handbook* 8: 1–73 (natural history).

Litoria alboguttata (Günther, 1867)

Chiroleptes alboguttatus Günther, A. (1867). Additions to the knowledge of Australian reptiles and fishes. *Ann. Mag. Nat. Hist. (3)* **20**: 45–68 [54]. Type data: syntypes, BMNH 1947.2.18.50–51, BMNH 1947.2.20.6–7, from Bowen (as Port Denison) and Cape York, Qld.

Chiroleptes albopunctatus Fletcher, J.J. (1894). Contributions to a more exact knowledge of the geographical distribution of Australian Batrachia. 4. *Proc. Linn. Soc. N.S.W. (2)* **8**: 524–533 [525] [*errore pro Chiroleptes alboguttatus* Günther, 1867].

Synonymy that of Cogger, H.G., this work.

Distribution: N Gulf, NE coastal, Lake Eyre basin, Bulloo River basin, Murray-Darling basin, Qld., N.S.W., N.T. Ecology: open scrub, low open woodland, low woodland, woodland, lentic freshwater, swamp, terrestrial, noctidiurnal, predator; free-living tadpole, anuriphagous, arthropod-feeder. Biological references: Parker, H.W. (1940). The Australasian frogs of the family Leptodactylidae. *Novit. Zool.* **42**: 1–106 (taxonomy); Moore, J.A. (1961). The frogs of eastern New South Wales. *Bull. Am. Mus. Nat. Hist.* **121**: 149–386 (taxonomy, natural history); Tyler, M.J. (1974). The systematic position and geographic distribution of the Australian frog *Chiroleptes alboguttatus* Günther. *Proc. R. Soc. Qd.* **85**: 27–32; Dawson, A.B. (1951). Functional and degenerate or rudimentary glomeruli in the kidney of two species of Australian frog, *Cyclorana (Chiroleptes) platycephalus* and *alboguttatus* (Günther). *Anat. Rec.* **109**: 417–430.

Litoria aurea (Lesson, 1829)

Rana aurea Lesson, R.P. (1829–31). Description de quelques reptiles nouveaux ou peu connus. *in* Duperrey, L.I. (ed.) (1826–1838). *Voyage autour du monde, executé par ordre du Roi, sur la corvette de Sa Majesté,* La Coquille, *pendant les années 1822, 1823, 1824 et 1825.* Zoologie 2(1): 34–65 and Atlas (Reptiles plates 1–7) Paris : Arthus Bertrand [60 Reptiles pl 7 fig 2] [Publication dates established from Sherborn, C.D. & Woodward, B.B. (1906). Notes on the dates of publication of the natural history portions of some French voyages. "Voyage autour du Monde .... sur .... la Coquille pendant .... 1822-25 .... Par L.J. Duperrey, "etc. A correction. *Ann. Mag. Nat. Hist. (7)* **17**: 335–336 (date of livraison and text); Anon. (1829). [Review of] *Voyage autour du monde*. Livraison 11. *Bull. Sci. Nat. Géol.* (Férussac, ed.) **17**: 267 (identifies livraison in which plate appears); 1830 on title page]. Type data: syntypes (probable), from Macquarie River at Bathurst, N.S.W., but MNHP 4857 from Port Jackson is cited as lectotype by Roux-Estève, R. (1979). Liste des amphibiens et reptiles des collections du Muséum National d'Histoire Naturelle de Paris, récoltés par Lesueur (1778–1846). *Bull. Trim. Soc. Géol. Normandie et Amis Museum du Harve* **66**: 25–29 [26].

Ranoidea jacksoniensis Tschudi, J.J. (1838). *Classification der Batrachier, mit Berücksichtigung der fossilen Tiere dieser Abteilung der Reptilien.* Neuchâtel : Petitpierre 102 pp. [35,76]. Type data: syntypes, MNHP 537?, MNHP 4857-8; MNHP 4858 regarded as holotype by Guibé, J. (1950). *Catalogue des types d'Amphibiens du Muséum National d'Histoire Naturelle.* Paris : Imprimerie Nationale 71 pp., from Port Jackson, N.S.W., MNHP 537 cited as a specimen of *Litoria adelaidensis* from Perth, W.A., by Roux-Estève, R. (1979). Liste des amphibiens et reptiles des collections du Muséum National d'Historie Naturelle de Paris, récoltés par Lesueur (1778–1846). *Bull. Trim. Soc. Géol. Normandie et Amis Museum du Harve* **66**: 25–29 [26].

Ranoidea resplendens Girard, C. (1853). Descriptions of new species of reptiles, collected by the U.S. Exploring Expedition, under the command of Capt. Charles Wilkes, U.S.N. Second Part. Including the species of batrachians, exotic to North America. *Proc. Acad. Nat. Sci. Philad.* **6**: 420–424 [422] [1854 on title page]. Type data: syntypes (probable), not found, from Wollongong, Illawarra, N.S.W.

H[yla] jacksonii Duméril, A.M.C., Bibron, G. & Duméril, A. (1854). *Erpétologie Générale* ou Histoire Naturelle Complète des Reptiles. Paris : Roret 9: xx 440 pp. [407] [emend. *pro Hyla jacksoniensis* Duméril and Bibron, 1841, itself only a new combination for *Ranoidea jacksoniensis* Tschudi, 1838].

Fanchonia elegans Werner, F. (1893). Herpetologische nova. *Zool. Anz.* **16**: 81–84 [82] [synonymy *vide* Werner, F. (1897). Über einige noch unbeschriebene Reptilian und Batrachier. *Zool. Anz.* **20**: 261–267]. Type data: holotype, not found, from unknown locality (as Tropical Africa).

Hyla blandsuttoni Procter, J.B. (1924). Unrecorded characters seen in living snakes, and description of a new tree-frog. *Proc. Zool. Soc. Lond.* **1924**: 1125–1129 [1128 pl 3]. Type data: holotype, presumed lost, from ?Qld.

Hyla aurea ulongae Loveridge, A. (1950). New frogs of the genera *Cyclorana* and *Hyla* from south-eastern Australia. *Proc. Biol. Soc. Wash.* **63**: 131–138 [133]. Type data: holotype, AM R13817 (*ex* CAS 83235), from Ulong, N.S.W.

[Hyla] [aurea] ulongensis Loveridge, A. (1950). New frogs of the genera *Cyclorana* and *Hyla* from south-eastern Australia. *Proc. Biol. Soc. Wash.* **63**: 131–138 [134] [?*lapsus pro Hyla aurea ulongae* Loveridge, 1950].

Synonymy that of Cogger, H.G., this work.

Distribution: SE coastal, Murray-Darling basin, Vic., N.S.W., A.C.T. Ecology: low open woodland, woodland, open forest, lentic freshwater, swamp, amphibious, noctidiurnal, predator; seasonal breeder, free-living tadpole, general carnivore. Biological references: Copland, S.J. (1957). Presidential address. Australian tree frogs of the genus *Hyla. Proc. Linn. Soc. N.S.W.* **82**: 9–108 (taxonomy); Moore, J.A. (1961). The frogs of eastern New South Wales. *Bull. Am. Mus. Nat. Hist.* **121**: 149–386 (taxonomy, natural history); Parker, H.W. (1938). The races of the Australian frog *Hyla aurea* Lesson.' *Ann. Mag. Nat. Hist. (11)* **2**: 302–305; Harrison, L. (1922). On the breeding habits of some Australian frogs. *Aust. Zool.* **3**: 17–34; Marriner, G.R. (1905). On the anatomy of *Hyla aurea. Trans. Proc. N.Z. Inst.* **38**: 257–265; Courtice, G.P. & Grigg, G.C. (1975). A taxonomic revision of the *Litoria aurea* complex (Anura : Hylidae) in south-eastern Australia. *Aust. Zool.* **18**: 149–163 (taxonomy).

Litoria bicolor (Gray, 1842)

Eucnemis bicolor Gray, J.E. (1842). Description of some hitherto unrecorded species of Australian reptiles and batrachians. *in* Gray, J.E. (ed.) *Zoological Miscellany.* London : Treuttel, Würtz & Co. pp. 51–57 [57]. Type data: holotype, BMNH 1947.2.22.59, from Port Essington, N.T.

Distribution: N coastal, N Gulf, NE coastal, W.A., N.T., Qld. Ecology: open scrub, low open woodland, low woodland, woodland, open forest, tall forest, closed forest, swamp, supralittoral, terrestrial, arboreal, noctidurnal, predator; seasonal breeder, free-living tadpole, riparian, arthropod -feeder. Biological references: Copland, S.J. (1957). Presidential address. Australian tree frogs of the genus *Hyla*. *Proc. Linn. Soc. N.S.W.* **82**: 9–108 (taxonomy); Moore, J.A. (1961). The frogs of eastern New South Wales. *Bull. Am. Mus. Nat. Hist.* **121**: 149–386 (taxonomy, natural history); Straughan, I.R. (1968). The *Hyla bicolor* complex (Anura : Hylidae) in north Queensland. *Proc. R. Soc. Qd.* **80**: 43–54 (taxonomy).

Litoria booroolongensis (Moore, 1961)

Hyla booroolongensis Moore, J.A. (1961). The frogs of eastern New South Wales. *Bull. Am. Mus. Nat. Hist.* **121**: 149–386 [292 figs 59–60, pl 45 fig 1]. Type data: holotype, AM R16006, from Guy Fawkes Creek, Ebor, N.S.W.

Distribution: SE coastal, N.S.W., (A.C.T.). Ecology: woodland, open forest, tall forest, lotic freshwater, amphibious, nocturnal, predator; seasonal breeder, free-living tadpole, riparian, arthropod-feeder. Biological references: Copland, S.J. (1957). Presidential address. Australian tree frogs of the genus *Hyla*. *Proc. Linn. Soc. N.S.W.* **82**: 9–108 (taxonomy, as *Hyla* X); Moore, J.A. (1961). The frogs of eastern New South Wales. *Bull. Am. Mus. Nat. Hist.* **121**: 149–386 (taxonomy, natural history).

Litoria brevipalmata Tyler, Martin and Watson, 1972

Litoria brevipalmata Tyler, M.J., Martin, A.A. & Watson, G.F. (1972). A new species of hylid frog from New South Wales. *Proc. Linn. Soc. N.S.W.* **97**: 82–86 [82 figs 1–2 pl 2]. Type data: holotype, SAMA R11236, from Ourimbah Creek, about 5 mi NW of Gosford, N.S.W.

Distribution: SE coastal, NE coastal, N.S.W., Qld. Ecology: tall forest, closed forest, swamp, terrestrial, nocturnal, predator; seasonal breeder, free-living tadpole, arthropod-feeder, breeds in swamps. Biological references: Czechura, G.V. (1978). A new locality for *Litoria brevipalmata* (Anura : Pelodryadidae) from south-east Queensland. *Vict. Nat.* **95**: 150–151.

Litoria burrowsi (Scott, 1942)

Hyla burrowsi Scott, E.O.G. (1942). A new *Hyla* from Cradle Valley, Tasmania. *Rec. Queen Victoria Mus., Launceston* **1**: 5–11 [7 pls 1–3]. Type data: holotype, QVM 1941. 41, from Cradle Valley, Tas.

Distribution: Tas. Ecology: alpine, tall forest, closed forest, lentic freshwater, arboreal, nocturnal, predator; winter dormant, seasonal breeder, free-living tadpole, arthropod-feeder, breeds in lentic freshwater. Biological references: Copland, S.J. (1957). Presidential address. Australian tree frogs of the genus *Hyla*. *Proc. Linn. Soc. N.S.W.* **82**: 9–108 (taxonomy); Hewer, A.M. (1965). Distribution of the species *Hyla burrowsi* in Tasmania. *Tas. Nat. Bull. Suppl.* **2**: 1–2.

Litoria caerulea (White, 1790)

Rana caerulea White, J. (1790). *Journal of a Voyage to New South Wales,* with sixty five plates of non-descript animals, birds, lizards, serpents, curious cones of trees and other natural productions. Appendix. London : Debrett 299 pp. [248 pl 33]. Type data: syntypes (probable), from N.S.W., presumed lost, see Tyler, M.J. & Dobson, J. (1973). On the identity, authorship and fate of the type specimens of *Rana caerulea*. *Herpetologica* **29**: 373–375.

Rana austrasiae Schneider, J.G. (1799). *Historiae Amphibiorum naturalis et literariae.* fasc. prim. Jena : Frommani xvi 266 pp. [150] [*nom. substit. pro Rana caerulea* White, 1790].

Hyla cyanea Daudin, F.M. (1803). *Histoire Naturelle, générale et particulière des Reptiles.* Paris : Dufart Tom. 8 439 pp. [43] [?*nom. substit. pro Rana austrasiae* Schneider, 1799].

R[ana] coerulea Daudin, F. M. (1803). *Histoire Naturelle des Rainettes, des Grenouilles et des Crapauds.* Paris : Levrault 70 pp. [70] [*pro Rana caerulea* White, 1790].

Hyla irrorata De Vis, C.W. (1884). On a new species of *Hyla*. *Proc. R. Soc. Qd.* **1**: 128–130 [128]. Type data: neotype, QM J9255, designation by Copland, S.J. (1962). Rediscovery of a little known Victorian frog. *Proc. Linn. Soc. N.S.W.* **86**: 258–261, from Gympie, Qld.

Synonymy that of Cogger, H.G., this work.

Distribution: SE coastal, NE coastal, N Gulf, N coastal, W plateau, Murray-Darling basin, Bulloo River basin, Lake Eyre basin, S.A., N.S.W., Qld., W.A., N.T., A.C.T. Ecology: tall open shrubland, open scrub, low open woodland, low woodland, woodland, open forest, tall forest, closed forest, swamp, arboreal, nocturnal, predator; seasonal breeder, free-living tadpole, arthropod-feeder, breeds in swamps. Biological references: Copland, S.J. (1957). Presidential address. Australian tree frogs of the genus *Hyla*. *Proc. Linn. Soc. N.S.W.* **82**: 9–108 (taxonomy); Moore, J.A. (1961). The frogs of eastern New South Wales. *Bull. Am. Mus. Nat. Hist.* **121**: 149–386 (taxonomy, natural history); Gillies, C.D. & Peberdy, E.F. (1917). The anatomy of *Hyla caerulea* White. 1. The pectoral girdle and vertebral column. *Proc. R. Soc. Qd.* **29**: 45–54 (as *Hyla caerulea*); Harrison, L. (1922). On the breeding habits of some Australian frogs. *Aust. Zool.* **3**: 17–34; Johnson, C.R. (1971). Daily variation in the thermal tolerance of *Litoria caerulea* (Anura : Hylidae). *Comp. Biochem. Physiol.* **40**: 1109–1111.

Litoria castanea (Steindachner, 1867)

Hyla castanea Steindachner, F. (1867). Amphibien. *in, Reise der österreichischen Fregatte* Novara *um die Erde in den Jahren 1857, 1858, 1859.* Zoologie 1(4): 1–70 Vienna : State Printer [62 pl 3 figs 9–13] [1869 on title page]. Type data: holotype, not found, from unknown locality.

Litoria flavipunctata Courtice, G.P. & Grigg, G.C. (1975). A taxonomic revision of the *Litoria aurea* complex (Anura : Hylidae) in south-eastern Australia. *Aust. Zool.* **18**: 149–163 [159]. Type data: holotype, AM R40676, from Booralong Creek Road, 12.8 km W of Guyra, N.S.W. in 30°16′S, 151°33′E.

Synonymy that of Cogger, H.G., this work.

Distribution: SE coastal, (Murray-Darling basin), N.S.W. Ecology: woodland, open forest, swamp, amphibious, noctidiurnal, predator; winter dormant, seasonal breeder, free-living tadpole, arthropod-feeder.

Litoria cavernicola Tyler and Davies, 1979

Litoria cavernicola Tyler, M.J. & Davies, M. (1979). A new species of cave-dwelling, hylid frog from Mitchell Plateau, Western Australia. *Trans. R. Soc. S. Aust.* **103**: 149–153 [149 figs 1–3]. Type data: holotype, WAM R43228, from approx. 3 km W Surveyor's Pool, Mitchell Plateau, Kimberley Division, W.A.

Distribution: N coastal, W.A. Ecology: caves, woodland, nocturnal, predator; arthropod-feeder.

Litoria chloris (Boulenger, 1893)

Hyla chloris Boulenger, G.A. (1893). Description of a new tree frog from New South Wales. *Proc. Linn. Soc. N.S.W. (2)* **7**: 403 [403]. Type data: holotype, BMNH 1947.2.23.86, from Dunoon, Richmond River, N.S.W.

Distribution: SE coastal, NE coastal, N.S.W., Qld. Ecology: tall forest, closed forest, swamp, arboreal, nocturnal, predator; seasonal breeder, free-living tadpole, arthropod-feeder, breeds in swamps. Biological references: Copland, S.J. (1957). Presidential address. Australian tree frogs of the genus *Hyla. Proc. Linn. Soc. N.S.W.* **82**: 9–108 (taxonomy); Moore, J.A. (1961). The frogs of eastern New South Wales. *Bull. Am. Mus. Nat. Hist.* **121**: 149–386 (taxonomy, natural history); Davies, M. & McDonald, K.R. (1979). A study of intraspecific variation in the green tree frog *Litoria chloris* (Boulenger) (Hylidae). *Aust. Zool.* **20**: 347–359; Watson, G.F. & Martin, A.A. (1979). Early development of the Australian green hylid frogs, *Litoria chloris, L. fallax* and *L. gracilenta. Aust. Zool.* **20**: 259–268.

Litoria citropa (Duméril and Bibron, 1841)

Dendrohyas citropa Tschudi, J.J. (1838). *Classification der Batrachier, mit Berücksichtigung der fossilen Tiere dieser Abteilung der Reptilien.* Neuchâtel : Petitpierre 102 pp. [34,75] [*nom. nud.*].

Hyla citropa Duméril, A.M.C. & Bibron, G. (1841). *Erpétologie Générale* ou Histoire Naturelle Complète des Reptiles. Paris : Roret 8: ii 792 pp. [600] [*ex* Péron and Lesueur ms]. Type data: holotype, MNHP 4854, from Port Jackson, N.S.W.

Hyla jenolanensis Copland, S.J. (1957). Presidential address. Australian tree frogs of the genus *Hyla. Proc. Linn. Soc. N.S.W.* **82**: 9–108 [97]. Type data: holotype, AM R14412, from Bottomless Pit, Jenolan Caves, N.S.W.

Synonymy that of Cogger, H.G., this work.

Distribution: SE coastal, Murray-Darling basin, Vic., N.S.W. Ecology: (woodland), open forest, tall forest, terrestrial, arboreal, nocturnal, predator; winter dormant, seasonal breeder, free-living tadpole, arthropod-feeder, breeds in swamps. Biological references: Copland, S.J. (1957). Presidential address. Australian tree frogs of the genus *Hyla. Proc. Linn. Soc. N.S.W.* **82**: 9–108 (taxonomy); Moore, J.A. (1961). The frogs of eastern New South Wales. *Bull. Am. Mus. Nat. Hist.* **121**: 149–386 (taxonomy, natural history); Tyler, M.J. & Anstis, M. (1975). Taxonomy and biology of frogs of the *Litoria citropa* complex (Anura : Hylidae). *Rec. S. Aust. Mus.* **17**: 41–50.

Litoria cooloolensis Liem, 1974

Litoria cooloolensis Liem, D.S. (1974). A new species of the *Litoria bicolor* species group from southeast Queensland, Australia (Anura : Hylidae). *Mem. Qd. Mus.* **17**: 169–174 [169 fig 1 pl 5]. Type data: holotype, QM J22646, from Lake Coolamera, Cooloola, SE Qld.

Distribution: NE coastal, Qld. Ecology: woodland, tall forest, closed forest, lentic freshwater, nocturnal, predator; free-living tadpole, arthropod-feeder. Biological references: Ingram, G.J. & Corben, C.J. (1975). The frog fauna of North Stradbroke Island with comments on the acid frogs of the wallum. *Proc. R. Soc. Qd.* **86**: 49–54 (natural history).

Litoria coplandi (Tyler, 1968)

***Hyla latopalmata watjulumensis* (part.)** Copland, S.J. (1957). Presidential address. Australian tree frogs of the genus *Hyla. Proc. Linn. Soc. N.S.W.* **82**: 9–108 [96] [Tyler, M.J. (1968). A taxonomic study of hylid frogs of the *Hyla lesueuri* complex occurring in north-western Australia. *Rec. S. Aust. Mus.* **15**: 711–727 cited 29 syntypes. Of these, those listed below he assigned to *Litoria coplandi* (Tyler, 1968). For his assignment of the other syntypes see *Litoria peronii* (Tschudi, 1838) and *Litoria wotjulumensis* (Copland, 1957)]. Type data: syntypes, WAM R11197, WAM R11638, WAM R11939, from Watjulum [*sic*] Mission, W.A.

Hyla coplandi Tyler, M.J. (1968). A taxonomic study of hylid frogs of the *Hyla lesueuri* complex occurring in north-western Australia. *Rec. S. Aust. Mus.* **15**: 711–727 [716]. Type data: holotype, WAM R13722G, from Inverway Station, N.T.

Synonymy that of Cogger, H.G., this work.

Distribution: N Gulf, N coastal, W.A., N.T. Ecology: low open woodland, woodland, open forest, terrestrial, nocturnal, predator; free-living tadpole, saxicoline, arthropod-feeder.

Litoria cyclorhyncha (Boulenger, 1882)

Litoria punctata Duméril, A. (1853). Mémoire sur les batraciens anoures, de la famille des hylaeformes ou rainettes. *Ann. Sci. Nat. (3)* **19** Zool.: 135–179 [149] [*non Hyla punctata* Daudin, 1802]. Type data: holotype, MNHP 758, from unknown locality (as environs of Sydney).

Litoria marmorata Duméril, A. (1853). Mémoire sur les batraciens anoures, de la famille des hylaeformes ou rainettes. *Ann. Sci. Nat. (3)* **19** Zool.: 135–179 [150] [*non Hyla marmorata* Sonnini and Latreille, 1801]. Type data: syntypes, MNHP 757 (2 specimens), from Australia.

Hyla hyposticta Cope, E.D. (1866). On the structure and distribution of the genera of arciferous Anura. *J. Acad. Nat. Sci. Philad. (2)* **6**: 67–112 [85] [*nom. nud.*].

Hyla dimolops Cope, E.D. (1866). On the structure and distribution of the genera of arciferous Anura. *J. Acad. Nat. Sci. Philad. (2)* **6**: 67–112 [85] [*nom. nud.*].

Hyla dimolops Cope, E.D. (1869). Seventh contribution to the herpetology of tropical America. *Proc. Am. Phil. Soc.* **11**: 147–169 [163] [*nom. nov. pro Litoria punctata* Duméril, 1853, at that time a junior secondary homonym of *Hyla punctata* Daudin, 1802; *nom. oblit.*].

Hyla thyposticta Cope, E.D. (1869). Seventh contribution to the herpetology of tropical America. *Proc. Am. Phil. Soc.* **11**: 147–169 [163] [*nom. nov. pro Litoria marmorata* Duméril, 1853, at that time a junior secondary homonym of *Hyla marmorata* Sonnini and Latreille, 1801; *nom. oblit.*].

Hyla aurea cyclorhynchus Boulenger, G.A. (1882). *Catalogue of the Batrachia Salientia s. Ecaudata in the collection of the British Museum.* 2nd edn. London : British Museum xvi 503 pp. [411]. Type data: syntypes, BMNH 69.7.22.3–7, AMNH 62493, from Western Australia.

Synonymy that of Cogger, H.G., this work.

Distribution: SW coastal, W plateau, W.A. Ecology: open heath, swamp, amphibious, noctidiurnal, predator; seasonal breeder, free-living tadpole, general carnivore. Biological references: Copland, S.J. (1957). Presidential address. Australian tree frogs of the genus *Hyla. Proc. Linn. Soc. N.S.W.* **82**: 9–108 (taxonomy); Main, A.R. (1965). Frogs of southern Western Australia. *West. Aust. Nat. Club Handbook* 8: 1–73 (natural history).

Litoria dahlii (Boulenger, 1896)

Chiroleptes dahlii Boulenger, G.A. (1896). Descriptions of a new snake and a new frog from north Australia. *Proc. Zool. Soc. Lond.* **1895**: 867 [867 pl 49 fig 2]. Type data: syntypes, BMNH 1947.2.18.49, UZMO K2002, from Daly River, N.T.

Distribution: N Gulf, N coastal, Qld., W.A., N.T. Ecology: tall open shrubland, low woodland, woodland, lentic freshwater, swamp, aquatic, terrestrial, noctidiurnal, predator; free-living tadpole, anuriphagous, arthropod-feeder. Biological references: Parker, H.W. (1940). The Australasian frogs of the family Leptodactylidae. *Novit. Zool.* **42**: 1–106 (taxonomy); Tyler, M.J., Davies, M. & King, M. (1978). The Australian frog *Chiroleptes dahlii* Boulenger: its systematic position, morphology, chromosomes and distribution. *Trans. R. Soc. S. Aust.* **102**: 17–24.

Litoria dentata (Keferstein, 1868)

Hyla dentata Keferstein, W. (1868). Über die Batrachier Australiens. *Arch. Naturgesch.* **34**: 253–290 [284] [holotype length differs from that given in original description but inconsistencies in other measurements suggest that stated length was erroneous]. Type data: holotype (probable), ZFMK (*ex* ZMG 123a), from N.S.W.

Distribution: SE coastal, NE coastal, (Murray-Darling basin), N.S.W., Qld. Ecology: woodland, open forest, tall forest, closed forest, lentic freshwater, arboreal, nocturnal, predator; seasonal breeder, free-living tadpole, arthropod-feeder, breeds in lentic freshwater. Biological references: Copland, S.J. (1957). Presidential address. Australian tree frogs of the genus *Hyla. Proc. Linn. Soc. N.S.W.* **82**: 9–108 (taxonomy); Moore, J.A. (1961). The frogs of eastern New South Wales. *Bull. Am. Mus. Nat. Hist.* **121**: 149–386 (taxonomy, natural history).

Litoria dorsalis Macleay, 1877

Litoria dorsalis Macleay, W. (1877). The batrachians of the *Chevert* Expedition. *Proc. Linn. Soc. N.S.W.* **2**: 135–138 [138]. Type data: holotype, presumed lost, from Mawatta, Binaturi River (as Katow), Papua New Guinea.

Hyla dorsalis microbelos Cogger, H.G. (1966). A new hylid frog from Australia. *Aust. Zool.* **13**: 220–227 [223 fig 1 pl 12]. Type data: holotype, AM R25836, from Cairns, Qld.

Synonymy that of Cogger, H.G., this work.

Distribution: NE coastal, N Gulf, N coastal, Qld., W.A., N.T. Ecology: low open woodland, woodland, open forest, swamp, mangrove, terrestrial, noctidiurnal, predator; free-living tadpole, arthropod-feeder. Biological references: Tyler, M.J. (1968). Papuan hylid frogs of the genus *Hyla. Zool. Verh.* **96**: 1–203 (taxonomy); Cogger,

H.G. & Lindner, D.A. (1974). Frogs and reptiles in fauna survey of the Port Essington District, Cobourg Peninsula, Northern Territory of Australia. Frith, H.J. & Calaby, J.H. (eds.) *CSIRO Div. Wildl. Res. Tech. Pap.* **28**: 63–107 (taxonomy).

Litoria ewingii (Duméril and Bibron, 1841)

Hyla ewingii Duméril, A.M.C. & Bibron, G. (1841). *Erpétologie Générale* ou Histoire Naturelle Complète des Reptiles. Paris : Roret 8: ii 792 pp. [597]. Type data: holotype, MNHP 4851 *vide* Guibé, J. (1950). *Catalogue des types d'Amphibiens du Muséum National d'Histoire Naturelle.* Paris : Imprimerie Nationale 71 pp., from Tas.

Hyla parvidens Peters, W. (1874). Ueber neue Amphibien. *Mber. K. Preuss. Akad. Wiss. Berl.* **1874**: 616–624 [620] [1875 on title page]. Type data: holotype, ZMB 8252, from Port Phillip, Vic.

Hyla calliscelis Peters, W. (1874). Ueber neue Amphibien. *Mber. K. Preuss. Akad. Wiss. Berl.* **1874**: 616–624 [620] [1875 on title page]. Type data: holotype, ZMB 8251 (not found), from Adelaide, S.A.

Hyla inguinalis Ahl, E. (1935). Beschreibung eines neuen Laubfrösche aus Südaustralien. *Zool. Anz.* **109**: 252–253 [252]. Type data: holotype, ZMB 14080, from S.A. (probably vicinity of Adelaide).

Hyla ewingii iuxtaewingii Copland, S.J. (1957). Presidential address. Australian tree frogs of the genus *Hyla. Proc. Linn. Soc. N.S.W.* **82**: 9–108 [60]. Type data: holotype, AM R85610, from 18 mi W of Bairnsdale, Vic.

Synonymy that of Cogger, H.G., this work.

Distribution: SE coastal, S Gulfs, Murray-Darling basin, S.A., Tas., Vic., N.S.W., A.C.T. Ecology: alpine, tussock grassland, woodland, open forest, tall forest, lentic freshwater, swamp, terrestrial, arboreal, nocturnal, predator; seasonal breeder, free-living tadpole, arthropod-feeder. Biological references: Copland, S.J. (1957). Presidential address. Australian tree frogs of the genus *Hyla. Proc. Linn. Soc. N.S.W.* **82**: 9–108 (taxonomy); Moore, J.A. (1961). The frogs of eastern New South Wales. *Bull. Am. Mus. Nat. Hist.* **121**: 149–386 (taxonomy, natural history); Littlejohn, M.J. (1965). Premating isolation in the *Hyla ewingi* complex (Anura : Hylidae). *Evolution* **19**: 234–243; Warburg, M.R. (1965). Studies on the water economy of some Australian frogs. *Aust. J. Zool.* **13**: 317–330; Watson, G.F. & Littlejohn, M.J. (1978). The *Litoria ewingi* complex (Anura : Hylidae) in south-eastern Australia. V. Interactions between northern *L. ewingi* and adjacent taxa. *Aust. J. Zool.* **26**: 175–195; Gartside, D.F. (1982). The *Litoria ewingi* complex (Anura : Hylidae) in south-eastern Australia. VI. Geographic variation in transferrins of four taxa. *Aust. J. Zool.* **30**. 103–113.

Litoria fallax (Peters, 1880)

Hylomantis fallax Peters, W. (1880). Mitteilung über neue oder weniger bekannte Amphibien des Berliner Zoologischen Museums. *Mber. K. Preuss. Akad. Wiss. Berl.* **1880**: 217–224 [224 fig 4] [1881 on title page]. Type data: syntypes, ZMB 5887 (2 specimens), ZMB 5925 (2 specimens), ZMB 6633 (2 specimens), ZMB 9650, from Port Clinton (as Port Bowen), Mackay and Rockhampton, Qld.

Hyla bicolor glauerti Copland, S.J. (1957). Presidential address. Australian tree frogs of the genus *Hyla. Proc. Linn. Soc. N.S.W.* **82**: 9–108 [16]. Type data: holotype, AM R18585, from Colo, near Sydney, N.S.W.

Synonymy that of Cogger, H.G., this work.

Distribution: SE coastal, NE coastal, (Murray-Darling basin), N.S.W., Qld. Ecology: woodland, open forest, tall forest, lentic freshwater, lotic freshwater, terrestrial, arboreal, noctidiurnal, predator, (closed forest); free-living tadpole, arthropod-feeder. Biological references: Moore, J.A. (1961). The frogs of eastern New South Wales. *Bull. Am. Mus. Nat. Hist.* **121**: 149–386 (taxonomy, as *Hyla bicolor*); Cogger, H.G. & Lindner, D.A. (1974). Frogs and reptiles in fauna survey of the Port Essington District, Cobourg Peninsula, Northern Territory of Australia. Frith, H.J. & Calaby, J.H. (eds.) *CSIRO Div. Wildl. Res. Tech. Pap.* **28**: 63–107 (taxonomy); Straughan, I.R. (1968). The *Hyla bicolor* complex (Anura : Hylidae) in north Queensland. *Proc. R. Soc. Qd.* **80**: 43–54 (taxonomy, as *Hyla glauerti*); Watson, G.F. & Martin, A.A. (1979). Early development of the Australian green hylid frogs, *Litoria chloris, L. fallax* and *L. gracilenta. Aust. Zool.* **20**: 259–268.

Litoria freycineti Tschudi, 1838

Litoria freycineti Tschudi, J.J. (1838). *Classification der Batrachier, mit Berücksichtigung der fossilen Tiere dieser Abteilung der Reptilien.* Neuchâtel : Petitpierre 102 pp. [36, 77]. Type data: syntypes, MNHP 246 (2 specimens), from Australia.

Hyla (Litoria) freycinetii verruculata Keferstein, W. (1867). Ueber einige neue oder seltene Batrachier aus Australien und dem tropischen Amerika. *Nachr. Ges. Wiss. Göttingen* **1867**: 341–361 [355]. Type data: syntypes (probable), presumed lost, from Australia.

Litoria copei Steindachner, F. (1867). Amphibien. *in, Reise der österreichischen Fregatte* Novara *um die Erde in den Jahren 1857, 1858, 1859.* Zoologie 1(4): 1–70 Vienna : State Printer [56] [1869 on title page]. Type data: holotype, NHMW 16381, from N.S.W.

Synonymy that of Cogger, H.G., this work.

Distribution: SE coastal, NE coastal, (Murray-Darling basin), N.S.W., Qld. Ecology: woodland, open forest, tall forest, terrestrial, nocturnal, predator; seasonal breeder, free-living tadpole, arthropod-feeder. Biological references: Copland, S.J. (1957). Presidential address.

Australian tree frogs of the genus *Hyla*. *Proc. Linn. Soc. N.S.W.* **82**: 9–108 (taxonomy); Moore, J.A. (1961). The frogs of eastern New South Wales. *Bull. Am. Mus. Nat. Hist.* **121**: 149–386 (taxonomy, natural history).

Litoria gilleni (Spencer, 1896)

Hyla gilleni Spencer, B. (1896). *Report on the work of the Horn Scientific Expedition to central Australia.* Pt. 2 Zoology: Amphibia. 152–175 London : Dulau [173]. Type data: holotype, NMV D9827, from Alice Springs, N.T.

Distribution: Lake Eyre basin, (S.A.), N.T. Ecology: low open shrubland, lentic freshwater, tall shrubland, arboreal, nocturnal, predator; opportunistic breeder, free-living tadpole, saxicoline, arthropod-feeder, breeds in lentic freshwater. Biological references: Copland, S.J. (1957). Presidential address. Australian tree frogs of the genus *Hyla*. *Proc. Linn. Soc. N.S.W.* **82**: 9–108 (taxonomy); Mertens, R. (1964). Die systematische Stellung des zentral-australischen Laubfrösches *Hyla gilleni*. *Senckenberg. Biol.* **45**: 15–21.

Litoria gracilenta (Peters, 1869)

Hyla gracilenta Peters, W. (1869). Mitteilung über Saurier. *Mber. K. Preuss. Akad. Wiss. Berl.* **1869**: 786–790 [789]. Type data: holotype, ZMB 6618, from Port Mackay, Qld.

Pelodryas granulata Peters, W. (1873). Über zwei Giftschlangen aus Afrika und über neue oder weniger bekannte Gattungen und Arten von Batrachiern. *Mber. K. Preuss. Akad. Wiss. Berl.* **1873**: 411–418 [417 pl 1 fig 4] [1874 on title page]. Type data: syntypes, ZMB 7708 (2 specimens), from Port Clinton (as Port Bowen), Qld.

Hyla fordii Günther, A. (1876). Descriptions of new species of reptiles from Australia. *J. Mus. Godeffroy* **5**: 45–47 [47]. Type data: syntypes, BMNH 1947.2.23.89–91, from Rockhampton, Qld.

Hyla luteiventris Ogilby, J.D. (1907). A new tree frog from Brisbane. *Proc. R. Soc. Qd.* **20**: 31–32 [31]. Type data: holotype, AM R4379, from Brisbane, Qld.

Synonymy that of Cogger, H.G., this work.

Distribution: SE coastal, NE coastal, N Gulf, N.S.W., Qld. Ecology: woodland, open forest, tall forest, closed forest, swamp, arboreal, nocturnal, predator; seasonal breeder, free-living tadpole, arthropod-feeder, breeds in swamps. Biological references: Copland, S.J. (1957). Presidential address. Australian tree frogs of the genus *Hyla*. *Proc. Linn. Soc. N.S.W.* **82**: 9–108 (taxonomy); Moore, J.A. (1961). The frogs of eastern New South Wales. *Bull. Am. Mus. Nat. Hist.* **121**: 149–386 (taxonomy, natural history); Johnson, C.R. (1972). Diel variation in the thermal tolerance of *Litoria gracilenta* (Anura : Hylidae). *Comp. Biochem. Physiol.* **41**: 727–730; Watson, G.F. & Martin, A.A. (1979). Early development of the Australian green hylid frogs, *Litoria chloris*, *L. fallax* and *L. gracilenta*. *Aust. Zool.* **20**: 259–268.

Litoria inermis (Peters, 1867)

Chiroleptes inermis Peters, W. (1867). Herpetologische Notizen. *Mber. K. Preuss. Akad. Wiss. Berl.* **1867**: 13–37 [30] [1868 on title page]. Type data: holotype, ZMB 5589, from Rockhampton, Qld. [4 other specimens have been variously regarded as syntypes, apparently in error: RMNH 1888, NHMW 16384 (2 specimens) and AMNH 23582].

Distribution: NE coastal, N Gulf, N coastal, Qld., W.A., N.T. Ecology: low open woodland, woodland, open forest, swamp, terrestrial, nocturnal, predator; seasonal breeder, free-living tadpole, arthropod-feeder. Biological references: Straughan, I.R. (1969). *Hyla inermis* (Peters), a species hitherto erroneously referred to the leptodactylid genus *Cyclorana* (Anura : Hylidae/Leptodactylidae). *Zool. Meded.* **43**: 207–212 (taxonomy); Parker, H.W. (1940). The Australasian frogs of the family Leptodactylidae. *Novit. Zool.* **42**: 1–106 (taxonomy, as *Cyclorana inermis*).

Litoria infrafrenata (Günther, 1867)

Hyla infrafrenata Günther, A. (1867). Additions to the knowledge of Australian reptiles and fishes. *Ann. Mag. Nat. Hist. (3)* **20**: 45–68 [56]. Type data: holotype, BMNH 1947.2.24.11, from Cape York, Qld.

Calamita dolichopsis Cope, E.D. (1867). On the families of raniform Anura. *J. Acad. Nat. Sci. Philad. (2)* **6**: 189–206 [204]. Type data: syntypes (probable), ANSP (presumed lost), from Amboina, Molucca Ils.

Pelodryas militarius Ramsay, E.P. (1877). Description of a new species of *Pelodryas* from New Ireland. *Proc. Linn. Soc. N.S.W.* **2**: 28–30 [28] [1878 on title page]. Type data: holotype, AM R5250, from New Ireland.

Litoria guttata Macleay, W. (1877). The batrachians of the *Chevert* Expedition. *Proc. Linn. Soc. N.S.W.* **2**: 135–138 [137]. Type data: holotype, AM R29956, from Mawatta, Binaturi River, near Daru, Papua New Guinea (as Katow).

Hyla infrataeniata Boulenger, G.A. (1885). Remarks on Mr. C.W. De Vis's recent contributions to the herpetology of Australia. *Ann. Mag. Nat. Hist. (5)* **16**: 386–387 [387] [*lapsus pro Hyla infrafrenata* Günther, 1867].

Hyla dolichopsis tenuigranulata Boettger, O. (1895). Liste der Reptilien und Batrachier der Insel Halmaheira nach den Sammlunge Prof. Dr. W. Kukenthal's. *Zool. Anz.* **18**(472): 129–140 [136]. Type data: lectotype, SMF 2117, designation by Mertens, R. (1967). Die herpetologische Sektion des Natur-Museums und Forschungs-Institutes Senckenberg in Frankfurt-a.-M. nebst einem Verzeichnis ihrer Typen. 1. *Senckenberg. Biol.* **48**: 1–106, from Halmahera and Ternate Ils., Molucca Ils.

Hyla dolichopsis pollicaris Werner, F. (1898). Vorläufige Mitteilung über die von Herrn Prof. F. Dahl im

Bismarck-Archipel gesammelten Reptilien und Batrachier. *Zool. Anz.* **21**: 552–556 [554]. Type data: syntypes, NHMW 16864, whereabouts of remaining syntype(s) unknown, from Ralum, New Britain, Bismarck Archipelago, see Haupl, M. & Tiedemann, F. (1978). Typenkatalog der herpetologischen Sammlung. Amphibia. *Kat. Wiss. Samml. Naturhist. Mus. Wien* **2**(Vertebrata 1): 1–35.

Hyla dolichopsis calcarifera Vogt, T. (1912). Beitrag zur Reptilien- und Amphibienfauna der Südseeinseln. *Sber. Ges. Naturf. Freunde Berl.* **1912**: 1–13 [10] [*nom. nud.*; name attributed to Werner].

Hyla spengeli Boulenger, G.A. (1912). On some tree frogs allied to *Hyla caerulea*, with remarks on noteworthy secondary sexual characteristics in the family Hylidae. *Zool. Jb. Suppl. 15* **1**: 211–218 [215]. Type data: holotype, BMNH 1947.2.23.15, from Dinawa, Owen Stanley Mts., Papua New Guinea.

Hyla trinilensis Ahl, E. (1929). Beschreibung eines neuen Laubfrösches der Gattung *Hyla* von Java. *Zool. Anz.* **85**: 269–271 [270]. Type data: syntypes, ZMB, from Trinil, Central Java, Indonesia.

Hyla spengleri Loveridge, A. (1948). New Guinean reptiles and amphibians in the Museum of Comparative Zoology and the United States National Museum. *Bull. Mus. Comp. Zool.* **101**: 305–430 [402] [*lapsus pro Hyla spengeli* Boulenger, 1912].

Synonymy that of Cogger, H.G., this work.

Distribution: NE coastal, N Gulf, Qld. Ecology: woodland, open forest, closed forest, swamp, arboreal, nocturnal, predator; free-living tadpole, arthropod-feeder, breeds in swamps. Biological references: Copland, S.J. (1957). Presidential address. Australian tree frogs of the genus *Hyla*. *Proc. Linn. Soc. N.S.W.* **82**: 9–108 (taxonomy); Tyler, M.J. (1968). Papuan hylid frogs of the genus *Hyla*. *Zool. Verh.* **96**: 1–203 (taxonomy); Davies, M. (1978). Variation in the cranial osteology of the Australo-Papuan hylid frog *Litoria infrafrenata*. *Rec. S. Aust. Mus.* **17**: 337–345.

Litoria jervisiensis (Duméril and Bibron, 1841)

Hyla jervisiensis Duméril, A.M.C. & Bibron, G. (1841). *Erpétologie Générale* ou Histoire Naturelle Complète des Reptiles. Paris : Roret 8: ii 792 pp. [580]. Type data: holotype, MNHP 4826, from Jervis Bay, N.S.W.

Hyla oculata Fitzinger, L. (1860). Ausbeute der österreichischen Naturforscher an Säugethiere und Reptilien während der Weltumsegelung Sr. Majestat Fregatte *Novara*. *Sber. Akad. Wiss. Wien* **42**: 384–416 [413] [*nom. nud.* introduced in synonymy, *ex* Péron ms].

Hyla krefftii Günther, A. (1863). On new species of batrachians from Australia. *Ann. Mag. Nat. Hist. (3)* **11**: 26–28 [28 pl 4 fig C]. Type data: holotype, BMNH 1947.2.22.78, from Sydney, N.S.W.

Synonymy that of Cogger, H.G., this work.

Distribution: SE coastal, Vic., N.S.W. Ecology: open forest, tall forest, lentic freshwater, terrestrial, arboreal, predator; seasonal breeder, free-living tadpole, arthropod-feeder, breeds in lentic freshwater. Biological references: Copland, S.J. (1957). Presidential address. Australian tree frogs of the genus *Hyla*. *Proc. Linn. Soc. N.S.W.* **82**: 9–108 (taxonomy); Moore, J.A. (1961). The frogs of eastern New South Wales. *Bull. Am. Mus. Nat. Hist.* **121**: 149–386 (taxonomy, natural history); Martin, A.A. & Littlejohn, M.J. (1966). The breeding biology and larval development of *Hyla jervisiensis* (Anura : Hylidae). *Proc. Lin. Soc. N.S.W.* **91**: 47–57; White, A.W., Whitford, D. & Watson, G.F. (1980). Re-description of the Jervis Bay tree frog *Litoria jervisiensis* (Anura : Hylidae), with notes on the identity of Krefft's frog (*Litoria krefftii*). *Aust. Zool.* **20**: 375–390.

Litoria latopalmata Günther, 1867

Litoria latopalmata Günther, A. (1867). Additions to the knowledge of Australian reptiles and fishes. *Ann. Mag. Nat. Hist. (3)* **20**: 45–68 [55]. Type data: syntypes, BMNH 1947.2.24.27–28, from Bowen (as Port Denison), Qld.

Hyla palmata Slevin, J.R. (1955). Notes on Australian amphibians. *Proc. Calif. Acad. Sci. (4)* **28**: 355–392 figs 28 [383] [*lapsus pro Litoria latopalmata* Günther, 1867].

Synonymy that of Cogger, H.G., this work.

Distribution: SE coastal, NE coastal, N Gulf, Murray-Darling basin, Bulloo River basin, Lake Eyre basin, N.S.W., Qld., W.A., N.T., (S.A.). Ecology: tussock grassland, open scrub, low open woodland, low woodland, woodland, open forest, tall forest, swamp, tall shrubland, terrestrial, nocturnal, predator; seasonal breeder, free-living tadpole, arthropod-feeder, breeds in swamps. Biological references: Copland, S.J. (1957). Presidential address. Australian tree frogs of the genus *Hyla*. *Proc. Linn. Soc. N.S.W.* **82**: 9–108 (taxonomy); Moore, J.A. (1961). The frogs of eastern New South Wales. *Bull. Am. Mus. Nat. Hist.* **121**: 149–386 (taxonomy, natural history); Tyler, M.J. (1968). A taxonomic study of hylid frogs of the *Hyla lesueuri* complex occurring in north-western Australia. *Rec. S. Aust. Mus.* **15**: 711–727 (taxonomy); Tyler, M.J. (1969). A synopsis of the frogs of the genus *Hyla* of northwestern Australia, with the description of a new species. *Rec. S. Aust. Mus.* **16**: 1–11 (taxonomy).

Litoria lesueuri (Duméril and Bibron, 1841)

Hyla lesueuri Duméril, A.M.C. & Bibron, G. (1841). *Erpétologie Générale* ou Histoire Naturelle Complète des Reptiles. Paris : Roret 8: ii 792 pp. [595]. Type data: syntypes, MNHP 4845–6, from Sydney, N.S.W. (as Port Jackson).

Ranoidea flavoviridus Girard, C. (1853). Descriptions of new species of reptiles, collected by the U.S. Exploring Expedition, under the command of Capt. Charles Wilkes, U.S.N. Second Part. Including the species of batrachians, exotic to North America. *Proc. Acad. Nat. Sci. Philad.* **6**: 420–424 [422] [1854 on title page]. Type data: syntypes (probable), not found, from Wollongong, Illawarra, N.S.W.

Euscelis lessonii Fitzinger, L. (1860). Ausbeute der österreichischen Naturforscher an Säugethiere und Reptilien während der Weltumsegelung Sr. Majestat Fregatte *Novara. Sber. Akad. Wiss. Wien* **42**: 384–416 [413] [*pro Hyla lesueuri* Duméril and Bibron, 1841]

Litoria wilcoxii Günther, A. (1864). Third contribution to our knowledge of batrachians from Australia. *Proc. Zool. Soc. Lond.* **1864**: 46–49 pl 7 [48]. Type data: syntypes, BMNH 1947.2.22.76–77, from Clarence River, N.S.W.

Hyla (Litoria) freycinetii unicolor Keferstein, W. (1867). Ueber einige neue oder seltene Batrachier aus Australien und dem tropischen Amerika. *Nachr. Ges. Wiss. Göttingen* **1867**: 341–361 [355]. Type data: holotype, ZMG 102c (presumed lost), from Australia.

Hyla vinosa Lamb, J. (1911). Description of three new batrachians from southern Queensland. *Ann. Qd. Mus.* **10**: 26–28 [27]. Type data: holotype, QM J74, from Ithaca Creek, Brisbane, Qld.

Hyla kinghorni Loveridge, A. (1950). New frogs of the genera *Cyclorana* and *Hyla* from south-eastern Australia. *Proc. Biol. Soc. Wash.* **63**: 131–138 [132]. Type data: holotype, AM R13818 (*ex* CAS 83234), from Ulong, N.S.W.

Synonymy that of Cogger, H.G., this work.

Distribution: SE coastal, NE coastal, Murray-Darling basin, Vic., N.S.W., Qld. Ecology: woodland, open forest, tall forest, closed forest, lentic freshwater, lotic freshwater, terrestrial, nocturnal, predator; seasonal breeder, free-living tadpole, arthropod-feeder. Biological references: Copland, S.J. (1957). Presidential address. Australian tree frogs of the genus *Hyla. Proc. Linn. Soc. N.S.W.* **82**: 9–108 (taxonomy); Moore, J.A. (1961). The frogs of eastern New South Wales. *Bull. Am. Mus. Nat. Hist.* **121**: 149–386 (taxonomy, natural history).

Litoria longirostris Tyler and Davies, 1977

Litoria longirostris Tyler, M.J. & Davies, M. (1977). A new species of hylid frog from northern Australia. *Copeia* **1977**: 620–623 [620 figs 1–3, 4A]. Type data: holotype, QM J26930, from McIlwraith Range, Cape York Peninsula, Qld.

Distribution: NE coastal, Qld. Ecology: closed forest.

Litoria lorica Davies and McDonald, 1979

Litoria lorica Davies, M. & McDonald, K.R. (1979). A new species of stream-dwelling hylid frog from northern Queensland. *Trans. R. Soc. S. Aust.* **103**: 169–176 [170 figs 1, 2A, B, C, 3]. Type data: holotype, QM J36090, from Alexandra Creek, near Thornton Peak (16°07'S, 145°20'E), Qld.

Distribution: NE coastal, Qld. Ecology: closed forest, lotic freshwater, amphibious, nocturnal; torrent dweller, riparian.

Litoria maculata (Spencer, 1901)

Hyla maculata Spencer, B. (1901). Two new species of frogs from Victoria. *Proc. R. Soc. Vict. (ns)* **13**: 175–178 [177]. Type data: holotype, NMV D8498, from Poowong, Vic.

Distribution: SE coastal, Vic., N.S.W. Ecology: (open forest), tall forest, lotic freshwater, noctidiurnal, predator; free-living tadpole, riparian, arthropod-feeder. Biological references: Copland, S.J. (1957). Presidential address. Australian tree frogs of the genus *Hyla. Proc. Linn. Soc. N.S.W.* **82**: 9–108 (taxonomy); Copland, S.J. (1962). Rediscovery of a little known Victorian frog. *Proc. Linn. Soc. N.S.W.* **86**: 258–261 (natural history).

Litoria meiriana (Tyler, 1969)

Hyla meiriana Tyler, M.J. (1969). A synopsis of the frogs of the genus *Hyla* of northwestern Australia, with the description of a new species. *Rec. S. Aust. Mus.* **16**: 1–11 [2 figs 1–2]. Type data: holotype, SAMA R9082, from 98 mi N of Mainoru, N.T.

Distribution: N Gulf, N coastal, W.A., N.T. Ecology: low open woodland, woodland, open forest, lotic freshwater, terrestrial, amphibious, noctidiurnal, predator; free-living tadpole, saxicoline, riparian, arthropod-feeder.

Litoria moorei (Copland, 1957)

Hyla moorei Copland, S.J. (1957). Presidential address. Australian tree frogs of the genus *Hyla. Proc. Linn. Soc. N.S.W.* **82**: 9–108 [83]. Type data: lectotype, WAM R5981, from Pemberton, W.A., designation by Cogger, H.G., this work.

Distribution: (NW coastal), SW coastal, W.A. Ecology: (open heath), woodland, open forest, tall forest, swamp, amphibious, noctidiurnal, predator; seasonal breeder, free-living tadpole, general carnivore. Biological references: Main, A.R. (1965). Frogs of southern Western Australia. *West. Aust. Nat. Club Handbook* 8: 1–73 (natural history).

Litoria nannotis (Andersson, 1916)

Hyla nannotis Andersson, L.G. (1916). Results of Dr. E. Mjöberg's Swedish Scientific Expeditions to Australia 1910–1913. 9. Batrachians from Queensland. *K. Svenska Vetensk.-Akad. Handl. (ns)* **52**: 1–20 [16 figs 5a–e]. Type data: holotype, NHRM 1645, from Tully River, N Qld.

Distribution: NE coastal, Qld. Ecology: closed forest, lotic freshwater, amphibious, nocturnal, predator; free-living tadpole, torrent dweller, riparian, arthropod-feeder. Biological references: Copland, S.J. (1957). Presidential address. Australian tree frogs of the genus *Hyla*. *Proc. Linn. Soc. N.S.W.* **82**: 9–108 (taxonomy); Tyler, M.J. (1965). Taxonomic studies of some hylid frogs of Australia and New Guinea. *Proc. Zool. Soc. Lond.* **145**: 91–106 figs 1–5 pl 1 (taxonomy); Liem, D.S. (1974). A review of the *Litoria nannotis* species group, and a description of a new species of *Litoria* from northern Queensland, Australia (Anura : Hylidae). *Mem. Qd. Mus.* **17**: 151–168 (taxonomy); Davies, M. & McDonald, K.R. (1979). A new species of stream-dwelling hylid frog from northern Queensland. *Trans. R. Soc. S. Aust.* **103**: 169–176 (taxonomy).

Litoria nasuta (Gray, 1842)

Pelodytes nasutus Gray, J.E. (1842). Description of some hitherto unrecorded species of Australian reptiles and batrachians. *in* Gray, J.E. (ed.) *Zoological Miscellany.* London : Treuttel, Würtz & Co. pp. 51–57 [56]. Type data: holotype, BMNH 1947.2.22.81, from Port Essington, N.T.

Hyla peninsulae De Vis, C.W. (1884). On a new species of *Hyla*. *Proc. R. Soc. Qd.* **1**: 128–130 [130]. Type data: syntypes (probable), presumed lost, from Cape York, Qld.

Hyla semoni Boettger, O. (1894). Lurche (Batrachia). *in* Semon, R. (ed.) *Zoologische Forschungsreisen in Australien und dem Malayischen Archipel.* Jena : Gustav Fischer 5: 107–114 [112]. Type data: holotype, SMF 2625 (cited as SMF 1414,2a in original description), from Bou, Milne Bay, SE New Guinea, see Semon, R. (1899). *In the Australian Bush and on the coast of the Coral Sea.* London : Macmillan Co. 552 pp. [385].

Synonymy that of Cogger, H.G., this work.

Distribution: SE coastal, NE coastal, N Gulf, N coastal, Murray-Darling basin, N.S.W., Qld., W.A., N.T. Ecology: tussock grassland, open scrub, low open woodland, low woodland, woodland, open forest, tall forest, swamp, tall shrubland, terrestrial, nocturnal, predator; seasonal breeder, free-living tadpole, arthropod-feeder, breeds in swamps. Biological references: Copland, S.J. (1957). Presidential address. Australian tree frogs of the genus *Hyla*. *Proc. Linn. Soc. N.S.W.* **82**: 9–108 (taxonomy); Moore, J.A. (1961). The frogs of eastern New South Wales. *Bull. Am. Mus. Nat. Hist.* **121**: 149–386 (taxonomy, natural history); Tyler, M.J. (1968). A taxonomic study of hylid frogs of the *Hyla lesueuri* complex occurring in north-western Australia. *Rec. S. Aust. Mus.* **15**: 711–727 (taxonomy); Peile, A.R. (1978). Gugadja aborigines and frogs. *Herpetofauna* **10**: 9–14 (ethnobiology).

Litoria nigrofrenata (Günther, 1867)

Hyla nigrofrenata Günther, A. (1867). Additions to the knowledge of Australian reptiles and fishes. *Ann. Mag. Nat. Hist. (3)* **20**: 45–68 [56]. Type data: syntypes, BMNH 1947.2.23.46–47, from Cape York, Qld.

Distribution: NE coastal, N Gulf, Qld. Ecology: woodland, open forest, closed forest, lentic freshwater, lotic freshwater, swamp, terrestrial, nocturnal, predator; seasonal breeder, free-living tadpole, riparian, arthropod-feeder. Biological references: Tyler, M.J. (1968). A taxonomic study of hylid frogs of the *Hyla lesueuri* complex occurring in north-western Australia. *Rec. S. Aust. Mus.* **15**: 711–727 (taxonomy).

Litoria nyakalensis Liem, 1974

Litoria nyakalensis Liem, D.S. (1974). A review of the *Litoria nannotis* species group, and a description of a new species of *Litoria* from northern Queensland, Australia (Anura : Hylidae). *Mem. Qd. Mus.* **17**: 151–168 [157 fig 2 pl 4]. Type data: holotype, QM J22624, from Henrietta Creek, Palmerston National Park, near Innisfail, Qld.

Distribution: NE coastal, Qld. Ecology: tall forest, closed forest, lotic freshwater, arboreal, nocturnal, predator; riparian, arthropod-feeder. Biological references: Davies, M. & McDonald, K.R. (1979). A new species of stream-dwelling hylid frog from northern Queensland. *Trans. R. Soc. S. Aust.* **103**: 169–176 (taxonomy).

Litoria olongburensis Liem and Ingram, 1977

Litoria olongburensis Liem, D.S. & Ingram, G.J. (1977). Two new species of frogs (Anura : Myobatrachidae, Pelodryadidae) from Queensland and New South Wales. *Vict. Nat.* **94**: 255–262 [259 figs 3–4]. Type data: holotype, QM J22652, from Coomboo Lake, Fraser Is., SE Qld.

Distribution: SE coastal, NE coastal, N.S.W., Qld. Ecology: woodland, open forest, tall forest, swamp, nocturnal, predator; free-living tadpole, arthropod-feeder.

Litoria paraewingi Watson, Loftus-Hills and Littlejohn, 1971

Litoria paraewingi Watson, G.F., Loftus-Hills, J.J. & Littlejohn, M.J. (1971). The *Litoria ewingi* complex (Anura : Hylidae) in south eastern Australia. 1. A new species from Victoria. *Aust. J. Zool.* **19**: 401–416 [414]. Type data: holotype, NMV D27601, from O'Connors Flat, 8.3 km SSE of Yea, Vic.

Distribution: SE coastal, Murray-Darling basin, Vic. Ecology: woodland, open forest, swamp, terrestrial, nocturnal, predator; free-living tadpole, arthropod-feeder, breeds in swamps. Biological references: Littlejohn, M.J. (1976). The *Litoria ewingi* complex (Anura : Hylidae) in south-eastern

Australia. 4. Variation in mating-call structure across a narrow hybrid zone between *L. ewingi* and *L. paraewingi*. *Aust. J. Zool.* **24**: 283–293.

Litoria peronii (Tschudi, 1838)

Dendrohyas peronii Tschudi, J.J. (1838). *Classification der Batrachier, mit Berücksichtigung der fossilen Tiere dieser Abteilung der Reptilien.* Neuchâtel : Petitpierre 102 pp. [34,75]. Type data: holotype, MNHP 4809, from Australia.

***Hyla latopalmata watjulumensis* (part.)** Copland, S.J. (1957). Presidential address. Australian tree frogs of the genus *Hyla*. *Proc. Linn. Soc. N.S.W.* **82**: 9–108 [96] [Tyler, M.J. (1968). A taxonomic study of hylid frogs of the *Hyla lesueuri* complex occurring in north-western Australia. *Rec. S. Aust. Mus.* **15**: 711–727 cited 29 syntypes. Of these, that listed below he assigned to *Litoria peronii* (Tschudi, 1838). For his assignment of the other syntypes see *Litoria coplandi* (Tyler, 1968) and *Litoria wotjulumensis* (Copland, 1957)]. Type data: syntype, WAM R11195, from Watjulum [*sic*] Mission, W.A.

Synonymy that of Tyler, M.J. (1968). A taxonomic study of hylid frogs of the *Hyla lesueuri* complex occurring in north-western Australia. *Rec. S. Aust. Mus.* **15**: 711–727.

Distribution: SE coastal, NE coastal, Murray-Darling basin, (Bulloo River basin), S.A., Vic., N.S.W., Qld., A.C.T. Ecology: low open woodland, low woodland, woodland, open forest, tall forest, lotic freshwater, swamp, arboreal, nocturnal, predator; seasonal breeder, free-living tadpole, riparian, arthropod-feeder, breeds in swamps. Biological references: Copland, S.J. (1957). Presidential address. Australian tree frogs of the genus *Hyla*. *Proc. Linn. Soc. N.S.W.* **82**: 9–108 (taxonomy); Moore, J.A. (1961). The frogs of eastern New South Wales. *Bull. Am. Mus. Nat. Hist.* **121**: 149–386 (taxonomy, natural history); Martin, A.A., Watson, G.F., Gartside, D.F., Littlejohn, M.J. & Loftus-Hills, J.J. (1979). A new species of the *Litoria peronii* complex (Anura : Hylidae) from eastern Australia. *Proc. Linn. Soc. N.S.W.* **103**: 23–35 (taxonomy, biology).

Litoria personata Tyler, Davies and Martin, 1978

Litoria personata Tyler, M.J., Davies, M. & Martin, A.A. (1978). A new species of hylid frog from the Northern Territory. *Trans. R. Soc. S. Aust.* **102**: 151–157 [151]. Type data: holotype, SAMA R16773, from Birndu (12°32'S, 132°08'E), SE of Cannon Hill Station, East Alligator River Region, N.T.

Distribution: N coastal, N.T. Ecology: woodland, open forest, lotic freshwater, terrestrial, arboreal, nocturnal, predator; free-living tadpole, riparian, arthropod-feeder.

Litoria phyllochroa (Günther, 1863)

Hyla phyllochroa Günther, A. (1863). Observations on Australian tree-frogs living in the Society's menagerie. *Proc. Zool. Soc. Lond.* **1863**: 249–251 [251 pl 30 figs C, c]. Type data: syntypes, BMNH 1947.2.24.3–5, ZMB 4929, from Sydney and Errumanga, N.S.W.

Hyla phyllochroa barringtonensis Copland, S.J. (1957). Presidential address. Australian tree frogs of the genus *Hyla*. *Proc. Linn. Soc. N.S.W.* **82**: 9–108 [45]. Type data: holotype, AM R9502, from Cutler's Pass, Williams River, N.S.W.

Hyla pearsoni Copland, S.J. (1960). A new tree-frog (genus *Hyla*) from Queensland. *Proc. Linn. Soc. N.S.W.* **85**: 154–156 [154] [*non Hyla pearsoni* Gaige, 1929; *Hyla pearsoni* Copland, 1960 permanently rejected as it is a junior primary homonym]. Type data: holotype, AM R18588, from Cedar Creek, E of Mt. Glorious, about 20 mi NW of Brisbane, Qld.

Hyla pearsoniana Copland, S.J. (1961). A new name for *Hyla pearsoni*, preoccupied. (Amphibia). *Proc. Linn. Soc. N.S.W.* **86**: 168 [168] [*nom. nov. pro Hyla pearsoni* Copland, 1960].

Hyla phyllochrous nudidigitus Copland, S.J. (1962). *Hyla phyllochrous* Günther (Amphibia) as an addition to the fauna of Victoria, with the description of a new race and a note on the name of the genus. *Proc. Linn. Soc. N.S.W.* **87**: 137–140 [137]. Type data: holotype, AM R51309 (*ex* S.J. Copland pers. coll. 7280), from Aberfeldy River, Vic.

Synonymy that of Cogger, H.G., this work.

Distribution: SE coastal, NE coastal, (Murray-Darling basin), Vic., N.S.W., Qld. Ecology: open forest, tall forest, closed forest, lotic freshwater, arboreal, nocturnal, predator; seasonal breeder, free-living tadpole, arthropod-feeder. Biological references: Moore, J.A. (1961). The frogs of eastern New South Wales. *Bull. Am. Mus. Nat. Hist.* **121**: 149–386 (taxonomy, natural history); Copland, S.J. (1957). Presidential address. Australian tree frogs of the genus *Hyla*. *Proc. Linn. Soc. N.S.W.* **82**: 9–108 (taxonomy).

Litoria raniformis (Keferstein, 1867)

Chirodryas raniformis Keferstein, W. (1867). Ueber einige neue oder seltene Batrachier aus Australien und dem tropischen Amerika. *Nachr. Ges. Wiss. Göttingen* **1867**: 341–361 [358]. Type data: holotype, ZMG 93h (*ex* 111a), presumed lost, from Australia.

Hyla aurea major Copland, S.J. (1957). Presidential address. Australian tree frogs of the genus *Hyla*. *Proc. Linn. Soc. N.S.W.* **82**: 9–108 [79]. Type data: holotype, QVM 1939.120, from Collins Gap, Tas.

Synonymy that of Cogger, H.G., this work.

Distribution: SE coastal, (S Gulfs), Murray-Darling basin, S.A., Tas., Vic., N.S.W., A.C.T. Ecology: low open woodland, woodland, open forest, tall forest, swamp, terrestrial, noctidiurnal, predator; seasonal breeder, free-living tadpole, general carnivore. Biological references: Copland, S.J. (1957). Presidential address. Australian tree frogs of the genus *Hyla*. *Proc. Linn. Soc. N.S.W.* **82**: 9–108 (taxonomy); Moore, J.A. (1961). The frogs of eastern New South Wales. *Bull. Am. Mus. Nat. Hist.* **121**: 149–386 (taxonomy, natural history); Courtice, G.P. & Grigg, G.C. (1975). A taxonomic revision of the *Litoria aurea* complex (Anura : Hylidae) in south-eastern Australia. *Aust. Zool.* **18**: 149–163 (taxonomy); Hobbs, J.N. (1980). Frogs as a deterrent to breeding success in reedwarblers and grassbirds. *Aust. Birds* **15**: 27–28 (behaviour).

Litoria rheocola Liem, 1974

Litoria rheocolus Liem, D.S. (1974). A review of the *Litoria nannotis* species group, and a description of a new species of *Litoria* from northern Queensland, Australia (Anura : Hylidae). *Mem. Qd. Mus.* **17**: 151–168 [152 figs 1, 4C–D pl 3A]. Type data: holotype, QM J22631, from Kuranda, 50 km W of Cairns, Qld.

Distribution: NE coastal, Qld. Ecology: closed forest, lotic freshwater, amphibious, nocturnal, predator; free-living tadpole, torrent dweller, riparian, arthropod-feeder. Biological references: Copland, S.J. (1957). Presidential address. Australian tree frogs of the genus *Hyla*. *Proc. Linn. Soc. N.S.W.* **82**: 9–108 (as *Hyla dayi*); Davies, M. & McDonald, K.R. (1979). A new species of stream-dwelling hylid frog from northern Queensland. *Trans. R. Soc. S. Aust.* **103**: 169–176 (taxonomy).

Litoria rothii (De Vis, 1884)

Hyla rothii De Vis, C.W. (1884). On some new batrachians from Queensland. Proc. Linn. Soc. N.S.W. **9**: 65–68 [66]. Type data: syntypes, not found, from Mackay, Qld.

Distribution: NE coastal, N Gulf, N coastal, Qld., W.A., N.T. Ecology: tall open shrubland, low open woodland, woodland, open forest, tall forest, lentic freshwater, lotic freshwater, swamp, arboreal, nocturnal, predator; seasonal breeder, free-living tadpole, riparian, arthropod-feeder. Biological references: Cogger, H.G. & Lindner, D.A. (1974). Frogs and reptiles in fauna survey of the Port Essington District, Cobourg Peninsula, Northern Territory of Australia. Frith, H.J. & Calaby, J.H. (eds.) *CSIRO Div. Wildl. Res. Tech. Pap.* **28**: 63–107 (taxonomy); Martin, A.A., Watson, G.F., Gartside, D.F., Littlejohn, M.J. & Loftus-Hills, J.J. (1979). A new species of the *Litoria peronii* complex (Anura : Hylidae) from eastern Australia. *Proc. Linn. Soc. N.S.W.* **103**: 23–35 (taxonomy).

Litoria rubella (Gray, 1842)

Hyla rubella Gray, J.E. (1842). Description of some hitherto unrecorded species of Australian reptiles and batrachians. *in* Gray, J.E. (ed.) *Zoological Miscellany*. London : Treuttel, Würtz & Co. pp. 51–57 [56]. Type data: syntypes, BMNH 1947.2.24.7–9, from Port Essington, N.T.

Hyla (Litoria) mystacina Keferstein, W. (1867). Ueber einige neue oder seltene Batrachier aus Australien und dem tropischen Amerika. *Nachr. Ges. Wiss. Göttingen* **1867**: 341–361 [356]. Type data: holotype, ZFMK (*ex* ZMG 102e), from Australia.

H[yla] nigrogularis Keferstein, W. (1867). Ueber einige neue oder seltene Batrachier aus Australien und dem tropischen Amerika. *Nachr. Ges. Wiss. Göttingen* **1867**: 341–361 [358] [*ex* Krefft ms; *nom. nud.* introduced in synonymy].

Synonymy that of Cogger, H.G., this work.

Distribution: SE coastal, NE coastal, N Gulf, N coastal, NW coastal, W plateau, S Gulfs, Murray-Darling basin, Bulloo River basin, Lake Eyre basin, S.A., N.S.W., Qld., W.A., N.T. Ecology: low open shrubland, low shrubland, tall open shrubland, open scrub, low open woodland, low woodland, woodland, open forest, tall forest, closed forest, lentic freshwater, swamp, tall shrubland, arboreal, nocturnal, predator; free-living tadpole, riparian, arthropod-feeder, breeds in swamps. Biological references: Copland, S.J. (1957). Presidential address. Australian tree frogs of the genus *Hyla*. *Proc. Linn. Soc. N.S.W.* **82**: 9–108 (taxonomy); Moore, J.A. (1961). The frogs of eastern New South Wales. *Bull. Am. Mus. Nat. Hist.* **121**: 149–386 (taxonomy, natural history); Main, A.R. (1965). Frogs of southern Western Australia. *West. Aust. Nat. Club Handbook* 8: 1–73 (natural history); Tyler, M.J. (1978). *Amphibians of South Australia.* Adelaide : South Australian Government 84 pp. (natural history); Warburg, M.R. (1965). Studies on the water economy of some Australian frogs. *Aust. J. Zool.* **13**: 317–330.

Litoria serrata (Andersson, 1916)

Hyla serrata Andersson, L.G. (1916). Results of Dr. E. Mjöberg's Swedish Scientific Expeditions to Australia 1910–1913. 9. Batrachians from Queensland. *K. Svenska Vetensk.-Akad. Handl. (ns)* **52**: 1–20 [17 fig 6]. Type data: syntypes, NHRM 1646 (2 specimens), NHRM 1647 (3 specimens), NHRM 1648, from Carrington, Malanda and Atherton, N Qld.

Distribution: NE coastal, Qld. Ecology: closed forest, lotic freshwater, arboreal, nocturnal, predator; free-living tadpole, riparian, arthropod-feeder. Biological references: Copland, S.J. (1957). Presidential address. Australian tree frogs of the genus *Hyla*. *Proc. Linn. Soc. N.S.W.* **82**: 9–108 (taxonomy, as *Hyla eucnemis*); Tyler, M.J. (1965). Taxonomic studies of some hylid frogs of Australia and New Guinea. *Proc. Zool. Soc. Lond.* **145**: 91–106 figs 1–5 pl 1 (taxonomy, as *Hyla eucnemis*).

Litoria splendida Tyler, Davies and Martin, 1977

Litoria splendida Tyler, M.J., Davies, M. & Martin, A.A. (1977). A new species of large, green tree frog from northern Western Australia. *Trans. R. Soc. S. Aust.* **101**: 133–138 [133 figs 1–3]. Type data: holotype, WAM R56840, from Lake Argyle Tourist Village, Kimberley Division, W.A.

Distribution: N coastal, W.A. Ecology: woodland, arboreal, nocturnal, predator; free-living tadpole, general carnivore.

Litoria subglandulosa Tyler and Anstis, 1983

Litoria glandulosa Tyler, M.J. & Anstis, M. (1975). Taxonomy and biology of frogs of the *Litoria citropa* complex (Anura : Hylidae). *Rec. S. Aust. Mus.* **17**: 41–50 [46 figs 2b, 3b–d, 4b, 6] [*non Litoria glandulosa* Bell, 1843] Type data: holotype, SAMA R13504, from Barwick Creek, Point Lookout, near Ebor, N.S.W.

Litoria subglandulosa Tyler, M.J. & Anstis, M. (1983). Replacement name for *Litoria glandulosa* Tyler & Anstis, 1975 (Anura : Hylidae). *Trans. R. Soc. S. Aust.* **107**: 130 [*nom. nov. pro Litoria glandulosa* Tyler and Anstis, 1975].

Distribution: SE coastal, (Murray-Darling basin), N.S.W., (Qld.); Ecology: woodland, open forest, tall forest, terrestrial, nocturnal, predator; seasonal breeder, free-living tadpole, arthropod-feeder.

Litoria tornieri (Nieden, 1923)

Pelodytes affinis Gray, J.E. (1842). Description of some hitherto unrecorded species of Australian reptiles and batrachians. *in* Gray, J.E. (ed.) *Zoological Miscellany.* London : Treuttel, Würtz & Co. pp. 51–57 [56] [*non Hyla affinis* Spix, 1824]. Type data: holotype, BMNH 1947.2.22.79, from Port Essington, N.T. (as North coast of Australia), see Boulenger, G.A. (1882). *Catalogue of the Batrachia Salientia s. Ecaudata in the collection of the British Museum.* 2nd edn. London : British Museum xvi 503 pp.

H[yla] tornieri Nieden, F. (1923). Amphibia. Anura I. Subordo Aglossa und Phaneroglossa. Sectio 1 Arcifera. *Das Tierreich* **46**: 1–584 [228] [*nom. nov. pro Pelodytes affinis* Gray, 1842, at that time a junior secondary homonym of *Hyla affinis* Spix, 1824].

Synonymy that of Cogger, H.G., this work.

Distribution: N coastal, N Gulf, (Qld.), (W.A.), N.T. Ecology: tall open shrubland, woodland, open forest, swamp, terrestrial, nocturnal, predator; seasonal breeder, free-living tadpole, arthropod-feeder, breeds in swamps. Biological references: Andersson, L.G. (1913). Results of Dr. E. Mjöberg's Swedish Scientific Expeditions to Australia 1910–1913. 4. Batrachians. *K. Svenska Vetensk.-Akad. Handl. (ns)* **52**: 1–26 (taxonomy); Cogger, H.G. & Lindner, D.A. (1974). Frogs and reptiles in fauna survey of the Port Essington District, Cobourg Peninsula, Northern Territory of Australia. Frith, H.J. & Calaby, J.H. (eds.) *CSIRO Div. Wildl. Res. Tech. Pap.* **28**: 63–107 (taxonomy).

Litoria tyleri Martin, Watson, Gartside, Littlejohn and Loftus-Hills, 1979

Litoria tyleri Martin, A.A., Watson, G.F., Gartside, D.F., Littlejohn, M.J. & Loftus-Hills, J.J. (1979). A new species of the *Litoria peronii* complex (Anura : Hylidae) from eastern Australia. *Proc. Linn. Soc. N.S.W.* **103**: 23–35 [25]. Type data: holotype, AM R64754, from Ryan's Swamp, Caves Beach Reserve, Jervis Bay Territory (as A.C.T.), 14 km S of Huskisson, Shire of Shoalhaven, N.S.W.

Distribution: SE coastal, NE coastal, N.S.W., Qld. Ecology: open forest, tall forest, swamp, arboreal, nocturnal, predator; seasonal breeder, free-living tadpole, arthropod-feeder, breeds in swamps.

Litoria verreauxii (Duméril, 1853)

Hyla verreauxii Duméril, A. (1853). Mémoire sur les batraciens anoures, de la famille des hylaeformes ou rainettes. *Ann. Sci. Nat. (3)* **19** Zool.: 135–179 [171]. Type data: syntypes, MNHP 4849 (5 specimens), MNHP 4850 (9 specimens), from Australia.

Hyla ewingii orientalis Fletcher, J.J. (1898). Contributions to a more exact knowledge of the geographical distribution of Australian Batrachia. 5. A. Batrachia of Tasmania. *Proc. Linn. Soc. N.S.W. (2)* **12**: 660–684 [670]. Type data: none designated, from N.S.W.

Hyla ewingii alpina Fry, D.B. (1915). Herpetological notes. *Proc. R. Soc. Qd.* **27**: 60–95 [79]. Type data: holotype, AM R4644, from Mt. Kosciusko, N.S.W.

Hyla ewingii loveridgei Copland, S.J. (1957). Presidential address. Australian tree frogs of the genus *Hyla. Proc. Linn. Soc. N.S.W.* **82**: 9–108 [65] [*non Nyctimystes loveridgei* Neill, 1954]. Type data: holotype, AM R18586, from near Porter's Retreat, N.S.W.

Hyla ewingii oberonensis Copland, S.J. (1963). Size at metamorphosis of the frog *Hyla aureus raniformis* (Keferstein). *Proc. Linn. Soc. N.S.W.* **88**: 105–107 [107] [*nom. nov. pro Hyla ewingii loveridgei* Copland, 1957; at that time a junior secondary homonym of *Hyla loveridgei* (Neill, 1954)].

Synonymy that of Cogger, H.G., this work.

Distribution: SE coastal, NE coastal, (Murray-Darling basin), Vic., N.S.W., Qld., A.C.T. Ecology: alpine, woodland, open forest, tall forest, swamp, terrestrial, nocturnal, predator; seasonal breeder, free-living tadpole, arthropod-feeder. Biological references: Copland, S.J. (1957). Presidential address. Australian tree frogs of the genus *Hyla. Proc. Linn. Soc. N.S.W.* **82**: 9–108 (taxonomy); Moore, J.A. (1961). The frogs of eastern New South Wales. *Bull. Am. Mus. Nat. Hist.* **121**: 149–386 (taxonomy, natural history); Watson, G.F., Loftus-Hills, J.J. & Littlejohn, M.J. (1971). The *Litoria ewingi* complex (Anura : Hylidae) in south eastern Australia. 1. A new species from Victoria. *Aust. J. Zool.* **19**: 401–416 (taxonomy); Anstis, M. (1976). Breeding biology and larval development of *Litoria verreauxi* (Anura : Hylidae). *Trans. R. Soc. S. Aust.* **100**: 193–202; Anstis, M. (1974). Australian frogs. Part 2. The behaviour and life history of *Litoria verreauxi* (Hylidae). *Bull. Herpet. (R. Zool. Soc. N.S.W.)* **1**: 19–21; Pengilley, R.K. (1971). The food of some Australian anurans (Amphibia). *J. Zool.* **163**: 93–103 (diet).

Litoria wotjulumensis (Copland, 1957)

***Hyla latopalmata watjulumensis* (part.)** Copland, S.J. (1957). Presidential address. Australian tree frogs of the genus *Hyla*. *Proc. Linn. Soc. N.S.W.* **82**: 9–108 [96] [Tyler, M.J. (1968). A taxonomic study of hylid frogs of the *Hyla lesueuri* complex occurring in north-western Australia. *Rec. S. Aust. Mus.* **15**: 711–727 cited 29 syntypes. Of these, those listed below he assigned to *Litoria wotjulumensis* (Copland, 1957). For his assignment of the other syntypes see *Litoria coplandi* (Tyler, 1968) and *Litoria peronii* (Tschudi, 1838)]. Type data: syntypes, WAM R11196, WAM R11198–99, WAM R11633–35, WAM R11896–907, WAM R11932–38, from Watjulum [*sic*] Mission, W.A.

Hyla spaldingi Hosmer, W. (1964). A new frog of the genus *Hyla* from Northern Territory, Australia. *Am. Mus. Novit.* 2182: 1–7 [2 figs 3–5]. Type data: holotype, AMNH 67835, from Elizabeth River, 50 mi S Darwin, N.T.

Hyla wotjulumensis Tyler, M.J. (1968). A taxonomic study of hylid frogs of the *Hyla lesueuri* complex occurring in north-western Australia. *Rec. S. Aust. Mus.* **15**: 711–727 [713] [valid emend. *pro Hyla latopalmata watjulumensis* Copland, 1957].

Synonymy that of Cogger, H.G., this work.

Distribution: N Gulf, N coastal, (Qld.), W.A., N.T. Ecology: tall open shrubland, woodland, open forest, swamp, terrestrial, nocturnal, predator; seasonal breeder, free-living tadpole, arthropod-feeder. Biological references: Tyler, M.J. (1968). A taxonomic study of hylid frogs of the *Hyla lesueuri* complex occurring in north-western Australia. *Rec. S. Aust. Mus.* **15**: 711–727 (taxonomy); Cogger, H.G. & Lindner, D.A. (1974). Frogs and reptiles in fauna survey of the Port Essington District, Cobourg Peninsula, Northern Territory of Australia. Frith, H.J. & Calaby, J.H. (eds.) *CSIRO Div. Wildl. Res. Tech. Pap.* **28**: 63–107 (taxonomy).

Nyctimystes Stejneger, 1916

Nyctimystes Stejneger, L. (1916). New generic name for a tree-toad from New Guinea. *Proc. Biol. Soc. Wash.* **29**: 85–86 [85]. Type species *Nyctimantis papua* Boulenger, 1897 by original designation.

This group is also found in New Guinea, see Zweifel, R.G. (1958). Results of the Archbold Expeditions. 78. Frogs of the Papuan hylid genus *Nyctimystes*. *Am. Mus. Novit.* 1896: 1–51.

Nyctimystes hosmeri Tyler, 1964

Nyctimystes hosmeri Tyler, M.J. (1964). Two new species of Australian frogs of the hylid genus *Nyctimystes*. *Trans. R. Soc. S. Aust.* **88**: 111–114 [111 fig 1 pl 1]. Type data: holotype, AMNH 65538, from Tully Falls, Cape York Peninsula, Qld.

Distribution: NE coastal, Qld. Ecology: closed forest, arboreal, nocturnal, predator; riparian, arthropod-feeder. Biological references: Tyler, M.J. & Davies, M. (1979). Redefinition and evolutionary origin of the Australopapuan hylid frog genus *Nyctimystes* Stejneger. *Aust. J. Zool.* **27**: 755–772.

Nyctimystes tympanocryptis (Andersson, 1916)

Hyla tympanocryptis Andersson, L.G. (1916). Results of Dr. E. Mjöberg's Swedish Scientific Expeditions to Australia 1910–1913. 9. Batrachians from Queensland. *K. Svenska Vetensk.-Akad. Handl. (ns)* **52**: 1–20 [19 figs 7a–e]. Type data: holotype, NHRM 1649, from Malanda, N Qld.

Distribution: NE coastal, Qld. Ecology: closed forest, arboreal, nocturnal, predator; riparian, arthropod-feeder. Biological references: Tyler, M.J. (1968). An additional Australian species of the hylid frog genus *Nyctimystes* with notes on *Hyla dayi* Günther. *Ark. Zool. (2)* **20**: 501–504 (taxonomy); Tyler, M.J. & Davies, M. (1979). Redefinition and evolutionary origin of the Australopapuan hylid frog genus *Nyctimystes* Stejneger. *Aust. J. Zool.* **27**: 755–772.

Nyctimystes vestigia Tyler, 1964

Nyctimystes vestigia Tyler, M.J. (1964). Two new species of Australian frogs of the hylid genus *Nyctimystes*. *Trans. R. Soc. S. Aust.* **88**: 111–114 [113]. Type data: holotype, NHMW 17187, from Mt. Bartle Frere, Cape York Peninsula, Qld.

Distribution: NE coastal, Qld. Ecology: closed forest, arboreal, nocturnal, predator; riparian, arthropod-feeder. Biological references: Tyler, M.J. & Davies, M. (1979). Redefinition and evolutionary origin of the Australopapuan hylid frog genus *Nyctimystes* Stejneger. *Aust. J. Zool.* **27**: 755–772.

MICROHYLIDAE

INTRODUCTION

A family of frogs occurring throughout the tropical and warm temperate regions of the world and containing about 250 species in 54 genera. Australia's two genera and 8 species are members of the Papuan regional subfamily Sphenophryinae, and are restricted to the wet forests of north-eastern Queensland, with one species also occurring in drier monsoonal forests and swamps from Cape York to Arnhem Land.

Characterised in Australia by: firmisternal pectoral girdle; intercalary cartilages absent; tips of fingers and toes with or without expanded pads, but with lateral grooves; sacral diapophyses dilated; maxillary teeth usually absent; toes not webbed; no dorso-lateral skin fold.

References

Barker, J. & Grigg, G. (1977). *A Field Guide to Australian Frogs.* Adelaide : Rigby Ltd. 229 pp.

Cogger, H.G. (1979). *Reptiles and Amphibians of Australia.* Sydney : A.H. & A.W. Reed 608 pp.

Parker, H.W. (1934). *A monograph of the frogs of the family Microhylidae.* London : British Museum viii 208 pp.

Tyler, M.J. (1982). *Frogs.* 2nd edn. Sydney : William Collins 256 pp.

Cophixalus Boettger, 1892

Cophixalus Boettger, O. (1892). *Katalog der Batrachier-Sammlung im Museum der Senckenbergischen Naturforschenden Gesellschaft in Frankfurt-am-Main.* Frankfurt-am-Main : Knauer x 73 pp. [24]. Type species *Cophixalus geislerorum* Boettger, 1892 by monotypy.

Phrynixalus Boettger, O. (1895). Liste der Reptilien und Batrachier der Insel Halmaheira nach den Sammlunge Prof. Dr. W. Kukenthal's. *Zool. Anz.* **18**(472): 129–140 [132]. Type species *Phrynixalus montanus* Boettger, 1895 by monotypy.

***Austrochaperina* (part.)** Fry, D.B. (1912). Description of *Austrochaperina*, a new genus of Engystomatidae from north Australia. *Rec. Aust. Mus.* **9**: 87–106 [87] [synonymy based on *Austrochaperina ornata* Fry, 1912, one of three originally included species]. Type species *Austrochaperina robusta* Fry, 1912 by original designation.

Aphantophryne Fry, D.B. (1917). Description of *Aphantophryne*, a new batrachian genus from New Guinea; with comparative notes on the pectoral musculature. *Proc. Linn. Soc. N.S.W.* **41**: 770–785 [772]. Type species *Aphantophryne pansa* Fry, 1917 by original designation. Compiled from secondary source: Zweifel, R.G. & Allison, A. (1982). A new montane microhylid frog from Papua New Guinea, and comments on the status of the genus *Aphantophryne*. *Am. Mus. Novit.* 2723: 1–14.

This group is also found in the Papuan biogeographical region, see Zweifel, R.G. (1962). Results of the Archbold Expeditions. 83. Frogs of the microhylid genus *Cophixalus* from the mountains of New Guinea. *Am. Mus. Novit.* 2087: 1–26; Zweifel, R.G. (1979). A new cryptic species of microhylid frog (genus *Cophixalus*) from Papua New Guinea, with notes on related forms. *Am. Mus. Novit.* 2678: 1–14.

Cophixalus concinnus Tyler, 1979

Cophixalus concinnus Tyler, M.J. (1979). A new species of *Cophixalus* (Anura : Microhylidae) from Queensland, Australia. *Copeia* **1979**: 118–121 [119 fig 1]. Type data: holotype, QM J30743, from Thornton Peak, Qld., in 16°12'S, 145°20'E.

Distribution: NE coastal, Qld. Ecology: closed forest, terrestrial, arboreal, nocturnal, predator; direct development, cryptozoic, arthropod-feeder.

Cophixalus exiguus Zweifel and Parker, 1969

Cophixalus exiguus Zweifel, R.G. & Parker, F. (1969). A new species of microhylid frog (genus *Cophixalus*) from Australia. *Am. Mus. Novit.* 2390: 1–10 [2 figs 1–5]. Type data: holotype, SAMA R10311, from between 1800 and 2000 ft on Mt. Hartley, 23 mi S and 5 mi E of Cooktown, Qld.

Distribution: NE coastal, Qld. Ecology: closed forest, terrestrial, (arboreal), nocturnal, predator; direct development, cryptozoic, arthropod-feeder. Biological references: Tyler, M.J. (1979). A new species of *Cophixalus* (Anura : Microhylidae) from Queensland, Australia. *Copeia* **1979**: 118–121 (taxonomy).

Cophixalus neglectus Zweifel, 1962

Cophixalus neglectus Zweifel, R.G. (1962). A systematic review of the microhylid frogs of Australia. *Am. Mus. Novit.* 2113: 1–40 [15 fig 5]. Type data: holotype, MCZ 18505, from Bellenden Ker Range, Qld.

Distribution: NE coastal, Qld. Ecology: closed forest, terrestrial, nocturnal, predator; direct development, cryptozoic, arthropod-feeder. Biological references: Tyler, M.J. (1979). A new species of *Cophixalus* (Anura : Microhylidae) from Queensland, Australia. *Copeia* **1979**: 118–121 (taxonomy).

Cophixalus ornatus (Fry, 1912)

Austrochaperina ornata Fry, D.B. (1912). Description of *Austrochaperina*, a new genus of Engystomatidae from north Australia. *Rec. Aust. Mus.* **9**: 87–106 [91 fig 38]. Type data: holotype, AM R222, from 25 mi inland from Cairns, Qld.

Phrynixalus reginae Andersson, L.G. (1916). Results of Dr. E. Mjöberg's Swedish Scientific Expeditions to Australia 1910–1913. 9. Batrachians from Queensland. *K. Svenska Vetensk.-Akad. Handl. (ns)* **52**: 1–20 [4 figs 1a–d]. Type data: syntypes, NHRM 1615 (4 specimens), NHRM 1616, MCZ 17805, BMNH 1929.5.17.1, 4 types not found, from Malanda and Cedar Creek, Qld.

Synonymy that of Cogger, H.G., this work.

Distribution: NE coastal, Qld. Ecology: closed forest, arboreal, nocturnal, predator; direct development, arthropod-feeder. Biological references: Zweifel, R.G. (1962). A systematic review of the microhylid frogs of Australia. *Am. Mus. Novit.* 2113: 1–40 (taxonomy); Zweifel, R.G. & Parker, F. (1977). A new species of frog from Australia (Microhylidae : *Cophixalus*). *Am. Mus. Novit.* 2614: 1–10 (taxonomy).

Cophixalus saxatilis Zweifel and Parker, 1977

Cophixalus saxatilis Zweifel, R.G. & Parker, F. (1977). A new species of frog from Australia (Microhylidae : *Cophixalus*). *Am. Mus. Novit.* 2614: 1–10 [2 figs 2–4]. Type data: holotype, MCZ 90205, from Black Mt., 13 mi S and 2 mi W of Cooktown, Qld.

Distribution: NE coastal, Qld. Ecology: terrestrial, nocturnal, predator; direct development, saxicoline, cryptozoic, arthropod-feeder.

Sphenophryne Peters and Doria, 1878

Sphenophryne Peters, W. & Doria, G. (1878). Catalogo dei Rettili e dei Batraci raccolti da O. Beccari, L.M. d'Albertis e A.A. Bruijn nella sotto-regione Austro-Malese. *Ann. Mus. Civ. Stor. Nat. Genova* **13**: 323–450 [430]. Type species *Sphenophryne cornuta* Peters and Doria, 1878 by monotypy.

Liophryne Boulenger, G.A. (1897). Descriptions of new lizards and frogs from Mount Victoria, Owen Stanley Range, New Guinea, collected by Mr. A.S. Anthony. *Ann. Mag. Nat. Hist. (6)* **19**: 6–13 [11]. Type species *Liophryne rhododactyla* Boulenger, 1897 (=*Sphenophryne rhododactyla*) by subsequent designation, see Parker, H.W. (1934). *A monograph of the frogs of the family Microhylidae.* London : British Museum viii 208 pp.

***Austrochaperina* (part.)** Fry, D.B. (1912). Description of *Austrochaperina*, a new genus of Engystomatidae from north Australia. *Rec. Aust. Mus.* **9**: 87–106 [87]. Type species *Austrochaperina robusta* Fry, 1912 by original designation.

Oxydactyla van Kampen, P.N. (1913). Amphibien, gesammelt von der niederländischen Süd-Neu-Guinea-Expedition von 1909–10. *Nova Guinea* **9**(Zoologie): 453–465 [464]. Type species *Oxydactyla brevicrus* van Kampen, 1913 (=*Sphenophryne brevicrus*) by monotypy. Compiled from secondary source: Parker, H.W. (1934). *A monograph of the frogs of the family Microhylidae.* London : British Museum viii 208 pp.

Synonymy that of Cogger, H.G., this work; Parker, H.W. (1934). *A monograph of the frogs of the family Microhylidae.* London : British Museum viii 208 pp.

This group is also found in the Papuan biogeographical region, see Zweifel, R.G. (1962). A systematic review of the microhylid frogs of Australia. *Am. Mus. Novit.* 2113: 1–40.

Sphenophryne fryi Zweifel, 1962

*[**Austrochaperina robusta** **(part.)**]* Fry, D.B. (1912). Description of *Austrochaperina*, a new genus of Engystomatidae from north Australia. *Rec. Aust. Mus.* **9**: 87–106 [91 pl 8 fig 2] [this paratype assigned to Var. B]. Type data: paratype, AM R2285, from Bloomfield River near Cooktown, Qld.

Austrochaperina brevipes Fry, D.B. (1915). Herpetological notes. *Proc. R. Soc. Qd.* **27**: 60–95 [61] [*non Liophryne brevipes* Boulenger, 1892 =*Sphenophryne brevipes*]. Type data: holotype, AM R2285, from Bloomfield River, near Cooktown, Qld.

Sphenophryne robustsa Parker, H.W. (1934). *A monograph of the frogs of the family Microhylidae.* London : British Museum viii 208 pp. [157] [invalid emend.; *lapsus pro Sphenophryne robusta* Fry, 1912].

Sphenophryne fryi Zweifel, R.G. (1962). A systematic review of the microhylid frogs of Australia. *Am. Mus.*

Novit. 2113: 1–40 [26] [*nom. nov. pro Austrochaperina brevipes* Fry, 1915].

Synonymy that of Cogger, H.G., this work.

Distribution: NE coastal, Qld. Ecology: closed forest, terrestrial, nocturnal, predator; direct development, cryptozoic, arthropod-feeder. Biological references: Zweifel, R.G. (1962). A systematic review of the microhylid frogs of Australia. *Am. Mus. Novit.* 2113: 1–40 (taxonomy); Czechura, G.V. (1978). An apparent case of altitudinal displacement between two frogs of the genus *Sphenophryne* (Anura : Microhylidae) in north-east Queensland. *N. Qd. Nat.* **44**: 4–6.

Sphenophryne pluvialis Zweifel, 1965

Sphenophryne pluvialis Zweifel, R.G. (1965). Revisionary notes on Australian microhylid frogs of the genus *Sphenophryne. Am. Mus. Novit.* 2214: 1–9 [6]. Type data: holotype, AMNH 54195, from Speewah, elevation 1500 ft, Qld.

Distribution: NE coastal, Qld. Ecology: closed forest, terrestrial, nocturnal, predator; direct development, cryptozoic, arthropod-feeder. Biological references: Czechura, G.V. (1978). An apparent case of altitudinal displacement between two frogs of the genus *Sphenophryne* (Anura : Microhylidae) in north-east Queensland. *N. Qd. Nat.* **44**: 4–6.

Sphenophryne robusta (Fry, 1912)

***Austrochaperina robusta* (part.)** Fry, D.B. (1912). Description of *Austrochaperina*, a new genus of Engystomatidae from north Australia. *Rec. Aust. Mus.* **9**: 87–106 [89] [holotype assigned to Var. A]. Type data: holotype, AM R5295, from Russell River, NE Qld.

Austrochaperina gracilipes Fry, D.B. (1912). Description of *Austrochaperina*, a new genus of Engystomatidae from north Australia. *Rec. Aust. Mus.* **9**: 87–106 [93]. Type data: holotype, AM R4536, from Somerset, Cape York Peninsula, Qld.

Synonymy that of Cogger, H.G., this work.

Distribution: NE coastal, N Gulf, N coastal, Qld., N.T.; extralimital in southern New Guinea. Ecology: tall open shrubland, woodland, open forest, tall forest, closed forest, swamp, terrestrial, nocturnal, predator; direct development, cryptozoic, arthropod-feeder, riparian. Biological references: Zweifel, R.G. (1962). A systematic review of the microhylid frogs of Australia. *Am. Mus. Novit.* 2113: 1–40 (taxonomy); Zweifel, R.G. (1965). Revisionary notes on Australian microhylid frogs of the genus *Sphenophryne. Am. Mus. Novit.* 2214: 1–9 (taxonomy); Cogger, H.G. & Lindner, D.A. (1974). Frogs and reptiles in fauna survey of the Port Essington District, Cobourg Peninsula, Northern Territory of Australia. Frith, H.J. & Calaby, J.H. (eds.) *CSIRO Div. Wildl. Res. Tech. Pap.* **28**: 63–107 (taxonomy).

RANIDAE

INTRODUCTION

A large cosmopolitan family containing more than 560 species in about 43 genera. A single species of the nominate genus *Rana* has reached Australia, where it is confined to Cape York Peninsula.

Characterised in Australia by: firmisternal pectoral girdle; intercalary cartilages absent; small finger and toe discs present, with lateral grooves; sacral diapophyses cylindrical; maxillary teeth present; toes fully webbed; a dorso-lateral skin fold on either side.

References

Barker, J. & Grigg, G. (1977). *A Field Guide to Australian Frogs.* Adelaide : Rigby Ltd. 229 pp.

Cogger, H.G. (1979). *Reptiles and Amphibians of Australia.* Sydney : A.H. & A.W. Reed 608 pp.

Tyler, M.J. (1982). *Frogs.* 2nd edn. Sydney : William Collins 256 pp.

Rana Linnaeus, 1758

Rana Linnaeus, C. (1758). *Systema Naturae.* Classis III. Amphibia. Holmiae : Laurentii Salvii 824 pp. 10th edn. 1: 194–238 [210] [all generic synonyms are based on extralimital species]. For extra-limital synonyms see Gorham, S.W. (1974). *Checklist of World Amphibians.* Saint John : New Brunswick Museum 172 pp. [139]. Type species *Rana temporaria* Linnaeus, 1758 by subsequent designation, see Mertens, R. & Wermuth, H. (1960). *Die Amphibien und Reptilien Europas.* Frankfurt-am-Main : Verl. Waldemar Kramer xii 264 pp.

This group is cosmopolitan.

Rana daemeli (Steindachner, 1868)

Hylorana daemeli Steindachner, F. (1868). Über eine neue *Hylorana*-Art von Cap-York in Australien. *Sber. Akad. Wiss. Wien* **57**: 532–536 [532]. Type data: syntypes, NHMW 3244 (7 specimens), NHMW 16382 (2 specimens), from Cape York, Qld.

Hylarana nebulosa Macleay, W. (1877). The batrachians of the *Chevert* Expedition. *Proc. Linn. Soc. N.S.W.* **2**: 135–138 [137]. Type data: holotype, presumed lost; incorrectly attributed to QM by Moore, J.A. (1961). The frogs of eastern New South Wales. *Bull. Am. Mus. Nat. Hist.* **121**: 149–386, see Covacevich, J. (1971). Amphibian and reptile type specimens in the Queensland Museum. *Mem. Qd. Mus.* **16**: 49–68, from Cape York, Qld.

Hyla nobilis De Vis, C.W. (1884). On a new species of *Hyla. Proc. R. Soc. Qd.* **1**: 128–130 [129]. Type data: syntypes (probable), presumed lost, from Cape York, Qld.

Synonymy that of Cogger, H.G., this work.

Distribution: NE coastal, N Gulf, Qld.; extralimital in New Guinea. Ecology: woodland, open forest, closed forest, terrestrial, nocturnal, predator; free-living tadpole, direct development, riparian, general carnivore. Biological references: Zweifel, R.G. (1980). Results of the Archbold expeditions. No. 130. Frogs and lizards from the Huon Peninsula, Papua New Guinea. *Bull. Am. Mus. Nat. Hist.* **165**: 387–434 (taxonomy).

BUFONIDAE

INTRODUCTION

A large cosmopolitan family containing about 240 species in 18 genera. The only representative in Australia was introduced into Queensland in 1935 in an attempt to combat several insect pests of sugar cane.

Characterised in Australia by: arciferal pectoral girdle; digits without discs or intercalary cartilages; supraocular region warty, separated from the interorbital region by a thick fold or ridge of skin; maxillary teeth absent; toes fully webbed.

References

Barker, J. & Grigg, G. (1977). *A Field Guide to Australian Frogs.* Adelaide : Rigby Ltd. 229 pp.

Cogger, H.G. (1979). *Reptiles and Amphibians of Australia.* Sydney : A.H. & A.W. Reed 608 pp.

Tyler, M.J. (1982). *Frogs.* 2nd edn. Sydney : William Collins 256 pp.

Bufo Laurenti, 1768

Bufo Laurenti, J.N. (1768). *Synopsin Reptilium.* Vienna : J. Thomae 214 pp. [25] For full generic synonymy and extralimital distribution see Gorham, S.W. (1974). *Checklist of World Amphibians.* Saint John : New Brunswick Museum 172 pp. [76] Type species *Bufo vulgaris* Laurenti, 1768 [=*Bufo bufo* (Linnaeus, 1758)] by subsequent designation, see Stejneger, L. (1907). Herpetology of Japan and adjacent territory. *Bull. U.S. Natl. Mus.* **58**: 1–577. Compiled from secondary source: Mertens, R. (1971). Die Typisierung der Krötengattung *Bufo. Senckenberg. Biol.* **52**: 425–427.

This group is also found in South America, Central America, North America, Europe, Asia and Africa; introduced to Australia in 1935.

Bufo marinus (Linnaeus, 1758)

Rana marina Linnaeus, C. (1758). *Systema Naturae.* Classis III. Amphibia. Holmiae : Laurentii Salvii 824 pp. 10th edn. 1: 194–238 [211]. Type data: whereabouts unknown, from America.
For species level synonymy see Gorham, S.W. (1974). *Checklist of World Amphibians.* Saint John : New Brunswick Museum 172 pp. [81] and Cochran, D.M. & Goin, C.J. (1970). Frogs of Colombia. *Bull. U.S. Natl. Mus.* **288**: i–xii 1–655.

Distribution: SE coastal, NE coastal, N Gulf, Murray-Darling basin, (Lake Eyre basin), N.S.W., Qld.; introduced species, native to far southern U.S., Mexico, Central America and northern South America. Ecology: open scrub, low open woodland, low woodland, woodland, open forest, tall forest, closed forest, lentic freshwater, swamp, mangrove, terrestrial, nocturnal, crepuscular, predator; opportunistic breeder, free-living tadpole, breeds in lentic freshwater. Biological references: Covacevich, J. & Archer, M. (1975). The distribution of the cane toad, *Bufo marinus*, in Australia and its effects on indigenous vertebrates. *Mem. Qd. Mus.* **17**: 305–310; van Beurden, E. & Grigg, G.C. (1980). An isolated and expanding population of the introduced toad *Bufo marinus* in New South Wales. *Aust. Wildl. Res.* **7**: 305–310; Broughton, W.C. & Sabath, M.D. (1980). The distribution records of the marine toad *Bufo marinus.* Part 1: Australia. *Griffith University, Brisbane, A.E.S. Working Pap.* **2/80**: 1–81; Easteal, S. (1981). The history of introductions of *Bufo marinus* (Amphibia : Anura); a natural experiment in evolution. *Biol. J. Linn. Soc.* **16**: 93–113.

CROCODYLIDAE

INTRODUCTION

A tropicopolitan family which is regarded either as embracing the crocodiles, alligators and caimans (totalling 22 living species in 7 genera) or containing only the true crocodiles (13 living species in 3 genera). Most crocodylids are fierce predators, although only a few species, including *Crocodylus porosus*, are known to prey on man. Two species occur in Australia.

Crocodiles are semi-aquatic reptiles which hunt in the water but emerge to bask on mud and sand banks. They are oviparous, laying their eggs in specially-constructed nest mounds. The female usually guards the nest and assists the young to emerge and reach the safety of water.

The principal characteristics of crocodiles are their large heavily-built bodies, the development of a thick skin containing a body armour of osteoderms and gastralia, a four-chambered heart, a longitudinal cloacal aperture and a single penis; their adaptations to an aquatic existence include a secondary palate and a pair of transverse fleshy throat valves isolating the air passages from the mouth, valvular nostrils and a powerful muscular tail for aquatic locomotion.

References

Cogger, H.G. (1979). *Reptiles and Amphibians of Australia.* Sydney : A.H. & A.W. Reed 608 pp.

Wermuth, H. & Mertens, R. (1961). *Schildkröten, Krokodile, Brückenechsen.* Jena : Gustav Fischer 422 pp.

Wermuth, H. & Mertens, R. (1977). Liste der rezenten Amphibien und Reptilien; Testudines, Crocodylia, Rhynchocephalia. *Das Tierreich* **100**: i–xxvii, 1–174

Worrell, E. (1970). *Reptiles of Australia.* 2nd edn. Sydney : Angus & Robertson xv 169 pp.

Crocodylus Laurenti, 1768

Crocodylus Gronovius, L.T. (1763). *Zoophylacium Gronovianum, exhibens animalia Quadrupeda, Amphibia atque Pisces, quae in Museo suo adservavit.* Lugduno - Batavia : T. Haak Vol. 1 [10] [*nom. nud.*; Gronovius' work rejected for nomenclatural purposes under International Commission on Zoological Nomenclature Opinion 89, 1925].

Crocodylus Laurenti, J.N. (1768). *Synopsin Reptilium.* Vienna : J. Thomae 214 pp. [53]. Type species *Crocodylus niloticus* Laurenti, 1768 by subsequent designation, see Brown, A.E. (1908). Generic types of nearctic Reptilia and Amphibia. *Proc. Acad. Nat. Sci. Philad.* **60**: 112–127.

Crocodillus Scopoli, J.A. (1777). *Introductio ad hist. naturalem, sistentes genera lapidum, plantarum et animalium, hactenus detecta, characteribus essentialibus donata, in tribus divisa, subinde ad leges naturae.* Prague : Gerle [463 Index An. 3] [*pro Crocodylus* or *Crocodilus* Bertrand, 1763 (based on fossil material)].

Crocodilus Bonnaterre, l'Abbé (1789). *Tableau Encyclopédique et Méthodique des trois règnes de la nature. Erpétologie.* Paris : Panckoucke xxviii 70 pp. [32]. Type species *Crocodilus caudiverbera* Bonnaterre, 1789 [? = *Crocodylus acutus* (Cuvier, 1807)] by subsequent designation, see Smith, H.M. & Smith, R.B. (1977). *Synopsis of the Herpetofauna of Mexico.* Vol. 5. Guide to Mexican Amphisbaenians and Crocodilians. Bibliographic Addendum 2. North Bennington, Vermont : John Johnson 187 pp.

Crocodilus Gmelin, J.F. (1789). Amphibia, Pisces. *in* Linnaeus, C. *Systema Naturae in quo naturae regna tria, secundum classes, ordines, genera, species, systematice proponuntur.* Lipsiae : G.E. Beer 13th edn. 1(3): 1033–1516 [1057] [*nom. substit. pro Crocodylus*

Laurenti, 1768 *vide* Wermuth, H. & Mertens, R. (1977). Liste der rezenten Amphibien und Reptilien; Testudines, Crocodylia, Rhynchocephalia. *Das Tierreich* **100**: i–xxvii, 1–174].

Mecistops Gray, J.E. (1844). *Catalogue of the Tortoises, Crocodiles, and Amphisbaenians in the Collection of the British Museum.* London : British Museum viii 80 pp. [57]. Type species *Mecistops bennettii* Gray, 1844 (= *Crocodylus cataphractus* Cuvier, 1825) by subsequent designation, see Wermuth, H. & Mertens, R. (1977). Liste der rezenten Amphibien und Reptilien; Testudines, Crocodylia, Rhynchocephalia. *Das Tierreich* **100**: i–xxvii, 1–174; Smith, H.M. & Smith, R.B. (1977). *Synopsis of the Herpetofauna of Mexico.* Vol. 5. Guide to Mexican Amphisbaenians and Crocodilians. Bibliographic Addendum 2. North Bennington, Vermont : John Johnson 187 pp.

Oopholis Gray, J.E. (1844). *Catalogue of the Tortoises, Crocodiles, and Amphisbaenians in the Collection of the British Museum.* London : British Museum viii 80 pp. [58] [described as subgenus of *Crocodilus*]. Type species *Crocodilus oopholis* Schneider, 1801 (= *Crocodylus porosus* Schneider, 1801) by tautonymy, but Smith, H.M. & Smith, R.B. (1977). *Synopsis of the Herpetofauna of Mexico.* Vol. 5. Guide to Mexican Amphisbaenians and Crocodilians. Bibliographic Addendum 2. North Bennington, Vermont : John Johnson 187 pp. erroneously designate *Crocodylus porosus* Schneider as type species.

Motinia Gray, J.E. (1844). *Catalogue of the Tortoises, Crocodiles, and Amphisbaenians in the Collection of the British Museum.* London : British Museum viii 80 pp. [60] [described as subgenus of *Crocodilus*]. Type species *Crocodilus americanus* Laurenti, 1768 (=? *Crocodylus acutus* Cuvier, 1807) by monotypy.

Palinia Gray, J.E. (1844). *Catalogue of the Tortoises, Crocodiles, and Amphisbaenians in the Collection of the British Museum.* London : British Museum viii 80 pp. [60] [described as subgenus of *Crocodilus*]. Type species *Crocodilus rhombifer* Cuvier, 1807 by monotypy.

Bombifrons Gray, J.E. (1862). A synopsis of the species of crocodiles. *Ann. Mag. Nat. Hist. (3)* **10**: 265–274 [269]. Type species *Crocodilus bombifrons* Gray, 1844 (= *Crocodylus palustris* Lesson, 1831) by tautonymy, but Smith, H.M. & Smith, R.B. (1977). *Synopsis of the Herpetofauna of Mexico.* Vol. 5. Guide to Mexican Amphisbaenians and Crocodilians. Bibliographic Addendum 2. North Bennington, Vermont : John Johnson 187 pp. erroneously designate *Crocodylus siamensis* Schneider as type species.

Molinia Gray, J.E. (1862). A synopsis of the species of crocodiles. *Ann. Mag. Nat. Hist. (3)* **10**: 265–274 [272] [almost certainly an emend. *pro Motinia* Gray, 1844, although Smith, H.M. & Smith, R.B. (1977). *Synopsis of the Herpetofauna of Mexico.* Vol. 5. Guide to Mexican Amphisbaenians and Crocodilians. Bibliographic Addendum 2. North Bennington, Vermont : John Johnson 187 pp. treat it as a new genus with type species *Crocodylus americanus* Laurenti, 1768].

Temsacus Gray, J.E. (1862). A synopsis of the species of crocodiles. *Ann. Mag. Nat. Hist. (3)* **10**: 265–274 [272] [described as subgenus of *Molinia* Gray, 1862]. Type species *Crocodylus intermedius* Graves, 1819 by monotypy.

Philas Gray, J.E. (1874). On *Crocodilus johnstoni. Proc. Zool. Soc. Lond.* **1874**: 177–178 [178 pl 27] [described as subgenus of *Crocodilus*]. Type species *Crocodylus johnstoni* Krefft, 1873 by monotypy.

Molina Romer, A.S. (1956). *Osteology of the Reptiles.* Chicago : Univ. Chicago Press xxii 772 pp. [604] [*errore pro Molinia* Gray, 1862].

Synonymy that of Cogger, H.G., this work; Wermuth, H. & Mertens, R. (1977). Liste der rezenten Amphibien und Reptilien; Testudines, Crocodylia, Rhynchocephalia. *Das Tierreich* **100**: i–xxvii, 1–174; Smith, H.M. & Smith, R.B. (1977). *Synopsis of the Herpetofauna of Mexico.* Vol. 5. Guide to Mexican Amphisbaenians and Crocodilians. Bibliographic Addendum 2. North Bennington, Vermont : John Johnson 187 pp.

This group is also found in Africa, Madagascar, southern and south-east Asia, Indo-Australian Archipelago, Oceania, southern and south-eastern North America, Central America and northern South America, see Wermuth, H. & Mertens, R. (1977). Liste der rezenten Amphibien und Reptilien; Testudines, Crocodylia, Rhynchocephalia. *Das Tierreich* **100**: i–xxvii, 1–174.

Crocodylus johnstoni Krefft, 1873

Crocodilus johnsoni Krefft, G. (1873). Remarks on Australian crocodiles, and description of a new species. *Proc. Zool. Soc. Lond.* **1873**: 334–335 [335]. Type data: holotype, AM 4629 skeleton, AM 4627 stuffed skin minus head from same specimen, from Cashmere, upper Herbert River, Qld.

Tomistoma krefftii Gray, J.E. *in* Krefft, G. (1873). Remarks on Australian crocodiles, and description of a new species. *Proc. Zool. Soc. Lond.* **1873**: 334–335 [334] [*nom. nud.*, introduced in synonymy].

Crocodilus (Philas) johnstoni Gray, J.E. (1874). On *Crocodilus johnstoni. Proc. Zool. Soc. Lond.* **1874**: 177–178 [177 pl 27] [justified emend. *pro Crocodilus johnsoni* Krefft, 1873].

Synonymy that of Cogger, H.G., this work.

Distribution: NE coastal, N Gulf, N coastal, Qld., W.A., N.T. Ecology: lentic freshwater, lotic freshwater, swamp, amphibious, noctidiurnal, predator; seasonal breeder, oviparous, riparian, general carnivore. Biological references: Grigg, G.C. & Alchin, J. (1976). The role of the cardiovascular system in thermoregulation of *Crocodylus johnstoni. Physiol. Zool.* **49**: 24–36; Johnson, C.R., Webb, G.J.W. & Tanner, C. (1976). Thermoregulation in crocodilians. II. A telemetric study of body temperature in the Australian crocodiles, *Crocodylus johnstoni* and *Crocodylus porosus. Comp. Biochem. Physiol.* **53A**: 143–146; Johnson, C.R. (1973). Behaviour of the Australian crocodiles, *Crocodylus johnstoni* and *C. porosus. Zool. J. Linn. Soc.* **52**: 315–336; Webb, G.J.W. (1981). Nesting biology of *Crocodylus johnstoni* in the Northern Territory. p. 170 *in* Banks, C.B. &

Martin, A.A. (eds.) *Proceedings of the Melbourne Herpetological Symposium, 1980.* Melbourne : Zoological Board of Victoria.

Crocodylus porosus Schneider, 1801

Crocodylus natans Meyer, F.A.A. (1795). *Synopsis Reptilium, sistematis novam generum methodum, nec non Gottingae huj. ordinis animalium enumerationem.* Göttingen : Vandenhoeck u. Ruprecht [20] [*nom. oblit.*]. Type data: status unknown, from Sri Lanka.

Crocodilus porosus Schneider, J.G. (1801). *Historiae Amphibiorum naturalis et literariae.* fasc. sec. Jena : Frommani vi 374 pp. [159]. Type data: lectotype, ZMB 278, designation by Mertens, R. (1960). Zur Systematik und Nomenklatur der Ceylon-Krokodile. *Senckenberg. Biol.* **41**: 267–272, from unknown locality; lectotype labelled "India", but type locality restricted to Sri Lanka by Mertens.

Crocodilus oopholis Schneider, J.G. (1801). *Historiae Amphibiorum naturalis et literariae.* fasc. sec. Jena : Frommani vi 374 pp. [165]. Type data: none designated, from unknown locality but restricted to Sri Lanka, see Mertens, R. (1960). Zur Systematik und Nomenklatur der Ceylon-Krokodile. *Senckenberg. Biol.* **41**: 267–272.

Crocodilus biporcatus Cuvier, G. (1807). Sur les differentes espèces de crocodiles vivans et sur leurs caractères distinctifs. *Ann. Mus. Natl. Hist. Nat. Paris* **10**: 8–86 [48]. Type data: syntypes, ?MNHP, from Java and Timor, Indonesia; Sri Lanka; restricted to Java, Indonesia, see Mertens, R. & Wermuth, H. (1955). Die rezenten Schildkröten, Krokodile und Brückenechsen. Eine kritische Liste der heute lebenden Arten und Rassen. *Zool. Jb. Syst.* **83**: 323–440.

Crocodilus biporcatus raninus Müller, S. & Schlegel, H. (1844). Over de Krokodillen van den Indischen Archipel. *in* Temminck, C.J. *Verhandelingen over de Natuurlijke Geschiedenis der Nederlandsche Overzeesche Bezittingen.* Zoologie : Reptilia Leiden : Natuurkundige Commissie in Indie 28 pp. [28, pl 3 figs 7–8]. Type data: syntypes, RMNH, from Borneo.

Oopholis pondicherianus Gray, J.E. (1862). A synopsis of the species of crocodiles. *Ann. Mag. Nat. Hist. (3)* **10**: 265–274 [268 pl 7]. Type data: holotype, BMNH, from Pondicherry, India.

Crocodylus porosus australis Deraniyagala, P.E.P. (1953). *A Coloured Atlas of Some Vertebrates from Ceylon.* Vol. 2. Tetrapod Reptilia. Colombo : Govt. Press vii 101 pp. [34] [type citation refers only to p.12 in Barrett, C. (1950). *Reptiles of Australia.* London : Cassell 168 pp., which lacks any description, but type citation probably refers to figure on facing page 13]. Type data: none designated, from Australia.

Crocodylus porosus minikanna Deraniyagala, P.E.P. (1955). A new race of estuarine crocodile from Ceylon. *Spolia Zeylan.* **27**: 277–279 [277]. Type data: holotype, COM ID (mounted head and neck), from Bolgoda, Sri Lanka.

Synonymy that of Cogger, H.G., this work; Wermuth, H. & Mertens, R. (1977). Liste der rezenten Amphibien und Reptilien; Testudines, Crocodylia, Rhynchocephalia. *Das Tierreich* **100**: i–xxvii, 1–174.

Distribution: NE coastal, N Gulf, N coastal, Qld., W.A., N.T.; extralimital in the western Pacific and Indo-Malaysian regions. Ecology: lentic freshwater, lotic freshwater, swamp, estuarine, inshore, amphibious, noctidiurnal, predator; seasonal breeder, oviparous, general carnivore. Biological references: Johnson, C.R., Webb, G.J.W. & Tanner, C. (1976). Thermoregulation in crocodilians. II. A telemetric study of body temperature in the Australian crocodiles, *Crocodylus johnstoni* and *Crocodylus porosus. Comp. Biochem. Physiol.* **53A**: 143–146; Drane, C.R., Webb, G.J.W. & Heuer, P. (1977). Patterns of heating in the body, trunk and tail of *Crocodylus porosus. J. Thermal Biol.* **2**: 127–130; Webb, G.J.W. & Messel, H. (1978). Morphometric analysis of *Crocodylus porosus* from the north coast of Arnhem Land, northern Australia. *Aust. J. Zool.* **26**: 1–27; Webb, G.J.W., Messel, H., Crawford, J. & Yerbury, M.J. (1978). Growth rates of *Crocodylus porosus* (Reptilia : Crocodilia) from Arnhem Land. *Aust. Wildl. Res.* **5**: 401–415; Johnson, C.R. (1973). Behaviour of the Australian crocodiles, *Crocodylus johnstoni* and *C. porosus. Zool. J. Linn. Soc.* **52**: 315–336; Johnson, C.R. (1974). Thermoregulation in crocodilians - I. Head-body temperature control in the Papuan - New Guinean crocodiles, *Crocodylus novaeguineae* and *Crocodylus porosus. Comp. Biochem. Physiol.* **49A**: 3–28; Magnusson, W.E. (1980). Hatching and crèche formation by *Crocodylus porosus. Copeia* **1980**: 359–362; Magnusson, W.E. (1979). Dispersal of hatchling crocodiles (*Crocodylus porosus*) (Reptilia, Crocodilidae). *J. Herpet.* **13**: 227–231; Magnusson, W.E. & Taylor, J.A. (1980). A description of developmental stages of *Crocodylus porosus*, for use in aging eggs in the field. *Aust. Wildl. Res.* **7**: 479–485; Messel, H., Vorlicek, G.C., Wells, A.G. & Green, W.J. (1981). The status and dynamics of *Crocodylus porosus* and *C. johnstoni*. pp. 78–103 *in* Banks, C.B. & Martin, A.A. (eds.) *Proceedings of the Melbourne Herpetological Symposium, 1980.* Melbourne : Zoological Board of Victoria; Magnusson, W.E. & Taylor, J.A. (1981). Growth of juvenile *Crocodylus porosus* as effected by season of hatching. *J. Herpet.* **15**: 242–245; Grigg, G.C., Taplin, L.E., Harlow, P. & Wright, J. (1980). Survival and growth of hatchling *Crocodylus porosus* in saltwater without access to fresh drinking water. *Oecologia* **47**: 264–266.

CHELIDAE

INTRODUCTION

A family primarily of aquatic turtles containing about 30 species in 8 genera, of which 5 genera are endemic to Australia and New Guinea; the remaining members of the family are restricted to South America. The taxonomic arrangement of the 15 Australian species presented below is a traditional one which is currently undergoing a much-needed review and can be expected to undergo major changes in the near future.

Characterised in Australia by: limbs jointed, not paddle-shaped, with 4-or 5-clawed, webbed feet; barbels well developed in most species; pleurodirous.

References

Cann, J. (1978). *Tortoises of Australia.* Sydney : Angus & Robertson 79 pp.

Cogger, H.G. (1979). *Reptiles and Amphibians of Australia.* Sydney : A.H. & A.W. Reed 608 pp.

Goode, J. (1964). *Freshwater Tortoises of Australia and New Guinea.* Melbourne : Landsdowne Press x 154 pp.

Wermuth, H. & Mertens, R. (1961). *Schildkröten, Krokodile, Brückenechsen.* Jena : Gustav Fischer 422 pp.

Wermuth, H. & Mertens, R. (1977). Liste der rezenten Amphibien und Reptilien; Testudines, Crocodylia, Rhynchocephalia. *Das Tierreich* **100**: i–xxvii, 1–174

Chelodina Fitzinger, 1826

Chelodina Fitzinger, L.J. (1826). *Neue Classification der Reptilien nach ihren naturlichen Verwandtschaften.* Vienna : Heubner 66 pp. [6]. Type species *Emys longicollis* (= *Testudo longicollis* Shaw, 1794) by original designation.

Hydraspis Bell, T. (1828). On *Hydraspis*, a new genus of freshwater tortoises, of the family Emydidae. *Zool. J.* **3**: 511–513 [512]. Type species *Testudo longicollis* Shaw, 1794 by original designation.

Chelyodina Agassiz, L. (1846). *Nomenclatoris Zoologici Index Universalis.* Soloduri : Jent et Gassman 393 pp. [79] [emend. *pro Chelodina* Fitzinger, 1826].

Chelydura Oudemans, J.T. (1894). Eidechsen und Schildkröten. *in* Semon, R. *Zoologische Forschungsreisen in Australien und den Malayischen Archipel.* Jena : Gustav Fischer **5**: 127–146 [145] [?*lapsus pro Chelodina* Fitzinger, 1826].

Chelidura Oudemans, J.T. (1894). Eidechsen und Schildkröten. *in* Semon, R. *Zoologische Forschungsreisen in Australien und den Malayischen Archipel.* Jena : Gustav Fischer **5**: 127–146 [130] [?*lapsus pro Chelodina* Fitzinger, 1826].

Synonymy that of Cogger, H.G., this work.

This group is also found in southern New Guinea.

Chelodina expansa Gray, 1857

Chelodina expansa Gray, J.E. (1857). Description of a new species of *Chelodina* from Australia. *Proc. Zool. Soc. Lond.* **1856**: 369–371 [370 pl 12] [1856 on title page]. Type data: syntypes, BMNH 1947.3.4.21, BMNH 1947.3.5.88, from Australia.

Distribution: NE coastal, Murray-Darling basin, Vic., N.S.W., Qld., S.A. Ecology: lotic freshwater, aquatic, noctidiurnal, predator, (lentic freshwater); seasonal breeder, oviparous, general carnivore. Biological references: Goode, J. (1968). Morphological variations in Victorian tortoises with special reference to the intermediate form of *Chelodina expansa. Vict. Nat.* **85**: 263–267; Goode, J. & Russell, J. (1968). Incubation of eggs in the species of chelid tortoises, and notes on their embryological development. *Aust. J. Zool.* **16**: 749–761; Legler, J.M. (1978). Observations on behaviour and ecology in an Australian turtle, *Chelodina expansa* (Testudines : Chelidae). *Can. J. Zool.* **56**: 2449–2453; Goode, J. (1965). Nesting behaviour of freshwater tortoises in Victoria. *Vict. Nat.* **82**: 212–222.

Chelodina longicollis (Shaw, 1794)

Testudo longicollis Shaw, G. (1794). *The Zoology of New Holland*. Illustrated by J. Sowerby. London 33 pp. [19 pl 7]. Type data: holotype, not specified but presumed to be BMNH 1947.3.5.86, from unspecified locality. Holotype presumed to be that specimen on which Shaw based a subsequent description of this species *vide* Shaw, G. (1802). *General Zoology*, or Systematic Natural History. Vol. 3 Pt. 1. Amphibia. London : Kearsley 312 pp.

Chelodina novaehollandiae Duméril, A.M.C. & Bibron, G. (1835). *Erpétologie Générale* ou Histoire Naturelle Complète des Reptiles. Paris : Roret 2: 680 pp. [443] [*nom. substit. pro Testudo longicollis* Shaw, 1802; Roux-Estève, R. (1979). Liste des Amphibiens et Reptiles des collections du Muséum National d'Histoire Naturelle de Paris, récoltés par Lesueur (1778–1846). *Bull. Trim. Soc. Géol. Normandie et Amis Muséum du Havre* **66**: 25–29 regards this as a distinct species with its own holotype, MNHP 6969, from Australia].

Chelodina sulcifera Gray, J.E. (1855). *Catalogue of Shield Reptiles in the Collection of the British Museum.* Part I. Testudinata (Tortoises). London : British Museum 79 pp. [59 pl 25 fig 2]. Type data: holotype, BMNH 1947.3.5.87, from Australia.

Chelodina sulcata Gray, J.E. (1856). On some new species of freshwater tortoises from North America, Ceylon and Australia, in the collection of the British Museum. *Proc. Zool. Soc. Lond.* **1855**: 197–202 [201] [*pro Chelodina sulcifera* Gray, 1855; 1855 on title page but p. 201 issued 5 Feb., 1856].

Synonymy that of Cogger, H.G., this work.

Distribution: SE coastal, NE coastal, Murray-Darling basin, S.A., Vic., N.S.W., Qld., A.C.T. Ecology: lentic freshwater, swamp, amphibious, noctidiurnal, predator, (lotic freshwater); seasonal breeder, oviparous, general carnivore. Biological references: Vestjens, W.J.M. (1969). Nesting, egg-laying and hatching of the snake-necked tortoise at Canberra, A.C.T. *Aust. Zool.* **15**: 141–148; Eisner, T., Jones, T.H., Meinwald, J. & Legler, J.M. (1978). Chemical composition of the odorous secretion of the Australian turtle, *Chelodina longicollis. Copeia* **1978**: 714–715; Kool, K. (1981). Is the musk of the long-necked turtle, *Chelodina longicollis*, a deterrent to predators? *Aust. J. Herp.* **1**: 45–53; Armstrong, G. (1982). Notes on feeding and growth rates in juvenile *Chelodina longicollis. Herpetofauna* **13**: 27; Siebenrock, F. (1915). Die Schildkröten Gattung *Chelodina* Fitz. *Sber. Akad. Wiss. Wien* **124**: 13–35 (taxonomy); Mertens, R. (1972). Eine oft verkannte Unterart der australischen Schlangenhals-Schildkröte *Chelodina longicollis. Salamandra* **8**: 27–31.

Chelodina novaeguineae Boulenger, 1888

Chelodina novaeguineae Boulenger, G.A. (1888). On the chelydoid chelonians of New Guinea. *Ann. Mus. Civ. Stor. Nat. Genova (2)* **6**: 449–452 [450]. Type data: syntypes, MCG C.E. 8407, BMNH 1946.1.22.36, from Mawatta, Binaturi River (as Katow), Papua New Guinea.

Distribution: NE coastal, N Gulf, Qld.; extralimital in New Guinea. Ecology: lotic freshwater, aquatic, noctidiurnal, predator, (lentic freshwater); seasonal breeder, oviparous, general carnivore. Biological references: Siebenrock, F. (1915). Die Schildkröten Gattung *Chelodina* Fitz. *Sber. Akad. Wiss. Wien* **124**: 13–35 (taxonomy).

Chelodina oblonga Gray, 1841

Chelodina oblonga Gray, J.E. (1841). A catalogue of the species of reptiles and Amphibia hitherto described as inhabiting Australia, with a description of some new species from Western Australia. Appendix E *in* Grey, G. *Journals of Two Expeditions of Discovery in North-west and Western Australia, during the years 1837, 38, and 39, under the authority of Her Majesty's Government.* London : T. & W. Boone 2: 422–449 [446 pl 7]. Type data: holotype, BMNH 1947.3.5.89, from Western Australia.

Chelodina colliei Gray, J.E. (1856). On some new species of freshwater tortoises from North America, Ceylon and Australia, in the collection of the British Museum. *Proc. Zool. Soc. Lond.* **1855**: 197–202 [200] [1855 on title page]. Type data: syntypes, BMNH 1946.1.22.5, BMNH 1947.3.5.90–91, from Swan River, Australia.

Synonymy that of Cogger, H.G., this work.

Distribution: SW coastal, W.A. Ecology: lentic freshwater, lotic freshwater, swamp, aquatic, noctidiurnal, predator; seasonal breeder, oviparous, general carnivore. Biological references: Siebenrock, F. (1915). Die Schildkröten Gattung *Chelodina* Fitz. *Sber. Akad. Wiss. Wien* **124**: 13–35 (taxonomy).

Chelodina rugosa Ogilby, 1890

Chelodina rugosa Ogilby, J.D. (1890). Description of a new Australian tortoise. *Rec. Aust. Mus.* **1**: 56–59 [56 pl 7]. Type data: holotype, AM R6256, from Cape York, Qld.

Chelodina intergularis Fry, D.B. (1915). Herpetological notes. *Proc. R. Soc. Qd.* **27**: 60–95 [88 pl 4]. Type data: holotype, AM R6255, from "Australia?".

Synonymy that of Cogger, H.G., this work.

Distribution: NE coastal, N Gulf, N coastal, Qld., W.A., N.T. Ecology: lentic freshwater, swamp, amphibious, noctidiurnal, predator, (lotic freshwater); seasonal breeder, oviparous, general carnivore. Biological references: Cogger, H.G. & Lindner, D.A. (1974). Frogs and reptiles. *in* Fauna Survey of the Port Essington District, Cobourg Peninsula, Northern Territory of Australia. Frith, H.J. & Calaby, J.H. (eds.) *CSIRO Div. Wildl. Res. Tech. Pap.* **28**: 63–107 (taxonomy).

Chelodina steindachneri Siebenrock, 1914

Chelodina steindachneri Siebenrock, F. (1914). Eine neue *Chelodina* Art aus Westaustralien. *Anz. Akad. Wiss. Wien* **17**: 386–387 [386]. Type data: holotype, NHMW 19798, from Marloo Station, Grey River, W.A.

Chelodina millymillyensis Glauert, L. (1923). A new freshwater tortoise from the Murchison River. *J. R. Soc. West. Aust.* **9**: 53–56 [53]. Type data: lectotype, WAM R1000, from Milly Milly Creek, Milly Milly Station, Murchison River, W.A., designation by Cogger, H.G., this work.

Synonymy that of Cogger, H.G., this work.

Distribution: NW coastal, W.A. Ecology: lentic freshwater, lotic freshwater, aquatic, noctidiurnal, predator; seasonal breeder, oviparous, general carnivore.

Elseya Gray, 1867

Elseya Gray, J.E. (1867). Description of a new Australian tortoise (*Elseya latisternum*). *Ann. Mag. Nat. Hist. (3)* **20**: 43–45 [44]. Type species *Chelymys dentata* Gray, 1863 by subsequent designation, see Lindholm, W.A. (1929). Revidiertes Verzeichnis der Gattungen der rezenten Schildkröten nebst Notizen zur Nomencklatur einiger Arten. *Zool. Anz.* **81**: 275–295

[Euchelymys*(part.)*] Gray, J.E. (1871). On *Euchelymys*, a new genus and two new species of Australian freshwater tortoises. *Ann. Mag. Nat. Hist. (4)* **8**: 117–118 [118]. Type species *Euchelymys sulcifera* Gray, 1871 by subsequent designation, Cogger, H.G., this work. [Synonymy based on *Euchelymys spinosa* Gray, 1871 (=*Elseya latisternum*), one of two species originally included in genus].

Synonymy that of Cogger, H.G., this work.

This group is also found in New Guinea.

Elseya dentata (Gray, 1863)

Chelymys dentata Gray, J.E. (1863). On the species of *Chelymys* from Australia, with the description of a new species. *Ann. Mag. Nat. Hist. (3)* **12**: 98–99, 246 [98; 246]. Type data: syntypes, BMNH 1947.3.6.3 (probable holotype), BMNH 1947.3.6.2, from Beagle's Valley, upper Victoria River, N.T.

C[helymys] elseyi Gray, J.E. (1864). On the genera of Chelydidae and the characters furnished by the study of their skulls. *Proc. Zool. Soc. Lond.* **1864**: 128–135 [132] [*nom. nud.*introduced in synonymy].

Elseya ? intermedia Gray, J.E. (1872). *Appendix to the Catalogue of Shield Reptiles in the Collection of the British Museum.* Part I. Testudinata (Tortoises). London : British Museum 28 pp. [23]. Type data: holotype, BMNH 1947.3.4.14, from upper part of Victoria River, N.T.

Synonymy that of Cogger, H.G., this work.

Distribution: NE coastal, N Gulf, N coastal, Lake Eyre basin, Qld., W.A., N.T. Ecology: lentic freshwater, lotic freshwater, aquatic, noctidiurnal, predator; seasonal breeder, oviparous, sexually dimorphic, general carnivore. Biological references: Coventry, A.J. & Tanner, C. (1973). Notes on the short-necked tortoises *Emydura australis* (Gray) and *Elseya dentata* (Gray) in the Victoria River system, Northern Territory. *Vict. Nat.* **90**: 351–353.

Elseya latisternum Gray, 1867

Elseya latisternum Gray, J.E. (1867). Description of a new Australian tortoise (*Elseya latisternum*). *Ann. Mag. Nat. Hist. (3)* **20**: 43–45 [44]. Type data: holotype, BMNH 1947.3.4.13, from "North Australia".

Euchelymys spinosa Gray, J.E. (1871). On *Euchelymys*, a new genus and two new species of Australian freshwater tortoises. *Ann. Mag. Nat. Hist. (4)* **8**: 117–118 [118]. Type data: holotype, BMNH 1946.1.22.77, from "North Australia".

Elseya latisternon Gray, J.E. (1871). Notes on Australian freshwater tortoises. *Ann. Mag. Nat. Hist. (4)* **8**: 291–292 [292] [*lapsus pro Elseya latisternum* Gray, 1867].

Synonymy that of Cogger, H.G., this work.

Distribution: SE coastal, NE coastal, Murray-Darling basin, N.S.W., Qld. Ecology: lotic freshwater, aquatic, noctidiurnal, predator, (lentic freshwater); seasonal breeder, oviparous, sexually dimorphic, general carnivore.

Emydura Bonaparte, 1836

Emydura Bonaparte, C.L. (1836). *Cheloniorum Tabula Analytica.* Rome pp. 1–9 [7] [a *nom. nud.* because its type species is a *nom. nud.*; *Emydura* should be conserved, however, on the grounds of general usage]. Type species *Emys macquaria* Cuvier, 1829 by monotypy.

Chelymys Gray, J.E. (1844). *Catalogue of the Tortoises, Crocodiles, and Amphisbaenians in the Collection of the British Museum.* London : British Museum viii 80 pp. [42]. Type species *Chelymys macquaria* (=*Chelys (Hydraspis) macquarii* Gray, 1830) by monotypy.

***Euchelymys* (part.)** Gray, J.E. (1871). On *Euchelymys*, a new genus and two new species of Australian freshwater tortoises. *Ann. Mag. Nat. Hist. (4)* **8**: 117–118 [8] [also in synonymy of *Elseya*]. Type species *Euchelymys sulcifera* Gray, 1871 by subsequent designation, Cogger, H.G., this work.

Synonymy that of Cogger, H.G., this work.

This group is also found in New Guinea.

Emydura krefftii (Gray, 1871)

Chelymys krefftii Gray, J.E. (1871). Notes on Australian freshwater tortoises. *Ann. Mag. Nat. Hist. (4)* **8**: 366 [366]. Type data: holotype, BMNH 1947.3.6.1, from Burnett River, Qld.

Chelymys victoriae* Var. 2. *marmorata Gray, J.E. (1872). On the genus *Chelymys* and its allies from Australia. *Proc. Zool. Soc. Lond.* **1872**: 504–514 [506]. Type data: holotype, BMNH 71.9.25.5 (Stimson, *pers. comm.*), from Burnett River, Qld.

Chelymys victoriae* Var. 3. *sulcata Gray, J.E. (1872). On the genus *Chelymys* and its allies from Australia. *Proc. Zool. Soc. Lond.* **1872**: 504–514 [506]. Type data: syntypes, BMNH 71.9.25.3–4 (Stimson, *pers. comm.*), from Burnett River, Qld.

Synonymy that of Cogger, H.G., this work.

Distribution: NE coastal, N Gulf, Murray-Darling basin, Bulloo River basin, Lake Eyre basin, Qld.; (S.A.). Ecology: lotic freshwater, aquatic, noctidiurnal, predator, lentic freshwater; seasonal breeder, oviparous, sexually dimorphic, general carnivore. Biological references: McNicol, K. & Georges, A. (1980). Observations on the eggs and hatchlings of *Emydura krefftii* from Fraser Island. *Herpetofauna* **12**: 10–12; Georges, A. (1982). Diet of the Australian freshwater turtle *Emydura krefftii* (Chelonia : Chelidae), in an unproductive lentic environment. *Copeia* **1982**: 331–336.

Emydura macquarii (Gray, 1830)

Emys macquaria Cuvier, G.L.C.F.D. (1829). *Le Règne Animal.* 2nd edn. Paris : Déterville Vol. 2 xv 406 pp. [11] [*nom. nud.*].

Ch[elys] (Hydraspis) macquarii Gray, J.E. (1830). A synopsis of the species of the class Reptilia, pp. 1–110 *in* Griffith, E. *The Animal Kingdom Arranged in Conformity with its Organization by the Baron Cuvier.* London : Whittaker, Treacher & Co. 9: 481+110 pp. [Synopsis,15] [*ex* Cuvier ms; 1831 on title page]. Type data: holotype, MNHP 9409, from Australia; subsequently corrected to Macquarie River, N.S.W., see Gray, J.E. (1831). *Synopsis Reptilium;* or Short Descriptions of the Species of Reptiles. Part I. Cataphracta. Tortoises, Crocodiles, and Enaliosaurians. London : Treuttel, Würtz & Co. viii 85 pp.; Gray, J.E. (1872). On the genus *Chelymys* and its allies from Australia. *Proc. Zool. Soc. Lond.* **1872**: 504–514.

Hydraspis macquarrii Gray, J.E. (1831). *Synopsis Reptilium;* or Short Descriptions of the Species of Reptiles. Part I. Cataphracta. Tortoises, Crocodiles, and Enaliosaurians. London : Treuttel, Würtz & Co. viii 85 pp. [40] [*pro Emys macquarii* (*sic*) Cuvier ms].

Hydraspis australis Gray, J.E. (1841). A catalogue of the species of reptiles and Amphibia hitherto described as inhabiting Australia, with a description of some new species from Western Australia. Appendix E *in* Grey, G. *Journals of Two Expeditions of Discovery in North-west and Western Australia, during the years 1837, 38, and 39, under the authority of Her Majesty's Government.* London : T. & W. Boone 2: 422–449 [445 pl 6]. Type data: holotype, BMNH 1947.3.4.36, from "Western Australia?", but from Macquarie River, N.S.W. *vide* corrected type locality provided by Gray, J.E. (1872). On the genus *Chelymys* and its allies from Australia. *Proc. Zool. Soc. Lond.* **1872**: 504–514, referring to his original description.

Euchelymys sulcifera Gray, J.E. (1871). On *Euchelymys*, a new genus and two new species of Australian freshwater tortoises. *Ann. Mag. Nat. Hist. (4)* **8**: 117–118 [118]. Type data: lectotype, BMNH 1947.3.5.97, from North Australia. designation by Cogger, H.G., this work.

Synonymy that of Cogger, H.G., this work.

Distribution: Murray-Darling basin, S.A., Vic., N.S.W., Qld., A.C.T. Ecology: lotic freshwater, aquatic, noctidiurnal, predator, (lentic freshwater); seasonal breeder, oviparous, sexually dimorphic, general carnivore. Biological references: Thompson, M. (1981). Gas tensions in natural nests and eggs of the tortoise *Emydura macquarii*. pp. 74–77 *in* Banks, C.B. & Martin, A.A. (eds.) *Proceedings of the Melbourne Herpetological Symposium.* Melbourne : Zoological Board of Victoria; Goode, J. (1965). Nesting behaviour of freshwater tortoises in Victoria. *Vict. Nat.* **82**: 212–222.

Emydura signata Ahl, 1932

Emydura signata Ahl, E. (1932). Beschreibung einer neuen Schildkröte aus Australien. *Sber. Ges. Naturf. Freunde Berl.* **1932**: 127–129 [127]. Type data: holotype, ZMB 34102, from Brisbane district, Qld.

Emydura canni Worrell, E. (1970). *Reptiles of Australia.* 2nd edn. Sydney : Angus & Robertson xv 169 pp. [pl 6] [*nom. nud.*].

Synonymy that of Cogger, H.G., this work.

Distribution: SE coastal, NE coastal, N.S.W., Qld. Ecology: lotic freshwater, aquatic, noctidiurnal, predator; seasonal breeder, oviparous, sexually dimorphic, general carnivore.

Emydura subglobosa (Krefft, 1876)

Euchelymys subglobosa Krefft, G. (1876). Notes on Australian animals in New Guinea with description of a new species of fresh water tortoise belonging to the genus *Euchelymys* (Gray). *Ann. Mus. Civ. Stor. Nat. Genova* **1**: 390–394 [390]. Type data: holotype, MCG C.E.2320, from Naiabui, on Amama River, SE Papua New Guinea.

Emydura albertisii Boulenger, G.A. (1888). On the chelydoid chelonians of New Guinea. *Ann. Mus. Civ. Stor. Nat. Genova (2)* **6**: 449–452 [449]. Type data: syntypes, MCG C.E. 8430–1, from Mawatta, Binaturi River (as Katow), Papua New Guinea.

Synonymy that of Cogger, H.G., this work.

Distribution: N Gulf, Qld.; extralimital in New Guinea. Ecology: lotic freshwater, aquatic, noctidiurnal, predator, lentic freshwater; seasonal breeder, oviparous, sexually dimorphic, general carnivore. Biological references: Covacevich, J.,

Ingram, G.J. & Czechura, G.V. (1982). Rare frogs and reptiles of Cape York Peninsula, Australia. *Biol. Conserv.* **22**: 283–294 (status in Australia).

Emydura victoriae (Gray, 1842)

Hydraspis victoriae Gray, J.E. (1842). Description of some hitherto unrecorded species of Australian reptiles and batrachians. *in* Gray, J.E. (ed.) *Zoological Miscellany*. London : Treuttel, Würtz & Co. pp. 51–57 [55]. Type data: syntypes, BMMH 1947.3.5.95–6, from Victoria River, N.T.

Distribution: N coastal, N Gulf, W.A., N.T.; (Qld.). Ecology: lotic freshwater, aquatic, noctidiurnal, predator, lentic freshwater; seasonal breeder, oviparous, sexually dimorphic, general carnivore.

Pseudemydura Siebenrock, 1901

Pseudemydura Siebenrock, F. (1901). Beschreibung einer neuen Schildkrötengattung aus der Familie Chelydidae aus Australien: *Pseudemydura*. *Anz. Akad. Wiss. Wien* **38**: 248–251 [248]. Type species *Pseudemydura umbrina* Siebenrock, 1901 by monotypy.

Pseudemydura umbrina Siebenrock, 1901

Pseudemydura umbrina Siebenrock, F. (1901). Beschreibung einer neuen Schildkrötengattung aus der Familie Chelydidae aus Australien: *Pseudemydura*. *Anz. Akad. Wiss. Wien* **38**: 248–251 [249]. Type data: holotype, NHMW 89 (8450) = 1296, from Australia.

Emydura inspectata Glauert, L. (1954). Herpetological miscellanea. IV. A new swamp tortoise from the Swan River district. *West. Aust. Nat.* **4**: 125–127 figs 2 [125 figs]. Type data: holotype, WAM R11092, from Warbrook, *ca.* 24 mi N of Perth, W.A.

Synonymy that of Cogger, H.G., this work.

Distribution: SW coastal, W.A. Ecology: swamp, amphibious, noctidiurnal, predator; seasonal breeder, oviparous, general carnivore. Biological references: Williams, E.E. (1958). Rediscovery of the Australian chelid genus *Pseudemydura* Siebenrock (Chelidae : Testudines). *Breviora* 84: 1–8; Burbidge, A.A. (1981). The ecology of the western swamp tortoise *Pseudemydura umbrina* (Testudines : Chelidae). *Aust. Wildl. Res.* **8**: 203–223.

Rheodytes Legler and Cann, 1980

Rheodytes Legler, J.M. & Cann, J. (1980). A new genus and species of chelid turtle from Queensland, Australia. *Los Angeles Co. Mus. Contrib. Sci.* 324: 1–18 [2]. Type species *Rheodytes leukops* Legler and Cann, 1980 by monotypy.

Rheodytes leukops Legler and Cann, 1980

Rheodytes leukops Legler, J.M. & Cann, J. (1980). A new genus and species of chelid turtle from Queensland, Australia. *Los Angeles Co. Mus. Contrib. Sci.* 324: 1–18 [2 figs 1–7]. Type data: holotype, QM J31701, from Fitzroy River, 63 km N and 25 km E of Duaringa, elevation 40m, 23°09'S, 149°55'E, Qld.

Distribution: NE coastal, Qld. Ecology: lotic freshwater, aquatic, noctidiurnal, predator; seasonal breeder, oviparous, general carnivore.

CARETTOCHELYDIDAE

INTRODUCTION

A cryptodirous family with a single living aquatic species which occurs in the rivers of southern New Guinea and Australia's Northern Territory.

Characterised in Australia by: bony shell covered by a fleshy, pitted skin without dermal scutes; paddle-shaped limbs, each with two claws; nostrils at the end of a prominent, fleshy proboscis.

References

Cann, J. (1978). *Tortoises of Australia.* Sydney : Angus & Robertson 79 pp.

Cogger, H.G. (1979). *Reptiles and Amphibians of Australia.* Sydney : A.H. & A.W. Reed 608 pp.

Pritchard, P.C.H. (1979). *Encyclopedia of Turtles.* Neptune, N.J. : T.F.H. Publications 895 pp.

Wermuth, H. & Mertens, R. (1961). *Schildkröten, Krokodile, Brückenechsen.* Jena: Gustav Fischer 422 pp.

Wermuth, H. & Mertens, R. (1977). Liste der rezenten Amphibien und Reptilien; Testudines, Crocodylia, Rhynchocephalia. *Das Tierreich* **100**: i–xxvii, 1–174

Carettochelys Ramsay, 1886

Carettochelys Ramsay, E.P. (1886). On a new genus and species of fresh water tortoise from the Fly River, New Guinea. *Proc. Linn. Soc. N.S.W. (2)* **1**: 158–162 [158] [1887 on title page of bound volume]. Type species *Carettochelys insculptus* Ramsay, 1886 by monotypy.

Carettocchelys Ramsay, E.P. (1886). On a new genus and species of fresh water tortoise from the Fly River, New Guinea. *Proc. Linn. Soc. N.S.W. (2)* **1**: 158–162 [158] [*errore pro Carettochelys* Ramsay, 1886; 1887 on title page of bound volume].

Carretochelys Ramsay, E.P. (1886). On a new genus and species of fresh water tortoise from the Fly River, New Guinea. *Proc. Linn. Soc. N.S.W. (2)* **1**: 158–162 [161] [*errore pro Carettochelys* Ramsay, 1886; 1887 on title page of bound volume].

Synonymy that of Cogger, H.G., this work.

This group is also found in southern New Guinea.

Carettochelys insculpta Ramsay, 1886

Carettocchelys* [*sic*] *insculptus Ramsay, E.P. (1886). On a new genus and species of fresh water tortoise from the Fly River, New Guinea. *Proc. Linn. Soc. N.S.W. (2)* **1**: 158–162 [158]. Type data: holotype, AM R3677, from Fly River, Papua New Guinea.

Carettochelys insculpta Boulenger, G.A. (1889). *Catalogue of the Chelonians, Rhynchocephalians, and Crocodiles in the British Museum (Natural History).* new edn. London : British Museum x 311 pp. [236] [valid emend. *pro Carettochelys insculptus* Ramsay, 1886].

Synonymy that of Cogger, H.G., this work.

Distribution: N coastal, (N Gulf), N.T.; extralimital in New Guinea. Ecology: lentic freshwater, lotic freshwater, aquatic, noctidiurnal, predator; seasonal breeder, oviparous, general carnivore. Biological references: Cogger, H.G. (1970). First record of the pitted-shelled turtle, *Carettochelys insculpta*, from Australia. *Search* **1**: 41; Schodde, R., Mason, I. & Wolfe, T.O. (1972). Further records of the pitted-shelled turtle (*Carettochelys insculpta*) from Australia. *Trans. R. Soc. S. Aust.* **96**: 115–117; Waite, E.R. (1905). The osteology of the New Guinea turtle (*Carettochelys insculpta*, Ramsay). *Rec. Aust. Mus.* **6**: 110–118.

CHELONIIDAE

INTRODUCTION

A family containing all but one of the large sea turtles. The six living species in four genera are largely confined to the tropics, but regularly range into warm temperate and occasionally cold temperate waters. Successful breeding occurs only in the tropics. Gravid females emerge on to beaches to lay their eggs in egg cavities which they first excavate, then later fill in and obscure. Minor deviations from an optimum incubation temperature recently have been shown to result in massively-skewed sex ratios of the hatchlings. Five species, representing all genera, occur in Australian waters; only one is endemic.

Characterised in Australia by: cryptodirous withdrawal of the head into the shell; strong, paddle-shaped limbs with one or more claws on the leading edge; nostrils level with surface of snout; shell consisting of a fused bony shell covered by large juxtaposed or imbricate dermal scutes and sometimes a soft, thin skin.

References

Cogger, H.G. (1979). *Reptiles and Amphibians of Australia.* Sydney : A.H. & A.W. Reed 608 pp.

Ernst, C.H. & Barbour, R.W. (1972). *Turtles of the United States.* Lexington : University Press of Kentucky x 347 pp.

Pritchard, P.C.H. (1979). *Encyclopedia of Turtles.* Neptune, N.J. : T.F.H. Pulbications 895 pp.

Wermuth, H. & Mertens, R. (1961). *Schildkröten, Krokodile, Brückenechsen.* Jena : Gustav Fischer 422 pp.

Wermuth, H. & Mertens, R. (1977). Liste der rezenten Amphibien und Reptilien; Testudines, Crocodylia, Rhynchocephalia. *Das Tierreich* **100**: i–xxvii 1–174

Caretta Rafinesque, 1814

Caretta Rafinesque, C.S. (1814). *Specchio della Science, o Giornale enciclopedico di Sicilia. Deposito letterario delle moderne cognizioni, scopperte et osservazioni sopra le scienze ed arti.* Palermo Vol. 2 Cahier 9 [66]. Type species *Caretta nasuta* Rafinesque, 1814 by monotypy. Compiled from secondary source: Wermuth, H. & Mertens, R. (1977). Liste der rezenten Amphibien und Reptilien; Testudines, Crocodylia, Rhynchocephalia. *Das Tierreich* **100**: i–xxvii, 1–174.

Thalassochelys Fitzinger, L.J. (1835). Entwurf einer systematischen Anordnung der Schildkröten nach den Grundsätzen der naturlichen Methode. *Ann. Mus. Wien* **1**: 103–108 [121] [described as subgenus of *Chelonia*]. Type species *Testudo caouana* Lacépède, 1788 by subsequent designation, see Fitzinger, L.J. (1843). *Systema Reptilium.* Vienna : Braümüller u. Seidel vi 106 pp.

Caouana Cocteau, J.T. (1838). Reptiles de l'île de Cuba. *in* Sagra, R. de la *Histoire physique, politique et naturelle de l'île de Cuba.* Paris : Arthus Bertrand [35]. Type species *Testudo caouana* Lacépède, 1788 by tautonymy. Compiled from secondary source: Wermuth, H. & Mertens, R. (1977). Liste der rezenten Amphibien und Reptilien; Testudines, Crocodylia, Rhynchocephalia. *Das Tierreich* **100**: i–xxvii, 1–174.

Halichelys Fitzinger, L.J. (1843). *Systema Reptilium.* Vienna : Braümüller u. Seidel vi 106 pp. [30]. Type species *Caretta atra* Merrem, 1820 by original designation.

***Caouana* (part.)** Gray, J.E. (1844). *Catalogue of the Tortoises, Crocodiles, and Amphisbaenians in the Collection of the British Museum.* London : British Museum viii 80 pp. [52] [?*non Caouana* Cocteau, 1838]. Type species none designated, *Testudo caretta* Linnaeus,

1758 and *Caouana elongata* Gray, 1844 are two of three included species.

Thalassiochelis Nardo, G.D. (1864). Nuova rarissima specie di Cheloniano pescato nelle nostre spiaggie: *Chelonia Mydas. Atti R. Istit. Veneto (3)* **9**: 1418–1427 [1421] [invalid emend. *pro Thalassochelys* Fitzinger, 1835, *vide* Smith, H.M. & Smith, R.B. (1979). *Synopsis of the Herpetofauna of Mexico.* Vol. 6. Guide to Mexican Turtles. Bibliographic Addendum 3. North Bennington, Vermont : John Johnson 1044 pp.].

Eremonia Gray, J.E. (1873). Notes on the genera of turtles (Oiacopodes), and especially on their skeletons and skulls. *Proc. Zool. Soc. Lond.* **1873**: 395–411 [408]. Type species *Caouana elongata* Gray, 1844 by monotypy.

Thalassiochelys Philippi, R.A. (1887). Vorläufige Nachricht über die chilenischen Seeschildkröten und einige Fische der chilenischen Küste. *Zool. Garten, Frankfurt-am-Main* **28**: 84–85 [85] [invalid emend. *pro Thalassochelys* Fitzinger, 1835, *vide* Smith, H.M. & Smith, R.B. (1979). *Synopsis of the Herpetofauna of Mexico.* Vol. 6. Guide to Mexican Turtles. Bibliographic Addendum 3. North Bennington, Vermont : John Johnson 1044 pp.].

Thallasochelys Barbour, T. & Cole, L.J. (1906). Vertebrata from Yucatán. Reptilia, Amphibia and Pisces. *Bull. Mus. Comp. Zool.* **50**: 146–155 [148] [*errore pro Thalassochelys* Fitzinger, 1835]. Compiled from secondary source: Smith, H.M. & Smith, R.B. (1979). *Synopsis of the Herpetofauna of Mexico.* Vol. 6. Guide to Mexican Turtles. Bibliographic Addendum 3. North Bennington, Vermont : John Johnson 1044 pp.

Synonymy that of Wermuth, H. & Mertens, R. (1977). Liste der rezenten Amphibien und Reptilien; Testudines, Crocodylia, Rhynchocephalia. *Das Tierreich* **100**: i–xxvii, 1–174; Smith, H.M. & Smith, R.B. (1979). *Synopsis of the Herpetofauna of Mexico.* Vol. 6. Guide to Mexican Turtles. Bibliographic Addendum 3. North Bennington, Vermont : John Johnson 1044 pp. [93; 294].

This group is pan-tropical, extending into temperate regions.

Caretta caretta (Linnaeus, 1758)

Testudo caretta Linnaeus, C. (1758). *Systema Naturae.* Classis III. Amphibia. Holmiae : Laurentii Salvii 824 pp. 10th edn. 1: 194–238 [197]. Type data: none designated, based on published and unpublished notes, from "About the American Islands"; restricted to Bermuda Ils. by Smith, H.M. & Taylor, E.H. (1950). An annotated checklist and key to the reptiles of Mexico exclusive of the snakes. *Bull. U.S. Natl. Mus.* **199**: i–iv 1–253 and to Bimini, British Bahamas by Schmidt, K.P. (1953). *A Checklist of North American Amphibians and Reptiles.* 6th edn. Chicago : Univ. Chicago Press viii 280 pp.

Testudo cephalo Schneider, J.G. (1783). *Allgemeine Naturgeschichte der Schildkröten, nebst einem systematischen Verzeichnisse der einzelnen Arten.* Leipzig : Müller xlviii 364 pp. [303]. Type data: none designated, from unspecified locality; restricted to Charleston, South Carolina, U.S.A., see Smith, H.M. & Taylor, E.H. (1950). An annotated checklist and key to the reptiles of Mexico exclusive of the snakes. *Bull. U.S. Natl. Mus.* **199**: i–iv 1–253.

Testudo caouana Lacépède, B.G.E. (1788). Histoire naturelle des Quadrupèdes ovipares et serpens. *in* Buffon, G.L.L. (ed.) *Histoire Naturelle, générale et particulière,* avec la description du cabinet du roi. Paris : Royal Printer 1 (Synops. Méth): 95–104 [95] [*nom. substit. pro Testudo caretta* Linnaeus, 1758].

Testudo nasicornis Lacépède, B.G.E. (1788). Histoire naturelle des Quadrupèdes ovipares et serpens. *in* Buffon, G.L.L. (ed.) *Histoire Naturelle, générale et particulière,* avec la description du cabinet du roi. Paris : Royal Printer 1 (Synops. Méth): 95–104 [103]. Type data: status unknown, from "les mers du nouveau Continent, voisines de l'équator"; restricted to Ascension Is., see Smith, H.M. & Smith, R.B. (1979). *Synopsis of the Herpetofauna of Mexico.* Vol. 6. Guide to Mexican Turtles. Bibliographic Addendum 3. North Bennington, Vermont : John Johnson 1044 pp.

Caretta nasuta Rafinesque, C.S. (1814). *Specchio della Science, o Giornale enciclopedico di Sicilia. Deposito letterario delle moderne cognizioni, scopperte et osservazioni sopra le scienze ed arti.* Palermo Vol. 2 Cahier 9 [66] [*nom. substit. pro Testudo caretta* Linnaeus, 1758, but Wermuth, H. & Mertens, R. (1977). Liste der rezenten Amphibien und Reptilien; Testudines, Crocodylia, Rhynchocephalia. *Das Tierreich* **100**: i–xxvii, 1–174 regard it as *sp. nov.* with type locality Sicily]. Type data: that of *Testudo caretta* Linnaeus, 1758.

Chelonia cavanna Oken, L. (1816). *Lehrbuch der Naturgeschichte. Part 2. Abtheilung Fleischthiere.* Leipzig : Reclam xvi 1270 pp. [350] [*errore pro Chelonia caouana* (Lacépède, 1788)]. Compiled from secondary source: Smith, H.M. & Smith, R.B. (1979). *Synopsis of the Herpetofauna of Mexico.* Vol. 6. Guide to Mexican Turtles. Bibliographic Addendum 3. North Bennington, Vermont : John Johnson 1044 pp..

Caretta atra Merrem, B. (1820). *Versuch eines Systems Amphibien. Tentamen Systematis Amphibiorum.* Marburg : Krieger xv 199 pp. [17]. Type data: status unknown, from Ascension Is.

Testudo corianna Gray, J.E. (1831). *Synopsis Reptilium;* or Short Descriptions of the Species of Reptiles. Part I. Cataphracta. Tortoises, Crocodiles, and Enaliosaurians. London : Treuttel, Würtz & Co. viii 85 pp. [53] [*errore pro Testudo caouana* Lacépède, 1788].

Chelonia pelasgorum Bory de Saint-Vincent, J.B.G.M. (1833). *Expédition scientifique en Morée, entérise et publée par ordre du gouvernement français. Travaux de la section des sciences physiques, sous la direction de M. le colonel Bory de Saint-Vincent.* Paris : P. Bertrand 3(1): 1–209 [64 pl 6]. Type data: status unknown, from between Arcadia and the mouth of the Neda River, Greece. Compiled from secondary source: Smith, H.M. & Smith, R.B. (1979). *Synopsis of the Herpetofauna of Mexico.* Vol. 6. Guide to Mexican Turtles. Bibliographic Addendum 3. North Bennington, Vermont : John Johnson 1044 pp.

Caouana elongata Gray, J.E. (1844). *Catalogue of the Tortoises, Crocodiles, and Amphisbaenians in the Collection of the British Museum.* London : British Museum viii 80 pp. [53]. Type data: holotype, BMNH

1947.3.5.41, from unknown locality; restricted to Ascension Is., see Smith, H.M. & Smith, R.B. (1979). *Synopsis of the Herpetofauna of Mexico.* Vol. 6. Guide to Mexican Turtles. Bibliographic Addendum 3. North Bennington, Vermont : John Johnson 1044 pp.

Thalassochelys corticata Girard, C. (1858). *United States Exploring Expedition during the years 1838, 1839, 1840, 1841, 1842.* Herpetology. 20: xvi 496 pp. Philadelphia : J.B. Lippincott [431 pl 29]. Type data: holotype, USNM 7778, from Funchal, Madeira. Compiled from secondary source: Smith, H.M. & Smith, R.B. (1979). *Synopsis of the Herpetofauna of Mexico.* Vol. 6. Guide to Mexican Turtles. Bibliographic Addendum 3. North Bennington, Vermont : John Johnson 1044 pp.

Thalassiochelys tarapacona Philippi, R.A. (1887). Vorläufige Nachricht über die chilenischen Seeschildkröten und einige Fische der chilenischen Küste. *Zool. Garten, Frankfurt-am-Main* **28**: 84–85 [85] [incorrect original spelling *vide* Smith, H.M. & Smith, R.B. (1979). *Synopsis of the Herpetofauna of Mexico.* Vol. 6. Guide to Mexican Turtles. Bibliographic Addendum 3. North Bennington, Vermont : John Johnson 1044 pp.]. Type data: holotype (probable), MNHN (presumed), from Iquique, Chile, see Smith, H.M. & Smith, R.B. (1979). *Synopsis of the Herpetofauna of Mexico.* Vol. 6. Guide to Mexican Turtles. Bibliographic Addendum 3. North Bennington, Vermont : John Johnson 1044 pp.

Thalassochelys tarapacana Philippi, R.A. (1899). Las tortugas Chilenas. *An. Univ. Chile* **104**: 725–736 [731] [valid emend. *pro Thalassiochelys tarapacona* Philippi, 1887, *vide* Smith, H.M. & Smith, R.B. (1979). *Synopsis of the Herpetofauna of Mexico.* Vol. 6. Guide to Mexican Turtles. Bibliographic Addendum 3. North Bennington, Vermont : John Johnson 1044 pp.].

Caretta gigas Deraniyagala, P.E.P. (1933). The loggerhead turtles (Carettidae) of Ceylon. *Ceylon J. Sci. (B)* **18**: 61–72 [66 figs 4–6 pl 5]. Type data: holotype, BMNH, from Gulf of Mannar, Sri Lanka.

Chelonia cahuano Tamayo, J.L. (1962). *Geografía general de México. Tomo III. Geografía biológica y humana.* México : Inst. Méx. Inv. Económicas 633 pp. [373] [*errore pro Chelonia caouana* (Lacépède, 1788), *vide* Smith, H.M. & Smith, R.B. (1979). *Synopsis of the Herpetofauna of Mexico.* Vol. 6. Guide to Mexican Turtles. Bibliographic Addendum 3. North Bennington, Vermont : John Johnson 1044 pp.].

Caretta careta Tamayo, J.L. (1962). *Geografía general de México. Tomo III. Geografía biológica y humana.* México : Inst. Méx. Inv. Económicas 633 pp. [373] [*errore pro Caretta caretta* (Linnaeus, 1758), *vide* Smith, H.M. & Smith, R.B. (1979). *Synopsis of the Herpetofauna of Mexico.* Vol. 6. Guide to Mexican Turtles. Bibliographic Addendum 3. North Bennington, Vermont : John Johnson 1044 pp.].

Synonymy that of Cogger, H.G., this work; Brongersma, L.D. (1961). Notes upon some sea turtles. *Zool. Verh. Leiden* **51**: 1–46; Wermuth, H. & Mertens, R. (1977). Liste der rezenten Amphibien und Reptilien; Testudines, Crocodylia, Rhynchocephalia. *Das Tierreich* **100**: i–xxvii, 1–174; Smith, H.M. & Smith, R.B. (1979). *Synopsis of the Herpetofauna of Mexico.* Vol. 6. Guide to Mexican Turtles. Bibliographic Addendum 3. North Bennington, Vermont : John Johnson 1044 pp.

Distribution: (SE coast), NE coast, Great Barrier Reef, Gulf of Carpentaria, N coast, NW Shelf, (W coast), Coral Sea, Arafura Sea, Timor Sea, Indian Ocean, (N.S.W.), Qld., W.A., N.T.; also pan-tropical, extending into temperate regions. Ecology: continental shelf, oceanic, aquatic, noctidiurnal, migratory, predator; seasonal breeder, oviparous, coral reefs, piscivore. Biological references: Cogger, H.G. & Lindner, D.L. (1969). Marine turtles in northern Australia. *Aust. Zool.* **15**: 150–159; Limpus, C.J. (1973). Loggerhead turtles (*Caretta caretta*) in Australia: food source while nesting. *Herpetologica* **29**: 42–44; Limpus, C.J. (1979). Notes on growth rates of wild turtles. *IUCN/SSC Marine Turtle Newsletter* **10**: 3–5.

Chelonia Sonnini and Latreille, 1802

Chelonia Brongniart, A. (1800). Essai d'une classification naturelle des reptiles. *Bull. Sci. Soc. Philomath. Paris* **2**: 89–91 [89] [*nom. nud., vide* Deraniyagala, P.E.P. (1953). *A Coloured Atlas of Some Vertebrates from Ceylon.* Vol. 2. Tetrapod Reptilia. Colombo : Govt. Press vii 101 pp.; for full discussion on the first available use of the name *Chelonia*, see Smith, H.M. & Smith, R.B. (1979). *Synopsis of the Herpetofauna of Mexico.* Vol. 6. Guide to Mexican Turtles. Bibliographic Addendum 3. North Bennington, Vermont : John Johnson 1044 pp. (259–260)].

Chelonia Sonnini de Manoncourt, C.S. & Latreille, P.A. (1802). *Histoire Naturelle des Reptiles*, avec Figures Déssinnées d'après Nature *in* Buffon, G.L.L. de *Histoire Naturelle de Buffon, classée d'après le système de Linné par R.R. Castel.* Nouvelle édition. (Suites). Paris : Déterville Vol. 4 [22]. Publication date established from Anon (1903). *Catalogue of the Books, Manuscripts, Maps and Drawings in the British Museum (Natural History). A—D.* London : British Museum 1: viii 500 pp. Type species *Testudo mydas* Linnaeus, 1758 by subsequent designation, see Fitzinger, L.J. (1843). *Systema Reptilium.* Vienna : Braümüller u. Seidel vi 106 pp.

Chelone Brongniart, A. (1805). Essai d'une classification naturelle des Reptiles. *Mém. Sav. Étrang.* **1**: 587–637 [610] [emend. *pro Chelonia* Brongniart, 1800]. Publication date established from Wermuth, H. & Mertens, R. (1977). Liste der rezenten Amphibien und Reptilien; Testudines, Crocodylia, Rhynchocephalia. *Das Tierreich* **100**: i–xxvii, 1–174 but Smith, H.M. & Smith, R.B. (1979). *Synopsis of the Herpetofauna of Mexico.* Vol. 6. Guide to Mexican Turtles. Bibliographic Addendum 3. North Bennington, Vermont : John Johnson 1044 pp. [257] cite 1806 as publication date. Type species *Testudo mydas* Linnaeus, 1758 by subsequent designation, see Fitzinger, L.J. (1843). *Systema Reptilium.* Vienna : Braümüller u. Seidel vi 106 pp.

Chelona Burmeister, H.K. (1807). *Handbuch der Naturgeschichte. Zum Gebrauch bei Vorlesungen entworfen von Herman Burmeister.* Berlin : T.C.F. Enslin viii xxvi xxi 858 pp. [731] [emend. *pro Chelonia* Sonnini and Latreille, 1802].

Chelonias Rafinesque, C.S. (1814). *Specchio della Science, o Giornale enciclopedico di Sicilia. Deposito letterario delle moderne cognizioni, scopperte et osservazioni sopra le scienze ed arti.* Palermo Vol. 2 Cahier 9 [66] [emend. *pro Chelonia* Sonnini and Latreille, 1802].

Mydas Cocteau, J.T. (1838). Reptiles de l'île de Cuba. *in* Sagra, R. de la *Histoire physique, politique et naturelle de l'île de Cuba.* Paris : Arthus Bertrand [19–20] [described as subgenus of *Chelonia* Sonnini and Latreille, 1802; *non Mydas* Fabricius, 1794; *non Mydas* Bonaparte, 1831]. Type species *Testudo mydas* Linnaeus, 1758 by tautonymy.

Mydasea Gervais, P. (1843). Chélonée. *in* Orbigny, A.C.V.D. d's (ed.) *Dictionnaire Universel d'Histoire Naturelle, résumant et complétant tous les faits présentées par les anciens dictionnaires scientifiques, les ouevres complètes de Buffon, de Lacépède, de Cuvier.* Paris : Langlois et Leclerq 3: 455–459 [457] [described as subgenus of *Chelonia* Sonnini and Latreille, 1802]. Publication date established from Sherborn, C.D. & Palmer, T.S. (1899). Dates of Charles d'Orbigny's "Dictionnaire Universel d'Histoire Naturelle". *Ann. Mag. Nat. Hist. (7)* **3**: 350–352. Type species *Testudo mydas* Linnaeus, 1758 by monotypy.

Euchelonia Tschudi, J.J. von (1846). *Untersuchungen über die Fauna Peruana auf einer Reise in Peru während der Jahre 1838, 39, 40, 41 und 42.* Herpetologie. St. Gallen : Scheitlin u. Zollikofer 80 pp. [22] [described as subgenus of *Chelonia* Sonnini and Latreille, 1802]. Publication date established from Peters, J.A. (1951). The proper citation for certain species described by Tschudi. *Copeia* **1951**: 70. Type species "*Chelonia midas* Schweigger, 1812" (= *Testudo mydas* Linnaeus, 1758) by monotypy.

Megemys Gistel, J. (1848). *Naturgeschichte des Thierreichs.* Stuttgart : Scheitlin & Krais xvi 216 pp. [viii] [*nom. nov. pro Chelonia* Sonnini and Latreille, 1802].

Euchelys Girard, C. (1858). *United States Exploring Expedition during the years 1838, 1839, 1840, 1841, 1842.* Herpetology. 20: xvi 496 pp. Philadelphia : J.B. Lippincott [447]. Type species *Testudo macropus* Walbaum, 1782 by monotypy.

Midas Herrera, A.L. (1901). *Nouvelle nomenclature des êtres organisés et des mineraux.* México : Govt. Fed. 88 pp. [68] [*nom. nov. pro Mydas* Cocteau, 1838, suppressed in 1922 by the International Commission on Zoological Nomenclature under Opinion 72].

Natator McCulloch, A.R. (1908). A new genus and species of turtle from north Australia. *Rec. Aust. Mus.* **7**: 126–128 [127 pls 26–27] [126]. Type species *Natator tessellatus* McCulloch, 1908 by monotypy.

Synonymy that of Cogger, H.G., this work; Wermuth, H. & Mertens, R. (1977). Liste der rezenten Amphibien und Reptilien; Testudines, Crocodylia, Rhynchocephalia. *Das Tierreich* **100**: i–xxvii, 1–174; Smith, H.M. & Smith, R.B. (1979). *Synopsis of the Herpetofauna of Mexico.* Vol. 6. Guide to Mexican Turtles. Bibliographic Addendum 3. North Bennington, Vermont : John Johnson 1044 pp.

This group is pan-tropical, extending into temperate regions.

Chelonia depressa Garman, 1880

***Chelonia depressa* (part.)** Garman, S. (1880). On certain species of Chelonioidae. *Bull. Mus. Comp. Zool.* **6**: 123–126 [124] [for paralectotype see *Chelonia mydas* (Linnaeus, 1758)]. Type data: lectotype, MCZ 4473, from northern Australia, see Barbour, T. (1914). On some Australian reptiles. *Proc. Biol. Soc. Wash.* **27**: 201–206, designation by Loveridge, A. (1934). Australian reptiles in the Museum of Comparative Zoology, Cambridge, Massachusetts. *Bull. Mus. Comp. Zool.* **77**: 243–383 who cited one of the "cotypes" as "type", which we regard as a lectotype designation.

Natator tessellatus McCulloch, A.R. (1908). A new genus and species of turtle from north Australia. *Rec. Aust. Mus.* **7**: 126–128 [127 pls 26–27]. Type data: holotype, AM R4158, from near Port Darwin, N.T.

Synonymy that of Cogger, H.G., this work.

Distribution: (SE coast), NE coast, Great Barrier Reef, Gulf of Carpentaria, N coast, NW Shelf, (W coast), Timor Sea, Indian Ocean, Coral Sea, Arafura Sea, (N.S.W.), Qld., W.A., N.T. Ecology: continental shelf, aquatic, (algal-feeder), noctidiurnal, migratory, predator; seasonal breeder, oviparous, coral reefs, piscivore. Biological references: Cogger, H.G. & Lindner, D.L. (1969). Marine turtles in northern Australia. *Aust. Zool.* **15**: 150–159; Fry, D.B. (1913). On the status of *Chelonia depressa*, Garman. *Rec. Aust. Mus.* **10**: 159–185 (taxonomy); Limpus, C.J., Parmenter, C.J., Parker, R. & Ford, N. (1981). The flatback turtle *Chelonia depressa* in Queensland: the Peak Island rookery. *Herpetofauna* **13**: 14–17; Williams, E.E., Grandison, A.G.C. & Carr, A.F. (1967). *Chelonia depressa* Garman reinvestigated. *Breviora* 271: 1–15.

Chelonia mydas (Linnaeus, 1758)

Testudo mydas Linnaeus, C. (1758). *Systema Naturae.* Classis III. Amphibia. Holmiae : Laurentii Salvii 824 pp. 10th edn. 1: 194–238 [197]. Type data: holotype, ZIUU 20, restricted to Ascension Is., see Mertens, R. & Müller, L. (1928). Liste der Amphibien und Reptilien Europas. *Abh. Senckenb. Naturf. Ges.* **41**: 1–62.

Testudo macropus Walbaum, J.J. (1782). *Chelonographia oder Beschreibung einiger Schildkröten nach natürlichen urbildern.* Lübeck : J.F. Gleditsch vi 132 pp. [122] [*nom. nud.*]. Compiled from secondary source: Wermuth, H. & Mertens, R. (1977). Liste der rezenten Amphibien und Reptilien; Testudines, Crocodylia, Rhynchocephalia. *Das Tierreich* **100**: i–xxvii, 1–174.

Testudo viridis Schneider, J.G. (1783). *Allgemeine Naturgeschichte der Schildkröten, nebst einem systematischen Verzeichnisse der einzelnen Arten.* Leipzig : Müller xlviii 364 pp. [299 pl 2]. Type data: none designated, from unspecified locality; restricted to Charleston, South Carolina, U.S.A., see Smith, H.M. &

Taylor, E.H. (1950). An annotated checklist and key to the reptiles of Mexico exclusive of the snakes. *Bull. U.S. Natl. Mus.* **199**: i–iv 1–253.

Testudo japonica Thunberg, C.P. (1787). Beskrifning pa Trenne Sköld-paddor. *K. Vet. Acad. Nya Handl. (2)* **8**: 178–180 [178 pl 7 fig 1]. Type data: status unknown, ?NHRM, from Japan.

Testudo marina vulgaris Lacépède, B.G.E. (1788). Histoire naturelle des Quadrupèdes ovipares et serpens. *in* Buffon, G.L.L. (ed.) *Histoire Naturelle, générale et particulière,* avec la description du cabinet du roi. Paris : Royal Printer 1 (Synops. Méth): 95–104 [54 pl 1] [*nom. substit. pro Testudo mydas* Linnaeus, 1758]. Type data: that of *Testudo mydas* Linnaeus, 1758.

Testudo viridisquamosa Lacépède, B.G.E. (1788). Histoire naturelle des Quadrupèdes ovipares et serpens. *in* Buffon, G.L.L. (ed.) *Histoire Naturelle, générale et particulière,* avec la description du cabinet du roi. Paris : Royal Printer 1 (Synops. Méth): 95–104 [92] [name suppressed in 1963 by the International Commission on Zoological Nomenclature under Opinion 660]. Type data: none designated, based on published and unpublished notes, from unspecified locality; restricted to Bocas del Toro by Wermuth, H. (1956). Versuch der Deutung einiger bisher übersehener Schildkröten-Namen. *Zool. Beitr. Berl.* **2**: 399–423 and later to Isla Blanquilla, Caribbean Sea by Brongersma, L.D. (1961). Notes upon some sea turtles. *Zool. Verh. Leiden* **51**: 1–46. Compiled from secondary source: Brongersma, L.D. (1961). Notes upon some sea turtles. *Zool. Verh. Leiden* **51**: 1–46.

Testudo chloronotus Bechstein, J.M. (1800). *Herrn de la Cépède's Naturgeschichte der Amphibien, aus dem französischen ubersetzt und mit Anmerkungen und Zusätzen versehen von J.M. Bechstein.* Weimar : Landes-Industrie-Comptoir 1: xxxii 524 pp. [107] [*nom. substit. pro Testudo viridisquamosa* Lacépède, 1788, *vide* Brongersma, L.D. (1961). Notes upon some sea turtles. *Zool. Verh. Leiden* **51**: 1–46]. Compiled from secondary source: Brongersma, L.D. (1961). Notes upon some sea turtles. *Zool. Verh. Leiden* **51**: 1–46.

Testudo cepediana Daudin, F.M. (1802). *Histoire Naturelle, générale et particulière des Reptiles.* Paris : Dufart Tom. 2 [50 pl 17 fig 1]. Type data: status unknown, ?MNHP, from unknown locality.

Testudo rugosa Daudin, F.M. (1802). *Histoire Naturelle, générale et particulière des Reptiles.* Paris : Dufart Tom. 2 [37] [*non Testudo rugosa* Shaw, 1802]. Type data: status unknown, ?MNHP, from Maldive Ils., Indian Ocean.

Chelonia virgata Schweigger, A.F. (1812). Prodromus monographiae Cheloniorum. *Königsberger Archiv.* **1**: 271–405. [291]. Type data: status unknown, from "Seas of Torrid Zone"; restricted to Bermuda Ils., see Smith, H.M. & Taylor, E.H. (1950). An annotated checklist and key to the reptiles of Mexico exclusive of the snakes. *Bull. U.S. Natl. Mus.* **199**: i–iv 1–253.

Caretta cepedii Merrem, B. (1820). *Versuch eines Systems Amphibien. Tentamen Systematis Amphibiorum.* Marburg : Krieger xv 199 pp. [18] [*nom. substit. pro Testudo cepediana* Daudin, 1802].

Caretta esculenta Merrem, B. (1820). *Versuch eines Systems Amphibien. Tentamen Systematis Amphibiorum.* Marburg : Krieger xv 199 pp. [18]. Type data: status unknown, from Atlantic Ocean.

Caretta thunbergii Merrem, B. (1820). *Versuch eines Systems Amphibien. Tentamen Systematis Amphibiorum.* Marburg : Krieger xv 199 pp. [19] [*nom. substit. pro Testudo japonica* Thunberg, 1787].

Chelonia lachrymata Cuvier, G.L.C.F.D. (1829). *Le Règne Animal.* 2nd edn. Paris : Déterville Vol. 2 xv 406 pp. [13]. Type data: status unknown, from unknown locality.

Chelonia maculosa Cuvier, G.L.C.F.D. (1829). *Le Règne Animal.* 2nd edn. Paris : Déterville Vol. 2 xv 406 pp. [13]. Type data: status unknown, from unknown locality; restricted to Ascension Is., see Smith, H.M. & Taylor, E.H. (1950). An annotated checklist and key to the reptiles of Mexico exclusive of the snakes. *Bull. U.S. Natl. Mus.* **199**: i–iv 1–253.

Chelonia bicarinata Lesson, R.-P. (1834). Reptiles, pp. 289–336 *in* Bélanger, C. *Voyage aux Indes-Orientales, par le Nord de l'Europe pendant les Années 1825, 1826, 1827, 1828 et 1829.* Paris : Arthus Bertrand xxxix 535 pp. [301]. Type data: status unknown, from Atlantic Ocean.

Chelonia marmorata Duméril, A.M.C. & Bibron, G. (1835). *Erpétologie Générale* ou Histoire Naturelle Complète des Reptiles. Paris : Roret 2: 680 pp. [546 pl 23 fig 1]. Type data: syntypes (2), ?MNHP, from Ascension Is.

Chelonia formosa Girard, C. (1858). *United States Exploring Expedition during the years 1838, 1839, 1840, 1841, 1842.* Herpetology. 20: xvi 496 pp. Philadelphia : J.B. Lippincott [456 pl 31 figs 1–4]. Type data: holotype, USNM 012386, from Fiji Ils.

Chelonia tenuis Girard, C. (1858). *United States Exploring Expedition during the years 1838, 1839, 1840, 1841, 1842.* Herpetology. 20: xvi 496 pp. Philadelphia : J.B. Lippincott [459 pl 31 fig 8]. Type data: syntypes, USNM 012390, from Honden Is., Tahiti, Eimeo and Rosa Is., Tuamotu Ils., South Pacific.

Chelonia albiventer Nardo, G.D. (1864). Nuova rarissima specie di Cheloniano pescato nelle nostre spiaggie: *Chelonia Mydas. Atti R. Istit. Veneto (3)* **9**: 1418–1427 [1420 pl 35]. Type data: status unknown, from Malamocco Harbour, Adriatic Sea, Italy, see Brongersma, L.D. (1961). Notes upon some sea turtles. *Zool. Verh. Leiden* **51**: 1–46.

Chelonia midas Bocage, J.V.B. (1866). Lista dos reptis possessoes Portuguezas d'Africa occidental que existem no Museu de Lisboa. *Jorn. Sci. Math. Lisboa* **1**: 37–56 [41] [*errore pro Testudo mydas* Linnaeus, 1758].

Chelonia agassizii Bocourt, F. (1868). Description de quelques Chélonians nouveaux appartenant à la faune Mexicaine. *Ann. Sci. Nat. Paris (5)* **10**: 121–122 [122]. Type data: syntypes, MNHP 1871–2, from mouth of Rio Nagualate, Pacific coast of Guatamala. Compiled from secondary source Smith, H.M. & Taylor, E.H. (1950). An annotated checklist and key to the reptiles of Mexico exclusive of the snakes. *Bull. U.S. Natl. Mus.* **199**: i–iv 1–253.

[Chelonia depressa* (part.)*] Garman, S. (1880). On certain species of Chelonioidae. *Bull. Mus. Comp. Zool.* **6**: 123–126 [124] [described as "cotype" of *Chelonia depressa* Garman, 1880 by Barbour, T. (1914). On some Australian reptiles. *Proc. Biol. Soc. Wash.* **27**: 201–206 but Loveridge, A. (1934). Australian reptiles in the Museum of Comparative Zoology, Cambridge, Massachusetts. *Bull. Mus. Comp. Zool.* **77**: 243–383 cited one of the "cotypes" as "type", which we have regarded as a lectotype designation for *Chelonia depressa* Garman, 1880]. Type data: paralectotype, MCZ 1413, from Penang, Malaysia.

Chelonia lata Philippi, R.A. (1887). Vorläufige Nachricht über die chilenischen Seeschildkröten und einige Fische der chilenischen Küste. *Zool. Garten, Frankfurt-am-Main* **28**: 84–85 [84]. Type data: status unknown, from Valparaiso, Chile.

Chelonia mydas carrinegra Caldwell, D.K. (1962). Sea turtles in Baja Californian waters (with special reference to those of the Gulf of California), and the description of a new subspecies of north-eastern Pacific green turtle. *Los Angeles Co. Mus. Contrib. Sci.* **61**: 1–31 [4 fig 1]. Type data: holotype, LACM 1696, from waters adjacent to Isla Angel de la Guarda, central Gulf of California, Mexico.

Testudo nigrita Tamayo, J.L. (1962). *Geografía general de México. Tomo III. Geografía biológica y humana.* México : Inst. Méx. Inv. Económicas 633 pp. [358] [*nom. nud.*].

Synonymy that of Cogger, H.G., this work; Brongersma, L.D. (1961). Notes upon some sea turtles. *Zool. Verh. Leiden* **51**: 1–46; Wermuth, H. & Mertens, R. (1977). Liste der rezenten Amphibien und Reptilien; Testudines, Crocodylia, Rhynchocephalia. *Das Tierreich* **100**: i–xxvii, 1–174; Smith, H.M. & Smith, R.B. (1979). *Synopsis of the Herpetofauna of Mexico.* Vol. 6. Guide to Mexican Turtles. Bibliographic Addendum 3. North Bennington, Vermont : John Johnson 1044 pp.

Distribution: (SE coast), NE coast, Great Barrier Reef, Gulf of Carpentaria, N coast, NW Shelf, (W coast), Timor Sea, Indian Ocean, Coral Sea, Arafura Sea, (N.S.W.), Qld., W.A., N.T.; also pan-tropical extending into temperate regions. Ecology: continental shelf, oceanic, aquatic, noctidiurnal, migratory; seasonal breeder, oviparous, coral reefs, general carnivore (juveniles), algal-feeder (adults). Biological references: Bustard, H.R. & Greenham, P.M. (1969). Nesting behaviour of the green sea turtle on a Great Barrier Reef island. *Herpetologica* **25**: 93–102; Cogger, H.G. & Lindner, D.L. (1969). Marine turtles in northern Australia. *Aust. Zool.* **15**: 150–159; Moorhouse, F.W. (1933). Notes on the green turtle (*Chelonia mydas*). *Rep. Gt. Barrier Reef Comm.* **4**: 1–2; Mrosovsky, N. & Pritchard, P.C.H. (1971). Body temperatures of *Dermochelys coriacea* and other sea turtles. *Copeia* **1971**: 624–631; Parker, W.K. (1880). Report on the development of the green turtle (*Chelone viridis*, Schneid.). Part 1. The cranium, face and cranial nerves. *in* Thomson, C.W. & Murray, J. (eds.) *Report of the Scientific Results of the Voyage of H.M.S.* Challenger *during the years 1873–1876, under the Command of Captain G.S. Nares and Captain F.T. Thomson.* Zoology **15**: 1–58 London (as *Chelone viridis*); Bjorndal, K.A. (1979). Cellulose digestion and volatile fatty acid production in the green turtle. *Chelonia mydas. Comp. Biochem. Physiol.* **63A**: 127–133; Bjorndal, K.A. (1979). Urine concentrations of ammonia, urea and uric acid in the green turtle. *Chelonia mydas. Comp. Biochem. Physiol.* **63A**: 509–510; Parmenter, C.J. (1980). Incubation of the eggs of the green sea turtle, *Chelonia mydas*, in Torres Strait, Australia : the effect of movement on hatchability. *Aust. Wildl. Res.* **7**: 479–485; Limpus, C.J. (1980). The green turtle, *Chelonia mydas* (L.) in eastern Australia. *James Cook Univ. N. Qd Res. Monogr.* **1**: 5–22; Limpus, C.J. & Walter, D.G. (1980). The growth of immature green turtles (*Chelonia mydas*) under natural conditions. *Herpetologica* **36**: 162–165; Bjorndal, K.A. (1980). Nutrition and grazing behavior of the green turtle *Chelonia mydas. Mar. Biol.* **56**: 147–154; Garnett, S.T. & Murray, R.M. (1981). Farm management and nutrition of the green turtle (*Chelonia mydas*). pp. 60–65 *in* Banks, C.B. & Martin, A.A. (eds.) *Proceedings of the Melbourne Herpetological Symposium, 1980.* Melbourne : Zoological Board of Victoria; Miller, J.D. & Limpus, C.J. (1981). Incubation period and sexual differentiation in the green turtle *Chelonia mydas* L. pp. 66–73 *in* Banks, C.B. & Martin, A.A. (eds.) *Proceedings of the Melbourne Herpetological Symposium, 1980.* Melbourne : Zoological Board of Victoria; Hirth, H. (1980). *Chelonia mydas* (Reptilia : Testudines : Cheloniidae). *Cat. Am. Amphib. Rept.* 249: 1–4.

Eretmochelys Fitzinger, 1843

Caretta Ritgen, F.A. von (1828). Versuch einer natürlichen Eintheilung der Amphibien. *Nova Acta Acad. Caes. Leop.-Carol. Halle* **14**: 245–284 [270] [*non Caretta* Rafinesque, 1814]. Type species *Testudo imbricata* Linnaeus, 1766 by subsequent designation see Lindholm, W.A. (1929). Revidiertes Verzeichnis der Gattungen der rezenten Schildkröten nebst Notizen zur Nomencklatur einiger Arten. *Zool. Anz.* **81**: 275–295.

Eretmochelys Fitzinger, L.J. (1843). *Systema Reptilium.* Vienna : Braümüller u. Seidel vi 106 pp. [30]. Type species *"Chelonia imbricata* Cuvier*"* (= *Testudo imbricata* Linnaeus, 1766) by original designation.

Eretmochelis Tschudi, J.J. von (1846). *Untersuchungen über die Fauna Peruana auf einer Reise in Peru während der Jahre 1838, 39, 40, 41 und 42.* Herpetologie. St. Gallen : Scheitlin u. Zollikofer 80 pp. [7] [emend. *pro Eretmochelys* Fitzinger, 1843].

Eritmochelys Tschudi, J.J. von (1846). *Untersuchungen über die Fauna Peruana auf einer Reise in Peru während der Jahre 1838, 39, 40, 41 und 42.* Herpetologie. St. Gallen : Scheitlin u. Zollikofer 80 pp. [22] [*pro Eretmochelys* Fitzinger, 1843].

Herpysmostes Gistel, J. (1868). *Blicke in das Leben der Natur und des Menschen.* Leipzig pp. 144–167 [145]. Type species *Testudo imbricata* Linnaeus, 1766 by monotypy. Compiled from secondary source: Smith, H.M. & Smith, R.B. (1979). *Synopsis of the Herpetofauna of Mexico.* Vol. 6. Guide to Mexican Turtles. Bibliographic Addendum 3. North Bennington, Vermont : John Johnson 1044 pp.

Onychochelys Gray, J.E. (1873). Notes on the genera of turtles (Oiacopodes), and especially on their skeletons and skulls. *Proc. Zool. Soc. Lond.* **1873**: 395–411 [397]. Type species *Onychochelys kraussi* Gray, 1873 by monotypy.

Erethmochelys Maldonado-Koerdell, M. (1953). Reptiles *in* Beltrán, E. (ed.) *Vida Silvestre y Recursos Naturales a lo largo de la Carretera Pan Americana.* México : Inst. Méxicano de Recursos Naturales Renovables, A.C. viii 228 pp. [130] [*errore pro Eretmochelys* Fitzinger, 1843, *vide* Smith, H.M. & Smith, R.B. (1979). *Synopsis of the Herpetofauna of Mexico.* Vol. 6. Guide to Mexican Turtles. Bibliographic Addendum 3. North Bennington, Vermont : John Johnson 1044 pp.].

Synonymy that of Cogger, H.G., this work; Wermuth, H. & Mertens, R. (1977). Liste der rezenten Amphibien und Reptilien; Testudines, Crocodylia, Rhynchocephalia. *Das Tierreich* **100**: i–xxvii, 1–174; Smith, H.M. & Smith, R.B. (1979). *Synopsis of the Herpetofauna of Mexico.* Vol. 6. Guide to Mexican Turtles. Bibliographic Addendum 3. North Bennington, Vermont : John Johnson 1044 pp.

This group is pan-tropical, extending into temperate regions.

Eretmochelys imbricata (Linnaeus, 1766)

Testudo imbricata Linnaeus, C. (1766). *Systema Naturae.* Amphibia. 12th edn. Holmiae : Laurenti Salvii 1: 347–393 [350]. Type data: holotype (probable), ZIUU 130, from "American and Asiatic seas"; restricted to Bermuda Ils., see Smith, H.M. & Taylor, E.H. (1950). An annotated checklist and key to the reptiles of Mexico exclusive of the snakes. *Bull. U.S. Natl. Mus.* **199**: i–iv 1–253. Compiled from secondary source: Smith, H.M. & Smith, R.B. (1979). *Synopsis of the Herpetofauna of Mexico.* Vol. 6. Guide to Mexican Turtles. Bibliographic Addendum 3. North Bennington, Vermont : John Johnson 1044 pp.

Chelonia radiata Cuvier, G.L.C.F.D. (1829). *Le Règne Animal.* 2nd edn. Paris : Déterville Vol. 2 xv 406 pp. [14]. Type data: pl 16 fig B (dated 1793) in Schoepff, J.D. (1801). *Historia Testudinum*, from unknown locality.

Chelonia pseudomydas Lesson, R.-P. (1834). Reptiles, pp. 289–336 *in* Bélanger, C. *Voyage aux Indes-Orientales, par le Nord de l'Europe pendant les Années 1825, 1826, 1827, 1828 et 1829.* Paris : Arthus Bertrand xxxix 535 pp. [299]. Type data: status unknown, from Atlantic Ocean; restricted to Bermuda Ils. by Smith, H.M. & Taylor, E.H. (1950). An annotated checklist and key to the reptiles of Mexico exclusive of the snakes. *Bull. U.S. Natl. Mus.* **199**: i–iv 1–253.

Chelonia pseudocaretta Lesson, R.-P. (1834). Reptiles, pp. 289–336 *in* Bélanger, C. *Voyage aux Indes-Orientales, par le Nord de l'Europe pendant les Années 1825, 1826, 1827, 1828 et 1829.* Paris : Arthus Bertrand xxxix 535 pp. [302]. Type data: status unknown, from Atlantic Ocean; restricted to Bermuda Ils. by Smith, H.M. & Taylor, E.H. (1950). An annotated checklist and key to the reptiles of Mexico exclusive of the snakes. *Bull. U.S. Natl. Mus.* **199**: i–iv 1–253.

Caretta bissa Rüppell, E. (1835). *Neue Wirbelthiere zu der Fauna von Abyssinien gehorig. Amphibien.* Frankfurt-am-Main : Schermber 18 pp. [4 pl 2]. Type data: lectotype, SMF 7886 M, from Red Sea, designation by Mertens, R. (1967). Die herpetologische Sektion des Natur-Museums und Forschungs-Institutes Senckenberg in Frankfurt-a.-M. nebst einem Verzeichnis ihrer Typen. 1. *Senckenberg. Biol.* **48**: 1–106,

Eretmochelys squamata Agassiz, J.L.R. (1857). *Contributions to the Natural History of the United States. North American Testudinata.* Boston : Little, Brown & Co. 1(2): 233–450 [382]. Type data: lectotype, MCZ 1416 from Singapore (original type localities were Singapore and Bengal, India), designation by Smith, H.M. & Taylor, E.H. (1950). An annotated checklist and key to the reptiles of Mexico exclusive of the snakes. *Bull. U.S. Natl. Mus.* **199**: i–iv 1–253.

Caretta squamosa Girard, C. (1858). *United States Exploring Expedition during the years 1838, 1839, 1840, 1841, 1842.* Herpetology. 20: xvi 496 pp. Philadelphia : J.B. Lippincott [442 pl 30 figs 1–7] [*nom. substit. pro Eretmochelys squamata* Agassiz, 1857].

Caretta rostrata Girard, C. (1858). *United States Exploring Expedition during the years 1838, 1839, 1840, 1841, 1842.* Herpetology. 20: xvi 496 pp. Philadelphia : J.B. Lippincott [446 pl 30 figs 8–13]. Type data: syntypes, USNM 012387–88, from Fiji Ils.

Onychochelys kraussi Gray, J.E. (1873). Notes on the genera of turtles (Oiacopodes), and especially on their skeletons and skulls. *Proc. Zool. Soc. Lond.* **1873**: 395–411 [398 figs 1–2]. Type data: syntypes (2), BMNH, from French Guiana.

Eretmochelys squammata Dugès, A.A.D. (1884). *Elementos de Zoologôa.* México : Secr. Fomento viii 4709 pp. [332] [*errore pro Eretmochelys squamata* Agassiz, 1857].

Eretmochelys imbricata bibba Pritchard, P.C.H. (1967). Scientific and common names of turtles of the world. *Bull. Ross Allen Rept. Inst.* **1967**: 1–10 [1] [*errore pro Caretta bissa* Rüppell, 1835].

Synonymy that of Cogger, H.G., this work; Wermuth, H. & Mertens, R. (1977). Liste der rezenten Amphibien und Reptilien; Testudines, Crocodylia, Rhynchocephalia. *Das Tierreich* **100**: i–xxvii, 1–174; Smith, H.M. & Smith, R.B. (1979). *Synopsis of the Herpetofauna of Mexico.* Vol. 6. Guide to Mexican Turtles. Bibliographic Addendum 3. North Bennington, Vermont : John Johnson 1044 pp.

Distribution: (SE coast), NE coast, Great Barrier Reef, Gulf of Carpentaria, N coast, NW Shelf, (W coast), Timor Sea, Indian Ocean, Coral Sea, Arafura Sea, (N.S.W.), Qld., W.A., N.T.; also pan-tropical, extending into temperate regions. Ecology: continental shelf, oceanic, aquatic, noctidiurnal, migratory, predator; seasonal breeder, oviparous, coral reefs, piscivore. Biological references: Cogger, H.G. & Lindner, D.L. (1969). Marine turtles in northern Australia. *Aust. Zool.* **15**: 150–159; Limpus, C.J. (1979). Notes on growth rates of wild turtles. *IUCN/SSC Marine Turtle Newsletter* **10**: 3–5; Limpus, C.J. (1980). Observations on the hawksbill turtle (*Eretmochelys imbricata*) nesting along the Great Barrier Reef. *Herpetologica* **36**: 265–271.

Lepidochelys Fitzinger, 1843

Lepidochelys Fitzinger, L.J. (1843). *Systema Reptilium.* Vienna : Braümüller u. Seidel vi 106 pp. [30]. Type species "*Thalass[ochelys] olivacea* Fitzinger" (= *Chelonia olivacea* Eschscholtz, 1829) by original designation.

***Caouana* (part.)** Gray, J.E. (1844). *Catalogue of the Tortoises, Crocodiles, and Amphisbaenians in the Collection of the British Museum.* London : British Museum viii 80 pp. [52] [*non Caouana* Cocteau, 1838]. Type species none designated; *Chelonia olivacea* Eschscholtz, 1829 is one of three included species.

Cephalochelys Gray, J.E. (1873). Notes on the genera of turtles (Oiacopodes), and especially on their skeletons and skulls. *Proc. Zool. Soc. Lond.* **1873**: 395–411 [408]. Type species *Cephalochelys oceanica* Gray, 1873 by monotypy.

Colpochelys Garman, S. (1880). On certain species of Chelonioidae. *Bull. Mus. Comp. Zool.* **6**: 123–126 [124] [described as subgenus of *Thalassochelys* Fitzinger, 1835]. Type species *Thalassochelys (Colpochelys) kempii* Garman, 1880 by monotypy.

Synonymy that of Cogger, H.G., this work; Wermuth, H. & Mertens, R. (1977). Liste der rezenten Amphibien und Reptilien; Testudines, Crocodylia, Rhynchocephalia. *Das Tierreich* **100**: i–xxvii, 1–174; Smith, H.M. & Smith, R.B. (1979). *Synopsis of the Herpetofauna of Mexico.* Vol. 6. Guide to Mexican Turtles. Bibliographic Addendum 3. North Bennington, Vermont : John Johnson 1044 pp.; Brongersma, L.D. (1961). Notes upon some sea turtles. *Zool. Verh. Leiden* **51**: 1–46.

This group is pan-tropical, extending into temperate regions.

Lepidochelys olivacea (Eschscholtz, 1829)

Testudo mydas minor Suckow, G.A. (1798). *Anfangsgründe der theoretischen und angewandten Naturgeschichte der Thiere.* Vol. 3. Von den Amphibien. Leipzig : Weidmann [30] [*nom. oblit.*]. Type data: status unknown, from Cap Blanco, Mexico; corrected to Isla Blanquilla, Caribbean Sea, see Brongersma, L.D. (1961). Notes upon some sea turtles. *Zool. Verh. Leiden* **51**: 1–46. Compiled from secondary source: Smith, H.M. & Smith, R.B. (1979). *Synopsis of the Herpetofauna of Mexico.* Vol. 6. Guide to Mexican Turtles. Bibliographic Addendum 3. North Bennington, Vermont : John Johnson 1044 pp.

Chelonia multiscutata Kuhl, H. (1820). *Beiträge zur Zoologie und vergleichenden Anatomie.* Frankfurt 130 pp. [78] [*nom. oblit.*]. Type data: status unknown, from unknown locality.

Chelonia olivacea Eschscholtz, J.F. von (1829). *Zoologischer Atlas, enhaltend Abbildungen und Beschreibungen neuer Thierarten, während des Flottcapitains von Kotzbue zweiter Reise um die Welt in den Jahren 1823, 24, 25 und 26.* Berlin : G. Reimer Vol. 1 [3 pl 3]. Type data: status unknown but possibly in the Museum of Tartu (formerly Dorpat), Estonia, U.S.S.R. *vide* Smith, H.M. & Smith, R.B. (1979). *Synopsis of the Herpetofauna of Mexico.* Vol. 6. Guide to Mexican Turtles. Bibliographic Addendum 3. North Bennington, Vermont : John Johnson 1044 pp., from Manila Bay, Philippines.

Chelonia dussumierii Duméril, A.M.C. & Bibron, G. (1835). *Erpétologie Générale* ou Histoire Naturelle Complète des Reptiles. Paris : Roret 2: 680 pp. [557] [*nom. substit. pro Chelonia olivacea* Eschscholtz, 1829].

Chelonia subcarinata Gray, J.E. (1844). *Catalogue of the Tortoises, Crocodiles, and Amphisbaenians in the Collection of the British Museum.* London : British Museum viii 80 pp. [53] [*ex* Rüppell ms; *nom. nud.* introduced in synonymy].

Caouana ruppellii Gray, J.E. (1844). *Catalogue of the Tortoises, Crocodiles, and Amphisbaenians in the Collection of the British Museum.* London : British Museum viii 80 pp. [53] [*nom. nud.* introduced in synonymy; Smith, H.M. & Smith, R.B. (1979). *Synopsis of the Herpetofauna of Mexico.* Vol. 6. Guide to Mexican Turtles. Bibliographic Addendum 3. North Bennington, Vermont : John Johnson 1044 pp., regarded this as a valid name applied to one of five varieties of *Caouana caretta* (the others unnamed); we believe that they are in error and that Gray simply indicated that his manuscript name had been applied to a particular specimen].

Caouana dessumierii Smith, A. (1849). *Illustrations of the Zoology of South Africa.* Reptilia. London : Smith, Elder & Co. [Appendix 2] [*errore pro Chelonia dussumierii* Duméril and Bibron, 1835].

Chelonia polyaspis Bleeker, P. (1857). Opsomming der soorten van Reptiliën, tot dus verre van het eiland Java bekend geworden. *Natuurk. Tijdschr. Ned.-Indie* **14**: 235–244 [239] [*nom. nud.*]. from Java, Indonesia (as Batavia). Compiled from secondary source: Brongersma, L.D. (1961). Notes upon some sea turtles. *Zool. Verh. Leiden* **51**: 1–46.

Chelonia dubia Gray, J.E. (1864). Notes on certain species of tortoises from the Asiatic islands transmitted to the British Museum by Dr Bleeker. *Proc. Zool. Soc. Lond.* **1864**: 11–13 [13] [*ex* Bleeker ms; *nom. nud.*]. from Borneo, see Wermuth, H. & Mertens, R. (1977). Liste der rezenten Amphibien und Reptilien; Testudines, Crocodylia, Rhynchocephalia. *Das Tierreich* **100**: i–xxvii, 1–174. Compiled from secondary source: Brongersma, L.D. (1961). Notes upon some sea turtles. *Zool. Verh. Leiden* **51**: 1–46.

Cephalochelys oceanica Gray, J.E. (1873). Notes on the genera of turtles (Oiacopodes), and especially on their skeletons and skulls. *Proc. Zool. Soc. Lond.* **1873**: 395–411 [408]. Type data: holotype, BMNH 1947.3.5.40, from "West coast of America (Mexico)?"; restricted to San Jose del Cabo, Baja California, Mexico. Compiled from secondary source: Smith, H.M. & Smith, R.B. (1979). *Synopsis of the Herpetofauna of Mexico.* Vol. 6. Guide to Mexican Turtles. Bibliographic Addendum 3. North Bennington, Vermont : John Johnson 1044 pp.

Chelonia olivaccea Velasco, A.L. (1892). *Geografía y estadística del estado de Tamaulipas. Geografía y estadística de la República Méxicana. (Reptiles and Amphibias).* México : Secr. Fomento 12: 78–80 [75] [invalid emend. *pro Chelonia olivacea* Eschscholtz, 1820].

Thalassochelys controversa Philippi, R.A. (1899). Las tortugas Chilenas. *An. Univ. Chile* **104**: 725–736 [731]. Type data: holotype, MNHN (presumed), from Quinteros, Chile. Compiled from secondary source: Smith, H.M. & Smith, R.B. (1979). *Synopsis of the Herpetofauna of Mexico.* Vol. 6. Guide to Mexican Turtles. Bibliographic Addendum 3. North Bennington, Vermont : John Johnson 1044 pp.; Loveridge, A. & Williams, E.D. (1957). Revision of the African tortoises and turtles of the suborder Cryptodira. *Bull. Mus. Comp. Zool.* **115**: 161–557.

Caretta remivaga Hay, O.P. (1908). On three existing species of sea turtles, one of them (*Caretta remivaga*) new. *Proc. U.S. Natl. Mus.* **34**: 183–198 [194 pl 10 figs 1–3 pl 11 fig 5]. Type data: holotype, USNM 09973, from Ventosa Bay, Gulf of Tehuantepec, Oaxaca, Mexico.

Synonymy that of Cogger, H.G., this work; Brongersma, L.D. (1961). Notes upon some sea turtles. *Zool. Verh. Leiden* **51**: 1–46; Smith, H.M. & Smith, R.B. (1979). *Synopsis of the Herpetofauna of Mexico.* Vol. 6. Guide to Mexican Turtles. Bibliographic Addendum 3. North Bennington, Vermont : John Johnson 1044 pp.; Loveridge, A. & Williams, E.D. (1957). Revision of the African tortoises and turtles of the suborder Cryptodira. *Bull. Mus. Comp. Zool.* **115**: 161–557.

Distribution: NE coast, Great Barrier Reef, (Gulf of Carpentaria), N coast, (NW Shelf), Qld., (W.A.), N.T.; also pan-tropical extending into temperate regions. Ecology: continental shelf, oceanic, aquatic, noctidiurnal, migratory, predator; seasonal breeder, oviparous, coral reefs, piscivore. Biological references: Cogger, H.G. & Lindner, D.L. (1969). Marine turtles in northern Australia. *Aust. Zool.* **15**: 150–159; Mrosovsky, N. & Pritchard, P.C.H. (1971). Body temperatures of *Dermochelys coriacea* and other sea turtles. *Copeia* **1971**: 624–631; Limpus, C.J. (1975). The Pacific Ridley, *Lepidochelys olivacea* (Eschscholtz) and other sea turtles in northeastern Australia. *Herpetologica* **31**: 444–445.

DERMOCHELYIDAE

INTRODUCTION

A monotypic family of sea turtles whose single species is the largest of all living turtles. The leatherback turtle occurs throughout tropical and subtropical seas, but more than any of the other sea turtles it wanders regularly into the colder waters of higher latitudes. The species breeds only rarely in Australia, the females coming ashore to lay their eggs in the same manner as other sea turtles (Cheloniidae).

Characterised by: cryptodirous withdrawal of the head into the shell; strong, paddle-shaped limbs, without claws on the leading edge; nostrils level with surface of snout; shell not a fused bony shield but a series of mostly small, free, polygonal plates forming a bony mosaic within a thick leathery skin.

References

Cogger, H.G. (1979). *Reptiles and Amphibians of Australia.* Sydney : A.H. & A.W. Reed 608 pp.

Ernst, C.H. & Barbour, R.W. (1972). *Turtles of the United States.* Lexington : University Press of Kentucky x 347 pp.

Pritchard, P.C.H. (1979). *Encyclopedia of Turtles.* Neptune, N.J. : T.F.H. Publications 895 pp.

Wermuth, H. & Mertens, R. (1961). *Schildkröten, Krokodile, Brückenechsen.* Jena : Gustav Fischer 422 pp.

Wermuth, H. & Mertens, R. (1977). Liste der rezenten Amphibien und Reptilien; Testudines, Crocodylia, Rhynchocephalia. *Das Tierreich* **100**: i–xxvii, 1–174

Dermochelys Blainville, 1816

Dermochelys Blainville, H.D. de (1816). Prodrome d'une nouvelle distribution systématique du Règne Animal. *Bull. Sci. Soc. Philomath. Paris* **1816**: 105–124 [119 (as 111)]. Type species *Testudo coriacea* Linnaeus, 1766 by subsequent designation, see Cuvier, G.L.C.F.D. (1829). *Le Règne Animal.* 2nd edn. Paris : Déterville Vol. 2 xv 406 pp.

Sphargis Merrem, B. (1820). *Versuch eines Systems Amphibien. Tentamen Systematis Amphibiorum.* Marburg : Krieger xv 199 pp. [19] [name suppressed in 1956 by the International Commission on Zoological Nomenclature under Direction 56]. Type species *Sphargis mercurialis* Merrem, 1820 by monotypy.

Coriudo Fleming, J. (1822). *The Philosophy of Zoology;* or a General View of the Structure, Functions, and Classification of Animals. Edinburgh : Constable 2: 1–618 [271] [name suppressed in 1956 by the International Commission on Zoological Nomenclature under Direction 56]. Type species *Testudo coriacea* Linnaeus, 1766 by monotypy. Compiled from secondary source: Smith, H.M. & Smith, R.B. (1979). *Synopsis of the Herpetofauna of Mexico.* Vol. 6. Guide to Mexican Turtles. Bibliographic Addendum 3. North Bennington, Vermont : John Johnson 1044 pp.

Siphargis Risso, A. (1826). *Histoire Naturelle des Principales Productions de l'Europe Méridionale.* Paris : F.-G. Levrault 3: 1–480 [85] [invalid emend. *pro Sphargis* Merrem, 1820].

Scytina Wagler, J.G. (1828). Vorläufige übersicht des Gerüstes seines Systema Amphibiorum. *Isis Von Oken, Jena* **21**: 859–861 [861] [*nom. substit. pro Sphargis* Merrem, 1820].

Dermochelis Cuvier, G.L.C.F.D. (1829). *Le Règne Animal.* 2nd edn. Paris : Déterville Vol. 2 xv 406 pp. [14] [*ex* Lesueur ms; emend. *pro Dermochelys* Blainville, 1816].

Dermatochelys Wagler, J.G. (1830). *Natürliches System der Amphibien, mit vorangehender Classification der Säugethiere und Vögel.* München : Cotta'schen vi 354 pp. [133] [*nom. substit. pro Dermochelys* Blainville, 1816].

Chelyra Rafinesque, C.S. (1832). Description of two new genera of soft shell turtles of North America. *Atlantic J. and Friend of Knowledge* **1**: 64–65 [64]. Type species *Testudo coriacca* [*sic*] Linnaeus, 1766 by original designation.

Scytena Gray, J.E. (1844). *Catalogue of the Tortoises, Crocodiles, and Amphisbaenians in the Collection of the British Museum.* London : British Museum viii 80 pp. [51] [*nom. substit. pro Scytina* Wagler, 1828].

Sphragis Agassiz, L. (1846). *Nomenclatoris Zoologici Index Universalis.* Soloduri : Jent et Gassman 393 pp. [347] [invalid emend. *pro Sphargis* Merrem, 1820].

Seytinia Neave, S.A. (ed.) (1940). *Nomenclator Zoologicus.* London : Zoological Society of London 4: 1–758 [186] [*nom. nud.*, incorrectly attributed to Wagler, 1828].

Seytina Romer, A.S. (1956). *Osteology of the Reptiles.* Chicago : Univ. Chicago Press xxii 772 pp. [511] [incorrectly attributed to Wagler, J.G. (1830). *Natürliches System der Amphibien, mit vorangehender Classification der Säugethiere und Vögel.* München : Cotta'schen vi 354 pp.; invalid emend. *pro Seytinia* Neave (=*Scytina* Wagler, 1828)].

Scytine Romer, A.S. (1956). *Osteology of the Reptiles.* Chicago : Univ. Chicago Press xxii 772 pp. [511] [incorrectly attributed to Wagler, J.G. (1830). *Natürliches System der Amphibien, mit vorangehender Classification der Säugethiere und Vögel.* München : Cotta'schen vi 354 pp.; invalid emend. *pro Scytina* Wagler, 1828].

Dermatochelis Romer, A.S. (1956). *Osteology of the Reptiles.* Chicago : Univ. Chicago Press xxii 772 pp. [511] [incorrectly attributed to "Lesueur, 1829"; invalid emend. *pro Dermatochelys* Wagler, 1830].

Dermotochelys Battersby, J.C. (1959). Reptiles. *Zool. Rec. 1956* **93**: 1–122 [115] [*errore pro Dermatochelys* Wagler, 1830].

Synonymy that of Cogger, H.G., this work; Wermuth, H. & Mertens, R. (1977). Liste der rezenten Amphibien und Reptilien; Testudines, Crocodylia, Rhynchocephalia. *Das Tierreich* **100**: i–xxvii, 1–174; Smith, H.M. & Smith, R.B. (1979). *Synopsis of the Herpetofauna of Mexico.* Vol. 6. Guide to Mexican Turtles. Bibliographic Addendum 3. North Bennington, Vermont : John Johnson 1044 pp.

This group is also pan-tropical, extending into temperate regions, see Wermuth, H. & Mertens, R. (1977). Liste der rezenten Amphibien und Reptilien; Testudines, Crocodylia, Rhynchocephalia. *Das Tierreich* **100**: i–xxvii, 1–174.

Dermochelys coriacea (Vandelli, 1761)

Testudo coriacea Vandelli, D. (1761). *Epistola de Holothuria, et* Testudine coriacea *ad Celeberrinum Carolum Linnaeum.* Padua : Conzetti 12 pp. [2, 4, 7, 12, pl 2]. Type data: holotype, MZUP unnumbered specimen, from Tyrrhenian coast near Rome, Italy. Compiled from secondary source: Fretey, J. & Bour, R. (1980). Redécouverte du type de *Dermochelys coriacea* (Vandelli) (Testudinata, Dermochelyidae). *Boll. Zool.* **47**: 193–205.

Testudo arcuata Catesby, M. (1771). *The Natural History of Carolina, Florida and the Bahama Islands.* (Revised by G. Edwards). To which is now added a Linnaean index of the animals and plants. London : B. White 2: 1–100 [40 pl 40]. Type data: none designated, from unknown locality, restricted to coastal regions of Carolina and Florida, see Mertens, R. & Wermuth, H. (1955). Die rezenten Schildkröten, Krokodile und Brückenechsen. Eine kritische Liste der heute lebenden Arten und Rassen. *Zool. Jb. Syst.* **83**: 323–440.

Testudo lyra Lacépède, B.G.E. (1788). Histoire naturelle des Quadrupèdes ovipares et serpens. *in* Buffon, G.L.L. (ed.) *Histoire Naturelle, générale et particulière,* avec la description du cabinet du roi. Paris : Royal Printer 1 (Synops. Méth): 95–104 [111 pl 3] [*nom. substit. pro Testudo coriacea* Vandelli, 1761].

Testudo tuberculata Pennant, T. *in* Schoepff, J.D. (1801). *Historia Testudinum iconibus illustrata.* Erlangen : J.J. Palm xii 136 pp. [123 pl 29a]. Type data: none designated, from unknown locality, but restricted to Palermo, Italy by Smith, H.M. & Smith, R.B. (1979). *Synopsis of the Herpetofauna of Mexico.* Vol. 6. Guide to Mexican Turtles. Bibliographic Addendum 3. North Bennington, Vermont : John Johnson 1044 pp.

Chelonias lutaria Rafinesque, C.S. (1814). *Specchio della Science, o Giornale enciclopedico di Sicilia. Deposito letterario delle moderne cognizioni, scopperte et osservazioni sopra le scienze ed arti.* Palermo Vol. 2 Cahier 9 [66]. Type data: none designated, from Sicily.

Sphargis mercurialis Merrem, B. (1820). *Versuch eines Systems Amphibien. Tentamen Systematis Amphibiorum.* Marburg : Krieger xv 199 pp. [19] [*nom. substit. pro Testudo coriacea* Linnaeus, 1766 (=*Testudo coriacea* Vandelli, 1761); name suppressed in 1956 by the International Commission on Zoological Nomenclature under Direction 56]. Type data: that of *Testudo coriacea* Linnaeus, 1766.

Dermochelis atlantica Cuvier, G.L.C.F.D. (1829). *Le Règne Animal.* 2nd edn. Paris : Déterville Vol. 2 xv 406 pp. [14] [*ex* Lesueur ms; *nom. nud.* introduced in synonymy].

Dermatochelys porcata Wagler, J.G. (1830). *Natürliches System der Amphibien, mit vorangehender Classification der Säugethiere und Vögel.* München : Cotta'schen vi 354 pp. [pl 1] [*nom. substit. pro Testudo coriacea* Linnaeus, 1766 (=*Testudo coriacea* Vandelli, 1761)]. Compiled from secondary source: Wermuth, H. & Mertens, R. (1961). *Schildkröten, Krokodile, Brückenechsen.* Jena : Gustav Fischer 422 pp.

Sphargis coriacea schlegelii Garman, S. (1884). The reptiles of Bermuda (and) list of sea turtles (Chelonioidae) with synonymy. *Bull. U.S. Natl. Mus.* **25**: 285–303 [303]. Type data: pls 1–3 in Temminck, C.J. & Schlegel, H. (1835–1838). *Fauna Japonica.* Reptilia. Leiden : Lalau xxi 144 pp., see Smith, H.M. & Smith, R.B. (1979). *Synopsis of the Herpetofauna of Mexico.* Vol. 6. Guide to Mexican Turtles. Bibliographic Addendum 3. North Bennington, Vermont : John Johnson 1044 pp., from tropical Pacific and Indian Oceans but restricted to Guaymas, Sonora, Mexico by Smith, H.M. & Taylor, E.H. (1950). An annotated checklist and key to the reptiles of Mexico exclusive of the snakes. *Bull. U.S. Natl. Mus.* **199**: i–iv 1–253.

Sphargis angusta Philippi, R.A. (1899). Las tortugas Chilenas. *An. Univ. Chile* **104**: 725–736 [728 pl]. Type data: holotype, MNHN, from Tocopilla, Chile. Compiled from secondary source: Smith, H.M. & Smith, R.B. (1979). *Synopsis of the Herpetofauna of Mexico.* Vol. 6. Guide to Mexican Turtles. Bibliographic Addendum 3. North Bennington, Vermont : John Johnson 1044 pp. Synonymy that of Cogger, H.G., this work; Wermuth, H. & Mertens, R. (1977). Liste der rezenten Amphibien und Reptilien; Testudines, Crocodylia, Rhynchocephalia. *Das Tierreich* **100**: i–xxvii, 1–174; Smith, H.M. & Smith, R.B. (1979). *Synopsis of the Herpetofauna of Mexico.* Vol. 6. Guide to Mexican Turtles. Bibliographic Addendum 3. North Bennington, Vermont : John Johnson 1044 pp.

Distribution: SE coast, NE coast, Great Barrier Reef, Gulf of Carpentaria, N coast, NW Shelf, W coast, Coral Sea, Arafura Sea, Timor Sea, Indian Ocean, N.S.W., Qld., W.A., N.T.; also pan-tropical, extending into temperate regions. Ecology: estuarine, inshore, continental shelf, oceanic, aquatic, noctidiurnal, predator; seasonal breeder, oviparous, feeds principally on coelenterates. Biological references: Mrosovsky, N. & Pritchard, P.C.H. (1971). Body temperatures of *Dermochelys coriacea* and other sea turtles. *Copeia* **1971**: 624–631; Mrosovsky, N. & Shettleworth, S.J. (1975). On the orientation circle of the leatherback turtle, *Dermochelys coriacea. Anim. Behav.* **23**: 568–591; Limpus, C.J. & McLachlan, N.C. (1979). Observations on the leatherback turtle, *Dermochelys coriacea* (L.), in Australia. *Aust. Wildl. Res.* **6**: 105–116; Pritchard, P.C.H. (1980). *Dermochelys coriacea* (Reptilia : Testudines : Dermochelyidae). *Cat. Am. Amphib. Rept.* 238: 1–4.

GEKKONIDAE

INTRODUCTION

A family primarily of small nocturnal lizards occurring throughout tropical and temperate regions. Of the 88 genera and 750 species currently recognised, some 77 species in 16 genera occur in Australia. Of these, about 70 per cent belong to the regionally-endemic subfamily, the Diplodactylinae.

Characterised in Australia by: well-developed pentadactyl limbs; eyes without movable eyelids, each covered by a fixed transparent spectacle; pupil (in daylight) a narrow vertical slit; external ear-opening present; body scales juxtaposed; tongue short, flat and fleshy, slightly notched in front.

References

Cogger, H.G. (1979). *Reptiles and Amphibians of Australia.* Sydney : A.H. & A.W. Reed 608 pp.

Stephenson, N.G. (1960). The comparative osteology of Australian geckos and its bearing on their morphological status. *Zool. J. Linn. Soc.* **44**: 278–299

Wermuth, H. (1965). Liste der rezenten Amphibien und Reptilien; Gekkonidae, Pygopodidae, Xantusiidae. *Das Tierreich* **80**: i–xxii 1–246

Worrell, E. (1970). *Reptiles of Australia.* 2nd edn. Sydney : Angus & Robertson xv 169 pp.

Carphodactylus Günther, 1897

Carphodactylus Günther, A. (1897). Description of new species of lizards and of a tree frog from north-eastern Queensland. *Novit. Zool.* **4**: 403–406 [403]. Type species *Carphodactylus laevis* Günther, 1897 by monotypy.

Carphodactylus laevis Günther, 1897

Carphodactylus laevis Günther, A. (1897). Description of new species of lizards and of a tree frog from north-eastern Queensland. *Novit. Zool.* **4**: 403–406 [403 pl 11]. Type data: holotype, presumed lost, from Mt. Bartle Frere, Qld.

Distribution: NE coastal, Qld. Ecology: litter, closed forest, terrestrial, nocturnal; oviparous, arthropod-feeder. Biological references: Russell, A.P. (1979). Parallelism and integrated design in the foot structure of gekkonine and diplodactyline geckos. *Copeia* **1979**: 1–21 (foot structure).

Crenadactylus Dixon and Kluge, 1964

Crenadactylus Dixon, J.R. & Kluge, A.G. (1964). A new gekkonid lizard genus from Australia. *Copeia* **1964**: 174–180 [174]. Type species *Diplodactylus ocellatus* Gray, 1845 by original designation.

Crenadactylus ocellatus (Gray, 1845)

Diplodactylus ocellatus Gray, J.E. (1845). *Catalogue of the Specimens of Lizards in the Collection of the British Museum.* London : British Museum xxviii 289 pp. [149]. Type data: holotype, BMNH 1947.3.6.68, from W.A.

Diplodactylus bilineatus Gray, J.E. (1845). *Catalogue of the Specimens of Lizards in the Collection of the British Museum.* London : British Museum xxviii 289 pp. [149]. Type data: syntypes, BMNH 1947.3.6.69 (2 specimens), from Houtman's Abrolhos, W.A.

Ebenavia horni Lucas, A.H.S. & Frost, C. (1895). Preliminary notice of a certain species of lizard from central Australia. *Proc. R. Soc. Vict. (ns)* **7**: 264–269 [264]. Type data: holotype, NMV D7533, from Charlotte Waters, N.T.

Crenadactylus ocellatus naso Storr, G.M. (1978). Seven new gekkonid lizards from Western Australia. *Rec. West. Aust. Mus.* **6**: 337–352 [337]. Type data: holotype, WAM R56206, from Crystal Creek, 14°30′S, 125°47′E, W.A.

Crenadactylus ocellatus rostralis Storr, G.M. (1978). Seven new gekkonid lizards from Western Australia. *Rec. West. Aust. Mus.* **6**: 337–352 [339 pl 1]. Type data:

holotype, WAM R32154, from Geikie Gorge, 18°07'S, 125°39'E, W.A.

Synonymy that of Cogger, H.G., this work.

Distribution: N Gulf, N coastal, NW coastal, SW coastal, W plateau, Lake Eyre basin, S.A., Qld., W.A., N.T. Ecology: desert, hummock grassland, low open shrubland, low woodland, open forest, tall shrubland, terrestrial, nocturnal; oviparous, arthropod-feeder, in clumps of spinifex. Biological references: Kluge, A.G. (1967). Systematics, phylogeny, and zoogeography of the lizard genus *Diplodactylus* Gray (Gekkonidae). *Aust. J. Zool.* **15**: 1007–1108.

Cyrtodactylus Gray, 1827

Cyrtodactylus Gray, J.E. (1827). A synopsis of the genera of saurian reptiles, in which some new genera are indicated, and the others reviewed by actual examination. *Phil. Mag. (2)* **3**: 54–58 [56]. Type species *Cyrtodactylus pulchella* Gray, 1827 by monotypy.

Cyrtopodion Fitzinger, L.J. (1843). *Systema Reptilium.* Vienna : Braümüller u. Seidel vi 106 pp. [93]. Type species *Stenodactylus scaber* Heyden, 1827 by original designation. Two species originally included: *G. scaber* and *G. cyprius* Fitzinger; the latter may be a *nom. nud.*, see Wermuth, H. (1965). Liste der rezenten Amphibien und Reptilien; Gekkonidae, Pygopodidae, Xantusiidae. *Das Tierreich* **80**: i–xxii 1–246.

Puellula Blyth, E. (1860). Report of Curator, Zoological Department. *J. Asiat. Soc. Beng.* **29**: 107–111 [109]. Type species *Puellula rubida* Blyth, 1860 (= *Cyrtodactylus rubidus*) by monotypy.

Geckoella Gray, J.E. (1867). On a new geckoid lizard from Ceylon. *Proc. Zool. Soc. Lond.* **1867**: 98–99 [98]. Type species *Geckoella punctata* Gray, 1867 (= *Cyrtodactylus triedrus* Günther, 1864) by monotypy.

Dinosaura Gistel, J. (1868). *Blicke in das Leben der Natur und des Menschen.* Leipzig pp. 144–167 [145] [*non Dinosaurus* Fischer-Waldheim, 1847]. Type species *Stenodactylus scaber* Heyden, 1827 by subsequent designation, see Loveridge, A. (1947). Revision of the African lizards of the family Gekkonidae. *Bull. Mus. Comp. Zool.* **8**: 1–469.

Crytodactylus Kluge, A.G. (1967). Higher taxonomic categories of gekkonid lizards and their evolution. *Bull. Am. Mus. Nat. Hist.* **135**: 1–59 [27] [*err. typ. pro Cyrtodactylus* Gray, 1827].

Synonymy that of Cogger, H.G., this work; Wermuth, H. (1965). Liste der rezenten Amphibien und Reptilien; Gekkonidae, Pygopodidae, Xantusiidae. *Das Tierreich* **80**: i–xxii 1–246; Brown, W.C. & Parker, F. (1973). A new species of *Cyrtodactylus* (Gekkonidae) from New Guinea with a key to species from the island. *Breviora* 417: 1–7.

This group is also found in Africa, Asia, Indo-malayan Archipelago, New Guinea and islands of western Pacific.

Cyrtodactylus galgajuga Ingram, 1978

Cyrtodactylus galgajuga Ingram, G.J. (1978). A new species of gecko, genus *Cyrtodactylus*, from Cape York Peninsula, Queensland, Australia. *Vict. Nat.* **95**: 142–146 [142 pl 1 fig 2 pl 2]. Type data: holotype, QM J29474, from near Black Mountain, 15°40'S, 145°14'E, Trevethan Range, 22 km S of Cooktown, Qld.

Distribution: NE coastal, Qld. Ecology: terrestrial, nocturnal; oviparous, saxicoline, arthropod-feeder.

Cyrtodactylus louisiadensis (De Vis, 1892)

Gymnodactylus louisiadensis De Vis, C.W. (1892). Zoology of British New Guinea. Part I. Vertebrata. *Ann. Qd. Mus.* **2**: 1–12 [11]. Type data: holotype, presumed lost, from Tagula (as Südest) Is., Papua New Guinea.

Hoplodactylus tuberculatus Lucas, A.H.S. & Frost, C. (1900). Description of a new lizard from northern Queensland. *Proc. R. Soc. Vict. (ns)* **12**: 145–146 [145]. Type data: holotype, NMV D7874, from Endeavour River, Qld.

Gymnodactylus olivii Garman, S. (1901). Some reptiles and batrachians from Australasia. *Bull. Mus. Comp. Zool.* **39**: 1–14 [1 pl 1 figs 1–1d]. Type data: holotype, MCZ 6470, from near Cooktown, Qld.

Synonymy that of Cogger, H.G., this work.

Distribution: NE coastal, Qld. Ecology: (caves), woodland, closed forest, terrestrial, arboreal, nocturnal; oviparous, (saxicoline), arthropod-feeder. Biological references: Brown, W.C. & Parker, F. (1973). A new species of *Cyrtodactylus* (Geckonidae) from New Guinea with a key to species from the island. *Breviora* 417: 1–7 (taxonomy).

[Cyrtodactylus marmoratus] auctorum

Gymnodactylus marmoratus Boulenger, G.A. (1887). Reptiles. *in* Günther, A. (ed.) Report on a zoological collection made by the officers of H.M.S. *Flying Fish* at Christmas Island, Indian Ocean. *Proc. Zool. Soc. Lond.* **1887**: 516–517 [516] [*non [Cyrtodactylus] marmoratus* Gray, 1831; although specimens of *Cyrtodactylus* from Christmas Is. have been assigned to *C. marmoratus*, they represent an undescribed endemic species currently being named by L.D. Brongersma].

Distribution: Christmas Is. Ecology: closed forest, (terrestrial), arboreal, nocturnal; secondary growth following phosphate mining, oviparous, arthropod-feeder. Biological references: Brongersma, L.D. (1953). *Gymnodactylus marmoratus* (Gray). *Verslag. Gew. Verg. Afd. Natuurk., K. Ned. Akad. Wet. Amsterdam* **62**: 172–175.

Cyrtodactylus pelagicus (Girard, 1858)

Heteronota pelagica Girard, C. (1858). Descriptions of some new reptiles, collected by the United States Exploring Expedition, under the command of Capt. Charles Wilkes, U.S.N. Fourth Part. Including the species of saurians exotic to North America. *Proc. Acad. Nat. Sci. Philad.* **1857**: 195–199 [197]. Type data: syntypes, USNM 5626 (2 specimens), from Samoa (as Navigator Ils.) and Fiji.

Heteronota fasciata Macleay, W. (1877). The lizards of the *Chevert* Expedition. Second paper. *Proc. Linn. Soc. N.S.W.* **2**: 97–104 [100] [1878 on title page; *non Gymnodactylus fasciatus* Duméril and Bibron, 1836]. Type data: holotype, AM R31934, from Hall Sound, Papua New Guinea.

Hetoronota* [*sic*] *marmorata Macleay, W. (1877). The lizards of the *Chevert* Expedition. Second paper. *Proc. Linn. Soc. N.S.W.* **2**: 97–104 [100] [*non Cyrtodactylus marmoratus* Gray, 1831; 1878 on title page]. Type data: lectotype, AM R42733, from Fitzroy Is., SE of Cairns, Qld., designation by Kluge, A.G. (1963). The systematic status of certain Australian and New Guinean gekkonid lizards. *Mem. Qd. Mus.* **14**: 77–86 (original description also cites Endeavour River, Qld.).

Heteronota eboracensis Macleay, W. (1877). The lizards of the *Chevert* Expedition. Second paper. *Proc. Linn. Soc. N.S.W.* **2**: 97–104 [101] [1878 on title page]. Type data: lectotype, AM R29951, from Cape York, Qld., designation by Kluge, A.G. (1963). The systematic status of certain Australian and New Guinean gekkonid lizards. *Mem. Qd. Mus.* **14**: 77–86.

Gymnodactylus heteronotus Boulenger, G.A. (1885). *Catalogue of the Lizards in the British Museum (Natural History).* 1. London : British Museum xii 436 pp. pls 32 [41] [*nom. nov. pro Heteronota fasciata* Macleay, 1877, at that time a junior secondary homonym of *Gymnodactylus fasciatus* Duméril and Bibron, 1836].

Gymnodactylus cheverti Boulenger, G.A. (1885). *Catalogue of the Lizards in the British Museum (Natural History).* 1. London : British Museum xii 436 pp. pls 32 [41] [*nom. nov. pro Heteronota marmorata* Macleay, 1877].

Synonymy that of Cogger, H.G., this work.

Distribution: NE coastal, N Gulf, Qld. Ecology: open forest, woodland, closed forest, supralittoral, terrestrial, nocturnal; oviparous, cryptozoic, arthropod-feeder. Biological references: Ingram, G.J. (1978). A new species of gecko, genus *Cyrtodactylus*, from Cape York Peninsula, Queensland, Australia. *Vict. Nat.* **95**: 142–146 (taxonomy); Brown, W.C. & Parker, F. (1973). A new species of *Cyrtodactylus* (Geckonidae) from New Guinea with a key to species from the island. *Breviora* 417: 1–7 (taxonomy); Schwaner, T.D. (1980). Reproductive biology of lizards on the American Samoan islands. *Occ. Pap. Mus. Nat. Hist. Univ. Kansas* **86**: 1–53 (biology).

Diplodactylus Gray, 1832

Diplodactylus Gray, J.E. (1832). (Three new animals brought from New Holland by Mr Cunningham). *Proc. Zool. Soc. Lond.* **1832**: 39–40 [40]. Type species *Diplodactylus vittatus* Gray, 1832 by monotypy.

Strophurus Fitzinger, L.J. (1843). *Systema Reptilium.* Vienna : Braümüller u. Seidel vi 106 pp. [96]. Type species *Discodactylus dumerilii* Fitzinger, 1843 by monotypy.

Strophura Gray, J.E. (1845). *Catalogue of the Specimens of Lizards in the Collection of the British Museum.* London : British Museum xxviii 289 pp. [148]. Type species *Diplodactylus spinigerus* Gray, 1842 by monotypy *vide* Kluge, A.G. (1967). Systematics, phylogeny, and zoogeography of the lizard genus *Diplodactylus* Gray (Gekkonidae). *Aust. J. Zool.* **15**: 1007–1108, but Wermuth, H. (1965). Liste der rezenten Amphibien und Reptilien; Gekkonidae, Pygopodidae, Xantusiidae. *Das Tierreich* **80**: i–xxii 1–246 correctly cited *Diplodactylus strophurus* (Duméril and Bibron) as type species by tautonymy.

Stenodactylopsis Steindachner, F. (1870). Herpetologische Notizen (II). Reptilien gesammelt Während einer Reise in Sengambien. *Sber. Akad. Wiss. Wien* **62**: 326–349 [343]. Type species *Stenodactylopsis pulcher* Steindachner, 1870 by monotypy.

Oedurella Lönnberg, E. & Andersson, G.L. (1913). Results of Dr. E. Mjöberg's Swedish Scientific Expeditions to Australia 1910–1913. III. Reptiles. *K. Svenska Vetensk.-Akad. Handl.* **52**(3): 1–17 [5]. Type species *Oedurella taeniata* Lönnberg and Andersson, 1913 by monotypy.

Synonymy that of Cogger, H.G., this work; Kluge, A.G. (1967). Systematics, phylogeny, and zoogeography of the lizard genus *Diplodactylus* Gray (Gekkonidae). *Aust. J. Zool.* **15**: 1007–1108.

Diplodactylus alboguttatus Werner, 1910

Diplodactylus alboguttatus Werner, F. (1910). Reptilia (Geckonidae und Scincidae). *in* Michaelsen, W. & Hartmeyer, R. (eds.) *Die Fauna Südwest-Australiens.* Jena : Gustav Fischer 2: 451–493 [462 fig 4]. Type data: syntypes, from Denham, Shark Bay, W.A., SMF 8343, ZMB 21435, third syntype not found; types previously believed lost, *vide* Kluge, A.G. (1967). Systematics, phylogeny, and zoogeography of the lizard genus *Diplodactylus* Gray (Gekkonidae). *Aust. J. Zool.* **15**: 1007–1108.

Distribution: NW coastal, SW coastal, W.A. Ecology: low shrubland, woodland, open forest, terrestrial, nocturnal; oviparous, cryptozoic, arthropod-feeder. Biological references: Glauert, L. (1961). *A Handbook of the Lizards of Western Australia.* Perth : West. Aust. Naturalists' Club Handbook 6 100 pp. (natural history); Kluge, A.G. (1967). Systematics, phylogeny, and zoogeography of the lizard genus *Diplodactylus* Gray (Gekkonidae). *Aust. J. Zool.* **15**: 1007–1108 (taxonomy).

Diplodactylus byrnei Lucas and Frost, 1896

Diplodactylus byrnei Lucas, A.H.S. & Frost, C. (1896). Reptilia. *Rep. Horn Expedition to Central Australia.* 2: 112–151 [124 pl 12 fig 2]. Type data: holotype, NMV D7534, from Charlotte Waters, N.T.

Distribution: Murray-Darling basin, Bulloo River basin, Lake Eyre basin, S.A., N.S.W., Qld., N.T. Ecology: desert, low open shrubland, low shrubland, open scrub, tall shrubland, terrestrial, nocturnal; oviparous, cryptozoic, arthropod-feeder. Biological references: Kluge, A.G. (1967). Systematics, phylogeny, and zoogeography of the lizard genus *Diplodactylus* Gray (Gekkonidae). *Aust. J. Zool.* **15**: 1007–1108 (taxonomy).

Diplodactylus ciliaris Boulenger, 1885

Diplodactylus ciliaris Boulenger, G.A. (1885). *Catalogue of the Lizards in the British Museum (Natural History).* 1. London : British Museum xii 436 pp. pls 32 [98 pl 8 fig 2]. Type data: lectotype, BMNH 1946.9.6.87, from Port Darwin, N.T., designation by Kluge, A.G. (1967). Systematics, phylogeny, and zoogeography of the lizard genus *Diplodactylus* Gray (Gekkonidae). *Aust. J. Zool.* **15**: 1007–1108.

Diplodactylus spinigerus aberrans Glauert, L. (1952). Herpetological miscellanea. I. Notes on some forms of *Diplodactylus*. pp. 166–168 —and— Some new western Australian lizards. pp. 168–170. *West. Aust. Nat.* **3**: 166–170 [167]. Type data: holotype, WAM R2803, from Mt. Wynne, West Kimberley, W.A.

Synonymy that of Cogger, H.G., this work.

Distribution: N Gulf, N coastal, NW coastal, W plateau, Lake Eyre basin, Bulloo River basin, S.A., N.S.W., Qld., W.A., N.T. Ecology: desert, hummock grassland, low open shrubland, woodland, open forest, tall shrubland, (terrestrial), arboreal, nocturnal; oviparous, arthropod-feeder, (in clumps of spinifex). Biological references: Kluge, A.G. (1967). Systematics, phylogeny, and zoogeography of the lizard genus *Diplodactylus* Gray (Gekkonidae). *Aust. J. Zool.* **15**: 1007–1108 (taxonomy); Pianka, E.R. & Pianka, H.D. (1976). Comparative ecology of twelve species of nocturnal lizards (Gekkonidae) in the Western Australian desert. *Copeia* **1976**: 125–142 (ecology); Rosenberg, H.I. & Russell, A.P. (1980). Structural and functional aspects of tail squirting: a unique defense mechanism of *Diplodactylus* (Reptilia : Gekkonidae). *Can. J. Zool.* **58**: 865–881.

Diplodactylus conspicillatus Lucas and Frost, 1897

Diplodactylus conspicillatus Lucas, A.H.S. & Frost, C. (1897). Descriptions of two new species of lizards from central Australia. *Proc. R. Soc. Vict. (ns)* **9**: 54–56 [55]. Type data: lectotype, NMV D7535, from Charlotte Waters, N.T., designation by Kluge, A.G. (1967). Systematics, phylogeny, and zoogeography of the lizard genus *Diplodactylus* Gray (Gekkonidae). *Aust. J. Zool.* **15**: 1007–1108.

Diplodactylus hillii Longman, H.A. (1915). Reptiles from Queensland and the Northern Territory. *Mem. Qd. Mus.* **3**: 30–34 [32]. Type data: holotype, QM J1994, from Port Darwin, N.T.

Gymnodactylus laevis Sternfeld, R. (1924). Beiträge zur Herpetologie Inner-Australiens. *Abh. Senckenb. Naturf. Ges.* **38**: 221–251 [229]. Type data: lectotype, SMF 8242, from Hermannsburg Mission, N.T., designation by Mertens, R. (1967). Die herpetologische Sektion des Natur-Museums und Forschungs-Institutes Senckenberg in Frankfurt-a.-M. nebst einem Verzeichnis ihrer Typen. 1. *Senckenberg. Biol.* **48**: 1–106.

Diplodactylus platyurus Parker, H.W. (1926). New reptiles and a new frog from Queensland. *Ann. Mag. Nat. Hist. (9)* **17**: 665–670 [665]. Type data: holotype, BMNH 1946.8.11.38, from Torrens Creek, in 21°25′S, 145°14′E, N Qld.

Synonymy that of Cogger, H.G., this work.

Distribution: NE coastal, N Gulf, N coastal, NW coastal, W plateau, Lake Eyre basin, S.A., (N.S.W.), Qld., W.A., N.T. Ecology: desert, low open shrubland, low shrubland, low open woodland, low woodland, woodland, tall shrubland, terrestrial, nocturnal; oviparous, (saxicoline), arthropod-feeder. Biological references: Kluge, A.G. (1967). Systematics, phylogeny, and zoogeography of the lizard genus *Diplodactylus* Gray (Gekkonidae). *Aust. J. Zool.* **15**: 1007–1108 (taxonomy); Pianka, E.R. & Pianka, H.D. (1976). Comparative ecology of twelve species of nocturnal lizards (Gekkonidae) in the Western Australian desert. *Copeia* **1976**: 125–142 (ecology).

Diplodactylus elderi Stirling and Zietz, 1893

Diplodactylus elderi Stirling, E.C. & Zietz, A. (1893). Vertebrata; Reptilia (collected by the Elder Exploring Expedition). *Trans. R. Soc. S. Aust.* **16**: 159–176 [161 pl 6 figs 1–1a]. Type data: holotype, SAMA R2027, from Barrow Range, W.A.

Distribution: NW coastal, W plateau, Murray-Darling basin, Bulloo River basin, Lake Eyre basin, S.A., Vic., N.S.W., Qld., W.A., N.T. Ecology: desert, hummock grassland, low open shrubland, open scrub, tall shrubland, terrestrial, nocturnal; oviparous, arthropod-feeder, in clumps of spinifex. Biological references: Bustard, H.R. (1965). Observations on Australian geckos. *Herpetologica* **21**: 294–302 (natural history); Kluge, A.G. (1967). Systematics, phylogeny, and zoogeography of the lizard genus *Diplodactylus* Gray (Gekkonidae). *Aust. J. Zool.* **15**: 1007–1108 (taxonomy); Pianka, E.R. & Pianka, H.D. (1976). Comparative ecology of twelve species of nocturnal lizards (Gekkonidae) in the Western Australian desert. *Copeia* **1976**: 125–142 (ecology); Rosenberg, H.I. & Russell, A.P. (1980). Structural and functional aspects of tail squirting: a unique defense mechanism of *Diplodactylus*

(Reptilia : Gekkonidae). *Can. J. Zool.* **58**: 865–881; Bustard, H.R. (1979). Defensive mechanisms and predation in gekkonid lizards. *Br. J. Herpet.* **6**: 9–11.

Diplodactylus fulleri Storr, 1978

Diplodactylus fulleri Storr, G.M. (1978). Seven new gekkonid lizards from Western Australia. *Rec. West. Aust. Mus.* **6**: 337–352 [345 pl 4]. Type data: holotype, WAM R31331, from 5 km W of the mouth of Savoury Creek, 23°20′S, 122°40′E, W.A.

Distribution: W plateau, W.A. Ecology: (desert), low open shrubland, terrestrial, nocturnal; oviparous, arthropod-feeder.

Diplodactylus galeatus Kluge, 1963

Diplodactylus galeatus Kluge, A.G. (1963). Three new species of the gekkonid lizard *Diplodactylus* Gray from Australia. *Rec. S. Aust. Mus.* **14**: 545–553 [545 pl 34]. Type data: holotype, SAMA R973, from Stuart Range, S.A.

Distribution: Lake Eyre basin, S.A., N.T. Ecology: low open shrubland, tall shrubland, terrestrial, nocturnal; oviparous, (saxicoline), (in clumps of spinifex), arthropod-feeder. Biological references: Kluge, A.G. (1967). Systematics, phylogeny, and zoogeography of the lizard genus *Diplodactylus* Gray (Gekkonidae). *Aust. J. Zool.* **15**: 1007–1108 (taxonomy).

Diplodactylus granariensis Storr, 1979

Diplodactylus granariensis Storr, G.M. (1979). The *Diplodactylus vittatus* complex (Lacertilia : Gekkonidae) in Western Australia. *Rec. West. Aust. Mus.* **7**: 391–402 [397]. Type data: holotype, WAM R54239, from Walyahmoning Rock, 30°38′S, 118°43′E, W.A.

Distribution: NW coastal, SW coastal, W plateau, S Gulfs, S.A., W.A. Ecology: low shrubland, open heath, open scrub, woodland, open forest, tall shrubland, terrestrial, nocturnal; oviparous, arthropod-feeder.

Diplodactylus intermedius Ogilby, 1892

Diplodactylus intermedius Ogilby, J.D. (1892). Descriptions of three new Australian lizards. *Rec. Aust. Mus.* **2**: 6–11 [10]. Type data: lectotype, AM R285, from interior of N.S.W., designation by Kluge, A.G. (1967). Systematics, phylogeny, and zoogeography of the lizard genus *Diplodactylus* Gray (Gekkonidae). *Aust. J. Zool.* **15**: 1007–1108, who also restricted the type locality to Nymagee, N.S.W.

Distribution: W plateau, S Gulfs, Murray-Darling basin, (Bulloo River basin), Lake Eyre basin, S.A., Vic., N.S.W., N.T., W.A. Ecology: desert, tussock grassland, low open shrubland, low shrubland, open scrub, tall shrubland, (terrestrial), arboreal, nocturnal; oviparous, arthropod-feeder, (in clumps of spinifex). Biological references: Kluge, A.G. (1967). Systematics, phylogeny, and zoogeography of the lizard genus *Diplodactylus* Gray (Gekkonidae). *Aust. J. Zool.* **15**: 1007–1108 (taxonomy); Warburg, M.R. (1966). Water economy of several Australian geckos, agamids, and water skinks. *Copeia* **1966**: 230–235 (as *D. strophurus intermedius*); Brazenor, C.W. (1951). On the Victorian species of tuberculated *Diplodactylus*. *Mem. Natl. Mus. Vict.* **17**: 215–222; Ehmann, H. (1980). Diurnal perching by the southern spiny-tailed gecko, *Diplodactylus intermedius*. *Herpetofauna* **12**: 37.

Diplodactylus maini Kluge, 1962

Diplodactylus maini Kluge, A.G. (1962). A new species of gekkonid lizard, genus *Diplodactylus* (Gray) from southern interior of Western Australia. *West. Aust. Nat.* **8**: 97–101 [97 figs 1–2]. Type data: holotype, WAM R12242, from 10 mi S of Queen Victoria Spring (33 mi N of Zanthus), W.A.

Distribution: SW coastal, W plateau, W.A. Ecology: low shrubland, open heath, open scrub, terrestrial, nocturnal; oviparous, arthropod-feeder.

Diplodactylus mcmillani Storr, 1978

Diplodactylus mcmillani Storr, G.M. (1978). Seven new gekkonid lizards from Western Australia. *Rec. West. Aust. Mus.* **6**: 337–352 [341 pl 2]. Type data: holotype, WAM R43230, from Mitchell Plateau, 14°57′S, 126°00′E, W.A.

Distribution: N coastal, W.A. Ecology: woodland, terrestrial, nocturnal; oviparous, (saxicoline), arthropod-feeder, in clumps of spinifex.

Diplodactylus michaelseni Werner, 1910

Diplodactylus michaelseni Werner, F. (1910). Reptilia (Geckonidae und Scincidae). *in* Michaelsen, W. & Hartmeyer, R. (eds.) *Die Fauna Südwest-Australiens.* Jena : Gustav Fischer 2: 451–493 [460 fig 3]. Type data: holotype, ZMB 21430, from Denham, Shark Bay, W.A.

Distribution: NW coastal, SW coastal, W.A. Ecology: open heath, terrestrial, nocturnal; oviparous, arthropod-feeder, in clumps of spinifex. Biological references: Kluge, A.G. (1967). Systematics, phylogeny, and zoogeography of the lizard genus *Diplodactylus* Gray (Gekkonidae). *Aust. J. Zool.* **15**: 1007–1108 (taxonomy); Storr, G.M. & Ford, J.R. (1967). Rediscovery and taxonomic status of the Western Australian gecko *Diplodactylus michaelseni*. *West. Aust. Nat.* **10**: 160–162; Rosenberg, H.I. & Russell, A.P. (1980). Structural and functional aspects of tail squirting: a unique defense mechanism of *Diplodactylus* (Reptilia : Gekkonidae). *Can. J. Zool.* **58**: 865–881.

Diplodactylus mitchelli Kluge, 1963

Diplodactylus mitchelli Kluge, A.G. (1963). Three new species of the gekkonid lizard *Diplodactylus* Gray from Australia. *Rec. S. Aust. Mus.* **14**: 545–553 [548 pl 34]. Type data: holotype, WAM R14823, from Coolawanyah Homestead, Pilbara Division, W.A.

Distribution: NW coastal, W.A. Ecology: hummock grassland, low open shrubland, terrestrial, nocturnal; oviparous, arthropod-feeder, (saxicoline). Biological references: Kluge, A.G. (1967). Systematics, phylogeny, and zoogeography of the lizard genus *Diplodactylus* Gray (Gekkonidae). *Aust. J. Zool.* **15**: 1007–1108 (taxonomy).

Diplodactylus ornatus Gray, 1845

Diplodactylus ornatus Gray, J.E. (1845). *Catalogue of the Specimens of Lizards in the Collection of the British Museum.* London : British Museum xxviii 289 pp. [149]. Type data: holotype, BMNH 1946.9.7.26, from Houtman's Abrolhos, W.A., probably error for Champion Bay, *vide* Storr, G.M. (1979). The *Diplodactylus vittatus* complex (Lacertilia : Gekkonidae) in Western Australia. *Rec. West. Aust. Mus.* **7**: 391–402.

Distribution: NW coastal, W.A. Ecology: hummock grassland, low open shrubland, open heath, open scrub, terrestrial, nocturnal; oviparous, arthropod-feeder. Biological references: Kluge, A.G. (1967). Systematics, phylogeny, and zoogeography of the lizard genus *Diplodactylus* Gray (Gekkonidae). *Aust. J. Zool.* **15**: 1007–1108 (taxonomy); Storr, G.M. (1979). The *Diplodactylus vittatus* complex (Lacertilia : Gekkonidae) in Western Australia. *Rec. West. Aust. Mus.* **7**: 391–402 (taxonomy).

Diplodactylus polyophthalmus Günther, 1867

***Diplodactylus polyophthalmus* (part.)** Günther, A. (1867). Additions to the knowledge of Australian reptiles and fishes. *Ann. Mag. Nat. Hist. (3)* **20**: 45–68 [49]. Type data: lectotype, BMNH 67.2.19.16 from Champion Bay, W.A.; for the paralectotype (BMNH unnumbered specimen, Nickol Bay, W.A.) see under *Diplodactylus stenodactylus* Boulenger, 1896; designation by Kluge, A.G. (1967). Systematics, phylogeny, and zoogeography of the lizard genus *Diplodactylus* Gray (Gekkonidae). *Aust. J. Zool.* **15**: 1007–1108.

Distribution: SW coastal, W.A. Ecology: open heath, woodland, open forest, terrestrial, nocturnal; oviparous, arthropod-feeder. Biological references: Kluge, A.G. (1967). Systematics, phylogeny, and zoogeography of the lizard genus *Diplodactylus* Gray (Gekkonidae). *Aust. J. Zool.* **15**: 1007–1108 (taxonomy); Storr, G.M. (1979). The *Diplodactylus vittatus* complex (Lacertilia : Gekkonidae) in Western Australia. *Rec. West. Aust. Mus.* **7**: 391–402 (taxonomy).

Diplodactylus pulcher (Steindachner, 1870)

Stenodactylopsis pulcher Steindachner, F. (1870). Herpetologische Notizen (II). Reptilien gesammelt Während einer Reise in Sengambien *Sber. Akad. Wiss. Wien* **62**: 326–349 [343 pl 2 figs 3–5]. Type data: lectotype, NHMW 17572:3, from Swan River, W.A., designation by Kluge, A.G. (1967). Systematics, phylogeny, and zoogeography of the lizard genus *Diplodactylus* Gray (Gekkonidae). *Aust. J. Zool.* **15**: 1007–1108.

Diplodactylus bilineatus Lucas, A.H.S. & Frost, C. (1903). Description of two new Australian lizards, *Varanus spenceri* and *Diplodactylus bilineatus. Proc. R. Soc. Vict. (ns)* **15**: 145–147 [146] [*non Diplodactylus bilineatus* Gray, 1845]. Type data: holotype, NMV D7570, from Minilya Station, 80 mi inland from Carnarvon (as Caernarvon), W.A.

Diplodactylus pulcher dorsotaeniata Pellegrin, J. (1909). Sur une collection de lézards de l'Australie Occidentale. *Bull. Soc. Zool. Fr.* **34**: 241–245 [242]. Type data: holotype, MNHP 09–190, from Birrigrin, W.A.

Diplodactylus pulcher dorsalis Werner, F. (1910). Reptilia (Geckonidae und Scincidae). *in* Michaelsen, W. & Hartmeyer, R. (eds.) *Die Fauna Südwest-Australiens.* Jena : Gustav Fischer 2: 451–493 [462]. Type data: holotype, not found, from Eradu, W.A.

Diplodactylus lucasi Fry, D.B. (1914). On a collection of reptiles and batrachians from Western Australia. *Rec. West. Aust. Mus. and Art Gall.* **1**: 174–210 [177] [*nom. nov. pro Diplodactylus bilineatus* Lucas and Frost, 1903].

Synonymy that of Cogger, H.G., this work.

Distribution: NW coastal, SW coastal, W plateau, W.A. Ecology: desert, hummock grassland, low open shrubland, open heath, tall shrubland, terrestrial, nocturnal; oviparous, arthropod-feeder. Biological references: Glauert, L. (1961). *A Handbook of the Lizards of Western Australia.* Perth : West. Aust. Naturalists' Club Handbook 6 100 pp. (natural history); Kluge, A.G. (1967). Systematics, phylogeny, and zoogeography of the lizard genus *Diplodactylus* Gray (Gekkonidae). *Aust. J. Zool.* **15**: 1007–1108 (taxonomy); Pianka, E.R. & Pianka, H.D. (1976). Comparative ecology of twelve species of nocturnal lizards (Gekkonidae) in the Western Australian desert. *Copeia* **1976**: 125–142 (ecology).

Diplodactylus rankini Storr, 1979

Diplodactylus rankini Storr, G.M. (1979). Five new lizards from Western Australia. *Rec. West Aust. Mus.* **8**: 134–142 [134 pl 1]. Type data: holotype, WAM R61474, from mouth of Yardie Creek, W.A., in 22°20′S, 113°48′E.

Distribution: NW coastal, W.A. Ecology: hummock grassland, low open shrubland, terrestrial, nocturnal; (arboreal), oviparous, arthropod-feeder, in clumps of spinifex.

Diplodactylus savagei Kluge, 1963

Diplodactylus savagei Kluge, A.G. (1963). Three new species of the gekkonid lizard *Diplodactylus* Gray from Australia. *Rec. S. Aust. Mus.* **14**: 545–553 [550 pl 35]. Type data: holotype, WAM R14369, from Marble Bar, Pilbara Division, W.A.

Distribution: NW coastal, W.A. Ecology: hummock grassland, low open shrubland, terrestrial, nocturnal; oviparous, arthropod-feeder, (saxicoline). Biological references: Kluge, A.G. (1967). Systematics, phylogeny, and zoogeography of the lizard genus *Diplodactylus* Gray (Gekkonidae). *Aust. J. Zool.* **15**: 1007–1108 (taxonomy).

Diplodactylus spinigerus Gray, 1842

Diplodactylus spinigerus Gray, J.E. (1842). Description of some hitherto unrecorded species of Australian reptiles and batrachians. *in* Gray, J.E. (ed.) *Zoological Miscellany.* London : Treuttel, Würtz & Co. pp. 51–57 [53]. Type data: syntypes, from Houtman's Abrolhos, W.A. (as Van Diemen's Land), see Gray, J.E. (1845). *Catalogue of the Specimens of Lizards in the Collection of the British Museum.* London : British Museum xxviii 289 pp., presumed lost; BMNH 64.10.6.7 erroneously cited as holotype by Kluge, A.G. (1967). Systematics, phylogeny, and zoogeography of the lizard genus *Diplodactylus* Gray (Gekkonidae). *Aust. J. Zool.* **15**: 1007–1108; BMNH xxi.3b and xxi.5a erroneously cited by Boulenger, G.A. (1885). *Catalogue of the Lizards in the British Museum (Natural History).* 1. London : British Museum xii 436 pp. pls 32 as syntypes.

Distribution: NW coastal, SW coastal, W plateau, W.A. Ecology: open heath, woodland, open forest, arboreal, nocturnal; oviparous, arthropod-feeder. Biological references: Brazenor, C.W. (1951). On the Victorian species of tuberculated *Diplodactylus. Mem. Natl. Mus. Vict.* **17**: 215–222 (taxonomy); Kluge, A.G. (1967). Systematics, phylogeny, and zoogeography of the lizard genus *Diplodactylus* Gray (Gekkonidae). *Aust. J. Zool.* **15**: 1007–1108 (taxonomy); Rosenberg, H.I. & Russell, A.P. (1980). Structural and functional aspects of tail squirting: a unique defense mechanism of *Diplodactylus* (Reptilia : Gekkonidae). *Can. J. Zool.* **58**: 865–881; Russell, A.P. & Rosenberg, H.I. (1981). Self-grooming in *Diplodactylus spinigerus* (Reptilia : Gekkonidae), with a brief review of such behaviour in reptiles. *Can. J. Zool.* **59**: 564–566.

Diplodactylus squarrosus Kluge, 1962

Diplodactylus squarrosus Kluge, A.G. (1962). A new species of gekkonid lizard, genus *Diplodactylus*, from the Carnarvon region, Western Australia. *West. Aust. Nat.* **8**: 73–75 [73 figs 1–2]. Type data: holotype, WAM R13805, from Hamelin Pool, Shark Bay, W.A.

Distribution: NW coastal, W plateau, W.A. Ecology: hummock grassland, tussock grassland, tall shrubland, terrestrial, nocturnal; oviparous, arthropod-feeder. Biological references: Kluge, A.G. (1967). Systematics, phylogeny, and zoogeography of the lizard genus *Diplodactylus* Gray (Gekkonidae). *Aust. J. Zool.* **15**: 1007–1108 (taxonomy).

Diplodactylus steindachneri Boulenger, 1885

Diplodactylus steindachneri Boulenger, G.A. (1885). *Catalogue of the Lizards in the British Museum (Natural History).* 1. London : British Museum xii 436 pp. pls 32 [102 pl 8 fig 5]. Type data: holotype, BMNH 76.3.4.58, from Peak Downs, Qld. (as Sydney, New South Wales), see BMNH records.

Diplodactylus stenurus Werner, F. (1909). Über neue oder seltene Reptilien des Musée Royal d'Histoire Naturelle de Belgique in Brussell. *Zool. Jb. Abt. Syst.* **28**: 263–288 [267]. Type data: holotype, RIB Typ. cat. 2022, Gen. cat. 3981, from Qld.

Synonymy that of Cogger, H.G., this work.

Distribution: NE coastal, N Gulf, Murray-Darling basin, Bulloo River basin, Lake Eyre basin, (S.A.), N.S.W., Qld. Ecology: open scrub, low open woodland, low woodland, woodland, open forest, terrestrial, nocturnal; oviparous, arthropod-feeder. Biological references: Kluge, A.G. (1967). Systematics, phylogeny, and zoogeography of the lizard genus *Diplodactylus* Gray (Gekkonidae). *Aust. J. Zool.* **15**: 1007–1108 (taxonomy).

Diplodactylus stenodactylus Boulenger, 1896

*[**Diplodactylus polyophthalmus*** **(part.)***]* Günther, A. (1867). Additions to the knowledge of Australian reptiles and fishes. *Ann. Mag. Nat. Hist. (3)* **20**: 45–68 [49]. Type data: paralectotype, BMNH unnumbered specimen, from Nickol Bay; for the lectotype (BMNH 67.2.19.16, Champion Bay) see under *Diplodactylus polyophthalmus* Günther, 1867.

Diplodactylus stenodactylus Boulenger, G.A. (1896). Description of four new lizards from Roebuck Bay, north Western Australia obtained by Dr. Dahl, for the Christiania Museum. *Ann. Mag. Nat. Hist. (6)* **18**: 232–235 [232]. Type data: holotype, UZMO K2001, from Roebuck Bay, W.A.

Diplodactylus woodwardi Fry, D.B. (1914). On a collection of reptiles and batrachians from Western Australia. *Rec. West. Aust. Mus. and Art Gall.* **1**: 174–210 [175 pl 27 fig 1]. Type data: holotype, WAM R14370, from W.A.

Synonymy that of Cogger, H.G., this work.

Distribution: N Gulf, N coastal, NW coastal, W plateau, (Bulloo River basin), Lake Eyre basin, S.A., N.S.W., Qld., W.A., N.T. Ecology: desert, tussock grassland, low open shrubland, low shrubland, low woodland, tall shrubland, terrestrial, nocturnal; oviparous, arthropod-feeder. Biological references: Kluge, A.G. (1967). Systematics, phylogeny, and zoogeography of the lizard genus *Diplodactylus* Gray (Gekkonidae). *Aust. J. Zool.* **15**: 1007–1108 (taxonomy); Pianka, E.R. & Pianka, H.D. (1976). Comparative ecology of twelve species of nocturnal lizards (Gekkonidae) in the Western Australian desert. *Copeia* **1976**: 125–142 (ecology).

Diplodactylus strophurus (Duméril and Bibron, 1836)

Phyllodactylus strophurus Duméril, A.M.C. & Bibron, G. (1836). *Erpétologie Générale* ou Histoire Naturelle Complète des Reptiles. Paris : Roret 3: iv 517 pp. [397 pl 32 fig 1]. Type data: neotype, WAM R6535, from Yalgoo, W.A. (original description cites Shark Bay, W.A.), designation by Kluge, A.G. (1967). Systematics, phylogeny, and zoogeography of the lizard genus *Diplodactylus* Gray (Gekkonidae). *Aust. J. Zool.* **15**: 1007–1108.

Discodactylus (Strophurus) dumerilii Fitzinger, L.J. (1843). *Systema Reptilium.* Vienna : Braümüller u. Seidel vi 106 pp. [96] [*nom. substit. pro Phyllodactylus strophurus* Duméril and Bibron, 1836].

Synonymy that of Cogger, H.G., this work.

Distribution: NW coastal, W plateau, W.A. Ecology: desert, low open shrubland, tall shrubland, (terrestrial), arboreal, nocturnal; oviparous, arthropod-feeder. Biological references: Kluge, A.G. (1967). Systematics, phylogeny, and zoogeography of the lizard genus *Diplodactylus* Gray (Gekkonidae). *Aust. J. Zool.* **15**: 1007–1108 (taxonomy); Pianka, E.R. & Pianka, H.D. (1976). Comparative ecology of twelve species of nocturnal lizards (Gekkonidae) in the Western Australian desert. *Copeia* **1976**: 125–142 (ecology); Rosenberg, H.I. & Russell, A.P. (1980). Structural and functional aspects of tail squirting: a unique defense mechanism of *Diplodactylus* (Reptilia : Gekkonidae). *Can. J. Zool.* **58**: 865–881.

Diplodactylus taeniatus (Lönnberg and Andersson, 1913)

Oedurella taeniata Lönnberg, E. & Andersson, G.L. (1913). Results of Dr. E. Mjöberg's Swedish Scientific Expeditions to Australia 1910–1913. III. Reptiles. *K. Svenska Vetensk.-Akad. Handl.* **52**(3): 1–17 [5 figs 1–3]. Type data: holotype, NHRM 2999, from near Broome, W.A., restricted to Streeters Station, near Broome by Kluge, A.G. (1967). Systematics, phylogeny, and zoogeography of the lizard genus *Diplodactylus* Gray (Gekkonidae). *Aust. J. Zool.* **15**: 1007–1108.

Distribution: N Gulf, N coastal, NW coastal, W plateau, Lake Eyre basin, Qld., W.A., N.T. Ecology: desert, hummock grassland, low open shrubland, low open woodland, low woodland, tall shrubland, terrestrial, nocturnal; oviparous, arthropod-feeder, in clumps of spinifex. Biological references: Kluge, A.G. (1967). Systematics, phylogeny, and zoogeography of the lizard genus *Diplodactylus* Gray (Gekkonidae). *Aust. J. Zool.* **15**: 1007–1108 (taxonomy); Storr, G.M. & Ford, J.R. (1967). Rediscovery and taxonomic status of the Western Australian gecko *Diplodactylus michaelseni. West. Aust. Nat.* **10**: 160–162 (taxonomy).

Diplodactylus taenicauda De Vis, 1886

Diplodactylus taenicauda De Vis, C.W. (1886). On certain geckos in the Queensland Museum. *Proc. Linn. Soc. N.S.W. (2)* **1**: 168–170 [169] [1887 on title page]. Type data: lectotype, QM J11215, from Chinchilla, Qld., designation by Kluge, A.G. (1967). Systematics, phylogeny, and zoogeography of the lizard genus *Diplodactylus* Gray (Gekkonidae). *Aust. J. Zool.* **15**: 1007–1108; Covacevich, J. (1971). Amphibian nd reptile type specimens in the Queensland Museum. *Mem. Qd. Mus.* **16**: 49–68 could locate only the original syntype QM J232.

Diplodactylus taeniocauda Boulenger, G.A. (1887). *Catalogue of the Lizards in the British Museum (Natural History).* 3. London : British Museum xii 575 pp. pls 40 [483] [invalid emend. *pro Diplodactylus taenicauda* De Vis, 1886].

Synonymy that of Cogger, H.G., this work.

Distribution: NE coastal, Murray-Darling basin, Qld. Ecology: open scrub, low open woodland, woodland, arboreal, nocturnal; oviparous, arthropod-feeder. Biological references: Kluge, A.G. (1967). Systematics, phylogeny, and zoogeography of the lizard genus *Diplodactylus* Gray (Gekkonidae). *Aust. J. Zool.* **15**: 1007–1108 (taxonomy); Dale, D.F. (1973). *Forty Queensland Lizards.* Brisbane : Queensland Museum 64 pp. (natural history).

Diplodactylus tessellatus (Günther, 1875)

Stenodactylopsis tessellatus Günther, A. (1875). Reptiles. A list of saurians of Australia and New Zealand. pp. 9–19 *Zoology of the Voyage of H.M.S.* Erebus *and* Terror. Vol. 2 London : E.W. Janson [16]. Type data: holotype, BMNH 70.6.26.60, from Australia, restriction to Brewarrina, N.S.W. proposed by Kluge, A.G. (1967). Systematics, phylogeny, and zoogeography of the lizard genus *Diplodactylus* Gray (Gekkonidae). *Aust. J. Zool.* **15**: 1007–1108.

Diplodactylus pachyurus Werner, F. (1909). Über neue oder seltene Reptilien des Musée Royal d'Histoire Naturelle de Belgique in Brussell. *Zool. Jb. Abt. Syst.* **28**: 263–288 [267]. Type data: holotype, RIB Typ.cat.2021, Gen.cat.4515, from Australia.

Synonymy that of Cogger, H.G., this work.

Distribution: W plateau, Lake Eyre basin, Murray-Darling basin, Bulloo River basin, S.A., Vic., N.S.W., Qld., N.T. Ecology: desert, (hummock grassland), low shrubland, low open woodland, low woodland, tall shrubland, terrestrial, nocturnal; oviparous, arthropod-feeder. Biological references: Kluge, A.G. (1967). Systematics, phylogeny, and zoogeography of the lizard genus *Diplodactylus* Gray (Gekkonidae). *Aust. J. Zool.* **15**: 1007–1108 (taxonomy).

Diplodactylus vittatus Gray, 1832

Diplodactylus vittatus Gray, J.E. (1832). (Three new animals brought from New Holland by Mr Cunningham). *Proc. Zool. Soc. Lond.* **1832**: 39–40 [40]. Type data: holotype, BMNH 1946.9.7.43, from Australia (as Nova Hollandia); restriction of type locality to Darling Range, W.A. by Kluge, A.G. (1967). Systematics, phylogeny, and zoogeography of the lizard genus *Diplodactylus* Gray (Gekkonidae). *Aust. J. Zool.* **15**: 1007–1108, shown to be invalid by Storr, G.M. (1979). The *Diplodactylus vittatus* complex (Lacertilia : Gekkonidae) in Western Australia. *Rec. West. Aust. Mus.* **7**: 391–402.

*[**Phyllodactylus lesueurii** (part.)]* Duméril, A.M.C. & Bibron, G. (1836). *Erpétologie Générale* ou Histoire Naturelle Complète des Reptiles. Paris : Roret 3: iv 517 pp. [392] [for the lectotype MNHP 268 and remaining paralectotypes, see under *Oedura lesueurii* (Duméril and Bibron, 1836)]. Type data: paralectotype, MNHP 6726 (smaller of the two specimens), from Port Jackson, N.S.W.

Diplodactylus furcosus Peters, W. (1863). Übersicht der von Hrn. Richard Schomburgk an das zoologische Museum eingesandten Amphibien, aus Buchsfelde bei Adelaide in Südaustralien. *Mber. K. Preuss. Akad. Wiss. Berl.* **1863**: 228–236 [229] [1864 on title page]. Type data: lectotype, ZMB 39231 (*ex* ZMB 4711), from Loos, 4.5 km W Gawler (as Buchsfelde), S.A., see Praite, R. & Tolley, J.C. (1970). *Place Names of South Australia.* Adelaide : Rigby 203 pp., designation by Kluge, A.G. (1967). Systematics, phylogeny, and zoogeography of the lizard genus *Diplodactylus* Gray (Gekkonidae). *Aust. J. Zool.* **15**: 1007–1108.

Phyllodactylus barbouri Angel, M.F. (1936). Deux gekkos nouveaux de Madagascar appartenant au genre *Phyllodactylus. Bull. Soc. Zool. Fr.* **61**: 508–511 [508]. Type data: holotype, MNHP 5600, from unknown locality (as Madagascar, *errore*).

Synonymy that of Cogger, H.G., this work.

Distribution: SE coastal, NE coastal, S Gulfs, Murray-Darling basin, (Bulloo River basin), S.A., Vic., N.S.W., Qld., A.C.T. Ecology: low open shrubland, low shrubland, open heath, open scrub, low open woodland, low woodland, woodland, open forest, tall forest, terrestrial, nocturnal; oviparous, arthropod-feeder, saxicoline. Biological references: Kluge, A.G. (1967). Systematics, phylogeny, and zoogeography of the lizard genus *Diplodactylus* Gray (Gekkonidae). *Aust. J. Zool.* **15**: 1007–1108 (taxonomy); King, M. (1977). Chromosomal and morphometric variation in the gecko *Diplodactylus vittatus* Gray. *Aust. J. Zool.* **25**: 43–57; Storr, G.M. (1979). The *Diplodactylus vittatus* complex (Lacertilia : Gekkonidae) in Western Australia. *Rec. West. Aust. Mus.* **7**: 391–402 (taxonomy); Bustard, H.R. (1968). Temperature dependent activity in the Australian gecko *Diplodactylus vittatus. Copeia* **1968**: 606–612.

Diplodactylus williamsi Kluge, 1963

Diplodactylus williamsi Kluge, A.G. (1963). A new species of gekkonid lizard, genus *Diplodactylus* Gray, from eastern Australia. *Proc. Linn. Soc. N.S.W.* **88**: 230–234 [230 pl 14 text fig]. Type data: holotype, AM R14987, from Warrumbungle Mts., N.S.W.

Distribution: NE coastal, N Gulf, Murray-Darling basin, N.S.W., Qld. Ecology: low open shrubland, open scrub, low open woodland, low woodland, woodland, open forest, arboreal, nocturnal; oviparous, arthropod-feeder. Biological references: Kluge, A.G. (1967). Systematics, phylogeny, and zoogeography of the lizard genus *Diplodactylus* Gray (Gekkonidae). *Aust. J. Zool.* **15**: 1007–1108 (taxonomy); Bustard, H.R. (1969). The ecology of the Australian geckoes *Diplodactylus williamsi* and *Gehyra australis* in northern New South Wales. *Proc. K. Ned. Akad. Wet. (C)* **72**: 451–477; Bustard, H.R. (1964). Defensive behaviour shown by Australian geckos, genus *Diplodactylus. Herpetologica* **20**: 198–200 (tail squirting).

Diplodactylus wombeyi Storr, 1978

Diplodactylus wombeyi Storr, G.M. (1978). Seven new gekkonid lizards from Western Australia. *Rec. West. Aust. Mus.* **6**: 337–352 [344 pl 3]. Type data: holotype, WAM R36747, from Cockeraga River, 22°05′S, 118°48′E, W.A.

Distribution: NW coastal, W.A. Ecology: hummock grassland, low open shrubland, terrestrial, nocturnal; oviparous, arthropod-feeder, saxicoline.

Gehyra Gray, 1834

Gehyra Gray, J.E. (1834). (Specimens were exhibited of several reptiles, which were accompanied by notes by Mr Gray. A new genus of Geckotidae). *Proc. Zool. Soc. Lond.* **1834**: 99–101 [100]. Type species *Gehyra pacifica* Gray, 1834 by monotypy.

Peropus Wiegmann, A.F.A. (1835). Amphibien. *in* Meyen, F.J.F. (ed.) Beiträge zur Zoologie, Gesammelt auf einer Reise um die Erde. *Nova Acta Acad. Leop.-Carol.* **17**: 185–268 [238]. Type species *Hemidactylus (Peropus) mutilatus* Wiegmann, 1835 by monotypy.

Peroplus Gray, J.E. (1842). Description of some new species of reptiles, chiefly from the British Museum collection. *in* Gray, J.E. (ed.) *Zoological Miscellany* London : Treuttel, Würtz & Co. pp. 57–59 [57] [*errore pro Peropus* Wiegmann, 1835].

Phyria Gray, J.E. (1842). Description of some hitherto unrecorded species of Australian reptiles and batrachians *in* Gray, J.E. (ed.) *Zoological Miscellany.* London : Treuttel, Würtz & Co. pp. 51–57 [53]. Type species *Phyria punctulata* Gray, 1842 by monotypy.

Perodactylus Fitzinger, L.J. (1843). *Systema Reptilium.* Vienna : Braümüller u. Seidel vi 106 pp. [19, 103] [described as subgenus of *Peropus* Wiegmann, 1835]. Type species *Hemidactylus oualensis* Duméril and Bibron, 1836 by original designation.

Dactyloperus Fitzinger, L.J. (1843). *Systema Reptilium.* Vienna : Braümüller u. Seidel vi 106 pp. [19, 103] [described as subgenus of *Peropus* Wiegmann, 1835]. Type species *Hemidactylus variegatus* Duméril and Bibron, 1836 by original designation.

Peripia Gray, J.E. (1845). *Catalogue of the Specimens of Lizards in the Collection of the British Museum.* London : British Museum xxviii 289 pp. [158]. Type species *Hemidactylus peronii* Duméril and Bibron, 1836 [= *Gehyra mutilata* (Wiegmann, 1835)] by subsequent designation, see Smith, M.A. (1935). *The Fauna of British India; including Ceylon and Burma.* Reptilia and Amphibia. Vol. II. Sauria. London : Taylor & Francis xiv 440 pp.

Chalinocnemis Dugès, A. (1883). Spasmocnemis, Chalinocnemis. *La Naturaleza, México* **6**: 309–312 [312]. Type species *Hemidactylus navarri* Dugès, 1883 [= *Gehyra mutilata* (Wiegmann, 1835)] by original designation. Compiled from secondary source: Smith, H.M. & Necker, W.L. (1943). Alfredo Dugès' types of Mexican reptiles and amphibians. *An. Esc. Nac. Cienc. Biol. Méx.* **3**: 179–219.

Spasmocnemis Dugès, A. (1883). Spasmocnemis, Chalinocnemis. *La Naturaleza, México* **6**: 309–312 [312]. Type species *Hemidactylus navarri* Dugès, 1883 [= *Gehyra mutilata* (Wiegmann, 1835)] by original designation. Compiled from secondary source: Smith, H.M. & Necker, W.L. (1943). Alfredo Dugès' types of Mexican reptiles and amphibians. *An. Esc. Nac. Cienc. Biol. Méx.* **3**: 179–219.

Pepopus Okada, Y. (1936). Studies on the lizards of Japan. Contribution I. Gekkonidae. *Sci. Rep. Tokyo Bunrika Daig. (b)* **2**: 233–289 [280] [*errore pro Peropus* Wiegmann, 1835].

Gehyla Okada, Y. (1938). *A Catalogue of Vertebrates of Japan.* pp. 97–113 Reptilia and Amphibia. Tokyo 412 pp. [105] [*errore pro Gehyra* Gray, 1834].

Synonymy that of Cogger, H.G., this work; Wermuth, H. (1965). Liste der rezenten Amphibien und Reptilien; Gekkonidae, Pygopodidae, Xantusiidae. *Das Tierreich* **80**: i–xxii 1–246.

This group is also found in Madagascar, Asia, Indo-Malayan Archipelago, New Guinea and Pacific islands; introduced to Mexico.

Gehyra australis Gray, 1845

Phyria punctulata Gray, J.E. (1842). Description of some hitherto unrecorded species of Australian reptiles and batrachians. *in* Gray, J.E. (ed.) *Zoological Miscellany.* London : Treuttel, Würtz & Co. pp. 51–57 [53] [*nom. oblit.*]. Type data: presumed lost, from Port Essington, N.T.

***Gehyra australis* (part.)** Gray, J.E. (1845). *Catalogue of the Specimens of Lizards in the Collection of the British Museum.* London : British Museum xxviii 289 pp. [163]. Type data: lectotype, BMNH xxii.55b, from Port Essington, N.T.; for the paralectotype BMNH xxii.55a (Swan River) see under *Gehyra variegata* (Duméril and Bibron, 1836), designation by Cogger, H.G., this work.

Synonymy that of Cogger, H.G., this work.

Distribution: NE coastal, N Gulf, N coastal, Murray-Darling basin, Bulloo River basin, Lake Eyre basin, S.A., N.S.W., Qld., W.A., N.T. Ecology: low open shrubland, tall open shrubland, open scrub, low open woodland, low woodland, woodland, open forest, tall forest, under bark, tall shrubland, arboreal, nocturnal; oviparous, arthropod-feeder. Biological references: Bustard, H.R. (1965). The systematic status of the Australian geckos *Gehyra variegata* (Duméril and Bibron, 1836) and *Gehyra australis* Gray, 1845. *Herpetologica* **20**: 259–272; Mitchell, F.J. (1965). Australian geckos assigned to the genus *Gehyra* Gray. (Reptilia : Gekkonidae). *Senckenberg. Biol.* **46**: 287–319 (taxonomy); Bustard, H.R. (1969). The ecology of the Australian geckos *Diplodactylus williamsi* and *Gehyra australis* in northern New South Wales. *Proc. K. Ned. Akad. Wet. (C)* **72**: 451–477; King, M. (1982). Karyotypic evolution in *Gehyra* (Gekkonidae : Reptilia) II. A new species from the Alligator Rivers region of northern Australia. *Aust. J. Zool.* **30**: 93–101.

Gehyra baliola (Duméril and Duméril, 1851)

Hemidactylus baliolus Duméril, A.M.C. & Duméril, A. (1851). *Catalogue Méthodique de la Collection des Reptiles.* Paris : Mus. Hist. Nat. iv 224 pp. [38]. Type data: holotype, MNHP 6574, from New Guinea.

Peripia torresiana Günther, A. (1877). Descriptions of three new species of lizards from islands of Torres Strait. *Ann. Mag. Nat. Hist. (4)* **19**: 413–415 [415]. Type data: syntypes, BMNH 77.3.3.9–11, from "Somerset [Cape York, Qld.] and ... the islands of Torres Straits" [Qld.].

Peripia marmorata Macleay, W. (1877). The lizards of the *Chevert* Expedition. Second paper. *Proc. Linn. Soc. N.S.W.* **2**: 97–104 [99] [1878 on title page]. Type data: holotype, AM R29943, from Mawatta, Binaturi River, near Daru (as Katow), Papua New Guinea.

Peripia brevicaudis Macleay, W. (1877). The lizards of the *Chevert* Expedition. Second paper. *Proc. Linn. Soc. N.S.W.* **2**: 97–104 [99] [1878 on title page]. Type data: lectotype, AM R29947 (*ex* MMUS MR931), from Darnley Is., Torres Strait, Qld., designation by Kluge, A.G. (1963). The systematic status of certain Australian and New Guinean gekkonid lizards. *Mem. Qd. Mus.* **14**: 77–86.

Synonymy that of Cogger, H.G., this work; for synonymy of *Peripia marmorata* and *P. brevicaudis* see Kluge, A.G. (1963). The systematic status of certain Australian and New Guinean gekkonid lizards. *Mem. Qd. Mus.* **14**: 77–86.

Distribution: NE coastal, N Gulf, Qld. Ecology: woodland, open forest, closed forest, mangrove, arboreal, nocturnal; oviparous, arthropod-feeder.

Gehyra catenata Low, 1979

Gehyra catenata Low, T. (1979). A new species of gecko, genus *Gehyra* (Reptilia : Gekkonidae) from Queensland. *Victorian Nat.* **96**: 190–196 [190 figs 1,2b pl 1]. Type data: holotype, QM J15633, from Batheaston Station, Qld., in 22°26'S, 148°47'E.

Distribution: NE coastal, Murray-Darling basin, Qld. Ecology: open scrub, woodland, arboreal, nocturnal; oviparous, arthropod-feeder.

Gehyra fenestra Mitchell, 1965

Gehyra fenestra Mitchell, F.J. (1965). Australian geckos assigned to the enus *Gehyra* Gray. (Reptilia : Gekkonidae). *Senckenberg. Biol.* **46**: 287–319 [307]. Type data: holotype, WAM R20199, from Mt. Herbert, W.A.

Distribution: NW coastal, W.A. Ecology: hummock grassland low open shrubland, terrestrial, nocturnal; oviparous, saxicoline, arthropofeeder.

Gehyra mutilata (Wiegmann, 1835)

Hemidactylus (Peropus) mutilatus Wiegmann, A.F.A. (1835). Amphibien. *in* Meyen, F.J.F. (ed.) Beiträge zur Zoologie, Gesammelt auf einer Reise um die Erde. *Nova Acta Acad. Leop.-Carol.* **17**: 185–268 [238]. Type data: holotype (probable), ?ZMB, from Manila, Philippines.

Hemidactylus peronii Duméril, A.M.C. & Bibron, G. (1836). *Erpétologie Générale* ou Histoire Naturelle Complète des Reptiles. Paris : Roret 3: iv 517 pp. [352 pl 30 fig 1]. Type data: syntypes (3), MNHP 6557, MNHP 6582, from Mauritius (as Île de France).

Hemidactylus platurus Bleeker, P. (1858). Der heer Bleeker doet verslag van de verzameling visschen van Amboina, der Vereeniging aangeboden door het lid den heer Schröder. *Natuurk. Tijdschr. Ned.-Indie (4)* **16**: 28–31 [31]. Type data: syntypes (probable), whereabouts unknown, from Java, Indonesia, restricted by Smith, M.A. (1935). *The Fauna of British India; including Ceylon and Burma.* Reptilia and Amphibia. Vol. II. Sauria. London : Taylor & Francis xiv 440 pp.

Dactyloperus insulensis Girard, C. (1858). Descriptions of some new reptiles, collected by the United States Exploring Expedition, under the command of Capt. Charles Wilkes, U.S.N. Fourth Part. Including the species of saurians exotic to North America. *Proc. Acad. Nat. Sci. Philad.* **1857**: 195–199 [197]. Type data: syntypes, ?USNM 21219, whereabouts of remaining syntypes unknown, from Hawaiian Ils. (as Sandwich Ils.), U.S.A.

Gecko pardus Tytler, R.C. (1865). Observations on a few species of geckos alive in the possession of the author. *J. Asiat. Soc. Beng.* **33**: 535–548 [547]. Type data: holotype (probable), whereabouts unknown, from Moulmein, Burma, restricted by Smith, M.A. (1935). *The Fauna of British India; including Ceylon and Burma.* Reptilia and Amphibia. Vol. II. Sauria. London : Taylor & Francis xiv 440 pp.

Gecko harrieti Tytler, R.C. (1865). Observations on a few species of geckos alive in the possession of the author. *J. Asiat. Soc. Beng.* **33**: 535–548 [548]. Type data: syntypes (probable), whereabouts unknown, from Port Blair and Umgebung, Andaman Ils.

Peropus packardii Cope, E.D. (1869). Observations on reptiles of the Old World. Article II. *Proc. Acad. Nat. Sci. Philad.* **1868**: 316–323 [319]. Type data: syntypes, MCZ 5470, whereabouts of remaining syntypes unknown, from Penang, Malaysia.

Hemidactylus navarri Dugès, A. (1883). Spasmocnemis, Chalinocnemis. *La Naturaleza, Mexico* **6**: 309–312 [309, 311 pl 7]. Type data: holotype, MDUG, from San Blas, Nayarit, Mexico, restricted by Smith, H.M. & Taylor, E.H. (1950). An annotated checklist and key to the reptiles of Mexico exclusive of the snakes. *Bull. U.S. Natl. Mus.* **199**: i–253. Compiled from secondary source: Smith, H.M. & Necker, W.L. (1943). Alfredo Dugès' types of Mexican reptiles and amphibians. *An. Esc. Nac. Cienc. Biol. Mex.* **3**: 179–219.

Gehyra beebei Annandale, N. (1913). Some new and interesting Batrachia and lizards from India, Ceylon and Borneo. *Rec. Ind. Mus.* **9**: 301–307 [306 pl 15 fig 2]. Type data: holotype, IM 17402, from Kapit, Sarawak, Borneo.

Gehyra insulanus Underwood, G. (1954). On the classificaiton and evolution of geckos. *Proc. Linn. Soc. Lond.* **124**: 469–492 [478] [*errore pro Dactyloperus insulensis* Girard, 1858].

Synonymy that of Wermuth, H. (1965). Liste der rezenten Amphibien und Reptilien; Gekkonidae, Pygopodidae, Xantusiidae. *Das Tierreich* **80**: i–xxii 1–246 [35].

Distribution: Christmas Is., Cocos (Keeling) Ils.; ranges extralimitally from Africa through Asia and the Pacific to the Americas. Ecology: closed forest, arboreal, nocturnal, peridomestic; secondary growth following phosphate mining (Christmas Is.), coconut plantations (Cocos (Keeling) Ils.), oviparous, arthropod-feeder.

Gehyra nana Storr, 1978

Gehyra nana Storr, G.M. (1978). Seven new gekkonid lizards from Western Australia. *Rec. West. Aust. Mus.* **6**: 337–352 [347 pl 5]. Type data: holotype, WAM R28214, from King Edward River, "presumably 14°52'S, 126°12'E", W.A.

Distribution: NE coastal, N Gulf, N coastal, Qld., W.A., N.T. Ecology: terrestrial, nocturnal; oviparous, saxicoline, arthropod-feeder.

Gehyra oceanica (Lesson, 1830)

Gecko oceanicus Lesson, R.P. (1830). Observations générales sur les reptiles recueillis dans le voyage. *in* Duperrey, L.I. (ed.). *Voyage autour du monde, executé par ordre du Roi, sur la corvette de Sa Majesté,* La Coquille, *pendant les années 1822, 1823, 1824 et 1825.* Zoologie 2(1): 1–65 Paris : Arthus Bertrand [livraison 15 pl 2 fig 3 (1830); 42 (1831)] [Sherborn, C.D. & Woodward, B.B. (1901). Notes on the dates of publication of the natural history portions of some French voyages. Part 1 *Ann. Mag. Nat. Hist. (7)* **7**: 388–392 (date of livraison); Anon. (1830). [Review of] *Voyage autour du monde.* Livraisons 14–17. *Bull. Sci. Nat. Géol.* (Férussac, ed.) **21**: 309–311 (identifies livraison in which plate appears); Sherborn, C.D. &

Woodward, B.B. (1906). Notes on the dates of publication of the natural history portions of some French voyages. "Voyage autour du Monde .... sur .... la Coquille pendant .... 1822-25 .... Par L.J. Duperrey, "etc. A correction. *Ann. Mag. Nat. Hist. (7)* **17**: 335–336 (date of text)]. Type data: ? as for *Hemidactylus oualensis* Duméril and Bibron, 1836, from Tahiti, Borabora and islands of the Pacific Ocean.

Gehyra pacifica Gray, J.E. (1834). (Specimens were exhibited of several reptiles, which were accompanied by notes by Mr Gray. A new genus of Geckotidae). *Proc. Zool. Soc. Lond.* **1834**: 99–101 [100]. Type data: holotype, presumed lost, from islands of the Pacific Ocean.

Hemidactylus oualensis Duméril, A.M.C. & Bibron, G. (1836). *Erpétologie Générale* ou Histoire Naturelle Complète des Reptiles. Paris : Roret 3: iv 517 pp. [350 pl 28 fig 7]. Type data: syntypes, MNHP 1776, MNHP 6608 (Kusiae), MNHP 2293 (Tongatapu), from Kusiae (as Oualan) Is., Caroline Ils., Tahiti, Vanicoro and Tongatapu (as Tonga-Tabou), Tonga.

Gehyra gularis Gray, J.E. (1842). Description of some new species of reptiles, chiefly from the British Museum collection. *in* Gray, J.E. (ed.) *Zoological Miscellany.* London : Treuttel, Würtz & Co. pp. 57–59 [58] [*nom. nud.*].

Synonymy that of Cogger, H.G., this work.

Distribution: Qld; Torres Strait Ils. Ecology: closed forest, mangrove, arboreal, nocturnal; oviparous, arthropod-feeder. Biological references: Chrapliwy, P.S., Smith, H.M. & Grant, C. (1961). Systematic status of the geckonid lizard genera *Gehyra*, *Peropus*, *Hoplodactylus* and *Naultinus*. *Herpetologica* **17**: 5–12; Schwaner, T.D. (1980). Reproductive biology of lizards on the American Samoan islands. *Occ. Pap. Mus. Nat. Hist. Univ. Kansas* **86**: 1–53.

Gehyra pamela King, 1982

Gehyra pamela King, M. (1982). Karyotypic evolution in *Gehyra* (Gekkonidae : Reptilia). A new species from the Alligator River region of northern Australia. *Aust. J. Zool.* **30**: 93–101 [97 fig 4]. Type data: holotype, NTM R8656, from Nourlangie Rock, N.T.

Distribution: N coastal, N.T. Ecology: tall open shrubland, woodland, open forest, nocturnal; oviparous, saxicoline, arthropod-feeder.

Gehyra pilbara Mitchell, 1965

Gehyra pilbara Mitchell, F.J. (1965). Australian geckos assigned to the genus *Gehyra* Gray. (Reptilia : Gekkonidae). *Senckenberg. Biol.* **46**: 287–319 [303]. Type data: holotype, WAM R20046, from Pilbara Plateau, W.A.

Distribution: N coastal, NW coastal, W plateau, W.A., N.T. Ecology: desert, hummock grassland, low open shrubland, terrestrial, nocturnal; oviparous, saxicoline, arthropod-feeder.

Gehyra punctata (Fry, 1914)

Peropus variegatus punctatus Fry, D.B. (1914). On a collection of reptiles and batrachians from Western Australia. *Rec. West. Aust. Mus. and Art Gall.* **1**: 174–210 [178]. Type data: holotype, from Strelly River, Pilbara District, W.A., presumed lost, see Mitchell, F.J. (1965). Australian geckos assigned to the genus *Gehyra* Gray. (Reptilia : Gekkonidae). *Senckenberg. Biol.* **46**: 287–319.

Distribution: NW coastal, W plateau, Lake Eyre basin, S.A., (N.S.W.), Qld., W.A., N.T. Ecology: terrestrial, nocturnal; oviparous, saxicoline, arthropod-feeder. Biological references: Mitchell, F.J. (1965). Australian geckos assigned to the genus *Gehyra* Gray. (Reptilia : Gekkonidae). *Senckenberg. Biol.* **46**: 287–319 (taxonomy); King, M. (1979). Karyotypic evolution in *Gehyra* (Gekkonidae : Reptilia) I. The *Gehyra variegata-punctata* complex. *Aust. J. Zool.* **27**: 373–393.

Gehyra variegata (Duméril and Bibron, 1836)

Hemidactylus variegatus Duméril, A.M.C. & Bibron, G. (1836). *Erpétologie Générale* ou Histoire Naturelle Complète des Reptiles. Paris : Roret 3: iv 517 pp. [353]. Type data: syntypes, MNHP 254 (3 specimens, Tasmania); MNHP 2295 (Shark Bay), from Tasmania (in error) and Shark Bay, W.A.

*[**Gehyra australis** **(part.)**]* Gray, J.E. (1845). *Catalogue of the Specimens of Lizards in the Collection of the British Museum.* London : British Museum xxviii 289 pp. [163]. Type data: paralectotype, BMNH xxii.55a, from Swan River, W.A.; for the lectotype BMNH xxii.55b (Port Essington) see under *Gehyra australis* Gray, 1845.

Gecko grayi Steindachner, F. (1867). Reptilien. *in, Reise der österreichischen Fregatte* Novara *um die Erde in den Jahren 1857, 1858, 1859.* Vienna : State Printer. Zoologie 1(3): 1–98 [11 pl 1 fig 3] [1869 on title page]. Type data: holotype, NHMW (unnumbered), from "New South Wales".

Peripia longicaudis Macleay, W. (1877). The lizards of the *Chevert* Expedition. Second paper. *Proc. Linn. Soc. N.S.W.* **2**: 97–104 [98] [1878 on title page]. Type data: holotype, AM R31933, from Endeavour River, Qld.

Peripia dubia Macleay, W. (1877). The lizards of the *Chevert* Expedition. Second paper. *Proc. Linn. Soc. N.S.W.* **2**: 97–104 [98] [1878 on title page]. Type data: holotype, from Cape Grenville, Cape York Peninsula, Qld., presumed lost *vide* Kluge, A.G. (1963). The systematic status of certain Australian and New Guinean gekkonid lizards. *Mem. Qd. Mus.* **14**: 77–86.

Perochirus mestoni De Vis, C.W. (1890). Descriptions of two lizards of genera new to Australian herpetology. *Proc. Linn. Soc. N.S.W. (2)* **4**: 1034–1036 [1035]. Type data: holotype, QM J236, from Bellenden Ker, NE Qld.

Synonymy that of Gogger, H.G., this work; Kluge, A.G. (1963). The systematic status of certain Australian and New Guinean gekkonid lizards. *Mem. Qd. Mus.* **14**: 77–86.

Distribution: (SE coastal), NE coastal, N Gulf, (N coastal), NW coastal, (SW coastal), W plateau, S Gulfs, Murray-Darling basin, Bulloo River basin, Lake Eyre basin, S.A., Vic., N.S.W., Qld., W.A., N.T. Ecology: low open shrubland, low shrubland, open heath, tall open shrubland, open scrub, low open woodland, low woodland, woodland, open forest, tall forest, under bark, tall shrubland, (terrestrial), arboreal, nocturnal; oviparous, (saxicoline), arthropod-feeder. Biological references: Bustard, H.R. (1965). The systematic status of the Australian geckos *Gehyra variegata* (Duméril and Bibron, 1836) and *Gehyra australis* Gray, 1845. *Herpetologica* **20**: 259–272; Bustard, H.R. (1967). Activity cycle and thermoregulation in the Australian gecko *Gehyra variegata. Copeia* **1967**: 753–758; Bustard, H.R. (1968). The ecology of the Australian gecko *Gehyra variegata*, in northern New South Wales. *J. Zool.* **154**: 113–138; Bustard, H.R. (1969). Population ecology of the gekkonid lizard (*Gehyra variegata* (Duméril and Bibron)) in exploited forests in northern New South Wales. *J. Anim. Ecol.* **38**: 35–51; Mitchell, F.J. (1965). Australian geckos assigned to the genus *Gehyra* Gray. (Reptilia : Gekkonidae). *Senckenberg. Biol.* **46**: 287–319; King, M. (1979). Karyotypic evolution in *Gehyra* (Gekkonidae : Reptilia) I. The *Gehyra variegata-punctata* complex. *Aust. J. Zool.* **27**: 373–393; Pianka, E.R. & Pianka, H.D. (1976). Comparative ecology of twelve species of nocturnal lizards (Gekkonidae) in the Western Australian desert. *Copeia* **1976**: 125–142; Warburg, M.R. (1966). Water economy of several Australian geckos, agamids, and water skinks. *Copeia* **1966**: 230–235.

Gehyra xenopus Storr, 1978

Gehyra xenopus Storr, G.M. (1978). Seven new gekkonid lizards from Western Australia. *Rec. West. Aust. Mus.* **6**: 337–352 [349 pl 6]. Type data: holotype, WAM R56429, from near the Port Warrender road crossing of the King Edward River, 14°52'S, 126°12'E, W.A.

Distribution: N coastal, W.A. Ecology: caves, low open woodland, woodland, terrestrial, nocturnal; oviparous, saxicoline, arthropod-feeder.

Hemidactylus Gray, 1825

Hemidact(ylus) Oken, L. (1817). Hemidactylus, Phyllurus. *Isis von Oken, Jena* **2**(8): 1183 [1183] [*nom. nud.* without indication or included species; regarded by most authors as a valid latinized version of vernacular "Hemidactyles" of Cuvier, G. (1817). *Le Règne Animal distribué d'après son organisation.* Tome 2, contenant les Reptiles, les Poissons, les Mollusques et les Annelidès. Paris : Deterville xviii 432 pp., with type species "*Gecko touberculeux* Daudin" [=*Hemidactylus mabouia* (Moreau de Jonnès, 1818)], see Loveridge, A. (1947). Revision of the African lizards of the family Gekkonidae. *Bull. Mus. Comp. Zool.* **8**: 1–469.

Hemidactylus Gray, J.E. (1825). A synopsis of the genera of reptiles and Amphibia, with a description of some new species. *Annls. Philos. (2)* **10**: 193–217 [199]. Type species *Gecko tuberculosus* Daudin, 1802 [=*Hemidactylus mabouia* (Moreau de Jonnès, 1818)] by subsequent designation, see Gray, J.E. (1827). A synopsis of the genera of saurian reptiles, in which some new genera are indicated, and the others reviewed by actual examination. *Phil. Mag.* **2**: 54–58.

Pnoepus Fitzinger, L.J. (1843). *Systema Reptilium.* Vienna : Braümüller u. Seidel vi 106 pp. [19, 106]. Type species *Hemidactylus javanicus* Fitzinger, 1843 by original designation.

Synonymy that of Cogger, H.G., this work; for discussion of original description, type species and extralimital synonyms see Neave, A.S. (1939). *Nomenclator Zoologicus.* London : Zoolological Society of London Vol. 2 1025 pp.; Wermuth, H. (1965). Liste der rezenten Amphibien und Reptilien; Gekkonidae, Pygopodidae, Xantusiidae. *Das Tierreich* **80**: i–xxii 1–246; Peters, J.A. & Donoso-Barros, R. (1970). Catalogue of the Neotropical Squamata. Part 2. Lizards and amphisbaenians. *Bull. U.S. Natl. Mus.* **297**: 1–293; Sherborn, C.D. (1927). *Index Animalium 1801–1851. H-L* London : British Museum pp. 2885–3746.

This group is also found in Africa, Asia, Indo-Malayan Archipelago, New Guinea and Pacific islands; also American tropics.

Hemidactylus frenatus Duméril and Bibron, 1836

Hemidactylus frenatus Duméril, A.M.C. & Bibron, G. (1836). *Erpétologie Générale* ou Histoire Naturelle Complète des Reptiles. Paris : Roret 3: iv 517 pp. [366] [*ex* Schlegel ms]. Type data: syntypes, MNHP 5135 (2 specimens), RMNH, *vide* Smith, H.M. & Taylor, E.H. (1950). An annotated checklist and key to the reptiles of Mexico exclusive of the snakes. *Bull. U.S. Natl. Mus.* **199**: i–253, and Guibé, J. (1954). *Catalogue des Types de Lézards du Muséum National d'Histoire Naturelle.* Paris : Muséum National d'Histoire Naturelle 119 pp., from southern Africa; Madagascar; Mauritius; Sri Lanka; Ambon, Java and Timor, Indonesia; Bengal, India; Marianas Ils., restricted to Java, Indonesia by Loveridge, A. (1947). Revision of the African lizards of the family Gekkonidae. *Bull. Mus. Comp. Zool.* **8**: 1–469.

Hemidactylus (Pnoepus) bojeri Fitzinger, L.J. (1843). *Systema Reptilium.* Vienna : Braümüller u. Seidel vi 106 pp. [106]. Type data: syntypes, MNHP, RMNH, BMNH, NHMW, from Cape of Good Hope, Africa; Madagascar; Mauritius.

Hemidactylus (Pnoepus) javanicus Fitzinger, L.J. (1843). *Systema Reptilium.* Vienna : Braümüller u. Seidel vi 106 pp. [106]. Type data: syntypes, MNHP, NHMW, BMNH, RMNH, "Asia. India. Bengala. Ceylon. Java. Timor. Amboina. Ins. Mariannae.", restricted to Java by Loveridge, A. (1947). Revision of the African lizards of the family Gekkonidae. *Bull. Mus. Comp. Zool.* **8**: 1–469.

Hemidactylus vittatus Gray, J.E. (1845). *Catalogue of the Specimens of Lizards in the Collection of the British*

Museum. London : British Museum xxviii 289 pp. [155] [the original description is often erroneously cited as an illustration (pl 15 fig 5) in either Günther, A. (1875). Reptiles. A list of saurians of Australia and New Zealand. pp. 9–19 *Zoology of the Voyage of H.M.S.* Erebus *and* Terror. Vol. 2 London : E.W. Janson or Gray, J.E. (1845). Reptiles. *in, Zoology of the Voyage of H.M.S.* Erebus *and* Terror. London : E.W. Janson 2: 1–8 with type specimens from Borneo and Port Essington, N.T.; this illustration was first published in 1867 in Gray, J.E. (1867). *The Lizards of Australia and New Zealand in the Collection of the British Museum*. London : British Museum]. Type data: syntypes (2) BMNH, from Borneo.

Hemidactylus punctatus Jerdon, T.C. (1853). Catalogue of reptiles inhabiting the peninsula of India. *J. Asiat. Soc. Beng.* **22**: 462–479 [467]. Type data: holotype (probable), from Tellicherry, Malabar, India, specimen lost, see Smith, M.A. (1935). *The Fauna of British India; including Ceylon and Burma.* Reptilia and Amphibia. Vol. II. Sauria. London : Taylor & Francis xiv 440 pp. Compiled from secondary source: Wermuth, H. (1965). Liste der rezenten Amphibien und Reptilien; Gekkonidae, Pygopodidae, Xantusiidae. *Das Tierreich* **80**: i–xxii 1–246.

Hemidactylus inornatus Hallowell, E. (1861). Report upon the Reptilia of the North Pacific Exploring Expedition, under command of Capt. John Rogers, U.S.N. *Proc. Acad. Nat. Sci. Philad.* **1860**: 480–510 [492]. Type data: syntypes (3), USNM 7429, from Ryukyu Ils., Japan. Compiled from secondary source: Cochran, D.M. (1961). Type specimens of reptiles and amphibians in the United States National Museum. *Bull. U.S. Natl. Mus.* **220**: 1–291.

Hemidactylus pumilis Hallowell, E. (1861). Report upon the Reptilia of the North Pacific Exploring Expedition, under command of Capt. John Rogers, U.S.N. *Proc. Acad. Nat. Sci. Philad.* **1860**: 480–510 [502]. Type data: holotype (probable), from Hong Kong, USNM, *vide* Smith, M.A. (1935). *The Fauna of British India; including Ceylon and Burma.* Reptilia and Amphibia. Vol. II. Sauria. London : Taylor & Francis xiv 440 pp., specimens lost, see Loveridge, A. (1947). Revision of the African lizards of the family Gekkonidae. *Bull. Mus. Comp. Zool.* **8**: 1–469.

Gecko caracal Tytler, R.C. (1865). Observations on a few species of geckos alive in the possession of the author. *J. Asiat. Soc. Beng.* **33**: 535–548 [547]. Type data: holotype, not found, from Rangoon, Burma.

Gecko chaus Tytler, R.C. (1865). Observations on a few species of geckos alive in the possession of the author. *J. Asiat. Soc. Beng.* **33**: 535–548 [547]. Type data: syntypes, not found, from Moulmein and Rangoon, Burma, restricted to Rangoon by Loveridge, A. (1947). Revision of the African lizards of the family Gekkonidae. *Bull. Mus. Comp. Zool.* **8**: 1–469.

Hemidactylus hexaspis Cope, E.D. (1869). Observations on reptiles of the Old World. Article II. *Proc. Acad. Nat. Sci. Philad.* **1868**: 316–323 [320]. Type data: syntypes, MCZ 5732, from Madagascar.

Hemidactylus longiceps Cope, E.D. (1869). Observations on reptiles of the Old World. Article II. *Proc. Acad. Nat. Sci. Philad.* **1868**: 316–323 [320]. Type data: holotype, MCZ 5730, from Manila, Luzon, Philippine Ils.

Peripia papuensis Macleay, W. (1877). The lizards of the *Chevert* Expedition. Second paper. *Proc. Linn. Soc. N.S.W.* **2**: 97–104 [97] [1878 on title page]. Type data: holotype, AM R29939 (*ex* MMUS MR800), from Mawatta, Binaturu River (as Katow), Papua New Guinea.

Hemidactylus tristis Sauvage, H.E. (1878). Notice sur quelques reptiles nouveaux ou peu connus de la Nouvelle-Guinee. *Bull. Soc. Philomath. Paris (7)* **3**: 47–61 [49]. Type data: syntypes, MNHP 5258 (2 specimens), from New Guinea.

Hemidactylus fragilis Calabresi, E. (1915). Contributo alla conoscenza dei rettili della Somalia. *Monit. Zool. Ital.* **26**: 234–247 [236 figs 1a–b]. Type data: whereabouts unknown, from Bur Meldac, Italian Somaliland. Compiled from secondary source: Wermuth, H. (1965). Liste der rezenten Amphibien und Reptilien Gekkonidae, Pygopodidae, Xantusiidae. *Das Tierreich* **80**: i–xxii 1–246.

Hemidactylus punilis Okada, Y. (1936). Studies on the lizards of Japan. Contribution I. Gekkonidae. *Sci. Rep. Tokyo Bunrika Daig. (b)* **2**: 233–289 [270] [*errore pro Hemidactylus pumilis* Hallowell, 1861].

Synonymy that of Wermuth, H. (1965). Liste der rezenten Amphibien und Reptilien; Gekkonidae, Pygopodidae, Xantusiidae. *Das Tierreich* **80**: i–xxii 1–246; Cogger, H.G., this work.

Distribution: NE coastal, N Gulf, N coastal, Qld., N.T., Christmas Is., Cocos (Keeling) Ils., Ashmore Ils. Ecology: arboreal, nocturnal; oviparous, arthropod-feeder, usually closely associated with man-made structures. Biological references: Bustard, H.R. (1970). Activity cycle of the tropical house gecko, *Hemidactylus frenatus. Copeia* **1970**: 173–176; Marcellini, D.L. (1976). Some aspects of the thermal ecology of the gecko *Hemidactylus frenatus. Herpetologica* **32**: 341–345; Snyder, G.K. & Weaters, W.W. (1976). Physiological responses to temperature in the tropical lizard, *Hemidactylus frenatus* Sauria : Gekkonidae. *Herpetologica* **32**: 252–256; King, M. (1978). A new chromosome form of *Hemidactylus frenatus* (Duméril and Bibron). *Herpetologica* **34**: 216–218; Marcellini, D.L. (1977). The function of a vocal display of the lizard *Hemidactylus frenatus* (Sauria : Gekkonidae). *Anim. Behav.* **25**: 414–417; Schwaner, T.D. (1980). Reproductive biology of lizards on the American Samoan islands. *Occ. Pap. Mus. Nat. Hist. Univ. Kansas* **86**: 1–53 (reproduction); Fyfe, G. (1981). Range extension for *Hemidactylus frenatus*, the Asian house gecko. *Herpetofauna* **13**: 33 (distribution); Frankenberg, E. & Werner, Y.L. (1981). Adaptability of the daily activity pattern to changes in longitude, in a colonizing lizard, *Hemidactylus frenatus. J. Herpet.* **15**: 373–376.

Heteronotia Wermuth, 1965

Heteronota Gray, J.E. (1845). *Catalogue of the Specimens of Lizards in the Collection of the British*

Museum. London : British Museum xxviii 289 pp. [174] [*non Heteronota* Germar, 1835]. Type species *Heteronota binoei* Gray, 1845 by subsequent designation, see Boulenger, G.A. (1885). *Catalogue of the Lizards in the British Museum (Natural History)*. 1. London : British Museum xii 436 pp. pls 32.

Heteronotia Wermuth, H. (1965). Liste der rezenten Amphibien und Reptilien; Gekkonidae, Pygopodidae, Xantusiidae. *Das Tierreich* **80**: i–xxii 1–246 [x,90] [*nom. nov. pro Heteronota* Gray, 1845].

Synonymy that of Cogger, H.G., this work.

Heteronotia binoei (Gray, 1845)

Heteronota binoei Gray, J.E. (1845). *Catalogue of the Specimens of Lizards in the Collection of the British Museum*. London : British Museum xxviii 289 pp. [174]. Type data: holotype, BMNH xxii.93a, from Houtman's Abrolhos, W.A.

Eublepharis derbianus Gray, J.E. (1845). *Catalogue of the Specimens of Lizards in the Collection of the British Museum*. London : British Museum xxviii 289 pp. [274]. Type data: holotype, BMNH 1962.199, from Port Essington, N.T.

Phyllodactylus anomalus Peters, W. (1867). Herpetologische Notizen. *Mber. K. Preuss. Akad. Wiss. Berl.* **1867**: 13–37 [14] [1868 on title page]. Type data: holotype, ZMB 5718, from Rockhampton, Qld.

Hoplodactylus (Pentadactylus) australis Steindachner, F. (1867). Reptilien. *in, Reise der österreichischen Fregatte* Novara *um die Erde in den Jahren 1857, 1858, 1859*. Vienna : State Printer. Zoologie 1(3): 1–98 [18 pl 1 figs 2–2c] [1869 on title page]. Type data: holotype, not found, from "New South Wales".

Diplodactylus annulatus Macleay, W. (1877). The lizards of the *Chevert* Expedition. Second paper. *Proc. Linn. Soc. N.S.W.* **2**: 97–104 [97] [1878 on title page]. Type data: lectotype, AM R29932 (*ex* MMUS MR762), from Palm Is. [Qld.], designation by Kluge, A.G. (1963). The systematic status of certain Australian and New Guinean gekkonid lizards. *Mem. Qd. Mus.* **14**: 77–86.

Heteronota bynoei Boulenger, G.A. (1897). Reptilia and Batrachia. Geckonidae. *Zool. Rec.* **33**: 19–20 [19] [emend. *pro Heteronota binoei* Gray, 1845].

Synonymy that of Cogger, H.G., this work.

Distribution: NE coastal, N Gulf, N coastal, NW coastal, SW coastal, W plateau, S Gulfs, Murray-Darling basin, Bulloo River basin, Lake Eyre basin, S.A., N.S.W., Qld., W.A., N.T., Vic. Ecology: desert, tussock grassland, low open shrubland, low shrubland, open heath, tall open shrubland, open scrub, low open woodland, low woodland, woodland, open forest, tall forest, closed forest, tall shrubland, terrestrial, nocturnal; oviparous, arthropod-feeder. Biological references: Bustard, H.R. (1968). The ecology of the Australian gecko *Heteronotia binoei* in northern New South Wales. *J. Zool.* **156**: 483–497; Warburg, M.R. (1966). Water economy of several Australian geckos, agamids, and water skinks. *Copeia* **1966**: 230–235; Kluge, A.G. (1963). A review of the gekkonid lizard genus *Heteronota* Gray, with a description of a new species from Western Australia. *J. R. Soc. West. Aust.* **46**: 63–67; Pianka, E.R. & Pianka, H.D. (1976). Comparative ecology of twelve species of nocturnal lizards (Gekkonidae) in the Western Australian desert. *Copeia* **1976**: 125–142; Bustard, H.R. (1970). The population ecology of the Australian gekkonid lizard *Heteronotia binoei* in an exploited forest. *J. Zool.* **162**: 31–42.

Heteronotia spelea (Kluge, 1963)

Heteronota spelea Kluge, A.G. (1963). A review of the gekkonid lizard genus *Heteronota* Gray, with a description of a new species from Western Australia. *J. R. Soc. West. Aust.* **46**: 63–67 [64]. Type data: holotype, WAM R12638, from "Prophecy West" mine at Bamboo Creek, Marble Bar District, W.A.

Distribution: (N Gulf), N coastal, NW coastal, W plateau, W.A., N.T. Ecology: caves, desert, hummock grassland, low open shrubland, woodland, terrestrial, nocturnal; oviparous, saxicoline, arthropod-feeder.

Lepidodactylus Fitzinger, 1843

Lepidodactylus Fitzinger, L.J. (1843). *Systema Reptilium*. Vienna : Braümüller u. Seidel vi 106 pp. [19, 98]. Type species *Platydactylus lugubris* Duméril and Bibron, 1836 by original designation.

Amydosaurus Gray, J.E. (1845). *Catalogue of the Specimens of Lizards in the Collection of the British Museum*. London : British Museum xxviii 289 pp. [145]. Type species *Platydactylus lugubris* Duméril and Bibron, 1836 by monotypy.

Synonymy that of Cogger, H.G., this work; Wermuth, H. (1965). Liste der rezenten Amphibien und Reptilien; Gekkonidae, Pygopodidae, Xantusiidae. *Das Tierreich* **80**: i–xxii 1–246.

This group is also found in Asia, Indo-Malayan Archipelago, New Guinea and Pacific islands.

Lepidodactylus listeri (Boulenger, 1889)

Gecko listeri Boulenger, G.A. (1889). On the reptiles of Christmas Island. *Proc. Zool. Soc. Lond.* **1888**: 534–536 [535]. Type data: holotype, BMNH 1946.8.25.91 M, from Christmas Is., Indian Ocean.

Distribution: Christmas Is. Ecology: closed forest, arboreal, nocturnal; oviparous, arthropod-feeder. Biological references: Kluge, A.G. (1968). Phylogenetic relationships of the gekkonid lizard genera *Lepidodactylus* Fitzinger, *Hemiphyllodactylus* Bleeker, and *Pseudogekko* Taylor. *Philipp. J. Sci.* **95**: 331–352 (taxonomy); Brown, W.C. & Parker, F. (1977). Lizards of the genus *Lepidodactylus* (Gekkonidae) from the Indo-Australian Archipelago and the islands of the Pacific, with descriptions of new species. *Proc. Calif. Acad. Sci. (4)* **41**: 253–265 (taxonomy).

Lepidodactylus lugubris (Duméril and Bibron, 1836)

Platydactylus lugubris Duméril, A.M.C. & Bibron, G. (1836). *Erpétologie Générale* ou Histoire Naturelle Complète des Reptiles. Paris : Roret 3: iv 517 pp. [304]. Type data: syntypes (2), MNHP 5323, from Tahiti.

Peropus neglectus Girard, C. (1858). Descriptions of some new reptiles, collected by the United States Exploring Expedition, under the command of Capt. Charles Wilkes, U.S.N. Fourth Part. : Including the species of saurians exotic to North America. *Proc. Acad. Nat. Sci. Philad.* **1857**: 195–199 [197]. Type data: holotype, from Rio de Janeiro, Brazil, specimen lost, see Brown, W.C. & Parker, F. (1977). Lizards of the genus *Lepidodactylus* (Gekkonidae) from the Indo-Australian Archipelago and the islands of the Pacific, with descriptions of new species. *Proc. Calif. Acad. Sci. (4)* **41**: 253–265.

Hemidactylus meijeri Bleeker, P. (1858). De heer Bleeker berigt omtrent de vischoorten en Reptiliën. *Natuurk. Tijdschr. Ned.-Indie. (4)* **16**: 45–47 [47]. Type data: holotype, BMNH 63.12.4.47 (*vide* British Museum register), from Bintang, Rhio Archipelago, Indonesia.

Peripia cantoris Günther, A. (1864). *The Reptiles of British India.* London : Royal Society xxvii 444 pp. [110]. Type data: holotype, BMNH 60.3.19.1073, from Penang, Malaysia.

Gecko moestus Peters, W. (1867). Herpetologische Notizen. *Mber. K. Preuss. Akad. Wiss. Berl.* **1867**: 13–37 [13] [1868 on title page]. Type data: holotype, from Republic of Belau (as Pelew Is.), ZMB, *vide* Brown, W.C. & Parker, F. (1977). Lizards of the genus *Lepidodactylus* (Gekkonidae) from the Indo-Australian Archipelago and the islands of the Pacific, with descriptions of new species. *Proc. Calif. Acad. Sci. (4)* **41**: 253–265.

Gymnodactylus caudeloti Bavay, A. (1869). Catalogue des reptiles de la Nouvelle-Calédonie et description d'espèces nouvelles. *Mém. Soc. Linn. Normandie* **15**: 1–37 [13]. Type data: holotype, not found, from New Caledonia.

Peripia meyeri Günther, A. (1872). On the reptiles and amphibians of Borneo. *Proc. Zool. Soc. Lond.* **1872**: 586–600 [594] [*errore pro Hemidactylus meijeri* Bleeker, 1858].

Peripia mysorensis Meyer, A.B. (1874). Eine Mitteilung von Hrn. Dr. Adolf Bernhard Meyer über die von ihm auf Neu-Guinea den Inseln Jobi, Mysore und Mafoor im Jahr 1873 gesammelten Amphibien. *Mber. K. Preuss. Akad. Wiss. Berl.* **1874**: 128–140 [129] [1875 on title page]. Type data: holotype, from Mysore, India, ZMB, *vide* Brown, W.C. & Parker, F. (1977). Lizards of the genus *Lepidodactylus* (Gekkonidae) from the Indo-Australian Archipelago and the islands of the Pacific, with descriptions of new species. *Proc. Calif. Acad. Sci. (4)* **41**: 253–265.

Peripia ornata Macleay, W. (1877). The lizards of the *Chevert* Expedition. Second paper. *Proc. Linn. Soc. N.S.W.* **2**: 97–104 [98] [1878 on title page]. Type data: syntypes (3), MMUS, presumed lost, from Port Moresby, Papua New Guinea.

Lepidodactylus divergens Taylor, E.H. (1918). Reptiles of Sulu Archipelago. *Philipp. J. Sci.* **13d**: 233–268 [242 pl 1 figs 1–3]. Type data: holotype, from Great Govenene Is., Sulu Archipelago, Philippines; Philippine Bureau of Science specimen destroyed, see Brown, W.C. & Alcala, A.C. (1978). *Philippine Lizards of the Family Gekkonidae.* Dumaguete City : Silliman University 146 pp.

Gehyra variegata ogasawarasimae Okada, Y. (1930). Notes on the herpetology of Chichijima, one of the Bonin Islands. *Bull. Biogeogr. Soc. Japan* **1**: 187–194 [188 (Japanese), 193 (English) pl 12] [Shigei, M. (1971). Reptiles of the Bonin Islands. *J. Fac. Sci. Tokyo Univ. (Zool)* **12**: 145–166 has determined that this name was applied to specimens of *Lepidodactylus lugubris*]. Type data: holotype, Okada collection no. 580, from Chichijima, Bonin Ils., Japan.

Synonymy that of Wermuth, H. (1965). Liste der rezenten Amphibien und Reptilien; Gekkonidae, Pygopodidae, Xantusiidae. *Das Tierreich* **80**: i–xxii 1–246; Cogger, H.G., this work.

Distribution: NE coastal, (N Gulf), Qld. Ecology: woodland, arboreal, nocturnal; oviparous, parthenogenetic, arthropod-feeder. Biological references: Brown, W.C. & Parker, F. (1977). Lizards of the genus *Lepidodactylus* (Gekkonidae) from the Indo-Australian Archipelago and the islands of the Pacific, with descriptions of new species. *Proc. Calif. Acad. Sci. (4)* **41**: 253–265 (taxonomy); Jones, R.E., Fitzgerald, K.T. & Duvall, D. (1978). Quantitative analysis of the ovarian cycle of the lizard *Lepidodactylus lugubris. Gen. Comp. Endocrinol.* **35**: 70–76; Cuellar, O. & Kluge, A.G. (1972). Natural parthenogenesis in the gekkonid lizard *Lepidodactylus lugubris. J. Genet.* **61**: 14–26; Werner, Y.L. (1980). Apparent homosexual behaviour in an all-female population of a lizard, *Lepidodactylus lugubris*, and its probable interpretation. *Z. Tierpsychol.* **54**: 144–150; Feder, M.E. & Feder, J.H. (1981). Diel variation of oxygen consumption in three species of Philippine gekkonid lizards. *Copeia* **1981**: 204–209 (physiology).

Lepidodactylus pumilus (Boulenger, 1885)

Gecko pumilus Boulenger, G.A. (1885). Descriptions of three new species of geckos. *Ann. Mag. Nat. Hist. (5)* **16**: 473–475 [473]. Type data: holotype, BMNH 1946.8.25.93, from Murray Is., Torres Strait, Qld.

Distribution: Qld.; Torres Strait islands. Ecology: woodland, closed forest, arboreal, nocturnal; oviparous, arthropod-feeder. Biological references: Brown, W.C. & Parker, F. (1977). Lizards of the genus *Lepidodactylus* (Gekkonidae) from the Indo-Australian Archipelago and the islands of the Pacific, with descriptions of new species. *Proc. Calif. Acad. Sci. (4)* **41**: 253–265 (taxonomy).

Lucasium Wermuth, 1965

Lucasius Kinghorn, J.R. (1929). Herpetological notes. 1. *Rec. Aust. Mus.* **17**: 76–84 [77] [*non Lucasius* Kinahan, 1859 and *non Lucasius* Dours, 1872]. Type species *Ceramodactylus damaeus* Lucas and Frost, 1896 by original designation.

Lucasium Wermuth, H. (1965). Liste der rezenten Amphibien und Reptilien; Gekkonidae, Pygopodidae, Xantusiidae. *Das Tierreich* **80**: i–xxii 1–246 [x,100] [*nom. nov. pro Lucasius* Kinghorn, 1929].

Synonymy that of Cogger, H.G., this work.

Lucasium damaeum (Lucas and Frost, 1896)

Ceramodactylus damaeus Lucas, A.H.S. & Frost, C. (1896). Reptilia. *Rep. Horn Expedition to Central Australia.* 2: 112–151 [119 pl 9 fig 2] [Wermuth, H. (1965). Liste der rezenten Amphibien und Reptilien; Gekkonidae, Pygopodidae, Xantusiidae. *Das Tierreich* **80**: i–xxii 1–246 and Kluge, A.G. (1967). Systematics, phylogeny, and zoogeography of the lizard genus *Diplodactylus* Gray (Gekkonidae). *Aust. J. Zool.* **15**: 1007–1108 erroneously cite *Proc. R. Soc. Vict. (2)* **8**: 1 of Apr., 1896 for original description; the Horn Expedition Report appeared in Feb., 1896]. Type data: holotype, NMV D7528, from Charlotte Waters, N.T.

Distribution: W plateau, S Gulfs, Murray-Darling basin, Bulloo River basin, Lake Eyre basin, S.A., N.S.W., Qld., W.A., N.T. Ecology: desert, tussock grassland, low open shrubland, low shrubland, open scrub, tall shrubland, terrestrial, nocturnal; oviparous, arthropod-feeder. Biological references: Kluge, A.G. (1967). Systematics, phylogeny, and zoogeography of the lizard genus *Diplodactylus* Gray (Gekkonidae). *Aust. J. Zool.* **15**: 1007–1108 (taxonomy, as *Diplodactylus damaeum*); Bustard, H.R. (1965). Observations on Australian geckos. *Herpetologica* **21**: 294–302 (natural history); Russell, A.P. (1979). Parallelism and integrated design in the foot structure of gekkonine and diplodactyline geckos. *Copeia* **1979**: 1–21 (foot structure).

Nephrurus Günther, 1876

Nephrurus Günther, A. (1876). Descriptions of new species of reptiles from Australia. *J. Mus. Godeffroy* **5**: 45–47 [46]. Type species *Nephrurus asper* Günther, 1876 by monotypy.

Nephrurus asper Günther, 1876

Nephrurus asper Günther, A. (1876). Descriptions of new species of reptiles from Australia. *J. Mus. Godeffroy* **5**: 45–47 [46]. Type data: holotype, BMNH 1946.8.23.34, from Peak Downs, Qld.

Distribution: N Gulf, N coastal, NE coastal, W plateau, (Murray-Darling basin), (Bulloo River basin), Lake Eyre basin, Qld., W.A., N.T. Ecology: desert, hummock grassland, low open woodland, low woodland, woodland, open forest, tall shrubland, terrestrial, nocturnal; oviparous, saxicoline, arthropod-feeder. Biological references: Storr, G.M. (1963). The gekkonid genus *Nephrurus* in Western Australia, including a new species and three new sub-species. *J. R. Soc. West. Aust.* **46**: 85–90 (taxonomy); Bustard, H.R. (1967). Defensive display behaviour of the Australian gecko *Nephrurus asper. Herpetologica* **23**: 126–129; Gow, G. (1979). Notes on the biology of *Nephrurus asper. N. Territory Nat.* **1**: 19–20 (natural history).

Nephrurus laevissimus Mertens, 1958

Nephrurus laevissimus Mertens, R. (1958). Neue Eidechsen aus Australien. *Senckenberg. Biol.* **39**: 51–56 [51]. Type data: holotype, SMF 53201, from 2 km NW of Ayers Rock, N.T.

Distribution: W plateau, Lake Eyre basin, S.A., W.A., N.T. Ecology: desert, low open shrubland, tall shrubland, terrestrial, nocturnal; oviparous, arthropod-feeder. Biological references: Storr, G.M. (1963). The gekkonid genus *Nephrurus* in Western Australia, including a new species and three new sub-species. *J. R. Soc. West. Aust.* **46**: 85–90 (taxonomy); Delean, S. & Harvey, C. (1981). Some observations on the knob-tailed gecko, *Nephrurus laevissimus* in the wild. *Herpetofauna* **13**: 1–3 (natural history).

Nephrurus levis De Vis, 1886

Nephrurus levis De Vis, C.W. (1886). On certain geckos in the Queensland Museum. *Proc. Linn. Soc. N.S.W. (2)* **1**: 168–170 [168] [1887 on title page]. Type data: holotype, QM J246, from unstated locality, restriction to Chinchilla, Qld. proposed by Covacevich, J. (1971). Amphibian and reptile type specimens in the Queensland Museum. *Mem. Qd. Mus.* **16**: 49–68.

Nephrurus platyurus Boulenger, G.A. (1886). Description of a new gecko of the genus *Nephrurus. Ann. Mag. Nat. Hist. (5)* **18**: 91 [91]. Type data: holotype, BMNH 1946.8.23.42, from Adelaide, S.A.

Nephrurus laevis Boulenger, G.A. (1887). *Catalogue of the Lizards in the British Museum (Natural History).* 3. London : British Museum xii 575 pp. pls 140 [477] [emend. *pro Nephrurus levis* De Vis, 1886].

Nephrurus levis occidentalis Storr, G.M. (1963). The gekkonid genus *Nephrurus* in Western Australia, including a new species and three new sub-species. *J. R. Soc. West. Aust.* **46**: 85–90 [88]. Type data: holotype, WAM R13918, from Narryer, 26°34'S, 115°56'E, W.A.

Nephrurus levis pilbarensis Storr, G.M. (1963). The gekkonid genus *Nephrurus* in Western Australia, including a new species and three new sub-species. *J. R. Soc. West. Aust.* **46**: 85–90 [88]. Type data: holotype, WAM R14835, from 12 mi E of Mundabullangana, W.A.

Synonymy that of Cogger, H.G., this work.

Distribution: NW coastal, W plateau, (S Gulfs), (SW coastal), Murray-Darling basin, Bulloo River basin, Lake Eyre basin, S.A., N.S.W., Qld., W.A., N.T. Ecology: desert, hummock grassland, tussock grassland, low open shrubland, low shrubland, tall shrubland, terrestrial, nocturnal; oviparous, arthropod-feeder. Biological references: Storr, G.M. (1963). The gekkonid genus *Nephrurus* in Western Australia, including a new species and three new sub-species. *J. R. Soc. West. Aust.* **46**: 85–90 (taxonomy); Warburg, M.R. (1966). Water economy of several Australian geckos, agamids, and water skinks. *Copeia* **1966**: 230–235.

Nephrurus stellatus Storr, 1968

Nephrurus stellatus Storr, G.M. (1968). *Nephrurus stellatus*, a new knob-tailed gecko from southern Australia. *West. Aust. Nat.* **10**: 180–182 [180 fig 1]. Type data: holotype, WAM R28363, from 41 mi E of Southern Cross, 31°25′S, 120°00′E, W.A.

Distribution: W plateau, S Gulfs, (SW coastal), S.A., W.A. Ecology: low shrubland, open heath, open scrub, tall shrubland, terrestrial, nocturnal; oviparous, arthropod-feeder. Biological references: Galliford, M. (1981). Notes on the starred knob-tailed gecko, *Nephrurus stellatus*, caught spotlighting. *Herpetofauna* **12**: 33–34 (natural history).

Nephrurus vertebralis Storr, 1963

Nephrurus vertebralis Storr, G.M. (1963). The gekkonid genus *Nephrurus* in Western Australia, including a new species and three new sub-species. *J. R. Soc. West. Aust.* **46**: 85–90 [88 fig 2]. Type data: holotype, WAM R5231, from Jibberding, 29°58′S, 116°51′E, W.A.

Distribution: (NW coastal), W plateau, S.A., W.A. Ecology: desert, hummock grassland, low open shrubland, terrestrial, nocturnal; oviparous, arthropod-feeder.

Nephrurus wheeleri Loveridge, 1932

Nephrurus wheeleri Loveridge, A. (1932). New lizards of the genera *Nephrurus* and *Amphibolurus* from Western Australia. *Proc. New Engl. Zool. Club* **13**: 31–34 [31]. Type data: holotype, MCZ 32950, from Yandil, 30 mi NE of Wiluna, W.A.

Nephrurus wheeleri cinctus Storr, G.M. (1963). The gekkonid genus *Nephrurus* in Western Australia, including a new species and three new sub-species. *J. R. Soc. West. Aust.* **46**: 85–90 [86 fig 2]. Type data: holotype, WAM R4284, from Tambrey, in 21°38′S, 117°37′E, W.A.

Synonymy that of Cogger, H.G., this work.

Distribution: NW coastal, W plateau, W.A. Ecology: hummock grassland, low open shrubland, tall shrubland, terrestrial, nocturnal; oviparous, (saxicoline), arthropod-feeder.

Oedura Gray, 1842

Oedura Gray, J.E. (1842). Description of some hitherto unrecorded species of Australian reptiles and batrachians. *in* Gray, J.E. (ed.) *Zoological Miscellany.* London : Treuttel, Würtz & Co. pp. 51–57 [52]. Type species *Oedura marmorata* Gray, 1842 by monotypy.

Pachyurus Fitzinger, L.J. (1843). *Systema Reptilium.* Vienna : Braümüller u. Seidel vi 106 pp. [94] [described as subgenus of *Diplodactylus* Gray, 1832; *non Pachyurus* Agassiz, 1831 and *non Pachyurus* Swainson, 1839]. Type species *Phyllodactylus lesueurii* Duméril and Bibron, 1836 by original designation.

Synonymy that of Cogger, H.G., this work.

Oedura castelnaui (Thominot, 1889)

Phyllodactylus (Oedura) castelnaui Thominot, A. (1889). Observations sur quelques reptiles et batraciens de la collection du Muséum d'Histoire Naturelle de Paris. *Bull. Soc. Philomath. Paris* **8**: 21–30 [22]. Type data: lectotype, MNHP 6802, from Australia, designation by Bustard, H.R. (1970). *Oedura marmorata* a complex of Australian geckos (Reptilia : Gekkonidae). *Senckenberg. Biol.* **51**: 21–40.

Oedura mayeri Garman, S. (1901). Some reptiles and batrachians from Australasia. *Bull. Mus. Comp. Zool.* **39**: 1–14 [3 pl 2 figs 2–2c]. Type data: syntypes, MCZ 6469 and MCZ 6471 (5 specimens, tags not attached), AMNH 12859, from Qld.

Synonymy that of Cogger, H.G., this work.

Distribution: NE coastal, N Gulf, Qld. Ecology: tussock grassland, woodland, open forest, arboreal, nocturnal; oviparous, arthropod-feeder, often found in ground debris and litter. Biological references: Cogger, H.G. (1957). Investigations in the gekkonid genus *Oedura* Gray. *Proc. Linn. Soc. N.S.W.* **82**: 167–179 (taxonomy, as *Oedura marmorata*); Bustard, H.R. (1970). *Oedura marmorata* a complex of Australian geckos (Reptilia : Gekkonidae). *Senckenberg. Biol.* **51**: 21–40 (taxonomy).

Oedura coggeri Bustard, 1966

Oedura coggeri Bustard, H.R. (1966). The *Oedura tryoni* complex: east Australian rock dwelling geckos (Reptilia : Gekkonidae). *Bull. Brit. Mus. Nat. Hist.* **14**: 1–14 [9 pls 1,3]. Type data: holotype, AM R17791, from Lappa Junction, N Qld.

Distribution: NE coastal, (N Gulf), Qld. Ecology: woodland, terrestrial, nocturnal; oviparous, saxicoline, arthropod-feeder.

Oedura lesueurii (Duméril and Bibron, 1836)

***Phyllodactylus lesueurii* (part.)** Duméril, A.M.C. & Bibron, G. (1836). *Erpétologie Générale* ou Histoire Naturelle Complète des Reptiles. Paris : Roret 3: iv 517 pp. [392]. [three paralectotypes from Port Jackson, N.S.W., King George's Sound, W.A. and New Guinea; for remaining paralectotype see *Diplodactylus vittatus*]. Type data: lectotype, MNHP 268, from Kangaroo Is., S.A. (from Australia and New Guinea in original description), designation by Cogger, H.G., this work; see also synonymy of *Diplodactylus vittatus* Gray, 1832.

[Oedura rhombifer* (part.)*] Gray, J.E. (1845). *Catalogue of the Specimens of Lizards in the Collection of the British Museum.* London : British Museum xxviii 289 pp. [147]. Type data: paralectotype, BMNH xxii.3c from Australia (W Australia in original description); for the lectotype (BMNH xxii.3a) see under *Oedura rhombifer* Gray, 1845.

Synonymy that of Cogger, H.G., this work.

Distribution: SE coastal, NE coastal, (Murray-Darling basin), N.S.W., Qld. Ecology: (low open woodland), woodland, open forest, tall forest, terrestrial, nocturnal; oviparous, saxicoline, arthropod-feeder. Biological references: Cogger, H.G. (1957). Investigations in the gekkonid genus *Oedura* Gray. *Proc. Linn. Soc. N.S.W.* **82**: 167–179 (taxonomy).

Oedura marmorata Gray, 1842

Oedura marmorata Gray, J.E. (1842). Description of some hitherto unrecorded species of Australian reptiles and batrachians. *in* Gray, J.E. (ed.) *Zoological Miscellany.* London : Treuttel, Wurz & Co. pp. 51–57 [52]. Type data: lectotype, BMNH xxii.2b, from Port Essington, N.T., designation by Bustard, H.R. (1970). *Oedura marmorata:* a complex of Australian geckos (Reptilia : Gekkonidae). *Senckenberg. Biol.* **51**: 21–40.

Oedura fracticolor De Vis, C.W. (1884). Notes on the fauna of the Gulf of Carpentaria. *Proc. R. Soc. Qd.* **1**: 154–160 [160]. Type data: holotype, presumed lost, from Kimberley, mouth of Norman River, Gulf of Carpentaria, Qld.

Oedura cincta De Vis, C.W. (1888). A contribution to the herpetology of Queensland. *Proc. Linn. Soc. N.S.W. (2)* **2**: 811–826 [811]. Type data: syntypes, ?QM J226; remaining syntype presumed lost, from Charleville, Qld.

Synonymy that of Cogger, H.G., this work.

Distribution: N Gulf, N coastal, NW coastal, W plateau, Murray-Darling basin, Bulloo River basin, Lake Eyre basin, S.A., N.S.W., Qld., W.A., N.T. Ecology: desert, low open shrubland, tall open shrubland, open scrub, low open woodland, low woodland, woodland, open forest, tall shrubland, terrestrial, arboreal, nocturnal; oviparous, (saxicoline), arthropod-feeder. Biological references: Cogger, H.G. (1957). Investigations in the gekkonid genus *Oedura* Gray. *Proc. Linn. Soc. N.S.W.* **82**: 167–179 (taxonomy); Bustard, H.R. (1970). *Oedura marmorata:* a complex of Australian geckos (Reptilia : Gekkonidae). *Senckenberg. Biol.* **51**: 21–40 (taxonomy).

Oedura monilis De Vis, 1888

***Oedura ocellata* (part.)** Boulenger, G.A. (1885). *Catalogue of the Lizards in the British Museum (Natural History).* 1. London : British Museum xii 436 pp. pls 1–32 [105 pl 9 fig 1]. Type data: syntypes, from Australia, BMNH 64.7.22.4–6, BMNH 64.10.27.2–3; for BMNH 54.11.1.1 (figured specimen) see under *Oedura tryoni* De Vis, 1884. The designation by Bustard (1967) of BMNH 64.10.27.2 as lectotype is regarded as invalid, because in choosing a lectotype, he not only ignored all relevant Recommendations of the I.C.Z.N., but chose a specimen which, though listed in the type series, is not consistent with the original description. Further, his selection was contrary to the actions of Boulenger, who subsequently fixed his species by relegating it to the synonymy of *Oedura tryoni*, the only species to which his description and figured specimen can apply.

Oedura monilis De Vis, C.W. (1888). A contribution to the herpetology of Queensland. *Proc. Linn. Soc. N.S.W. (2)* **2**: 811–826 [811]. Type data: holotype, QM J228, from Qld.

Synonymy that of Cogger, H.G., this work.

Distribution: NE coastal, Murray-Darling basin, N.S.W., Qld. Ecology: open scrub, woodland, open forest, arboreal, nocturnal; oviparous, arthropod-feeder. Biological references: Bustard, H.R. (1967). The status of the gekkonid lizard name *Oedura ocellata* Boulenger, 1885. *Herpetologica* **23**: 312–313 (nomenclature); Cogger, H.G. (1964). The comparative osteology and systematic status of the gekkonid genera *Afroedura* Loveridge and *Oedura* Gray. *Proc. Linn. Soc. N.S.W.* **89**: 364–372 (osteology); Roesler, H. (1980). Ein australier, der Samtgecko *Oedura monilis* (De Vis, 1888). *Elaphe (G.D.R.)* **1980**: 53–56 (natural history).

Oedura reticulata Bustard, 1969

Oedura reticulata Bustard, H.R. (1969). *Oedura reticulata*, a new velvet gecko from south-west Western Australia. *West. Aust. Nat.* **11**: 82–86 [83 figs 1–2]. Type data: holotype, WAM R23350, from 9 mi N of Kelanning, in 33°22'S, 117°32'E, W.A.

Distribution: SW coastal, (W plateau), W.A. Ecology: woodland, arboreal, nocturnal; oviparous, arthropod-feeder. Biological references: Jenkins, R.W.G. (1979). The status of endangered Australian reptiles. pp. 169–176 *in* The Status of Endangered Australasian Wildlife. Adelaide : Royal Zoological Society of South Australia.

Oedura rhombifer Gray, 1845

***Oedura rhombifer* (part.)** Gray, J.E. (1845). *Catalogue of the Specimens of Lizards in the Collection of the British Museum.* London : British Museum xxviii 289 pp. [147]. Type data: lectotype, BMNH xxii.3a M, from N Australia (W Australia in original description), designation by Cogger, H.G., this work.

Oedura rhombifera Gray, J.E. (1867). *The Lizards of Australia and New Zealand in the Collection of the British Museum.* London : British Museum [pl 16 fig 6] [invalid emend. *pro Oedura rhombifer* Gray, 1845].

Synonymy that of Cogger, H.G., this work.

Distribution: NE coastal, N Gulf, N coastal, W plateau, Lake Eyre basin, Qld., W.A., N.T. Ecology: tussock grassland, low open shrubland, low open woodland, woodland, open forest, tall shrubland, arboreal, nocturnal; oviparous, arthropod-feeder. Biological references: Cogger, H.G. (1957). Investigations in the gekkonid genus *Oedura* Gray. *Proc. Linn. Soc. N.S.W.* **82**: 167–179 (taxonomy).

Oedura robusta Boulenger, 1885

Oedura robusta Boulenger, G.A. (1885). *Catalogue of the Lizards in the British Museum (Natural History).* 1. London : British Museum xii 436 pp. pls 132 [106 pl 10 fig 1]. Type data: syntypes, BMNH 54.11.1.2, BMNH 32.8.10.5, BMNH 63.6.16.35, BMNH 66.2.13.26, from Australia.

Distribution: SE coastal, NE coastal, Murray-Darling basin, N.S.W., Qld. Ecology: open scrub, low open woodland, woodland, open forest, tall forest, arboreal, nocturnal; oviparous, arthropod-feeder. Biological references: Cogger, H.G. (1957). Investigations in the gekkonid genus *Oedura* Gray. *Proc. Linn. Soc. N.S.W.* **82**: 167–179 (taxonomy).

Oedura tryoni De Vis, 1884

Oedura tryoni De Vis, C.W. (1884). On a new Australian lizard. *Proc. R. Soc. Qd.* **1**: 53–56 [54]. Type data: neotype, AM R21601, from Stanthorpe, Qld., designation by Bustard, H.R. (1966). The *Oedura tryoni* complex: east Australian rock dwelling geckos (Reptilia : Gekkonidae). *Bull. Brit. Mus. Nat. Hist.* **14**: 1–14.

***Oedura ocellata* (part.)** Boulenger, G.A. (1885). *Catalogue of the Lizards in the British Museum (Natural History).* 1. London : British Museum xii 436 pp. pls 132 [105 pl 9 fig 1]. Type data: syntypes, BMNH 54.11.1.1 (figured specimen) from Australia; for the other syntypes, BMNH 64.7.22.4–6 and BMNH 64.10.27.2–3 see under *Oedura monilis* De Vis, 1888.

Synonymy that of Cogger, H.G., this work.

Distribution: SE coastal, NE coastal, Murray-Darling basin, N.S.W., Qld. Ecology: low open woodland, woodland, open forest, tall forest, terrestrial, nocturnal; oviparous, saxicoline, arthropod-feeder. Biological references: Cogger, H.G. (1957). Investigations in the gekkonid genus *Oedura* Gray. *Proc. Linn. Soc. N.S.W.* **82**: 167–179 (taxonomy); Bustard, H.R. (1966). The *Oedura tryoni* complex: east Australian rock dwelling geckos (Reptilia : Gekkonidae). *Bull. Brit. Mus. Nat. Hist.* **14**: 1–14 (taxonomy); Milton, D. (1980). An example of community egglaying in *Oedura tryoni* (De Vis). *Herpetofauna* **11**: 19–23.

Phyllodactylus Gray, 1828

Phyllodactylus Gray, J.E. (1828). *Spicilegia Zoologica, or Original Figures and Short Systematic Descriptions of New and Unfigured Animals.* London : Treuttel, Würz & Co. 1–8 [3]. Type species *Phyllodactylus pulcher* Gray, 1828 by monotypy.

Euleptes Fitzinger, L.J. (1843). *Systema Reptilium.* Vienna : Braümüller u. Seidel vi 106 pp. [95]. Type species *Euleptes wagleri* Fitzinger, 1843 by original designation.

Discodactylus Fitzinger, L.J. (1843). *Systema Reptilium.* Vienna : Braümüller u. Seidel vi 106 pp. [95] [*non Discodactylus* Wagler, 1833]. Type species *Phyllodactylus tuberculosus* Wiegmann, 1835 by original designation.

Gerrhopygus Gray, J.E. (1845). *Catalogue of the Specimens of Lizards in the Collection of the British Museum.* London : British Museum xxviii 289 pp. [150]. Type species *Diplodactylus gerrhopygus* Wiegmann, 1835 by tautonymy.

Paroedura Günther, A. (1879). On mammals and reptiles from Johanna, Comoro Islands. Reptiles. *Ann. Mag. Nat. Hist. (5)* **3**: 217–219 [217]. Type species *Paroedura sanctijohannis* Günther, 1879 by monotypy.

Synonymy that of Cogger, H.G., this work; Wermuth, H. (1965). Liste der rezenten Amphibien und Reptilien; Gekkonidae, Pygopodidae, Xantusiidae. *Das Tierreich* **80**: i–xxii 1–246; Dixon, J.R. (1964). The systematics and distribution of lizards of the genus *Phyllodactylus* in north and central Africa. *New Mexico State Univ. Sci. Bull.* **64**: 1–139; Dixon, J.R. & Huey, R.B. (1970). Systematics of the lizards of the gekkonid genus *Phyllodactylus* of mainland South America. *Los Angeles Co. Mus. Contrib. Sci.* 192: 1–78.

This group is also found in Africa, Madagascar, Europe, west Asia and the Americas.

Phyllodactylus guentheri Boulenger, 1885

Phyllodactylus guentheri Boulenger, G.A. (1885). *Catalogue of the Lizards in the British Museum (Natural History).* 1. London : British Museum xii 436 pp. pls 132 [90 pl 7 fig 3]. Type data: syntypes, BMNH 67.2.19.18, BMNH 62.1.6.29–31, BMNH 56.7.7.19, from Norfolk Is., Lord Howe Is., Champion Bay, NW coast of Australia (the latter presumed to be in error, *vide* Cogger, H.G. (1971). The reptiles of Lord Howe Island. *Proc. Linn. Soc. N.S.W.* **96**: 23–38).

Distribution: Lord Howe Is., Norfolk Is. Ecology: terrestrial, arboreal, nocturnal; oviparous, saxicoline, arthropod-feeder. Biological references: Cogger, H.G. (1971). The reptiles of Lord Howe Island. *Proc. Linn. Soc. N.S.W.* **96**: 23–38 (taxonomy, natural history).

Phyllodactylus marmoratus (Gray, 1845)

Phyllodactylus peronii Fitzinger, L.J. (1843). *Systema Reptilium.* Vienna : Braümüller u. Seidel vi 106 pp. [95] [*nom. oblit.*; Fitzinger based his name on Australian specimens assigned to the extralimital species *Phyllodactylus porphyreus* (Daudin, 1802) by Duméril, A.M.C. & Bibron, G. (1836). *Erpétologie Générale* ou Histoire Naturelle Complète des Reptiles. Paris : Roret 3: iv 517 pp., but these specimens have not previously been identified as types]. Type data: syntypes, MNHP (probably including 6735, two specimens from Kangaroo

Is., S.A. collected by Péron and Lesueur and cited by Roux-Estève, R. (1979). Liste des Amphibiens et Reptiles des collections du Muséum National d'Histoire Naturelle de Paris, récoltés par Lesueur (1778–1846). *Bull. Trim. Soc. Géol. Normandie et Amis Muséum du Havre* **66**: 25–29).

Diplodactylus marmoratus Gray, J.E. (1845). *Catalogue of the Specimens of Lizards in the Collection of the British Museum.* London : British Museum xxviii 289 pp. [149]. Type data: syntypes, BMNH xxi.9.a–d, from Australia and Houtman's Abrolhos (?).

Goniodactylus australis Gray, J.E. (1845). *Catalogue of the Specimens of Lizards in the Collection of the British Museum.* London : British Museum xxviii 289 pp. [172]. Type data: holotype, BMNH xxii.86a, from Australia.

Phyllodactylus macrodactylus Boulenger, G.A. (1885). *Catalogue of the Lizards in the British Museum (Natural History).* 1. London : British Museum xii 436 pp. pls 132 [89 pl 7 fig 2]. Type data: syntypes, BMNH 70.6.26.60 (2 specimens), from unknown locality, but assumed to be Australia.

Phyllodactylus affinis Boulenger, G.A. (1885). *Catalogue of the Lizards in the British Museum (Natural History).* 1. London : British Museum xii 436 pp. pls 132 [89 pl 7 fig 4]. Type data: holotype, BMNH 55.8.16.15, from Aneiteum, Vanuatu, presumed to be in error, *vide* Cogger, H.G. (1971). The reptiles of Lord Howe Island. *Proc. Linn. Soc. N.S.W.* **96**: 23–38.

Synonymy that of Cogger, H.G., this work.

Distribution: SE coastal, SW coastal, W plateau, S Gulfs, Murray-Darling basin, S.A., W.A., Vic., N.S.W., A.C.T. Ecology: open heath, open scrub, low open woodland, woodland, open forest, terrestrial, arboreal, nocturnal; oviparous, arthropod-feeder, (saxicoline). Biological references: Cogger, H.G. (1971). The reptiles of Lord Howe Island. *Proc. Linn. Soc. N.S.W.* **96**: 23–38 (taxonomy); King, M. & Rofe, R. (1976). Karyotypic variation in the Australian gecko *Phyllodactylus marmoratus* (Gray) (Gekkonidae : Reptilia). *Chromosoma (Berl.)* **54**: 75–87; Bustard, H.R. (1965). Observations on Australian geckos. *Herpetologica* **21**: 294–302 (natural history); Warburg, M.R. (1966). Water economy of several Australian geckos, agamids, and water skinks. *Copeia* **1966**: 230–235; King, M. (1977). Reproduction in the Australian gekko *Phyllodactylus marmoratus* (Gray). *Herpetologica* **33**: 7–13; King, M. & King, D. (1977). An additional chromosome race of *Phyllodactylus marmoratus* (Gray)(Reptilia : Gekkonidae) and its phylogenetic implications. *Aust. J. Zool.* **25**: 667–672; King, M. & Hayman, D. (1980). Seasonal variation of chiasma frequency in *Phyllodactylus marmoratus* (Gray) (Gekkonidae : Reptilia). *Chromosoma (Berl.)* **69**: 131–154.

Phyllurus Schinz, 1822

Geckoides Péron, F. (1807). *Voyage de Découvertes aux Terres Australes, Fait par Ordre du Gouvernement, sur les Corvettes le* Géographe, *le* Naturaliste, *et la goülette le* Casuarina, *pendant les années 1800, 1801, 1802, 1803 et 1804. Historique.* Vol. 1 Paris : Arthus Bertrand [450] [*nom. oblit.*]. Type species "*Gecko platurus* Shaw" (=*Lacerta platura* White, 1790) by original designation.

Phyllurus Oken, L. (1817). Hemidactylus, Phyllurus. *Isis Von Oken, Jena* **2**(8): 1183 [1183] [*nom. nud.*].

Phyllurus Schinz, H.R. (1822). *Das Thierreich eingetheilt nach dem Bau der Thiere von Cuvier.* Stuttgart & Tübingen [vii,79]. Type species *Phyllurus novaehollandiae* Schinz, 1822 (*ex* Cuvier ms) by monotypy.

Phyllura Kaup, J.J. (1827). Zoologische Monographien. *Isis Von Oken, Jena* **20**: 610–625 [613] [*nom. substit. pro Phyllurus* Oken, 1817].

Synonymy that of Cogger, H.G., this work; Wermuth, H. (1965). Liste der rezenten Amphibien und Reptilien; Gekkonidae, Pygopodidae, Xantusiidae. *Das Tierreich* **80**: i–xxii 1–246.

Phyllurus caudiannulatus Covacevich, 1975

Phyllurus caudiannulatus Covacevich, J. (1975). A review of the genus *Phyllurus* (Lacertilia : Gekkonidae). *Mem. Qd. Mus.* **17**: 293–303 [297 figs 2–4 pls 36–40]. Type data: holotype, QM J15619, from Bulburin State Forest, via Many Peaks, Qld.

Distribution: NE coastal, Qld. Ecology: tall forest, closed forest, arboreal, nocturnal; oviparous, arthropod-feeder.

Phyllurus cornutus (Ogilby, 1892)

Gymnodactylus cornutus Ogilby, J.D. (1892). Descriptions of three new Australian lizards. *Rec. Aust. Mus.* **2**: 6–11 [8]. Type data: holotype, AM R749, from Bellenden-Ker Ranges, Qld.

Phyllurus lichenosus Günther, A. (1897). Description of new species of lizards and of a tree frog from north-eastern Queensland. *Novit. Zool.* **4**: 403–406 [404 pl 12]. Type data: holotype, presumed lost, from Mount Bartle Frere, Qld.

Gymnodactylus spyrurus Barrett, C. (1950). *Reptiles of Australia.* London : Cassell 168 pp. [31] [*non Gymnodactylus sphyrurus* Ogilby, 1892; *lapsus pro Gymnodactylus cornutus* Ogilby, 1892].

Synonymy that of Cogger, H.G., this work.

Distribution: SE coastal, NE coastal, N.S.W., Qld. Ecology: tall forest, closed forest, arboreal, nocturnal; oviparous, arthropod-feeder. Biological references: Covacevich, J. (1975). A review of the genus *Phyllurus* (Lacertilia : Gekkonidae). *Mem. Qd. Mus.* **17**: 293–303 (taxonomy); Rosler, H. (1981). De australische Bladstaartgekko, *Phyllurus cornutus. Lacerta* **39**: 128–129 (natural history).

Phyllurus platurus (White, 1790)

Lacerta platura White, J. (1790). *Journal of a Voyage to New South Wales,* with sixty five plates of non-descript animals, birds, lizards, serpents, curious cones of trees and other natural productions. Appendix. London : Debrett 299 pp. [246 pl 32 fig 2]. [*ex* Shaw ms]. Type data: holotype (probable), BMNH xxii.98.a, now an alizarin transparency, regarded as likely holotype by Covacevich, J. (1975). A review of the genus *Phyllurus* (Lacertilia : Gekkonidae). *Mem. Qd. Mus.* **17**: 293–303, from N.S.W.

Stellio phyllurus Schneider, J.G. (1797). *Amphibiorum Physiologiae Specimen Secundeum.* 2nd edn. Zurich [31] [*nom. substit. pro Lacerta platura* White, 1790].

Agama discosura Merrem, B. (1820). *Versuch eines Systems Amphibien. Tentamen Systematis Amphibiorum.* Marburg : Krieger xv 199 pp. [51] [*pro* vernacular "lézard discosure" of Lacépède, B.G.E. (1804). Mémoire sur plusieurs animaux de la Nouvelle-Hollande dont la description n'a pas encore été publiée. *Ann. Mus. Natl. Hist. Nat. Paris* **4**: 184–211]. Type data: none designated, MNHP, not found, from Australia.

Agama platyura Merrem, B. (1820). *Versuch eines Systems Amphibien. Tentamen Systematis Amphibiorum.* Marburg : Krieger xv 199 pp. [51] [*nom. substit. pro Lacerta platura* White, 1790].

Phyll[urus] novaehollandiae Schinz, H.R. (1822). *Das Thierreich eingetheilt nach dem Bau der Thiere von Cuvier.* Stuttgart & Tübingin [79] [*ex* Cuvier ms; *nom. substit. pro Lacerta platura* White, 1790].

Phyllurus cuvieri Bory de Saint-Vincent, J.B.G.M. (1825). *Dictionnaire Classique d'Histoire Naturelle.* Paris : Rey et Gravier 7: 183 [183] [also figured in Bory de Saint-Vincent, J.B.G.M. (1831). *Dictionnaire Classique d'Histoire Naturelle.* Atlas. *Phyllurus milii, Phyllurus cuvieri.* Paris : Rey et Gravier Tom. 17 (pl 120 fig 2)]. Type data: holotype, MNHP (presumed lost), from "environs of Port Jackson", N.S.W.

P[hyllurus] whitii Gray, J.E. (1825). A synopsis of the genera of reptiles and Amphibia, with a description of some new species. *Annls. Philos. (2)* **10**: 193–217 [198]. Type data: unknown, from unknown locality.

Gecko platycaudus Schinz, H.R. (1835). *Naturgeschichte und Abbildungen der Reptilien.* Leipzig iv 240 pp. [75 pl 17 fig 3] [*nom. substit. pro Stellio phyllurus* Schneider, 1797].

Phyllurus australis Swainson, W. (1839). The natural history and classification of fishes, amphibians, and reptiles. *in* Lardner, D. (ed.) *The Cabinet of Natural History.* London : Longman, Orme, Brown, Green & Longmans 2: 1–452 [370 fig] [?*nom. substit. pro Phyllurus platurus* White, 1790].

Phyllurus inermis Gray, J.E. (1845). *Catalogue of the Specimens of Lizards in the Collection of the British Museum.* London : British Museum xxviii 289 pp. [176]. Type data: holotype, BMNH xxii.100a, from Australia.

Synonymy that of Cogger, H.G., this work.

Distribution: SE coastal, N.S.W. Ecology: caves, open forest, tall forest, terrestrial, nocturnal; oviparous, saxicoline, arthropod-feeder. Biological references: Covacevich, J. (1975). A review of the genus *Phyllurus* (Lacertilia : Gekkonidae). *Mem. Qd. Mus.* **17**: 293–303 (taxonomy); Green, D. (1973). Observations on the southern leaf-tailed gecko *Phyllurus platurus* (Shaw). *Herpetofauna* **6**: 21–24 (natural history).

Phyllurus salebrosus Covacevich, 1975

Phyllurus salebrosus Covacevich, J. (1975). A review of the genus *Phyllurus* (Lacertilia : Gekkonidae). *Mem. Qd. Mus.* **17**: 293–303 [300 figs 3–4 pls 36–40]. Type data: holotype, QM J8142, from Monto, SE Qld.

Distribution: NE coastal, Murray-Darling basin, Qld. Ecology: caves, open scrub, woodland, tall forest, closed forest, arboreal, nocturnal; oviparous, saxicoline, arthropod-feeder.

Pseudothecadactylus Brongersma, 1936

Torresia Brongersma, L.D. (1934). Contributions to Indo-Australian herpetology. *Zool. Meded.* **17**: 161–251 [176] [*non Torresia* Castelnau, 1875]. Type species *Thecadactylus australis* Günther, 1877 by original designation.

Pseudothecadactylus Brongersma, L.D. (1936). Herpetological note. XIII. *Zool. Meded.* **19**: 136 [136] [*nom. nov. pro Torresia* Brongersma, 1934].

Synonymy that of Cogger, H.G., this work.

Pseudothecadactylus australis (Günther, 1877)

Thecadactylus australis Günther, A. (1877). Descriptions of three new species of lizards from islands of Torres Strait. *Ann. Mag. Nat. Hist. (4)* **19**: 413–415 [414]. Type data: holotype, BMNH 77.3.3.12, from Islands of Torres Strait, Qld.

Distribution: NE coastal, N Gulf, Qld. Ecology: woodland, open forest, closed forest, arboreal, nocturnal; oviparous, arthropod-feeder. Biological references: Cogger, H.G. (1975). New lizards of the genus *Pseudothecadactylus* (Lacertilia : Gekkonidae) from Arnhem Land and northwestern Australia. *Rec. Aust. Mus.* **30**: 87–97 (taxonomy).

Pseudothecadactylus lindneri Cogger, 1975

Pseudothecadactylus lindneri lindneri Cogger, H.G. (1975). New lizards of the genus *Pseudothecadactylus* (Lacertilia : Gekkonidae) from Arnhem Land and northwestern Australia. *Rec. Aust. Mus.* **30**: 87–97 [89 figs 1–5]. Type data: holotype, AM R38734, from Deaf Adder Gorge, Arnhem Land, N.T.

Pseudothecadactylus lindneri cavaticus Cogger, H.G. (1975). New lizards of the genus *Pseudothecadactylus* (Lacertilia : Gekkonidae) from Arnhem Land and northwestern Australia. *Rec. Aust. Mus.* **30**: 87–97 [93 figs 5–6]. Type data: holotype, WAM R43176, from near Mitchell River Falls, approx. 25 km SW of Crystal Head, Port Warrender (approx. 14°40'S, 125°42'E), W.A.

Synonymy that of Cogger, H.G., this work.

Distribution: N coastal, W.A., N.T. Ecology: caves, tall open shrubland, woodland, open forest, terrestrial, arboreal, nocturnal; oviparous, saxicoline, arthropod-feeder.

Rhynchoedura Günther, 1867

Rhynchoedura Günther, A. (1867). Additions to the knowledge of Australian reptiles and fishes. *Ann. Mag. Nat. Hist. (3)* **20**: 45–68 [50]. Type species *Rhynchoedura ornata* Günther, 1867 by monotypy.

Rhynchoedura ornata Günther, 1867

Rhynchoedura ornata Günther, A. (1867). Additions to the knowledge of Australian reptiles and fishes. *Ann. Mag. Nat. Hist. (3)* **20**: 45–68 [51]. Type data: holotype, BMNH 1946.8.20.50, from Nickol Bay, W.A.

Distribution: N Gulf, N coastal, NW coastal, W plateau, S Gulfs, Murray-Darling basin, Bulloo River basin, Lake Eyre basin, S.A., (Vic.), N.S.W., Qld., W.A., N.T. Ecology: desert, hummock grassland, tussock grassland, low open shrubland, low shrubland, open scrub, low open woodland, tall shrubland, terrestrial, nocturnal; oviparous, arthropod-feeder. Biological references: Kluge, A.G. (1967). Systematics, phylogeny, and zoogeography of the lizard genus *Diplodactylus* Gray (Gekkonidae). *Aust. J. Zool.* **15**: 1007–1108 (taxonomy); Pianka, E.R. & Pianka, H.D. (1976). Comparative ecology of twelve species of nocturnal lizards (Gekkonidae) in the Western Australian desert. *Copeia* **1976**: 125–142 (ecology).

Underwoodisaurus Wermuth, 1965

Anomalurus Fitzinger, L.J. (1843). *Systema Reptilium.* Vienna : Braümüller u. Seidel vi 106 pp. [18, 90] [described as subgenus of *Gymnodactylus* Spix, 1825; *non Anomalurus* Waterhouse, Jan. 1843]. Type species *Gymnodactylus miliusii* Duméril and Bibron, 1836 by monotypy.

Underwoodisaurus Wermuth, H. (1965). Liste der rezenten Amphibien und Reptilien; Gekkonidae, Pygopodidae, Xantusiidae. *Das Tierreich* **80**: i–xxii 1–246 [ix,47] [*nom. nov. pro Anomalurus* Fitzinger, 1843].

Synonymy that of Cogger, H.G., this work.

Underwoodisaurus milii (Bory de Saint-Vincent, 1825)

Phyllurus milii Bory de Saint-Vincent, J.B.G.M. (1825). *Dictionnaire Classique d'Histoire Naturelle.* Paris : Rey et Gravier 7: 183 [183] [also figured in Bory de Saint-Vincent (1831). *Dictionnaire Classique d'Histoire Naturelle.* Atlas. *Phyllurus milii, Phyllurus cuvieri.* Paris : Rey et Gravier 17 (pl 120 fig 1)]. Type data: presumed lost, from Shark Bay, W.A.

Cyrtodactylus nilii Gray, J.E. (1831). A synopsis of the species of the class Reptilia, pp. 1–110 *in* Griffith, E. *The Animal Kingdom Arranged in Conformity with its Organization by the Baron Cuvier.* London : Whittaker, Treacher & Co. 9: 481+110 pp. [52] [*lapsus pro Phyllurus milii* Bory de Saint-Vincent, 1825].

Gymnodactylus miliusii Duméril, A.M.C. & Bibron, G. (1836). *Erpétologie Générale* ou Histoire Naturelle Complète des Reptiles. Paris : Roret 3: iv 517 pp. [430 pl 33 fig 1] [*nom. substit. pro Phyllurus milii* Bory, 1825].

Synonymy that of Cogger, H.G., this work.

Distribution: SE coastal, NE coastal, NW coastal, SW coastal, W plateau, S Gulfs, Murray-Darling basin, Bulloo River basin, Lake Eyre basin, S.A., Vic., N.S.W., A.C.T., W.A., Qld. Ecology: low open shrubland, low shrubland, open heath, open scrub, low open woodland, low woodland, woodland, open forest, tall forest, terrestrial, nocturnal; oviparous, arthropod-feeder. Biological references: Waite, E.R. (1929). *The Reptiles and Amphibians of South Australia.* Adelaide : Harrison Weir 270 pp. (natural history); McPhee, D. (1979). *The Observer's Book of Snakes and Lizards of Australia.* Sydney : Methuen 157 pp. (natural history); Stephenson, N.G. (1960). The comparative osteology of Australian geckos and its bearing on their morphological status. *Zool. J. Linn. Soc.* **44**: 278–299.

Underwoodisaurus sphyrurus (Ogilby, 1892)

Gymnodactylus sphyrurus Ogilby, J.D. (1892). Descriptions of three new Australian lizards. *Rec. Aust. Mus.* **2**: 6–11 [6]. Type data: holotype, AM R3800, from interior of N.S.W. (Tumut?).

Heteronota walshi Kinghorn, J.R. (1931). Herpetological notes. 3. *Rec. Aust. Mus.* **18**: 267–269 [268 fig 2]. Type data: holotype, AM R10266, from Boggabri, N.S.W.

Synonymy that of Cogger, H.G., this work.

Distribution: Murray-Darling basin, N.S.W., Qld. Ecology: low open woodland, woodland, terrestrial, nocturnal; oviparous, arthropod-feeder.

PYGOPODIDAE

INTRODUCTION

A family of oviparous, limbless snake- or worm-like lizards endemic to the Australian region. Of the 31 species in 8 genera, only one species (confined to New Guinea) does not occur in Australia. They are most closely allied to the geckos (family Gekkonidae).

Known as snake lizards, legless lizards or flap-footed lizards, they are characterised in Australia by : forelimbs absent; hindlimbs vestigial, represented by a pair of scaly skin flaps, one on either side of the cloacal opening; eyes without movable lids and covered by a snake-like spectacle; external ear opening usually present; tail (unless regenerating) usually at least as long as the snout-vent length.

References

Cogger, H.G. (1979). *Reptiles and Amphibians of Australia.* Sydney : A.H. & A.W. Reed 608 pp.

Kinghorn, J.R. (1926). A brief review of the family Pygopodidae. *Rec. Aust. Mus.* **15**: 40–64

Kluge, A.G. (1974). A taxonomic revision of the lizard family Pygopodidae. *Misc. Publ. Mus. Zool. Univ. Mich.* **147**: 1–221

Kluge, A.G. (1976). Phylogenetic relationships in the lizard family Pygopodidae : an evaluation of theory, methods and data. *Misc. Publ. Mus. Zool. Univ. Mich.* **152**: 1–72

Stephenson, N.G. (1962). The comparative morphology of the head skeleton, girdles and hind limbs in the Pygopodidae. *Zool. J. Linn. Soc.* **44**: 627–644

Wermuth, H. (1965). Liste der rezenten Amphibien und Reptilien: Gekkonidae, Pygopodidae, Xanthusiidae. *Das Tierreich* **80**: i–xxii 1–246

Underwood, G. (1957). On lizards of the family Pygopodidae. A contribution to the morphology and phylogeny of the Squamata. *J. Morph.* **100**. 207–268

Aclys Kluge, 1974

Aclys Kluge, A.G. (1974). A taxonomic revision of the lizard family Pygopodidae. *Misc. Publ. Mus. Zool. Univ. Mich.* 147: 1–221 [46]. Type species *Aclys concinna* Kluge, 1974 by original designation.

Aclys concinna Kluge, 1974

Aclys concinna Kluge, A.G. (1974). A taxonomic revision of the lizard family Pygopodidae. *Misc. Publ. Mus. Zool. Univ. Mich.* 147: 1–221 [46 figs 16–17]. Type data: holotype, WAM R17312, from Sorrento, 11 mi NW of Perth, W.A.

Distribution: NW coastal, SW coastal, W.A.

Ecology: low shrubland, open heath, open forest, fossorial; oviparous.

Aprasia Gray, 1839

Aprasia Gray, J.E. (1839). Catalogue of the slender-tongued saurians, with descriptions of many new genera and species. (contd.). *Ann. Mag. Nat. Hist.* **2**: 331–337 [331]. Type species *Aprasia pulchella* Gray, 1839 by monotypy.

Ophioseps Bocage, J.V.B. (1873). Mélanges erpétologiques. III. Sur quelques sauriens nouveaux de la Nouvelle Calédonie et de l'Australie. *J. Sci. Nat. Lisbon* **4**: 228–232 [231]. Type species *Ophioseps nasutus* Bocage, 1873 by monotypy.

Ophiopsiseps Boulenger, G.A. (1887). *Catalogue of the Lizards in the British Museum (Natural History).* 3. London : British Museum xii 575 pp. pls 40 [436] [*nom. nov. pro Ophioseps* Bocage, 1873, regarded by Boulenger as a junior homonym of *Ophiseps* Blyth, 1853 (= *Ophisaurus* Daudin, 1803)].

Synonymy that of Cogger, H.G., this work; Kluge, A.G. (1974). A taxonomic revision of the lizard family Pygopodidae. *Misc. Publ. Mus. Zool. Univ. Mich.* 147: 1–221.

Aprasia aurita Kluge, 1974

Aprasia aurita Kluge, A.G. (1974). A taxonomic revision of the lizard family Pygopodidae. *Misc. Publ. Mus. Zool. Univ. Mich.* 147: 1–221 [49 figs 18–20]. Type data: holotype, NMV D599, from Ouyen, Vic.

Distribution: Murray-Darling basin, Vic., N.S.W. Ecology: open scrub, tall shrubland, fossorial, nocturnal; oviparous.

Aprasia haroldi Storr, 1978

Aprasia haroldi Storr, G.M. (1978). Taxonomic notes on the reptiles of the Shark Bay region, Western Australia. *Rec. West. Aust. Mus.* **6**: 303–318 [307]. Type data: holotype, WAM R54766, from Parrot Isle, 5 km N of Carrarang, in 26°20′S, 113°30′E, W.A.

Distribution: NW coastal, W.A. Ecology: fossorial; oviparous.

Aprasia inaurita Kluge, 1974

Aprasia inaurita Kluge, A.G. (1974). A taxonomic revision of the lizard family Pygopodidae. *Misc. Publ. Mus. Zool. Univ. Mich.* 147: 1–221 [51 fig 21]. Type data: holotype, NMV D11181, from Birthday Tank, Sunset County (N Cowangie), Vic.

Distribution: W plateau, S Gulfs, Murray-Darling basin, Vic., S.A., N.S.W., W.A. Ecology: low shrubland, open scrub, tall shrubland, fossorial, nocturnal; oviparous.

Aprasia parapulchella Kluge, 1974

Aprasia parapulchella Kluge, A.G. (1974). A taxonomic revision of the lizard family Pygopodidae. *Misc. Publ. Mus. Zool. Univ. Mich.* 147: 1–221 [53 figs 22–23]. Type data: holotype, WAM R41231, from Coppins Crossing, Molonglo River, A.C.T.

Distribution: Murray-Darling basin, N.S.W., A.C.T. Ecology: woodland, open forest, fossorial, nocturnal; oviparous. Biological references: Jenkins, R.W.G. (1979). The status of endangered Australian reptiles. pp. 169–176 *in, The Status of Endangered Australasian Wildlife.* Adelaide : Royal Zoological Society of South Australia (endangered status); Jenkins, R.W.G. & Bartell, R.J. (1980). *A Field Guide to Reptiles of the Australian High Country.* Melbourne : Inkata Press 187 pp. (natural history).

Aprasia pseudopulchella Kluge, 1974

Aprasia pseudopulchella Kluge, A.G. (1974). A taxonomic revision of the lizard family Pygopodidae. *Misc. Publ. Mus. Zool. Univ. Mich.* 147: 1–221 [56 fig 24]. Type data: holotype, SAMA R6360, from a few mi N of Burra, S.A.

Distribution: S Gulfs, (Lake Eyre basin), S.A. Ecology: low open shrubland, open scrub, fossorial, nocturnal; oviparous.

Aprasia pulchella Gray, 1839

Aprasia pulchella Gray, J.E. (1839). Catalogue of the slender-tongued saurians, with descriptions of many new genera and species. (contd.). *Ann. Mag. Nat. Hist.* **2**: 331–337 [332]. Type data: holotype, BMNH 1946.8.30.93, from Australia, restriction to Kalamunda National Park, 13 mi E of Perth, W.A. proposed by Kluge, A.G. (1974). A taxonomic revision of the lizard family Pygopodidae. *Misc. Publ. Mus. Zool. Univ. Mich.* 147: 1–221.

Ophioseps nasutus Bocage, J.V.B. (1873). Mélanges erpétologiques. III. Sur quelques sauriens nouveaux de la Nouvelle Calédonie et de l'Australie. *J. Sci. Nat. Lisbon* **4**: 228–232 [232]. Type data: holotype, MB T73–(439), from Australia.

Aprasia brevirostris Werner, F. (1909). Reptilia exkl. Geckonidae und Scincidae. *in* Michaelsen, W. & Hartmeyer, R. (eds.) *Die Fauna Südwest-Australiens.* Jena : Gustav Fischer 2: 251–278 [266 fig 2]. Type data: syntypes, (2 specimens, presumed lost), from Mt. Helena (as Lion Mill) and Donnybrook, W.A.

Synonymy that of Cogger, H.G., this work.

Distribution: SW coastal, W.A. Ecology: woodland, open forest, (tall forest), fossorial, nocturnal; oviparous. Biological references: Kluge, A.G. (1974). A taxonomic revision of the lizard family Pygopodidae. *Misc. Publ. Mus. Zool. Univ. Mich.* 147: 1–221; Parker, H.W. (1956). The lizard genus *Aprasia*, its taxonomy and temperature correlated variation. *Bull. Brit. Mus. Nat. Hist. (Zool.)* **3**: 365–385 (taxonomy).

Aprasia repens (Fry, 1914)

Ophioseps repens Fry, D.B. (1914). On a collection of reptiles and batrachians from Western Australia. *Rec. West. Aust. Mus. and Art Gall.* **1**: 174–210 [178 figs 2–3]. Type data: holotype, WAM R364, from W.A., restriction to Swan View National Park, in 35°51′S, 116°03′E, W.A. proposed by Kluge, A.G. (1974). A taxonomic revision of the lizard family Pygopodidae. *Misc. Publ. Mus. Zool. Univ. Mich.* 147: 1–221.

Distribution: NW coastal, SW coastal, W plateau, W.A. Ecology: open heath, open scrub, woodland, open forest, fossorial, nocturnal; oviparous. Biological references: Kluge, A.G. (1974). A taxonomic revision of the lizard family Pygopodidae. *Misc. Publ. Mus. Zool. Univ. Mich.* 147: 1–221; Stokely, P.S. (1947). The post-cranial skeleton of *Aprasia repens. Copeia* **1947**: 22–28;

Parker, H.W. (1956). The lizard genus *Aprasia*, its taxonomy and temperature correlated variation. *Bull. Brit. Mus. Nat. Hist. (Zool.)* **3**: 365–385 (taxonomy).

Aprasia rostrata Parker, 1956

Aprasia repens rostrata Parker, H.W. (1956). The lizard genus *Aprasia*, its taxonomy and temperature correlated variation. *Bull. Brit. Mus. Nat. Hist. (Zool.)* **3**: 365–385 [384]. Type data: holotype, WAM R13861, from Hermite Is., Monte Bello Group, W.A.

Aprasia rostrata fusca Storr, G.M. (1979). Five new lizards from Western Australia. *Rec. West. Aust. Mus.* **8**: 134–142 [136]. Type data: holotype, WAM R61077, from 3 km NW of Bullara, in 22°40′S, 114°02′E, W.A.

Synonymy that of Cogger, H.G., this work.

Distribution: NW coastal, W.A. Ecology: hummock grassland, low open shrubland, fossorial, nocturnal; oviparous. Biological references: Kluge, A.G. (1974). A taxonomic revision of the lizard family Pygopodidae. *Misc. Publ. Mus. Zool. Univ. Mich.* 147: 1–221; Hill, F.L. (1955). Notes on the natural history of the Monte Bello Islands. *Proc. Linn. Soc. Lond.* **165**: 113–124 pl 1 fig 1 (natural history, as *Aprasia* sp.).

Aprasia smithi Storr, 1970

Aprasia smithi Storr, G.M. (1970). *Aprasia smithi*, a new worm-lizard (Pygopodidae) from Western Australia. *West. Aust. Nat.* **11**: 141 [141]. Type data: holotype, WAM R34325, from Kalbarri townsite, in 27°43′S, 114°10′E, W.A.

Distribution: NW coastal, W.A. Ecology: fossorial, nocturnal; oviparous. Biological references: Kluge, A.G. (1974). A taxonomic revision of the lizard family Pygopodidae. *Misc. Publ. Mus. Zool. Univ. Mich.* 147: 1–221.

Aprasia striolata Lütken, 1863

Aprasia pulchella striolata Lütken, C. (1863). Nogle nye krybdyr og padder. *Vidensk. Meddr. Dansk Naturh. Foren.* **1862**: 292–331 [300 pls 1–2 figs 3a–d]. Type data: holotype, UZM R3505, from unknown locality, restriction to one mi NE of Tea Tree Gully, S.A. proposed by Kluge, A.G. (1974). A taxonomic revision of the lizard family Pygopodidae. *Misc. Publ. Mus. Zool. Univ. Mich.* 147: 1–221.

Aprasia pulchella lineolata Lütken, C. (1863). Nogle nye krybdyr og padder. *Vidensk. Meddr. Dansk Naturh. Foren.* **1862**: 292–331 [300]. Type data: holotype, UZM R3504, from unknown locality.

Aprasia octolineata Peters, W. (1863). Übersicht der von Hrn. Richard Schomburgk an das zoologische Museum eingesandten Amphibien, aus Buchsfelde bei Adelaide in Südaustralien. *Mber. K. Preuss. Akad. Wiss. Berl.* **1863**: 228–236 [233] [1864 on title page]. Type data: syntypes, ZMB 4720 (3 specimens), from Loos, 4.5 km W Gawler (as Buchsfelde), S.A., see Praite, R. & Tolley, J.C. (1970). *Place Names of South Australia.* Adelaide : Rigby 203 pp.

Aprasia striolata glauerti Parker, H.W. (1956). The lizard genus *Aprasia*, its taxonomy and temperature correlated variation. *Bull. Brit. Mus. Nat. Hist. (Zool.)* **3**: 365–385 [378]. Type data: holotype, WAM R10949, from Albany, W.A.

Synonymy that of Cogger, H.G., this work.

Distribution: (SE coastal), SW coastal, W plateau, S Gulfs, Murray-Darling basin, Lake Eyre basin, S.A., Vic., W.A., N.T. Ecology: low shrubland, open heath, open scrub, tall shrubland, fossorial, nocturnal; oviparous. Biological references: Parker, H.W. (1956). The lizard genus *Aprasia*, its taxonomy and temperature correlated variation. *Bull. Brit. Mus. Nat. Hist. (Zool.)* **3**: 365–385 (taxonomy); Kluge, A.G. (1974). A taxonomic revision of the lizard family Pygopodidae. *Misc. Publ. Mus. Zool. Univ. Mich.* 147: 1–221.

Delma Gray, 1831

Delma Gray, J.E. (1831). Description of a new genus of ophisaurean animal, discovered by the late James Hunter. *in* Gray, J.E. (ed.) *Zoological Miscellany.* London : Treuttel, Würtz & Co. [14]. Type species *Delma fraseri* Gray, 1831 by monotypy.

Nisara Gray, J.E. (1867). *The Lizards of Australia and New Zealand in the Collection of the British Museum.* London : British Museum [3]. Type species *Delma grayii* Smith, 1849 by monotypy.

Pseudodelma Fischer, J.G. (1882). Herpetologische Bemerkungen. *Arch. Naturgesch.* **48**: 281–302 [286]. Type species *Pseudodelma impar* Fischer, 1882 by monotypy.

Synonymy that of Cogger, H.G., this work; Kluge, A.G. (1974). A taxonomic revision of the lizard family Pygopodidae. *Misc. Publ. Mus. Zool. Univ. Mich.* 147: 1–221.

Delma australis Kluge, 1974

Delma australis Kluge, A.G. (1974). A taxonomic revision of the lizard family Pygopodidae. *Misc. Publ. Mus. Zool. Univ. Mich.* 147: 1–221 [77 figs 41–42]. Type data: holotype, WAM R27359, from Port Lincoln, S.A.

Distribution: NW coastal, SW coastal, W plateau, S Gulfs, Murray-Darling basin, Lake Eyre basin, S.A., Vic., N.S.W., W.A., N.T. Ecology: open heath, open scrub, tall shrubland, terrestrial, noctidiurnal; oviparous, arthropod-feeder.

Delma borea Kluge, 1974

Delma borea Kluge, A.G. (1974). A taxonomic revision of the lizard family Pygopodidae. *Misc. Publ. Mus. Zool. Univ. Mich.* 147: 1–221 [81 figs 44–46]. Type data: holotype, WAM R37131, from 48 mi S of Darwin, N.T.

Distribution: N Gulf, N coastal, NW coastal, W plateau, Lake Eyre basin, S.A., Qld., W.A., N.T. Ecology: desert, low open shrubland, low woodland, woodland, open forest, terrestrial, noctidiurnal; oviparous.

Delma elegans Kluge, 1974

Delma elegans Kluge, A.G. (1974). A taxonomic revision of the lizard family Pygopodidae. *Misc. Publ. Mus. Zool. Univ. Mich.* 147: 1–221 [82 figs 48–50]. Type data: holotype, WAM R20070, from Mt. Herbert, northern foothills of Hamersley Range, W.A.

Distribution: NW coastal, W.A. Ecology: hummock grassland, terrestrial; oviparous.

Delma fraseri Gray, 1831

Delma fraseri Gray, J.E. (1831). Description of a new genus of ophisaurean animal, discovered by the late James Hunter. *in* Gray, J.E. (ed.) *Zoological Miscellany.* London : Treuttel, Würtz & Co. [14]. Type data: syntypes, BMNH 1946.8.26.98–9, from Australia, restriction to John Forrest National Park, approximately 17 mi E of Perth, W.A. proposed by Kluge, A.G. (1974). A taxonomic revision of the lizard family Pygopodidae. *Misc. Publ. Mus. Zool. Univ. Mich.* 147: 1–221.

Distribution: SW coastal, W plateau, S.A., W.A. Ecology: open heath, open scrub, (woodland), open forest, tall shrubland, terrestrial, noctidiurnal; oviparous. Biological references: Kluge, A.G. (1974). A taxonomic revision of the lizard family Pygopodidae. *Misc. Publ. Mus. Zool. Univ. Mich.* 147: 1–221.

Delma grayii Smith, 1849

Delma grayii Smith, A. (1849). *Illustrations of the Zoology of South Africa.* Reptilia. London : Smith, Elder & Co. [pl 76 fig 2]. Type data: holotype, BMNH 1946.8.30.81, from interior of Southern Africa (in error), restriction to Australia proposed by Gray, J.E. (1867). *The Lizards of Australia and New Zealand in the Collection of the British Museum.* London : British Museum; further restriction to Kalamunda National Park, 13 mi E Perth, W.A. proposed by Kluge, A.G. (1974). A taxonomic revision of the lizard family Pygopodidae. *Misc. Publ. Mus. Zool. Univ. Mich.* 147: 1–221.

Distribution: NW coastal, SW coastal, W.A. Ecology: hummock grassland, open heath, open scrub, woodland, terrestrial, noctidiurnal; oviparous. Biological references: Kluge, A.G. (1974). A taxonomic revision of the lizard family Pygopodidae. *Misc. Publ. Mus. Zool. Univ. Mich.* 147: 1–221.

Delma impar (Fischer, 1882)

Ps[eudodelma] impar Fischer, J.G. (1882). Herpetologische Bemerkungen. *Arch. Naturgesch.* **48**: 281–302 [287 pl 16 figs 1–4] [holotype MTKD D158 from Melbourne, Vic., presumed destroyed *vide* Obst, F.J. (1977). Die herpetologische Sammlung des Staatlichen Museums für Tierkunde Dresden und ihre Typusexemplare. *Zool. Abh. Mus. Tierk. Dresden* **34**: 171–186]. Type data: neotype, NMV D10883, designation by Kluge, A.G. (1974). A taxonomic revision of the lizard family Pygopodidae. *Misc. Publ. Mus. Zool. Univ. Mich.* 147: 1–221, from Hamilton, 37°45′S, 142°02′E, Vic.

Delma lineata Rosén, N. (1905)

Delma lineata Rosén, N. (1905). List of lizards in the Zoological Museum of Lund, with descriptions of new species. *Ann. Mag. Nat. Hist. (7)* **16**: 129–142 [131 pl 8(1) figs 2a–c]. Type data: holotype, ZMLU Cat. Typ. No. 295, from Vic.

Synonymy that of Cogger, H.G., this work.

Distribution: SE coastal, Murray-Darling basin, Vic., N.S.W., A.C.T. Ecology: low open woodland, woodland, open forest, terrestrial, noctidiurnal; oviparous. Biological references: Kluge, A.G. (1974). A taxonomic revision of the lizard family Pygopodidae. *Misc. Publ. Mus. Zool. Univ. Mich.* 147: 1–221; Jenkins, R.W.G. & Bartell, R.J. (1980). *A Field Guide to Reptiles of the Australian High Country.* Melbourne : Inkata Press 187 pp. (natural history).

Delma inornata Kluge, 1974

Delma inornata Kluge, A.G. (1974). A taxonomic revision of the lizard family Pygopodidae. *Misc. Publ. Mus. Zool. Univ. Mich.* 147: 1–221 [101 figs 62–64]. Type data: holotype, NMV D3492, from Kewell, Vic.

Distribution: SE coastal, NE coastal, (S Gulfs), Murray-Darling basin, S.A., Vic., N.S.W., Qld., A.C.T. Ecology: open scrub, low open woodland, woodland, terrestrial, noctidiurnal; oviparous. Biological references: Thompson, M.B. (1980). *Delma inornata* Kluge (Reptilia, Pygopodidae) in South Australia. *S. Aust. Nat.* **84**: 42–43 (natural history).

Delma molleri Lütken, 1863

Delma molleri Lütken, C. (1863). Nogle nye krybdyr og padder. *Vidensk. Meddr. Dansk Naturh. Foren.* **1862**: 292–331 [296 pl 1/2 figs 2a–c]. Type data: holotype, UZM R3503, from unknown locality, restriction to one mi NE of Tea Tree Gully, S.A. proposed by Kluge, A.G. (1974). A taxonomic revision of the lizard family Pygopodidae. *Misc. Publ. Mus. Zool. Univ. Mich.* 147: 1–221.

Distribution: S Gulfs, (Murray-Darling basin), (Lake Eyre basin), S.A. Ecology: terrestrial; oviparous.

Delma nasuta Kluge, 1974

Delma nasuta Kluge, A.G. (1974). A taxonomic revision of the lizard family Pygopodidae. *Misc. Publ. Mus. Zool. Univ. Mich.* 147: 1–221 [109 figs 70–72]. Type data: holotype, WAM R40178, from Pollock Hills, 144 mi W of Sandy Blight Junction, W.A.

Distribution: N Gulf, N coastal, NW coastal, W plateau, S Gulfs, Lake Eyre basin, S.A., (N.S.W.), Qld., W.A., N.T. Ecology: desert, hummock grassland, low open shrubland, low shrubland, open scrub, low open woodland, low woodland, tall shrubland, terrestrial, noctidiurnal; oviparous.

Delma pax Kluge, 1974

Delma pax Kluge, A.G. (1974). A taxonomic revision of the lizard family Pygopodidae. *Misc. Publ. Mus. Zool. Univ. Mich.* 147: 1–221 [113 figs 74–77]. Type data: holotype, WAM R14804, from Jones River, 21 mi SE of Roeburne, in 20°58′S, 117°23′E, W.A.

Distribution: NW coastal, W.A. Ecology: terrestrial; oviparous.

Delma plebeia De Vis, 1888

Delma plebeia De Vis, C.W. (1888). A contribution to the herpetology of Queensland. *Proc. Linn. Soc. N.S.W. (2)* **2**: 811–826 [824]. Type data: lectotype, QM J254, designation by Kluge, A.G. (1974). A taxonomic revision of the lizard family Pygopodidae. *Misc. Publ. Mus. Zool. Univ. Mich.* 147: 1–221, from Brisbane and Gympie, Qld.; restricted to Brisbane by choice of lectotype.

Distribution: SE coastal, NE coastal, Murray-Darling basin, N.S.W., Qld. Ecology: woodland, open forest, terrestrial; oviparous. Biological references: Kluge, A.G. (1974). A taxonomic revision of the lizard family Pygopodidae. *Misc. Publ. Mus. Zool. Univ. Mich.* 147: 1–221.

Delma tincta De Vis, 1888

Delma tincta De Vis, C.W. (1888). A contribution to the herpetology of Queensland. *Proc. Linn. Soc. N.S.W. (2)* **2**: 811–826 [824]. Type data: lectotype, QM J241, designation by Kluge, A.G. (1974). A taxonomic revision of the lizard family Pygopodidae. *Misc. Publ. Mus. Zool. Univ. Mich.* 147: 1–221, from Normanton and Springsure, Qld.; restricted to Normanton by choice of lectotype.

Delma reticulata Garman, S. (1901). Some reptiles and batrachians from Australasia. *Bull. Mus. Comp. Zool.* **39**: 1–14 [5 pl 2 figs 1–1f]. Type data: holotype, MCZ 6486, from Qld., restriction to Cooktown, Qld. proposed by Barbour, T. & Loveridge, A. (1929). Typical reptiles and amphibians in the Museum of Comparative Zoology. *Bull. Mus. Comp. Zool.* **69**: 205–360.

Synonymy that of Cogger, H.G., this work.

Distribution: NE coastal, N Gulf, N coastal, NW coastal, W plateau, Murray-Darling basin, Bulloo River basin, Lake Eyre basin, S.A., N.S.W., Qld., W.A., N.T. Ecology: desert, hummock grassland, tussock grassland, low woodland, woodland, open forest, tall shrubland, terrestrial, noctidiurnal; oviparous. Biological references: Kluge, A.G. (1974). A taxonomic revision of the lizard family Pygopodidae. *Misc. Publ. Mus. Zool. Univ. Mich.* 147: 1–221.

Delma torquata Kluge, 1974

Delma torquata Kluge, A.G. (1974). A taxonomic revision of the lizard family Pygopodidae. *Misc. Publ. Mus. Zool. Univ. Mich.* 147: 1–221 [125 figs 85 86]. Type data: holotype, QM J14365, from Upper Brookfield, 15 mi WNW of Brisbane, Qld.

Distribution: NE coastal, Qld. Ecology: woodland, terrestrial, (fossorial); oviparous.

Lialis Gray, 1835

Lialis Gray, J.E. (1835). Characters of a new genus of reptiles (*Lialis*) from N.S.W. *Proc. Zool. Soc. Lond.* **1834**: 134–135 [134] [1834 on title page]. Type species *Lialis burtonis* Gray, 1835 by monotypy.

Ophiophthalmus Fitzinger, L.J. (1843). *Systema Reptilium.* Vienna : Braümüller u. Seidel vi 106 pp. [23]. Type species *Lialis burtonis* Gray, 1835 by original designation.

Alopecosaurus Lindholm, W.A. (1905). Über einige Eidechsen und Schlangen aus Deutsch-Neuguinea. *Jb. Nassau Ver. Naturk.* **58**: 227–240 [230]. Type species *Alopecosaurus cuneirostris* Lindholm, 1905 (=*Lialis jicari* Boulenger, 1903) by original designation.

Synonymy that of Cogger, H.G., this work; Kluge, A.G. (1974). A taxonomic revision of the lizard family Pygopodidae. *Misc. Publ. Mus. Zool. Univ. Mich.* 147: 1–221.

This group is also found in New Guinea.

Lialis burtonis Gray, 1835

Lialis burtonis Gray, J.E. (1835). Characters of a new genus of reptiles (*Lialis*) from N.S.W. *Proc. Zool. Soc. Lond.* **1834**: 134–135 [134] [1834 on title page]. Type data: neotype, AM R27914, designation by Kluge, A.G. (1974). A taxonomic revision of the lizard family Pygopodidae. *Misc. Publ. Mus. Zool. Univ. Mich.* 147: 1–221, from Round Hill Fauna Reserve, between Lake Cargelligo and Mt. Hope, in 32°58′S, 146°10′E, N.S.W.; original type locality N.S.W. (as Nova Cambria Australi).

Lialis bicatenata Gray, J.E. (1842). Description of some hitherto unrecorded species of Australian reptiles and batrachians. *in* Gray, J.E. (ed.) *Zoological Miscellany.* London : Treuttel, Würtz & Co. pp. 51–57 [52]. Type data: holotype, BMNH 1946.8.26.71, from Port Essington, N.T.

Lialis punctulata Gray, J.E. (1842). Description of some hitherto unrecorded species of Australian reptiles and batrachians. *in* Gray, J.E. (ed.) *Zoological Miscellany.* London : Treuttel, Würtz & Co. pp. 51–57 [52]. Type data: holotype, BMNH 1946.8.30.89, from Port Essington, N.T.

Lialis leptorhyncha Peters, W. (1873). Über eine neue Schildkrötenart, *Cinosternon effeldtii* und einige andere neue oder weniger bekannte Amphibien. *Mber. K. Preuss. Akad. Wiss. Berl.* **1873**: 603–618 [605] [1874 on title page]. Type data: holotype, from Port Mackay, Qld., original description cites ZMB 5947; Kluge, A.G. (1974). A taxonomic revision of the lizard family Pygopodidae. *Misc. Publ. Mus. Zool. Univ. Mich.* 147: 1–221 cites ZMB 5948.

Lialis punctulata concolor Peters, W. (1873). Über eine neue Schildkrötenart, *Cinosternon effeldtii* und einige andere neue oder weniger bekannte Amphibien. *Mber. K. Preuss. Akad. Wiss. Berl.* **1873**: 603–618 [606] [1874 on title page]. Type data: syntypes, ZMB 5135 (not found), ZMB 5323 (Clarence River, N.S.W.), remaining two syntypes not found, from Sydney, N.S.W. and Perth, W.A. (and presumably Clarence River, N.S.W.).

Synonymy that of Cogger, H.G., this work.

Distribution: SE coastal, NE coastal, N Gulf, N coastal, NW coastal, SW coastal, W plateau, S Gulfs, Murray-Darling basin, Bulloo River basin, Lake Eyre basin, S.A., Vic., N.S.W., Qld., W.A., N.T., A.C.T. Ecology: desert, hummock grassland, tussock grassland, low open shrubland, low shrubland, open heath, tall open shrubland, open scrub, low open woodland, low woodland, woodland, open forest, tall forest, closed forest, tall shrubland, terrestrial, noctidiurnal; oviparous, lacertiphagous, (arthropod-feeder). Biological references: Kinghorn, J.R. (1924). A short review of lizards belonging to the genus *Lialis* Gray. *Rec. Aust. Mus.* **14**: 184–188 (taxonomy); Kluge, A.G. (1974). A taxonomic revision of the lizard family Pygopodidae. *Misc. Publ. Mus. Zool. Univ. Mich.* 147: 1–221; Wever, E.G. (1974). The ear of *Lialis burtonis* (Sauria : Pygopodidae), its structure and function. *Copeia* **1974**: 297–305; Gorman, G.C. & Gress, F. (1970). Sex chromosomes of a pygopodid, *Lialis burtonis*. *Experientia* **26**: 206–207; Neill, W.T. (1957). Notes on the pygopodid lizards, *Lialis burtoni* and *L. jicari*. *Copeia* **1957**: 230–232; Bradshaw, S.D., Gans, C. & Saint Girons, H. (1980). Behavioural thermoregulation in a pygopodid lizard, *Lialis burtonis*. *Copeia* **1980**: 738–743 (thermoregulation).

Ophidiocephalus Lucas and Frost, 1897

Ophidiocephalus Lucas, A.H.S. & Frost, C. (1897). Descriptions of two new species of lizards from central Australia. *Proc. R. Soc. Vict. (ns)* **9**: 54–56 [54]. Type species *Ophidiocephalus taeniatus* Lucas and Frost, 1897 by monotypy.

Ophidiocephalus taeniatus Lucas and Frost, 1897

O[phidiocephalus] taeniatus Lucas, A.H.S. & Frost, C. (1897). Descriptions of two new species of lizards from central Australia. *Proc. R. Soc. Vict. (ns)* **9**: 54–56 [54]. Type data: holotype, NMV D11761, from Charlotte Waters, N.T.

Distribution: Lake Eyre basin, S.A., N.T. Ecology: tussock grassland, low shrubland, fossorial, crepuscular, nocturnal; oviparous. Biological references: Kluge, A.G. (1974). A taxonomic revision of the lizard family Pygopodidae. *Misc. Publ. Mus. Zool. Univ. Mich.* 147: 1–221; Ehmann, H. (1981). The natural history and conservation of the bronzeback (*Ophidiocephalus taeniatus* Lucas and Frost) (Lacertilia, Pygopodidae). pp. 7–13 *in* Banks, C.B. & Martin, A.A. (eds.) *Proceedings of the Melbourne Herpetological Symposium, 1980.* Melbourne : Zoological Board of Victoria.

Paradelma Kinghorn, 1926

Paradelma Kinghorn, J.R. (1926). A brief review of the family Pygopodidae. *Rec. Aust. Mus.* **15**: 40–64 [48]. Type species *Delma orientalis* Günther, 1876 by monotypy.

Paradelma orientalis (Günther, 1876)

Delma orientalis Günther, A. (1876). Descriptions of new species of reptiles from Australia. *J. Mus. Godeffroy* **5**: 45–47 [45]. Type data: lectotype, BMNH 1946.8.13.48 M, from Peak Downs, Qld., designation by Kluge, A.G. (1974). A taxonomic revision of the lizard family Pygopodidae. *Misc. Publ. Mus. Zool. Univ. Mich.* 147: 1–221.

Distribution: NE coastal, Murray-Darling basin, Qld. Ecology: terrestrial; oviparous. Biological references: Kluge, A.G. (1974). A taxonomic revision of the lizard family Pygopodidae. *Misc. Publ. Mus. Zool. Univ. Mich.* 147: 1–221; Jenkins, R.W.G. (1979). The status of endangered Australian reptiles. pp. 169–176 *in, The Status of Endangered Australasian Wildlife.* Adelaide : Royal Zoological Society of South Australia (endangered status).

Pletholax Cope, 1864

Pletholax Cope, E.D. (1864). On the characters of the higher groups of Reptilia-Squamata—and especially of the Diploglossa. *Proc. Acad. Nat. Sci. Philad.* **1864**: 224–231 [229]. Type species *Pletholax gracilis* Cope, 1864 by original designation.

Pletholax gracilis Cope, 1864

Pletholax gracilis Cope, E.D. (1864). On the characters of the higher groups of Reptilia-Squamata —and especially of the Diploglossa. *Proc. Acad. Nat. Sci. Philad.* **1864**: 224–231 [229] [*ex Pygopus gracilis* Schlegel ms; specific name attributed to Schlegel by Kluge, A.G. (1974). A taxonomic revision of the lizard family Pygopodidae. *Misc. Publ. Mus. Zool. Univ. Mich.* 147: 1–221, although there seems little doubt that its attribution to Cope by all previous authors was correct, as Cope states simply that "... I am indebted (to Schlegel) for the opportunity of making this description."]. Type data: holotype, RMNH 3670, from "South West Australia".

Pletholax gracilis edelensis Storr, G.M. (1978). Taxonomic notes on the reptiles of the Shark Bay region, Western Australia. *Rec. West. Aust. Mus.* **6**: 303–318 [304 fig 1]. Type data: holotype, WAM R54627, from 4 km S of Useless Loop, in 26°10'S, 113°25'E, W.A.

Synonymy that of Cogger, H.G., this work.

Distribution: NW coastal, SW coastal, W.A. Ecology: open heath, woodland, open forest, fossorial; oviparous. Biological references: Kluge, A.G. (1974). A taxonomic revision of the lizard family Pygopodidae. *Misc. Publ. Mus. Zool. Univ. Mich.* 147: 1–221.

Pygopus Merrem, 1820

Pygopus Merrem, B. (1820). *Versuch eines Systems Amphibien. Tentamen Systematis Amphibiorum.* Marburg : Krieger xv 199 pp. [77]. Type species *Bipes lepidopus* [*sic*] Lacépède, 1804 by monotypy.

Hysteropus Duméril, A.M.C. & Bibron, G. (1839). *Erpétologie Générale* ou Histoire Naturelle Complète des Reptiles. Paris : Roret 5: viii 854 pp. [826] [*ex* Duméril ms]. Type species *Sheltopusik novaehollandiae* Oppel, 1811 by monotypy.

Cryptodelma Fischer, J.G. (1882). Herpetologische Bemerkungen. *Arch. Naturgesch.* **48**: 281–302 [289]. Type species *Cryptodelma nigriceps* Fischer, 1882 by monotypy.

Synonymy that of Cogger, H.G., this work; Kluge, A.G. (1974). A taxonomic revision of the lizard family Pygopodidae. *Misc. Publ. Mus. Zool. Univ. Mich.* 147: 1–221.

Pygopus lepidopodus (Lacépède, 1804)

Bipes lepidopodus Lacépède, B.G.E. (1804). Mémoire sur plusieurs animaux de la Nouvelle-Hollande dont la description n'a pas encore été publiée. *Ann. Mus. Natl. Hist. Nat. Paris* **4**: 184–211 [209 pl 55 fig 1]. Type data: holotype, MNHP 7154, from Australia.

Sheltopusik novaehollandiae Oppel, M. (1811). *Die Ordnungen, Familien und Gattungen der Reptilien, als prodrom einer Naturgeschichte Derselben.* München : Lindauer xii 86 pp. [40]. Type data: type(s) not located, from Australia.

Pygopus lepidopus Merrem, B. (1820). *Versuch eines Systems Amphibien. Tentamen Systematis Amphibiorum.* Marburg : Krieger xv 199 pp. [77] [invalid emend., *lapsus pro Bipes lepidopodus* Lacépède, 1804].

Pygopus squamiceps Gray, J.E. (1845). *Catalogue of the Specimens of Lizards in the Collection of the British Museum.* London : British Museum xxviii 289 pp. [68]. Type data: holotype, BMNH xii.2a, from Australia.

Pygopus longicaudatus Tepper, J.C.O. (1882). Notes on some South Australian lizards. *Trans. Proc. R. Soc. S. Aust.* **5**: 32–33 [32] [*nom. oblit.*, *vide* Kluge, A.G. (1974). A taxonomic revision of the lizard family Pygopodidae. *Misc. Publ. Mus. Zool. Univ. Mich.* 147: 1–221]. Type data: not known, from near Kangarilla, S.A.

Synonymy that of Cogger, H.G., this work.

Distribution: SE coastal, NE coastal, NW coastal, SW coastal, W plateau, S Gulfs, Murray-Darling basin, S.A., Vic., N.S.W., Qld., N.T., A.C.T., W.A. Ecology: low open shrubland, low shrubland, open heath, open scrub, low open woodland, woodland, open forest, terrestrial, noctidiurnal; oviparous, lacertiphagous, arthropod-feeder. Biological references: Kluge, A.G. (1974). A taxonomic revision of the lizard family Pygopodidae. *Misc. Publ. Mus. Zool. Univ. Mich.* 147: 1–221; Wells, R. & Husband, G. (1979). Comments on the reproduction of *Pygopus lepidopodus* (Lacépède). *Herpetofauna* **11**: 22–25; May, E. (1978). Einige Beobachtungen zum Verhalten von *Pygopus lepidopodus* (Reptilia : Sauria : Pygopodidae). *Salamandra* **14**: 1–8 (natural history).

Pygopus nigriceps (Fischer, 1882)

Cryptodelma nigriceps Fischer, J.G. (1882). Herpetologische Bemerkungen. *Arch. Naturgesch.* **48**: 281–302 [290 pl 16 figs 5–9]. Type data: holotype, SMNS 2259, from Nickol Bay, W.A.

Delma (Cryptodelma) baileyi Günther, A. (1897). Description of a new species of *Delma* from Western Australia. *Ann. Mag. Nat. Hist. (6)* **19**: 170–172 [170 fig]. Type data: holotype, presumed lost, from neighbourhood of Cue, W.A.

Pygopus schraderi Boulenger, G.A. (1913). Description of new lizards in the collection of the British Museum. *Ann. Mag. Nat. Hist. (8)* **12**: 563–566 [564]. Type data: holotype, BMNH 1946.8.27.2, from Milparinka (as Milparinha), N.S.W.

Synonymy that of Cogger, H.G., this work.

Distribution: NE coastal, N Gulf, N coastal, NW coastal, SW coastal, W plateau, S Gulfs, Murray-Darling basin, Bulloo River basin, Lake Eyre basin, S.A., Vic., N.S.W., Qld., W.A., N.T. Ecology: desert, hummock grassland, tussock grassland, low open shrubland, tall open shrubland, open scrub, low open woodland, low woodland, woodland, tall shrubland, noctidiurnal; oviparous, lacertiphagous, arthropod-feeder. Biological references: Kluge, A.G. (1974). A taxonomic revision of the lizard family Pygopodidae. *Misc. Publ. Mus. Zool. Univ. Mich.* 147: 1–221.

AGAMIDAE

INTRODUCTION

A family of moderate to large, primarily diurnal lizards ranging from Africa through southern Europe, Asia, and the Indo-Australian Archipelago to New Guinea and Australia. Of the 36 genera and 325 species currently recognised, some 64 species in 10 genera occur in Australia. The current generic arrangement is conservative and many genera are clearly polyphyletic.

Characterised in Australia by: well-developed pentadactyl limbs; eyes with movable eyelids; pupil more or less circular; external ear opening either obvious or hidden; body scales juxtaposed; tongue broad, flat and fleshy, slightly notched in front; scales on top of head small and irregular.

References

Cogger, H.G. (1979). *Reptiles and Amphibians of Australia.* Sydney : A.H. & A.W. Reed 608 pp.

Wermuth, H. (1967). Liste der rezenten Amphibien und Reptilien; Agamidae. *Das Tierreich* **86**: 1–127

Worrell, E. (1970). *Reptiles of Australia.* 2nd edn. Sydney : Angus & Robertson xv 169 pp.

Amphibolurus Wagler, 1830

Gemmatophora Kaup, J.J. (1827). Zoologische Monographien. *Isis Von Oken, Jena* **20**: 610–625 [621] [*nom. oblit.*; described as subgenus of *Calotes* Cuvier, 1817]. Type species *Lacerta muricata* White, 1790 by original designation.

Amphibolurus Wagler, J.G. (1830). *Natürliches System der Amphibien, mit vorangehender Classification der Säugethiere und Vögel.* München : Cotta'schen vi 354 pp. [145] [*nom. substit. pro Gemmatophora* Kaup, 1827, which Wagler rejected as an invalid hybrid name ("*vox hybrida*")].

Grammatophora Duméril, A.M.C. & Bibron, G. (1837). *Erpétologie Générale* ou Histoire Naturelle Complète des Reptiles. Paris : Roret 4: ii 571 pp. [468] [*non Grammatophora* Stephens, 1829 (*nom. nud.*); emend. *pro Gemmatophora* Kaup, 1827].

Homalonotus Fitzinger, L.J. (1843). *Systema Reptilium.* Vienna : Braümüller u. Seidel vi 106 pp. [18,83] [*non Homalonotus* Koenig, 1825 and *non Homalonotus* Schoenherr, 1836]. Type species *Grammatoph[ora] gaimardii* Duméril and Bibron, 1837 by original designation.

Ctenophorus Fitzinger, L.J. (1843). *Systema Reptilium.* Vienna : Braümüller u. Seidel vi 106 pp. [18,83]. Type species *Grammatoph[ora] decresii* Duméril and Bibron, 1837 by original designation.

Petroplanis Fitzinger, L.J. (1843). *Systema Reptilium.* Vienna : Braümüller u. Seidel vi 106 pp. [83,84] [*nom. nud.* introduced in synonymy of *Amphibolurus* Wagler, 1830]. Type species *Petroplanis jacksoniensis* Fitzinger, 1843 (=?*Agama jacksoniensis* Cloquet, 1816) by monotypy.

Polylophus Fitzinger, L.J. (1843). *Systema Reptilium.* Vienna : Braümüller u. Seidel vi 106 pp. [83,84] [*nom. nud.* introduced in synonymy of *Amphibolurus* Wagler, 1830]. Type species *Polylophus jacksoniensis* Fitzinger, 1843 (=?*Agama jacksoniensis* Cloquet, 1816) by monotypy.

Macrops De Vis, C.W. (1884). On new species of Australian lizards. *Proc. R. Soc. Qd.* **1**: 97–100 [97] [*non Macrops* Wagler, 1830; *non Macrops* Burmeister, 1835; *non Macrops* Kirby, 1837; *non Macrops* Duméril, 1856]. Type species *Macrops nuchalis* De Vis, 1884 by monotypy.

Synonymy that of Cogger, H.G., this work; Wermuth, H. (1967). Liste der rezenten Amphibien und Reptilien; Agamidae. *Das Tierreich* **86**: 1–127.

Amphibolurus adelaidensis (Gray, 1841)

Grammatophora muricata adelaidensis Gray, J.E. (1841). Description of some new species and four new genera of reptiles from Western Australia, discovered by John Gould, Esq. *Ann. Mag. Nat. Hist.* **7**: 86–91 [88]. Type data: syntypes, BMNH 1946.9.4.45–57 from Swan River

[W.A.] and "W. Australia", original description cites Adelaide, Western Australia. [type specimens cited from Gray, J.E. (1845). *Catalogue of the Specimens of Lizards in the Collection of the British Museum.* London : British Museum xxviii 289 pp. and Boulenger, G.A. (1885). *Catalogue of the Lizards in the British Museum (Natural History).* 1. London : British Museum xii 436 pp. pls 32, but these specimens are not currently regarded as types by BMNH authorities].

***Grammatophora angulifera* (part.)** Gray, J.E. (1845). *Catalogue of the Specimens of Lizards in the Collection of the British Museum.* London : British Museum xxviii 289 pp. [252] [var. 2 and var. 3; for remaining syntypes see *Amphibolurus diemensis* (Gray, 1841)]. Type data: syntypes, BMNH xxiv. 49.n,o,p,q,r, BMNH 69.5.25.15, from Western Australia and Perth, W.A.

Amphibolurus pulcherrimus Boulenger, G.A. (1885). *Catalogue of the Lizards in the British Museum (Natural History).* 1. London : British Museum xii 436 pp. pls 32 [388 pl 30 fig 2]. Type data: syntypes, BMNH 69.5.25.11–14, from "West Australia".

Amphibolurus pallidus Boulenger, G.A. (1885). *Catalogue of the Lizards in the British Museum (Natural History).* 1. London : British Museum xii 436 pp. pls 32 [388 pl 30 fig 3]. Type data: syntypes, BMNH xxiv.49.n,o,p,q,r, BMNH 69.5.25.15, from "West Australia" and Perth, W.A.

Amphibolurus adelaidensis tasmaniensis Boulenger, G.A. (1885). *Catalogue of the Lizards in the British Museum (Natural History).* 1. London : British Museum xii 436 pp. pls 1–32 [388]. Type data: syntypes, BMNH xxiii.50.j,l–o (9 specimens), from Tas. (apparently in error, BMNH records leave little doubt that correct localities for these syntypes are "Swan River, W.A." and "West Australia").

Amphibolurus adelaidensis chapmani Storr, G.M. (1977). The *Amphibolurus adelaidensis* species group (Lacertilia, Agamidae) in Western Australia. *Rec. West. Aust. Mus.* **5**: 73–81 [80]. Type data: holotype, WAM R24657, from 11 km SSE of Cocklebiddy, W.A., in 32°08′S, 126°08′E.

Synonymy that of Cogger, H.G., this work.

Distribution: NW coastal, SW coastal, (W plateau), S Gulfs, S.A., W.A. Ecology: low shrubland, open heath, open scrub, woodland, open forest, terrestrial, diurnal; oviparous, arthropod-feeder. Biological references: Tyler, M.J. (1960). Observations on the diet and size variation of *Amphibolurus adelaidensis* (Gray)(Reptilia : Agamidae) on the Nullarbor Plain. *Trans. R. Soc. S. Aust.* **83**: 111–117 figs 2; Storr, G.M. (1977). The *Amphibolurus adelaidensis* species group (Lacertilia, Agamidae) in Western Australia. *Rec. West. Aust. Mus.* **5**: 73–81 (taxonomy); Houston, T.F. (1978). *Dragon Lizards and Goannas of South Australia.* Adelaide : South Australian Museum 84 pp.

Amphibolurus auritus (Storr, 1981)

Tympanocryptis aurita Storr, G.M. (1981). Three new agamid lizards from Western Australia. *Rec. West. Aust. Mus.* **8**: 599–607 [599]. Type data: holotype, WAM R66296 M, from 27 km SSE of Halls Creek, W.A., in 18°27′S, 127°45′E.

Distribution: N coastal, W.A. Ecology: hummock grassland, terrestrial, diurnal; oviparous, arthropod-feeder.

Amphibolurus barbatus (Cuvier, 1829)

Ag[ama] barbata Cuvier, G.L.C.F.D. (1829). *Le Règne Animal.* 2nd edn. Paris : Déterville Vol. 2 xv 406 pp. [35]. Type data: syntypes, MNHP 7375, MNHP 7489, from Australia.

Distribution: SE coastal, NE coastal, (W plateau), S Gulfs, Murray-Darling basin, S.A., N.S.W., Qld., A.C.T., Vic. Ecology: open scrub, low open woodland, woodland, open forest, tall forest, terrestrial, arboreal, diurnal; oviparous, omnivore. Biological references: Badham, J.A. (1976). The *Amphibolurus barbatus* species-group (Lacertilia : Agamidae). *Aust. J. Zool.* **24**: 423–443 (taxonomy); Warburg, M.R. (1965). The influence of ambient temperature and humidity on the body temperature and water loss from two Australian lizards, *Tiliqua rugosa* (Gray) (Scincidae) and *Amphibolurus barbatus* Cuvier (Agamidae). *Aust. J. Zool.* **13**: 331–350; Badham, J.A. (1971). Albumin formation in eggs of the agamid *Amphibolurus barbatus barbatus. Copeia* **1971**: 543–545; Brattstrom, B.H. (1971). Social and thermoregulatory behavior of the bearded dragon, *Amphibolurus barbatus. Copeia* **1971**: 484–497 (behaviour, may be referring to *A. vitticeps*); Bartholomew, G.A. & Tucker, V.A. (1963). Control of changes in body temperature, metabolism, and circulation by the agamid lizard, *Amphibolurus barbatus. Physiol. Zool.* **36**: 199–218; Carpenter, C.C., Badham, J.A. & Kimble, B. (1970). Behavior patterns of three species of *Amphibolurus* (Agamidae). *Copeia* **1970**: 497–505 (behaviour, may be referring to *A. vitticeps*).

Amphibolurus caudicinctus (Günther, 1875)

Grammatophora caudicincta Günther, A. (1875). Reptiles. A list of saurians of Australia and New Zealand. pp. 9–19 *Zoology of the Voyage of H.M.S.* Erebus *and* Terror. Vol. 2 London : E.W. Janson [19]. Type data: holotype, BMNH 1946.9.4.43, from Nickol (as Nicol) Bay, W.A.

Amphibolurus imbricatus Peters, W. (1876). Über die von S.M.S. *Gazelle* mitgebrachten Amphibien. *Mber. K. Preuss. Akad. Wiss. Berl.* **1876**: 528–535 [529] [1877 on title page]. Type data: syntypes, ZMB 8933 (2 specimens), from Mermaid Strait, W.A.(as Nova Hollandia borealis, Meermaidstraat).

Amphibolurus caudicinctus graafi Storr, G.M. (1967). Geographic races of the agamid lizard *Amphibolurus caudicinctus. J. R. Soc. West. Aust.* **50**: 49–56 [51]. Type data: holotype, WAM R25914, from Mt. Eveline, in 26°10′S, 127°06′E, W.A.

Amphibolurus caudicinctus infans Storr, G.M. (1967). Geographic races of the agamid lizard *Amphibolurus caudicinctus*. *J. R. Soc. West. Aust.* **50**: 49–56 [51]. Type data: holotype, WAM R25945, from Deeba Rock-Hole, 25 mi NE of Laverton, in 28°22′S, 122°35′E, W.A.

Amphibolurus caudicinctus mensarum Storr, G.M. (1967). Geographic races of the agamid lizard *Amphibolurus caudicinctus*. *J. R. Soc. West. Aust.* **50**: 49–56 [51]. Type data: holotype, WAM R19486, from 5 mi S of Meekatharra, in 26°40′S, 118°27′E, W.A.

Amphibolurus caudicinctus macropus Storr, G.M. (1967). Geographic races of the agamid lizard *Amphibolurus caudicinctus*. *J. R. Soc. West. Aust.* **50**: 49–56 [53]. Type data: holotype, SAMA R3229, from 4.5 mi SSE of Oenpelli, N.T.

Amphibolurus caudicinctus slateri Storr, G.M. (1967). Geographic races of the agamid lizard *Amphibolurus caudicinctus*. *J. R. Soc. West. Aust.* **50**: 49–56 [52]. Type data: holotype, WAM R26793, from Hermannsburg, in 23°58′S, 132°46′E, N.T.

Synonymy that of Cogger, H.G., this work.

Distribution: N Gulf, N coastal, NW coastal, W plateau, Lake Eyre basin, Qld., W.A., N.T., (S.A.). Ecology: desert, hummock grassland, low open shrubland, tall open shrubland, low open woodland, woodland, open forest, terrestrial, diurnal; oviparous, sexually dimorphic, saxicoline, arthropod-feeder. Biological references: Storr, G.M. (1967). Geographic races of the agamid lizard *Amphibolurus caudicinctus*. *J. R. Soc. West. Aust.* **50**: 49–56 (taxonomy); Bradshaw, S.D. & Main, A.R. (1968). Behavioural attitudes and regulation of temperature in *Amphibolurus* lizards. *J. Zool.* **154**: 193–221; Bradshaw, S.D. (1970). Seasonal changes in the water and electrolyte metabolism of *Amphibolurus* lizards in the field. *Comp. Biochem. Physiol.* **36**: 689–718; Houston, T.F. (1978). *Dragon Lizards and Goannas of South Australia*. Adelaide : South Australian Museum 84 pp. (taxonomy, natural history); Houston, T.F. (1974). Revision of the *Amphibolurus decresii* complex (Lacertilia : Agamidae) of South Australia. *Trans. R. Soc. S. Aust.* **98**: 49–60 (taxonomy).

Amphibolurus clayi Storr, 1966

Amphibolurus clayi Storr, G.M. (1966). The *Amphibolurus reticulatus* species-group (Lacertilia : Agamidae) in Western Australia. *J. R. Soc. West. Aust.* **49**: 17–25 [24]. Type data: holotype, WAM R14462, from 3 mi S of Learmonth, in 22°16′S, 114°06′E, W.A.

Distribution: NW coastal, W plateau, Lake Eyre basin, S.A., W.A., N.T. Ecology: desert, hummock grassland, low open shrubland, open scrub, tall shrubland, terrestrial, diurnal, (low woodland); oviparous, arthropod-feeder.

Amphibolurus cristatus (Gray, 1841)

Grammatophora cristata Gray, J.E. (1841). Description of some new species and four new genera of reptiles from Western Australia, discovered by John Gould, Esq. *Ann. Mag. Nat. Hist.* **7**: 86–91 [87]. Type data: holotype, BMNH 1946.9.4.41, from W.A.

Distribution: SW coastal, W plateau, S Gulfs, S.A., W.A. Ecology: desert, low open shrubland, open scrub, tall shrubland, terrestrial, diurnal; oviparous, sexually dimorphic, arthropod-feeder. Biological references: Warburg, M.R. (1965). Studies on the environmental physiology of some Australian lizards from arid and semiarid habitats. *Aust. J. Zool.* **13**: 563–575; Pianka, E.R. (1971). Notes on the biology of *Amphibolurus cristatus* and *Amphibolurus scutulatus*. *West. Aust. Nat.* **12**: 36–41; Pianka, E.R. (1972). Zoogeography and speciation of Australian desert lizards: an ecological perspective. *Copeia* **1972**: 127–144.

Amphibolurus decresii (Duméril and Bibron, 1837)

Agama decresiensis Cloquet, H. (1816). *Dictionnaire des Sciences Naturelles,* dans lequel on Traité Méthodiquement des Differens Etres de la Nature. G.L.C.F.D. Cuvier (ed.) Paris & Strasbourg [75] [*nom. nud.*].

Grammatophora decresii Duméril, A.M.C. & Bibron, G. (1837). *Erpétologie Générale* ou Histoire Naturelle Complète des Reptiles. Paris : Roret 4: ii 571 pp. [472]. Type data: syntypes, MNHP 6545 (2 specimens), from Kangaroo Is. (as Île de Decrès), S.A.

Amphibolurus modestus Ahl, E. (1926). Neue Eidechsen und Amphibien. *Zool. Anz.* **67**: 186–192 [187]. Type data: syntypes, ZMB 30092 (2 specimens), from Australia.

Synonymy that of Cogger, H.G., this work.

Distribution: S Gulfs, Lake Eyre basin, S.A., N.S.W. Ecology: tussock grassland, low open shrubland, tall shrubland, terrestrial, diurnal, (low shrubland); oviparous, sexually dimorphic, saxicoline, arthropod-feeder. Biological references: Warburg, M.R. (1965). Studies on the environmental physiology of some Australian lizards from arid and semiarid habitats. *Aust. J. Zool.* **13**: 563–575; Houston, T.F. (1978). *Dragon Lizards and Goannas of South Australia*. Adelaide : South Australian Museum 84 pp. (natural history); Houston, T.F. (1974). Revision of the *Amphibolurus decresii* complex (Lacertilia : Agamidae) of South Australia. *Trans. R. Soc. S. Aust.* **98**: 49–60 (taxonomy); Gibbons, J.R.H. (1979). The hind leg pushup display of the *Amphibolurus decresii* species complex (Lacertilia : Agamidae). *Copeia* **1979**: 29–40 (behaviour); Gibbons, J.R.H. & Lillywhite, H.B. (1981). Ecological segregation, color matching, and speciation in lizards of the *Amphibolurus decresii* species complex (Lacertilia : Agamidae). *Ecology* **62**: 1573–1584.

Amphibolurus diemensis (Gray, 1841)

[Agama jacksoniensis **(part.)***]* Cloquet, H. (1816). *Dictionnaire des Sciences Naturelles,* dans lequel on Traité Méthodiquement des Differens Etres de la Nature. G.L.C.F.D. Cuvier (ed.) Paris & Strasbourg [74]. Type data: paralectotype, from Australia, RMNH 3117 (smaller of 2 specimens) from Australia; for other syntype designated lectotype in this work see under *Amphibolurus muricatus* (White, 1790).

Agama diemenensis Cloquet, H. (1816). *Dictionnaire des Sciences Naturelles,* dans lequel on Traité Méthodiquement des Differens Etres de la Nature. G.L.C.F.D. Cuvier (ed.) Paris & Strasbourg [75] [*nom. nud.*].

Grammatophora muricata diemensis Gray, J.E. (1841). Description of some new species and four new genera of reptiles from Western Australia, discovered by John Gould, Esq. *Ann. Mag. Nat. Hist.* **7**: 86–91 [88]. Type data: whereabouts unknown, from Tas.

Grammatophora angulifera **(part.)** Gray, J.E. (1845). *Catalogue of the Specimens of Lizards in the Collection of the British Museum.* London : British Museum xxviii 289 pp. [252] [var. 1; for remaining syntypes see *Amphibolurus adelaidensis* (Gray, 1841)]. Type data: syntypes, BMNH 1946.9.4.29–30, BMNH 1946.9.4.58–66, BMNH xxiv.49.f, from Tas.

Agama caelaticeps Smith, A. (1849). *Illustrations of the Zoology of South Africa.* Reptilia. London : Smith, Elder & Co. [pl 74 and accompanying text]. Type data: syntypes, BMNH 1946.9.4.33–34, from "towards the Tropic of Capricorn, South Africa" (*errore*).

Synonymy that of Cogger, H.G., this work.

Distribution: SE coastal, Murray-Darling basin, Tas., Vic., N.S.W., A.C.T. Ecology: alpine, woodland, open forest, terrestrial, arboreal, diurnal, (tall forest); oviparous, arthropod-feeder. Biological references: Baehr, M. (1976). Beiträge zur Verbreitung und Ökologie tasmanischer Reptilien. *Stuttgart Beitr. Naturk. (A)* **292**: 1–24 (natural history); Jenkins, R.W.G. & Bartell, R.J. (1980). *A Field Guide to Reptiles of the Australian High Country.* Melbourne : Inkata Press 187 pp. (natural history).

Amphibolurus femoralis Storr, 1965

Amphibolurus femoralis Storr, G.M. (1965). The *Amphibolurus maculatus* species-group (Lacertilia : Agamidae) in Western Australia. *J. R. Soc. West. Aust.* **48**: 45–54 [49]. Type data: holotype, WAM R9008, from 23 mi NE of Giralia, in 22°29′S, 114°33′E, W.A.

Distribution: NW coastal, W.A. Ecology: hummock grassland, low open shrubland, terrestrial, diurnal; oviparous, arthropod-feeder.

Amphibolurus fionni Procter, 1923

Amphibolurus fionni Procter, J.B. (1923). On new and rare reptiles and batrachians from the Australian region. *Proc. Zool. Soc. Lond.* **1923**: 1069–1077 [1075 figs 4a–c]. Type data: holotype, BMNH 1946.9.4.67, from Port Lincoln, S.A. (original description cites S.A., "coast of the mainland", S.A.).

Distribution: W plateau, S Gulfs, S.A. Ecology: open scrub, low woodland, terrestrial, diurnal; oviparous, saxicoline, arthropod-feeder. Biological references: Houston, T.F. (1974). Revision of the *Amphibolurus decresii* complex (Lacertilia : Agamidae) of South Australia. *Trans. R. Soc. S. Aust.* **98**: 49–60 (taxonomy); Warburg, M.R. (1966). Water economy of several Australian geckos, agamids, and water skinks. *Copeia* **1966**: 230–235; Houston, T.F. (1978). *Dragon Lizards and Goannas of South Australia.* Adelaide : South Australian Museum 84 pp. (natural history); Gibbons, J.R.H. (1979). The hind leg pushup display of the *Amphibolurus decresii* species complex (Lacertilia : Agamidae). *Copeia* **1979**: 29–40 (behaviour); Gibbons, J.R.H. & Lillywhite, H.B. (1981). Ecological segregation, color matching, and speciation in lizards of the *Amphibolurus decresii* species complex (Lacertilia : Agamidae). *Ecology* **62**: 1573–1584.

Amphibolurus fordi Storr, 1965

Amphibolurus fordi Storr, G.M. (1965). The *Amphibolurus maculatus* species-group (Lacertilia : Agamidae) in Western Australia. *J. R. Soc. West. Aust.* **48**: 45–54 [48]. Type data: holotype, WAM R19151, from 12 mi NW of Coolgardie, in 30°49′S, 121°02′E, W.A.

Distribution: W plateau, S Gulfs, Murray-Darling basin, Bulloo River basin, Lake Eyre basin, S.A., Vic., N.S.W., W.A. Ecology: open scrub, tall shrubland, terrestrial, diurnal; oviparous, arthropod-feeder, saxicoline. Biological references: Cogger, H.G. (1974). Thermal relations of the mallee dragon, *Amphibolurus fordi* (Lacertilia : Agamidae). *Aust. J. Zool.* **22**: 319–339; Cogger, H.G. (1978). Reproductive cycles, fat body cycles and socio-sexual behaviour in the mallee dragon, *Amphibolurus fordi* (Lacertilia : Agamidae). *Aust. J. Zool.* **26**: 653–672; Bradshaw, S.D. (1981). Ecophysiology of Australian desert lizards: studies on the genus *Amphibolurus*. pp. 1393–1434 *in* Keast, A. (ed.) *Ecological Biogeography of Australia.* The Hague : Junk.

Amphibolurus gibba Houston, 1974

Amphibolurus gibba Houston, T.F. (1974). *Amphibolurus gibba*, a new dragon lizard (Lacertilia : Agamidae) from northern South Australia. *Trans. R. Soc. S. Aust.* **98**: 209–212 [209 figs 2–4]. Type data: holotype, from 5.5 km NNW of Alberrie Creek Railway Siding, in 29°35′S,

137°31'E, S.A., SAMA R13954, although Houston, T.F. (1976). Vertebrate type-specimens in the South Australian Museum. III. Reptiles. *Rec. S. Aust. Mus.* **17**: 181–187 lists SAMA R13954A.

Distribution: Lake Eyre basin, S.A. Ecology: desert, terrestrial, diurnal; oviparous, arthropod-feeder.

Amphibolurus isolepis (Fischer, 1881)

Grammatophora isolepis Fischer, J.G. (1881). Beschreibung neuer Reptilien. *Arch. Naturgesch.* **47**: 225–238 [232 pl 12 figs 10–12]. Type data: syntypes, SMNS 2051, ZMB 10072 (2 specimens), from W.A. [Storr, G.M. (1965). The *Amphibolurus maculatus* species-group (Lacertilia : Agamidae) in Western Australia. *J. R. Soc. West. Aust.* **48**: 45–54 records 2 specimens registered under SMNS 2051; Storr also restricted the type locality to Nickol Bay, W.A.].

Amphibolurus maculatus gularis Sternfeld, R. (1924). Beiträge zur Herpetologie Inner-Australiens. *Abh. Senckenb. Naturf. Ges.* **38**: 221–251 [231]. Type data: lectotype, SMF 10294, from Hermannsburg Mission, upper Finke River, N.T., designation by Wermuth, H. (1967). Liste der rezenten Amphibien und Reptilien; Agamidae. *Das Tierreich* **86**: 1–127.

Amphibolurus isolepis rubens Storr, G.M. (1965). The *Amphibolurus maculatus* species-group (Lacertilia : Agamidae) in Western Australia. *J. R. Soc. West. Aust.* **48**: 45–54 [50]. Type data: holotype, WAM R18740, from Yanrey, in 22°30'S, 114°48'E, W.A.

Amphibolurus isolepis citrinus Storr, G.M. (1965). The *Amphibolurus maculatus* species-group (Lacertilia : Agamidae) in Western Australia. *J. R. Soc. West. Aust.* **48**: 45–54 [51]. Type data: holotype, WAM R21706, from 2 mi W of Boorabbin, in 31°12'S, 120°17'E, W.A.

Synonymy that of Cogger, H.G., this work.

Distribution: N coastal, NW coastal, W plateau, Lake Eyre basin, S.A., Qld., W.A., N.T., (N Gulf). Ecology: desert, hummock grassland, tussock grassland, low open shrubland, woodland, tall shrubland, terrestrial, diurnal; oviparous, sexually dimorphic, arthropod-feeder, in clumps of spinifex. Biological references: Storr, G.M. (1965). The *Amphibolurus maculatus* species-group (Lacertilia : Agamidae) in Western Australia. *J. R. Soc. West. Aust.* **48**: 45–54 (taxonomy); Pianka, E.R. (1971). Ecology of the agamid lizard *Amphibolurus isolepis* in Western Australia. *Copeia* **1971**: 527–536; Warburg, M.R. (1966). Water economy of several Australian geckos, agamids, and water skinks. *Copeia* **1966**: 230–235 (as *A.m. maculatus*).

Amphibolurus maculatus (Gray, 1831)

Uromastyx maculatus Gray, J.E. (1831). A synopsis of the species of the class Reptilia, pp. 1–110 *in* Griffith, E. *The Animal Kingdom Arranged in Conformity with its Organization by the Baron Cuvier.* London : Whittaker, Treacher & Co. 9: 481+110 pp. [62] [Gray's description referred to Cuvier's ms name applied to a specimen, almost certainly MNHP 1259, on which the name *Grammatophora gaimardii* is also based]. Type data: holotype (probable), MNHP 1259, from unspecified locality, but probable holotype from Shark Bay, W.A.

Grammatophora gaimardii Duméril, A.M.C. & Bibron, G. (1837). *Erpétologie Générale* ou Histoire Naturelle Complète des Reptiles. Paris : Roret 4: ii 571 pp. [470]. Type data: holotype, MNHP 1259, from vicinity of Shark Bay, W.A. [treated as a distinct species with its own type specimen *vide* Guibé, J. (1954). *Catalogue des Types de Lézards du Muséum National d'Histoire Naturelle.* Paris : Muséum National d'Histoire Naturelle 119 pp., but is more properly regarded as a replacement name for *Uromastyx maculatus* Gray, 1831].

Amphibolurus maculatus badius Storr, G.M. (1965). The *Amphibolurus maculatus* species-group (Lacertilia : Agamidae) in Western Australia. *J. R. Soc. West. Aust.* **48**: 45–54 [46]. Type data: holotype, WAM R18624, from 15 mi SE of Cardabia, in 23°15'S, 113°55'E, W.A.

Amphibolurus maculatus dualis Storr, G.M. (1965). The *Amphibolurus maculatus* species-group (Lacertilia : Agamidae) in Western Australia. *J. R. Soc. West. Aust.* **48**: 45–54 [47]. Type data: holotype, WAM R23637, from 7 mi SSE of Cocklebiddy, in 32°08'S, 126°08'E, W.A.

Amphibolurus maculatus griseus Storr, G.M. (1965). The *Amphibolurus maculatus* species-group (Lacertilia : Agamidae) in Western Australia. *J. R. Soc. West. Aust.* **48**: 45–54 [47]. Type data: holotype, WAM R21708, from 14 mi W of Boorabbin, in 31°15'S, 120°05'E, W.A.

Synonymy that of Cogger, H.G., this work.

Distribution: NW coastal, SW coastal, W plateau, W.A. Ecology: open heath, open scrub, woodland, tall shrubland, terrestrial, diurnal; oviparous, sexually dimorphic, arthropod-feeder.

Amphibolurus maculosus (Mitchell, 1948)

Tympanocryptis maculosa Mitchell, F.J. (1948). A revision of the lacertilian genus *Tympanocryptis. Rec. S. Aust. Mus.* **9**: 57–86 [78 fig 9 pl 5]. Type data: holotype, SAMA R2220a, from Lake Eyre North, S.A.

Distribution: Lake Eyre basin, S.A. Ecology: desert, terrestrial, diurnal; oviparous, arthropod-feeder, found on salt-encrusted margins of salt lakes. Biological references: Mitchell, F.J. (1965). The affinities of *Tympanocryptis maculosa* Mitchell (Lacertilia : Agamidae). *Rec. S. Aust.* **15**: 179–191 (taxonomy); Mitchell, F.J. (1973). Studies on the ecology of the agamid lizard *Amphibolurus maculosus* (Mitchell). *Trans. R. Soc. S. Aust.* **97**: 47–76; Braysher, M.L. (1976). The excretion of hyperosmotic urine and other aspects of the electrolyte balance of the lizard *Amphibolurus maculosus*. *Comp. Biochem. Physiol.* **54A**: 341–345.

Amphibolurus mckenziei Storr, 1981

Amphibolurus mckenziei Storr, G.M. (1981). Three new agamid lizards from Western Australia. *Rec. West. Aust.*

Mus. **8**: 599–607 [605]. Type data: holotype, WAM R59754 M, from 8 km SW of Ponier Rock, W.A., in 32°59'S, 123°26'E.

Distribution: W plateau, W.A. Ecology: low shrubland, open scrub, terrestrial, diurnal; oviparous, arthropod-feeder.

Amphibolurus microlepidotus Glauert, 1952

Amphibolurus barbatus microlepidotus Glauert, L. (1952). Herpetological miscellanea. I. Notes on some forms of *Diplodactylus*. pp. 166–168; Some new western Australian lizards. pp. 168–170. *West. Aust. Nat.* **3**: 166–170 [168]. Type data: syntypes, WAM R591-2, from Drysdale River Mission, North Kimberley, W.A.

Distribution: N coastal, W.A. Ecology: woodland, terrestrial, arboreal, diurnal; oviparous, arthropod-feeder. Biological references: Badham, J.A. (1976). The *Amphibolurus barbatus* species-group (Lacertilia : Agamidae). *Aust. J. Zool.* **24**: 423–443 (taxonomy).

Amphibolurus minimus Loveridge, 1933

Amphibolurus barbatus minimus Loveridge, A. (1933). New agamid lizards of the genera *Amphibolurus* and *Physignathus* from Australia. *Proc. New. Engl. Zool. Club* **13**: 69–72 [69]. Type data: holotype, MCZ 32972, from West Wallaby Is., Houtman's Abrolhos, W.A.

Distribution: NW coastal, SW coastal, W plateau, W.A. Ecology: open heath, open scrub, woodland, open forest, terrestrial, arboreal, diurnal, (hummock grassland); oviparous, arthropod-feeder. Biological references: Bradshaw, S.D. & Main, A.R. (1968). Behavioural attitudes and regulation of temperature in *Amphibolurus* lizards. *J. Zool.* **154**: 193–221; Badham, J.A. (1976). The *Amphibolurus barbatus* species-group (Lacertilia : Agamidae). *Aust. J. Zool.* **24**: 423–443 (taxonomy); Bradshaw, S.D. (1970). Seasonal changes in the water and electrolyte metabolism of *Amphibolurus* lizards in the field. *Comp. Biochem. Physiol.* **36**: 689–718.

Amphibolurus minor Sternfeld, 1919

Amphibolurus barbatus minor Sternfeld, R. (1919). Neue Schlangen und Echsen aus Zentralaustralien. *Senckenbergiana* **1**: 76–83 [78]. Type data: lectotype, SMF 10339, from Hermannsburg Mission, upper Finke River, N.T., designation by Mertens, R. (1967). Die herpetologische Sektion des Natur-Museums und Forschungs-Institutes Senckenberg in Frankfurt-a.-M. nebst einem Verzeichnis ihrer Typen. 1. *Senckenberg. Biol.* **48**: 1–106.

Distribution: NW coastal, W plateau, Lake Eyre basin, S.A., W.A., N.T. Ecology: desert, hummock grassland, low open shrubland, open scrub, tall shrubland, terrestrial, arboreal, diurnal; oviparous, arthropod-feeder. Biological references: Bradshaw, S.D. & Main, A.R. (1968). Behavioural attitudes and regulation of temperature in *Amphibolurus* lizards. *J. Zool.* **154**: 193–221; Badham, J.A. (1976). The *Amphibolurus barbatus* species-group (Lacertilia : Agamidae). *Aust. J. Zool.* **24**: 423–443 (taxonomy); Bradshaw, S.D. (1981). Ecophysiology of Australian desert lizards: studies on the genus *Amphibolurus*. pp. 1393–1434 *in* Keast, A. (ed.) *Ecological Biogeography of Australia.* The Hague : Junk.

Amphibolurus mitchelli Badham, 1976

Amphibolurus mitchelli Badham, J.A. (1976). The *Amphibolurus barbatus* species-group (Lacertilia : Agamidae). *Aust. J. Zool.* **24**: 423–443 [435 fig 4c]. Type data: holotype, WAM R26828, from Derby, W.A.

Distribution: N coastal, NW coastal, W plateau, W.A., N.T. Ecology: desert, low open shrubland, low open woodland, terrestrial, arboreal, diurnal; oviparous, arthropod-feeder.

Amphibolurus muricatus (White, 1790)

Lacerta muricata White, J. (1790). *Journal of a Voyage to New South Wales,* with sixty five plates of non-descript animals, birds, lizards, serpents, curious cones of trees and other natural productions. Appendix. London : Debrett 299 pp. [244 pl 31 fig 2] [*ex* Shaw ms]. Type data: holotype, BMNH 1946.9.4.44, from N.S.W.

***Agama jacksoniensis* (part.)** Cloquet, H. (1816). *Dictionnaire des Sciences Naturelles,* dans lequel on Traité Méthodiquement des Differens Etres de la Nature. G.L.C.F.D. Cuvier (ed.) Paris & Strasbourg [74] [for the paralectotype see under *Amphibolurus diemensis* Gray, 1841]. Type data: lectotype, RMNH 3117 (larger of 2 specimens), from Australia, designation by Cogger, H.G., this work.

Agama jaksoniensis Guérin-Méneville, F.E. (1838). *Iconographiae du Règne Animal de G.L.C.F.D. Cuvier,* ou représentation d'après nature de l'une des espèces les plus remarquables, et souvent non encore figurées, de chaque genre animaux. 1. Reptiles. Paris [pl 7 fig 1] [invalid emend. *pro Agama jacksoniensis* Cloquet, 1816].

Synonymy that of Cogger, H.G., this work.

Distribution: SE coastal, NE coastal, W plateau, S Gulfs, Murray-Darling basin, S.A., Vic., N.S.W., W.A., A.C.T., Qld. Ecology: low open woodland, woodland, open forest, terrestrial, arboreal, diurnal, (open scrub), (tall forest); oviparous, arthropod-feeder. Biological references: Chong, G., Heatwole, H. & Firth, B.T. (1973). Panting thresholds of lizards. II. Diel variation in the panting threshold of *Amphibolurus muricatus. Comp. Biochem. Physiol.* **46A**: 827–829; Witten, G.J. (1972). A new species of *Amphibolurus* from eastern Australia. *Herpetologica* **28**: 191–195 (taxonomy); Firth, B.T. & Heatwole, H. (1976). Panting thresholds of lizards: the role of the pineal complex in panting responses in an agamid, *Amphibolurus muricatus. Gen. Comp. Endocrinol.*

29: 388–401; Heatwole, H., Firth, B.T. & Stoddart, H. (1975). Influence of season, photoperiod and thermal acclimation on the panting threshold of *Amphibolurus muricatus*. *J. Exp. Zool.* **191**: 183–192; Carpenter, C.C., Badham, J.A. & Kimble, B. (1970). Behavior patterns of three species of *Amphibolurus* (Agamidae). *Copeia* **1970**: 497–505; Groom, S. (1973). Notes on the keeping and distribution of the jacky lizard (*Amphibolurus muricatus*). *Herpetofauna* **5**: 3–5 (natural history).

Amphibolurus nobbi Witten, 1972

Amphibolurus nobbi nobbi Witten, G.J. (1972). A new species of *Amphibolurus* from eastern Australia. *Herpetologica* **28**: 191–195 [191 fig 1]. Type data: holotype, AM R33440, from 24 mi ENE Guyra, N.S.W.

Amphibolurus nobbi coggeri Witten, G.J. (1972). A new species of *Amphibolurus* from eastern Australia. *Herpetologica* **28**: 191–195 [192 fig 2]. Type data: holotype, AM R17660, from Warrumbungle Mts., N.S.W.

Synonymy that of Cogger, H.G., this work.

Distribution: SE coastal, NE coastal, S Gulfs, Lake Eyre basin, Murray-Darling basin, S.A., Vic., N.S.W., Qld. Ecology: open scrub, low open woodland, low woodland, woodland, open forest, tall shrubland, terrestrial, arboreal, diurnal; oviparous, arthropod-feeder. Biological references: Bustard, H.R. (1968). The reptiles of Merriwindi State Forest, Pilliga West, northern New South Wales. *Herpetologica* **24**: 131–140 (natural history, as *Amphibolurus* sp.); Witten, G.J. (1974). Population movements of the agamid lizard, *Amphibolurus nobbi*. *Aust. Zool.* **18**: 129–132.

Amphibolurus nuchalis (De Vis, 1884)

Macrops nuchalis De Vis, C.W. (1884). On new species of Australian lizards. *Proc. R. Soc. Qd.* **1**: 97–100 [97]. Type data: syntypes, QM J1405–9; ?QM J1410, from Delta Station, Bogantungan, Qld.

Grammatophora inermis De Vis, C.W. (1888). A contribution to the herpetology of Queensland. *Proc. Linn. Soc. N.S.W. (2)* **2**: 811–826 [812]. Type data: none designated, whereabouts unknown, from central Qld.

Amphibolurus reticulatus major Sternfeld, R. (1919). Neue Schlangen und Echsen aus Zentralaustralien. *Senckenberg. Biol.* **1**: 76–83 [78]. Type data: lectotype, SMF 10316, from Hermannsburg Mission, upper Finke River, N.T., designation by Mertens, R. (1967). Die herpetologische Sektion des Natur-Museums und Forschungs-Institutes Senckenberg in Frankfurt-a.-M. nebst einem Verzeichnis ihrer Typen. 1. *Senckenberg. Biol.* **48**: 1–106.

Synonymy that of Cogger, H.G., this work.

Distribution: (N Gulf), N coastal, NW coastal, (SW coastal), W plateau, Murray-Darling basin, Bulloo River basin, Lake Eyre basin, S.A., N.S.W., Qld., W.A., N.T. Ecology: desert, hummock grassland, tussock grassland, low open shrubland, low shrubland, open heath, open scrub, low open woodland, woodland, tall shrubland, terrestrial, diurnal, burrowing; oviparous, arthropod-feeder. Biological references: Storr, G.M. (1966). The *Amphibolurus reticulatus* species-group (Lacertilia : Agamidae) in Western Australia. *J. R. Soc. West. Aust.* **49**: 17–25 (taxonomy, as *A. inermis*); Heatwole, H. (1970). Thermal ecology of the desert dragon *Amphibolurus inermis*. *Ecol. Monogr.* **40**: 425–457 (as *A. inermis*); Warburg, M.R. (1965). Studies on the environmental physiology of some Australian lizards from arid and semiarid habitats. *Aust. J. Zool.* **13**: 563–575 (as *A. reticulatus inermis*); Pianka, E.R. (1971). Comparative ecology of two lizards. *Copeia* **1971**: 129–138 (as *A. inermis*); Bradshaw, S.D. & Main, A.R. (1968). Behavioural attitudes and regulation of temperature in *Amphibolurus* lizards. *J. Zool.* **154**: 193–221 (as *A. inermis*); Bradshaw, S.D. (1970). Seasonal changes in the water and electrolyte metabolism of *Amphibolurus* lizards in the field. *Comp. Biochem. Physiol.* **36**: 689–718 (as *A. inermis*); Bradshaw, S.D. (1975). Osmoregulation and pituitary-adrenal function in desert reptiles. *Gen. Comp. Endocrinol.* **25**: 230–248 (physiology, as *A. inermis*); Bradshaw, S.D. (1981). Ecophysiology of Australian desert lizards: studies on the genus *Amphibolurus*. pp. 1393–1434 *in* Keast, A. (ed.) *Ecological Biogeography of Australia*. The Hague : Junk.

Amphibolurus nullarbor Badham, 1976

Amphibolurus nullarbor Badham, J.A. (1976). The *Amphibolurus barbatus* species-group (Lacertilia : Agamidae). *Aust. J. Zool.* **24**: 423–443 [440 fig 4g]. Type data: holotype, WAM R29667, from 16 km NW Naretha Railway Station, W.A.

Distribution: W plateau, W.A., S.A. Ecology: low shrubland, tall shrubland, terrestrial, (arboreal), diurnal, omnivore; oviparous. Biological references: Houston, T.F. (1978). *Dragon Lizards and Goannas of South Australia*. Adelaide : South Australian Museum 84 pp. (natural history).

Amphibolurus ornatus (Gray, 1845)

Grammatophora ornata Gray, J.E. (1845). *Catalogue of the Specimens of Lizards in the Collection of the British Museum*. London : British Museum xxviii 289 pp. [253]. Type data: holotype, BMNH 1946.4.42, from W.A.

Amphibolurus tibialis Ahl, E. (1926). Neue Eidechsen und Amphibien. *Zool. Anz.* **67**: 186–192 [188]. Type data: holotype, ZMB 30100, from Australia.

Synonymy that of Cogger, H.G., this work.

Distribution: (NW coastal), SW coastal, W plateau, W.A. Ecology: open heath, open scrub, woodland, open forest, terrestrial, diurnal; oviparous, saxicoline, arthropod-feeder. Biological references: Bradshaw, S.D. & Main, A.R. (1968). Behavioural attitudes and regulation of temperature in *Amphibolurus* lizards. *J. Zool.* **154**: 193–221; Bradshaw, S.D. & Shoemaker, V.H. (1967). Aspects of water and electrolyte changes in a field population of *Amphibolurus* lizards. *Comp. Biochem. Physiol.* **20**: 855–865; Baverstock, P.R. (1975). Effect of variations in rate of growth on physiological parameters in the lizard *Amphibolurus ornatus. Comp. Biochem. Physiol.* **51A**: 619–632; Bradshaw, S.D. (1970). Seasonal changes in the water and electrolyte metabolism of *Amphibolurus* lizards in the field. *Comp. Biochem. Physiol.* **36**: 689–718; Bradshaw, S.D. (1971). Growth and mortality in a field population of *Amphibolurus* lizards exposed to seasonal cold and aridity. *J. Zool.* **165**: 1–25; Baverstock, P.R. & Bradshaw, S.D. (1975). Variation in rate of growth and adrenal cortico steroidogenesis in field and laboratory populations of the lizard *Amphibolurus ornatus. Comp. Biochem. Physiol.* **52A**: 557–565; Bentley, P.J. & Bradshaw, S.D. (1972). Electrical potential difference across the cloaca and colon of the Australian lizards *Amphibolurus ornatus* and *A. inermis. Comp. Biochem. Physiol.* **42A**: 465–471 (physiology); Bradshaw, S.D. (1981). Ecophysiology of Australian desert lizards: studies on the genus *Amphibolurus*. pp. 1393–1434 *in* Keast, A. (ed.) *Ecological Biogeography of Australia.* The Hague : Junk.

Amphibolurus pictus Peters, 1866

Amphibolurus pictus Peters, W. (1866). Eine Mitteilung über neue Amphibien (*Amphibolurus*, *Lygosoma*, *Cyclodus*, *Masticophis*, *Crotaphopeltis*) und Fische (*Diagramma*: *Hapalogenys*) des Kgl. zoologischen Museums. *Mber. K. Preuss. Akad. Wiss. Berl.* **1866**: 86–96 [88] [1867 on title page]. Type data: syntypes, ZMB 4738 (3 specimens), ZMB 4739 (3 specimens), from S.A. (Adelaide, S.A. according to ZMB records).

Distribution: W plateau, S Gulfs, Murray-Darling basin, Bulloo River basin, Lake Eyre basin, S.A., Vic., N.S.W., Qld., W.A., N.T. Ecology: desert, tussock grassland, low open shrubland, low shrubland, open scrub, tall shrubland, terrestrial, diurnal, burrowing; oviparous, arthropod-feeder, sexually dimorphic. Biological references: Storr, G.M. (1966). The *Amphibolurus reticulatus* species-group (Lacertilia : Agamidae) in Western Australia. *J. R. Soc. West. Aust.* **49**: 17–25; Warburg, M.R. (1965). Studies on the environmental physiology of some Australian lizards from arid and semiarid habitats. *Aust. J. Zool.* **13**: 563–575; Houston, T.F. (1978). *Dragon Lizards and Goannas of South Australia.* Adelaide : South Australian Museum 84 pp.; Hudson, P. (1979). Notes on the behavioural antics of the painted dragon *Amphibolurus pictus* Peters. *Herpetofauna* **11**: 25–26.

Amphibolurus reticulatus (Gray, 1845)

Grammatophora reticulata Gray, J.E. (1845). *Catalogue of the Specimens of Lizards in the Collection of the British Museum.* London : British Museum xxviii 289 pp. [252]. Type data: syntypes, BMNH 1946.9.4.35–36, from W.A.

Grammatophora laevis Günther, A. (1867). Additions to the knowledge of Australian reptiles and fishes. *Ann. Mag. Nat. Hist. (3)* **20**: 45–68 [52]. Type data: syntypes, BMNH 1946.9.4.37–40, from Champion Bay, W.A.

Amphibolurus darlingtoni Loveridge, A. (1932). New lizards of the genera *Nephrurus* and *Amphibolurus* from Western Australia. *Proc. New Engl. Zool. Club* **13**: 31–34 [33]. Type data: holotype, MCZ 32958, from Mullewa, W.A.

Synonymy that of Cogger, H.G., this work.

Distribution: NW coastal, SW coastal, W plateau, Lake Eyre basin, S.A., W.A., N.T. Ecology: hummock grassland, tussock grassland, low open shrubland, open heath, open scrub, woodland, tall shrubland, terrestrial, diurnal; oviparous, arthropod-feeder. Biological references: Storr, G.M. (1966). The *Amphibolurus reticulatus* species-group (Lacertilia : Agamidae) in Western Australia. *J. R. Soc. West. Aust.* **49**: 17–25; Bradshaw, S.D. & Main, A.R. (1968). Behavioural attitudes and regulation of temperature in *Amphibolurus* lizards. *J. Zool.* **154**: 193–221.

Amphibolurus rufescens Stirling and Zietz, 1893

Amphibolurus rufescens Stirling, E.C. & Zietz, A. (1893). Vertebrata; Reptilia (collected by the Elder Exploring Expedition). *Trans. R. Soc. S. Aust.* **16**: 159–176 [164 pl 6 figs 2–2a]. Type data: lectotype, SAMA R1423, from Mt. Sir Thomas, Birksgate Range, S.A., designation by Houston, T.F. (1976). Vertebrate type-specimens in the South Australian Museum. III. Reptiles. *Rec. S. Aust. Mus.* **17**: 181–187.

Distribution: W plateau, (Lake Eyre basin), S.A., W.A., N.T. Ecology: low open shrubland, (low woodland), tall shrubland, terrestrial, diurnal; oviparous, saxicoline, arthropod-feeder. Biological references: Storr, G.M. (1967). Geographic races of the agamid lizard *Amphibolurus caudicinctus. J. R. Soc. West. Aust.* **50**: 49–56 (taxonomy); Houston, T.F. (1978). *Dragon Lizards and Goannas of South Australia.* Adelaide : South Australian Museum 84 pp. (natural history); Houston, T.F. (1974). Revision of the *Amphibolurus decresii* complex (Lacertilia : Agamidae) of South Australia. *Trans. R. Soc. S. Aust.* **98**: 49–60 (taxonomy); Gibbons, J.R.H. (1979). The hind leg pushup display of the *Amphibolurus decresii* species complex (Lacertilia : Agamidae). *Copeia* **1979**: 29–40.

Amphibolurus salinarum Storr, 1966

Amphibolurus pictus salinarum Storr, G.M. (1966). The *Amphibolurus reticulatus* species-group (Lacertilia : Agamidae) in Western Australia. *J. R. Soc. West. Aust.* **49**: 17–25 [18]. Type data: holotype, WAM R17649, from Norseman, in 32°10'S, 121°46'E, W.A.

Distribution: NW coastal, SW coastal, W plateau, W.A. Ecology: low open shrubland, low shrubland, salt lake, tall shrubland, terrestrial, diurnal; oviparous, arthropod-feeder. Biological references: Storr, G.M., Hanlon, T.M.S. & Harold, G. (1981). Herpetofauna of the shores and hinterland of the Great Australian Bight, Western Australia. *Rec. West. Aust. Mus.* **9**: 23–39.

Amphibolurus scutulatus Stirling and Zietz, 1893

Amphibolurus scutulatus Stirling, E.C. & Zietz, A. (1893). Vertebrata; Reptilia (collected by the Elder Exploring Expedition). *Trans. R. Soc. S. Aust.* **16**: 159–176 [165 pl 7 figs 1–2]. Type data: syntypes, from between Fraser Range and Queen Victoria Springs, W.A., SAMA R1459 (9 specimens), SAMA R3024 (2 specimens), SAMA R4814 (12 specimens), *vide* Houston, T.F. (1976). Vertebrate type-specimens in the South Australian Museum. III. Reptiles. *Rec. S. Aust. Mus.* **17**: 181–187.

Amphibolurus websteri Boulenger, G.A. (1904). Description of a new lizard from Western Australia. *Ann. Mag. Nat. Hist. (7)* **14**: 414–415 [414 pl 11]. Type data: syntypes, BMNH 1946.9.4.31–32, from Coolgardie District, 400 mi inland of Perth, W.A.

Amphibolurus holsti Rosén, N. (1905). List of lizards in the Zoological Museum of Lund, with descriptions of new species. *Ann. Mag. Nat. Hist. (7)* **16**: 129–142 [134 and corrigendum 141 pl 9]. Type data: holotype, ZMLU Cat.Typ.Nr. 284, from W.A.

Synonymy that of Cogger, H.G., this work.

Distribution: NW coastal, SW coastal, W plateau, (Lake Eyre basin), S.A., W.A. Ecology: desert, open scrub, woodland, tall shrubland, terrestrial, diurnal; oviparous, arthropod-feeder. Biological references: Houston, T.F. (1978). *Dragon Lizards and Goannas of South Australia.* Adelaide : South Australian Museum 84 pp.; Pianka, E.R. (1971). Notes on the biology of *Amphibolurus cristatus* and *Amphibolurus scutulatus. West. Aust. Nat.* **12**: 36–41; Pianka, E.R. (1972). Zoogeography and speciation of Australian desert lizards: an ecological perspective. *Copeia* **1972**: 127–144.

Amphibolurus vadnappa Houston, 1974

Amphibolurus vadnappa Houston, T.F. (1974). Revision of the *Amphibolurus decresii* complex (Lacertilia : Agamidae) of South Australia. *Trans. R. Soc. S. Aust.* **98**: 49–60 [55]. Type data: holotype, SAMA R3416B, from Aroona Waters (138°21'E, 30°35'S), Flinders Ranges, S.A.

Distribution: S Gulfs, Lake Eyre basin, S.A. Ecology: low open shrubland, low shrubland, terrestrial, diurnal; oviparous, saxicoline, arthropod-feeder, sexually dimorphic. Biological references: Houston, T.F. (1978). *Dragon Lizards and Goannas of South Australia.* Adelaide : South Australian Museum 84 pp. (natural history); Gibbons, J.R.H. (1979). The hind leg pushup display of the *Amphibolurus decresii* species complex (Lacertilia : Agamidae). *Copeia* **1979**: 29–40 (behaviour); Gibbons, J.R.H. & Lillywhite, H.B. (1981). Ecological segregation, color matching, and speciation in lizards of the *Amphibolurus decresii* species complex (Lacertilia : Agamidae). *Ecology* **62**: 1573–1584.

Amphibolurus vitticeps Ahl, 1926

Amphibolurus vitticeps Ahl, E. (1926). Neue Eidechsen und Amphibien. *Zool. Anz.* **67**: 186–192 [189]. Type data: holotype, ZMB 30101, from Australia.

Distribution: W plateau, S Gulfs, Murray-Darling basin, (N Gulf), Bulloo River basin, Lake Eyre basin, S.A., Vic., N.S.W., Qld., N.T. Ecology: desert, tussock grassland, low open shrubland, low shrubland, open scrub, low open woodland, low woodland, woodland, tall shrubland, terrestrial, arboreal, diurnal, omnivore; oviparous. Biological references: Badham, J.A. (1976). The *Amphibolurus barbatus* species-group (Lacertilia : Agamidae). *Aust. J. Zool.* **24**: 423–443 (taxonomy); Brattstrom, B.H. (1971). Social and thermoregulatory behavior of the bearded dragon, *Amphibolurus barbatus. Copeia* **1971**: 484–497 (behaviour of *Amphibolurus barbatus* may be referring to *A. vitticeps*); Carpenter, C.C., Badham, J.A. & Kimble, B. (1970). Behavior patterns of three species of *Amphibolurus* (Agamidae). *Copeia* **1970**: 497–505 (behaviour of *Amphibolurus barbatus* may be referring to *A. vitticeps*); Johnston, G.R. (1979). The eggs, incubation and young of the bearded dragon *Amphibolurus vitticeps* Ahl. *Herpetofauna* **11**: 5–8.

Amphibolurus yinnietharra Storr, 1981

Amphibolurus yinnietharra Storr, G.M. (1981). Three new agamid lizards from Western Australia. *Rec. West. Aust. Mus.* **8**: 599–607 [602]. Type data: holotype, WAM R51675 M, from 5 km E of Yinnietharra, W.A. in 24°41'S, 116°13'E.

Distribution: NW coastal, W.A. Ecology: tall shrubland, terrestrial, diurnal; oviparous, arthropod-feeder.

Caimanops Storr, 1974

Caimanops Storr, G.M. (1974). Agamid lizards of the genera *Caimanops*, *Physignathus* and *Diporiphora* in Western Australia and Northern Territory. *Rec. West.*

Aust. Mus. **3**: 121–146 [125]. Type species *Diporophora amphiboluroides* Lucas and Frost, 1902 by original designation.

Caimanops amphiboluroides (Lucas and Frost, 1902)

Diporophora amphiboluroides Lucas, A.H.S. & Frost, C. (1902). Descriptions of some new lizards from Western Australia. *Proc. R. Soc. Vict.* **15**: 76–79 [76]. Type data: holotype, NMV D8813, from W.A.

Distribution: NW coastal, W plateau, W.A. Ecology: hummock grassland, low open shrubland, tall shrubland, terrestrial, diurnal; oviparous, arthropod-feeder. Biological references: Storr, G.M. (1974). Agamid lizards of the genera *Caimanops*, *Physignathus* and *Diporiphora* in Western Australia and Northern Territory. *Rec. West. Aust. Mus.* **3**: 121–146 (taxonomy).

Chelosania Gray, 1845

Chelasonia Gray, J.E. (1845). *Catalogue of the Specimens of Lizards in the Collection of the British Museum.* London : British Museum xxviii 289 pp. [231] [this spelling, which appears in Gray's generic synopsis and in his index, has long been regarded as a *lapsus* for the spelling *Chelosania* which is used several times, and consistently, in the formal descriptions of the genus and its sole included species, *brunnea*].

Chelosania Gray, J.E. (1845). *Catalogue of the Specimens of Lizards in the Collection of the British Museum.* London : British Museum xxviii 289 pp. [245]. Type species *Chelosania brunnea* Gray, 1845 by monotypy.

Synonymy that of Cogger, H.G., this work.

Chelosania brunnea Gray, 1845

Chelosania brunnea Gray, J.E. (1845). *Catalogue of the Specimens of Lizards in the Collection of the British Museum.* London : British Museum xxviii 289 pp. [245]. Type data: holotype, BMNH 1946.8.11.43, from W.A.

Distribution: N Gulf, N coastal, Qld., W.A., N.T. Ecology: tall open shrubland, woodland, terrestrial, arboreal, diurnal; oviparous, arthropod-feeder.

Chlamydosaurus Gray, 1825

Clamydosaurus Gray, J.E. (1825). A synopsis of the genera of reptiles and Amphibia, with a description of some new species. *Annls. Philos. (2)* **10**: 193–217 [197] [*nom. oblit.*; regarded as an *errore pro Chlamydosaurus* by Wermuth, H. (1967). Liste der rezenten Amphibien und Reptilien; Agamidae. *Das Tierreich* **86**: 1–127, although as the name appears only once there is no indication of an error in the original publication; the subsequent spelling *Chlamydosaurus*, however, is used consistently by Gray and consequently we have regarded it as a valid emend.]. Type species *Clamydosaurus kingii* Gray, 1825 by monotypy.

Chlamydosaurus Gray, J.E. (1826). Reptilia. Appendix *in* King, P.P. *Narrative of a survey of the Intertropical and Western Coasts of Australia performed between the years 1818 and 1822.* London : John Murray 2: 424–434 [424] [valid emend. *pro Clamydosaurus* Gray, 1825; 1827 on title page].

Chlamydophorus Gray, J.E. (1842). Description of some hitherto unrecorded species of Australian reptiles and batrachians. *in* Gray, J.E. (ed.) *Zoological Miscellany.* London : Treuttel, Würtz & Co. pp. 51–57 [53] [?*lapsus pro Chlamydosaurus* Gray, 1826].

Synonymy that of Cogger, H.G., this work.

This group is also found in southern New Guinea.

Chlamydosaurus kingii Gray, 1825

Clamydosaurus kingii Gray, J.E. (1825). A synopsis of the genera of reptiles and Amphibia, with a description of some new species. *Annls. Philos. (2)* **10**: 193–217 [197]. Type data: holotype, presumed lost, from Australia, restricted to Port Nelson, NW coast of Australia by Gray, J.E. (1826). Reptilia. Appendix *in* King, P.P. *Narrative of a survey of the Intertropical and Western Coasts of Australia performed between the years 1818 and 1822.* London : John Murray 2: 424–434.

Distribution: NE coastal, N Gulf, N coastal, (Murray-Darling basin), (Lake Eyre basin), Qld., W.A., N.T. Ecology: open scrub, low open woodland, low woodland, woodland, open forest, terrestrial, arboreal, diurnal; oviparous, arthropod-feeder. Biological references: Beddard, F.E. (1905). A contribution to the anatomy of the frilled lizard (*Chlamydosaurus kingi*) and some other agamids. *Proc. Zool. Soc. Lond.* **1905**: 9–22; Kent, W.S. (1895). Observations on the frilled lizard, *Chlamydosaurus kingi. Proc. Zool. Soc. Lond.* **1895**: 712–719 (natural history).

Diporiphora Gray, 1842

Diporiphora Gray, J.E. (1842). Description of some hitherto unrecorded species of Australian reptiles and batrachians. *in* Gray, J.E. (ed.) *Zoological Miscellany.* London : Treuttel, Würtz & Co. pp. 51–57 [53]. Type species *Diporiphora bilineata* Gray, 1842 by monotypy.

Diporophora Gray, J.E. (1845). *Catalogue of the Specimens of Lizards in the Collection of the British Museum.* London : British Museum xxviii 289 pp. [232,250] [emend. *pro Diporiphora* Gray, 1842].

Gindalia Gray, J.E. (1845). *Catalogue of the Specimens of Lizards in the Collection of the British Museum.* London : British Museum xxviii 289 pp. [231,246]. Type species *Gindalia bennettii* Gray, 1845 by monotypy.

Calotella Steindachner, F. (1867). Reptilien. *in, Reise der österreichischen Fregatte* Novara *um die Erde in den Jahren 1857, 1858, 1859.* Vienna : State Printer. Zoologie 1(3): 1–98 [28] [1869 on title page]. Type species *Calotella australis* Steindachner, 1867 by monotypy.

Synonymy that of Cogger, H.G., this work.

This group is also found in southern New Guinea.

Diporiphora albilabris Storr, 1974

Diporiphora albilabris albilabris Storr, G.M. (1974). Agamid lizards of the genera *Caimanops*, *Physignathus* and *Diporiphora* in Western Australia and Northern Territory. *Rec. West. Aust. Mus.* **3**: 121–146 [133]. Type data: holotype, WAM R43517, from Mitchell Plateau, in 14°48'S, 125°50'E, W.A.

Diporiphora albilabris sobria Storr, G.M. (1974). Agamid lizards of the genera *Caimanops*, *Physignathus* and *Diporiphora* in Western Australia and Northern Territory. *Rec. West. Aust. Mus.* **3**: 121–146 [135]. Type data: holotype, WAM R23180, from Pine Creek, in 14°04'S, 131°58'E, N.T.

Synonymy that of Cogger, H.G., this work.

Distribution: N coastal, W.A., N.T. Ecology: woodland, terrestrial, (arboreal), diurnal; oviparous, arthropod-feeder.

Diporiphora australis (Steindachner, 1867)

Calotella australis Steindachner, F. (1867). Reptilien. *in, Reise der österreichischen Fregatte* Novara *um die Erde in den Jahren 1857, 1858, 1859.* Vienna : State Printer. Zoologie 1(3): 1–98 [29 pl 1 fig 9] [1869 on title page]. Type data: syntypes (probable), NHMW 19821 (2 specimens), from Australia (Cape York, Qld. on type label).

Grammatophora macrolepis Günther, A. (1867). Additions to the knowledge of Australian reptiles and fishes. *Ann. Mag. Nat. Hist. (3)* **20**: 45–68 [51]. [synonymy with *Diporiphora australis* (Steindachner, 1867), in doubt]. Type data: holotype, BMNH 1946.8.12.74, from Australia (by implication).

Grammatophora calotella Günther, A. (1867). Additions to the knowledge of Australian reptiles and fishes. *Ann. Mag. Nat. Hist. (3)* **20**: 45–68 [52] [*nom. substit. pro Calotella australis* Steindachner, 1867].

Diporophora nuchalis De Vis, C.W. (1884). On new species of Australian lizards. *Proc. R. Soc. Qd.* **1**: 97–100 [98]. Type data: syntypes, whereabouts unknown, from central and south coast district, Qld.

Diporophora ornata De Vis, C.W. (1884). On new species of Australian lizards. *Proc. R. Soc. Qd.* **1**: 97–100 [99]. Type data: holotype (probable), whereabouts unknown, from unknown locality.

Physignatus* [*sic*] *nigricollis Lönnberg, E. & Andersson, G.L. (1915). Results of Dr. E. Mjöberg's Swedish Scientific Expeditions to Australia 1910–1913. VII. Reptiles collected in northern Queensland. *K. Svenska Vetensk.-Akad. Handl.* **52**(7): 1–9 [4]. Type data: holotype, NHRM 3209, from Cooktown, Qld.

Synonymy that of Cogger, H.G., this work.

Distribution: NE coastal, N Gulf, Qld. Ecology: open scrub, (low open woodland), low woodland, woodland, open forest, terrestrial, diurnal; oviparous, arthropod-feeder.

Diporiphora bennettii (Gray, 1845)

Gindalia bennettii Gray, J.E. (1845). *Catalogue of the Specimens of Lizards in the Collection of the British Museum.* London : British Museum xxviii 289 pp. [247]. Type data: holotype, BMNH 1946.8.12.77, from "N.W. Coast of Australia".

Diporiphora bennettii arnhemica Storr, G.M. (1974). Agamid lizards of the genera *Caimanops*, *Physignathus* and *Diporiphora* in Western Australia and Northern Territory. *Rec. West. Aust. Mus.* **3**: 121–146 [137]. Type data: holotype, ANWC R740, from near upper Katherine River, in 14°13'S, 132°36'E, N.T.

Synonymy that of Cogger, H.G., this work.

Distribution: N Gulf, N coastal, W.A., N.T. Ecology: tall open shrubland, woodland, terrestrial, diurnal; oviparous, arthropod-feeder. Biological references: Storr, G.M. (1974). Agamid lizards of the genera *Caimanops*, *Physignathus* and *Diporiphora* in Western Australia and Northern Territory. *Rec. West. Aust. Mus.* **3**: 121–146 (taxonomy).

Diporiphora bilineata Gray, 1842

Diporiphora bilineata Gray, J.E. (1842). Description of some hitherto unrecorded species of Australian reptiles and batrachians. *in* Gray, J.E. (ed.) *Zoological Miscellany.* London : Treuttel, Würtz & Co. pp. 51–57 [54]. Type data: syntypes, BMNH 1946.8.12.75–76, from Port Essington, N.T.

Grammatophora jugularis Macleay, W. (1877). The lizards of the *Chevert* Expedition. Second paper. *Proc. Linn. Soc. N.S.W.* **2**: 97–104 [104] [1878 on title page] Type data: syntypes, AM R40672–4, from Cape Grenville, Cape York Peninsula, Qld.

Diporophora brevicauda De Vis, C.W. (1884). On new species of Australian lizards. *Proc. R. Soc. Qd.* **1**: 97–100 [99]. Type data: holotype (probable), whereabouts unknown, from Cape York, Qld.

Diporophora pentalineata De Vis, C.W. (1884). On new species of Australian lizards. *Proc. R. Soc. Qd.* **1**: 97–100 [99]. Type data: holotype (probable), whereabouts unknown, from Cape York, Qld.

Synonymy that of Cogger, H.G., this work.

Distribution: NE coastal, N Gulf, N coastal, Qld. W.A., N.T. Ecology: tussock grassland, low open shrubland, tall open shrubland, (low open woodland), low woodland, woodland, terrestrial, (arboreal), diurnal; oviparous, arthropod-feeder. Biological references: Cogger, H.G. & Lindner, D.A. (1974). Frogs and reptiles in fauna survey of the Port Essington District, Cobourg Peninsula, Northern Territory of Australia. Frith, H.J. & Calaby, J.H. (eds.) *CSIRO Div. Wildl. Res. Tech. Pap.* **28**: 63–107; Storr, G.M. (1974). Agamid lizards of the genera *Caimanops*, *Physignathus* and *Diporiphora* in Western Australia and Northern Territory. *Rec. West. Aust. Mus.* **3**: 121–146 (taxonomy).

Diporiphora convergens Storr, 1974

Diporiphora convergens Storr, G.M. (1974). Agamid lizards of the genera *Caimanops*, *Physignathus* and

Diporiphora in Western Australia and Northern Territory. *Rec. West. Aust. Mus.* **3**: 121–146 [133]. Type data: holotype, WAM R42931, from Crystal Creek, in 14°30′S, 125°47′E, W.A.

Distribution: N coastal, W.A. Ecology: woodland, terrestrial; oviparous, arthropod-feeder.

Diporiphora lalliae Storr, 1974

Diporiphora lalliae Storr, G.M. (1974). Agamid lizards of the genera *Caimanops*, *Physignathus* and *Diporiphora* in Western Australia and Northern Territory. *Rec. West. Aust. Mus.* **3**: 121–146 [138]. Type data: holotype, WAM R23030, from Langey Crossing, in 17°39′S, 123°34′E, W.A.

Distribution: N Gulf, N coastal, W plateau, Lake Eyre basin, S.A., W.A., N.T. Ecology: desert, hummock grassland, low open shrubland, (low woodland), tall shrubland, terrestrial, diurnal; oviparous, arthropod-feeder.

Diporiphora linga Houston, 1977

Diporiphora linga Houston, T.F. (1977). A new species of *Diporiphora* from South Australia and geographic variation in *D. winneckei* Lucas and Frost (Lacertilia : Agamidae). *Trans. R. Soc. S. Aust.* **101**: 199–206 [203 figs 6–7]. Type data: holotype, SAMA R15020 F, from 23 km N of Koonibba Mission, in 31°42′S, 133°26′E, S.A.

Distribution: W plateau, S.A. Ecology: desert, low shrubland, tall shrubland, terrestrial, diurnal; oviparous, arthropod-feeder. Biological references: Houston, T.F. (1978). *Dragon Lizards and Goannas of South Australia.* Adelaide : South Australian Museum 84 pp. (natural history).

Diporiphora magna Storr, 1974

Diporiphora magna Storr, G.M. (1974). Agamid lizards of the genera *Caimanops*, *Physignathus* and *Diporiphora* in Western Australia and Northern Territory. *Rec. West. Aust. Mus.* **3**: 121–146 [137]. Type data: holotype, WAM R42786, from Old Lissadell, in 16°30′S, 128°41′E (now submerged by Lake Argyle), W.A.

Distribution: N Gulf, N coastal, Qld., W.A., N.T. Ecology: low open woodland, (low woodland), woodland, terrestrial, diurnal; oviparous, arthropod-feeder.

Diporiphora margaretae Storr, 1974

Diporiphora bilineata margaretae Storr, G.M. (1974). Agamid lizards of the genera *Caimanops*, *Physignathus* and *Diporiphora* in Western Australia and Northern Territory. *Rec. West. Aust. Mus.* **3**: 121–146 [143]. Type data: holotype, WAM R27648, from Kalumburu, in 14°18′S, 126°30′E, W.A.

Distribution: N Gulf, N coastal, N.T., W.A. Ecology: woodland, terrestrial, diurnal; oviparous, arthropod-feeder. Biological references: Storr, G.M. (1979). Two new *Diporiphora* (Lacertilia : Agamidae) from Western Australia. *Rec. West. Aust. Mus.* **7**: 255–263 (taxonomy).

Diporiphora pindan Storr, 1979

Diporiphora pindan Storr, G.M. (1979). Two new *Diporiphora* (Lacertilia : Agamidae) from Western Australia. *Rec. West. Aust. Mus.* **7**: 255–263 [256 pl 1]. Type data: holotype, WAM R58402, from 5 km N of Coulomb Point, in 17°18′S, 122°10′E, W.A.

Distribution: N coastal, W.A. Ecology: (low open shrubland), woodland, terrestrial, diurnal; oviparous, arthropod-feeder.

Diporiphora reginae Glauert, 1959

Diporiphora reginae Glauert, L. (1959). A new agamid lizard from Queen Victoria Springs, Western Australia. *Diporiphora reginae*, sp. nov. *Proc. R. Zool. Soc. N.S.W.* **1957–58**: 10 [10]. Type data: holotype, WAM R12961, from Karin Rock, 14 mi from Cunderlee Mission in the Queen Victoria Springs area, W.A.

Distribution: W plateau, W.A. Ecology: low open shrubland, low shrubland, (open scrub), tall shrubland, terrestrial, diurnal; oviparous, arthropod-feeder. Biological references: Storr, G.M. (1974). Agamid lizards of the genera *Caimanops*, *Physignathus* and *Diporiphora* in Western Australia and Northern Territory. *Rec. West. Aust. Mus.* **3**: 121–146 (taxonomy).

Diporiphora superba Storr, 1974

Diporiphora superba Storr, G.M. (1974). Agamid lizards of the genera *Caimanops*, *Physignathus* and *Diporiphora* in Western Australia and Northern Territory. *Rec. West. Aust. Mus.* **3**: 121–146 [144]. Type data: holotype, WAM R43178, from Mitchell River, in 14°25′S, 125°50′E, W.A.

Distribution: N coastal, W.A. Ecology: woodland, terrestrial, (arboreal), diurnal; oviparous, arthropod-feeder.

Diporiphora valens Storr, 1979

Diporiphora valens Storr, G.M. (1979). Two new *Diporiphora* (Lacertilia : Agamidae) from Western Australia. *Rec. West. Aust. Mus.* **7**: 255–263 [259 pl 2]. Type data: holotype, WAM R31009, from near Tom Price, in *ca.* 22°45′S, 117°45′E, W.A.

Distribution: NW coastal, (W plateau), W.A. Ecology: hummock grassland, low open shrubland, terrestrial, diurnal; oviparous, arthropod-feeder.

Diporiphora winneckei Lucas and Frost, 1896

Diporophora winneckei Lucas, A.H.S. & Frost, C. (1896). Reptilia. *Rep. Horn Expedition to Central Australia.* 2: 112–151 [132 pl 11 fig 5]. Type data:

lectotype, NMV D10156, from Charlotte Waters, N.T., designation by Coventry, A.J. (1970). Reptile and amphibian type specimens housed in the National Museum of Victoria. *Mem. Natl. Mus. Vic.* **31**: 115–124.

Distribution: N coastal, NW coastal, W plateau, Lake Eyre basin, S.A., N.S.W., Qld., W.A., N.T. Ecology: desert, hummock grassland, low open shrubland, (open scrub), low open woodland, tall shrubland, terrestrial, diurnal; oviparous, arthropod-feeder. Biological references: Houston, T.F. (1977). A new species of *Diporiphora* from South Australia and geographic variation in *D. winneckei* Lucas and Frost (Lacertilia : Agamidae). *Trans. R. Soc. S. Aust.* **101**: 199–206; Storr, G.M. (1974). Agamid lizards of the genera *Caimanops*, *Physignathus* and *Diporiphora* in Western Australia and Northern Territory. *Rec. West. Aust. Mus.* **3**: 121–146 (taxonomy); Houston, T.F. (1978). *Dragon Lizards and Goannas of South Australia.* Adelaide : South Australian Museum 84 pp. (natural history); Francis, M. (1981). Observations on heat regulating behaviour in captive specimens of *Diporiphora winneckei* (Lucas and Frost). *Herpetofauna* **12**: 35–36.

Gonocephalus Kaup, 1825

Lophyrus Duméril, A.M.C. (1806). *Zoologie Analytique,* ou Méthode Naturelle de Classification des Animaux. Paris : Allais xxxii 344 pp. [80] [*non Lophyrus* Poli, 1791; *non Lophyrus* Latreille, 1802; 1805 on title page]. Without included species; type species presumably *Iguana chamaeleontina* Laurenti, 1768, *vide* Wermuth, H. (1967). Liste der rezenten Amphibien und Reptilien; Agamidae. *Das Tierreich* **86**: 1–127.

Lophirus Rafinesque, C.S. (1815). *Analyse de la Nature* ou Tableau de l'Univers et des Corps Organises. Palermo [75] [emend. *pro Lophyrus* Duméril, 1806; *nom. oblit.*].

Gonocephalus Kaup, J.J. (1825). Einige Bemerkungen zu Merrem's Handbuch. *Isis Von Oken, Jena* **1825**: 589–590 [590]. Type species *Ag[ama] tigrina* Kaup, 1825 [= *Gonocephalus chamaeleontinus* (Laurenti, 1768)] by monotypy .

Goniocephalus Kaup, J.J. (1827). Zoologische Monographien. *Isis Von Oken, Jena* **20**: 610–625 [610,614] [emend. *pro Gonocephalus* Kaup, 1825].

Gonyocephalus Wagler, J.G. (1830). *Natürliches System der Amphibien, mit vorangehender Classification der Säugethiere und Vögel.* München : Cotta'schen vi 354 pp. [132,150] [emend. *pro Gonocephalus* Kaup, 1825].

Tiaris Duméril, A.M.C. & Bibron, G. (1837). *Erpétologie Générale* ou Histoire Naturelle Complète des Reptiles. Paris : Roret 4: ii 571 pp. [419] [*non Tiaris* Swainson, 1827]. Type species *Lophyrus dilophus* Duméril and Bibron, 1837 (= *Gonocephalus dilophus* Duméril and Bibron, 1837) by monotypy.

Lophosaurus Fitzinger, L.J. (1843). *Systema Reptilium.* Vienna : Braümüller u. Seidel vi 106 pp. [15,45] [*nom. substit. pro Tiaris* Duméril and Bibron, 1837].

Dilophyrus Gray, J.E. (1845). *Catalogue of the Specimens of Lizards in the Collection of the British Museum.* London : British Museum xxviii 289 pp. [231, 238]. Type species *Dilophyrus grandis* Gray, 1845 (= *Gonocephalus grandis* Gray, 1845) by monotypy.

Gonycephalus Agassiz, L. (1846). *Nomenclatoris Zoologici Index Universalis.* Soloduri : Jent et Gassman 393 pp. [166] [emend. *pro Gonocephalus* Kaup, 1825].

Coryphophylax Fitzinger, L. (1861). Ausbeute der österreichischen Naturforscher an Säugethiere und Reptilien während der Weltumsegelung Sr. Majestat Fregatte *Novara. Sber. Akad. Wiss. Wien* **42**: 384–416 [387,397] [*nom. nud.*].

Hypsilurus Peters, W. (1867). Über Flederthiere und Amphibien. *Mber K. Preuss. Akad. Wiss. Berl.* **1867**: 703–713 [707] [1868 on title page; described as subgenus of *Lophura* Gray, 1825]. Type species *Lophura (Hypsilurus) godeffroyi* Peters, 1867 by monotypy.

Coryphophylax Steindachner, F. (1867). Reptilien. *in, Reise der österreichischen Fregatte* Novara *um die Erde in den Jahren 1857, 1858, 1859.* Vienna : State Printer. Zoologie 1(3): 1–98 [30] [1869 on title page; *ex* Fitzinger ms]. Type species *Coryphophylax maximiliani* Steindachner, 1867 [= *Gonocephalus subcristatus* (Blyth, 1860)] by monotypy.

Arua Doria, G. (1874). Enumerazione dei rettili raccolti dal Dott. O. Beccari in Amboina alle Isole Aru ed alle Isole Kei durante gli anni. *Ann. Mus. Civ. Stor. Nat. Giacomo Doria* **6**: 325–357 [345] [described as subgenus of *Gonyocephalus* (*sic*) Kaup, 1825]. Type species *Gonyocephalus (Arua) inornatus* Doria, 1874 (= *Gonocephalus modestus* Meyer, 1874) by monotypy.

Lophosteus Peters, W. & Doria, G. (1878). Catalogo dei Rettili e dei Batraci raccolti da O. Beccari, L.M. d'Albertis e A.A. Bruijn nella sotto-regione Austro-Malese. *Ann. Mus. Civ. Stor. Nat. Genova* **13**: 323–450 [377] [described as subgenus of *Gonyocephalus* (*sic*) Kaup, 1825]. Type species *Gonyocephalus (Lophosteus) albertisii* Peters and Doria, 1878 [= *Gonocephalus papuensis* (Macleay, 1877)] by monotypy.

Diptychodera Boettger, O. (1893). *Katalog der Reptilien-Sammlung im Museum der Senckenbergischen Naturforschenden Gesellschaft in Frankfurt-am-Main.* I. Teil (Rhynchocephalen, Schildkröten, Krokodile, Eidechsen, Chamäleons). Frankfurt-am-Main : Knauer x 140 pp. [45]. Type species *Diptychodera lobata* Boettger, 1893 (= *Gonocephalus modestus* Meyer, 1874) by monotypy.

Diptychodlia Romer, A.S. (1956). *Osteology of the Reptiles.* Chicago : Univ. Chicago Press xxii 772 pp. [537] [*lapsus pro Diptychodera* Boettger, 1893].

Synonymy that of Cogger, H.G., this work; Wermuth, H. (1967). Liste der rezenten Amphibien und Reptilien; Agamidae. *Das Tierreich* **86**: 1–127.

This group is also found in India through Indo-Australian Archipelago to New Guinea and Solomon Ils.

Gonocephalus boydii (Macleay, 1884)

Tiaris boydii Macleay, W. (1884). Notes on some reptiles from the Herbert River, Queensland. *Proc. Linn. Soc. N.S.W.* **8**: 432–436 [432] [1883 on title page]. Type data: syntypes, AM R31884–5, from Ripple Creek district, Herbert River, Qld.

Distribution: NE coastal, Qld. Ecology: closed forest, arboreal, diurnal; oviparous, arthropod-feeder. Biological references: Zwinenberg, A.J. (1974). Boyd's forest dragon, *Goniocephalus boydii* (Macleay). *Bull. Maryland Herpet. Soc.* **10**: 99–101 (natural history); Wells, R. (1972). Notes on *Goniocephalus boydii* (Macleay). *Herpetofauna* **5**: 24 (natural history).

Gonocephalus godeffroyi (Peters, 1867)

Lophura (Hypsilurus) godeffroyi Peters, W. (1867). Über Flederthiere und Amphibien. *Mber K. Preuss. Akad. Wiss. Berl.* **1867**: 703–713 [707 fig 1] [1868 on title page]. Type data: syntypes, (2 specimens) not found, from Republic of Belau (as Pelew Ils.).

Hypsilurus macrolepis Peters, W. (1872). Über neue oder weniger bekannte Saurier. *Mber. K. Preuss. Akad. Wiss. Berl.* **1872**: 774–776 [775] [1873 on title page]. Type data: holotype, ?ZMB (not found), from Solomon Ils.

Tiaris longii Macleay, W. (1877). The lizards of the *Chevert* Expedition. Second paper. *Proc. Linn. Soc. N.S.W.* **2**: 97–104 [103] [1878 on title page]. Type data: holotype, AM R31935, from N Qld.

Synonymy that of Cogger, H.G., this work.

Distribution: NE coastal, Qld.; Australian record based on holotype of *Tiaris longii* Macleay, 1877, the provenance of which is doubtful, *vide* Cogger, H.G. (1979). Type specimens of reptiles and amphibians in The Australian Museum. *Rec. Aust. Mus.* **32**: 163–210.

Gonocephalus spinipes (Duméril and Duméril, 1851)

Lophyrus spinipes Duméril, A.M.C. & Duméril, A. (1851). *Catalogue Méthodique de la Collection des Reptiles.* Paris : Mus. Hist. Nat. iv 224 pp. [90]. Type data: holotype, MNHP 2560, from Australia.

Gonyocephalus cristipes Longley, G. (1943). Notes on *Goniocephalus cristipes. Proc. R. Zool. Soc. N.S.W.* **1942–43**: 30–31 [30] [*lapsus pro Gonocephalus spinipes* (Duméril and Duméril, 1851)].

Synonymy that of Cogger, H.G., this work.

Distribution: SE coastal, NE coastal, N.S.W., Qld. Ecology: (tall forest), closed forest, arboreal, diurnal; oviparous, arthropod-feeder. Biological references: Longley, G. (1943). Notes on *Goniocephalus cristipes. Proc. R. Zool. Soc. N.S.W.* **1942–43**: 30–31 (natural history); Bustard, H.R. (1970). *Australian Lizards.* Sydney : Collins 162 pp. (behaviour).

Lophognathus Gray, 1842

Lophognathus Gray, J.E. (1842). Description of some hitherto unrecorded species of Australian reptiles and batrachians. *in* Gray, J.E. (ed.) *Zoological Miscellany.* London : Treuttel, Wärtz & Co. pp. 51–57 [53]. Type species *Lophognathus gilberti* Gray, 1842 by monotypy.

Redtenbacheria Steindachner, F. (1867). Reptilien. *in, Reise der österreichischen Fregatte* Novara *um die Erde in den Jahren 1857, 1858, 1859.* Vienna : State Printer. Zoologie 1(3): 1–98 [31] [*non Redtenbacheria* Schiner, 1861; 1869 on title page]. Type species *Redtenbacheria fasciata* Steindachner, 1867 by monotypy.

Synonymy that of Cogger, H.G., this work.

This group is also found in Tanimbar Ils. and southern New Guinea.

Lophognathus gilberti Gray, 1842

Lophognathus gilberti Gray, J.E. (1842). Description of some hitherto unrecorded species of Australian reptiles and batrachians. *in* Gray, J.E. (ed.) *Zoological Miscellany.* London : Treuttel, Wärtz & Co. pp. 51–57 [53]. Type data: holotype, BMNH 1946.8.28.69, from Port Essington, N.T.

*[**Grammatophora temporalis*** **(part.)**]* Günther, A. (1867). Additions to the knowledge of Australian reptiles and fishes. *Ann. Mag. Nat. Hist. (3)* **20**: 45–68 [52] [for lectotype designation see *Grammatophora temporalis* Günther, 1867]. Type data: paralectotype, BMNH 1946.8.12.73 from Nickol Bay, W.A.

Redtenbacheria fasciata Steindachner, F. (1867). Reptilien. *in, Reise der österreichischen Fregatte* Novara *um die Erde in den Jahren 1857, 1858, 1859.* Zoologie 1(3): 1–98 Vienna : State Printer [31] [1869 on title page]. Type data: holotype, not found, from Australia.

Physignathus incognitus Ahl, E. (1926). Neue Eidechsen und Amphibien. *Zool. Anz.* **67**: 186–192 [190]. Type data: holotype, ZMB 30086, from Australia.

Physignathus gilberti centralis Loveridge, A. (1933). New agamid lizards of the genera *Amphibolurus* and *Physignathus* from Australia. *Proc. New. Engl. Zool. Club* **13**: 69–72 [71]. Type data: holotype, MCZ 35207, from Anningie 30 mi W of Teatree Well, N.T.

Synonymy that of Cogger, H.G., this work.

Distribution: NE coastal, N Gulf, N coastal, NW coastal, W plateau, Lake Eyre basin, (Murray-Darling basin), Qld., W.A., N.T. Ecology: desert, (tussock grassland), low open shrubland, tall open shrubland, low open woodland, woodland, open forest, tall shrubland, terrestrial, arboreal, diurnal; oviparous, arthropod-feeder. Biological references: Storr, G.M. (1974). Agamid lizards of the genera *Caimanops*, *Physignathus* and *Diporiphora* in Western Australia and Northern Territory. *Rec. West. Aust. Mus.* **3**: 121–146 (taxonomy); Cogger, H.G. & Lindner, D.A. (1974). Frogs and reptiles in fauna survey of the Port Essington District, Cobourg Peninsula, Northern Territory of Australia. Frith, H.J. & Calaby, J.H. (eds.) *CSIRO Div. Wildl. Res. Tech.*

Pap. **28**: 63–107; Houston, T.F. (1978). *Dragon Lizards and Goannas of South Australia.* Adelaide : South Australian Museum 84 pp. (natural history).

Lophognathus longirostris Boulenger, 1883

Lophognathus longirostris Boulenger, G.A. (1883). Remarks on the lizards of the genus *Lophognathus. Ann. Mag. Nat. Hist. (5)* **12**: 225–226 [225]. Type data: syntypes, BMNH 1946.8.12.64–65 (Nickol Bay), BMNH 1946.8.28.73 (Champion Bay); original description cites Champion Bay, W.A. but two syntypes from Nickol Bay, W.A.

Physignathus eraduensis Werner, F. (1909). Reptilia exkl. Geckonidae und Scincidae. *in* Michaelsen, W. & Hartmeyer, R. (eds.) *Die Fauna Südwest-Australiens.* Jena : Gustav Fischer 2: 251–278 [275]. Type data: holotype, status unknown, from Eradu, W.A.

Ph[ysignathus] longirostris quattuorfasciatus Sternfeld, R. (1924). Beiträge zur Herpetologie Inner-Australiens. *Abh. Senckenb. Naturf. Ges.* **38**: 221–251 [236]. Type data: lectotype, SMF 10366, from Hermannsburg Misson, N.T., designation by Mertens, R. (1967). Die herpetologische Sektion des Natur-Museums und Forschungs-Institutes Senckenberg in Frankfurt-a.-M. nebst einem Verzeichnis ihrer Typen. 1. *Senckenberg. Biol.* **48**: 1–106.

Synonymy that of Cogger, H.G., this work.

Distribution: (N Gulf), N coastal, NW coastal, W plateau, Lake Eyre basin, S.A., Qld., W.A., N.T. Ecology: desert, hummock grassland, low open shrubland, low open woodland, (low woodland), tall shrubland, terrestrial, arboreal, diurnal; oviparous, arthropod-feeder. Biological references: Storr, G.M. (1974). Agamid lizards of the genera *Caimanops*, *Physignathus* and *Diporiphora* in Western Australia and Northern Territory. *Rec. West. Aust. Mus.* **3**: 121–146 (taxonomy); Houston, T.F. (1978). *Dragon Lizards and Goannas of South Australia.* Adelaide : South Australian Museum 84 pp. (natural history); Baehr, M. (1976). Beobachtungen zur bipeden Fortbewegung bei der australischen Agame *Physignathus longirostris* (Boulenger). *Stuttgarter Beitr. Naturk. (A)* **291**: 1–7 (as *Physignathus longirostris*).

Lophognathus temporalis (Günther, 1867)

***Grammatophora temporalis* (part.)** Günther, A. (1867). Additions to the knowledge of Australian reptiles and fishes. *Ann. Mag. Nat. Hist. (3)* **20**: 45–68 [52] [for the other paralectotype BMNH 1946.8.12.73 (Nickol Bay) see under *Lophognathus gilberti* (Gray, 1842)]. Type data: lectotype, BMNH 1946.8.28.72 paralectotype BMNH 1946.8.12.63, designation by Cogger, H.G., this work, from Port Essington, N.T. and Nickol Bay, W.A. but restricted to Port Essington by lectotype designation.

Lophognathus lateralis Macleay, W. (1877). The lizards of the *Chevert* Expedition. Second paper. *Proc. Linn. Soc. N.S.W.* **2**: 97–104 [103] [1878 on title page]. Type data: holotype, AM R31882, from Mawatta, Binaturi River (as Katow), Papua New Guinea.

Lophognathus labialis Boulenger, G.A. (1883). Remarks on the lizards of the genus *Lophognathus. Ann. Mag. Nat. Hist. (5)* **12**: 225–226 [225]. Type data: syntypes, BMNH 1946.8.12.72, BMNH 1946.8.12.63, from Port Essington, N.T.

Synonymy that of Cogger, H.G., this work.

Distribution: NE coastal, N Gulf, N coastal, Qld., W.A., N.T. Ecology: tall open shrubland, woodland, open forest, terrestrial, arboreal, diurnal; oviparous, arthropod-feeder. Biological references: Cogger, H.G. & Lindner, D.A. (1974). Frogs and reptiles in fauna survey of the Port Essington District, Cobourg Peninsula, Northern Territory of Australia. Frith, H.J. & Calaby, J.H. (eds.) *CSIRO Div. Wildl. Res. Tech. Pap.* **28**: 63–107 (taxonomy); Storr, G.M. (1974). Agamid lizards of the genera *Caimanops*, *Physignathus* and *Diporiphora* in Western Australia and Northern Territory. *Rec. West. Aust. Mus.* **3**: 121–146 (taxonomy).

Moloch Gray, 1841

Moloch Gray, J.E. (1841). Description of some new species and four new genera of reptiles from Western Australia, discovered by John Gould, Esq. *Ann. Mag. Nat. Hist.* **7**: 86–91 [88]. Type species *Moloch horridus* Gray, 1841 by monotypy.

Molochus Steindachner, F. (1867). Amphibien. *in, Reise der österreichischen Fregatte* Novara *um die Erde in den Jahren 1857, 1858, 1859.* Vienna : State Printer. Zoologie 1(4): 1–70 [26] [emend. *pro Moloch* Gray, 1841; 1869 on title page].

Synonymy that of Cogger, H.G., this work.

Moloch horridus Gray, 1841

Moloch horridus Gray, J.E. (1841). Description of some new species and four new genera of reptiles from Western Australia, discovered by John Gould, Esq. *Ann. Mag. Nat. Hist.* **7**: 86–91 [89]. Type data: syntypes (2 specimens), whereabouts unknown, from W.A.

Distribution: N Gulf, N coastal, NW coastal, SW coastal, W plateau, S Gulfs, Lake Eyre basin, S.A., Qld., W.A., N.T. Ecology: desert, hummock grassland, (tussock grassland), low open shrubland, low woodland, (woodland), tall shrubland, terrestrial, diurnal; oviparous, formiciphagous. Biological references: Sporn, C.C. (1955). The breeding of the mountain devil in captivity. *West. Aust. Nat.* **5**: 1–5; Sporn, C.C. (1958). Further observations on the mountain devil in captivity. *West. Aust. Nat.* **6**: 136–137; Sporn, C.C. (1965). Additional observations on the life history of the mountain devil, *Moloch horridus*, in captivity. *West. Aust. Nat.* **9**: 157–159; Pianka, E.R. & Pianka, H.D. (1970). The ecology of *Moloch* *horridus* (Lacertilia : Agamidae) in Western Australia. *Copeia* **1970**: 90–103; Bentley, P.J. & Blumer, W.F.C. (1962). Uptake of water by the lizard, *Moloch horridus*. *Nature* **194**: 699–700.

Physignathus Cuvier, 1829

Physignathus Cuvier, G. (1829). *Le Règne Animal.* 2nd edn. Paris : Déterville Vol. 2 xv 406 pp. [41]. Type species *Phyhignat,us* [*sic*] *cocincinus* Cuvier, 1829 by monotypy.

Phrysignathus Gray, J.E. (1845). *Catalogue of the Specimens of Lizards in the Collection of the British Museum.* London : British Museum xxviii 289 pp. [231] [*lapsus pro Physignathus* Cuvier, 1829].

Physognathus Agassiz, L. (1846). *Nomenclatoris Zoologici Index Universalis.* Soloduri : Jent et Gassman 393 pp. [290] [emend. *pro Physignathus* Cuvier, 1829].

Synonymy that of Cogger, H.G., this work; Wermuth, H. (1967). Liste der rezenten Amphibien und Reptilien; Agamidae. *Das Tierreich* **86**: 1–127 [91].

This group is also found in New Guinea and south-east Asia.

Physignathus lesueurii (Gray, 1831)

***Lacerta muricata* var.** White, J. (1790). *Journal of a Voyage to New South Wales,* with sixty five plates of non-descript animals, birds, lizards, serpents, curious cones of trees and other natural productions. Appendix. London : Debrett 299 pp. [255 pl 40] [*ex* Shaw ms]. Type data: none designated, whereabouts unknown, from N.S.W.

L[ophura] lesueurii Gray, J.E. (1831). A synopsis of the species of the class Reptilia, pp. 1–110 *in* Griffith, E. *The Animal Kingdom Arranged in Conformity with its Organization by the Baron Cuvier.* London : Whittaker, Treacher & Co. 9: 481+110 pp. [60]. Type data: holotype, MNHP 392, from Paramatta [*sic*], N.S.W.

Iguana paramatensis Fitzinger, L.J. (1843). *Systema Reptilium.* Vienna : Braümüller u. Seidel vi 106 pp. [49] *nom. nud.*, Péron ms name introduced in synonymy of *Lophura lesueurii* Gray, 1831].

Amphibolurus maculiferus Girard, C. (1857). Descriptions of some new reptiles, collected by the United States Exploring Expedition, under the command of Capt. Charles Wilkes, U.S.N. Fourth Part : including the species of saurians exotic to North America. *Proc. Acad. Nat. Sci. Philad.* **1857**: 195–199 [199]. Type data: not found, from N.S.W.

Amphibolurus heterurus Peters, W. (1866). Eine Mitteilung über neue Amphibien (*Amphibolurus, Lygosoma, Cyclodus, Masticophis, Crotaphopeltis*) und Fische (*Diagramma*: *Hapalogenys*) des Kgl. zoologischen Museums. *Mber. K. Preuss. Akad. Wiss. Berl.* **1866**: 86–96 [86] [1867 on title page]. Type data: syntypes, ZMB 5321–2, from Clarence River, N.S.W.

Amphibolurus branchialis De Vis, C.W. (1884). On a new Australian lizard. *Proc. R. Soc. Qd.* **1**: 53–56 [55]. Type data: syntypes, from neighbourhood of Brisbane, Qld., QM, presumed lost, see Covacevich, J. (1971). Amphibian and reptile type specimens in the Queensland Museum. *Mem. Qd. Mus.* **16**: 49–68.

Physignathus lesueuri howitti McCoy, F. (1884). The Gippsland water lizard. *Prod. Zool. Vict.* **9**: 7–10 [7 pl 81]. Type data: lectotype, NMV D1822, from Upper Buchan River, Vic., designation by Coventry, A.J. (1970). Reptile and amphibian type specimens housed in the National Museum of Victoria. *Mem. Natl. Mus. Vict.* **31**: 115–124.

Synonymy that of Cogger, H.G., this work.

Distribution: SE coastal, NE coastal, Murray-Darling basin, Vic., N.S.W., Qld., A.C.T. Ecology: low open woodland, woodland, open forest, tall forest, closed forest, arboreal, diurnal; oviparous, riparian, general carnivore, arthropod-feeder, saxicoline. Biological references: Australian Herpetological Society Members (1976). (Observations on the eastern water dragon *Physignathus lesueurii* in the natural state and in captivity). *Herpetofauna* **8**: 20–22; Smith, J. (1979). Notes on incubation and hatching of eggs of the eastern water dragon. *Herpetofauna* **10**: 12–14; Zwinenberg, A.J. (1977). Die australische Wasseragame (*Physignathus lesueurii*). *Aquaria* **24**: 93–97 (natural history); Giddings, S. (1980). Breeding water dragons (*Physignathus lesueurii lesueurii* and *Physignathus lesueurii howittii*) in captivity. *S. Aust. Herpet. Gp. Newsletter* **Feb. 1980**: 2–3; Hay, M. (1972). The breeding of *Physignathus lesueurii l.* in captivity. *Herpetofauna* **5**: 4–5.

Tympanocryptis Peters, 1863

Tympanocryptis Peters, W. (1863). Übersicht der von Hrn. Richard Schomburgk an das zoologische Museum eingesandten Amphibien, aus Buchsfelde bei Adelaide in Südaustralien. *Mber. K. Preuss. Akad. Wiss. Berl.* **1863**: 228–236 [230] [1864 on title page]. Type species *Tympanocryptis lineata* Peters, 1863 by monotypy.

Tympanocryptis cephalus Günther, 1867

Tympanocryptis cephalus Günther, A. (1867). Additions to the knowledge of Australian reptiles and fishes. *Ann. Mag. Nat. Hist. (3)* **20**: 45–68 [52]. Type data: lectotype, BMNH 1946.8.12.78, from Nickol (as Nicol) Bay, W.A., designation by Cogger, H.G., this work.

Tympanocryptis cephalus gigas Mitchell, F.J. (1948). A revision of the lacertilian genus *Tympanocryptis*. *Rec. S. Aust. Mus.* **9**: 57–86 [65 fig 4 pl 5]. Type data: holotype, SAMA R2434, from between Ashburton and Gascoyne Rivers, W.A.

Synonymy that of Cogger, H.G., this work.

Distribution: N Gulf, NW coastal, (SW coastal), W plateau, Lake Eyre basin, S.A., Qld., W.A., N.T. Ecology: desert, hummock grassland, low open shrubland, (open scrub), tall shrubland, terrestrial, diurnal; oviparous, arthropod-feeder. Biological references: Mitchell, F.J. (1948). A revision of the lacertilian genus *Tympanocryptis*. *Rec. S. Aust. Mus.* **9**: 57–86 (taxonomy); Storr, G.M. (1964).

The agamid lizards of the genus *Tympanocryptis* in Western Australia. *J. R. Soc. West. Aust.* **47**: 43–50 (taxonomy).

Tympanocryptis intima Mitchell, 1948

Tympanocryptis intima Mitchell, F.J. (1948). A revision of the lacertilian genus *Tympanocryptis*. *Rec. S. Aust. Mus.* **9**: 57–86 [60 fig 2 pl 4]. Type data: holotype, SAMA R2331, from Oodnadatta, S.A.

Distribution: W plateau, Lake Eyre basin, (Bulloo River basin), S.A., N.S.W., Qld., W.A., N.T. Ecology: desert, low open shrubland, tall shrubland, terrestrial, diurnal; oviparous, arthropod-feeder. Biological references: Storr, G.M. (1964). The agamid lizards of the genus *Tympanocryptis* in Western Australia. *J. R. Soc. West. Aust.* **47**: 43–50 (taxonomy); Houston, T.F. (1978). *Dragon Lizards and Goannas of South Australia.* Adelaide : South Australian Museum 84 pp. (natural history).

Tympanocryptis lineata Peters, 1863

Tympanocryptis lineata Peters, W. (1863). Übersicht der von Hrn. Richard Schomburgk an das zoologische Museum eingesandten Amphibien, aus Buchsfelde bei Adelaide in Südaustralien. *Mber. K. Preuss. Akad. Wiss. Berl.* **1863**: 228–236 [230] [1864 on title page]. Type data: syntypes, ZMB 4714 (2 specimens), ZMB 740, from Loos, 4.5 km W Gawler (as Buchsfelde), S.A., see Praite, R. & Tolley, J.C. (1970). *Place names of South Australia.* Adelaide : Rigby 203 pp.

Tympanocryptis lineata centralis Sternfeld, R. (1924). Beiträge zur Herpetologie Inner-Australiens. *Abh. Senckenb. Naturf. Ges.* **38**: 221–251 [234]. Type data: lectotype, SMF 10349, from Hermannsburg Mission, upper Finke River, N.T., designation by Mertens, R. (1967). Die herpetologische Sektion des Natur-Museums und Forschungs-Institutes Senckenberg in Frankfurt-a.-M. nebst einem Verzeichnis ihrer Typen. 1. *Senckenberg. Biol.* **48**: 1–106.

Tympanocryptis lineata pinguicolla Mitchell, F.J. (1948). A revision of the lacertilian genus *Tympanocryptis*. *Rec. S. Aust. Mus.* **9**: 57–86 [70 fig 6 pl 4]. Type data: holotype, SAMA R2468a, from S Vic.

Synonymy that of Cogger, H.G., this work.

Distribution: SE coastal, N Gulf, (N coastal), W plateau, S Gulfs, Murray-Darling basin, Bulloo River basin, Lake Eyre basin, S.A., Vic., N.S.W., Qld., W.A., N.T., A.C.T. Ecology: desert, hummock grassland, tussock grassland, low open shrubland, low shrubland, (open scrub), low open woodland, low woodland, woodland, tall shrubland, terrestrial, diurnal; oviparous, arthropod-feeder. Biological references: Mitchell, F.J. (1948). A revision of the lacertilian genus *Tympanocryptis*. *Rec. S. Aust. Mus.* **9**: 57–86 (taxonomy); Storr, G.M. (1964). The agamid lizards of the genus *Tympanocryptis* in Western Australia. *J. R. Soc. West. Aust.* **47**: 43–50 (taxonomy); Warburg, M.R. (1966). Water economy of several Australian geckos, agamids, and water skinks. *Copeia* **1966**: 230–235.

Tympanocryptis parviceps Storr, 1964

Tympanocryptis parviceps Storr, G.M. (1964). The agamid lizards of the genus *Tympanocryptis* in Western Australia. *J. R. Soc. West. Aust.* **47**: 43–50 [46 figs 2–3]. Type data: holotype, WAM R16984, from 11 mi SE of Ningaloo, in 22°48′S, 113°49′E, W.A.

Amphibolurus parviceps butleri Storr, G.M. (1977). The *Amphibolurus adelaidensis* species group (Lacertilia, Agamidae) in Western Australia. *Rec. West. Aust. Mus.* **5**: 73–81 [75]. Type data: holotype, WAM R54728, from near east shore of Useless Inlet, 29 km NW of Carrarang, W.A. in 26°18′S, 113°21′E.

Synonymy that of Cogger, H.G., this work.

Distribution: NW coastal, W.A. Ecology: hummock grassland, low open shrubland, tall shrubland, terrestrial, diurnal; oviparous, arthropod-feeder. Biological references: Storr, G.M. (1977). The *Amphibolurus adelaidensis* species group (Lacertilia, Agamidae) in Western Australia. *Rec. West. Aust. Mus.* **5**: 73–81 (taxonomy).

Tympanocryptis tetraporophora Lucas and Frost, 1895

Tympanocryptis tetraporophora Lucas, A.H.S. & Frost, C. (1895). Preliminary notice of a certain species of lizard from central Australia. *Proc. R. Soc. Vict. (ns)* **7**: 264–269 [265]. Type data: lectotype, NMV D7701, from either Adminga or Dalhousie, S.A., designation by Coventry, A.J. (1970). Reptile and amphibian type specimens housed in the National Museum of Victoria. *Mem. Natl. Mus. Vic.* **31**: 115–124.

Distribution: (N Gulf), N coastal, W plateau, S Gulfs, Murray-Darling basin, Bulloo River basin, Lake Eyre basin, S.A., Vic., N.S.W., Qld., W.A., N.T. Ecology: desert, (hummock grassland), tussock grassland, low open shrubland, low open woodland, low woodland, woodland, terrestrial, diurnal; oviparous, arthropod-feeder. Biological references: Mitchell, F.J. (1948). A revision of the lacertilian genus *Tympanocryptis*. *Rec. S. Aust. Mus.* **9**: 57–86 (taxonomy); Storr, G.M. (1964). The agamid lizards of the genus *Tympanocryptis* in Western Australia. *J. R. Soc. West. Aust.* **47**: 43–50 (taxonomy); Houston, T.F. (1978). *Dragon Lizards and Goannas of South Australia.* Adelaide : South Australian Museum 84 pp. (natural history).

Tympanocryptis uniformis Mitchell, 1948

Tympanocryptis uniformis Mitchell, F.J. (1948). A revision of the lacertilian genus *Tympanocryptis*. *Rec. S. Aust. Mus.* **9**: 57–86 [76 fig 8 pl 4]. Type data: holotype, SAMA R705, from near Darwin, N.T.

Distribution: N Gulf, N coastal, W.A., N.T. Ecology: tall open shrubland, woodland, open forest, terrestrial, diurnal; oviparous, arthropod-feeder. Biological references: Storr, G.M. (1964). The agamid lizards of the genus *Tympanocryptis* in Western Australia. *J. R. Soc. West. Aust.* **47**: 43–50 (taxonomy).

VARANIDAE

INTRODUCTION

A family of lizards in which all 31 living species are placed in a single genus, *Varanus*. The family ranges from Africa to southern Asia, the Indo-Australian Archipelago, New Guinea and Australia. The 24 Australian species are primarily active, diurnal predators, although many also feed on carrion.

Characterised in Australia by: well-developed pentadactyl limbs; movable eyelids; pupil more or less circular; external ear-opening present; body scales mostly juxtaposed; scales on top of head mostly small and irregular; tongue long, slender and deeply-forked like that of a snake.

References

Cogger, H.G. (1979). *Reptiles and Amphibians of Australia.* Sydney : A.H. & A.W. Reed 608 pp.

Mertens, R. (1942). Die Familie der Warane (Varanidae). Erster Teil : Allgemeines. *Abh. Senckenb. Naturf. Ges.* 462: 1–116

Mertens, R. (1942). Die Familie der Warane (Varanidae). Zweiter Teil : der Schädel. *Abh. Senckenb. Naturf. Ges.* 465: 117–234

Mertens, R. (1942). Die Familie der Warane (Varanidae). Dritter Teil : Taxonomie. *Abh. Senckenb. Naturf. Ges.* 466: 235–391

Mertens, R. (1958). Bemerkungen über die Warane Australiens. *Senckenberg. Biol.* **39**: 229–264

Mertens, R. (1963). Liste der rezenten Amphibien und Reptilien; Helodermatidae, Varanidae, Lanthanotidae. *Das Tierreich* **79**: i–x 1–26

Varanus Merrem, 1820

Varanus Rafinesque, C.S. (1815). *Analyse de la Nature* ou Tableau de l'Univers et des Corps Organises. Palermo [75] [*nom. nud.*].

Varanus Merrem, B. (1820). *Versuch eines Systems Amphibien. Tentamen Systematis Amphibiorum.* Marburg : Krieger xv 199 pp. [58]. Type species *Lacerta varia* White, 1790 by subsequent designation, see Gray, J.E. (1827). A synopsis of the genera of saurian reptiles, in which some new genera are indicated, and the others reviewed by actual examination. *Phil. Mag. (2)* **3**: 54–58.

Odatria Gray, J.E. (1838). A catalogue of the slender-tongued saurians, with descriptions of many new genera and species. *Ann. Mag. Nat. Hist.* **1**: 388–394 [394]. Type species *Odatria punctata* Gray, 1838 by monotypy.

Cylindrurus Fitzinger, L.J. (1843). *Systema Reptilium.* Vienna : Braümüller u. Seidel vi 106 pp. [19] [described as subgenus of *Hydrosaurus* Wagler, 1830]. Type species *Odatria punctata* Gray, 1838 by original designation.

Pantherosaurus Fitzinger, L.J. (1843). *Systema Reptilium.* Vienna : Braümüller u. Seidel vi 106 pp. [19] [described as subgenus of *Euprepiosaurus* Fitzinger, 1843]. Type species *Hydrosaurus gouldii* Gray, 1838 by original designation.

Agalmatosaurus Fitzinger, L.J. (1843). *Systema Reptilium.* Vienna : Braümüller u. Seidel vi 106 pp. [19] [described as subgenus of *Euprepiosaurus* Fitzinger, 1843]. Type species *Varanus timoriensis* Duméril and Bibron, 1836 by original designation.

Euprepiosaurus Fitzinger, L.J. (1843). *Systema Reptilium.* Vienna : Braümüller u. Seidel vi 106 pp. [19] [described as new genus and subgenus]. Type species *Varanus chlorostigma* Duméril and Bibron, 1836 by original designation.

Synonymy that of Cogger, H.G., this work; for synonyms based on extra-limital species see Mertens, R. (1963).

Liste der rezenten Amphibien und Reptilien Helodermatidae, Varanidae, Lanthanotidae. *Das Tierreich* **79**: i–x 1–26.

This group is also found in Africa, S Asia, Indo-Australian Archipelago, New Guinea, Micronesia and Solomon Islands. Species now known not to occur in Australia: *Varanus salvator* (Laurenti, 1768), see Cogger, H.G. (1979). *Reptiles and Amphibians of Australia.* Sydney : A.H. & A.W. Reed 608 pp.; Australian record based on BMNH 78.1.31.14, purportedly from Somerset, Qld., cited in Boulenger, G.A. (1885). *Catalogue of the Lizards in the British Museum (Natural History).* 2. London : British Museum xiii 497 pp.

Varanus acanthurus Boulenger, 1885

Odatria ocellata Gray, J.E. (1845). *Catalogue of the Specimens of Lizards in the Collection of the British Museum.* London : British Museum xxviii 289 pp. [8] [*non Varanus ocellatus* Heyden, 1830]. Type data: holotype, BMNH 1945.8.30.97, from NW coast of Australia.

Varanus acanthurus Boulenger, G.A. (1885). *Catalogue of the Lizards in the British Museum (Natural History).* 2. London : British Museum xiii 497 pp. pls 12 [324] [*nom. nov. pro Odatria ocellata* Gray, 1845].

Varanus acanthurus brachyurus Sternfeld, R. (1919). Neue Schlangen und Echsen aus Zentralaustralien. *Senckenberg. Biol.* **1**: 76–83 [78]. Type data: lectotype, SMF 11640, from Hermannsburg Mission, upper Finke River, N.T., designation by Mertens, R. (1967). Die herpetologische Sektion des Natur-Museums und Forschungs-Institutes Senckenberg in Frankfurt-a.-M. nebst einem Verzeichnis ihrer Typen. 1. *Senckenberg.Biol.* **48**: 1–106.

Varanus (Odatria) acanthurus insulanicus Mertens, R. (1958). Bemerkungen über die Warane Australiens. *Senckenberg. Biol.* **39**: 229–264 [233]. Type data: holotype, AM R11037, from Groote Eylandt, Gulf of Carpentaria.

Synonymy that of Cogger, H.G., this work.

Distribution: NE coastal, N Gulf, N coastal, NW coastal, W plateau, Lake Eyre basin, S.A., Qld., W.A., N.T. Ecology: desert, hummock grassland, tussock grassland, low open shrubland, (tall open shrubland), low open woodland, low woodland, woodland, tall shrubland, terrestrial, diurnal, predator; oviparous, saxicoline. Biological references: Mertens, R. (1958). Bemerkungen über die Warane Australiens. *Senckenberg. Biol.* **39**: 229–264 (taxonomy); Husband, G.A. (1979). Notes on a nest and hatchlings of *Varanus acanthurus. Herpetofauna* **11**: 29–30; Mertens, R. (1942). Die Familie der Warane (Varanidae). Dritter Teil : Taxonomie. *Abh. Senckenb. Naturf. Ges.* 466: 235–391 (taxonomy); Storr, G.M. (1980). The monitor lizards (genus *Varanus* Merrem, 1820) of Western Australia. *Rec. West. Aust. Mus.* **8**: 237–293 (taxonomy); Swanson, S. (1979). Some rock-dwelling reptiles of the Arnhem Land escarpment. *N. Territory Nat.* **1**: 14–18 (natural history); Saint Girons, H., Rice, G.E. & Bradshaw, S.D. (1981). Histologie comparée et ultrastructure de la glande nasale externe de quelques Varanidae (Reptilia : Lacertilia). *Ann. Sci. Nat. Zool. (12)* **3**: 15–21.

Varanus brevicauda Boulenger, 1898

Varanus brevicauda Boulenger, G.A. (1898). Third report on additions to the collection of lizards in the British Museum. *Proc. Zool. Soc. Lond.* **1898**: 912–923 [916,920 pl 56 fig 2]. Type data: syntypes, BMNH 1946.8.30.46–47, from Sherlock River, Nickol (as Nicol) Bay, W.A.

Distribution: N coastal, NW coastal, W plateau, Lake Eyre basin, S.A., Qld., W.A., N.T. Ecology: desert, hummock grassland, low open shrubland, low open woodland, tall shrubland, terrestrial, diurnal, predator; oviparous. Biological references: Mertens, R. (1958). Bemerkungen über die Warane Australiens. *Senckenberg. Biol.* **39**: 229–264 (taxonomy); Pianka, E.R. (1970). Notes on *Varanus brevicauda. West. Aust. Nat.* **11**: 113–116 (natural history); Mertens, R. (1942). Die Familie der Warane (Varanidae). Dritter Teil : Taxonomie. *Abh. Senckenb. Naturf. Ges.* 466: 235–391 (taxonomy); Storr, G.M. (1980). The monitor lizards (genus *Varanus* Merrem, 1820) of Western Australia. *Rec. West. Aust. Mus.* **8**: 237–293 (taxonomy).

Varanus caudolineatus Boulenger, 1885

Varanus caudolineatus Boulenger, G.A. (1885). *Catalogue of the Lizards in the British Museum (Natural History).* 2. London : British Museum xiii 497 pp. [324 pl 18]. Type data: holotype, BMNH 64.7.22.3, from Champion Bay, W.A.

Varanus laudolineatus Lucas, A.H.S. & Frost, C. (1901). A census of Australian lizards. *Rep. Australas. Ass. Adv. Sci.* **8**: 256–261 [259] [*lapsus pro Varanus caudolineatus* Boulenger, 1885].

Synonymy that of Cogger, H.G., this work.

Distribution: NW coastal, SW coastal, W plateau, W.A. Ecology: hummock grassland, low open shrubland, (woodland), tall shrubland, (terrestrial), arboreal, diurnal, predator; oviparous. Biological references: Mertens, R. (1958). Bemerkungen über die Warane Australiens. *Senckenberg. Biol.* **39**: 229–264 (taxonomy); Pianka, E.R. (1969). Notes on the biology of *Varanus caudolineatus* and *Varanus gilleni. West. Aust. Nat.* **11**: 76–82 (natural history); Mertens, R. (1942). Die Familie der Warane (Varanidae). Dritter Teil : Taxonomie. *Abh. Senckenb. Naturf. Ges.* 466: 235–391 (taxonomy); Storr, G.M. (1980). The monitor lizards (genus *Varanus* Merrem, 1820) of Western Australia. *Rec. West. Aust. Mus.* **8**: 237–293 (taxonomy).

Varanus eremius Lucas and Frost, 1895

Varanus eremius Lucas, A.H.S. & Frost, C. (1895). Preliminary notice of a certain species of lizard from central Australia. *Proc. R. Soc. Vict. (ns)* **7**: 264–269 [267]. Type data: holotype, NMV D9136, from Idracowra, N.T., see Mertens, R. (1963). Liste der rezenten Amphibien und Reptilien; Helodermatidae, Varanidae, Lanthanotidae. *Das Tierreich* **79**: i–x 1–26.

Distribution: NW coastal, W plateau, Lake Eyre basin, S.A., (Qld.), W.A., N.T. Ecology: desert, hummock grassland, low open shrubland, tall shrubland, terrestrial, diurnal, predator; oviparous. Biological references: Mertens, R. (1958). Bemerkungen über die Warane Australiens. *Senckenberg. Biol.* **39**: 229–264 (taxonomy); Pianka, E.R. (1968). Notes on the biology of *Varanus eremius. West. Aust. Nat.* **11**: 39–44 (natural history); Mertens, R. (1942). Die Familie der Warane (Varanidae). Dritter Teil : Taxonomie. *Abh. Senckenb. Naturf. Ges.* 466: 235–391 (taxonomy); Storr, G.M. (1980). The monitor lizards (genus *Varanus* Merrem, 1820) of Western Australia. *Rec. West. Aust. Mus.* **8**: 237–293 (taxonomy); Pianka, E.R. (1982). Observations on the ecology of *Varanus* in the Great Victoria Desert. *West. Aust. Nat.* **15**: 1–8.

Varanus giganteus (Gray, 1845)

Hydrosaurus giganteus Gray, J.E. (1845). *Catalogue of the Specimens of Lizards in the Collection of the British Museum.* London : British Museum xxviii 289 pp. [13]. Type data: holotype, BMNH I.22a, from N coast of Australia [type has Gray's Catalogue number I.21a].

Distribution: NW coastal, W plateau, Lake Eyre basin, S.A., Qld., W.A., N.T. Ecology: desert, hummock grassland, low open shrubland, tall shrubland, terrestrial, diurnal, predator; oviparous, (saxicoline). Biological references: Mertens, R. (1958). Bemerkungen über die Warane Australiens. *Senckenberg. Biol.* **39**: 229–264; (taxonomy); Houston, T.F. (1978). *Dragon Lizards and Goannas of South Australia.* Adelaide : South Australian Museum 84 pp. (natural history); King, M. & King, D. (1975). Chromosomal evolution in the lizard genus *Varanus* (Reptilia). *Aust. J. Biol. Sci.* **28**: 89–108; Mertens, R. (1942). Die Familie der Warane (Varanidae). Dritter Teil : Taxonomie. *Abh. Senckenb. Naturf. Ges.* 466: 235–391 (taxonomy); Storr, G.M. (1980). The monitor lizards (genus *Varanus* Merrem, 1820) of Western Australia. *Rec. West. Aust. Mus.* **8**: 237–293 (taxonomy); Saint Girons, H., Rice, G.E. & Bradshaw, S.D. (1981). Histologie comparée et ultrastructure de la glande nasale externe de quelques Varanidae (Reptilia : Lacertilia). *Ann. Sci. Nat. Zool. (12)* **3**: 15–21; Pianka, E.R. (1982). Observations on the ecology of *Varanus* in the Great Victoria Desert. *West. Aust. Nat.* **15**: 1–8.

Varanus gilleni Lucas and Frost, 1895

Varanus gilleni Lucas, A.H.S. & Frost, C. (1895). Preliminary notice of a certain species of lizard from central Australia. *Proc. R. Soc. Vict. (ns)* **7**: 264–269 [266]. Type data: holotype, NMV D11759, from between Glen Edith and Deering Creek, N.T.

Distribution: W plateau, S Gulfs, Lake Eyre basin, NW coastal, S.A., Qld., W.A., N.T. Ecology: desert, hummock grassland, low open shrubland, low woodland, tall shrubland, arboreal, diurnal, predator; oviparous. Biological references: Mertens, R. (1958). Bemerkungen über die Warane Australiens. *Senckenberg. Biol.* **39**: 229–264 (taxonomy); Mitchell, M. & Mitchell, L.A. (1976). A further analysis of the combat ritual of the pigmy mulga monitor, *Varanus gilleni* (Reptilia : Varanidae). *Herpetologica* **32**: 35–40; Pianka, E.R. (1969). Notes on the biology of *Varanus caudolineatus* and *Varanus gilleni. West. Aust. Nat.* **11**: 76–82 (natural history); Murphy, J.B. & Mitchell, L.A. (1974). Ritualized combat behavior of the pigmy mulga monitor lizard, *Varanus gilleni* (Sauria : Varanidae). *Herpetologica* **30**: 90–97; Houston, T.F. (1978). *Dragon Lizards and Goannas of South Australia.* Adelaide : South Australian Museum 84 pp. (natural history); Mertens, R. (1942). Die Familie der Warane (Varanidae). Dritter Teil : Taxonomie. *Abh. Senckenb. Naturf. Ges.* 466: 235–391 (taxonomy); Storr, G.M. (1980). The monitor lizards (genus *Varanus* Merrem, 1820) of Western Australia. *Rec. West. Aust. Mus.* **8**: 237–293 (taxonomy); Pianka, E.R. (1982). Observations on the ecology of *Varanus* in the Great Victoria Desert. *West. Aust. Nat.* **15**: 1–8.

Varanus glauerti Mertens, 1957

Varanus (Odatria) timorensis glauerti Mertens, R. (1957). Two new goannas from Australia. *West. Aust. Nat.* **5**: 183–185 [183]. Type data: holotype, WAM R12337, from Wotjulum, West Kimberley, W.A.

Distribution: N coastal, W.A. Ecology: woodland, diurnal, predator; oviparous. Biological references: Mertens, R. (1958). Bemerkungen über die Warane Australiens. *Senckenberg. Biol.* **39**: 229–264 (taxonomy); Storr, G.M. (1980). The monitor lizards (genus *Varanus* Merrem, 1820) of Western Australia. *Rec. West. Aust. Mus.* **8**: 237–293 (taxonomy).

Varanus glebopalma Mitchell, 1955

Varanus (Odatria) glebopalma Mitchell, F.J. (1955). Preliminary account of the Reptilia and Amphibia collected by the National Geographic Society—Commonwealth Government—Smithsonian Institution Expedition to Arnhem Land (April to November, 1948). *Rec. S. Aust. Mus.* **11**: 373–408 [389]. Type data: holotype, SAMA R3222, from S end of Lake Hubert, N.T.

Distribution: N Gulf, N coastal, Qld., W.A., N.T. Ecology: tall open shrubland, low open woodland, low woodland, woodland, open forest, terrestrial, diurnal, predator; oviparous, saxicoline. Biological references: Mertens, R. (1958). Bemerkungen über die Warane Australiens. *Senckenberg. Biol.* **39**: 229–264 (taxonomy); Horn, H.-G. & Schürer, U. (1978). Bemerkungen zu *Varanus (Odatria) glebopalma* Mitchell, 1955 (Reptilia : Sauria : Varanidae). *Salamandra* **14**: 105–116; Storr, G.M. (1980). The monitor lizards (genus *Varanus* Merrem, 1820) of Western Australia. *Rec. West. Aust. Mus.* **8**: 237–293 (taxonomy); Swanson, S. (1979). Some rock-dwelling reptiles of the Arnhem Land escarpment. *N. Territory Nat.* **1**: 14–18 (natural history).

Varanus gouldii (Gray, 1838)

Hydrosaurus gouldii Gray, J.E. (1838). A catalogue of the slender-tongued saurians, with descriptions of many new genera and species. *Ann. Mag. Nat. Hist.* **1**: 388–394 [394]. Type data: lectotype, BMNH 1946.9.7.61, from Australia (restricted to north-western Australia by choice of lectotype), designation by Mertens, R. (1958). Bemerkungen über die Warane Australiens. *Senckenberg. Biol.* **39**: 229–264.

Hydrosaurus ocellarius Blyth, E. (1878). *Hydrosaurus ocellarius*, *in* Theobald, W. *J. Asiat. Soc. Beng.* **47** extra number: 21 [21] [regarded as a *nom. dub.* by Mertens, R. (1963). Liste der rezenten Amphibien und Reptilien; Helodermatidae, Varanidae, Lanthanotidae. *Das Tierreich* **79**: 1–x 1–26]. Type data: status unknown, from Australia.

Varanus gouldiae Berney, F.L. (1936). Gould's monitor (*Varanus gouldiae*). *Qd. Nat.* **10**: 12–14 [12] [*errore pro Varanus gouldii* (Gray, 1838)].

Varanus (Varanus) gouldii flavirufus Mertens, R. (1958). Bemerkungen über die Warane Australiens. *Senckenberg. Biol.* **39**: 229–264 [250 pl 25 fig 3, pl 29 fig 11]. Type data: holotype, SMF 53271, from Bat Caves S of Alice Springs, N.T.

Synonymy that of Cogger, H.G., this work.

Distribution: SE coastal, NE coastal, N Gulf, N coastal, NW coastal, SW coastal, W plateau, S Gulfs, Murray-Darling basin, Bulloo River basin, Lake Eyre basin, S.A., Vic., N.S.W., Qld., W.A., N.T., A.C.T. Ecology: desert, hummock grassland, tussock grassland, low open shrubland, low shrubland, open heath, tall open shrubland, open scrub, low open woodland, low woodland, woodland, open forest, tall shrubland, terrestrial, (arboreal), diurnal, predator; oviparous. Biological references: Mertens, R. (1958). Bemerkungen über die Warane Australiens. *Senckenberg. Biol.* **39**: 229–264; Pianka, E.R. (1982). Observations on the ecology of *Varanus* in the Great Victoria Desert. *West. Aust. Nat.* **15**: 1–8 (taxonomy); Warburg, M.R. (1965). Studies on the environmental physiology of some Australian lizards from arid and semiarid habitats. *Aust. J. Zool.* **13**: 563–575; Pianka, E.R. (1970). Notes on the biology of *Varanus gouldi flavirufus. West. Aust. Nat.* **11**: 141–144 (natural history); Johnson, C.R. (1976). Some behavioural observations on wild and captive sand monitors, *Varanus gouldii* (Sauria : Varanidae). *Zool. J. Linn. Soc.* **59**: 377–380; Green, B. (1972). Water loss of the sand goanna (*Varanus gouldii*) in its natural environment. *Ecology* **53**: 452–457; Johnson, C.R. (1972). Head-body temperature differences in *Varanus gouldii* (Sauria : Varanidae). *Comp. Biochem. Physiol.* **43A**: 1025–1029; King, M. & King, D. (1975). Chromosomal evolution in the lizard genus *Varanus* (Reptilia). *Aust. J. Biol. Sci.* **28**: 89–108; Mertens, R. (1942). Die Familie der Warane (Varanidae). Dritter Teil : Taxonomie. *Abh. Senckenb. Naturf. Ges.* 466: 235–391 (taxonomy); Bennett, A.F. (1973). Blood physiology and oxygen transport during activity in two lizards, *Varanus gouldii* and *Sauromalus hispidus. Comp. Biochem. Physiol.* **46A**: 673–690; Storr, G.M. (1980). The monitor lizards (genus *Varanus* Merrem, 1820) of Western Australia. *Rec. West. Aust. Mus.* **8**: 237–293 (taxonomy).

Varanus indicus (Daudin, 1802)

Tupinambis indicus Daudin, F.M. (1802). *Histoire Naturelle, générale et particulière des Reptiles.* Tom. 3 Paris : Dufart 452 pp. [46 pl 30]. Type data: none designated, whereabouts unknown, from Ambon, Moluccas, Indonesia.

Varanus guttatus Merrem, B. (1820). *Versuch eines Systems Amphibien. Tentamen Systematis Amphibiorum.* Marburg : Krieger xv 199 pp. [58] [we have followed Mertens, R. (1963). Liste der rezenten Amphibien und Reptilien; Helodermatidae, Varanidae, Lanthanotidae. *Das Tierreich* **79**: i–x 1–26 in regardiing *Varanus guttatus* as a replacement name for *Tupinambis indicus* Daudin, 1802]. Type data: that of *Tupinambis indicus* Daudin, 1802, from Ambon, Indonesia (Bengal in original description).

Monitor kalabek Lesson, R.P. (1830). Observations générales sur les reptiles recueillis dans le voyage. *in* L.I. Duperrey (ed.) *Voyage autour du monde, executé par ordre du Roi, sur la corvette de La Majesté,* La Coquille, *pendant les années 1822, 1823, 1824 et 1825.* Zoologie 2(1): 1–65 Paris : Arthus Bertrand [52]. Type data: status unknown, from Fofak (as Offack) Bay, N coast of Waigeu, Indonesia, see Mertens, R. (1963). Liste der rezenten Amphibien und Reptilien; Helodermatidae, Varanidae, Lanthanotidae. *Das Tierreich* **79: i–x 1–26.**

Monitor douarrha Lesson, R.P. (1830). Observations générales sur les reptiles recueillis dans le voyage. *in* L.I.

Duperrey (ed.) *Voyage autour du monde, executé par ordre du Roi, sur la corvette de Sa Majesté,* La Coquille, *pendant les années 1822, 1823, 1824 et 1825.* Zoologie 2(1): 1–65 Paris : Arthus Bertrand [53]. Type data: status unknown, from Praslin Harbour, New Ireland (as New Mecklenberg), Papua New Guinea.

Monitor chlorostigma Gray, J.E. (1831). A synopsis of the species of the class Reptilia, pp. 1–110 *in* Griffith, E. *The Animal Kingdom Arranged in Conformity with its Organization by the Baron Cuvier.* London : Whittaker, Treacher & Co. 9: 481+110 pp. [26]. Type data: syntypes, MNHP presumed lost, from Rawack Is., N of Waigeu, Indonesia.

Monitor doreanus Meyer, A.B. (1874). Eine Mitteilung von Hrn. Dr. Adolf Bernhard Meyer über die von ihm auf Neu-Guinea den Inseln Jobi, Mysore und Mafoor im Jahr 1873 gesammelten Amphibien. *Mber. K. Preuss. Akad. Wiss. Berl.* **1874**: 128–140 [130]. Type data: none designated, ZMB, from Doreh, Berou Peninsula, NW New Guinea.

Varanus indicus rouxi Mertens, R. (1926). Über die Rassen einiger indo-australischer Reptilien. *Senckenberg. Biol.* **8**: 272–279 [276]. Type data: holotype, SMF 11583, from Durdjela, Wammer, Aru Ils., Indonesia.

Varanus tsukamotoi Kishida, K. (1929). A new monitor from the island of Saipan, South Sea Islands. *Lansania* **1**: 13–16 [13]. Type data: whereabouts unknown, from Saipan, Caroline Ils.

Varanus indicus jobiensis Ahl, E. (1932). Eine neue Eidechse und zwei neue Frösche von der Insel Jobi. *Mitt. Zool. Mus. Berlin* **17**: 892–899 [892 fig]. Type data: holotype, ZMB, from Japen (as Jobi) Is., Indonesia.

Varanus leucostigma Hediger, H. (1934). Beitrag zur Herpetologie und Zoogeographiae New Britaniens. *Zool. Jb. Abt. Syst.* **65**: 441–582 [497] [*errore pro Monitor chlorostigma* Gray, 1831].

Varanus indicus spinulosus Mertens, R. (1941). Zwei neue Warane des australischen Fauenengebietes. *Senckenbergiana* **23**: 266–272 [269 fig 2]. Type data: holotype, NHMW 3709, from George's Is., Solomon Ils.

Synonymy that of Cogger, H.G., this work.

Distribution: NE coastal, N Gulf, N coastal, Qld., N.T. Ecology: closed forest, mangrove, terrestrial, arboreal, diurnal, predator; oviparous. Biological references: Mertens, R. (1958). Bemerkungen über die Warane Australiens. *Senckenberg. Biol.* **39**: 229–264 (taxonomy); Mertens, R. (1959). Liste der Warane Asiens und der indo-australischen Inselwelt mit systematischen Bemerkungen. *Senckenberg. Biol.* **40**: 221–240 (taxonomy); King, M. & King, D. (1975). Chromosomal evolution in the lizard genus *Varanus* (Reptilia). *Aust. J. Biol. Sci.* **28**: 89–108; Mertens, R. (1942). Die Familie der Warane (Varanidae). Dritter Teil : Taxonomie. *Abh. Senckenb. Naturf. Ges.* 466: 235–391 (taxonomy).

Varanus kingorum Storr, 1980

Varanus kingorum Storr, G.M. (1980). The monitor lizards (genus *Varanus* Merrem, 1820) of Western Australia. *Rec. West. Aust. Mus.* **8**: 237–293 [268]. Type data: holotype, WAM R60374, from 10 km WNW of Timber Creek, N.T., in 15°37'S, 130°23'E.

Distribution: N coastal, W.A., N.T. Ecology: woodland, terrestrial, diurnal, predator; oviparous, saxicoline.

Varanus mertensi Glauert, 1951

Varanus mertensi Glauert, L. (1951). A new *Varanus* from East Kimberley, *Varanus mertensi* sp. n. *West. Aust. Nat.* **3**: 14–16 [14 fig]. Type data: holotype, WAM R5819, from Moola Bulla Cattle Station, near Hall's Creek, East Kimberley, W.A.

Varanus bulliwallah Worrell, E. (1956). A new water-monitor from northern Australia. *Aust. Zool.* **12**: 201–202 [201 pls 27–29]. Type data: holotype, AM R14810, from Bulliwallah Station, on the Belyando River, Qld.

Synonymy that of Cogger, H.G., this work.

Distribution: N Gulf, N coastal, Qld., W.A., N.T. Ecology: tussock grassland, low open shrubland, tall open shrubland, low open woodland, low woodland, woodland, open forest, lentic freshwater, lotic freshwater, swamp, terrestrial, arboreal, diurnal, predator; oviparous, semi-aquatic. Biological references: Mertens, R. (1958). Bemerkungen über die Warane Australiens. *Senckenberg. Biol.* **39**: 229–264 (taxonomy); Bustard, H.R. (1970). *Australian Lizards.* Sydney : Collins 162 pp. (natural history); Schürer, U. & Horn, H-G. (1976). Freiland- und Gefangenschafts beobachtungen am australischen Wasserwaran, *Varanus mertensi. Salamandra* **12**: 176–188 (natural history); Murphy, J.B. & Lamoreaux, W.E. (1978). Threatening behaviour in Merten's water monitor *Varanus mertensi* (Sauria : Varanidae). *Herpetologica* **34**: 202–205; Storr, G.M. (1980). The monitor lizards (genus *Varanus* Merrem, 1820) of Western Australia. *Rec. West. Aust. Mus.* **8**: 237–293 (taxonomy); Hermes, N. (1981). Mertens water monitor feeding on trapped fish. *Herpetofauna* **13**: 34.

Varanus mitchelli Mertens, 1958

Varanus (Varanus) mitchelli Mertens, R. (1958). Bemerkungen über die Warane Australiens. *Senckenberg Biol.* **39**: 229–264 [256 pls 27,31]. Type data: holotype, SAMA R3230, from 5 mi W of Oenpelli (as Ocupelli) N.T.

Distribution: N Gulf, N coastal, W.A., N.T. Ecology: tall open shrubland, woodland, open forest, lentic freshwater, lotic freshwater, swamp terrestrial, arboreal, diurnal, predator; oviparous semi-aquatic. Biological references: Bustard, H.R (1970). *Australian Lizards.* Sydney : Collins 162 pp. (natural history); Swanson, S. (1976). *Lizards of Australia.* Sydney : Angus & Robertson 80 pp (natural history); Storr, G.M. (1980). The monitor lizards (genus *Varanus* Merrem, 1820) of Western Australia. *Rec. West. Aust. Mus.* **8**: 237–293 (taxonomy).

Varanus panoptes Storr, 1980

Varanus panoptes panoptes Storr, G.M. (1980). The monitor lizards (genus *Varanus* Merrem, 1820) of Western Australia. *Rec. West. Aust. Mus.* **8**: 237–293 [273]. Type data: holotype, WAM R44792, from Lake Argyle, W.A., in 16°03'S, 128°47'E.

Varanus panoptes rubidus Storr, G.M. (1980). The monitor lizards (genus *Varanus* Merrem, 1820) of Western Australia. *Rec. West. Aust. Mus.* **8**: 237–293 [276 pl 8]. Type data: holotype, WAM R19132, from Wilgie Mia, 60 km NNW of Cue, W.A., in 26°0'S, 117°42'E.

Synonymy that of Cogger, H.G., this work.

Distribution: N Gulf, N coastal, NW coastal, SW coastal, W plateau, W.A., N.T. Ecology: hummock grassland, low open shrubland, tall open shrubland, woodland, open forest, tall shrubland, terrestrial, diurnal, predator; oviparous.

Varanus pilbarensis Storr, 1980

Varanus pilbarensis Storr, G.M. (1980). The monitor lizards (genus *Varanus* Merrem, 1820) of Western Australia. *Rec. West. Aust. Mus.* **8**: 237–293 [278 pl 8]. Type data: holotype, WAM R39782, from Chichester Range, W.A., in 22°03'S, 118°48'E.

Distribution: NW coastal, W.A. Ecology: hummock grassland, low open shrubland, terrestrial, diurnal, predator; oviparous, saxicoline.

Varanus prasinus (Schlegel, 1839)

Monitor viridis Gray, J.E. (1831). A synopsis of the species of the class Reptilia, pp. 1–110 *in* Griffith, E. *The Animal Kingdom Arranged in Conformity with its Organization by the Baron Cuvier.* London : Whittaker, Treacher & Co. 9: 481+110 pp. [synopsis 26] [we have followed Mertens, R. (1963). Liste der rezenten Amphibien und Reptilien; Helodermatidae, Varanidae, Lanthanotidae. *Das Tierreich* **79**: i–x 1–26 in regarding *Monitor viridis* as a *nom. dub.*]. Type data: none designated, RMNH.

Monitor prasinus Schlegel, H. (1839). *Abbildungen neuer oder unvollständig bekannter Amphibien, nach der Natur der dem Leben entworfen.* Düsseldorf : Arnz xiv 141 pp. [78 pl 22 fig 5]. Type data: none designated, whereabouts unknown, from Fort Du Bus and Oeroe-Langoeroe Bay (as W coast of New Guinea), Irian Jaya, Indonesia, see Mertens, R. (1963). Liste der rezenten Amphibien und Reptilien; Helodermatidae, Varanidae, Lanthanotidae, *Das Tierreich* **79**: i–x 1–26.

Monitor beccarii Doria, G. (1874). Enumerazione dei rettili raccolti dal Dott. O. Beccari in Amboina alle Isole Aru ed alle Isole Kei durante gli anni. *Ann. Mus. Civ. Stor. Nat. Giacomo Doria* **6**: 325–357 [331 fig 11a]. Type data: syntypes (3), MCG C.E. 28723, other syntypes not found, from Wokam (as Wokan), Aru Ils., Indonesia.

Monitor kordensis Meyer, A.B. (1874). Eine Mitteilung von Hrn. Dr. Adolf Bernhard Meyer über die von ihm auf Neu-Guinea den Inseln Jobi, Mysore und Mafoor im Jahr 1873 gesammelten Amphibien. *Mber. K. Preuss. Akad. Wiss. Berl.* **1874**: 128–140 [131]. Type data: none designated, whereabouts unknown, from Biak Is., Schouten Group (as Kordo, Mysore), Irian Jaya, Indonesia.

Synonymy that of Cogger, H.G., this work; Mertens, R. (1963). Liste der rezenten Amphibien und Reptilien; Helodermatidae, Varanidae, Lanthanotidae. *Das Tierreich* **79**: i–x 1–26.

Distribution: NE coastal, Qld. Ecology: closed forest, mangrove, arboreal, diurnal, predator; oviparous. Biological references: Mertens, R. (1959). Liste der Warane Asiens und der indo-australischen Inselwelt mit systematischen Bemerkungen. *Senckenberg. Biol.* **40**: 221–240 (taxonomy); Mertens, R. (1942). Die Familie der Warane (Varanidae). Dritter Teil : Taxonomie. *Abh. Senckenb. Naturf. Ges.* 466: 235–391 (taxonomy); Czechura, G.V. (1980). The emerald monitor *Varanus prasinus* (Schlegel): an addition to the Australian mainland herpetofauna. *Mem. Qd. Mus.* **20**: 103–109.

Varanus primordius Mertens, 1942

Varanus acanthurus primordius Mertens, R. (1942). Ein weiterer neuer Waran aus Australien. *Zool. Anz.* **137**: 41–44 [41 fig 1]. Type data: holotype, ZMB 6204, from North Australia.

Distribution: N coastal, N.T. Ecology: low open woodland, low woodland, woodland, terrestrial, diurnal, predator; oviparous, saxicoline. Biological references: Storr, G.M. (1966). Rediscovery and taxonomic status of the Australian lizard, *Varanus primordius. Copeia* **1966**: 583–584; Mertens, R. (1966). Ein neuer Zwergwaran aus Australien. *Senckenberg. Biol.* **47**: 437–441; Mertens, R. (1958). Bemerkungen über die Warane Australiens. *Senckenberg. Biol.* **39**: 229–264 (taxonomy).

Varanus rosenbergi Mertens, 1957

Varanus gouldii rosenbergi Mertens, R. (1957). Ein neuer melanistischer Waran aus dem südlichen Australien. (*V. gouldii rosenbergi*, subsp. nov.). *Zool. Anz.* **159**: 17–20 [18]. Type data: holotype, WAM R822, from Monigup Pass, Stirling Range, W.A.

Distribution: SW coastal, W plateau, S Gulfs, S.A., W.A., Vic. Ecology: open heath, open scrub, woodland, open forest, tall shrubland, terrestrial, diurnal, predator; oviparous. Biological references: Storr, G.M. (1980). The monitor lizards (genus *Varanus* Merrem, 1820) of Western Australia. *Rec. West. Aust. Mus.* **8**: 237–293 (taxonomy); Green, B. & King, D. (1978). Home range and activity patterns of the sand goanna, *Varanus gouldii* (Reptilia : Varanidae). *Aust. Wildl. Res.* **5**:

417–424 (as *V. gouldii*); King, D. & Green, B. (1979). Notes on the diet and reproduction of the sand goanna, *Varanus gouldii rosenbergi*. *Copeia* **1979**: 64–70; King, M. & King, D. (1975). Chromosomal evolution in the lizard genus *Varanus* (Reptilia). *Aust. J. Biol. Sci.* **28**: 89–108; King, D. (1980). The thermal biology of free-living sand goannas (*Varanus gouldii*) in southern Australia. *Copeia* **1980**: 755–767 (as *V. gouldii*); Saint Girons, H., Rice, G.E. & Bradshaw, S.D. (1981). Histologie comparée et ultrastructure de la glande nasale externe de quelques Varanidae (Reptilia : Lacertilia). *Ann. Sci. Nat. Zool. (12)* **3**: 15–21.

Varanus semiremex Peters, 1869

Varanus (Odatria) semiremex Peters, W. (1869). Über neue Gattungen und Arten von Eidechsen. *Mber. K. Preuss. Akad. Wiss. Berl.* **1869**: 57–66 [65]. Type data: holotype, ZMB 5776, from Cape York, NE Australia.

Odatria semireme Günther, A. (1875). Reptiles. A list of saurians of Australia and New Zealand. pp. 9–19 *Zoology of the Voyage of H.M.S.* Erebus *and* Terror. Vol. 2 London : E.W. Janson [9] [*lapsus pro Varanus (Odatria) semiremex* Peters, 1869].

Varanus boulengeri Kinghorn, J.R. (1923). A new *Varanus* from Coquet Island, Queensland. *Rec. Aust. Mus.* **14**: 135–137 [135 pls 17–18]. Type data: holotype, AM R8083, from Coquet Is., Howick Group, Qld.

Synonymy that of Cogger, H.G., this work.

Distribution: NE coastal, N Gulf, Qld. Ecology: woodland, open forest, swamp, mangrove, arboreal, diurnal, predator; oviparous. Biological references: Mertens, R. (1958). Bemerkungen über die Warane Australiens. *Senckenberg. Biol.* **39**: 229–264 (taxonomy); Bustard, H.R. (1970). *Australian Lizards.* Sydney : Collins 162 pp. (natural history); King, M. & King, D. (1975). Chromosomal evolution in the lizard genus *Varanus* (Reptilia). *Aust. J. Biol. Sci.* **28**: 89–108; Dunson, W.A. (1974). Salt gland secretion in a mangrove monitor lizard. *Comp. Biochem. Physiol.* **47A**: 1245–1255; Polleck, R. (1980). Der Rostkopfwaran (*Varanus semiremex*). *Aqu.-Mag. (Stuttgart)* **14**: 39.

Varanus spenceri Lucas and Frost, 1903

Varanus spenceri Lucas, A.H.S. & Frost, C. (1903). Description of two new Australian lizards, *Varanus spenceri* and *Diplodactylus bilineatus*. *Proc. R. Soc. Vict. (ns)* **15**: 145–147 [145]. Type data: holotype, NMV D12350, from Tablelands, 50 mi to NE of Tennant's Creek, N.T.

Varanus ingrami Boulenger, G.A. (1906). Description of a new lizard and a new snake from Australia. *Ann. Mag. Nat. Hist. (7)* **18**: 440–441 [440]. Type data: holotype, BMNH 1946.8.31.6, from Alexandria, N.T.

Synonymy that of Cogger, H.G., this work.

Distribution: N Gulf, W plateau, Lake Eyre basin, Qld., N.T. Ecology: tussock grassland, terrestrial, diurnal, predator; oviparous. Biological references: Mertens, R. (1958). Bemerkungen über die Warane Australiens. *Senckenberg. Biol.* **39**: 229–264 (taxonomy); Christian, T. (1979). Notes on Spencer's monitor (*Varanus spenceri*). *Vict. Herp Soc. Newsletter* **14**: 13–14; Mertens, R. (1942). Die Familie der Warane (Varanidae). Dritter Teil : Taxonomie. *Abh. Senckenb. Naturf. Ges.* 466: 235–391 (taxonomy); Pengilley, R. (1981). Notes on the biology of *Varanus spenceri* and *V. gouldii*, Barkly Tablelands, Northern Territory. *Aust. J. Herp.* **1**: 23–26.

Varanus storri Mertens, 1966

Varanus (Odatria) storri Mertens, R. (1966). Ein neuer Zwergwaran aus Australien. *Senckenberg. Biol.* **47**: 437–441 [437 figs 1–3]. Type data: holotype, SMF 59011, from Charters Towers, Qld.

Varanus storri ocreatus Storr, G.M. (1980). The monitor lizards (genus *Varanus* Merrem, 1980) of Western Australia. *Rec. West. Aust. Mus.* **8**: 237–293 [283 pl 11]. Type data: holotype, WAM R42717, from Argyle Downs airstrip, W.A., in 16°20'S, 128°46'E.

Synonymy that of Cogger, H.G., this work.

Distribution: NE coastal, N Gulf, N coastal, (Lake Eyre basin), Qld., W.A., N.T. Ecology: hummock grassland, (tussock grassland), low open shrubland, low woodland, woodland, terrestrial, diurnal, predator; oviparous, saxicoline. Biological references: Storr, G.M. (1980). The monitor lizards (genus *Varanus* Merrem, 1820) of Western Australia. *Rec. West. Aust. Mus.* **8**: 237–293 (taxonomy); Storr, G.M. (1966). Rediscovery and taxonomic status of the Australian lizard, *Varanus primordius*. *Copeia* **1966**: 583–584; Peters, U. (1973). A contribution to the ecology of *Varanus* (*Odatria*) *storri*. *Koolewong* **2**: 12–13; King, M. & King, D. (1975). Chromosomal evolution in the lizard genus *Varanus* (Reptilia). *Aust. J. Biol. Sci.* **28**: 89–108; Sprackland, R.G. (1980). Some notes on Storr's dwarf spiny-tailed monitor. *Kansas Herp. Soc. Newsletter* **40**: 7–9.

Varanus timorensis (Gray, 1831)

Monitor timorensis Gray, J.E. (1831). A synopsis of the species of the class Reptilia, pp. 1–110 *in* Griffith, E. *The Animal Kingdom Arranged in Conformity with its Organization by the Baron Cuvier.* London : Whittaker, Treacher & Co. 9: 481+110 pp. [26]. Type data: none designated, MNHP (not found), from Timor, Indonesia.

Tup[inambis] viridomaculatus Gray, J.E. (1831). A synopsis of the species of the class Reptilia, pp. 1–110 *in* Griffith, E. *The Animal Kingdom Arranged in Conformity with its Organization by the Baron Cuvier.* London : Whittaker, Treacher & Co. 9: 481+110 pp. [26] [*nom. nud.*, introduced in synonymy].

Varanus timoriensis Duméril, A.M.C. & Bibron, G. (1836). *Erpétologie Générale* ou Histoire Naturelle Complète des Reptiles. Paris : Roret 3: iv 517 pp. [473] [emend. *pro Monitor timorensis* Gray, 1831].

Varanus timerensis Martens, E. von (1876). *Die Preussische Expedition nach Ost-asien.* Zoologischer Teil. Berlin : Decker Vol. 1 412 pp. [374] [*err. typ. pro Monitor timorensis* Gray, 1831].

Varanus timorensis scalaris Mertens, R. (1941). Zwei neue Warane des australischen Fauenengebietes. *Senckenbergiana* **23**: 266–272 [266 fig]. Type data: holotype, SMF 32806, from Mission Beagle Bay, Dampier Land, NW Australia.

Varanus (Odatria) timorensis similis Mertens, R. (1958). Bemerkungen über die Warane Australiens. *Senckenberg. Biol.* **39**: 229–264 [239 pl 24 fig 2 pl 28 fig 9]. Type data: holotype, AM R10207, from Groote Eylandt, N.T.

Synonymy that of Cogger, H.G., this work.

Distribution: NE coastal, N Gulf, N coastal, Qld., W.A., N.T. Ecology: low open shrubland, tall open shrubland, low open woodland, low woodland, woodland, open forest, arboreal, diurnal, predator; oviparous. Biological references: Mertens, R. (1958). Bemerkungen über die Warane Australiens. *Senckenberg. Biol.* **39**: 229–264 (taxonomy); Storr, G.M. (1980). The monitor lizards (genus *Varanus* Merrem, 1820) of Western Australia. *Rec. West. Aust. Mus.* **8**: 237–293 (taxonomy); Cogger, H.G. & Lindner, D.A. (1974). Frogs and reptiles in fauna survey of the Port Essington District, Cobourg Peninsula, Northern Territory of Australia. Frith, H.J. & Calaby, J.H. (eds.) *CSIRO Div. Wildl. Res. Tech. Pap.* **28**: 63–107 (taxonomy); King, M. & King, D. (1975). Chromosomal evolution in the lizard genus *Varanus* (Reptilia). *Aust. J. Biol. Sci.* **28**: 89–108; Mertens, R. (1942). Die Familie der Warane (Varanidae). Dritter Teil : Taxonomie. *Abh. Senckenb. Naturf. Ges.* 466: 235–391; Ruegg, R. (1974). Nachzucht beim Timor-Baumwaran, *Varanus timorensis similis* Mertens, 1958. *Aquarium mit Aquaterra* **8**: 360–363 (natural history).

Varanus tristis (Schlegel, 1839)

Odatria punctata Gray, J.E. (1838). A catalogue of the slender-tongued saurians, with descriptions of many new genera and species. *Ann. Mag. Nat. Hist.* **1**: 388–394 [394] [*non Varanus punctatus* Merrem, 1820]. Type data: holotype, BMNH i.3a, from Shark Bay, W.A.

Monitor tristis Schlegel, H. (1839). *Abbildungen neuer oder unvollständig bekannter Amphibien, nach der Natur der dem Leben entworfen.* Düsseldorf : Arnz xiv 141 pp. [73]. Type data: whereabouts unknown, from Swan River, W.A.

Varanus punctatus orientalis Fry, D.B. (1913). On a *Varanus* and a frog from Burnett River, Queensland, and a revision of the variations in *Limnodynastes dorsalis* Gray. *Rec. Aust. Mus.* **10**: 17–34 [18 figs 7,9]. Type data: holotype, AM R5313, from Eidsvold, upper Burnett River, Qld.

Varanus (Odatria) tristis centralis Mertens, R. (1957). Two new goannas from Australia. *West. Aust. Nat.* **5**: 183–185 [185]. Type data: holotype, SMF 11632, from Hermannsburg, Finke River, N.T.

Varanus occidentalis Zinniker, W. (1961). Australische Freunde. *Aquaria* (St. Gallen) **2**: 15–17 [16] [*nom. dub.*]. Type data: none designated, whereabouts unknown, from unknown locality, restricted to W.A. by Mertens, R. (1963). Liste der rezenten Amphibien und Reptilien; Helodermatidae, Varanidae, Lanthanotidae. *Das Tierreich* **79**: i–x 1–26.

Synonymy that of Cogger, H.G., this work.

Distribution: NE coastal, N Gulf, N coastal, NW coastal, (SW coastal), W plateau, Murray-Darling basin, Bullo River basin, Lake Eyre basin, S.A., N.S.W., Qld., W.A., N.T. Ecology: desert, hummock grassland, tussock grassland, low open shrubland, low shrubland, tall shrubland, tall open shrubland, open scrub, low open woodland, low woodland, woodland, terrestrial, arboreal, diurnal, predator; oviparous, (saxicoline). Biological references: Mertens, R. (1958). Bemerkungen über die Warane Australiens. *Senckenberg. Biol.* **39**: 229–264 (taxonomy); Pianka, E.R. (1971). Notes on the biology of *Varanus tristis. West. Aust. Nat.* **11**: 180–182 (natural history); Fyfe, G. (1979). Notes on the black-headed monitor (*Varanus tristis*). *Vict. Herp. Soc. Newsletter* **13**: 17–18 (natural history); Mertens, R. (1942). Die Familie der Warane (Varanidae). Dritter Teil : Taxonomie. *Abh. Senckenb. Naturf. Ges.* 466: 235–391 (taxonomy); Storr, G.M. (1980). The monitor lizards (genus *Varanus* Merrem, 1820) of Western Australia. *Rec. West. Aust. Mus.* **8**: 237–293 (taxonomy); Christian, T. (1981). *Varanus tristis* — a variable monitor. *Herpetofauna* **12**: 7–12 (natural history); Pianka, E.R. (1982). Observations on the ecology of *Varanus* in the Great Victoria Desert. *West. Aust. Nat.* **15**: 1–8.

Varanus varius (White, 1790)

Lacerta varia White, J. (1790). *Journal of a Voyage to New South Wales,* with sixty five plates of non-descript animals, birds, lizards, serpents, curious cones of trees and other natural productions. Appendix. London : Debrett 299 pp. [253 pl 38] [*ex* Shaw ms; junior primary homonym of *Lacerta varia* Wyttenbach, 1789, itself a *nom. oblit.*; conservation of *Lacerta varia* Shaw will require use of the Plenary powers of the International Commission on Zoological Nomenclature]. Type data: holotype, presumed lost, from N.S.W.

Tupinambis variegatus Daudin, F.M. (1802). *Histoire Naturelle, générale et particulière des Reptiles.* Tom. 3 Paris : Dufart 452 pp. [76]. Type data: holotype, MNHP 8287, from Sydney, N.S.W. (as Port Jackson).

Varanus bellii Duméril, A.M.C. & Bibron, G. (1836). *Erpétologie Générale* ou Histoire Naturelle Complète des Reptiles. Paris : Roret 3: iv 517 pp. [493 pl 35 fig 1]. Type data: holotype, MNHP 8283, from Australia.

Varanus (Hydrosaurus) mustelinus Borre, A.P. de (1870). Description d'une nouvelle espèce africaine du genre varan (*Varanus*). *Bull. Acad. R. Belg. Cl. Sci. (2)* **29**: 122–130 [124]. Type data: holotype, RIB 9422 (Typ. Cat. 2002), from "Côte de Guinée".

Varanus various Berney, F.L. (1936). Gould's monitor (*Varanus gouldiae*). *Qd. Nat.* **10**: 12–14 [13] [*errore pro Varanus varius* (White, 1790)].

Synonymy that of Cogger, H.G., this work.

Distribution: SE coastal, NE coastal, Murray-Darling basin, S Gulfs, (N Gulf), S.A., Vic., N.S.W., Qld., A.C.T. Ecology: open scrub, low open woodland, woodland, open forest, tall forest, arboreal, diurnal, predator; oviparous. Biological references: Longley, G. (1945). Notes on the lace monitor (*Varanus varius*). *Proc. R. Zool. Soc. N.S.W.* **1944–45**: 20–22 (natural history); Mertens, R. (1958). Bemerkungen über die Warane Australiens. *Senckenberg. Biol.* **39**: 229–264 (taxonomy); Bustard, H.R. (1970). *Australian Lizards.* Sydney : Collins 162 pp. (natural history); Horn, J.-G. (1980). Bisher unbekannte Details zur kenntnis von *Varanus varius* auf Grund von feldherpetologischen und terraristischen Beobachtungen (Reptilia : Sauria : Varanidae). *Salamandra* **16**: 1–18 (natural history); Copland, S.J. (1953). Recent Australian herpetology. The blood vascular system of the trunk of the goanna *Varanus varius* (Shaw). *Proc. Linn. Soc. N.S.W.* **78**: v–xviii; Stebbins, R.C. & Barwick, R.E. (1968). Radiotelemetric study of thermoregulation in a lace monitor. *Copeia* **1968**: 541–547; King, M. & King, D. (1975). Chromosomal evolution in the lizard genus *Varanus* (Reptilia). *Aust. J. Biol. Sci.* **28**: 89–108; Mertens, R. (1942). Die Familie der Warane (Varanidae). Dritter Teil : Taxonomie. *Abh. Senckenb. Naturf. Ges.* 466: 235–391 (taxonomy).

SCINCIDAE

INTRODUCTION

This largest and most diverse family of lizards contains more than 1,275 species in 85 genera; of these, some 250 species and 22 genera occur in Australia and make up about 50% of our total lizard fauna. The family is represented in all tropical and temperate regions, although it is most speciose in the region encompassing Africa, southern Asia, the Indo-Australian Archipelago, New Guinea and Australia.

Characterised in Australia by: limbs variable, ranging from well-developed pentadactyl forms through various stages of limb and digit reduction to snake- or worm-like forms in which limbs are entirely absent; eyelids either movable or fixed, the latter either forming a complete, sealed spectacle or with a slit-like opening above; pupil usually circular, sometimes vertically elliptical; external ear opening either present or absent; body scales usually imbricate, rarely juxtaposed; tongue broad, flat and fleshy, slightly notched in front; scales on top of head usually enlarged to form a series of symmetrical shields.

References

Cogger, H.G. (1979). *Rcptilcs and Amphibians of Australia.* Sydney : A.H. & A.W. Reed 608 pp.

Worrell, E. (1970). *Reptiles of Australia.* 2nd edn. Sydney : Angus & Robertson xv 169 pp.

Anomalopus Duméril and Duméril, 1851

Anomalopus Duméril, A.M.C. & Duméril, A. (1851). *Catalogue Méthodique de la Collection des Reptiles.* Paris : Mus. Hist. Nat. iv 224 pp. [185]. Type species *Anomalopus verreauxii* Duméril and Duméril, 1851 by monotypy.

Ophioscincus Peters, W. (1873). Mitteilung über neue Saurier (*Spoeriodactylus*, *Anolis*, *Phrynosoma*, *Tropidolcpisma*, *Lygosoma*, *Ophioscincus*) aus Centralamerica, Mexico und Australien. *Mber. K. Preuss. Akad. Wiss. Berl.* **1873**: 738–747 [746] [1874 on title page]. Type species *Ophioscincus australis* Peters, 1873 by monotypy.

Anomalpus Günther, A. (1875). Reptiles. A list of saurians of Australia and New Zealand. pp. 9–19 *Zoology of the Voyage of H.M.S.* Erebus *and* Terror. Vol. 2 London : E.W. Janson [14] [*lapsus pro Anomalopus* Duméril and Duméril, 1851].

Coloscincus Peters, W. (1876). Über die von S.M.S. *Gazelle* mitgebrachten Amphibien. *Mber. K. Preuss. Akad. Wiss. Berl.* **1876**: 528–535 [532] [1877 on title page]. Type species *Coloscincus truncatus* Peters, 1876 by monotypy.

Synonymy that of Cogger, H.G., this work; Mittleman, M.B. (1952). A generic synopsis of the lizards of the subfamily Lygosominae. *Smithson. Misc. Coll.* **117**(4069): 1–35.

Anomalopus frontalis (De Vis, 1888)

Ophioscincus frontalis De Vis, C.W. (1888). A contribution to the herpetology of Queensland. *Proc. Linn. Soc. N.S.W. (2)* **2**: 811–826 [823]. Type data: syntypes, QM J243, QM J11499, QM J19737–41, from Innisfail (as Geraldton), Qld.

Distribution: NE coastal, Qld. Ecology: tall forest, closed forest, fossorial; oviparous, arthropod-feeder.

Anomalopus lentiginosus De Vis, 1888

Brachymeles leuckartii Weinland, D.F. (1862). Beschreibung und Abbildung von drei neuen Sauriern. *Abh. Senckenb. Naturf. Ges.* **4**: 131–143 [140 pl 5 fig 3] [*nom. oblit.*]. Type data: syntypes (2), GZM, from Australia.

Anomalopus lentiginosus De Vis, C.W. (1888). A contribution to the herpetology of Queensland. *Proc. Linn. Soc. N.S.W. (2)* **2**: 811–826 [823]. Type data: syntypes, from Brisbane, Qld., QM, not found *vide* Covacevich, J. (1971). Amphibian and reptile type specimens in the Queensland Museum. *Mem. Qd. Mus.* **16**: 49–68.

Lygosoma verreauxii biunguiculata Oudemans, J.T. (1894). Eidechsen und Schildkröten. *in* Semon, R. *Zoologische Forschungsreisen in Australien und den Malayischen Archipel.* Jena : Gustav Fischer **5**: 127–146 [145]. Type data: syntypes, ZMA 11398–9, from Burnett River, Qld.

Lygosoma bancrofti Longman, H.A. (1916). Snakes and lizards from Queensland and the Northern Territory. *Mem. Qd. Mus.* **5**: 46–51 [49]. Type data: holotype, QM J2560, from Gyranda, Dawson River (as Upper Dawson River), Qld.

Lygosoma verreauxii biungulata Zietz, F.R. (1920). Catalogue of Australian lizards. *Rec. S. Aust. Mus.* **1**: 181–228 [219] [*lapsus pro Lygosoma verreauxii biunguiculata* Oudemans, 1894].

Lygosoma bankrofti Zietz, F.R. (1920). Catalogue of Australian lizards. *Rec. S. Aust. Mus.* **1**: 181–228 [219] [*lapsus pro Lygosoma bancrofti* Longman, 1916].

Synonymy that of Cogger, H.G., this work.

Distribution: SE coastal, NE coastal, Murray-Darling basin, N.S.W., Qld. Ecology: low open woodland, low woodland, woodland, open forest, fossorial; oviparous, arthropod-feeder. Biological references: Greer, A.E. (1980). Critical thermal maximum temperatures in Australian scincid lizards: their ecological and evolutionary significance. *Aust. J. Zool.* **28**: 91–102.

Anomalopus ophioscincus (Boulenger, 1887)

Ophioscincus australis Peters, W. (1873). Mitteilung über neue Saurier (*Spoeriodactylus*, *Anolis*, *Phrynosoma*, *Tropidolepisma*, *Lygosoma*, *Ophioscincus*) aus Centralamerica, Mexico und Australien. *Mber. K. Preuss. Akad. Wiss. Berl.* **1873**: 738–747 [747] [1874 on title page; *non Tiliqua australis* Gray, 1838 and *non Lygosoma australis* Gray, 1839]. Type data: holotype, ZMB 8046, from Port Clinton, Qld. (as Port Bowen).

Lygosoma ophioscincus Boulenger, G.A. (1887). *Catalogue of the Lizards in the British Museum (Natural History).* 3. London : British Museum xii 575 pp. pls 40 [343] [*nom. nov. pro Ophioscincus australis* Peters, 1873, at that time a junior secondary homonym within the genus *Lygosoma*].

Synonymy that of Cogger, H.G., this work.

Distribution: NE coastal, Qld. Ecology: (woodland), tall forest, closed forest; oviparous. Biological references: Heyer, W.R. (1972). A new limbless skink (Reptilia : Scincidae) from Thailand with comments on the generic status of the limbless skinks of Southeast Asia. *Fieldiana Zool.* **58**: 109–129 (taxonomy).

Anomalopus pluto Ingram, 1977

Anomalopus pluto Ingram, G.J. (1977). A new species of legless skink *Anomalopus pluto* from Cape York, Queensland. *Vict. Nat.* **94**: 52–53 [52 fig 1]. Type data: holotype, QM J26261, from McDonald Crossing, Cockatoo Creek, 115 km S of Bamaga, Cape York, in 11°33′S, 142°26′E, Qld.

Distribution: NE coastal, N Gulf, Qld. Ecology: woodland, closed forest, open heath, fossorial; oviparous.

Anomalopus reticulatus (Günther, 1873)

Chelomeles reticulatus Günther, A. (1873). Notes on and descriptions of some lizards with rudimentary limbs, in the British Museum. *Ann. Mag. Nat. Hist. (4)* **12**: 145–146 [146]. Type data: holotype, BMNH 1946.8.3.25, from Clarence River, N.S.W.

Distribution: SE coastal, NE coastal, N.S.W., Qld. Ecology: (tall forest), closed forest, fossorial; oviparous. Biological references: McDonald, K.R. (1977). Observations on the skink *Anomalopus reticulatus* (Günther) (Lacertilia : Scincidae). *Vict. Nat.* **94**: 99–103 (natural history).

Anomalopus truncatus (Peters, 1876)

Coloscincus truncatus Peters, W. (1876). Über die von S.M.S. *Gazelle* mitgebrachten Amphibien. *Mber. K. Preuss. Akad. Wiss. Berl.* **1876**: 528–535 [532 pl fig 1] [1877 on title page]. Type data: holotype, ZMB 8932, from Peale Is., Moreton Bay, Qld.

Lygosoma truncatum monswilsonensis Copland, S.J. (1952). A mainland race of the scincid lizard *Lygosoma truncatum* (Peters). *Proc. Linn. Soc. N.S.W.* **77**: 126–131 [128 figs 2–3 pl 6]. Type data: holotype, AM R18587, from Wilson's Peak, at the junction of the Macpherson Range with the Great Dividing Range, Qld.

Synonymy that of Cogger, H.G., this work.

Distribution: SE coastal, NE coastal, N.S.W., Qld. Ecology: woodland, tall forest, closed forest, fossorial; oviparous.

Anomalopus verreauxii Duméril and Duméril, 1851

Anomalopus verreauxii Duméril, A.M.C. & Duméril, A. (1851). *Catalogue Méthodique de la Collection des Reptiles.* Paris : Mus. Hist. Nat. iv 224 pp. [185]. Type data: holotype, MNHP 3047, from Tas. (in error).

S[iaphus] simplex Cope, E.D. (1864). On the characters of the higher groups of Reptilia-Squamata—and especially of the Diploglossa. *Proc. Acad. Nat. Sci. Philad.* **1864**: 224–231 [229]. Type data: status and whereabouts unknown, from Australia.

Anomalopus godeffroyi Peters, W. (1867). Herpetologische Notizen. *Mber. K. Preuss. Akad. Wiss. Berl.* **1867**: 13–37 [24]. Type data: syntypes, ZMB 5296–7, from East Australia.

Chelomeles pseudopus Günther, A. (1873). Notes on and descriptions of some lizards with rudimentary limbs, in the British Museum. *Ann. Mag. Nat. Hist. (4)* **12**: 145–146 [145]. Type data: holotype, BMNH 1946.8.3.59, from unknown locality.

Synonymy that of Cogger, H.G., this work.

Distribution: SE coastal, NE coastal, N.S.W., Qld. Ecology: tall forest, closed forest, fossorial; oviparous. Biological references: Dale, D.F. (1973). *Forty Queensland Lizards.* Brisbane : Queensland Museum 64 pp. (natural history).

Carlia Gray, 1845

Heteropus Duméril, A.M.C. & Bibron, G. (1839). *Erpétologie Générale* ou Histoire Naturelle Complète des Reptiles. Paris : Roret 5: viii 854 pp. [757] [*non Heteropus* De Beauvois, 1820 and other authors, *vide* Neave, A.S. (1939). *Nomenclator Zoologicus.* London : Zoological Society of London Vol. 2 1025 pp.]. Type species *Heteropus fuscus* Duméril and Bibron, 1839 by subsequent designation, see Fitzinger, L.J. (1843). *Systema Reptilium.* Vienna : Braümüller u. Seidel vi 106 pp.

Carlia Gray, J.E. (1845). *Catalogue of the Specimens of Lizards in the Collection of the British Museum.* London : British Museum xxviii 289 pp. [271]. Type species *Mocoa melanopogon* Gray, 1845 by monotypy.

Myophila De Vis, C.W. (1884). New Queensland lizards. *Proc. R. Soc. Qd.* **1**: 77–78 [77]. Type species *Myophila vivax* De Vis, 1884 by monotypy.

Lygisaurus De Vis, C.W. (1884). New Queensland lizards. *Proc. R. Soc. Qd.* **1**: 77–78 [77]. Type species *Lygisaurus foliorum* De Vis, 1884 by monotypy.

Synonymy that of Cogger, H.G., this work.

This group is also found in E Indonesia, New Guinea and islands of W Pacific, including Micronesia, see Greer, A.E. (1974). The generic relationships of the scincid lizard genus *Leiolopisma* and its relatives. *Aust. J. Zool. Suppl. Ser.* **31**: 1–67. [The species *Carlia fusca* (Duméril and Bibron, 1839) has long been regarded as occurring in Australia, but recent work (Ingram & Covacevich, *in litt.*) has shown that Australian populations are referable to *Carlia longipes* (Macleay, 1877)].

Carlia amax Storr, 1974

Carlia amax Storr, G.M. (1974). The genus *Carlia* (Lacertilia : Scincidae) in Western Australia and Northern Territory. *Rec. West. Aust. Mus.* **3**: 151–165 [160]. Type data: holotype, WAM R43350, from Mitchell Plateau, in 14°52'S, 125°50'E, W.A.

Carlia johnstonei grandensis Storr, G.M. (1974). The genus *Carlia* (Lacertilia : Scincidae) in Western Australia and Northern Territory. *Rec. West. Aust. Mus.* **3**: 151–165 [164]. Type data: holotype, AM R13464, from Groote Eylandt, N.T.

Synonymy that of Cogger, H.G., this work.

Distribution: N Gulf, N coastal, Qld., W.A., N.T. Ecology: tall open shrubland, woodland, open forest, terrestrial, diurnal; oviparous, arthropod-feeder. Biological references: Storr, G.M., Smith, L.A. & Johnstone, R.E. (1981). *Lizards of Western Australia. I. Skinks.* Perth : University of Western Australia Press and Western Australian Museum 200 pp.

Carlia bicarinata (Macleay, 1877)

Heteropus bicarinatus Macleay, W. (1877). The lizards of the *Chevert* Expedition. *Proc. Linn. Soc. N.S.W.* **2**: 60–69 [68] [as 1878 on title page]. Type data: holotype, MMUS (presumed lost), from Hall Sound, Papua New Guinea.

Heteropus albertisii Peters, W. & Doria, G. (1878). Catalogo dei Rettili e dei Batraci raccolti da O. Beccari, L.M. d'Albertis e A.A. Bruijn nella sotto-regione Austro-Malese. *Ann. Mus. Civ. Stor. Nat. Genova* **13**: 323–450 [362]. Type data: syntypes (9), MCG C.E. 28052–3, from Yule Is. and coast near Mt. Epa, Papua New Guinea.

Synonymy that of Cogger, H.G., this work.

Distribution: NE coastal, N Gulf, Qld. Ecology: open forest, woodland, terrestrial, diurnal; oviparous, arthropod-feeder. Biological references: Mitchell, F.J. (1953). A brief revision of the four-fingered members of the genus *Leiolopisma* (Lacertilia). *Rec. S. Aust. Mus.* **11**: 75–90.

Carlia burnetti (Oudemans, 1894)

Ablepharus burnetti Oudemans, J.T. (1894). Eidechsen und Schildkröten. *in* Semon, R. *Zoologische Forschungsreisen in Australien und den Malayischen Archipel.* Jena : Gustav Fischer **5**: 127–146 [145]. Type data: syntypes, from Burnett River, Qld., ZMA 11345–6, remaining 2 syntypes not located.

Ablepharus heteropus Garman, S. (1901). Some reptiles and batrachians from Australasia. *Bull. Mus. Comp. Zool.* **39**: 1–14 [9]. Type data: holotype, MCZ 6484, from Great Barrier Reef, Qld.

Lygosoma aeratum Garman, S. (1901). Some reptiles and batrachians from Australasia. *Bull. Mus. Comp. Zool.* **39**: 1–14 [7]. Type data: holotype, MCZ 6476, from near Cooktown, Qld.

Ablepharus burnetti sydneyensis Copland, S.J. (1949). Taxonomic notes on the genus *Ablepharus* (Sauria : Scincidae). II. The races of *Ablepharus burnetti* Oudemans. *Proc. Linn. Soc. N.S.W.* **73**: 362–371 [362 figs 2–8 pl 22] [1948 on title page]. Type data: holotype, AM R18589, from Mt. Riverview Lookout, near Blaxland, on the eastern scarp of the Blue Mts., *ca.* 33°44'S, 150°39'E, N.S.W.

Synonymy that of Cogger, H.G., this work.

Distribution: SE coastal, NE coastal, Murray-Darling basin, N.S.W., Qld. Ecology: closed forest, tall forest, open forest, woodland, terrestrial, diurnal; oviparous, arthropod-feeder. Biological references: Greer, A.E. (1980). Critical thermal maximum temperatures in Australian scincid lizards: their ecological and evolutionary significance. *Aust. J. Zool.* **28**: 91–102.

Carlia coensis (Mitchell, 1953)

Leiolopisma coense Mitchell, F.J. (1953). A brief revision of the four-fingered members of the genus *Leiolopisma* (Lacertilia). *Rec. S. Aust. Mus.* **11**: 75–90 [82 fig 2]. Type data: holotype, MCZ 37171, from Coen, N Qld.

Distribution: NE coastal, Ql/d. Ecology: woodland, open forest, closed forest, terrestrial, diurnal; oviparous, saxicoline, arthropod-feeder.

Carlia dogare Covacevich and Ingram, 1975

Carlia dogare Covacevich, J. & Ingram, G. (1975). Three new species of rainbow skinks of the genus *Carlia* from northern Queensland. *Vict. Nat.* **92**: 19–22 [21 pl 1b]. Type data: holotype, QM J20557, from 5–6 km N mouth McIvor River, in 15°08′S, 145°15′E, NE Qld.

Distribution: NE coastal, Qld. Ecology: woodland, terrestrial, diurnal; oviparous, arthropod-feeder.

Carlia foliorum (De Vis, 1884)

Mocoa melanopogon Gray, J.E. (1845). *Catalogue of the Specimens of Lizards in the Collection of the British Museum.* London : British Museum xxviii 289 pp. [81] [junior secondary homonym rejected by Boulenger, G.A. (1887). *Catalogue of the Lizards in the British Museum (Natural History).* 3. London : British Museum xii 575 pp. pls 40; *non Lygosoma melanopogon* Duméril and Bibron, 1839]. Type data: holotype, BMNH 1946.8.16.42, from Port Essington, N.T.

Lygisaurus foliorum De Vis, C.W. (1884). New Queensland lizards. *Proc. R. Soc. Qd.* **1**: 77–78 [77]. Type data: holotype, from Brisbane, Qld., QM presumed lost, see Covacevich, J. (1971). Amphibian and reptile type specimens in the Queensland Museum. *Mem. Qd. Mus.* **16**: 49–68.

Heteropus mundus De Vis, C.W. (1885). A conspect of the genus, *Heteropus. Proc. R. Soc. Qd.* **1**: 166–173 [172]. Type data: syntypes, from Warro, Port Curtis, Qld., QM (2 specimens) presumed lost, see Covacevich, J. (1971). Amphibian and reptile type specimens in the Queensland Museum. *Mem. Qd. Mus.* **16**: 49–68.

Heteropus blackmanni De Vis, C.W. (1885). A conspect of the genus, *Heteropus. Proc. R. Soc. Qd.* **1**: 166–173 [168]. Type data: syntypes, QM J7773, QM J19968–90, from Port Curtis, Qld.

Synonymy that of Cogger, H.G., this work.

Distribution: NE coastal, N Gulf, N coastal, NW coastal, W plateau, Lake Eyre basin, Qld., W.A., N.T. Ecology: open forest, woodland, low woodland, tall open shrubland, low open shrubland, tall shrubland, terrestrial, diurnal; oviparous, arthropod-feeder. Biological references: Greer, A.E. (1975). Notes on the systematics of the genus *Carlia* (Lacertilia : Scincidae). I. *Carlia melanopogon* Gray, 1845. *Herpetologica* **31**: 70–75 (as *Carlia melanopogon*); Storr, G.M., Smith, L.A. & Johnstone, R.E. (1981). *Lizards of Western Australia. I. Skinks.* Perth : University of Western Australia Press and Western Australian Museum 200 pp. (taxonomy).

Carlia gracilis Storr, 1974

Carlia gracilis Storr, G.M. (1974). The genus *Carlia* (Lacertilia : Scincidae) in Western Australia and Northern Territory. *Rec. West. Aust. Mus.* **3**: 151–165 [158]. Type data: holotype, WAM R43219, from Mitchell Plateau, in 14°52′S, 125°50′E, W.A.

Distribution: N Gulf, N coastal, W.A., N.T. Ecology: tall open shrubland, woodland, open forest, terrestrial, diurnal; oviparous, arthropod-feeder. Biological references: Storr, G.M., Smith, L.A. & Johnstone, R.E. (1981). *Lizards of Western Australia. I. Skinks.* Perth : University of Western Australia Press and Western Australian Museum 200 pp.

Carlia jarnoldae Covacevich and Ingram, 1975

Carlia jarnoldae Covacevich, J. & Ingram, G. (1975). Three new species of rainbow skinks of the genus *Carlia* from northern Queensland. *Vict. Nat.* **92**: 19–22 [19 pl 1a]. Type data: holotype, QM J20739, from Wakooka Outstation, Starcke Station, in 14°33′S, 144°33′E, NE Qld.

Distribution: NE coastal, N Gulf, Qld. Ecology: woodland, terrestrial, diurnal; oviparous, arthropod-feeder, sexually dimorphic.

Carlia johnstonei Storr, 1974

Carlia johnstonei johnstonei Storr, G.M. (1974). The genus *Carlia* (Lacertilia : Scincidae) in Western Australia and Northern Territory. *Rec. West. Aust. Mus.* **3**: 151–165 [162]. Type data: holotype, WAM R43170, from Mitchell Plateau, in 14°52′S, 125°50′E, W.A.

Distribution: N coastal, W.A. Ecology: low open woodland, woodland, terrestrial, diurnal; oviparous, arthropod-feeder. Biological references: Storr, G.M., Smith, L.A. & Johnstone, R.E. (1981). *Lizards of Western Australia. I. Skinks.* Perth : University of Western Australia Press and Western Australian Museum 200 pp.

Carlia longipes (Macleay, 1877)

Heteropus longipes Macleay, W. (1877). The lizards of the *Chevert* Expedition. *Proc. Linn. Soc. N.S.W.* **2**: 60–69 [66] [1878 on title page]. Type data: holotype, AM R31878, from Endeavour River, Qld.

Heteropus variegatus Macleay, W. (1877). The lizards of the *Chevert* Expedition. *Proc. Linn. Soc. N.S.W.* **2**: 60–69 [66] [1878 on title page]. Type data: syntypes, AM R31868–70, from Darnley Is., Torres Strait, Qld.

Heteropus cheverti Macleay, W. (1877). The lizards of the *Chevert* Expedition. *Proc. Linn. Soc. N.S.W.* **2**: 60–69 [67] [1878 on title page]. Type data: syntypes, AM R31875–6, from Barrow Is., Qld.

Heteropus quinquecarinatus Macleay, W. (1877). The lizards of the *Chevert* Expedition. *Proc. Linn. Soc.*

N.S.W. **2**: 60–69 [67] [1878 on title page]. Type data: syntypes, AM R31871–5, from Darnley Is., Torres Strait, Qld.

Heteropus sexdentatus Macleay, W. (1877). The lizards of the *Chevert* Expedition. *Proc. Linn. Soc. N.S.W.* **2**: 60–69 [67] [1878 on title page]. Type data: syntypes, AM R31879–81, from Cape Grenville, Qld.

Heteropus maculatus De Vis, C.W. (1885). A conspect of the genus *Heteropus. Proc. R. Soc. Qd.* **1**: 166–173 [169]. Type data: holotype (probable), from Cape York, Qld., QM presumed lost, see Covacevich, J. (1971). Amphibian and reptile type specimens in the Queensland Museum. *Mem. Qd. Mus.* **16**: 49–68.

Heteropus rubricatus De Vis, C.W. (1885). A conspect of the genus *Heteropus. Proc. R. Soc. Qd.* **1**: 166–173 [170]. Type data: holotype (probable), from Cape York, Qld., QM presumed lost, see Covacevich, J. (1971). Amphibian and reptile type specimens in the Queensland Museum. *Mem. Qd. Mus.* **16**: 49–68.

Heteropus rostralis De Vis, C.W. (1885). A conspect of the genus *Heteropus. Proc. R. Soc. Qd.* **1**: 166–173 [171]. Type data: holotype, QM J230, from Cardwell, Qld.

Synonymy that of Cogger, H.G., this work.

Distribution: NE coastal, N Gulf, N coastal, Qld., N.T. Ecology: woodland, open forest, closed forest, terrestrial, diurnal; oviparous, arthropod-feeder. Biological references: Mitchell, F.J. (1953). A brief revision of the four-fingered members of the genus *Leiolopisma* (Lacertilia). *Rec. S. Aust. Mus.* **11**: 75–90 (as *Leiolopisma fuscum*); Wilhoft, D.C. & Reiter, E.O. (1965). Sexual cycle of the lizard, *Leiolopisma fuscum*, a tropical Australian skink. *J. Morph.* **116**: 379–388 (as *Leiolopisma fuscum*); Greer, A.E. (1980). Critical thermal maximum temperatures in Australian scincid lizards: their ecological and evolutionary significance. *Aust. J. Zool.* **28**: 91–102 (as *Carlia fusca*).

Carlia mundivensis (Broom, 1898)

Heteropus vertebralis De Vis, C.W. (1888). A contribution to the herpetology of Queensland. *Proc. Linn. Soc. N.S.W. (2)* **2**: 811–826 [821] [*non Lygosoma vertebrale* Hallowell, 1860]. Type data: lectotype, QM J248, from Chinchilla, Darling Downs, Qld., designation by Covacevich, J. (1971). Amphibian and reptile type specimens in the Queensland Museum. *Mem. Qd. Mus.* **16**: 49–68.

Lygosoma mundivense Broom, R. (1898). On the lizards of the Chillagoe District, N Queensland. *Proc. Linn. Soc. N.S.W.* **22**: 639–645 [643]. Type data: holotype, BMNH 1946.8.17.81, from Muldiva, N Qld.

Lygosoma waitei Zietz, F.R. (1920). Catalogue of Australian lizards. *Rec. S. Aust. Mus.* **1**: 181–228 [211] [*nom. nov. pro Heteropus vertebralis* De Vis, 1888; preoccupied in *Lygosoma*].

Synonymy that of Cogger, H.G., this work.

Distribution: NE coastal, (Murray-Darling basin), Qld. Ecology: open scrub, low open woodland, woodland, terrestrial, diurnal; oviparous, saxicoline, arthropod-feeder.

Carlia novaeguineae (Meyer, 1874)

Lygosoma (Carlia) novaeguineae Meyer, A.B. (1874). Eine Mitteilung von Hrn. Dr. Adolf Bernhard Meyer über die von ihm auf Neu-Guinea den Inseln Jobi, Mysore und Mafoor im Jahr 1873 gesammelten Amphibien. *Mber. K. Preuss. Akad. Wiss. Berl.* **1874**: 128–140 [132] [1875 on title page]. Type data: syntypes (probable), ZMB (not found), from New Guinea.

Carlia macfarlani Günther, A. (1877). Descriptions of three new species of lizards from islands of Torres Strait. *Ann. Mag. Nat. Hist. (4)* **19**: 413–415 [413]. Type data: syntypes, BMNH 1946.8.16.49–51, from Islands of Torres Straits, Qld.

Lygosoma laeve Oudemans, J.T. (1894). Eidechsen und Schildkröten *in* Semon, R. *Zoologische Forschungsreisen in Australien und den Malayischen Archipel.* Jena : Gustav Fischer **5**: 127–146 [144]. Type data: holotype, ZMA 10994, from Cooktown, Qld.

Lygosoma curtum Boulenger, G.A. (1897). Descriptions of new lizards and frogs from Mount Victoria, Owen Stanley Range, New Guinea, collected by Mr. A.S. Anthony. *Ann. Mag. Nat. Hist. (6)* **19**: 6–13 [9]. Type data: syntypes, BMNH 1946.8.17.85–88, from Mt. Victoria, Owen Stanley Range, Papua New Guinea.

Synonymy that of Cogger, H.G., this work.

Distribution: NE coastal, Qld. Ecology: woodland, closed forest, terrestrial, diurnal; oviparous, arthropod-feeder. Biological references: Mitchell, F.J. (1953). A brief revision of the four-fingered members of the genus *Leiolopisma* (Lacertilia). *Rec. S. Aust. Mus.* **11**: 75–90.

Carlia pectoralis (De Vis, 1885)

Heteropus pectoralis De Vis, C.W. (1885). A conspect of the genus, *Heteropus. Proc. R. Soc. Qd.* **1**: 166–173 [169]. Type data: holotype, QM J1414, from Warro, Port Curtis, Qld.

Distribution: NE coastal, Murray-Darling basin, N.S.W., Qld. Ecology: open scrub, woodland, terrestrial, diurnal; oviparous, arthropod-feeder. Biological references: Mitchell, F.J. (1953). A brief revision of the four-fingered members of the genus *Leiolopisma* (Lacertilia). *Rec. S. Aust. Mus.* **11**: 75–90; Greer, A.E. (1980). Critical thermal maximum temperatures in Australian scincid lizards: their ecological and evolutionary significance. *Aust. J. Zool.* **28**: 91–102.

Carlia rhomboidalis (Peters, 1869)

Heteropus rhomboidalis Peters, W. (1869). Neue Gattungen und neue oder weniger bekannte Arten von Amphibien; Nachtrag. *Mber. K. Preuss. Akad. Wiss. Berl.* **1869**: 445–447 [446]. Type data: syntypes, BMNH

1946.8.16.57 (Port Mackay), ZMB 6509 (5 specimens, Port Mackay), ZMH 931 (probable, from Australia), from Port Mackay, Qld. and Australia.

Distribution: NE coastal, Qld. Ecology: woodland, terrestrial, diurnal; oviparous, arthropod-feeder. Biological references: Mitchell, F.J. (1953). A brief revision of the four-fingered members of the genus *Leiolopisma* (Lacertilia). *Rec. S. Aust. Mus.* **11**: 75–90; Wilhoft, D.C. (1963). Reproduction in the tropical Australian skink, *Leiolopisma rhomboidalis*. *Am. Midl. Nat.* **70**: 442–461; Wilhoft, D.C. (1963). Gonadal histology and seasonal changes in the tropical Australian lizard, *Leiolopisma rhomboidalis*. *J. Morph.* **113**: 185–204.

Carlia rimula Ingram and Covacevich, 1980

Carlia rimula Ingram, G.J. & Covacevich, J. (1980). Two new lygosomine skinks endemic to Cape York Peninsula. pp. 45–48 *in* Stevens, N.C. & Bailey, A. (eds.) *Contemporary Cape York*. Brisbane : Royal Society of Queensland [46 pl 1B]. Type data: holotype, QM J24603, from second Claudie River crossing, Iron Range road, via Coen, in 12°44'S, 143°13'E, NE Qld.

Distribution: NE coastal, Qld. Ecology: woodland, terrestrial, diurnal; oviparous, saxicoline, arthropod-feeder.

Carlia rufilatus Storr, 1974

Carlia rufilatus Storr, G.M. (1974). The genus *Carlia* (Lacertilia : Scincidae) in Western Australia and Northern Territory. *Rec. West. Aust. Mus.* **3**: 151–165 [157]. Type data: holotype, WAM R23271, from Tumbling Waters, in 12°46'S, 130°57'E, N.T.

Distribution: N coastal, W.A., N.T. Ecology: woodland, terrestrial, diurnal; oviparous, arthropod-feeder. Biological references: Storr, G.M., Smith, L.A. & Johnstone, R.E. (1981). *Lizards of Western Australia. I. Skinks*. Perth : University of Western Australia Press and Western Australian Museum 200 pp. (taxonomy).

Carlia schmeltzii (Peters, 1867)

Heteropus schmeltzii Peters, W. (1867). Herpetologische Notizen. *Mber. K. Preuss. Akad. Wiss. Berl.* **1867**: 13–37 [23] [1868 on title page]. Type data: holotype, ZMB 5750, from Rockhampton, Qld.

Carlia prava Covacevich, J. & Ingram, G. (1975). Three new species of rainbow skinks of the genus *Carlia* from northern Queensland. *Vict. Nat.* **92**: 19–22 [22 pl 1c]. Type data: holotype, QM J14101, from Magnificent Creek, Kowanyama (Mitchell River Mission), in 15°18'S, 141°44'E, Qld.

Synonymy that of Cogger, H.G., this work.

Distribution: (SE coastal), NE coastal, N Gulf, N.S.W., Qld. Ecology: open scrub, woodland, terrestrial, diurnal; oviparous, arthropod-feeder.

Carlia scirtetis Ingram and Covacevich, 1980

Carlia scirtetis Ingram, G.J. & Covacevich, J. (1980). Two new lygosomine skinks endemic to Cape York Peninsula. pp. 45–48 *in* Stevens, N.C. & Bailey, A. (eds.) *Contemporary Cape York*. Brisbane : Royal Society of Queensland [45 pl 1A]. Type data: holotype, NMV D12092, from Black Mt., Black Trevethan Range, approximately 20 km S of Cooktown, in 15°40'S, 145°24'E, NE Qld.

Distribution: NE coastal, Qld. Ecology: woodland, terrestrial, diurnal; oviparous, saxicoline, arthropod-feeder.

Carlia tetradactyla (O'Shaughnessy, 1879)

Mocoa tetradactyla O'Shaughnessy, A.W.E. (1879). Description of new species of lizards in the collection of the British Museum. *Ann. Mag. Nat. Hist. (5)* **4**: 300–301 [300]. Type data: holotype, BMNH 1946.8.17.43, from unknown locality (labelled "Queensland?").

Lygosoma maccooeyi Ramsay, E.P. & Ogilby, J.D. (1890). Redescription of an Australian skink. *Rec. Aust. Mus.* **1**: 8–9 [8]. Type data: syntypes, AM R672, AM R676–7, AM R683–6, BMNH 1946.8.17.76–78, MCZ 6304, from Brawlin near Cootamundra, N.S.W.

Synonymy that of Cogger, H.G., this work.

Distribution: SE coastal, NE coastal, Murray-Darling basin, Vic., N.S.W., Qld., A.C.T. Ecology: low open woodland, woodland, open forest, terrestrial, diurnal; oviparous, arthropod-feeder. Biological references: Mitchell, F.J. (1953). A brief revision of the four-fingered members of the genus *Leiolopisma* (Lacertilia). *Rec. S. Aust. Mus.* **11**: 75–90.

Carlia triacantha (Mitchell, 1953)

Leiolopisma triacantha Mitchell, F.J. (1953). A brief revision of the four-fingered members of the genus *Leiolopisma* (Lacertilia). *Rec. S. Aust. Mus.* **11**: 75–90 [88]. Type data: holotype, SAMA R2697, from Adelaide River, N.T.

Distribution: N Gulf, N coastal, NW coastal, W plateau, Lake Eyre basin, S.A., W.A., N.T. Ecology: desert, hummock grassland, low open shrubland, woodland, tall shrubland, terrestrial, diurnal; oviparous, (saxicoline), arthropod-feeder. Biological references: Storr, G.M., Smith, L.A. & Johnstone, R.E. (1981). *Lizards of Western Australia. I. Skinks*. Perth : University of Western Australia Press and Western Australian Museum 200 pp. (taxonomy).

Carlia vivax (De Vis, 1884)

Myophila vivax De Vis, C.W. (1884). New Queensland lizards. *Proc. R. Soc. Qd.* **1**: 77–78 [77]. Type data: syntypes (probable), from Brisbane and neighbourhood,

Qld., QM presumed lost, see Covacevich, J. (1971). Amphibian and reptile type specimens in the Queensland Museum. *Mem. Qd. Mus.* **16**: 49–68.

Distribution: SE coastal, NE coastal, N Gulf, (Murray-Darling basin), N.S.W., Qld. Ecology: low open woodland, low woodland, woodland, open forest, terrestrial, diurnal; oviparous, arthropod-feeder. Biological references: Greer, A.E. (1980). Critical thermal maximum temperatures in Australian scincid lizards: their ecological and evolutionary significance. *Aust. J. Zool.* **28**: 91–102.

Cryptoblepharus Wiegmann, 1834

Cryptoblepharus Wiegmann, A.F.A. (1834). *Herpetologia Mexicana,* seu Descriptio Amphibiorum Novae Hispaniae. Berlin : Lüderitz [12]. Type species *Ablepharus poeciliopleurus* Wiegmann, 1834 by subsequent designation, see Stejneger, L. (1899). The land reptiles of the Hawaiian Islands. *Proc. U.S. Natl. Mus.* **21**: 78–813.

Cryptoblepharis Cocteau, J.T. (1836). *Études sur les Scincoïdes.* Paris [8] [emend. *pro Cryptoblepharus* Wiegmann, 1834].

Petia Gray, J.E. (1839). Catalogue of the slender-tongued saurians, with descriptions of many new genera and species. (contd.). *Ann. Mag. Nat. Hist.* **2**: 331–337 [335] [*nom. nud.*; ms name introduced in synonymy of *Cryptoblepharus* Wiegmann, 1834].

Synonymy that of Cogger, H.G., this work; Mertens, R. (1931). *Ablepharus boutonii* (Desjardin) und seine geographische Variation. *Zool. Jb., Jena* **61**: 63–210.

This group is also found in Madagascar, East Africa, eastern Indonesia, New Guinea, Micronesia, Melanesia and Polynesia. Species now known not to occur in Australia: *Cryptoblepharus boutonii* (Desjardin, 1831), excluded by Storr, G.M. (1976). The genus *Cryptoblepharus* (Lacertilia : Scincidae) in Western Australia. *Rec. West. Aust. Mus.* **4**: 53–63.

Cryptoblepharus carnabyi Storr, 1976

Cryptoblepharus carnabyi Storr, G.M. (1976). The genus *Cryptoblepharus* (Lacertilia : Scincidae) in Western Australia. *Rec. West. Aust. Mus.* **4**: 53–63 [60]. Type data: holotype, WAM R21182, from 11 km WSW of Youanmi, in 28°37′S, 118°43′E, W.A.

Distribution: NE coastal, N Gulf, N coastal, NW coastal, SW coastal, Murray-Darling basin, Bulloo River basin, S.A., Vic., N.S.W., Qld., W.A., N.T. Ecology: hummock grassland, low open shrubland, tall open shrubland, open scrub, low open woodland, low woodland, woodland, tall shrubland, arboreal, diurnal; oviparous, arthropod-feeder. Biological references: Börner, A.-R. & Schüttler, B. (1981). Über die australischen Skinke des *Cryptoblepharus boutonii*-Komplexes. *Miscellaneous Articles in Saurology* 8: 1–10 (taxonomy); Storr, G.M., Smith, L.A. & Johnstone, R.E. (1981). *Lizards of Western Australia. I. Skinks.* Perth : University of Western Australia Press and Western Australian Museum 200 pp. (taxonomy).

Cryptoblepharus egeriae (Boulenger, 1889)

Ablepharus egeriae Boulenger, G.A. (1889). On the reptiles of Christmas Island. *Proc. Zool. Soc. Lond.* **1888**: 534–536 [535]. Type data: syntypes (3) BMNH 1946.8.15.86–88 from Christmas Is., Indian Ocean.

Distribution: Christmas Is. Ecology: closed forest, peridomestic, terrestrial, arboreal, diurnal; arthropod-feeder, oviparous.

Cryptoblepharus fuhni Covacevich and Ingram, 1978

Cryptoblepharus fuhni Covacevich, J. & Ingram, G.J. (1978). An undescribed species of rock dwelling *Cryptoblepharus* (Lacertilia : Scincidae). *Mem. Qd. Mus.* **18**: 151–154 [151]. Type data: holotype, QM J20566, from Melville Range, Cape Melville, Cape York, 14°16′S, 144°30′E, Qld.

Distribution: NE coastal, Qld. Ecology: woodland, terrestrial, diurnal; oviparous, saxicoline, arthropod-feeder. Biological references: Börner, A.-R. & Schüttler, B. (1981). Über die australischen Skinke des *Cryptoblepharus boutonii*-Komplexes. *Miscellaneous Articles in Saurology* 8: 1–10 (taxonomy).

Cryptoblepharus litoralis (Mertens, 1958)

Ablepharus boutonii litoralis Mertens, R. (1958). Neue Eidechsen aus Australien. *Senckenberg. Biol.* **39**: 51–56 [54]. Type data: holotype, SMF 53219, from Flying Fish Point, 6 mi E Innisfail, Qld.

Distribution: NE coastal, Qld. Ecology: mangrove, supralittoral, terrestrial, diurnal; oviparous, (saxicoline), arthropod-feeder. Biological references: Covacevich, J. & Ingram, G.J. (1978). An undescribed species of rock dwelling *Cryptoblepharus* (Lacertilia : Scincidae). *Mem. Qd. Mus.* **18**: 151–154 (taxonomy); Greer, A.E. (1980). Critical thermal maximum temperatures in Australian scincid lizards: their ecological and evolutionary significance. *Aust. J. Zool.* **28**: 91–102; Börner, A.-R. & Schüttler, B. (1981). Über die australischen Skinke des *Cryptoblepharus boutonii*-Komplexes. *Miscellaneous Articles in Saurology* 8: 1–10 (taxonomy).

Cryptoblepharus megastictus Storr, 1976

Cryptoblepharus megastictus Storr, G.M. (1976). The genus *Cryptoblepharus* (Lacertilia : Scincidae) in Western Australia. *Rec. West. Aust. Mus.* **4**: 53–63 [61]. Type data: holotype, WAM R43245, from Mitchell Plateau, in 14°52′S, 125°50′E, W.A.

Distribution: N Gulf, N coastal, W.A., N.T. Ecology: tall open shrubland, woodland, terrestrial, diurnal; oviparous, saxicoline, arthropod-feeder. Biological references: Storr, G.M., Smith, L.A. & Johnstone, R.E. (1981). *Lizards of Western Australia. I. Skinks.* Perth : University of Western Australia Press and Western Australian Museum 200 pp. (taxonomy).

Cryptoblepharus plagiocephalus (Cocteau, 1836)

***Scincus plagiocephalus* (part.)** Cocteau, J.T. (1836). *Études sur les Scincoïdes.* Paris [pl] [according to Mertens, R. (1931). *Ablepharus boutonii* (Desjardin) und seine geographische Variation. *Zool. Jb., Jena* **61**: 63–210 the type series of *Scincus plagiocephalus* Cocteau is composite, containing representatives of two currently recognised species, *Cryptoblepharus plagiocephalus* and the extra limital *Cryptoblepharus poecilopleurus*; in order to fix the name to the taxon to which it has been widely applied, it has been necessary to designate as lectotype the MNHP syntype from Shark Bay, W.A.; because the syntypes are extant (for details see Guibé, J. (1954). *Catalogue des Types de Lézards du Muséum National d'Histoire Naturelle.* Paris : Muséum National d'Histoire Naturelle 119 pp.), the action of Storr, G.M. (1961). *Ablepharus boutonii clarus*, a new skink from the Esperance District, Western Australia. *West. Aust. Nat.* **7**: 176–178 in restricting the type locality to Swan River, W.A. is inappropriate]. Type data: lectotype, MNHP 7150, from Shark Bay, W.A. (paralectotypes from Tasmania (in error), Tahiti, Hawaiian Ils. and Puña Is., Ecuador), designation by Cogger, H.G., this work.

Scincus plagyocephalus Cocteau, J.T. (1836). *Études sur les Scincoïdes.* Paris [7] [*ex* Péron ms; *lapsus pro Scincus plagiocephalus* Cocteau, 1836].

Tiliqua buchananii Gray, J.E. (1838). Catalogue of the slender-tongued saurians, with descriptions of many new genera and species. (contd.) *Ann. Mag. Nat. Hist.* **2**: 287–293 [291] [1839 on title page]. Type data: syntypes, BMNH 1946.8.19.73–74, from Australia, (W.A. *vide* Boulenger, G.A. (1887). *Catalogue of the Lizards in the British Museum (Natural History).* 3. London : British Museum xii 575 pp. pls 40).

***Ablepharus peronii* (part.)** Duméril, A.M.C. & Bibron, G. (1839). *Erpétologie Générale* ou Histoire Naturelle Complète des Reptiles. Paris : Roret 5: viii 854 pp. [813] [composite syntype series containing representatives of three currently recognised species, *C. plagiocephalus*, and the extralimital species *Cryptoblepharus leschnault* and *Cryptoblepharus poecilopleurus*]. Type data: syntypes, MNHP 7150 (Shark Bay), MNHP 3088 (Tas.), from Tas. (in error) and Shark Bay, W.A.; for extralimital syntypes see Guibé, J. (1954). *Catalogue des Types de Lézards du Muséum National d'Histoire Naturelle.* Paris : Muséum National d'Histoire Naturelle 119 pp.

Cryptoblepharus burtonii Macleay, W. (1877). The lizards of the *Chevert* Expedition. *Proc. Linn. Soc. N.S.W.* **2**: 60–69 [61] [1878 on title page; probable *lapsus pro Scincus boutonii* Desjardins, 1831 but referring to specimens of *C. plagiocephalus*].

Ablepharus boutonii metallicus Boulenger, G.A. (1887). *Catalogue of the Lizards in the British Museum (Natural History).* 3. London : British Museum xii 575 pp. pls 40 [347]. Type data: syntypes, BMNH 57.10.24.38–40, from "North Australian Exped.".

Ablepharus boutoni australis Sternfeld, R. (1918). Zur Tiergeographie Papuasiens und der pazifischen Inselwelt. *Abh. Senckenb. Naturf. Ges.* **36**: 375–436 [424]. Type data: lectotype, SMF 15683, from Hermannsburg, N.T. (as West Central Australia), designation by Mertens, R. (1967). Die herpetologische Sektion des Natur-Museums und Forschungs-Institutes Senckenberg in Frankfurt-a.-M. nebst einem Verzeichnis ihrer Typen. 1. *Senckenberg. Biol.* **48**: 1–106.

Ablepharus boutoni punctatus Sternfeld, R. (1918). Zur Tiergeographie Papuasiens und der pazifischen Inselwelt. *Abh. Senckenb. Naturf. Ges.* **36**: 375–436 [424]. Type data: lectotype, SMF 15685, from Yalgoo, W.A. ("West-Australien" in original description), designation by Mertens, R. (1967). Die herpetologische Sektion des Natur-Museums und Forschungs-Institutes Senckenberg in Frankfurt-a.-M. nebst einem Verzeichnis ihrer Typen. 1. *Senckenberg. Biol.* **48**: 1–106.

Cryptoblepharus plagiocephalus ruber Börner, A.-R. & Schüttler, B. (1981). Über die australischen Skinke des *Cryptoblepharus boutonii* Komplexes. *Misc. Articles in Saurology* **8**: 1–10 [4]. Type data: holotype, SMF 32823, from Kalindi Grotto, Bachsten Creek, NW Australia.

Synonymy that of Cogger, H.G., this work.

Distribution: NE coastal, N Gulf, N coastal, NW coastal, SW coastal, W plateau, (S Gulfs), Murray-Darling basin, Bulloo River basin, Lake Eyre basin, S.A., Vic., N.S.W., Qld., W.A., N.T. Ecology: desert, hummock grassland, low open shrubland, low shrubland, open heath, tall open shrubland, open scrub, low open woodland, low woodland, woodland, open forest, tall forest, arboreal, diurnal; oviparous, arthropod-feeder. Biological references: Mertens, R. (1931). *Ablepharus boutonii* (Desjardin) und seine geographische Variation. *Zool. Jb., Jena* **61**: 63–210 (taxonomy); Storr, G.M. (1976). The genus *Cryptoblepharus* (Lacertilia : Scincidae) in Western Australia. *Rec. West. Aust. Mus.* **4**: 53–63 (taxonomy); Waite, E.R. (1929). *The Reptiles and Amphibians of South Australia.* Adelaide : Harrison Weir 270 pp. (as *Ablepharus boutonii metallicus*); Storr, G.M., Smith, L.A. & Johnstone, R.E. (1981). *Lizards of Western Australia. I. Skinks.* Perth : University of Western Australia Press and Western Australian Museum 200 pp. (taxonomy).

Cryptoblepharus virgatus (Garman, 1901)

Ablepharus virgatus Garman, S. (1901). Some reptiles and batrachians from Australasia. *Bull. Mus. Comp. Zool.* **39**: 1–14 [10]. Type data: holotype, MCZ 6485, from Cooktown, Qld.

Ablepharus boutoni pulcher Sternfeld, R. (1918). Zur Tiergeographie Papuasiens und der pazifischen Inselwelt. *Abh. Senckenb. Naturf. Ges.* **36**: 375–436 [423]. Type data: lectotype, SMF 15680, from Australia, designation by Mertens, R. (1967). Die herpetologische Sektion des Natur-Museums und Forschungs-Institutes Senckenberg in Frankfurt-a.-M. nebst einem Verzeichnis ihrer Typen. 1. *Senckenberg. Biol.* **48**: 1–106.

Ablepharus boutonii clarus Storr, G.M. (1961). *Ablepharus boutonii clarus*, a new skink from the Esperance District, Western Australia. *West. Aust. Nat.* **7**: 176–178 [177 fig]. Type data: syntypes, WAM R18228–31, from lower Dalyup River, in 33°45′S, 121°32′E (20 mi WNW of Esperance), W.A.

Synonymy that of Cogger, H.G., this work.

Distribution: SE coastal, NE coastal, N Gulf, (SW coastal), W plateau, S Gulfs, Murray-Darling basin, S.A., Vic., N.S.W., A.C.T., Qld., W.A. Ecology: low shrubland, open heath, open scrub, low open woodland, woodland, open forest, tall shrubland, arboreal, diurnal; oviparous, arthropod-feeder. Biological references: Storr, G.M. (1976). The genus *Cryptoblepharus* (Lacertilia : Scincidae) in Western Australia. *Rec. West. Aust. Mus.* **4**: 53–63 (taxonomy); Mertens, R. (1931). *Ablepharus boutonii* (Desjardin) und seine geographische Variation. *Zool. Jb., Jena* **61**: 63–210 (taxonomy); Covacevich, J. & Ingram, G.J. (1978). An undescribed species of rock dwelling *Cryptoblepharus* (Lacertilia : Scincidae). *Mem. Qd. Mus.* **18**: 151–154 (taxonomy); Cook, R. (1973). The wall lizard, *Cryptoblepharus boutonii virgatus*. *Herpetofauna* **6**: 15–16 (natural history); Storr, G.M., Smith, L.A. & Johnstone, R.E. (1981). *Lizards of Western Australia. I. Skinks.* Perth : University of Western Australia Press and Western Australian Museum 200 pp. (taxonomy).

Ctenotus Storr, 1964

Ctenotus Storr, G.M. (1964). *Ctenotus*, a new generic name for a group of Australian skinks. *West. Aust. Nat.* **9**: 84–85 [84]. Type species *Lacerta taeniolata* Shaw, 1790 [=*Ctenotus taeniolatus* (White, 1790)], by original designation.

This group is also found in S New Guinea.

Ctenotus alacer Storr, 1970

Ctenotus alacer Storr, G.M. (1970). The genus *Ctenotus* (Lacertilia : Scincidae) in the Northern Territory. *J. R. Soc. West. Aust.* **52**: 97–108 [104] [1969 on title page]. Type data: holotype, WAM R20903, from Alice Springs, in 23°38′S, 133°52′E, N.T.

Distribution: N coastal, W plateau, Lake Eyre basin, Qld., W.A., N.T. Ecology: desert, hummock grassland, low open shrubland, (low woodland), tall shrubland, terrestrial, diurnal; oviparous, (saxicoline), arthropod-feeder.

Ctenotus alleni Storr, 1974

Ctenotus alleni Storr, G.M. (1974). The genus *Ctenotus* (Lacertilia : Scincidae) in the South-west and Eucla Divisions of Western Australia. *J. R. Soc. West. Aust.* **56**: 86–93 [89] [1973 on title page]. Type data: holotype, WAM R33602, from 11 mi N Galena, 27°41′S, 114°39′E, W.A.

Distribution: NW coastal, W.A. Ecology: terrestrial, diurnal; oviparous, arthropod-feeder. Biological references: Storr, G.M., Smith, L.A. & Johnstone, R.E. (1981). *Lizards of Western Australia. I. Skinks.* Perth : University of Western Australia Press and Western Australian Museum 200 pp. (taxonomy).

Ctenotus allotropis Storr, 1981

Ctenotus allotropis Storr, G.M. (1981). Ten new *Ctenotus* (Lacertilia : Scincidae) from Australia. *Rec. West. Aust. Mus.* **9**: 125–146 [142 fig. 11]. Type data: holotype, AM R27832, from Round Hill Fauna Reserve, near Euabalong, N.S.W.

Distribution: Murray-Darling basin, (Bulloo River basin), N.S.W., Qld. Ecology: low shrubland, open scrub, low open woodland, low woodland, woodland, terrestrial, diurnal; oviparous, arthropod-feeder.

Ctenotus ariadnae Storr, 1969

Ctenotus ariadnae Storr, G.M. (1969). The genus *Ctenotus* (Lacertilia : Scincidae) in the Eastern Division of Western Australia. *J. R. Soc. West. Aust.* **51**: 97–109 [104] [1968 on title page]. Type data: holotype, WAM R30006, from 24 mi ENE of Laverton, in 28°31′S, 122°45′E, W.A.

Distribution: W plateau, W.A. Ecology: desert, low open shrubland, tall shrubland, terrestrial, diurnal; oviparous, arthropod-feeder. Biological references: Pianka, E.R. (1969). Sympatry of desert lizards (*Ctenotus*) in Western Australia. *Ecology* **50**: 1012–1030 (ecology); Storr, G.M., Smith, L.A. & Johnstone, R.E. (1981). *Lizards of Western Australia. I. Skinks.* Perth : University of Western Australia Press and Western Australian Museum 200 pp. (taxonomy).

Ctenotus arnhemensis Storr, 1981

Ctenotus arnhemensis Storr, G.M. (1981). Ten new *Ctenotus* (Lacertilia : Scincidae) from Australia. *Rec. West. Aust. Mus.* **9**: 125–146 [127 fig. 2]. Type data: holotype, AM R88613, from Jabiluka, in 12°33′S, 132°55′E, N.T.

Distribution: N coastal, N.T. Ecology: woodland, terrestrial, diurnal; oviparous, arthropod-feeder.

Ctenotus atlas Storr, 1969

Ctenotus atlas Storr, G.M. (1969). The genus *Ctenotus* (Lacertilia : Scincidae) in the Eastern Division of Western Australia. *J. R. Soc. West. Aust.* **51**: 97–109 [105] [1968 on title page]. Type data: holotype, WAM R30001, from 17 mi S of Atley, in 28°27′S, 119°05′E, W.A.

Distribution: NW coastal, W plateau, Murray-Darling basin, (S Gulfs), S.A., Vic., N.S.W., W.A. Ecology: low shrubland, open scrub, tall shrubland, terrestrial, diurnal; oviparous, arthropod-feeder. Biological references: Storr, G.M. (1971). The genus *Ctenotus* (Lacertilia : Scincidae) in South Australia. *Rec. S. Aust. Mus.* **16**: 1–15 (taxonomy); Storr, G.M. (1974). The genus *Ctenotus* (Lacertilia : Scincidae) in the South-west and Eucla Divisions of Western Australia. *J. R. Soc. West. Aust.* **56**: 86–93 (taxonomy); Pianka, E.R. (1969). Sympatry of desert lizards (*Ctenotus*) in Western Australia. *Ecology* **50**: 1012–1030; Storr, G.M., Smith, L.A. & Johnstone, R.E. (1981). *Lizards of Western Australia. I. Skinks.* Perth : University of Western Australia Press and Western Australian Museum 200 pp. (taxonomy).

Ctenotus australis (Gray, 1838)

***Tiliqua australis* (part.)** Gray, J.E. (1838). Catalogue of the slender-tongued saurians, with descriptions of many new genera and species. (contd.) *Ann. Mag. Nat. Hist.* **2**: 287–293 [291] [1839 on title page; *ex* Cocteau ms]. Type data: syntypes (2), BMNH 38.3.10.9, from Australia. These specimens are unlabelled and share a jar with two unlabelled non-types (BMNH 44.4.13.111); these four specimens belong to at least three distinct species and until the name is fixed by appropriate lectotype designation, synonymy follows that of previous authors.

Lygosoma lesueurii Duméril, A.M.C. & Bibron, G. (1839). *Erpétologie Générale* ou Histoire Naturelle Complète des Reptiles. Paris : Roret 5: viii 854 pp. [733]. Type data: syntypes, MNHP 2982 (2 specimens), from Australia.

Synonymy that of Cogger, H.G., this work.

Distribution: NW coastal, SW coastal, W.A. Ecology: open heath, (woodland), terrestrial, diurnal; oviparous, arthropod-feeder. Biological references: Storr, G.M. (1974). The genus *Ctenotus* (Lacertilia : Scincidae) in the South-west and Eucla Divisions of Western Australia. *J. R. Soc. West. Aust.* **56**: 86–93 (taxonomy, as *Ctenotus lesueurii*); Storr, G.M., Smith, L.A. & Johnstone, R.E. (1981). *Lizards of Western Australia. I. Skinks.* Perth : University of Western Australia Press and Western Australian Museum 200 pp. (taxonomy, as *Ctenotus lesueurii*).

Ctenotus brachyonyx Storr, 1971

Ctenotus brachyonyx Storr, G.M. (1971). The genus *Ctenotus* (Lacertilia : Scincidae) in South Australia. *Rec. S. Aust. Mus.* **16**: 1–15 [6]. Type data: holotype, NMV D13944, from Annuello (34°52′S, 142°49′E), Vic.

Distribution: Murray-Darling basin, S.A., Vic., N.S.W. Ecology: open scrub, tall shrubland, terrestrial, diurnal; oviparous, arthropod-feeder.

Ctenotus brooksi (Loveridge, 1933)

Sphenomorphus leae brooksi Loveridge, A. (1933). New scincid lizards of the genera *Sphenomorphus*, *Rhodona* and *Lygosoma* from Australia. *Occ. Pap. Bost. Soc. Nat. Hist.* **8**: 95–100 [95]. Type data: holotype, MCZ 25055, from Perth, W.A.

Lygosoma (Sphenomorphus) taeniata Mitchell, F.J. (1949). A new species of *Lygosoma* (*Lygosoma (Sphenomorphus) taeniata sp. nov.*). *Rec. S. Aust. Mus.* **9**: 180 [180]. Type data: holotype, SAMA R2803, from Andamooka Ranges, immediately W of Lake Torrens, S.A.

Ctenotus brooksi aranda Storr, G.M. (1970). The genus *Ctenotus* (Lacertilia : Scincidae) in the Northern Territory. *J. R. Soc. West. Aust.* **52**: 97–108 [108] [1969 on title page]. Type data: holotype, NTM 2931, from Ringwood, in 24°21′S, 135°05′E, N.T.

Ctenotus brooksi euclae Storr, G.M. (1971). The genus *Ctenotus* (Lacertilia : Scincidae) in South Australia. *Rec. S. Aust. Mus.* **16**: 1–15 [14]. Type data: holotype, WAM R287, from Eucla (31°43′S, 128°53′E), W.A.

Ctenotus brooksi iridis Storr, G.M. (1981). Ten new *Ctenotus* (Lacertilia : Scincidae) from Australia. *Rec. West. Aust. Mus.* **9**: 125–146 [144 fig 12]. Type data: holotype, NMV D52837, from 6.6 km ENE of Chinaman Well, Vic.

Synonymy that of Cogger, H.G., this work.

Distribution: W plateau, Lake Eyre basin, Murray-Darling basin, S.A., Vic., N.S.W., Qld., W.A., N.T. Ecology: desert, low open shrubland, open scrub, tall shrubland, terrestrial, diurnal; oviparous, arthropod-feeder. Biological references: Storr, G.M. (1978). Notes on the *Ctenotus* (Lacertilia : Scincidae) of Queensland. *Rec. West. Aust. Mus.* **6**: 319–332 (taxonomy); Pianka, E.R. (1969). Sympatry of desert lizards (*Ctenotus*) in Western Australia. *Ecology* **50**: 1012–1030 (ecology); Storr, G.M. (1969). The genus *Ctenotus* (Lacertilia : Scincidae) in the Eastern Division of Western Australia. *J. R. Soc. West. Aust.* **51**: 97–109 (taxonomy); Pianka, E.R. (1972). Zoogeography and speciation of Australian desert lizards: an ecological perspective. *Copeia* **1972**: 127–144; Storr, G.M., Smith, L.A. & Johnstone, R.E. (1981). *Lizards of Western Australia. I. Skinks.* Perth : University of Western Australia Press and Western Australian Museum 200 pp. (taxonomy).

Ctenotus calurus Storr, 1969

Ctenotus calurus Storr, G.M. (1969). The genus *Ctenotus* (Lacertilia : Scincidae) in the Eastern Division of Western Australia. *J. R. Soc. West. Aust.* **51**: 97–109 [107] [1968 on title page]. Type data: holotype, WAM R30003, from 24 mi ENE of Laverton, in 28°31′S, 122°45′E, W.A.

Distribution: NW coastal, W plateau, S.A., W.A., N.T. Ecology: desert, hummock grassland, low open shrubland, tall shrubland, terrestrial, diurnal; oviparous, arthropod-feeder. Biological references: Pianka E.R. (1969). Sympatry of desert lizards (*Ctenotus*) in Western Australia. *Ecology* **50**: 1012–1030 (ecology); Storr, G.M. (1975). The genus *Ctenotus* (Lacertilia : Scincidae) in the Kimberley and North-west Divisions of Western Australia. *Rec. West. Aust. Mus.* **3**: 209–243 (taxonomy); Storr, G.M. (1971). The genus *Ctenotus* (Lacertilia : Scincidae) in South Australia. *Rec. S. Aust. Mus.* **16**: 1–15 (taxonomy); Storr, G.M., Smith, L.A. & Johnstone, R.E. (1981). *Lizards of Western Australia. I. Skinks.* Perth : University of Western Australia Press and Western Australian Museum 200 pp. (taxonomy).

Ctenotus capricorni Storr, 1981

Ctenotus capricorni Storr, G.M. (1981). Ten new *Ctenotus* (Lacertilia : Scincidae) from Australia. *Rec. West. Aust. Mus.* **9**: 125–146 [133 fig 6]. Type data: holotype, QM J39470, from 33 km W of Jericho, in 23°34′S, 145°48′E, Qld.

Distribution: Lake Eyre basin, (NE coastal), Qld. Ecology: (open scrub), woodland, terrestrial, diurnal; oviparous, arthropod-feeder.

Ctenotus catenifer Storr, 1974

Ctenotus catenifer Storr, G.M. (1974). The genus *Ctenotus* (Lacertilia : Scincidae) in the South-west and Eucla Divisions of Western Australia. *J. Proc. R. Soc. West. Aust.* **56**: 86–93 [92] [1973 on title page]. Type data: holotype, WAM R21819, from Cheyne Beach, in 34°52′S, 118°34′E, W.A.

Distribution: SW coastal, W.A. Ecology: open heath, woodland, open forest, terrestrial, diurnal; oviparous, arthropod-feeder. Biological references: Storr, G.M., Smith, L.A. & Johnstone, R.E. (1981). *Lizards of Western Australia. I. Skinks.* Perth : University of Western Australia Press and Western Australian Museum 200 pp. (taxonomy).

Ctenotus colletti (Boulenger, 1896)

Lygosoma colletti Boulenger, G.A. (1896). Description of four new lizards from Roebuck Bay, north Western Australia obtained by Dr. Dahl, for the Christiania Museum. *Ann. Mag. Nat. Hist. (6)* **18**: 232–235 [234]. Type data: holotype, UZMO K2008, from Roebuck Bay, W.A.

Ctenotus colletti nasutus Storr, G.M. (1969). The genus *Ctenotus* (Lacertilia : Scincidae) in the Eastern Division of Western Australia. *J. R. Soc. West. Aust.* **51**: 97–109 [107] [1968 on title page]. Type data: holotype, WAM R30250, from 5 mi NE of Dunges Table, in 28°08′S, 123°55′E, W.A.

Ctenotus colletti rufescens Storr, G.M. (1979). Five new lizards from Western Australia. *Rec. West. Aust. Mus.* **8**: 134–142 [137 pl 2]. Type data: holotype, WAM R61152, from 3 km E of Giralia, in 22°41′S, 114°23′E, W.A.

Synonymy that of Cogger, H.G., this work.

Distribution: N coastal, NW coastal, W plateau, W.A., N.T. Ecology: desert, hummock grassland, low open shrubland, terrestrial, diurnal; oviparous, arthropod-feeder. Biological references: Storr, G.M. (1970). The genus *Ctenotus* (Lacertilia : Scincidae) in the Northern Territory. *J. R. Soc. West. Aust.* **52**: 97–108 (taxonomy); Pianka, E.R. (1979). Sympatry of desert lizards (*Ctenotus*) in Western Australia. *Ecology* **50**: 1012–1030 (ecology); Storr, G.M., Smith, L.A. & Johnstone, R.E. (1981). *Lizards of Western Australia. I. Skinks.* Perth : University of Western Australia Press and Western Australian Museum 200 pp. (taxonomy).

Ctenotus decaneurus Storr, 1970

Ctenotus decaneurus Storr, G.M. (1970). The genus *Ctenotus* (Lacertilia : Scincidae) in the Northern Territory. *J. R. Soc. West. Aust.* **52**: 97–108 [104] [1969 on title page]. Type data: holotype, WAM R23130, from 21 mi WNW of Newry, in 15°59′S, 129°00′E, N.T.

Ctenotus decaneurus yampiensis Storr, G.M. (1975). The genus *Ctenotus* (Lacertilia : Scincidae) in the Kimberley and North-west Divisions of Western Australia. *Rec. West. Aust. Mus.* **3**: 209–243 [235]. Type data: holotype, WAM R11795, from Wotjulum Mission Station, in 16°11′S, 123°37′E, W.A.

Synonymy that of Cogger, H.G., this work.

Distribution: N coastal, W.A., N.T. Ecology: woodland, terrestrial, diurnal; oviparous, arthropod-feeder. Biological references: Rankin, P.R. (1978). A new species of lizard (Lacertilia : Scincidae) from the Northern Territory, closely allied to *Ctenotus decaneurus* Storr. *Rec. Aust. Mus.* **31**: 395–407 (taxonomy); Storr, G.M., Smith, L.A. & Johnstone, R.E. (1981). *Lizards of Western Australia. I. Skinks.* Perth : University of Western Australia Press and Western Australian Museum 200 pp. (taxonomy).

Ctenotus delli Storr, 1974

Ctenotus delli Storr, G.M. (1974). The genus *Ctenotus* (Lacertilia : Scincidae) in the South-west and Eucla Divisions of Western Australia. *J. R. Soc. West. Aust.* **56**: 86–93 [92] [1973 on title page]. Type data: holotype, WAM R37478, from 6 mi E of Kalamunda, in 31°57′S, 116°08′E, W.A.

Distribution: SW coastal, W.A. Ecology: open forest, terrestrial, diurnal; oviparous, arthropod-feeder. Biological references: Storr, G.M., Smith, L.A. & Johnstone, R.E. (1981). *Lizards of Western Australia. I. Skinks.* Perth : University of Western Australia Press and Western Australian Museum 200 pp. (taxonomy).

Ctenotus dux Storr, 1969

Ctenotus dux Storr, G.M. (1969). The genus *Ctenotus* (Lacertilia : Scincidae) in the Eastern Division of Western Australia. *J. R. Soc. West. Aust.* **51**: 97–109 [104] [1968 on title page]. Type data: holotype, WAM R30002, from 5 mi NE of Dunges Table, in 28°08′S, 123°55′E, W.A.

Distribution: W plateau, S.A., W.A., N.T. Ecology: desert, terrestrial, diurnal; oviparous, arthropod-feeder. Biological references: Pianka, E.R. (1969). Sympatry of desert lizards (*Ctenotus*) in Western Australia. *Ecology* **50**: 1012–1030 (ecology); Storr, G.M., Smith, L.A. & Johnstone, R.E. (1981). *Lizards of Western Australia. I. Skinks.* Perth : University of Western Australia Press and Western Australian Museum 200 pp. (taxonomy).

Ctenotus essingtonii (Gray, 1842)

Tiliqua essingtonii Gray, J.E. (1842). Description of some hitherto unrecorded species of Australian reptiles and batrachians. *in* Gray, J.E. (ed.) *Zoological Miscellany.* London : Treuttel, Würtz & Co. pp. 51–57 [51]. Type data: holotype, BMNH 1937.12.6.11, from Port Essington, N.T.

Ctenotus essingtonii brevipes Storr, G.M. (1981). Ten new *Ctenotus* (Lacertilia : Scincidae) from Australia. *Rec. West. Aust. Mus.* **9**: 125–146 [130]. Type data: holotype, QM J39469, from Venture Creek, 62 km E of Croyden, in 18°13′S, 142°49′E, Qld.

Synonymy that of Cogger, H.G., this work.

Distribution: NE coastal, N Gulf, N coastal, Qld., N.T. Ecology: (tussock grassland), low woodland, woodland, open forest, terrestrial, diurnal; oviparous, arthropod-feeder. Biological references: Storr, G.M. (1970). The genus *Ctenotus* (Lacertilia : Scincidae) in the Northern Territory. *J. R. Soc. West. Aust.* **52**: 97–108 (taxonomy); Cogger, H.G. & Lindner, D.A. (1974). Frogs and reptiles. *in* Fauna survey of the Port Essington District, Cobourg Peninsula, Northern Territory of Australia. Frith, H.J. & Calaby, J.H. (eds.) *CSIRO Div. Wildl. Res. Tech. Pap.* **28**: 63–107 (taxonomy).

Ctenotus eutaenius Storr, 1981

Ctenotus eutaenius Storr, G.M. (1981). Ten new *Ctenotus* (Lacertilia : Scincidae) from Australia. *Rec. West. Aust. Mus.* **9**: 125–146 [129 fig 3]. Type data: holotype, QM J39467, from Charters Towers, in 20°05′S, 146°16′E, Qld.

Distribution: NE coastal, Qld. Ecology: woodland, terrestrial, diurnal; oviparous, arthropod-feeder.

Ctenotus fallens Storr, 1974

Ctenotus fallens Storr, G.M. (1974). The genus *Ctenotus* (Lacertilia : Scincidae) in the South-west and Eucla Divisions of Western Australia. *J. R. Soc. West. Aust.* **56**: 86–93 [88] [1973 on title page]. Type data: holotype, WAM R33780, from 11 mi SSE of Kalbarri, in 27°52′S, 114°12′E, W.A.

Distribution: NW coastal, SW coastal, W.A. Ecology: open heath, open forest, terrestrial, diurnal; oviparous, arthropod-feeder. Biological references: Storr, G.M., Smith, L.A. & Johnstone, R.E. (1981). *Lizards of Western Australia. I. Skinks.* Perth : University of Western Australia Press and Western Australian Museum 200 pp. (taxonomy).

Ctenotus gemmula Storr, 1974

Ctenotus gemmula Storr, G.M. (1974). The genus *Ctenotus* (Lacertilia : Scincidae) in the South-west and Eucla Divisions of Western Australia. *J. R. Soc. West. Aust.* **56**: 86–93 [91] [1973 on title page]. Type data: holotype, WAM R29640, from South Perth, in 32°00′S, 115°49′E, W.A.

Distribution: SW coastal, W plateau, W.A. Ecology: open heath, (open scrub), woodland, open forest, terrestrial, diurnal; oviparous, arthropod-feeder. Biological references: Storr, G.M., Smith, L.A. & Johnstone, R.E. (1981). *Lizards of Western Australia. I. Skinks.* Perth : University of Western Australia Press and Western Australian Museum 200 pp. (taxonomy).

Ctenotus grandis Storr, 1969

Ctenotus grandis Storr, G.M. (1969). The genus *Ctenotus* (Lacertilia : Scincidae) in the Eastern Division of Western Australia. *J. R. Soc. West. Aust.* **51**: 97–109 [100] [1968 on title page]. Type data: holotype, WAM R30005, from 24 mi ENE of Laverton, in 28°31′S, 122°45′E, W.A.

Ctenotus grandis titan Storr, G.M. (1980). The *Ctenotus grandis* species-group (Lacertilia : Scincidae). *Rec. West. Aust. Mus.* **8**: 415–422 [419 pl 1]. Type data: holotype, WAM R61444, from 1 km S of Exmouth, W.A.

Synonymy that of Cogger, H.G., this work.

Distribution: N coastal, NW coastal, W plateau, W.A., N.T., (S.A.). Ecology: desert, hummock grassland, low open shrubland, low open woodland, terrestrial, diurnal; oviparous, arthropod-feeder. Biological references: Storr, G.M. (1980). The *Ctenotus grandis* species-group (Lacertilia : Scincidae). *Rec. West. Aust. Mus.* **8**: 415–422 (taxonomy); Pianka, E.R. (1969). Sympatry of desert lizards (*Ctenotus*) in Western Australia. *Ecology* **50**: 1012–1030 (ecology); Pianka, E.R. (1972). Zoogeography and speciation of Australian desert lizards: an ecological perspective. *Copeia* **1972**: 127–144; Storr, G.M., Smith, L.A. & Johnstone, R.E. (1981). *Lizards of Western Australia. I. Skinks.* Perth : University of Western Australia Press and Western Australian Museum 200 pp. (taxonomy).

Ctenotus greeri Storr, 1979

Ctenotus greeri Storr, G.M. (1979). *Ctenotus greeri*, a new scincid lizard from Western Australia. *Rec. West. Aust. Mus.* **8**: 143–146 [143 pl 1]. Type data: holotype, WAM R65980, from 9 km S of Mt. Linden, in 29°24'S, 122°20'E, W.A.

Distribution: W plateau, W.A. Ecology: desert, terrestrial, diurnal; oviparous, arthropod-feeder. Biological references: Storr, G.M., Smith, L.A. & Johnstone, R.E. (1981). *Lizards of Western Australia. I. Skinks.* Perth : University of Western Australia Press and Western Australian Museum 200 pp. (taxonomy).

Ctenotus hanloni Storr, 1980

Ctenotus hanloni Storr, G.M. (1980). The *Ctenotus grandis* species-group (Lacertilia : Scincidae). *Rec. West. Aust. Mus.* **8**: 415–422 [420 pl 2]. Type data: holotype, WAM R60952, from 22 km NNW of Giralia, in 22°30'S, 114°19'E, W.A.

Distribution: NW coastal, W plateau, S.A., W.A., N.T. Ecology: desert, hummock grassland, low open shrubland, terrestrial, diurnal; oviparous, arthropod-feeder. Biological references: Storr, G.M., Smith, L.A. & Johnstone, R.E. (1981). *Lizards of Western Australia. I. Skinks.* Perth : University of Western Australia Press and Western Australian Museum 200 pp. (taxonomy).

Ctenotus hebetior Storr, 1978

Ctenotus hebetior Storr, G.M. (1978). Notes on the *Ctenotus* (Lacertilia : Scincidae) of Queensland. *Rec. West. Aust. Mus.* **6**: 319–332 [328 pl 3]. Type data: holotype, QM J28374, from 5 km NW of Fermoy, in 23°09'S, 143°00'E, Qld.

Ctenotus hebetior schuettleri Börner, A.-R. (1981). A new subspecies of the *Ctenotus leonhardii* complex. *Misc. Articles in Saurology* **7**: 1–10 [3 figs 4–6]. Type data: holotype, Börner pers. coll. AU Scinc 6B, from 1.5 km E of the Barkly Hwy, *ca* 25 km N of Mt. Isa, Qld.

Distribution: N Gulf, Lake Eyre basin, Qld., N.T. Ecology: tussock grassland, low open shrubland, low open woodland, terrestrial, diurnal; oviparous, (saxicoline), arthropod-feeder.

Ctenotus helenae Storr, 1969

Lygosoma lesueurii concolor Glauert, L. (1952). Herpetological miscellanea. I. Notes on some forms of *Diplodactylus*. pp. 166–168; Some new Western Australian lizards. pp. 168–170. *West. Aust. Nat.* **3**: 166–170 [169] [junior primary homonym; *non Lygosoma (Rhodona) bipes concolor* Werner, 1910]. Type data: holotype, WAM R5049, from Marrilla Station, inland from Maud's Landing, W.A.

Ctenotus helenae Storr, G.M. (1969). The genus *Ctenotus* (Lacertilia : Scincidae) in the Eastern Division of Western Australia. *J. R. Soc. West. Aust.* **51**: 97–109 [100] [1968 on title page]. Type data: holotype, WAM R30004, from 24 mi ENE of Laverton, in 28°31'S, 122°45'E, W.A.

Synonymy that of Storr, G.M. (1975). The genus *Ctenotus* (Lacertilia : Scincidae) in the Kimberley and North-west Divisions of Western Australia. *Rec. West. Aust. Mus.* **3**: 209–243.

Distribution: N Gulf, N coastal, NW coastal, W plateau, Lake Eyre basin, S.A., W.A., N.T. Ecology: desert, hummock grassland, low open shrubland, tall shrubland, terrestrial, diurnal; oviparous, arthropod-feeder. Biological references: Storr, G.M. (1970). The genus *Ctenotus* (Lacertilia : Scincidae) in the Northern Territory. *J. R. Soc. West. Aust.* **52**: 97–108 (taxonomy); Storr, G.M. (1978). *Ctenotus rubicundus*, a new scincid lizard from Western Australia. *Rec. West. Aust. Mus.* **6**: 333–335 (taxonomy); Pianka, E.R. (1969). Sympatry of desert lizards (*Ctenotus*) in Western Australia. *Ecology* **50**: 1012–1030 (ecology); Pianka, E.R. (1972). Zoogeography and speciation of Australian desert lizards: an ecological perspective. *Copeia* **1972**: 127–144; Storr, G.M., Smith, L.A. & Johnstone, R.E. (1981). *Lizards of Western Australia. I. Skinks.* Perth : University of Western Australia Press and Western Australian Museum 200 pp. (taxonomy).

Ctenotus hilli Storr, 1970

Ctenotus hilli Storr, G.M. (1970). The genus *Ctenotus* (Lacertilia : Scincidae) in the Northern Territory. *J. R. Soc. West. Aust.* **52**: 97–108 [103] [1969 on title page]. Type data: holotype, WAM R23569, from Darwin, in 12°26'S, 130°55'E, N.T.

Distribution: N Gulf, N coastal, N.T. Ecology: tall open shrubland, woodland, open forest, terrestrial, diurnal; oviparous, arthropod-feeder. Biological references: Cogger, H.G. & Lindner, D.A. (1974). Frogs and reptiles. *in* Fauna survey of the Port Essington District, Cobourg Peninsula, Northern Territory of Australia. Frith, H.J. & Calaby, J.H. (eds.) *CSIRO Div. Wildl. Res. Tech. Pap.* **28**: 63–107 (taxonomy).

Ctenotus impar Storr, 1969

Ctenotus impar Storr, G.M. (1969). The genus *Ctenotus* (Lacertilia : Scincidae) in the Eastern Division of Western Australia. *J. R. Soc. West. Aust.* **51**: 97–109 [105] [1968 on title page]. Type data: holotype, WAM R1333, from Tambellup, in 34°02'S, 117°38'E, W.A.

Distribution: SW coastal, W plateau, W.A. Ecology: (open heath), (open scrub), woodland, open forest, terrestrial, diurnal; oviparous, arthropod-feeder. Biological references: Storr, G.M. (1974). The genus *Ctenotus* (Lacertilia : Scincidae) in the South-west and Eucla Divisions of Western Australia. *J. R. Soc. West. Aust.* **56**: 86–93 (taxonomy); Storr, G.M. (1971). The genus *Ctenotus* (Lacertilia : Scincidae) in South Australia. *Rec. S. Aust. Mus.* **16**: 1–15 (queries provenance of specimen in SAMA); Storr, G.M., Smith, L.A. & Johnstone, R.E. (1981). *Lizards of Western Australia. I. Skinks.* Perth : University of Western Australia Press and Western Australian Museum 200 pp. (taxonomy).

Ctenotus inornatus (Gray, 1845)

Hinulia inornata Gray, J.E. (1845). *Catalogue of the Specimens of Lizards in the Collection of the British Museum.* London : British Museum xxviii 289 pp. [78]. Type data: holotype, BMNH 1946.8.15.45, from Swan River, W.A. (in error) *vide* Storr, G.M. (1970). The genus *Ctenotus* (Lacertilia : Scincidae) in the Northern Territory. *J. R. Soc. West. Aust.* **52**: 97–108.

Distribution: N Gulf, N coastal, Qld., W.A., N.T. Ecology: tall open shrubland, (low open woodland), woodland, open forest, terrestrial, diurnal; oviparous, arthropod-feeder. Biological references: Storr, G.M. (1970). The genus *Ctenotus* (Lacertilia : Scincidae) in the Northern Territory. *J. R. Soc. West. Aust.* **52**: 97–108 (taxonomy); Storr, G.M. (1978). Notes on the *Ctenotus* (Lacertilia : Scincidae) of Queensland. *Rec. West. Aust. Mus.* **6**: 319–332 (taxonomy); Rankin, P.R. & Gillam, M.W. (1979). A new lizard in the genus *Ctenotus* (Lacertilia : Scincidae) from the Northern Territory with notes on its biology. *Rec. Aust. Mus.* **32**: 501–511 (taxonomy); Storr, G.M., Smith, L.A. & Johnstone, R.E. (1981). *Lizards of Western Australia. I. Skinks.* Perth : University of Western Australia Press and Western Australian Museum 200 pp. (taxonomy).

Ctenotus joanae Storr, 1970

Ctenotus joanae Storr, G.M. (1970). The genus *Ctenotus* (Lacertilia : Scincidae) in the Northern Territory. *J. R. Soc. West. Aust.* **52**: 97–108 [102] [1969 on title page]. Type data: holotype, NMV D5587, from Newcastle Waters, in 17°23′S, 123°25′E, N.T.

Distribution: N Gulf, N coastal, W plateau, Lake Eyre basin, Qld., N.T. Ecology: hummock grassland, tussock grassland, low open woodland, tall shrubland, terrestrial, diurnal; oviparous, arthropod-feeder. Biological references: Storr, G.M. (1978). Notes on the *Ctenotus* (Lacertilia : Scincidae) of Queensland. *Rec. West. Aust. Mus.* **6**: 319–332 (taxonomy).

Ctenotus labillardieri (Duméril and Bibron, 1839)

Tiliqua labillardii Gray, J.E. (1838). Catalogue of the slender-tongued saurians, with descriptions of many new genera and species. (contd.) *Ann. Mag. Nat. Hist.* **2**: 287–293 [289] [*nom. nud.*; 1839 on title page; *ex* Cocteau ms].

Lygosoma labillardieri Duméril, A.M.C. & Bibron, G. (1839). *Erpétologie Générale* ou Histoire Naturelle Complète des Reptiles. Paris : Roret 5: viii 854 pp. [731]. Type data: syntypes, from Australia and Waigeo (as Waigieu) and Rawack Ils., Indonesia, MNHP 2980, cited as "type" by Roux-Estève, R. (1979). Liste des Amphibiens et Reptiles des collections du Muséum National d'Histoire Naturelle de Paris, récoltés par Lesueur (1778–1846). *Bull. Trim. Soc. Géol. Normandie et Amis Muséum du Havre* **66**: 25–29, other syntypes not found.

Hinulia greyii Gray, J.E. (1845). *Catalogue of the Specimens of Lizards in the Collection of the British Museum.* London : British Museum xxviii 289 pp. [76]. Type data: holotype, BMNH 1946.8.19.57, from Swan River, W.A.

Synonymy that of Cogger, H.G., this work.

Distribution: SW coastal, W.A. Ecology: open heath, woodland, open forest, terrestrial, diurnal; oviparous, saxicoline, arthropod-feeder. Biological references: Ford, J. (1969). Distribution and variation of the skink *Ctenotus labillardieri* (Gray) of southwestern Australia. *J. R. Soc. West. Aust.* **51**: 68–75 (taxonomy); Storr, G.M. (1974). The genus *Ctenotus* (Lacertilia : Scincidae) in the South-west and Eucla Divisions of Western Australia. *J. R. Soc. West. Aust.* **56**: 86–93 (taxonomy); Storr, G.M., Smith, L.A. & Johnstone, R.E. (1981). *Lizards of Western Australia. I. Skinks.* Perth : University of Western Australia Press and Western Australian Museum 200 pp. (taxonomy).

Ctenotus lancelini Ford, 1969

Ctenotus labillardieri lancelini Ford, J. (1969). Distribution and variation of the skink *Ctenotus labillardieri* (Gray) of southwestern Australia. *J. R. Soc. West. Aust.* **51**: 68–75 [69] [1968 on title page]. Type data: holotype, WAM R18872, from Lancelin Is., W.A.

Distribution: SW coastal, W.A. Ecology: terrestrial, diurnal; oviparous, saxicoline, arthropod-feeder. Biological references: Storr, G.M. (1974). The genus *Ctenotus* (Lacertilia : Scincidae) in the South-west and Eucla Divisions of Western Australia. *J. Proc. R. Soc. West. Aust.* **56**: 86–93 (taxonomy); Storr, G.M., Smith, L.A. & Johnstone, R.E. (1981). *Lizards of Western Australia. I. Skinks.* Perth : University of Western Australia Press and Western Australian Museum 200 pp. (taxonomy).

Ctenotus lateralis Storr, 1978

Ctenotus lateralis Storr, G.M. (1978). Notes on the *Ctenotus* (Lacertilia : Scincidae) of Queensland. *Rec. West. Aust. Mus.* **6**: 319–332 [323 pl 1]. Type data: holotype, QM J28372, from 14 km N of Mt. Isa, in 20°37S, 139°32′E, Qld.

Distribution: (NE coastal), N Gulf, Lake Eyre basin, Qld. Ecology: tussock grassland, low open shrubland, low woodland, terrestrial, diurnal; oviparous, arthropod-feeder.

Ctenotus leae (Boulenger, 1887)

Lygosoma leae Boulenger, G.A. (1887). *Catalogue of the Lizards in the British Museum (Natural History).* 3. London : British Museum xii 575 pp. pls 40 [226 pl 12 fig 2]. Type data: holotype, BMNH 1946.8.15.33, from Adelaide, S.A.

Distribution: W plateau, S Gulfs, Lake Eyre basin, S.A., N.S.W., Qld., W.A., N.T. Ecology: desert, low shrubland, tall shrubland, terrestrial, diurnal; oviparous, arthropod-feeder. Biological references: Storr, G.M. (1969). The genus *Ctenotus* (Lacertilia : Scincidae) in the Eastern Division of Western Australia. *J. R. Soc. West. Aust.* **51**: 97–109 (taxonomy); Storr, G.M. (1971). The genus *Ctenotus* (Lacertilia : Scincidae) in South Australia. *Rec. S. Aust. Mus.* **16**: 1–15 (taxonomy); Pianka, E.R. (1969). Sympatry of desert lizards (*Ctenotus*) in Western Australia. *Ecology* **50**: 1012–1030 (ecology); Storr, G.M. (1969). The genus *Ctenotus* (Lacertilia : Scincidae) in the Eastern Division of Western Australia. *J. R. Soc. West. Aust.* **51**: 97–109 (taxonomy); Storr, G.M., Smith, L.A. & Johnstone, R.E. (1981). *Lizards of Western Australia. I. Skinks.* Perth : University of Western Australia Press and Western Australian Museum 200 pp. (taxonomy).

Ctenotus leonhardii (Sternfeld, 1919)

Lygosoma (Hinulia) taeniolatum maculata Rosén, N. (1905). List of lizards in the Zoological Museum of Lund, with descriptions of new species. *Ann. Mag. Nat. Hist. (7)* **16**: 129–142 [140 pl 8(2)] [junior primary homonym; *non Lygosoma maculatum* Blyth, 1853]. Type data: syntypes, ZMLU Cat. Typ. Nr. 281 (2 specimens), from "West Australia".

Lygosoma (Hinulia) leonhardii Sternfeld, R. (1919). Neue Schlangen und Echsen aus Zentralaustralien. *Senckenbergiana* **1**: 76–83 [79]. Type data: lectotype, SMF 14410, from Hermannsburg Mission, Upper Finke River, N.T., designation by Mertens, R. (1967). Die herpetologische Sektion des Natur-Museums und Forschungs-Institutes Senckenberg in Frankfurt-a.-M. nebst einem Verzeichnis ihrer Typen. 1. *Senckenb. Biol.* **48**: 1–106.

Synonymy that of Cogger, H.G., this work.

Distribution: NE coastal, NW coastal, W plateau, Murray-Darling basin, Bulloo River basin, Lake Eyre basin, S.A., N.S.W., Qld., W.A., N.T. Ecology: desert, hummock grassland, tussock grassland, low open shrubland, low woodland, woodland, tall shrubland, terrestrial, diurnal; oviparous, arthropod-feeder. Biological references: Storr, G.M. (1969). The genus *Ctenotus* (Lacertilia : Scincidae) in the Eastern Division of Western Australia. *J. R. Soc. West. Aust.* **51**: 97–109 (taxonomy); Storr, G.M. (1970). The genus *Ctenotus* (Lacertilia : Scincidae) in the Northern Territory. *J. R. Soc. West. Aust.* **52**: 97–108 (taxonomy); Storr, G.M. (1971). The genus *Ctenotus* (Lacertilia : Scincidae) in South Australia. *Rec. S. Aust. Mus.* **16**: 1–15 (taxonomy); Warburg, M.R. (1966). On the water economy of several Australian geckos, agamids, and skinks. *Copeia* **1966**: 230–235 (as *Sphenomorphus leonhardii*); Pianka, E.R. (1969). Sympatry of desert lizards (*Ctenotus*) in Western Australia. *Ecology* **50**: 1012–1030 (ecology); Pianka, E.R. (1972). Zoogeography and speciation of Australian desert lizards: an ecological perspective. *Copeia* **1972**: 127–144; Storr, G.M., Smith, L.A. & Johnstone, R.E. (1981). *Lizards of Western Australia. I. Skinks.* Perth : University of Western Australia Press and Western Australian Museum 200 pp. (taxonomy).

Ctenotus mastigura Storr, 1975

Ctenotus mastigura mastigura Storr, G.M. (1975). The genus *Ctenotus* (Lacertilia : Scincidae) in the Kimberley and North-west Divisions of Western Australia. *Rec. West. Aust. Mus.* **3**: 209–243 [228]. Type data: holotype, WAM R43210, from Mitchell Plateau, in 14°52'S, 125°50'E, W.A.

Ctenotus mastigura burbidgei Storr, G.M. (1975). The genus *Ctenotus* (Lacertilia : Scincidae) in the Kimberley and North-west Divisions of Western Australia. *Rec. West. Aust. Mus.* **3**: 209–243 [228]. Type data: holotype, WAM R41296, from Augustus Is., in 15°20'S, 124°35'E, W.A.

Synonymy that of Cogger, H.G., this work.

Distribution: N coastal, W.A. Ecology: woodland, terrestrial, diurnal; oviparous, arthropod-feeder. Biological references: Storr, G.M., Smith, L.A. & Johnstone, R.E. (1981). *Lizards of Western Australia. I. Skinks.* Perth : University of Western Australia Press and Western Australian Museum 200 pp. (taxonomy).

Ctenotus militaris Storr, 1975

Ctenotus militaris Storr, G.M. (1975) The genus *Ctenotus* (Lacertilia : Scincidae) in the Kimberley and North-west Divisions of Western Australia. *Rec. West. Aust. Mus.* **3**: 209–243 [231]. Type data: holotype, WAM R40779, from Lake Argyle, in 16°25'S, 128°40'E, W.A.

Distribution: N coastal, W.A., N.T. Ecology: woodland, terrestrial, diurnal; oviparous, arthropod-feeder. Biological references: Storr, G.M., Smith, L.A. & Johnstone, R.E. (1981). *Lizards of Western Australia. I. Skinks.* Perth : University of Western Australia Press and Western Australian Museum 200 pp. (taxonomy).

Ctenotus mimetes Storr, 1969

Ctenotus mimetes Storr, G.M. (1969). The genus *Ctenotus* (Lacertilia : Scincidae) in the Eastern Division of Western Australia. *J. R. Soc. West. Aust.* **51**: 97–109

[103] [1968 on title page]. Type data: holotype, WAM R17991, from 12 mi E of Paynes Find, in 29°13′S, 117°53′E, W.A.

Distribution: NW coastal, SW coastal, W plateau, W.A. Ecology: (open heath), open scrub, woodland, tall shrubland, terrestrial, diurnal; oviparous, arthropod-feeder. Biological references: Storr, G.M. (1974). The genus *Ctenotus* (Lacertilia : Scincidae) in the South-west and Eucla Divisions of Western Australia. *J. R. Soc. West. Aust.* **56**: 86–93 (taxonomy); Storr, G.M., Smith, L.A. & Johnstone, R.E. (1981). *Lizards of Western Australia. I. Skinks.* Perth : University of Western Australia Press and Western Australian Museum 200 pp. (taxonomy).

Ctenotus monticola Storr, 1981

Ctenotus monticola Storr, G.M. (1981). Ten new *Ctenotus* (Lacertilia : Scincidae) from Australia. *Rec. West. Aust. Mus.* **9**: 125–146 [132 fig 5]. Type data: holotype, QM J39468, from 11 km W of Mareeba, in 17°02′S, 145°20′E, Qld.

Distribution: NE coastal, Qld. Ecology: terrestrial, diurnal; oviparous, arthropod-feeder.

Ctenotus pallescens Storr, 1970

Ctenotus schomburgkii pallescens Storr, G.M. (1970). The genus *Ctenotus* (Lacertilia : Scincidae) in the Northern Territory. *J. R. Soc. West. Aust.* **52**: 97–108 [106] [1969 on title page]. Type data: holotype, WAM R24218, from Morphett Creek, 7 mi S of Banka Banka, in 18°53′S, 134°05′E, N.T.

Distribution: (N coastal), W plateau, N.T. Ecology: desert, low open shrubland, tall shrubland, terrestrial, diurnal; oviparous, arthropod-feeder. Biological references: Storr, G.M. (1974). The genus *Ctenotus* (Lacertilia : Scincidae) in the South-west and Eucla Divisions of Western Australia. *J. R. Soc. West. Aust.* **56**: 86–93 (elevation to specific rank).

Ctenotus pantherinus (Peters, 1866)

Lygosoma (Hinulia) pantherinum Peters, W. (1866). Eine Mitteilung über neue Amphibien (*Amphibolurus, Lygosoma, Cyclodus, Masticophis, Crotaphopeltis*) und Fische (*Diagramma*: *Hapalogenys*) des Kgl. zoologischen Museums. *Mber. K. Preuss. Akad. Wiss. Berl.* **1866**: 86–96 [89] [1867 on title page]. Type data: holotype, ZMB 5379, from Swan River, W.A.

Lygosoma ocellatum Boulenger, G.A. (1896). Description of four new lizards from Roebuck Bay, north Western Australia obtained by Dr. Dahl, for the Christiania Museum. *Ann. Mag. Nat. Hist. (6)* **18**: 232–235 [233] [*non Mocoa ocellata* Gray, 1845]. Type data: holotype, UZMO K2009, from Roebuck Bay, W.A.

Lygosoma ocelliferum Boulenger, G.A. (1896). *Erratum* (*Annals*, Sept. 1896, p. 233) for *Lygosoma ocellatum* read *Lygosoma ocelliferum. Ann. Mag. Nat. Hist. (6)* **18**: 342 [342] [*nom. nov. pro Lygosoma ocellatum* Boulenger, 1896, at that time a junior secondary homonym of *Mocoa ocellata* Gray, 1845].

Egernia whitei carnarae Kinghorn, J.R. (1931). Herpetological notes. 2. *Rec. Aust. Mus.* **18**: 85–91 [88]. Type data: holotype, AM R9981, from between Canara district and North West Cape, W.A.

Lygosoma (Hinulia) breviunguis Kinghorn, J.R. (1932). Description of a new species of *Lygosoma* from north-west Australia. *Rec. Aust. Mus.* **18**: 300–301 [300 fig 1]. Type data: holotype, AM R9981, from Carnarvon District, North West Cape, W.A.

Ctenotus pantherinus ocellifer Storr, G.M. (1969). The genus *Ctenotus* (Lacertilia : Scincidae) in the Eastern Division of Western Australia. *J. R. Soc. West. Aust.* **51**: 97–109 [99] [1968 on title page; emend. *pro Lygosoma ocelliferum* Boulenger, 1896].

Ctenotus pantherinus calx Storr, G.M. (1970). The genus *Ctenotus* (Lacertilia : Scincidae) in the Northern Territory. *J. R. Soc. West. Aust.* **52**: 97–108 [99] [1969 on title page]. Type data: holotype, NMV D10810, from Dunmarra, in 16°40′S, 133°23′E, N.T.

Ctenotus pantherinus acripes Storr, G.M. (1975). The genus *Ctenotus* (Lacertilia : Scincidae) in the Kimberley and North-west Divisions of Western Australia. *Rec. West. Aust. Mus.* **3**: 209–243 [216]. Type data: holotype, WAM R27843, from Barrow Is., in 20°45′S, 115°25′E, W.A.

Synonymy that of Cogger, H.G., this work.

Distribution: N Gulf, N coastal, NW coastal, SW coastal, W plateau, S Gulfs, Murray-Darling basin, Bulloo River basin, Lake Eyre basin, S.A., N.S.W., Qld., W.A., N.T. Ecology: desert, hummock grassland, tussock grassland, low open shrubland, (open heath), open scrub, low woodland, woodland, terrestrial, diurnal; oviparous, arthropod-feeder. Biological references: Pianka, E.R. (1969). Sympatry of desert lizards (*Ctenotus*) in Western Australia. *Ecology* **50**: 1012–1030 (ecology); Storr, G.M. (1975). The genus *Ctenotus* (Lacertilia : Scincidae) in the Kimberley and North-west Divisions of Western Australia. *Rec. West. Aust. Mus.* **3**: 209–243 (taxonomy); Storr, G.M. (1978). Notes on the *Ctenotus* (Lacertilia : Scincidae) of Queensland. *Rec. West. Aust. Mus.* **6**: 319–332 (taxonomy); Storr, G.M., Smith, L.A. & Johnstone, R.E. (1981). *Lizards of Western Australia. I. Skinks.* Perth : University of Western Australia Press and Western Australian Museum 200 pp. (taxonomy).

Ctenotus piankai Storr, 1969

Ctenotus piankai Storr, G.M. (1969). The genus *Ctenotus* (Lacertilia : Scincidae) in the Eastern Division of Western Australia. *J. R. Soc. West. Aust.* **51**: 97–109 [106] [1968 on title page]. Type data: holotype, WAM R30000, from 24 mi ENE of Laverton, 28°31′S, 122°45′E, W.A.

Ctenotus piankai duricola Storr, G.M. (1975). The genus *Ctenotus* (Lacertilia : Scincidae) in the Kimberley and

North-west Divisions of Western Australia. *Rec. West. Aust. Mus.* **3**: 209–243 [239]. Type data: holotype, WAM R17163, from Mt. Edgar, in 21°19'S, 120°02'E, W.A.

Synonymy that of Cogger, H.G., this work.

Distribution: N Gulf, N coastal, NW coastal, W plateau, Lake Eyre basin, S.A., Qld., W.A., N.T. Ecology: desert, hummock grassland, low open shrubland, tall shrubland, terrestrial, diurnal; oviparous, arthropod-feeder. Biological references: Pianka, E.R. (1969). Sympatry of desert lizards (*Ctenotus*) in Western Australia. *Ecology* **50**: 1012–1030 (ecology); Storr, G.M. (1978). Notes on the *Ctenotus* (Lacertilia : Scincidae) of Queensland. *Rec. West. Aust. Mus.* **6**: 319–332 (taxonomy).

Ctenotus pulchellus Storr, 1978

Ctenotus pulchellus Storr, G.M. (1978). Notes on the *Ctenotus* (Lacertilia : Scincidae) of Queensland. *Rec. West. Aust. Mus.* **6**: 319–332 [326 pl 2]. Type data: holotype, QM J28375, from 72 road km NNE of Mt. Isa, in 20°19'S, 139°47'E, Qld.

Distribution: N Gulf, W plateau, Lake Eyre basin, Qld., N.T. Ecology: hummock grassland, tussock grassland, low open shrubland, low open woodland, low woodland, terrestrial, diurnal; oviparous, arthropod-feeder.

Ctenotus quattuordecimlineatus (Sternfeld, 1919)

Lygosoma (Hinulia) quattuordecimlineatum Sternfeld, R. (1919). Neue Schlangen und Echsen aus Zentralaustralien. *Senckenbergiana* **1**: 76–83 [80]. Type data: holotype, SMF 14422, from Hermannsburg Mission, Upper Finke River, N.T.

Ctenotus quattuordecimlineatus iapetus Storr, G.M. (1975). The genus *Ctenotus* (Lacertilia : Scincidae) in the Kimberley and North-west Divisions of Western Australia. *Rec. West. Aust. Mus.* **3**: 209–243 [236]. Type data: holotype, WAM R27887, from 13 km S of North West Cape, in 21°54'S, 114°08'E, W.A.

Synonymy that of Cogger, H.G., this work.

Distribution: NW coastal, W plateau, Lake Eyre basin, S.A. W.A., N.T. Ecology: desert, hummock grassland, low open shrubland, tall shrubland, terrestrial, diurnal; oviparous, arthropod-feeder. Biological references: Pianka, E.R. (1969). Sympatry of desert lizards (*Ctenotus*) in Western Australia. *Ecology* **50**: 1012–1030 (ecology); Storr, G.M. (1971). The genus *Ctenotus* (Lacertilia : Scincidae) in South Australia. *Rec. S. Aust. Mus.* **16**: 1–15 (taxonomy); Pianka, E.R. (1972). Zoogeography and speciation of Australian desert lizards: an ecological perspective. *Copeia* **1972**: 127–144; Storr, G.M., Smith, L.A. & Johnstone, R.E. (1981). *Lizards of Western Australia. I. Skinks.* Perth : University of Western Australia Press and Western Australian Museum 200 pp. (taxonomy).

Ctenotus quinkan Ingram, 1979

Ctenotus quinkan Ingram, G.J. (1979). Two new species of skinks, genus *Ctenotus* (Reptilia : Lacertilia : Scincidae), from Cape York Peninsula, Queensland, Australia. *J. Herpet.* **13**: 279–282 [280 pl 1]. Type data: holotype, QM J28353, from 30.4 km SW of Laura (15°42'S, 144°36'E), Cape York Peninsula, Qld.

Distribution: NE coastal, Qld. Ecology: woodland, terrestrial, diurnal; oviparous, arthropod-feeder.

Ctenotus rawlinsoni Ingram, 1979

Ctenotus rawlinsoni Ingram, G.J. (1979). Two new species of skinks, genus *Ctenotus* (Reptilia : Lacertilia : Scincidae), from Cape York Peninsula, Queensland, Australia. *J. Herpet.* **13**: 279–282 [279 pl 1]. Type data: holotype, QM J32398, from 36 km N of Hopevale Community (15°07'S, 145°15'E), Cape York Peninsula, Qld.

Distribution: NE coastal, Qld. Ecology: woodland, terrestrial, diurnal; oviparous, arthropod-feeder.

Ctenotus regius Storr, 1971

Ctenotus regius Storr, G.M. (1971). The genus *Ctenotus* (Lacertilia : Scincidae) in South Australia. *Rec. S. Aust. Mus.* **16**: 1–15 [7]. Type data: holotype, WAM R24492, from Kingoonya (30°54'S, 135°19'E), S.A.

Distribution: W plateau, S Gulfs, Murray-Darling basin, Bulloo River basin, Lake Eyre basin, S.A., Vic., N.S.W., Qld., N.T. Ecology: desert, low open shrubland, low shrubland, (low open woodland), tall shrubland, terrestrial, diurnal; oviparous, arthropod-feeder. Biological references: Storr, G.M. (1978). Notes on the *Ctenotus* (Lacertilia : Scincidae) of Queensland. *Rec. West. Aust. Mus.* **6**: 319–332 (taxonomy).

Ctenotus robustus Storr, 1970

Ctenotus robustus Storr, G.M. (1970). The genus *Ctenotus* (Lacertilia : Scincidae) in the Northern Territory. *J. R. Soc. West. Aust.* **52**: 97–108 [100] [1969 on title page]. Type data: holotype, NMV D4957, from Barrow Creek, in 21°31'S, 133°53'E, N.T.

Distribution: SE coastal, NE coastal, N Gulf, N coastal, W plateau, S Gulfs, Murray-Darling basin, Lake Eyre basin, S.A., Vic., N.S.W., Qld., W.A., N.T., A.C.T. Ecology: low open shrubland, low shrubland, open scrub, low open woodland, woodland, open forest, supralittoral, tall shrubland, terrestrial, diurnal; oviparous, arthropod-feeder. Biological references: Storr, G.M. (1971). The genus *Ctenotus* (Lacertilia : Scincidae) in South Australia. *Rec. S. Aust. Mus.* **16**: 1–15 (taxonomy); Storr, G.M., Smith, L.A. & Johnstone, R.E. (1981). *Lizards of Western Australia. I.*

Skinks. Perth : University of Western Australia Press and Western Australian Museum 200 pp. (taxonomy).

Ctenotus rubicundus Storr, 1978

Ctenotus rubicundus Storr, G.M. (1978). *Ctenotus rubicundus*, a new scincid lizard from Western Australia. *Rec. West. Aust. Mus.* **6**: 333–335 [333 pl 1]. Type data: holotype, WAM R57257, from Mt. Herbert, in 21°20'S, 117°12'E, W.A.

Distribution: NW coastal, W.A. Ecology: hummock grassland, low open shrubland, terrestrial, diurnal; oviparous, arthropod-feeder. Biological references: Storr, G.M., Smith, L.A. & Johnstone, R.E. (1981). *Lizards of Western Australia. I. Skinks.* Perth : University of Western Australia Press and Western Australian Museum 200 pp. (taxonomy).

Ctenotus rutilans Storr, 1980

Ctenotus rutilans Storr, G.M. (1980). A new *Lerista* and two new *Ctenotus* (Lacertilia : Scincidae) from Western Australia. *Rec. West. Aust. Mus.* **8**: 441–447 [446 pl 3]. Type data: holotype, WAM R63628, from 27 km SE of Mt. Meharry, in 23°10'S, 118°45'E, W.A.

Distribution: NW coastal, W.A. Ecology: hummock grassland, low open shrubland, terrestrial, diurnal; oviparous, arthropod-feeder. Biological references: Storr, G.M., Smith, L.A. & Johnstone, R.E. (1981). *Lizards of Western Australia. I. Skinks.* Perth : University of Western Australia Press and Western Australian Museum 200 pp. (taxonomy).

Ctenotus saxatilis Storr, 1970

Ctenotus saxatilis Storr, G.M. (1970). The genus *Ctenotus* (Lacertilia : Scincidae) in the Northern Territory. *J. R. Soc. West. Aust.* **52**: 97–108 101] [1969 on title page]. Type data: holotype, WAM R24239, from 10 mi E of Tennant Creek, in 19°40'S, 134°20'E, N.T.

Distribution: N Gulf, N coastal, NW coastal, W plateau, Lake Eyre basin, S.A., W.A., N.T. Ecology: desert, hummock grassland, low open shrubland, tall shrubland, terrestrial, diurnal; oviparous, saxicoline, arthropod-feeder. Biological references: Storr, G.M. (1971). The genus *Ctenotus* (Lacertilia : Scincidae) in South Australia. *Rec. S. Aust. Mus.* **16**: 1–15 (taxonomy); Storr, G.M., Smith, L.A. & Johnstone, R.E. (1981). *Lizards of Western Australia. I. Skinks.* Perth : University of Western Australia Press and Western Australian Museum 200 pp. (taxonomy).

Ctenotus schevilli (Loveridge, 1933)

Sphenomorphus schevilli Loveridge, A. (1933). New scincid lizards of the genera *Sphenomorphus*, *Rhodona* and *Lygosoma* from Australia. *Occ. Pap. Bost. Soc. Nat. Hist.* **8**: 95–100 [96]. Type data: holotype, QM J5805, from Army Downs, near Richmond, Qld.

Distribution: N Gulf, Lake Eyre basin, Qld. Ecology: tussock grassland, low open shrubland, terrestrial, diurnal; oviparous, arthropod-feeder. Biological references: Storr, G.M. (1978). Notes on the *Ctenotus* (Lacertilia : Scincidae) of Queensland. *Rec. West. Aust. Mus.* **6**: 319–332 (taxonomy).

Ctenotus schomburgkii (Peters, 1863)

Lygosoma schomburgkii Peters, W. (1863). Übersicht der von Hrn. Richard Schomburgk an das zoologische Museum eingesandten Amphibien, aus Buchsfelde bei Adelaide in Südaustralien. *Mber. K. Preuss. Akad. Wiss. Berl.* **1863**: 228–236 [231] [1864 on title page]. Type data: lectotype, ZMB 4713, from Loos, 4.5 km W Gawler (as Buchsfelde), S.A., see Praite, R. & Tolley, J.C. (1970). *Place Names of South Australia.* Adelaide : Rigby 203 pp., designation by Storr, G.M. (1971). The genus *Ctenotus* (Lacertilia : Scincidae) in South Australia. *Rec. S. Aust. Mus.* **16**: 1–15.

Hinulia muelleri Fischer, J.G. (1882). Herpetologische Bemerkungen. *Arch. Naturgesch.* **48**: 281–302 [295 pl 16 figs 16–19] [*non Scincus mulleri* Schlegel, 1837]. Type data: syntypes, ZMB 10073 (not found), SMNS (not found), from Nickol Bay, W.A.

Lygosoma fischeri Boulenger, G.A. (1887). *Catalogue of the Lizards in the British Museum (Natural History).* 3. London : British Museum xii 575 pp. pls 40 [228] [*nom. nov. pro Hinulia muelleri* Fischer, 1882, at that time a junior secondary homonym of *Scincus mulleri* Schlegel, 1837].

Synonymy that of Cogger, H.G., this work.

Distribution: NW coastal, SW coastal, W plateau, S Gulfs, Murray-Darling basin, Bulloo River basin, Lake Eyre basin, S.A., Vic., N.S.W., Qld., W.A., N.T. Ecology: desert, hummock grassland, low open shrubland, (open heath), open scrub, low woodland, tall shrubland, terrestrial, diurnal; oviparous, arthropod-feeder. Biological references: Storr, G.M. (1974). The genus *Ctenotus* (Lacertilia : Scincidae) in the South-west and Eucla Divisions of Western Australia. *J. R. Soc. West. Aust.* **56**: 86–93 (taxonomy); Pianka, E.R. (1969). Sympatry of desert lizards (*Ctenotus*) in Western Australia. *Ecology* **50**: 1012–1030 (ecology); Storr, G.M., Smith, L.A. & Johnstone, R.E. (1981). *Lizards of Western Australia. I. Skinks.* Perth : University of Western Australia Press and Western Australian Museum 200 pp. (taxonomy).

Ctenotus serventyi Storr, 1975

Ctenotus serventyi Storr, G.M. (1975). The genus *Ctenotus* (Lacertilia : Scincidae) in the Kimberley and North-west Divisions of Western Australia. *Rec. West.*

Aust. Mus. **3**: 209–243 [233]. Type data: holotype, WAM R46000, from Christmas Creek, 18°53′S, 124°53′E, W.A.

Distribution: N coastal, NW coastal, W plateau, W.A. Ecology: tussock grassland, low open shrubland, low open woodland, terrestrial, diurnal; oviparous, arthropod-feeder. Biological references: Storr, G.M., Smith, L.A. & Johnstone, R.E. (1981). *Lizards of Western Australia. I. Skinks.* Perth : University of Western Australia Press and Western Australian Museum 200 pp. (taxonomy).

Ctenotus severus Storr, 1969

Ctenotus severus Storr, G.M. (1969). The genus *Ctenotus* (Lacertilia : Scincidae) in the Eastern Division of Western Australia. *J. R. Soc. West. Aust.* **51**: 97–109 [101] [1968 on title page]. Type data: holotype, WAM R25680, from Galena, W.A.

Distribution: NW coastal, SW coastal, W plateau, W.A. Ecology: low open shrubland, tall shrubland, terrestrial, diurnal; oviparous, arthropod-feeder. Biological references: Storr, G.M. (1974). The genus *Ctenotus* (Lacertilia : Scincidae) in the South-west and Eucla Divisions of Western Australia. *J. R. Soc. West. Aust.* **56**: 86–93 (taxonomy); Storr, G.M., Smith, L.A. & Johnstone, R.E. (1981). *Lizards of Western Australia. I. Skinks.* Perth : University of Western Australia Press and Western Australian Museum 200 pp. (taxonomy).

Ctenotus spaldingi (Macleay, 1877)

Hinulia spaldingi Macleay, W. (1877). The lizards of the *Chevert* Expedition. *Proc. Linn. Soc. N.S.W.* **2**: 60–69 [63] [1878 on title page]. Type data: syntypes, AM R31843–6, from Endeavour River, Qld.

Lygosoma dorsale Boulenger, G.A. (1887). *Catalogue of the Lizards in the British Museum (Natural History).* 3. London : British Museum xii 575 pp. pls 40 [226 pl 12 fig 1]. Type data: syntypes, BMNH 1946.8.15.51–52, from Fly River, Papua New Guinea.

Synonymy that of Cogger, H.G., this work.

Distribution: NE coastal, N Gulf, N coastal, (W plateau), Qld., W.A., N.T. Ecology: low woodland, woodland, open forest, supralittoral, terrestrial, diurnal; oviparous, arthropod-feeder. Biological references: Copland, S.J. (1947). Catalogue of reptiles in the Macleay Museum. II. *Sphenomorphus spaldingi* (Macleay). *Proc. Linn. Soc. N.S.W.* **71**: 136–144 (taxonomy, as *Sphenomorphus spaldingi*); Storr, G.M. (1970). The genus *Ctenotus* (Lacertilia : Scincidae) in the Northern Territory. *J. R. Soc. West. Aust.* **52**: 97–108 (taxonomy); Storr, G.M. (1978). Notes on the *Ctenotus* (Lacertilia : Scincidae) of Queensland. *Rec. West. Aust. Mus.* **6**: 319–332 (taxonomy).

Ctenotus storri Rankin, 1978

Ctenotus storri Rankin, P.R. (1978). A new species of lizard (Lacertilia : Scincidae) from the Northern Territory, closely allied to *Ctenotus decaneurus* Storr. *Rec. Aust. Mus.* **31**: 395–407 [395 figs 1–3]. Type data: holotype, ANWC R745, from Tapa Bay, Cox Peninsula, in 12°27′S, 130°37′E, N.T.

Distribution: N coastal, N.T. Ecology: woodland, terrestrial, diurnal; oviparous, arthropod-feeder.

Ctenotus strauchii (Boulenger, 1887)

Lygosoma strauchii Boulenger, G.A. (1887). *Catalogue of the Lizards in the British Museum (Natural History).* 3. London : British Museum xii 575 pp. pls 40 [229 pl 12 fig 3]. Type data: holotype, BMNH 1946.8.19.66, from Gayndah, Qld.

Ctenotus strauchii varius Storr, G.M. (1981). Ten new *Ctenotus* (Lacertilia : Scincidae) from Australia. *Rec. West. Aust. Mus.* **9**: 125–146 [139 fig 9]. Type data: holotype, WAM R49507, from Charlotte Waters, N.T.

Synonymy that of Cogger, H.G., this work.

Distribution: NE coastal, (N Gulf), Murray-Darling basin, Bulloo River basin, Lake Eyre basin, S.A., N.S.W., Qld., N.T. Ecology: desert, tussock grassland, low open shrubland, open scrub, low open woodland, woodland, tall shrubland, terrestrial, diurnal; oviparous, arthropod-feeder. Biological references: Storr, G.M. (1970). The genus *Ctenotus* (Lacertilia : Scincidae) in the Northern Territory. *J. R. Soc. West. Aust.* **52**: 97–108 (taxonomy); Storr, G.M. (1971). The genus *Ctenotus* (Lacertilia : Scincidae) in South Australia. *Rec. S. Aust. Mus.* **16**: 1–15 (taxonomy); Storr, G.M. (1981). Ten new *Ctenotus* (Lacertilia : Scincidae) from Australia. *Rec. West. Aust. Mus.* **9**: 125–146 (taxonomy).

Ctenotus striaticeps Storr, 1978

Ctenotus striaticeps Storr, G.M. (1978). Notes on the *Ctenotus* (Lacertilia : Scincidae) of Queensland. *Rec. West. Aust. Mus.* **6**: 319–332 [330 pl 4]. Type data: holotype, QM J28373, from 72 road km NNE of Mt. Isa, in 20°19′S, 139°47′E, Qld.

Distribution: N Gulf, Qld. Ecology: terrestrial, diurnal; oviparous, arthropod-feeder.

Ctenotus taeniolatus (White, 1790)

Lacerta taeniolata White, J. (1790). *Journal of a Voyage to New South Wales,* with sixty five plates of non-descript animals, birds, lizards, serpents, curious cones of trees and other natural productions. Appendix. London : Debrett 299 pp. [245 pl 32 fig 1] [*ex* Shaw ms to whom authorship of this species has also been ascribed, *e.g.* Gray, J.E. (1845). *Catalogue of the Specimens of Lizards in the Collection of the British Museum.* London : British Museum xxviii 289 pp.]. Type data: holotype, BMNH xv.16a (presumed lost), from N.S.W.

Scincus octolineatus Daudin, F. M. (1802). *Histoire Naturelle, générale et particulière des Reptiles.* Paris : Dufart Tom. 4 397 pp. [285] [*nom. substit. pro Lacerta taeniolata* White, 1790].

Scincus decemlineatus Lacépède, B.G.E. (1804). Mémoire sur plusieurs animaux de la Nouvelle-Hollande dont la description n'a pas encore été publiée. *Ann. Mus. Natl. Hist. Nat. Paris* **4**: 184–211 [208]. Type data: none designated, MNHP (not found), from Australia.

***Scincus undecimstriatus* (part.)** Kuhl, H. (1820). *Beiträge zur Zoologie und vergleichenden Anatomie.* Frankfurt 130 pp. [129] Type data: syntypes, from Australia; for only extant syntype see *Leiolopisma entrecasteauxii* (Duméril and Bibron, 1839); remaining syntypes not found, but unpublished Kuhl figure of *S. undecimstriatus* indicates that at least one syntype was *Ctenotus taeniolatus*.

Scincus multilineatus Lesson, R.P. (1829). Description de quelques reptiles nouveaux ou peu connus. *in* Duperrey, L.I. (ed.) (1826–1838). *Voyage autour du monde, executé par ordre du Roi, sur la corvette de Sa Majesté,* La Coquille, *pendant les années 1822, 1823, 1824 et 1825.* Zoologie 2(1): 34–65 and Atlas (Reptiles plates 1–7) Paris : Arthus Bertrand [livraison 11 pl 3 fig 2 (1829); 4 (1831)] [Publication dates established from Sherborn, C.D. & Woodward, B.B. (1901). Notes on the dates of publication of the natural history portions of some French voyages. Part 1 *Ann. Mag. Nat. Hist. (7)* **7**: 388–392 (date of livraison); Anon. (1829). [Review of] *Voyage autour du monde.* Livraison 11. *Bull. Sci. Nat. Géol.* (Férussac, ed.) **17**: 267 (identifies livraison in which plate appears); Sherborn, C.D. & Woodward, B.B. (1906). Notes on the dates of publication of the natural history portions of some French voyages. "Voyage autour du Monde .... sur .... la Coquille pendant .... 1822–25 .... Par L.J. Duperrey," etc. A correction. *Ann. Mag. Nat. Hist. (7)* **17**: 335–336 (date of text)]. Type data: syntypes, from N.S.W. and Bourou and Ambon Ils., Indonesia, MNHP 2983 (Australia), 2 specimens, only one of which, the figured specimen, has been found, see Roux-Estève, R. (1979). Liste des Amphibiens et Reptiles des collections du Muséum National d'Histoire Naturelle de Paris, récoltés par Lesueur (1778–1846). *Bull. Trim. Soc. Géol. Normandie et Amis Muséum du Havre* **66**: 25–29.

Synonymy that of Cogger, H.G., this work.

Distribution: SE coastal, NE coastal, Murray-Darling basin, Vic., N.S.W., Qld., A.C.T. Ecology: (open scrub), woodland, open forest, terrestrial, diurnal; oviparous, (saxicoline), arthropod-feeder. Biological references: Swanson, S. (1976). *Lizards of Australia.* Sydney : Angus & Robertson 80 pp. (natural history); Storr, G.M. (1978). Notes on the *Ctenotus* (Lacertilia : Scincidae) of Queensland. *Rec. West. Aust. Mus.* **6**: 319–332 (taxonomy); Greer, A.E. (1980). Critical thermal maximum temperatures in Australian scincid lizards: their ecological and evolutionary significance. *Aust. J. Zool.* **28**: 91–102.

Ctenotus tanamiensis Storr, 1970

Ctenotus tanamiensis Storr, G.M. (1970). The genus *Ctenotus* (Lacertilia : Scincidae) in the Northern Territory. *J. R. Soc. West. Aust.* **52**: 97–108 [103] [1969 on title page]. Type data: holotype, NTM 2079, from The Granites, in 20°43'S, 130°22'E, N.T.

Distribution: W plateau, (Lake Eyre basin), W.A., N.T. Ecology: desert, terrestrial, diurnal; oviparous, arthropod-feeder. Biological references: Storr, G.M., Smith, L.A. & Johnstone, R.E. (1981). *Lizards of Western Australia. I. Skinks.* Perth : University of Western Australia Press and Western Australian Museum 200 pp. (taxonomy).

Ctenotus tantillus Storr, 1975

Ctenotus tantillus Storr, G.M. (1975). The genus *Ctenotus* (Lacertilia : Scincidae) in the Kimberley and North-west Divisions of Western Australia. *Rec. West. Aust. Mus.* **3**: 209–243 [242]. Type data: holotype, WAM R45567, from Kununurra, in 15°42'S, 128°42'E, W.A.

Distribution: N coastal, W.A. Ecology: woodland, terrestrial, diurnal; oviparous, arthropod-feeder. Biological references: Storr, G.M., Smith, L.A. & Johnstone, R.E. (1981). *Lizards of Western Australia. I. Skinks.* Perth : University of Western Australia Press and Western Australian Museum 200 pp. (taxonomy).

Ctenotus uber Storr, 1969

Ctenotus uber Storr, G.M. (1969). The genus *Ctenotus* (Lacertilia : Scincidae) in the Eastern Division of Western Australia. *J. R. Soc. West. Aust.* **51**: 97–109 [102] [1968 on title page]. Type data: holotype, WAM R17654, from 22 mi SE of Yalgoo, in 28°34'S, 116°52'E, W.A.

Ctenotus uber orientalis Storr, G.M. (1971). The genus *Ctenotus* (Lacertilia : Scincidae) in South Australia. *Rec. S. Aust. Mus.* **16**: 1–15 [8]. Type data: holotype, NMV D825, from Ouyen (35°05'S, 142°19'E), Vic.

Ctenotus uber johnstonei Storr, G.M. (1980). A new *Lerista* and two new *Ctenotus* (Lacertilia : Scincidae) from Western Australia. *Rec. West. Aust. Mus.* **8**: 441–447 [444 pl 2]. Type data: holotype, WAM R63277, from Balgo Hill, in 20°07'S, 127°48'E, W.A.

Synonymy that of Cogger, H.G., this work.

Distribution: (SE coastal), (N coastal), NW coastal, SW coastal, W plateau, S Gulfs, Murray-Darling basin, Bulloo River basin, Lake Eyre basin, S.A., Vic., N.S.W., Qld., W.A., N.T., (A.C.T.). Ecology: desert, low open shrubland, low shrubland, (woodland), tall shrubland, terrestrial, diurnal; oviparous, arthropod-feeder. Biological references: Storr, G.M. (1974). The genus *Ctenotus* (Lacertilia : Scincidae) in the South-west and Eucla Divisions of Western Australia. *J. R. Soc. West. Aust.* **56**: 86–93 (taxonomy); Storr, G.M., Smith, L.A. & Johnstone, R.E. (1981).

Lizards of Western Australia. I. Skinks. Perth : University of Western Australia Press and Western Australian Museum 200 pp. (taxonomy).

Ctenotus vertebralis Rankin and Gillam, 1979

Ctenotus vertebralis Rankin, P.R. & Gillam, M.W. (1979). A new lizard in the genus *Ctenotus* (Lacertilia : Scincidae) from the Northern Territory with notes on its biology. *Rec. Aust. Mus.* **32**: 501–511 [501 figs 1–3]. Type data: holotype, CAM R410, from Arnold River, Cox River Station, in 15°43′S, 134°32′E, N.T.

Distribution: N coastal, N.T. Ecology: woodland, open forest, terrestrial, diurnal; oviparous, arthropod-feeder.

Ctenotus xenopleura Storr, 1981

Ctenotus xenopleura Storr, G.M. (1981). Ten new *Ctenotus* (Lacertilia : Scincidae) from Australia. *Rec. West. Aust. Mus.* **9**: 125–146 [135 fig 7]. Type data: holotype, WAM R72212, from 15 km NE Bungalbin Hill, in 30°17′S, 119°44′E, W.A.

Distribution: W plateau, (SW coastal), W.A. Ecology: terrestrial, diurnal; oviparous, arthropod-feeder.

Ctenotus youngsoni Storr, 1975

Ctenotus youngsoni Storr, G.M. (1975). The genus *Ctenotus* (Lacertilia : Scincidae) in the Kimberley and North-west Divisions of Western Australia. *Rec. West. Aust. Mus.* **3**: 209–243 [227]. Type data: holotype, WAM R42363, from Dirk Hartog Is., in 25°50′S, 113°05′E, W.A.

Distribution: NW coastal, W.A. Ecology: terrestrial, diurnal; oviparous, arthropod-feeder. Biological references: Storr, G.M., Smith, L.A. & Johnstone, R.E. (1981). *Lizards of Western Australia. I. Skinks.* Perth : University of Western Australia Press and Western Australian Museum 200 pp. (taxonomy).

Ctenotus zebrilla Storr, 1981

Ctenotus zebrilla Storr, G.M. (1981). Ten new *Ctenotus* (Lacertilia : Scincidae) from Australia. *Rec. West. Aust. Mus.* **9**: 125–146 [136 fig 8]. Type data: holotype, QM J39471, from Venture Creek, 62 km E of Croyden, in 18°13′S, 142°49′E, Qld.

Distribution: N Gulf, Qld. Ecology: low woodland, open forest, terrestrial, diurnal; oviparous, arthropod-feeder.

Egernia Gray, 1838

Egernia Gray, J.E. (1838). Catalogue of the slender-tongued saurians, with descriptions of many new genera and species. (contd.) *Ann. Mag. Nat. Hist.* **2**: 287–293 [288] [1839 on title page]. Type species *Tiliqua cunninghami* Gray, 1832 by monotypy.

Tropidolopisma Duméril, A.M.C. & Bibron, G. (1839). *Erpétologie Générale* ou Histoire Naturelle Complète des Reptiles. Paris : Roret 5: viii 854 pp. [744]. Type species *Tropidolopisma dumerilii* Duméril and Bibron, 1839 by monotypy.

Torpidolepisma Fitzinger, L.J. (1843). *Systema Reptilium.* Vienna : Braümüller u. Seidel vi 106 pp. [22] [*pro Tropidolopisma* Duméril and Bibron, 1839].

Tropidolepisma Peters, W. (1844). Eine neue Art von *Tropidolepisma. Ber. Verh. K. Preuss. Akad. Wiss. Berl.* **1844**: 36–37 [36] [?*pro Tropidolopisma* Duméril and Bibron, 1839].

Silubosaurus Gray, J.E. (1845). *Catalogue of the Specimens of Lizards in the Collection of the British Museum.* London : British Museum xxviii 289 pp. [104]. Type species *Silubosaurus stokesii* Gray, 1845 by monotypy.

Lissolepis Peters, W. (1872). Über neue oder weniger bekannte Saurier. *Mber. K. Preuss. Akad. Wiss. Berl.* **1872**: 774–776 [776] [1873 on title page]. Type species *Cyclodus (Omolepida) luctuosus* Peters, 1866 by original designation.

Synonymy that of Cogger, H.G., this work; Mitchell, F.J. (1950). The scincid genera *Egernia* and *Tiliqua* (Lacertilia). *Rec. S. Aust. Mus.* **9**: 275–308.

This group is also found in E Indonesia and New Guinea.

Egernia carinata Smith 1939

Egernia carinata Smith, H.M. (1939). A new Australian lizard, with a note on *Hemiergis. Field Mus. Nat. Hist. Zool. Ser.* **24**: 11–14 [11 fig 3]. Type data: holotype, FMNH 11729, from Toolbrunup, W.A.

Distribution: SW coastal, W plateau, W.A. Ecology: open heath, (open scrub), woodland, terrestrial, diurnal; viviparous, arthropod-feeder. Biological references: Storr, G.M. (1978). The genus *Egernia* (Lacertilia : Scincidae) in Western Australia. *Rec. West. Aust. Mus.* **6**: 147–187 (taxonomy).

Egernia coventryi Storr, 1978

Egernia coventryi Storr, G.M. (1978). The genus *Egernia* (Lacertilia : Scincidae) in Western Australia. *Rec. West. Aust. Mus.* **6**: 147–187 [175 fig 1]. Type data: holotype, NMV D18284, from Shipwreck Creek, approx. 20 km SW of Mallacoota, in 37°37′S, 149°39′E, Vic.

Distribution: SE coastal, S.A., Vic., (N.S.W.). Ecology: open heath, low open woodland, open forest, terrestrial, diurnal; viviparous, arthropod-feeder. Biological references: Smales, I. (1981). The herpetofauna of Yellingbo State Faunal Reserve. *Vict. Nat.* **98**: 234–246 (natural history).

Egernia cunninghami (Gray, 1832)

Tiliqua cunninghami Gray, J.E. (1832). (Three new animals brought from New Holland by Mr Cunningham). *Proc. Zool. Soc. Lond.* **1832**: 39–40 [40]. Type data: holotype, BMNH xv.85a (stuffed), from eastern Australia in lat. 29°S.

Egernia krefftii Peters, W. (1871). Mitteilung über neue Eidechsen (*Egernia krefftii, Euprepes (Riopa) punctatolineatus*). *Mber. K. Preuss. Akad. Wiss. Berl.* **1871**: 30–32 [30] [1872 on title page]. Type data: holotype, ZMB 341, from Sydney, N.S.W.

Egernia lohmanni Werner, F. (1917). Über einige neue Reptilien und einen neuen Frosch des zoologischen Museums in Hamburg. *Mitt. Zool. Mus. Hamb.* **34**: 31–36 [32]. Type data: holotype, whereabouts unknown, from unknown locality.

Synonymy that of Cogger, H.G., this work.

Distribution: SE coastal, S Gulfs, Murray-Darling basin, S.A., Vic., N.S.W., Qld., A.C.T. Ecology: low open woodland, woodland, open forest, terrestrial, diurnal, omnivore; viviparous, saxicoline, arthropod-feeder. Biological references: Mitchell, F.J. (1950). The scincid genera *Egernia* and *Tiliqua* (Lacertilia). *Rec. S. Aust. Mus.* **9**: 275–308 (taxonomy); Barwick, R.E. & Bryant, C. (1966). Physiological and biochemical aspects of hibernation in the scincid lizard *Egernia cunninghami* (Gray, 1832). *Physiol. Zool.* **39**: 1–20; Johnson, C.R. (1977). Thermoregulation in four Australian lizards of the genus *Egernia* (Sauria : Scincidae). *J. Linn. Soc. Zool.* **60**: 381–390; Niekisch, M. (1980). Terraristiche Beobachtungen zur Biologie von *Egernia cunninghami* (Reptilia : Sauria : Scincidae). *Salamandra* **16**: 162–176 (natural history).

Egernia depressa (Günther, 1875)

Silubosaurus depressus Günther, A. (1875). Reptiles. A list of saurians of Australia and New Zealand. pp. 9–19 *Zoology of the Voyage of H.M.S.* Erebus *and* Terror. Vol. 2 London : E.W. Janson [15]. Type data: syntypes, BMNH 1946.8.18.41–42, from S interior of W.A. (as Swan River), see Storr, G.M. (1978). The genus *Egernia* (Lacertilia : Scincidae) in Western Australia. *Rec. West. Aust. Mus.* **6**: 147–187.

Distribution: NW coastal, W plateau, S.A., W.A., N.T. Ecology: desert, hummock grassland, low open shrubland, tall shrubland, terrestrial, diurnal; viviparous, (saxicoline), arthropod-feeder. Biological references: Mitchell, F.J. (1950). The scincid genera *Egernia* and *Tiliqua* (Lacertilia). *Rec. S. Aust. Mus.* **9**: 275–308 (taxonomy); Storr, G.M. (1978). The genus *Egernia* (Lacertilia : Scincidae) in Western Australia. *Rec. West. Aust. Mus.* **6**: 147–187 (taxonomy); Day, K. (1980). Notes on the pygmy spiny tailed skink, *Egernia depressa* (Günther) in captivity. *Herpetofauna* **11**: 29; Storr, G.M., Smith, L.A. & Johnstone, R.E. (1981). *Lizards of Western Australia. I. Skinks.* Perth : University of Western Australia Press and Western Australian Museum 200 pp. (taxonomy).

Egernia douglasi Glauert, 1956

Egernia striolata douglasi Glauert, L. (1956). A new skink from West Kimberley. *Egernia striolata douglasi ssp. nov. West. Aust. Nat.* **5**: 117–119 [117]. Type data: syntypes, WAM R11793–4, from Wotjulum Mission, West Kimberley, W.A.

Distribution: N coastal, W.A. Ecology: woodland, terrestrial, diurnal, (arboreal); viviparous, arthropod-feeder. Biological references: Storr, G.M. (1978). The genus *Egernia* (Lacertilia : Scincidae) in Western Australia. *Rec. West. Aust. Mus.* **6**: 147–187 (taxonomy).

Egernia formosa Fry, 1914

Egernia formosa Fry, D.B. (1914). On a collection of reptiles and batrachians from Western Australia. *Rec. West. Aust. Mus. and Art Gall.* **1**: 174–210 [184 pl 27 fig 2 text fig 4]. Type data: holotype, AM R3058, from Perth, ("in error, probably for Boulder, W.A." *vide* Storr, G.M. (1978). The genus *Egernia* (Lacertilia : Scincidae) in Western Australia. *Rec. West. Aust. Mus.* **6**: 147–187).

Distribution: NW coastal, W plateau, W.A. Ecology: (hummock grassland), low open shrubland, tall shrubland, terrestrial, arboreal, diurnal; viviparous, arthropod-feeder. Biological references: Storr, G.M. (1978). The genus *Egernia* (Lacertilia : Scincidae) in Western Australia. *Rec. West. Aust. Mus.* **6**: 147–187 (taxonomy).

Egernia frerei Günther, 1897

Egernia frerei Günther, A. (1897). Description of new species of lizards and of a tree frog from north-eastern Queensland. *Novit. Zool.* **4**: 403–406 [405]. Type data: holotype, whereabouts unknown, from Mt. Bartle Frere, Qld.

Distribution: SE coastal, NE coastal, N Gulf, N coastal, N.S.W., Qld., N.T. Ecology: tall open shrubland, woodland, open forest, tall forest, closed forest, terrestrial, diurnal; viviparous, arthropod-feeder. Biological references: Johnson, C.R. (1977). Thermoregulation in four Australian lizards of the genus *Egernia* (Sauria : Scincidae). *J. Linn. Soc. Zool.* **60**: 381–390 (as *Egernia major*); Mitchell, F.J. (1950). The scincid genera *Egernia* and *Tiliqua* (Lacertilia). *Rec. S. Aust. Mus.* **9**: 275–308 (as *E. major*); Swanson, S. (1979). Some rock-dwelling reptiles of the Arnhem Land escarpment. *N. Territory Nat.* **1**: 14–18 (natural history).

Egernia hosmeri Kinghorn, 1955

Egernia hosmeri Kinghorn, J.R. (1955). Herpetological notes. No.5. *Rec. Aust. Mus.* **23**: 283–286 [286 pl 14]. Type data: holotype, AM R12927, from Kaban, Qld.

Distribution: NE coastal, N Gulf, Lake Eyre basin, Qld., N.T. Ecology: tussock grassland, low open woodland, low woodland, woodland, terrestrial, diurnal; viviparous, saxicoline, arthropod-feeder. Biological references: Zwinenberg, A.J. (1978).

Stekelskinken de "Zindelijke" hagedissen van Australie. *Blijdorp Geluiden* **26**: 2–5 (natural history).

Egernia inornata Rosén, 1905

Egernia inornata Rosén, N. (1905). List of lizards in the Zoological Museum of Lund, with descriptions of new species. *Ann. Mag. Nat. Hist. (7)* **16**: 129–142 [139 pl 7(2) fig 3]. Type data: holotype, ZMLU Cat. Typ. Nr. 279, from W Australia.

Distribution: NW coastal, SW coastal, W plateau, S Gulfs, Murray-Darling basin, Bulloo River basin, Lake Eyre basin, S.A., Vic., N.S.W., Qld., W.A., N.T. Ecology: desert, low open shrubland, low shrubland, low woodland, tall shrubland, terrestrial, burrowing, diurnal; viviparous, arthropod-feeder. Biological references: Storr, G.M. (1968). Revision of the *Egernia whitei* species-group (Lacertilia : Scincidae). *J. R. Soc. West. Aust.* **51**: 51–62 (taxonomy); Mitchell, F.J. (1950). The scincid genera *Egernia* and *Tiliqua* (Lacertilia). *Rec. S. Aust. Mus.* **9**: 275–308 (taxonomy); Storr, G.M. (1978). The genus *Egernia* (Lacertilia : Scincidae) in Western Australia. *Rec. West. Aust. Mus.* **6**: 147–187 (taxonomy); Webber, P. (1979). Burrow density, position and relationship of burrows to vegetation coverage shown by Rosén's desert skink *Egernia inornata* (Lacertilia : Scincidae). *Herpetofauna* **10**: 16–20; Storr, G.M., Smith, L.A. & Johnstone, R.E. (1981). *Lizards of Western Australia. I. Skinks.* Perth : University of Western Australia Press and Western Australian Museum 200 pp. (taxonomy); Pianka, E.R. & Giles, W.F. (1982). Notes on the biology of two species of nocturnal skinks, *Egernia inornata* and *Egernia striata*, in the Great Victoria Desert. *West. Aust. Nat.* **15**: 8–13.

Egernia kingii (Gray, 1838)

Tiliqua kingii Gray, J.E. (1838). Catalogue of the slender-tongued saurians, with descriptions of many new genera and species. (contd.) *Ann. Mag. Nat. Hist.* **2**: 287–293 [290] [1839 on title page]. Type data: holotype, BMNH 1946.8.9.86, from Australia.

Scinc[us] nicitensis Gray, J.E. (1838). Catalogue of the slender-tongued saurians, with descriptions of many new genera and species. (contd.) *Ann. Mag. Nat. Hist.* **2**: 287–293 [290] [*nom. nud.*; Cocteau ms name introduced in synonymy].

***Tropidolopisma dumerilii* (part.)** Duméril, A.M.C. & Bibron, G. (1839). *Erpétologie Générale* ou Histoire Naturelle Complète des Reptiles. Paris : Roret 5: viii 854 pp. [745 pl 50]. Type data: syntypes, MNHP 3001 (Kangaroo Island, presumably in error), MNHP 1242, 3003, MNHP 7126 (2 specimens, Australia), MNHP 3002, MNHP 7127, MNHP 7128 (4 specimens, King George's Sound), from Australia, Kangaroo Is., S.A., King George's Sound, W.A. (all as from "différents points de la Nouvelle-Hollande"), see Guibé, J. (1954). *Catalogue des Types de Lézards du Muséum National d'Histoire Naturelle.* Paris : Muséum National d'Histoire Naturelle 119 pp. [for remaining syntype see *Egernia pulchra* (Werner, 1910)].

Scincus aterrimus Duméril, A.M.C. & Bibron, G. (1839). *Erpétologie Générale* ou Histoire Naturelle Complète des Reptiles. Paris : Roret 5: viii 854 pp. [745] [*nom. nud.*, Péron ms name introduced in synonymy].

Scincus nuittensis Duméril, A.M.C. & Bibron, G. (1839). *Erpétologie Générale* ou Histoire Naturelle Complète des Reptiles. Paris : Roret 5: viii 854 pp. [745] [*nom. nud.*, Péron ms name introduced in synonymy].

Scincus trifasciatus Duméril, A.M.C. & Bibron, G. (1839). *Erpétologie Générale* ou Histoire Naturelle Complète des Reptiles. Paris : Roret 5: viii 854 pp. [745] [*nom. nud.*, Péron ms name introduced in synonymy].

Tropidolepisma nitida Gray, J.E. (1845). *Catalogue of the Specimens of Lizards in the Collection of the British Museum.* London : British Museum xxviii 289 pp. [106]. Type data: holotype, BMNH 43.5.11.2, from West Australia (original description cites Australia).

Synonymy that of Cogger, H.G., this work.

Distribution: NW coastal, SW coastal, W.A. Ecology: open heath, open forest, terrestrial, diurnal; viviparous, arthropod-feeder. Biological references: Mitchell, F.J. (1950). The scincid genera *Egernia* and *Tiliqua* (Lacertilia). *Rec. S. Aust. Mus.* **9**: 275–308 (taxonomy); Storr, G.M. (1978). The genus *Egernia* (Lacertilia : Scincidae) in Western Australia. *Rec. West. Aust. Mus.* **6**: 147–187 (taxonomy); Storr, G.M., Smith, L.A. & Johnstone, R.E. (1981). *Lizards of Western Australia. I. Skinks.* Perth : University of Western Australia Press and Western Australian Museum 200 pp. (taxonomy).

Egernia kintorei Stirling and Zietz, 1893

Egernia kintorei Stirling, E.C. & Zietz, A. (1893). Vertebrata; Reptilia (collected by the Elder Exploring Expedition). *Trans. R. Soc. S. Aust.* **16**: 159–176 [171]. Type data: lectotype, SAMA R2925, from south of Barrow Range, Victoria Desert, W.A., designation by Mitchell, F.J. (1950). The scincid genera *Egernia* and *Tiliqua* (Lacertilia). *Rec. S. Aust. Mus.* **9**: 275–308.

Egernia dahlii Boulenger, G.A. (1896). Description of four new lizards from Roebuck Bay, north Western Australia obtained by Dr. Dahl, for the Christiania Museum. *Ann. Mag. Nat. Hist. (6)* **18**: 232–235 [233]. Type data: syntypes, BMNH 1946.8.21.95, UZMO K2007, from Roebuck Bay, W.A.

Synonymy that of Cogger, H.G., this work.

Distribution: N coastal, W plateau, S.A., W.A., N.T. Ecology: desert, low open shrubland, terrestrial, burrowing, diurnal; viviparous, arthropod-feeder. Biological references: Storr, G.M. (1978). The genus *Egernia* (Lacertilia : Scincidae) in Western Australia. *Rec. West. Aust. Mus.* **6**: 147–187 (taxonomy); Storr, G.M., Smith, L.A. & Johnstone, R.E. (1981). *Lizards of Western*

Australia. I. Skinks. Perth : University of Western Australia Press and Western Australian Museum 200 pp. (taxonomy).

Egernia luctuosa (Peters, 1866)

Cyclodus (Omolepida) luctuosus Peters, W. (1866). Eine Mitteilung über neue Amphibien (*Amphibolurus, Lygosoma, Cyclodus, Masticophis, Crotaphopeltis*) und Fische (*Diagramma*: *Hapalogenys*) des Kgl. zoologischen Museums. *Mber. K. Preuss. Akad. Wiss. Berl.* **1866**: 86–96 [90] [1867 on title page]. Type data: holotype, ZMB 5380, from King George's Sound, W.A.

Egernia lauta De Vis, C.W. (1888). A contribution to the herpetology of Queensland. *Proc. Linn. Soc. N.S.W. (2)* **2**: 811–826 [813]. Type data: holotype, QM J249, from Qld. (in error).

Synonymy that of Cogger, H.G., this work.

Distribution: SW coastal, W.A. Ecology: open forest, terrestrial, diurnal; viviparous, arthropod-feeder. Biological references: Storr, G.M. (1978). The genus *Egernia* (Lacertilia : Scincidae) in Western Australia. *Rec. West. Aust. Mus.* **6**: 147–187 (taxonomy); Storr, G.M., Smith, L.A. & Johnstone, R.E. (1981). *Lizards of Western Australia. I. Skinks.* Perth : University of Western Australia Press and Western Australian Museum 200 pp. (taxonomy).

Egernia major (Gray, 1845)

Tropidolepisma major Gray, J.E. (1845). *Catalogue of the Specimens of Lizards in the Collection of the British Museum.* London : British Museum xxviii 289 pp. [107]. Type data: holotype, BMNH xv.88a, from Australia, but restricted to E Australia (Sydney and Rockhampton) by Günther, A. (1875). Reptiles. A list of saurians of Australia and New Zealand. pp. 9–19 *Zoology of the Voyage of H.M.S.* Erebus *and* Terror. Vol. 2 London : E.W. Janson.

Tropidolepisma majus Günther, A. (1875). Reptiles. A list of saurians of Australia and New Zealand. pp. 9–19 *Zoology of the Voyage of H.M.S.* Erebus *and* Terror. Vol. 2 London : E.W. Janson [15 pl 14] [*pro Tropidolepisma major* Gray, 1845].

Egernia bungana De Vis, C.W. (1888). A contribution to the herpetology of Queensland. *Proc. Linn. Soc. N.S.W. (2)* **2**: 811–826 [814]. Type data: holotype, from S Queensland, QM (not found), see Covacevich, J. (1971). Amphibian and reptile type specimens in the Queensland Museum. *Mem. Qd. Mus.* **16**: 49–68.

Synonymy that of Cogger, H.G., this work.

Distribution: SE coastal, NE coastal, N.S.W., Qld. Ecology: tall forest, closed forest, terrestrial, diurnal; viviparous, arthropod-feeder. Biological references: Johnson, C.R. (1977). Thermoregulation in four Australian lizards of the genus *Egernia* (Sauria : Scincidae). *J. Linn. Soc. Zool.* **60**: 381–390 (as *Egernia bungana*); Mitchell, F.J. (1950). The scincid genera *Egernia* and *Tiliqua* (Lacertilia). *Rec. S. Aust. Mus.* **9**: 275–308 (taxonomy); Longley, G. (1946). The giant skink (*Egernia major*) in the vivarium. *Proc. R. Zool. Soc. N.S.W.* **1945–46**: 33–34 (natural history).

Egernia margaretae Storr, 1968

Egernia margaretae margaretae Storr, G.M. (1968). Revision of the *Egernia whitei* species-group (Lacertilia : Scincidae). *J. R. Soc. West. Aust.* **51**: 51–62 [53]. Type data: holotype, WAM R29129, from Palm Valley, in 24°03′S, 132°42′E, N.T.

Egernia margaretae personata Storr, G.M. (1968). Revision of the *Egernia whitei* species-group (Lacertilia : Scincidae). *J. R. Soc. West. Aust.* **51**: 51–62 [53]. Type data: holotype, SAMA R3748, from Wilpena Gorge, S.A.

Synonymy that of Cogger, H.G., this work.

Distribution: W plateau, S Gulfs, Lake Eyre basin, S.A., W.A., N.T. Ecology: desert, low open shrubland, low woodland, tall shrubland, terrestrial, diurnal; viviparous, saxicoline, arthropod-feeder. Biological references: Storr, G.M. (1978). The genus *Egernia* (Lacertilia : Scincidae) in Western Australia. *Rec. West. Aust. Mus.* **6**: 147–187 (taxonomy); Storr, G.M., Smith, L.A. & Johnstone, R.E. (1981). *Lizards of Western Australia. I. Skinks.* Perth : University of Western Australia Press and Western Australian Museum 200 pp. (taxonomy).

Egernia modesta Storr, 1968

Egernia geophana Anon [Horton, D.] (1968). New lizard found in New England. *The Armidale Express* **7 Feb. 1968**: 5 [5 fig] [?*nom. nud.*; the status of the above description is equivocal. It would appear to fulfil the criteria of availability under Articles 11–14 of the International Code of Zoological Nomenclature, but the question of anonymity (Article 15) will probably require resolution by the International Commission on Zoological Nomenclature]. Type data: holotype (not specified in original description but designated by the author when presenting the type series to the Australian Museum), AM R26623, from Yarrowyck, 20 mi W Armidale, N.S.W. (not specified in original description).

Egernia whitei modesta Storr, G.M. (1968). Revision of the *Egernia whitei.* species-group (Lacertilia : Scincidae). *J. R. Soc. West. Aust.* **51**: 51–62 [55]. Type data: holotype, QM J464, from Chinchilla, Qld.

Distribution: SE coastal, NE coastal, Murray-Darling basin, N.S.W., Qld. Ecology: low open woodland, woodland, (tall shrubland), terrestrial, diurnal; viviparous, (saxicoline), arthropod-feeder.

Egernia multiscutata Mitchell and Behrndt, 1949

Egernia whitii multiscutata Mitchell, F.J. & Behrndt, A.C. (1949.). Fauna and flora of the Greenly Islands. Part I. Introductory narrative and vertebrate fauna. (Reptiles. 175–178). *Rec. S. Aust. Mus.* **9**: 167–179 [176 fig 2]. Type data: holotype, SAMA R2636, from Greenly Is., S.A.

Egernia bos Storr, G.M. (1960). *Egernia bos*, a new skink from the south coast of Western Australia. *West. Aust. Nat.* **7**: 99–103 fig 2 [99]. Type data: holotype, WAM R10751, from Cheyne Beach, 32 mi E of Albany, W.A.

Synonymy that of Cogger, H.G., this work.

Distribution: NW coastal, SW coastal, W plateau, S Gulfs, Murray-Darling basin, S.A., Vic., W.A. Ecology: open heath, open scrub, woodland, open forest, terrestrial, diurnal; arthropod-feeder. Biological references: Storr, G.M. (1968). Revision of the *Egernia whitei* species-group (Lacertilia : Scincidae). *J. R. Soc. West. Aust.* **51**: 51–62 (taxonomy); Storr, G.M. (1978). The genus *Egernia* (Lacertilia : Scincidae) in Western Australia. *Rec. West. Aust. Mus.* **6**: 147–187 (taxonomy); Storr, G.M., Smith, L.A. & Johnstone, R.E. (1981). *Lizards of Western Australia. I. Skinks.* Perth : University of Western Australia Press and Western Australian Museum 200 pp. (taxonomy).

Egernia napoleonis (Gray, 1838)

Tiliqua napoleonis Gray, J.E. (1838). Catalogue of the slender-tongued saurians, with descriptions of many new genera and species. (contd.) *Ann. Mag. Nat. Hist.* **2**: 287–293 [290] [*ex* Cuvier ms; 1839 on title page]. Type data: holotype, MNHP 1242 *vide* Roux-Estève, R. (1979). Liste des Amphibiens et Reptiles des collections du Muséum National d'Histoire Naturelle de Paris, récoltés par Lesueur (1778–1846). *Bull. Trim. Soc. Géol. Normandie et Amis Muséum du Havre* **66**: 25–29, from S coast of Australia (as Terre Napoléon; original description cites Australia).

Distribution: SW coastal, W plateau, W.A. Ecology: open heath, open forest, terrestrial, diurnal; viviparous, arthropod-feeder. Biological references: Storr, G.M. (1978). The genus *Egernia* (Lacertilia : Scincidae) in Western Australia. *Rec. West. Aust. Mus.* **6**: 147–187 (taxonomy); Storr, G.M., Smith, L.A. & Johnstone, R.E. (1981). *Lizards of Western Australia. I. Skinks.* Perth : University of Western Australia Press and Western Australian Museum 200 pp. (taxonomy).

Egernia pilbarensis Storr, 1978

Egernia pilbarensis Storr, G.M. (1978). The genus *Egernia* (Lacertilia : Scincidae) in Western Australia. *Rec. West. Aust. Mus.* **6**: 147–187 [177 fig 2]. Type data: holotype, WAM R31117, from 6 km NW of Pyramid, in 21°01'S, 117°24'E, W.A.

Distribution: NW coastal, W.A. Ecology: (hummock grassland), low open shrubland, terrestrial, diurnal; saxicoline, arthropod-feeder. Biological references: Storr, G.M., Smith, L.A. & Johnstone, R.E. (1981). *Lizards of Western Australia. I. Skinks.* Perth : University of Western Australia Press and Western Australian Museum 200 pp. (taxonomy).

Egernia pulchra Werner, 1910

***Tropidolopisma dumerilii* (part.)** Duméril, A.M.C. & Bibron, G. (1839). *Erpétologie Générale* ou Histoire Naturelle Complète des Reptiles. Paris : Roret 5: viii 854 pp. [745]. Type data: syntype, MNHP 3005, from Australia [for data on the other syntypes see under *Egernia kingii* (Gray, 1838)].

Egernia pulchra Werner, F. (1910). Reptilia (Geckonidae und Scincidae). *in* Michaelsen, W. & Hartmeyer, R. (eds.) *Die Fauna Südwest-Australiens.* Jena : Gustav Fischer 2: 451–493 [470 fig 6]. Type data: holotype, ZMB 21458, from Torbay, W.A.

Egernia pulchra longicauda Ford, J. (1963). The distribution and variation of the skinks *Egernia pulchra* and *E. bos* in Western Australia. *West. Aust. Nat.* **9**: 25–29 [26]. Type data: holotype, WAM R16772, from Favourite Is., Jurien Bay, W.A.

Synonymy that of Cogger, H.G., this work.

Distribution: SW coastal, W.A. Ecology: open heath, (woodland), open forest, terrestrial, diurnal; viviparous, arthropod-feeder. Biological references: Storr, G.M. (1978). The genus *Egernia* (Lacertilia : Scincidae) in Western Australia. *Rec. West. Aust. Mus.* **6**: 147–187 (taxonomy); Storr, G.M., Smith, L.A. & Johnstone, R.E. (1981). *Lizards of Western Australia. I. Skinks.* Perth : University of Western Australia Press and Western Australian Museum 200 pp. (taxonomy).

Egernia richardi (Peters, 1869)

Tropidolepisma richardi Peters, W. (1869). Mitteilung über Saurier. *Mber. K. Preuss. Akad. Wiss. Berl.* **1869**: 786–790 [787] [provenance of only known specimen (the holotype) is doubtful, as it is identical in most respects to *Egernia carinata* Smith, 1939 from southern W.A. and over which its name would have precedence if the two are shown to be conspecific]. Type data: holotype, ZMB 6613, from Alligator River, N Australia.

Egernia rugosa De Vis, 1888

Egernia rugosa De Vis, C.W. (1888). A contribution to the herpetology of Queensland. *Proc. Linn. Soc. N.S.W. (2)* **2**: 811–826 [815]. Type data: holotype, from Herbert Gorge, N Qld., QM (not found), see Covacevich, J. (1971). Amphibian and reptile type specimens in the Queensland Museum. *Mem. Qd. Mus.* **16**: 49–68.

Distribution: NE coastal, N Gulf, Murray-Darling basin, Qld. Ecology: open scrub, woodland, terrestrial, diurnal; viviparous, burrower, arthropod-feeder. Biological references: Mitchell, F.J. (1950). The scincid genera *Egernia* and *Tiliqua* (Lacertilia). *Rec. S. Aust. Mus.* **9**: 275–308 (as *Egernia dorsalis*).

Egernia saxatilis Cogger, 1960

Egernia saxatilis saxatilis Cogger, H.G. (1960). The ecology, morphology, distribution and speciation of a new species and subspecies of the genus *Egernia* (Lacertilia : Scincidae). *Rec. Aust. Mus.* **25**: 95–105 [96 pl 2]. Type data: holotype, AM R15282, from Warrumbungle Mts., N.S.W.

Egernia saxatilis intermedia Cogger, H.G. (1960). The ecology, morphology, distribution and speciation of a new species and subspecies of the genus *Egernia* (Lacertilia : Scincidae). *Rec. Aust. Mus.* **25**: 95–105 [96 pl 2]. Type data: holotype, AM R15273, from Kanangra Walls (3,400 ft), N.S.W.

Synonymy that of Cogger, H.G., this work.

Distribution: SE coastal, NE coastal, Murray-Darling basin, Vic., N.S.W., Qld., A.C.T. Ecology: low open woodland, woodland, open forest, terrestrial, diurnal; viviparous, saxicoline, arthropod-feeder. Biological references: Wells, R. (1972). On the occurrence of the skink *Egernia saxatilis intermedia* Cogger in the Blue Mountains west of Sydney, N.S.W. with additional notes on its behaviour in captivity. *Herpetofauna* **5**: 2–4; Greer, A.E. (1980). Critical thermal maximum temperatures in Australian scincid lizards: their ecological and evolutionary significance. *Aust. J. Zool.* **28**: 91–102.

Egernia slateri Storr, 1968

Egernia slateri slateri Storr, G.M. (1968). Revision of the *Egernia whitei* species-group (Lacertilia : Scincidae). *J. R. Soc. West. Aust.* **51**: 51–62 [59]. Type data: holotype, WAM R29130, from Alice Springs, in 23°46′S, 133°53′E, N.T.

Egernia slateri virgata Storr, G.M. (1968). Revision of the *Egernia whitei* species-group (Lacertilia : Scincidae). *J. R. Soc. West. Aust.* **51**: 51–62 [60]. Type data: holotype, SAMA R602, from between Oodnadatta and the Everard Range, S.A.

Synonymy that of Cogger, H.G., this work.

Distribution: W plateau, Lake Eyre basin, S.A., Qld., W.A., N.T. Ecology: desert, tall shrubland, terrestrial, burrowing, diurnal; viviparous, arthropod-feeder.

Egernia stokesii (Gray, 1845)

Silubosaurus stokesii Gray, J.E. (1845). *Catalogue of the Specimens of Lizards in the Collection of the British Museum.* London : British Museum xxviii 289 pp. [105]. Type data: syntypes, BMNH 1946.8.9.74–76 (Houtman Abrolhos), BMNH 1946.8.9.77–82 (W.A.), BMNH 44.12.16.3–4 (W.A.), from W.A. and Houtman Abrolhos, W.A.

Silubosaurus zellingi De Vis, C.W. (1884). On a new Australian lizard. *Proc. R. Soc. Qd.* **1**: 53–56 [53]. Type data: holotype, QM J253, from Barcoo, Qld.

Egernia stokesii aethiops Storr, G.M. (1978). The genus *Egernia* (Lacertilia : Scincidae) in Western Australia. *Rec. West. Aust. Mus.* **6**: 147–187 [156]. Type data: holotype, WAM R25731, from Baudin Is., Freycinet Estuary, in 26°31′S, 113°39′E, W.A.

Egernia stokesii badia Storr, G.M. (1978). The genus *Egernia* (Lacertilia : Scincidae) in Western Australia. *Rec. West. Aust. Mus.* **6**: 147–187 [158]. Type data: holotype, WAM R29590, from Rothsay, in 29°18′S, 116°54′E, W.A.

Synonymy that of Cogger, H.G., this work.

Distribution: NW coastal, SW coastal, W plateau, S Gulfs, Bulloo River basin, Lake Eyre basin, S.A., N.S.W., Qld., W.A., N.T. Ecology: low open shrubland, low shrubland, woodland, tall shrubland, terrestrial, diurnal; viviparous, saxicoline, arthropod-feeder. Biological references: Storr, G.M. (1978). The genus *Egernia* (Lacertilia : Scincidae) in Western Australia. *Rec. West. Aust. Mus.* **6**: 147–187 (taxonomy); Pianka, E.R. (1972). Zoogeography and speciation of Australian desert lizards: an ecological perspective. *Copeia* **1972**: 127–144 (ecology); Zwinenberg, A.J. (1978). Stekelskinken de "Zindelijke" hagedissen van Australie. *Blijdorp Geluiden* **26**: 2–5 (natural history); Storr, G.M., Smith, L.A. & Johnstone, R.E. (1981). *Lizards of Western Australia. I. Skinks.* Perth : University of Western Australia Press and Western Australian Museum 200 pp. (taxonomy).

Egernia striata Sternfeld, 1919

Egernia striata Sternfeld, R. (1919). Neue Schlangen und Echsen aus Zentralaustralien. *Senckenbergiana* **1**: 76–83 [79]. Type data: lectotype, SMF 13998, from Hermannsburg Mission, upper Finke River, N.T., designation by Mertens, R. (1967). Die herpetologische Sektion des Natur-Museums und Forschungs-Institutes Senckenberg in Frankfurt-a.-M. nebst einem Verzeichnis ihrer Typen. 1. *Senckenberg. Biol.* **48**: 1–106.

Distribution: W plateau, Lake Eyre basin, (NW coastal), S.A., W.A., N.T. Ecology: desert, low open shrubland, tall shrubland, terrestrial, burrowing, noctidiurnal; viviparous, arthropod-feeder. Biological references: Storr, G.M. (1968). Revision of the *Egernia whitei* species-group (Lacertilia : Scincidae). *J. R. Soc. West. Aust.* **51**: 51–62 (taxonomy); Storr, G.M. (1978). The genus *Egernia* (Lacertilia : Scincidae) in Western Australia. *Rec. West. Aust. Mus.* **6**: 147–187 (taxonomy); Pianka, E.R. (1972). Zoogeography and speciation of Australian desert lizards: an ecological perspective. *Copeia* **1972**: 127–144 (ecology); Storr, G.M., Smith, L.A. & Johnstone, R.E. (1981). *Lizards of Western Australia. I. Skinks.* Perth : University of Western Australia Press and Western Australian Museum 200 pp. (taxonomy); Pianka, E.R. & Giles, W.F. (1982). Notes on the biology of two species of nocturnal skinks, *Egernia inornata* and *Egernia striata*, in the Great Victoria Desert. *West. Aust. Nat.* **15**: 8–13.

Egernia striolata (Peters, 1870)

Tropidolepisma striolatum Peters, W. (1870). Eine Mitteilung über neue Amphibien (*Hemidactylus, Urosaura, Tropidolepisma* ...) des Königlich Zoologischen Museum. *Mber. K. Preuss. Akad. Wiss. Berl.* **1870**: 641–652 [642] [1871 on title page]. Type data: holotype, ZMB 6925, from Lake Elphinstone, Qld.

Tropidolepisma dorsale Peters, W. (1873). Mitteilung über neue Saurier (*Spoeriodactylus, Anolis, Phrynosoma, Tropidolepisma, Lygosoma, Ophioscincus*) aus Centralamerica, Mexico und Australien. *Mber. K. Preuss. Akad. Wiss. Berl.* **1873**: 738–747 [743] [1874 on title page]. Type data: holotype, ZMB 7512, from Port Clinton (as Port Bowen), Qld.

Synonymy that of Cogger, H.G., this work; Rawlinson, P. *in litt.*

Distribution: SE coastal, NE coastal, N Gulf, S Gulfs, Murray-Darling basin, Bulloo River basin, Lake Eyre basin, S.A., Vic., N.S.W., Qld., N.T., (A.C.T.). Ecology: low shrubland, open scrub, low open woodland, low woodland, woodland, open forest, tall shrubland, arboreal, diurnal; viviparous, arthropod-feeder. Biological references: Cogger, H.G. (1960). The ecology, morphology, distribution and speciation of a new species and subspecies of the genus *Egernia* (Lacertilia : Scincidae). *Rec. Aust. Mus.* **25**: 95–105 (taxonomy); Bustard, H.R. (1970). A population study of the scincid lizard *Egernia striolata* in northern New South Wales. *Proc. K. Ned. Akad. Wet.* **73C**: 186–213; Matz, G. (1979). Les scinques australians. 3. Observations sur la reproduction de *Egernia* Gray. *Aquarama* **13**: 43–44.

Egernia whitii (Lacépède, 1804)

Scineus* [*sic*] *whitii Lacépède, B.G.E. (1804). Mémoire sur plusieurs animaux de la Nouvelle-Hollande dont la description n'a pas encore été publiée. *Ann. Mus. Natl. Hist. Nat. Paris* **4**: 184–211 [209]. Type data: syntypes, MNHP 2988 (two specimens), from Kangaroo Is., S.A. (as Australia).

Scincus compressicauda Quoy, J.R.C. & Gaimard, J.P. (1824). Zoologie *in* Freycinet, L.C.D. *Voyage Autour du Monde Exécute sur l'*Uranie *et la* Physicienne *pendant les années 1817–1820.* Paris iv 712 pp. [180 pl 42 fig 2]. Type data: none designated, MNHP (not found, *vide* Guibé, J. (1954). *Catalogue des Types de Lézards du Muséum National d'Histoire Naturelle.* Paris : Muséum National d'Histoire Naturelle 119 pp.), from Port Jackson, N.S.W.

Tiliqua leucopsis Gray, J.E. (1838). Catalogue of the slender-tongued saurians, with descriptions of many new genera and species. (contd.) *Ann. Mag. Nat. Hist.* **2**: 287–293 [291] [1839 on title page]. Type data: holotype (probable), BMNH xv.17a, from Kangaroo Is., S.A. (original description cites Australia).

Lygosoma moniligera Duméril, A.M.C. & Bibron, G. (1839). *Erpétologie Générale* ou Histoire Naturelle Complète des Reptiles. Paris : Roret 5: viii 854 pp. [736]. Type data: syntypes, MNHP 2988 (2 specimens, Kangaroo Island), MNHP 2989 (Australia), MNHP 2992 (2 specimens, Australia), ?MNHP 2990 (Australia), ?MNHP 7123 (2 specimens, Hobart, Tas.), ?MCZ 2133 (Australia), from Australia and Kangaroo Is., S.A. (original description cites only Australia). Guibé, J. (1954). *Catalogue des Types de Lézards du Muséum National d'Histoire Naturelle.* Paris : Muséum National d'Histoire Naturelle 119 pp. cites only the first five specimens as types.

Scincus ocellatus Duméril, A.M.C. & Bibron, G. (1839). *Erpétologie Générale* ou Histoire Naturelle Complète des Reptiles. Paris : Roret 5: viii 854 pp. [73] [*nom. nud.*, name introduced in synonymy; *ex* Péron ms].

Scincus leuwinensis Duméril, A.M.C. & Bibron, G. (1839). *Erpétologie Générale* ou Histoire Naturelle Complète des Reptiles. Paris : Roret 5: viii 854 pp. [736] [*nom. nud.*, name introduced in synonymy; *ex* Péron ms].

Egernia whitei tenebrosa Condon, H.T. (1941). Further records of lizards and frogs from Kangaroo Island. *Rec. S. Aust. Mus.* **7**: 111–116 [111 fig 1]. Type data: holotype, SAMA R2161, from Flinders Chase, Kangaroo Is., S.A.

Synonymy that of Cogger, H.G., this work.

Distribution: SE coastal, NE coastal, S Gulfs, Murray-Darling basin, S.A., Tas., Vic., N.S.W., Qld., A.C.T. Ecology: low open woodland, woodland, open forest, tall forest, terrestrial, diurnal; viviparous, (saxicoline), arthropod-feeder. Biological references: Storr, G.M. (1968). Revision of the *Egernia whitei* species-group (Lacertilia : Scincidae). *J. R. Soc. West. Aust.* **51**: 51–62 (taxonomy); Johnson, C.R. (1977). Thermoregulation in four Australian lizards of the genus *Egernia* (Sauria : Scincidae). *J. Linn. Soc. Zool.* **60**: 381–390; Mitchell, F.J. (1950). The scincid genera *Egernia* and *Tiliqua* (Lacertilia). *Rec. S. Aust. Mus.* **9**: 275–308 (taxonomy); Hickman, J.L. (1960). Observations on the skink lizard *Egernia whitii* (Lacépède). *Pap. Proc. R. Soc. Tasm.* **94**: 111–118 (natural history); Storr, G.M. (1978). The genus *Egernia* (Lacertilia : Scincidae) in Western Australia. *Rec. West. Aust. Mus.* **6**: 147–187 (taxonomy).

Emoia Gray, 1845

Eusoma Fitzinger, L.J. (1843). *Systema Reptilium.* Vienna : Braümüller u. Seidel vi 106 pp. [22] [*non Eusoma* Germar, 1817]. Type species *Eumeces lessonii* Duméril and Bibron, 1839 by original designation.

Emoia Gray, J.E. (1845). *Catalogue of the Specimens of Lizards in the Collection of the British Museum.* London : British Museum xxviii 289 pp. [95]. Type species *Scincus atrocostatus* Lesson, 1826 by subsequent designation, see Mittleman, M.B. (1952). A generic synopsis of the lizards of the subfamily Lygosominae. *Smithson. Misc. Collect.* **117**(4069): 1–35.

Emoa Girard, C. (1857). Descriptions of some new reptiles, collected by the United States Exploring Expedition, under the command of Capt. Charles Wilkes, U.S.N. Fourth Part. Including the species of saurians exotic to North America. *Proc. Acad. Nat. Sci. Philad.* **1857**: 195–199 [197] [emend. *pro Emoia* Gray, 1845; 1858 on title page].

Synonymy that of Cogger, H.G., this work.

This group is also found in SE Asia through Indo-Australian Archipelago to New Guinea and Pacific islands, see Greer, A.E. (1974). The generic relationships of the scincid lizard genus *Leiolopisma* and its relatives. *Aust. J. Zool. Suppl. Ser.* **31**: 1–67. Australian records of the widely distributed species *Emoia cyanogaster* are now referable to *Emoia longicauda*.

Emoia atrocostata (Lesson, 1830)

Scincus atrocostatus Lesson, R.P. (1830). Description de quelques reptiles nouveaux ou peu connus. *in* Duperrey, L.I. (ed.) (1826–1838). *Voyage autour du monde, executé par ordre du Roi, sur la corvette de Sa Majesté,* La Coquille, *pendant les années 1822, 1823, 1824 et 1825.* Zoologie 2(1): 34–65 and Atlas (Reptiles plates 1–7) Paris : Arthus Bertrand [livraison 15 pl 4 fig 3 (1830); 50 (1831)] [Publication dates established from Sherborn, C.D. & Woodward, B.B. (1901). Notes on the dates of publication of the natural history portions of some French voyages. Part 1 *Ann. Mag. Nat. Hist. (7)* **7**: 388–392 (date of livraison); Anon. (1830). [Review of] Voyage autour du monde. Livraisons 14–17. *Bull. Sci. Nat. Géol.* (Férussac, ed.) **21**: 309–311 (identifies livraison in which plate appears); Sherborn, C.D. & Woodward, B.B. (1906). Notes on the dates of publication of the natural history portions of some French voyages. "Voyage autour du Monde .... sur .... la Coquille pendant .... 1822–25 .... Par L.J. Duperrey," etc. A correction. *Ann. Mag. Nat. Hist. (7)* **17**: 335–336 (date of text)]. Type data: none designated, whereabouts unknown, from Kusiae (as Oualan) Is., Caroline Ils.

Eumeces freycinetii Duméril, A.M.C. & Bibron, G. (1839). *Erpétologie Générale* ou Histoire Naturelle Complète des Reptiles. Paris : Roret 5: viii 854 pp. [648]. Type data: syntypes, MNHP 2905 (New Guinea), MNHP 2906–7 (Vanikoro), from New Guinea and Vanikoro Is., Solomon Ils. (original description cites only Vanicoro Is.).

Mocoa cumingii Gray, J.E. (1845). *Catalogue of the Specimens of Lizards in the Collection of the British Museum.* London : British Museum xxviii 289 pp. [81]. Type data: holotype, BMNH 1946.8.6.86, from Philippine Ils.

Euprepes bitaeniatus Peters, W. (1864). Die Eidechsenfamilie der Scincoiden, insbesondere über die Schneider'schen, Wiegmann'schen und neue Arten des zoologischen Museums. *Mber. K. Preuss. Akad. Wiss. Berl.* **1864**: 44–58 [53] [1865 on title page]. Type data: syntypes, ZMB, from Paracali (Luzon Is.) and Catbalong (Samar Is.), Philippine Ils.

Euprepes (Mabuia) microstictus Peters, W. (1874). Über einige neue Reptilien. *Mber. K. Preuss. Akad. Wiss. Berl.* **1874**: 368–377 [373] [1875 on title page]. Type data: holotype (probable), ZMB, from Republic of Belau (as Pellew Islands).

Mabouia marmorata Macleay, W. (1877). The lizards of the *Chevert* Expedition. *Proc. Linn. Soc. N.S.W.* **2**: 60–69 [65] [1878 on title page]. Type data: holotype, AM R31852, from Long Is., Torres Strait, Qld.

Mabouia irrorata Macleay, W. (1877). The lizards of the *Chevert* Expedition. *Proc. Linn. Soc. N.S.W.* **2**: 60–69 [66] [1878 on title page]. Type data: holotype, AM R31851, from Hall Sound, Papua New Guinea.

Lygosoma sinus Smith, M.A. (1929). Descriptions of a new skink from Christmas Island and a new frog from Annam. *Ann. Mag. Nat. Hist. (10)* **3**: 294–297 [294 fig 2]. Type data: holotype, BMNH 1946.8.10.94, from Christmas Is., Indian Ocean.

Synonymy that of Cogger, H.G., this work.

Distribution: NE coastal, (N Gulf), Qld., Christmas Is. Ecology: mangrove, littoral, supralittoral, terrestrial, (arboreal), diurnal; oviparous, arthropod-feeder. Biological references: Ingram, G.J. (1979). The occurrence of lizards of the genus *Emoia* (Lacertilia : Scincidae) in Australia. *Mem. Qd. Mus.* **19**: 431–437; Alcala, A.C. & Brown, W.A. (1967). Population ecology of the tropical scincoid lizard, *Emoia atrocostata*, in the Philippines. *Copeia* **1967**: 596–604.

Emoia longicauda (Macleay, 1877)

Euprepis longicaudis Macleay, W. (1877). The lizards of the *Chevert* Expedition. *Proc. Linn. Soc. N.S.W.* **2**: 60–69 [68] [1878 on title page]. Type data: lectotype, AM R31859, from Darnley Is., Torres Strait, Qld., designation by Ingram, G.J. (1979). The occurrence of lizards of the genus *Emoia* (Lacertilia : Scincidae) in Australia. *Mem. Qd. Mus.* **19**: 431–437.

Euprepis simillimus Macleay, W. (1877). The lizards of the *Chevert* Expedition. *Proc. Linn. Soc. N.S.W.* **2**: 60–69 [69] [1878 on title page]. Type data: lectotype, AM R31855, from Mawatta, Binaturi River (as Katow), Papua New Guinea. designation by Ingram, G.J. (1979). The occurrence of lizards of the genus *Emoia* (Lacertilia : Scincidae) in Australia. *Mem. Qd. Mus.* **19**: 431–437.

Emoa cuneiceps De Vis, C.W. (1890). Reptiles from New Guinea. *Proc. Linn. Soc. N.S.W. (2)* **5**: 497–500 [498]. Type data: syntypes, QM (not found), from St. Joseph River, Papua New Guinea.

Synonymy that of Cogger, H.G., this work.

Distribution: NE coastal, (N Gulf), Qld. Ecology: (woodland), closed forest, arboreal, diurnal; oviparous, arthropod-feeder. Biological references: Ingram, G.J. (1979). The occurrence of lizards of the genus *Emoia* (Lacertilia : Scincidae) in Australia. *Mem. Qd. Mus.* **19**: 431–437; Brown, W.C. (1954). Notes on several lizards of the genus

Emoia, with descriptions of new species from the Solomon Islands. *Fieldiana Zool.* **34**: 263–276 (taxonomy).

Emoia nativitatis (Boulenger, 1887)

Lygosoma nativitatis Boulenger, G.A. (1887). Reptiles. *in* Günther, A. (ed.) Report on a zoological collection made by the officers of H.M.S. *Flying Fish* at Christmas Island, Indian Ocean. *Proc. Zool. Soc. Lond.* **1887**: 516–517 [516]. Type data: holotype, BMNH 1946.8.10.59, from Christmas Is., Indian Ocean.

Distribution: Christmas Is. Ecology: closed forest, terrestrial, diurnal; oviparous, arthropod-feeder.

Eremiascincus Greer, 1979

Eremiascincus Greer, A.E. (1979). *Eremiascincus*, a new generic name for some Australian sand swimming skinks (Lacertilia : Scincidae). *Rec. Aust. Mus.* **32**: 321–338 [323]. Type species *Hinulia richardsonii* Gray, 1845 by original designation.

Eremiascincus fasciolatus (Günther, 1867)

Hinulia fasciolata Günther, A. (1867). Additions to the knowledge of Australian reptiles and fishes. *Ann. Mag. Nat. Hist. (3)* **20**: 45–68 [47]. Type data: syntypes, BMNH 1946.8.3.49 (Port Curtis), BMNH 1946.8.13.91 (Rockhampton), from Port Curtis and Rockhampton, Qld.

Euprepes striatulus Steindachner, F. (1867). Reptilien. *in, Reise der österreichischen Fregatte* Novara *um die Erde in den Jahren 1857, 1858, 1859.* Zoologie 1(3): 1–98 Vienna : State Printer [49] [1869 on title page]. Type data: holotype, NHMW 9720, from N.S.W.

Hinulia pallida Günther, A. (1875). Reptiles. A list of saurians of Australia and New Zealand. pp. 9–19 *Zoology of the Voyage of H.M.S.* Erebus *and* Terror. Vol. 2 London : E.W. Janson [12]. Type data: holotype, BMNH 1946.8.3.38, from Nickol Bay, W.A.

Lygosma* [*sic*] *(Hinulia) fasciolatum intermedium Sternfeld, R. (1919). Neue Schlangen und Echsen aus Zentralaustralien. *Senckenbergiana* **1**: 76–83 [81]. Type data: holotype, SMF 14446, from Hermannsburg Mission, upper Finke River, N.T.

Synonymy that of Cogger, H.G., this work.

Distribution: NE coastal, N Gulf, N coastal, NW coastal, W plateau, S Gulfs, Murray-Darling basin, Bulloo River basin, Lake Eyre basin, S.A., Vic., N.S.W., Qld., W.A., N.T. Ecology: desert, hummock grassland, tussock grassland, low open shrubland, low shrubland, open scrub, low open woodland, low woodland, woodland, tall shrubland, terrestrial, fossorial, nocturnal; sand swimming, oviparous (probable), arthropod-feeder. Biological references: Storr, G.M. (1974). Revision of the *Sphenomorphus richardsonii* species-group (Lacertilia : Scincidae). *Rec. West. Aust. Mus.* **3**: 66–70 (taxonomy); Greer, A.E. (1979). *Eremiascincus*, a new generic name for some Australian sand swimming skinks (Lacertilia : Scincidae). *Rec. Aust. Mus.* **32**: 321–338 (taxonomy).

Eremiascincus richardsonii (Gray, 1845)

Hinulia richardsonii Gray, J.E. (1845). *Catalogue of the Specimens of Lizards in the Collection of the British Museum.* London : British Museum xxviii 289 pp. [271]. Type data: holotype, BMNH 1946.8.3.14, from Houtman's Abrolhos, W.A.

Lygosoma monotropis Boulenger, G.A. (1887). *Catalogue of the Lizards in the British Museum (Natural History).* 3. London : British Museum xii 575 pp. pls 40 [237 pl 144 fig 2]. Type data: syntypes, BMNH 1946.8.26.12–13, BMNH 1946.8.5.45, from Champion Bay, W.A. and W.A.

Hinulia ambigua De Vis, C.W. (1888). A contribution to the herpetology of Queensland. *Proc. Linn. Soc. N.S.W. (2)* **2**: 811–826 [817]. Type data: holotype, QM J242, from Charleville, Qld.

Synonymy that of Cogger, H.G., this work.

Distribution: N Gulf, N coastal, NW coastal, SW coastal, W plateau, S Gulfs, Murray-Darling basin, Bulloo River basin, Lake Eyre basin, S.A., N.S.W., Qld., W.A., N.T. Ecology: desert, hummock grassland, tussock grassland, low open shrubland, low shrubland, open scrub, low open woodland, low woodland, woodland, tall shrubland, terrestrial, fossorial, nocturnal; sand swimming, oviparous (probable), arthropod-feeder. Biological references: Storr, G.M. (1974). Revision of the *Sphenomorphus richardsonii* species-group (Lacertilia : Scincidae). *Rec. West. Aust. Mus.* **3**: 66–70 (taxonomy); Greer, A.E. (1979). *Eremiascincus*, a new generic name for some Australian sand swimming skinks (Lacertilia : Scincidae). *Rec. Aust. Mus.* **32**: 321–338 (taxonomy).

Eugongylus Fitzinger, 1843

Eugongylus Fitzinger, L.J. (1843). *Systema Reptilium.* Vienna : Braümüller u. Seidel vi 106 pp. [23]. Type species *Eumeces oppelii* Duméril and Bibron, 1839 by original designation.

This group is also found in E Indonesia, New Guinea, Solomon Ils. and Santa Cruz Ils., see Greer, A.E. (1974). The generic relationships of the scincid lizard genus *Leiolopisma* and its relatives. *Aust. J. Zool. Suppl. Ser.* **31**: 1–67.

Eugongylus albofasciolatus (Günther, 1872)

Eumeces albofasciolatus Günther, A. (1872). Notice of a new species of lizard (*Eumeces albofasciolatus*) from north Australia. *Ann. Mag. Nat. Hist. (4)* **10**: 370–371 [370] [provenance of holotype doubtful; no subsequent Australian records of this widely distributed extralimital species]. Type data: holotype, BMNH 1946.8.13.74, from northern Australia.

Distribution: (NE coastal), (N Gulf), Qld.; status in Australia uncertain. Ecology: terrestrial, nocturnal; oviparous, arthropod-feeder. Biological references: Covacevich, J., Ingram, G.J. & Czechura, G.V. (1982). Rare frogs and reptiles of Cape York Peninsula, Australia. *Biol. Conserv.* **22**: 283–294 (reject Australian records).

Eugongylus rufescens (Shaw, 1802)

***Scincus aurata* (part.)** Schneider, J.G. (1801). *Historiae Amphibiorum naturalis et literariae.* fasc. sec. Jena : Frommani vi 374 pp. [176] [original description based in part on specimens from Indo-Australian region; New World syntypes have been referred to *Mabuya aurata* (Schneider, 1801) and *Mabuya sloanii* (Daudin, 1802); we are not aware of any lectotype or neotype designation]. Type data: syntypes, whereabouts unknown, from "Surinam".

***Lacerta rufescens* (part.)** Shaw, G. (1802). *General Zoology*, or Systematic Natural History. Vol. 3 Pt. 1. Amphibia. London : Kearsley 312 pp. [285] [composite type series based, at least in part, on literature records; for extralimital taxa included in type series see Boulenger, G.A. (1887). *Catalogue of the Lizards in the British Museum (Natural History).* 3. London : British Museum xii 575 pp. pls 40]. Type data: none designated, whereabouts unknown, from "Arabia, Egypt, and some European islands including Cyprus".

Eumeces oppelii Duméril, A.M.C. & Bibron, G. (1839). *Erpétologie Générale* ou Histoire Naturelle Complète des Reptiles. Paris : Roret 5: viii 854 pp. [656]. Type data: syntypes, MNHP 1243, MNHP 2918, from New Guinea.

Mabouia macrura Günther, A. (1867). Additions to the knowledge of Australian reptiles and fishes. *Ann. Mag. Nat. Hist. (3)* **20**: 45–68 [48]. Type data: holotype, BMNH 1946.8.13.73, from Cape York, Qld.

Eumeces uniformis Meyer, A.B. (1874). Eine Mitteilung von Hrn. Dr. Adolf Bernhard Meyer über die von ihm auf Neu-Guinea den Inseln Jobi, Mysore und Mafoor im Jahr 1873 gesammelten Amphibien. *Mber. K. Preuss. Akad. Wiss. Berl.* **1874**: 128–140 [133] [1875 on title page]. Type data: syntypes, from Mafoor, New Guinea, MTKD 242–243, destroyed *vide* Obst, F.J. (1977). Die herpetologische Sammlung des Staatlichen Museums für Tierkunde Dresden und ihre Typusexemplare. *Zool. Abh. Mus. Tierk. Dresden* **34**: 171–186.

Eumeces brunneus Macleay, W. (1877). The lizards of the *Chevert* Expedition. *Proc. Linn. Soc. N.S.W.* **2**: 60–69 [65] [1878 on title page]. Type data: syntypes, AM R31864–6, from Darnley Is., Torres Strait, Qld.

Mabouia uniformis Macleay, W. (1877). The lizards of the *Chevert* Expedition. *Proc. Linn. Soc. N.S.W.* **2**: 60–69 [65] [junior secondary homonym; *non Eumeces uniformis* Meyer, 1874]. Type data: holotype, AM R31867, from Coconut Is., Torres Strait, Qld.

Euprepes (Tiliqua) cingulatus Peters, W. & Doria, G. (1878). Catalogo dei Rettili e dei Batraci raccolti da O. Beccari, L.M. d'Albertis e A.A. Bruijn nella sotto-regione Austro-Malese. *Ann. Mus. Civ. Stor. Nat. Genova* **13**: 323–450 [352]. Type data: holotype, MCG C.E. 27904, from Mansinam, Irian Jaya, Indonesia.

Synonymy that of Cogger, H.G., this work.

Distribution: NE coastal, (N Gulf), Qld. Ecology: (woodland), open forest, closed forest, terrestrial, nocturnal; oviparous, arthropod-feeder.

Hemiergis Wagler, 1830

Tetradactylus Cuvier, G.L.C.F.D. (1829). *Le Règne Animal.* 2nd edn. Paris : Déterville Vol. 2 xv 406 pp. [64] [*ex* Péron ms; junior homonym; *non Tetradactylus* Merrem, 1820]. Type species *Tetradactylus decresiensis* Cuvier, 1829 by monotypy.

Tridactylus Cuvier, G.L.C.F.D. (1829). *Le Règne Animal.* 2nd edn. Paris : Déterville Vol. 2 xv 406 pp. [64] *ex* Péron ms; junior homonym, *non Tridactylus* Lacépède, 1799]. Type species *Tridactylus decresiensis* Cuvier, 1829 by monotypy.

Hemiergis Wagler, J.G. (1830). *Natürliches System der Amphibien, mit vorangehender Classification der Säugethiere und Vögel.* München : Cotta'schen vi 354 pp. [160]. Type species *Tridactylus decresiensis* Cuvier, 1829 (*ex* Péron ms) by monotypy.

Peromelis Wagler, J.G. (1830). *Natürliches System der Amphibien, mit vorangehender Classification der Säugethiere und Vögel.* München : Cotta'schen vi 354 pp. [160]. Type species *Tetradactylus decresiensis* Cuvier, 1829 (*ex* Péron ms) by monotypy.

Chelomeles Duméril, A.M.C. & Bibron, G. (1839). *Erpétologie Générale* ou Histoire Naturelle Complète des Reptiles. Paris : Roret 5: viii 854 pp. [774]. Type species *Chelomeles quadrilineatus* Duméril and Bibron, 1839 by monotypy.

Synonymy that of Cogger, H.G., this work.

Hemiergis decresiensis (Cuvier, 1829)

Zygnis decresiensis Fitzinger, L.J. (1826). *Neue Classification der Reptilien nach ihren natürlichen Verwandtschaften.* Vienna : Heubner 66 pp. [53] [*nom. nud.*].

Tridactylus decresiensis Cuvier, G.L.C.F.D. (1829). *Le Règne Animal.* 2nd edn. Paris : Déterville Vol. 2 xv 406 pp. [64] [*ex* Péron ms]. Type data: syntypes, MNHP 1601, MNHP 3041, NHMW 16629 (probable), from Kangaroo Is., S.A.

Hemiergis polylepis Günther, A. (1867). Additions to the knowledge of Australian reptiles and fishes. *Ann. Mag. Nat. Hist. (3)* **20**: 45–68 [48]. Type data: holotype, BMNH 1946.8.16.31, from S.A.

Hemiergis decresiensis continentis Copland, S.J. (1946). Geographic variation in the lizard *Hemiergis decresiensis* (Fitzinger). *Proc. Linn. Soc. N.S.W.* **70**: 62–92 [69 pl 6 fig 2] [1945 on title page]. Type data: holotype, SAMA R2190, from Myponga, in 35°25'S, 138°20'E, S.A.

Hemiergis decresiensis talbingoensis Copland, S.J. (1946). Geographic variation in the lizard *Hemiergis decresiensis* (Fitzinger). *Proc. Linn. Soc. N.S.W.* **70**: 62–92 [71 pl 6 fig 3] [1945 on title page]. Type data: holotype, AM R57654, from Talbingo, N.S.W.

Hemiergis decresiensis davisi Copland, S.J. (1946). Geographic variation in the lizard *Hemiergis decresiensis* (Fitzinger). *Proc. Linn. Soc. N.S.W.* **70**: 62–92 [79 pl 6 fig 4] [1945 on title page]. Type data: holotype, AM R58733, from Bendemeer, N.S.W.

Synonymy that of Cogger, H.G., this work.

Distribution: SE coastal, S Gulfs, Murray-Darling basin, S.A., Vic., N.S.W., A.C.T.; E islands of Great Aust. Bight. Ecology: open scrub, low open woodland, woodland, open forest, tall forest, fossorial, nocturnal; viviparous, arthropod-feeder. Biological references: Warburg, M.R. (1966). On the water economy of several Australian geckos, agamids, and skinks. *Copeia* **1966**: 230–235; Robertson, P. (1981). Comparative reproductive ecology of two south-eastern Australian skinks. pp. 25–37 *in* Banks, C.B. & Martin, A.A. (eds.) *Proceedings of the Melbourne Herpetological Symposium.* Melbourne : Zoological Board of Victoria; Crome, B. (1981). The diet of some ground-layer lizards in three woodlands of the New England Tableland of Australia. *Herpetofauna* **13**: 4–11.

Hemiergis graciloides (Lönnberg and Andersson, 1913)

Lygosoma graciloides Lönnberg, E. & Andersson, G.L. (1913). Results of Dr. E. Mjöberg's Swedish Scientific Expeditions to Australia 1910–1913. III. Reptiles. *K. Svenska Vetensk.-Akad. Handl.* **52**(3): 1–17 [10]. Type data: syntypes, NHRM 3097 (2 specimens), MCZ 93724, from Yandina, at the foot of Blackall Range, Qld.

Lygosoma scharffi Boulenger, G.A. (1915). Descriptions of two new lizards from Australia. *Ann. Mag. Nat. Hist. (8)* **16**: 64–66 [64]. Type data: holotype, BMNH 1915.3.30.1, from One Tree Hill, near Brisbane, Qld.

Synonymy that of Cogger, H.G., this work.

Distribution: NE coastal, Qld. Ecology: litter, tall forest, closed forest, nocturnal, crepuscular; oviparous, arthropod-feeder. Biological references: Czechura, G.V. (1981). The rare scincid lizard, *Nannoscincus graciloides*: a reappraisal. *J. Herpet.* **15**: 315–320 (taxonomy, natural history, as *Nannoscincus graciloides*).

Hemiergis initialis (Werner, 1910)

Lygosoma (Hemiergis) initiale Werner, F. (1910). Reptilia (Geckonidae und Scincidae). *in* Michaelsen, W. & Hartmeyer, R. (eds.) *Die Fauna Südwest-Australiens.* Jena : Gustav Fischer 2: 451–493 [480]. Type data: syntypes, NHMW 16637 (Lion Mill), ZMH 3503 (Lion Mill), ZMB 21465 (3 specimens, Lion Mill), 4 other syntypes not found, from Mt. Helena (as Lion Mill) and Jarrahdale, south W.A.

Hemiergis initialis brookeri Storr, G.M. (1975). The genus *Hemiergis* (Lacertilia : Scincidae) in Western Australia. *Rec. West. Aust. Mus.* **3**: 251–260 [254]. Type data: holotype, WAM R24677, from 40 km WSW of Caiguna, in 32°20'S, 125°00'E, W.A.

Synonymy that of Cogger, H.G., this work.

Distribution: SW coastal, W plateau, W.A. Ecology: open heath, woodland, open forest, fossorial, nocturnal; viviparous, arthropod-feeder. Biological references: Storr, G.M., Smith, L.A. & Johnstone, R.E. (1981). *Lizards of Western Australia. I. Skinks.* Perth : University of Western Australia Press and Western Australian Museum 200 pp. (taxonomy).

Hemiergis maccoyi (Lucas and Frost, 1894)

Siaphos maccoyi Lucas, A.H.S. & Frost, C. (1894). The lizards indigenous to Victoria. *Proc. R. Soc. Vict. (ns)* **6**: 24–92 [85 pl 2 figs 2 2a]. Type data: lectotype, NMV D1851, from Ringwood, Vic., designation by Coventry, A.J. (1970). Reptile and amphibian type specimens housed in the National Museum of Victoria. *Mem. Natl. Mus. Vict.* **31**: 115–124.

Distribution: SE coastal, Murray-Darling basin, Vic., N.S.W., A.C.T. Ecology: woodland, open forest, tall forest, alpine, terrestrial, noctidiurnal; oviparous, cryptozoic, arthropod-feeder. Biological references: Spellerberg, I.F. (1972). Temperature tolerances of southeast Australian reptiles examined in relation to reptile thermoregulation, behaviour and distribution. *Oecologia* **9**: 23–46; Greer, A.E. (1980). Critical thermal maximum temperatures in Australian scincid lizards: their ecological and evolutionary significance. *Aust. J. Zool.* **28**: 91–102; Robertson, P. (1981). Comparative reproductive ecology of two south-eastern Australian skinks. pp. 25–37 *in* Banks, C.B. & Martin, A.A. (eds.) *Proceedings of the Melbourne Herpetological Symposium.* Melbourne : Zoological Board of Victoria.

Hemiergis millewae Coventry, 1976

Hemiergis millewae Coventry, A.J. (1976). A new species of *Hemiergis* (Scincidae : Lygosominae) from Victoria. *Mem. Natl. Mus. Vict.* **37**: 23–26 [23 fig 1]. Type data: holotype, NMV D47410, from Millewa South bore, in 34°56'28"S, 141°4'E, Vic.

Distribution: W plateau, S Gulfs, Murray-Darling basin, Lake Eyre basin, W.A., S.A., Vic., N.S.W. Ecology: open scrub, fossorial; arthropod-feeder. Biological references: Storr, G.M., Smith, L.A. & Johnstone, R.E. (1981). *Lizards of Western Australia. I. Skinks.* Perth : University of Western Australia Press and Western Australian Museum 200 pp. (taxonomy).

Hemiergis peronii (Gray, 1831)

Seps peronii Fitzinger, L.J. (1826). *Neue Classification der Reptilien nach ihren natürlichen Verwandtschaften.* Vienna : Heubner 66 pp. [53] [*nom. nud.*, based on *Tetradactylus descresiensis* Cuvier ms].

Tetradactylus decresiensis Cuvier, G.L.C.F.D. (1829). *Le Règne Animal.* 2nd edn. Paris : Déterville Vol. 2 xv 406 pp. [64] [*ex* Péron ms; long regarded as a junior secondary homonym of *Tridactylus decresiensis* Cuvier,

1829]. Type data: syntypes (probable), MNHP 3040, MNHP 7135, MNHP 7136 (2 specimens), NHMW 9961, from Kangaroo Is., S.A.

Seps peronii Gray, J.E. (1831). A synopsis of the species of the class Reptilia, pp. 1–110 *in* Griffith, E. *The Animal Kingdom Arranged in Conformity with its Organization by the Baron Cuvier.* London : Whittaker, Treacher & Co. 9: 481+110 pp. [Synopsis:72] [*pro Seps peronii* Fitzinger, 1826 (*nom. nud.*), itself based on Cuvier's types of *Tetradactylus decresiensis*].

Chiamela duvaucellii Gray, J.E. (1839). Catalogue of the slender-tongued saurians, with descriptions of many new genera and species. (contd.). *Ann. Mag. Nat. Hist.* **2**: 331–337 [333] [*ex* Cocteau ms; published Jan., 1839]. Publication date established from Sherborn, C.D. (1925). *Index Animalium.* London : British Museum 2: 1773–2880. Type data: none designated, MNHP (presumed lost), from King George's Sound, W.A.

Chelomeles quadrilineatus Duméril, A.M.C. & Bibron, G. (1839). *Erpétologie Générale* ou Histoire Naturelle Complète des Reptiles. Paris : Roret 5: viii 854 pp. [774] [published Nov., 1839; Publication date established from Sherborn, C.D. (1930). *Index Animalium 1801–1851. Q–R.* London : British Museum pp. 5353–6357]. Type data: holotype, MNHP 7140, from Australia.

Hemiergis woodwardi Lucas, A.H.S. & Frost, C. (1902). Descriptions of some new lizards from Western Australia. *Proc. R. Soc. Vict.* **15**: 76–79 [77]. Type data: holotype, NMV D11772, from W.A.

Lygosoma (Hemiergis) quadridigitatum Werner, F. (1910). Reptilia (Geckonidae und Scincidae). *in* Michaelsen, W. & Hartmeyer, R. (eds.) *Die Fauna Südwest-Australiens.* Jena : Gustav Fischer 2: 451–493 [480] [*nom. nov. pro Seps peronii* Fitzinger, 1826].

Lygosoma peronii tridactylum Boulenger, G.A. (1915). Descriptions of two new lizards from Australia. *Ann. Mag. Nat. Hist. (8)* **16**: 64–66 [65]. Type data: syntypes, BMNH 1946.8.17.65–66, from Yallingup, SW Australia.

Synonymy that of Cogger, H.G., this work.

Distribution: SE coastal, SW coastal, W plateau, S Gulfs, Murray-Darling basin, S.A., Vic., W.A. Ecology: low shrubland, open heath, open scrub, woodland, open forest, fossorial, nocturnal; viviparous, arthropod-feeder. Biological references: Storr, G.M. (1975). The genus *Hemiergis* (Lacertilia : Scincidae) in Western Australia. *Rec. West. Aust. Mus.* **3**: 251–260 (taxonomy); Smyth, M. (1968). The distribution and life history of the skink, *Hemiergis peronii* (Fitzinger). *Trans. R. Soc. S. Aust.* **92**: 51–58; Smyth, M. & Smith, M.J. (1968). Obligatory sperm storage in the skink *Hemiergis peronii. Science* **161**: 575–576; Warburg, M.R. (1966). On the water economy of several Australian geckos, agamids, and skinks. *Copeia* **1966**: 230–235; Smyth, M. (1974). Changes in the fat stores of the skinks *Morethia boulengeri* and *Hemiergis peronii* (Lacertilia). *Aust. J. Zool.* **22**: 135–145.

Lampropholis Fitzinger, 1843

Lampropholis Fitzinger, L.J. (1843). *Systema Reptilium.* Vienna : Braümüller u. Seidel vi 106 pp. [22]. Type species *Lygosoma guichenoti* Duméril and Bibron, 1839 by original designation.

Mocoa Gray, J.E. (1845). *Catalogue of the Specimens of Lizards in the Collection of the British Museum.* London : British Museum xxviii 289 pp. [80]. Type species *Lygosoma guichenoti* Duméril and Bibron, 1839 by subsequent designation, see Mittleman, M.B. (1952). A generic synopsis of the lizards of the subfamily Lygosominae. *Smithson. Misc. Coll.* **117**(4069): 1–35.

Synonymy that of Cogger, H.G., this work.

This group is introduced to Hawaii, see Baker, J.K. (1979). The rainbow skink, *Lampropholis delicata*, in Hawaii. *Pac. Sci.* **33**: 207–212.

Lampropholis amicula Ingram and Rawlinson, 1981

Lampropholis amicula Ingram, G.J. & Rawlinson, P. (1981). Five new species of skinks (genus *Lampropholis*) from Queensland and New South Wales. *Mem. Qd. Mus.* **20**: 311–317 [314 figs 4,6]. Type data: holotype, QM J24333, from 3 km E of Lake Coolamera, in 26°02'S, 153°05'E, SE Qld.

Distribution: NE coastal, N.S.W., Qld. Ecology: open heath, open forest, terrestrial, diurnal; oviparous, arthropod-feeder.

Lampropholis basiliscus Ingram and Rawlinson, 1981

Lampropholis basiliscus Ingram, G.J. & Rawlinson, P. (1981). Five new species of skinks (genus *Lampropholis*) from Queensland and New South Wales. *Mem. Qd. Mus.* **20**: 311–317 [313 figs 3,7]. Type data: holotype, QM J34409, from Charmillin Creek, via Ravenshoe, in 17°43'S, 145°31'E, NE Qld.

Distribution: NE coastal, Qld. Ecology: closed forest, terrestrial, diurnal; oviparous, arthropod-feeder.

Lampropholis caligula Ingram and Rawlinson, 1981

Lampropholis caligula Ingram, G.J. & Rawlinson, P. (1981). Five new species of skinks (genus *Lampropholis*) from Queensland and New South Wales. *Mem. Qd. Mus.* **20**: 311–317 [315 figs 5–6]. Type data: holotype, AM R13313, from Pol Blue, Barrington Tops, in 31°59'S, 151°30'E, mid-E N.S.W.

Distribution: SE coastal, N.S.W. Ecology: woodland, tall forest, terrestrial, diurnal; oviparous, arthropod-feeder.

Lampropholis challengeri (Boulenger, 1887)

Lygosoma challengeri Boulenger, G.A. (1887). *Catalogue of the Lizards in the British Museum (Natural History).*

3. London : British Museum xii 575 pp. pls 40 [268 pl 19 fig 3]. Type data: syntypes, BMNH 1946.8.16.55–56, from Qld.

Mocoa spectabilis De Vis, C.W. (1888). A contribution to the herpetology of Queensland. *Proc. Linn. Soc. N.S.W. (2)* **2**: 811–826 [819]. Type data: syntypes, QM J244, QM J255, QM J19742–3, from Gympie, Qld.

Synonymy that of Cogger, H.G., this work.

Distribution: SE coastal, NE coastal, N.S.W., Qld. Ecology: tall forest, closed forest, terrestrial, diurnal; oviparous, arthropod-feeder. Biological references: Domrow, R. (1974). Miscellaneous mites from Australian vertebrates. *Proc. Linn. Soc. N.S.W.* **99**: 1–48; Dale, D.F. (1973). *Forty Queensland Lizards.* Brisbane : Queensland Museum 64 pp. (natural history).

Lampropholis czechurai Ingram and Rawlinson, 1981

Lampropholis czechurai Ingram, G.J. & Rawlinson, P. (1981). Five new species of skinks (genus *Lampropholis*) from Queensland and New South Wales. *Mem. Qd. Mus.* **20**: 311–317 [312 figs 2,6]. Type data: holotype, QM J34402, from Charmillin Creek, via Ravenshoe, in 17°43′S, 145°31′E, NE Qld.

Distribution: NE coastal, Qld. Ecology: closed forest, terrestrial, diurnal; oviparous, arthropod-feeder.

Lampropholis delicata (De Vis, 1888)

Mocoa delicata De Vis, C.W. (1888). A contribution to the herpetology of Queensland. *Proc. Linn. Soc. N.S.W. (2)* **2**: 811–826 [820]. Type data: holotype, QM (presumed lost), from Warro Station, Port Curtis (as Warro, Central Queensland), Qld.

Distribution: SE coastal, NE coastal, S Gulfs, (Murray-Darling basin), S.A., Tas., Vic., N.S.W., Qld., A.C.T. Ecology: open scrub, low open woodland, woodland, open forest, tall forest, terrestrial, diurnal; oviparous, arthropod-feeder. Biological references: Clarke, C.J. (1965). A comparison between some Australian five-fingered lizards of the genus *Leiolopisma* Duméril and Bibron (Lacertilia : Scincidae). *Aust. J. Zool.* **13**: 577–592 (taxonomy); Baker, J.K. (1979). The rainbow skink, *Lampropholis delicata*, in Hawaii. *Pac. Sci.* **33**: 207–212; Wells, R. (1981). Utilisation of the same site for communal egg-laying by *Lampropholis delicata* and *L. guichenoti. Aust. J. Herp.* **1**: 35–36; Humphreys, W.F. (1976). Spider induced egg mortality in a skink population. *Copeia* **1976**: 404; Crome, B. (1981). The diet of some ground-layer lizards in three woodlands of the New England Tableland of Australia. *Herpetofauna* **13**: 4–11.

Lampropholis guichenoti (Duméril and Bibron, 1839)

Lygosoma guichenoti Duméril, A.M.C. & Bibron, G. (1839). *Erpétologie Générale* ou Histoire Naturelle Complète des Reptiles. Paris : Roret 5: viii 854 pp. [713] Type data: lectotype, MNHP 5264, from Kangaroo Is.; original description cites "Australia", designation by Rawlinson, P.A. *in litt.* [lectotype designation not made here; indication only of work by Rawlinson in preparation].

***Lygosoma duperreyii* (part.)** Duméril, A.M.C. & Bibron, G. (1839). *Erpétologie Générale* ou Histoire Naturelle Complète des Reptiles. Paris : Roret 5: viii 854 pp. [715] [var. B; *ex* Cocteau ms; for other syntypes see *Leiolopisma trilineatum* (Gray, 1838)]. Type data: syntypes, MNHP 3014 (New Zealand), from New Zealand and Tas. (original description cites New Zealand and Australia).

Synonymy that of Cogger, H.G., this work.

Distribution: SE coastal, NE coastal, S Gulfs, Murray-Darling basin, S.A., Vic., N.S.W., Qld., A.C.T. Ecology: low open woodland, woodland, open forest, tall forest, terrestrial, diurnal; oviparous, arthropod-feeder. Biological references: Clarke, C.J. (1965). A comparison between some Australian five-fingered lizards of the genus *Leiolopisma* Duméril and Bibron (Lacertilia : Scincidae). *Aust. J. Zool.* **13**: 577–592 (taxonomy); Mitchell, F.J. (1959). Communal egg-laying in the lizard *Leiolopisma guichenoti* (Duméril and Bibron). *Trans. R. Soc. S. Aust.* **82**: 121–122; Milton, D. (1980). Some aspects of the population dynamics of *Lampropholis guichenoti* in Toohey Forest near Brisbane. *Herpetofauna* **11**: 19–23; Joss, J.M.P. (1978). A rhythm in hydroxyindole-o-methyl-transferase (HIOMT) activity in the scincid lizard, *Lampropholis guichenoti. Gen. Comp. Endocrin.* **36**: 521–525; Powell, H., Heatwole, H. & Heatwole, M. (1979). Winter aggregation of *Leiolopisma guichenoti. Br. J. Herpet.* **5**: 789–791; Wells, R. (1981). Utilisation of the same site for communal egg-laying by *Lampropholis delicata* and *L. guichenoti. Aust. J. Herp.* **1**: 35–36; Crome, B. (1981). The diet of some ground-layer lizards in three woodlands of the New England Tableland of Australia. *Herpetofauna* **13**: 4–11.

Lampropholis mirabilis Ingram and Rawlinson, 1981

Lampropholis mirabilis Ingram, G.J. & Rawlinson, P. (1981). Five new species of skinks (genus *Lampropholis*) from Queensland and New South Wales. *Mem. Qd. Mus.* **20**: 311–317 [311 figs 1,6]. Type data: holotype, QM J24439, from Magnetic Is., in 19°08′S, 146°50′E, NE Qld.

Distribution: NE coastal, Qld. Ecology: woodland, terrestrial, diurnal; oviparous, saxicoline, arthropod-feeder.

Lampropholis mustelina (O'Shaughnessy, 1874)

Mocoa mustelina O'Shaughnessy, A.W.E. (1874). A description of a new species of Scincidae in the collection of the British Museum. *Ann. Mag. Nat. Hist. (4)* **13**: 298–301 [299]. Type data: syntypes, BMNH 1946.8.16.78 (N.S.W.), BMNH 1946.8.16.86 (Sydney), BMNH 1946.8.17.16 (N.S.W.), from N.S.W. and Sydney, N.S.W. (original description cites only N.S.W.).

Lygosoma (Mocoa) orichalceum Boettger, O. (1878). Studien über neue oder wenig bekannte Eidechsen. I. *Offenbach Ver. Nat. Ber.* **17–18**: 1–12 [2 pl 1 fig 1]. Type data: holotype, SMF 14675, from Australia.

Lygosoma (Mocoa) sonderi Peters, W. (1878). (Über zwei Scincoiden aus Australien und einer neuen *Amphisbaena* von West Africa). *Sber. Ges. Naturf. Freunde Berl.* **1878**: 191–193 [191]. Type data: holotype, ZMB 9374, from Port Phillip, Vic.

Lygosoma (Mocoa) lacrymans Peters, W. & Doria, G. (1878). Catalogo dei Rettili e dei Batraci raccolti da O. Beccari, L.M. d'Albertis e A.A. Bruijn nella sotto-regione Austro-Malese. *Ann. Mus. Civ. Stor. Nat. Genova* **13**: 323–450 [348]. Type data: holotype, MCG C.E. 27967, from Mt. Victoria, N.S.W.

Lygosoma (Liolepisma) pseudotropis Werner, F. (1903). Neue Reptilien und Batrachier aus dem Naturhistorischen Museum in Brussel. *Zool. Anz.* **26**: 246–253 [247] [*non Mocoa pseudotropis* Günther, 1875]. Type data: holotype, RIB 1105, from N.S.W.

Lygosoma paraeneum Ahl, E. (1925). Herpetologische Notizen. *Zool. Anz.* **65**: 18–20 [20] [*nom. nov. pro Lygosoma (Liolepisma) pseudotropis* Werner, 1903].

Synonymy that of Cogger, H.G., this work.

Distribution: SE coastal, Murray-Darling basin, Vic., N.S.W., A.C.T. Ecology: low open woodland, open forest, tall forest, terrestrial, diurnal; oviparous, arthropod-feeder. Biological references: Spellerberg, I.F. (1972). Temperature tolerances of southeast Australian reptiles examined in relation to reptile thermoregulation behaviour and distribution. *Oecologia* **9**: 23–46; Wells, R. (1979). A large aggregation of skink eggs. *Herpetofauna* **11**: 19–20.

Lampropholis tetradactyla Greer and Kluge, 1980

Lampropholis tetradactyla Greer, A.E. & Kluge, A.G. (1980). A new species of *Lampropholis* (Lacertilia : Scincidae) from the rainforests of northeastern Queensland. *Occ. Pap. Mus. Zool. Univ. Mich.* 691: 1–12 [2 figs 1–2]. Type data: holotype, QM J29853, from vicinity of Smoko Creek at the Kennedy-Kirrama road, Kirrama State Forest, NE Qld.

Distribution: NE coastal, Qld. Ecology: closed forest, terrestrial, diurnal; oviparous, cryptozoic, arthropod-feeder.

Leiolopisma Duméril and Bibron, 1839

Leiolopisma Duméril, A.M.C. & Bibron, G. (1839). *Erpétologie Générale* ou Histoire Naturelle Complète des Reptiles. Paris : Roret 5: viii 854 pp. [742]. Type species *Scincus telfairii* Desjardin, 1831 by monotypy.

Leiolepisma Fitzinger, L.J. (1843). *Systema Reptilium.* Vienna : Braümüller u. Seidel vi 106 pp. [22] [emend. *pro Leiolopisma* Duméril and Bibron, 1839]. Type species that of *Leiolopisma* Duméril and Bibron, 1839.

Eulepis Fitzinger, L.J. (1843). *Systema Reptilium.* Vienna : Braümüller u. Seidel vi 106 pp. [22]. Type species *Lygosoma duperreyii* Duméril and Bibron, 1839 by original designation.

Liolopisma Agassiz, L. (1846). *Nomenclatoris Zoologici Index Universalis.* Soloduri : Jent et Gassman 393 pp. [212] [emend. *pro Leiolopisma* Duméril and Bibron, 1839].

Synonymy that of Cogger, H.G., this work; for synonyms based on extralimital species see Greer, A.E. (1974). The generic relationships of the scincid lizard genus *Leiolopisma* and its relatives. *Aust. J. Zool. Suppl. Ser.* **31**: 1–67.

This group is also found in New Zealand, New Caledonia, Vanuatu, Loyalty Ils. and Mauritius (Round Is.).

Leiolopisma coventryi Rawlinson, 1975

Leiolopisma coventryi Rawlinson, P.A. (1975). Two new lizard species from the genus *Leiolopisma* (Scincidae : Lygosominae) in southeastern Australia and Tasmania. *Mem. Natl. Mus. Vict.* **36**: 1–16 [2 fig 1, pl 1 fig 1]. Type data: holotype, NMV D40080, from 2 km N of Mt. St. Leonard, in 37°33′S, 145°32′E, Vic.

Distribution: SE coastal, Murray-Darling basin, Vic., N.S.W., A.C.T. Ecology: low open woodland, open forest, tall forest, terrestrial, diurnal; viviparous, arthropod-feeder.

Leiolopisma entrecasteauxii (Duméril and Bibron, 1839)

***Scincus undecimstriatus* (part.)** Kuhl, H. (1820). *Beiträge zur Zoologie und vergleichenden Anatomie.* Frankfurt 130 pp. [129]. Type data: syntype, RMNH 2479 (Kangaroo Is.); whereabouts of remaining syntypes unknown, from Kangaroo Is., S.A. (original description cites Australia), see also *Ctenotus taeniolatus* (White, 1790).

Tiliqua entrecasteaux Gray, J.E. (1838). Catalogue of the slender-tongued saurians, with descriptions of many new genera and species. (contd.) *Ann. Mag. Nat. Hist.* **2**: 287–293 [292] [*nom. nud.*].

Lygosoma entrecasteauxii Duméril, A.M.C. & Bibron, G. (1839). *Erpétologie Générale* ou Histoire Naturelle Complète des Reptiles. Paris : Roret 5: viii 854 pp. [717]. Type data: syntypes, MNHP 2473 (2 specimens), from Australia.

***Mocoa pseudocarinata* (part.)** O'Shaughnessy, A.W.E. (1874). A description of a new species of Scincidae in the collection of the British Museum. *Ann. Mag. Nat. Hist. (4)* **13**: 298–301 [300]. Type data: lectotype, BMNH 1946.8.16.21, from Tas., designation by Rawlinson, P.A., *in litt.* [lectotype designation not made in this work;

indication only of work by Rawlinson in preparation; for the paralectotype, BMNH 1946.8.16.22 see under *Leiolopisma metallicum*].

Mocoa pseudotropis Günther, A. (1875). Reptiles. A list of saurians of Australia and New Zealand. pp. 9–19 *Zoology of the Voyage of H.M.S.* Erebus *and* Terror. Vol. 2 London : E.W. Janson [13] [*nom. substit. pro Mocoa pseudocarinata* O'Shaughnessy, 1874, the latter regarded by Günther as a *vox hybrida*].

Lygosoma (Mocoa) guttulatum Peters, W. (1881). (Eine Mitteilung über drei neue Eidechsen, zu der Familie der Scincoiden gehörig, eine *Lipinia* (mit geckonenähnlicher Bildung der Zehen) aus Neu-Guinea und zwei *Mocoa* aus Neuholland). *Sber. Ges. Naturf. Freunde Berl.* **1881**: 81–85 [83]. Type data: syntypes, ZMB 9508 (2 specimens), from Adelaide, S.A.

Lygosoma (Leiolepisma) pagenstecheri Lindholm, W.A. (1901). Bemerkungen und Beschreibung einer neuen Eidechsensart versehen von W.A. Lindholm. *in* Lampe, E., Catalog der Reptilien-Sammlung (Schildkröten, Crocodile, Eidechsen und Chamaeleons) des Naturhist. Museums zu Wiesbaden. *Jb. Nassau Ver. Naturk.* **54**: 177–222 [214 pl 3 figs 3–5]. Type data: holotype, WNS, lost when forwarded to platemaker in Frankfurt (Rawlinson, P.A., *in litt.*), from "?Süd-Australien".

Synonymy that of Cogger, H.G., this work.

Distribution: SE coastal, S Gulfs, Murray-Darling basin, S.A., Tas., Vic., N.S.W., A.C.T. Ecology: alpine, open scrub, low open woodland, woodland, open forest, tall forest, terrestrial, diurnal; viviparous, arthropod-feeder. Biological references: Rawlinson, P.A. (1974). Biogeography and ecology of the reptiles of Tasmania and the Bass Strait area. Chapt. 11 pp. 291–338 *in* Williams, W.D. (ed.) *Biogeography and Ecology in Tasmania.* The Hague : Junk (biogeography); Harrison, L. & Weekes, H.C. (1925). On the occurrence of placentation in the scincid lizard *Lygosoma entrecasteauxii. Proc. Linn. Soc. N.S.W.* **50**: 470–486; Baehr, M. (1976). Beiträge zur Verbreitung und Ökologie tasmanischer Reptilien. *Stuttgart Beitr. Naturk. (A)* **292**: 1–24 (natural history).

Leiolopisma greeni Rawlinson, 1975

Leiolopisma greeni Rawlinson, P.A. (1975). Two new lizard species from the genus *Leiolopisma* (Scincidae : Lygosominae) in southeastern Australia and Tasmania. *Mem. Natl. Mus. Vict.* **36**: 1–16 [8 fig 2, pl 1 fig 2]. Type data: holotype, NMV D18243, from Mt. Barrow summit (1413 m), Tas., in 41°23′S, 147°25′E.

Distribution: Tas. Ecology: alpine, terrestrial, diurnal; arthropod-feeder.

Leiolopisma lichenigerum (O'Shaughnessy, 1874)

Mocoa lichenigera O'Shaughnessy, A.W.E. (1874). A description of a new species of Scincidae in the collection of the British Museum. *Ann. Mag. Nat. Hist. (4)* **13**: 298–301 [298]. Type data: holotype, BMNH 1946.8.16.27, from Lord Howe Is.

Distribution: Lord Howe Is., Norfolk Is. Ecology: tussock grassland, tall forest, closed forest, terrestrial, nocturnal; viviparous, arthropod-feeder. Biological references: Loveridge, A. (1934). Australian reptiles in the Museum of Comparative Zoology, Cambridge, Massachusetts. *Bull. Mus. Comp. Zool.* **77**: 243–383 (taxonomy as *Leiolopisma cuprea*); Cogger, H.G. (1971). The reptiles of Lord Howe Island. *Proc. Linn. Soc. N.S.W.* **96**: 23–38 (taxonomy, natural history).

Leiolopisma metallicum (O'Shaughnessy, 1874)

*[**Mocoa ocellata** **(part.)**]* Gray, J.E. (1845). *Catalogue of the Specimens of Lizards in the Collection of the British Museum.* London : British Museum xxviii 289 pp. [82] [for other specimens in type series see *Leiolopisma ocellata* (Gray, 1845)]. Type data: paralectotype, BMNH 1946.8.10.78, from Australia.

Mocoa metallica O'Shaughnessy, A.W.E. (1874). A description of a new species of Scincidae in the collection of the British Museum. *Ann. Mag. Nat. Hist. (4)* **13**: 298–301 [299]. Type data: syntypes, BMNH 1946.8.7.7–14, from Tas.

*[**Mocoa pseudocarinata** **(part.)**]* O'Shaughnessy, A.W.E. (1874). A description of a new species of Scincidae in the collection of the British Museum. *Ann. Mag. Nat. Hist. (4)* **13**: 298–301 [300] [for the lectotype, BMNH 1946.8.16.21 see *Leiolopisma entrecasteauxii*]. Type data: paralectotype BMNH 1946.8.16.22, from Tas.

Synonymy that of Cogger, H.G., this work.

Distribution: SE coastal, Tas., Vic. Ecology: alpine, tall forest, closed forest, terrestrial, diurnal; viviparous, arthropod-feeder. Biological references: Rawlinson, P.A. (1974). Biogeography and ecology of the reptiles of Tasmania and the Bass Strait area. Chapt. 11 pp. 291–338 *in* Williams, W.D. (ed.) *Biogeography and Ecology in Tasmania.* The Hague : Junk (biogeography); Baehr, M. (1976). Beiträge zur Verbreitung und Ökologie tasmanischer Reptilien. *Stuttgart Beitr. Naturk. (A)* **292**: 1–24 (natural history); Spellerberg, I.F. (1972). Temperature tolerances of southeast Australian reptiles examined in relation to reptile thermoregulation, behaviour and distribution. *Oecologia* **9**: 23–46.

Leiolopisma ocellatum (Gray, 1845)

***Mocoa ocellata* (part.)** Gray, J.E. (1845). *Catalogue of the Specimens of Lizards in the Collection of the British Museum.* London : British Museum xxviii 289 pp. [82]. Of the paralectotypes, BMNH 1946.8.16.44–46 are referable to this species, but for BMNH 1946.8.10.78 see *Leiolopisma metallicum* (O'Shaughnessy, 1874). Type data: lectotype, BMNH 1946.8.10.83, from Australia., designation by Rawlinson, P.A., *in litt.* [lectotype designation not made in this work; indication only of work by Rawlinson in preparation].

Lygosoma hieroglyphicum Duméril, A.M.C. & Duméril, A. (1851). *Catalogue Méthodique de la Collection des Reptiles.* Paris : Mus. Hist. Nat. iv 224 pp. [166] see also [pl 5 fig 1] in Jacquinot, H. & Guichenot, M.A. (1853). Reptiles et Poissons. *in* Dumont d'Urville, M.J. (ed.) *Voyage au Pole Sud et dans l'océanie sur les corvettes L'*Astrolabe *et La* Zélée, exécut pendant les années 1837–40. Atlas Sauriens. Zoologie 3: 1–28 Paris : Gide et Baudry Type data: syntypes, MNHP 3012 (3 specimens), from Hobart, Tas.

Synonymy that of Cogger, H.G., this work.

Distribution: Tas. Ecology: open forest, tall forest, closed forest, terrestrial, diurnal; viviparous, arthropod-feeder. Biological references: Rawlinson, P.A. (1974). Biogeography and ecology of the reptiles of Tasmania and the Bass Strait area. Chapt. 11 pp. 291–338 *in* Williams, W.D. (ed.) *Biogeography and Ecology in Tasmania.* The Hague : Junk (biogeography); Baehr, M. (1976). Beiträge zur Verbreitung und Ökologie tasmanischer Reptilien. *Stuttgart Beitr. Naturk. (A)* **292**: 1–24 (natural history).

Leiolopisma platynotum (Peters, 1881)

Lygosoma (Mocoa) platynotum Peters, W. (1881). (Eine Mitteilung über drei neue Eidechsen, zu der Familie der Scincoiden gehörig, eine *Lipinia* (mit geckonenähnlicher Bildung der Zehen) aus Neu-Guinea und zwei *Mocoa* aus Neuholland). *Sber. Ges. Naturf. Freunde Berl.* **1881**: 81–85 [84]. Type data: holotype, ZMB 9509, from Adelaide, S.A. (presumably in error).

Distribution: SE coastal, Murray-Darling basin, Vic., N.S.W., Qld., A.C.T. Ecology: woodland, open forest, tall forest, terrestrial, diurnal; viviparous, arthropod-feeder.

Leiolopisma pretiosum (O'Shaughnessy, 1874)

Mocoa pretiosa O'Shaughnessy, A.W.E. (1874). A description of a new species of Scincidae in the collection of the British Museum. *Ann. Mag. Nat. Hist. (4)* **13**: 298–301 [298]. Type data: holotype, BMNH 1946.8.16.54, from Tas.

Mocoa microlepidota O'Shaughnessy, A.W.E. (1874). A description of a new species of Scincidae in the collection of the British Museum. *Ann. Mag. Nat. Hist. (4)* **13**: 298–301 [299]. Type data: holotype, BMNH 1946.8.16.17, from Tas.

Synonymy that of Cogger, H.G., this work.

Distribution: Tas. Ecology: tussock grassland, woodland, open forest, tall forest, terrestrial, diurnal; viviparous, arthropod-feeder. Biological references: Rawlinson, P.A. (1974). Biogeography and ecology of the reptiles of Tasmania and the Bass Strait area. Chapt. 11 pp. 291–338 *in* Williams, W.D. (ed.) *Biogeography and Ecology in Tasmania.* The Hague : Junk (biogeography).

Leiolopisma trilineatum (Gray, 1838)

Tiliqua trilineata Gray, J.E. (1838). Catalogue of the slender-tongued saurians, with descriptions of many new genera and species. (contd.) *Ann. Mag. Nat. Hist.* **2**: 287–293 [291] [1839 on title page]. Type data: syntypes BMNH 1946.8.19.75–7, BMNH 1946.8.19.72, from Australia.

Tiliqua duperreyi Gray, J.E. (1838). Catalogue of the slender-tongued saurians, with descriptions of many new genera and species. (contd.) *Ann. Mag. Nat. Hist.* **2**: 287–293 [292]. [1839 on title page; *ex* Cocteau ms] Type data: lectotype, MNHP 7102 (total length 131. mm), from Tas. (original description cites Kangaroo Is. S.A.), designation by Rawlinson, P.A., *in litt.* [lectotype designation not made in this work; indication only of work by Rawlinson in preparation].

***Lygosoma duperreyii* (part.)** Duméril, A.M.C. & Bibron, G. (1839). *Erpétologie Générale* ou Histoire Naturelle Complète des Reptiles. Paris : Roret 5: viii 854 pp. [715] [var. A; *ex* Cocteau ms; *non Tiliqua duperreyi* Gray 1838]. Type data: syntypes, MNHP 7102 (Tas., two specimens), from New Zealand and Tas. (original description cites New Zealand and Australia).

Synonymy that of Cogger, H.G., this work.

Distribution: SE coastal, SW coastal, W plateau, S Gulfs, Murray-Darling basin, S.A., Tas., Vic. N.S.W., W.A., A.C.T. Ecology: alpine, low open woodland, woodland, open forest, tall forest terrestrial, diurnal; viviparous, arthropod-feeder Biological references: Rawlinson, P.A. (1974) Biogeography and ecology of the reptiles of Tasmania and the Bass Strait area. Chapt. 11 pp 291–338 *in* Williams, W.D. (ed.) *Biogeography and Ecology in Tasmania.* The Hague : Junk (biogeography); Baehr, M. (1976). Beiträge zur Verbreitung und Ökologie tasmanischer Reptilien *Stuttgart Beitr. Naturk. (A)* **292**: 1–24 (natural history); Storr, G.M., Smith, L.A. & Johnstone, R.E. (1981). *Lizards of Western Australia. I Skinks.* Perth : University of Western Australia Press and Western Australian Museum 200 pp. (taxonomy).

Leiolopisma zia Ingram and Ehmann, 1981

Leiolopisma zia Ingram, G.J. & Ehmann, H. (1981). A new species of scincid lizard of the genus *Leiolopisma* (Scincidae : Lygosominae) from south eastern Queensland and north eastern New South Wales. *Mem Qd. Mus.* **20**: 311–317 [307 fig 1]. Type data: holotype, QM J30563, from Mt. Superbus State Forest, in 28°13'S, 152°26'E, SE Qld.

Distribution: SE coastal, NE coastal, N.S.W., Qld. Ecology: tall forest, closed forest, terrestrial, diurnal; oviparous, arthropod-feeder.

Lerista Bell, 1833

Lerista Bell, T. (1833). Characters of two new genera of reptiles. *Proc. Zool. Soc. Lond.* **1**: 98–99 [99]. Type species *Lerista lineata* Bell, 1833 by monotypy.

Rhodona Gray, J.E. (1839). Catalogue of the slender-tongued saurians, with descriptions of many new genera and species. (contd.). *Ann. Mag. Nat. Hist.* **2**:

31–337 [335]. Type species *Rhodona punctata* Gray, 839 by monotypy.

'oridia Gray, J.E. (1839). Catalogue of the ender-tongued saurians, with descriptions of many new enera and species. (contd.). *Ann. Mag. Nat. Hist.* **2**: 31–337 [335]. Type species *Soridia lineata* Gray, 1839 y monotypy.

rachystopus Duméril, A.M.C. & Bibron, G. (1839). *:rpétologie Générale* ou Histoire Naturelle Complète des eptiles. Paris : Roret 5: viii 854 pp. [778] [*ex* Smith is]. Type species *Brachystopus lineopunctulatus* Duméril nd Bibron, 1839 by monotypy.

'raepeditus Duméril, A.M.C. & Bibron, G. (1839). *:rpétologie Générale* ou Histoire Naturelle Complète des eptiles. Paris : Roret 5: viii 854 pp. [787]. Type species *'oridia lineata* Gray, 1839 by monotypy.

Ronia Gray, J.E. (1841). Description of some new pecies and four new genera of reptiles from Western ustralia, discovered by John Gould, Esq. *Ann. Mag. Vat. Hist.* **7**: 86–91 [87]. Type species *Ronia catenulata* ;ray, 1841 by monotypy.

:eptosoma Fitzinger, L.J. (1843). *Systema Reptilium.* 'ienna : Braümüller u. Seidel vi 106 pp. [23] [junior omonym; *non Leptosoma* Desmarest and other authors, *'ide* Sherborn, C.D. (1927). *Index Animalium '801–1851.* H-L London : British Museum pp. 885–3746]. Type species *Riopa bougainvillii* Duméril nd Bibron, 1839 by original designation.

)phiopsis Fitzinger, L.J. (1843). *Systema Reptilium.* 'ienna : Braümüller u. Seidel vi 106 pp. [23]. Type pecies *Lerista lineata* Bell, 1833 by original designation.

Micalia Gray, J.E. (1845). *Catalogue of the Specimens)f Lizards in the Collection of the British Museum.* .ondon : British Museum xxviii 289 pp. [62,284] [*errore)ro Miculia* Gray, 1845].

Miculia Gray, J.E. (1845). *Catalogue of the Specimens)f Lizards in the Collection of the British Museum.* .ondon : British Museum xxviii 289 pp. [66]. Type pecies *Miculia elegans* Gray, 1845 by monotypy.

Pholeophilus Smith, A. (1849). *Illustrations of the Zoology of South Africa.* Reptilia. London : Smith, Elder & Co. [15]. Type species *Pholeophilus capensis* Smith, 849 by monotypy.

Phaneropis Fischer, J.G. (1881). Beschreibung neuer Reptilien. *Arch. Naturgesch.* **47**: 225–238 [236]. Type species *Phaneropis muelleri* Fischer, 1881 by monotypy.

Phaneropus Smith, M.A. (1935). *The Fauna of British India; including Ceylon and Burma.* Reptilia and Amphibia. Vol. II. Sauria. London : Taylor & Francis xiv 440 pp. [309] [*lapsus pro Phaneropis* Fischer, 1881].

Nodorha Mittleman, M.B. (1952). A generic synopsis of the lizards of the subfamily Lygosominae. *Smithson. Misc. Coll.* **117**(4069): 1–35 [14]. Type species *Riopa bougainvillii* Duméril and Bibron, 1839 by original designation.

Synonymy that of Cogger, H.G., this work; Greer, A.E. (1967). A new generic arrangement for some Australian scincid lizards. *Breviora* 267: 1–19.

Lerista allanae (Longman, 1937)

Rhodona allanae Longman, H.A. (1937). Herpetological notes. *Mem. Qd. Mus.* **11**: 165–168 [167]. Type data: holotype, QM J6180, from Retro Station, Capella, central Qld.

Distribution: NE coastal, (Lake Eyre basin), Qld. Ecology: open scrub, woodland, fossorial, nocturnal; (oviparous), arthropod-feeder.

Lerista ameles Greer, 1979

Lerista ameles Greer, A.E. (1979). A new species of *Lerista* (Lacertilia : Scincidae) from northern Queensland, with remarks on the origin of the genus. *Rec. Aust. Mus.* **32**: 383–388 [383 fig 1]. Type data: holotype, QM J30004, from 19.7 km W of the junction of the Kennedy and Gulf Highways along the Gulf Highway, NE Qld.

Distribution: NE coastal, Qld. Ecology: woodland, fossorial, nocturnal; (oviparous), arthropod-feeder.

Lerista apoda Storr, 1976

Lerista apoda Storr, G.M. (1976). Revisionary notes on the *Lerista* (Lacertilia : Scincidae) of Western Australia. *Rec. West. Aust. Mus.* **4**: 241–256 [255]. Type data: holotype, WAM R52163, from Cygnet Bay, 16°26'S, 123°01'E, W.A.

Distribution: N coastal, W.A. Ecology: woodland, fossorial, nocturnal; (oviparous), arthropod-feeder. Biological references: Storr, G.M., Smith, L.A. & Johnstone, R.E. (1981). *Lizards of Western Australia. I. Skinks.* Perth : University of Western Australia Press and Western Australian Museum 200 pp. (taxonomy).

Lerista bipes (Fischer, 1882)

Rhodona bipes Fischer, J.G. (1882). Herpetologische Bemerkungen. *Arch. Naturgesch.* **48**: 281–302 [292 pl 16 figs 10–15]. Type data: holotype, SMNS 1217, from Nickol Bay, W.A.

Distribution: N coastal, NW coastal, W plateau, Lake Eyre basin, S.A., Qld., W.A., N.T. Ecology: desert, hummock grassland, tussock grassland, tall shrubland, fossorial, nocturnal; oviparous, arthropod-feeder. Biological references: Storr, G.M. (1976). Revisionary notes on the *Lerista* (Lacertilia : Scincidae) of Western Australia. *Rec. West. Aust. Mus.* **4**: 241–256 (taxonomy); Storr, G.M. (1971). The genus *Lerista* (Lacertilia : Scincidae) in Western Australia. *J. R. Soc. West. Aust.* **54**: 59–75 (taxonomy); Storr, G.M., Smith, L.A. & Johnstone, R.E. (1981). *Lizards of Western Australia. I. Skinks.* Perth : University of Western Australia Press and Western Australian Museum 200 pp. (taxonomy).

Lerista borealis Storr, 1971

Lerista borealis Storr, G.M. (1971). The genus *Lerista* (Lacertilia : Scincidae) in Western Australia. *J. R. Soc. West. Aust.* **54**: 59–75 [64]. Type data: holotype, WAM R22363, from Thompson Spring, 16°02′S, 128°57′E, W.A.

Distribution: N coastal, W.A. Ecology: woodland, fossorial, nocturnal; (oviparous), arthropod-feeder. Biological references: Storr, G.M., Smith, L.A. & Johnstone, R.E. (1981). *Lizards of Western Australia. I. Skinks.* Perth : University of Western Australia Press and Western Australian Museum 200 pp. (taxonomy).

Lerista bougainvillii (Gray, 1839)

Riopa brougainvillii Gray, J.E. (1839). Catalogue of the slender-tongued saurians, with descriptions of many new genera and species. (contd.). *Ann. Mag. Nat. Hist.* **2**: 331–337 [332] [*ex* Cocteau ms; published Jan. 1839]. Type data: holotype, MNHP 3016 (by implication), from Australia.

Lygosoma bougainvillii Duméril, A.M.C. & Bibron, G. (1839). *Erpétologie Générale* ou Histoire Naturelle Complète des Reptiles. Paris : Roret 5: viii 854 pp. [716] [published Oct., 1839; unintentional but justified emend. *pro Riopa brougainvillii* Gray, 1839]. Type data: holotype, MNHP 3016, from Australia.

Lygosoma laterale Günther, A. (1867). Additions to the knowledge of Australian reptiles and fishes. *Ann. Mag. Nat. Hist. (3)* **20**: 45–68 [45] [*non Scincus lateralis* Say, 1823, at one time a senior secondary homonym of *Lygosoma laterale* Günther, 1867]. Type data: holotype, BMNH 1946.8.15.58, from S.A.

Synonymy that of Cogger, H.G., this work.

Distribution: SE coastal, (W plateau), S Gulfs, Murray-Darling basin, S.A., Tas., Vic., N.S.W., A.C.T. Ecology: open scrub, woodland, open forest, tall forest, fossorial, nocturnal; oviparous(part), viviparous(part), arthropod-feeder.

Lerista christinae Storr, 1979

Lerista christinae Storr, G.M. (1979). Five new lizards from Western Australia. *Rec. West. Aust. Mus.* **8**: 134–142 [139 pl 3]. Type data: holotype, WAM R59022, from 22 km S of Badgingarra, in 30°33′S, 115°26′E, W.A.

Distribution: SW coastal, W.A. Ecology: open heath, fossorial, nocturnal; (oviparous), arthropod-feeder. Biological references: Storr, G.M., Smith, L.A. & Johnstone, R.E. (1981). *Lizards of Western Australia. I. Skinks.* Perth : University of Western Australia Press and Western Australian Museum 200 pp. (taxonomy).

Lerista connivens Storr, 1971

Lerista connivens Storr, G.M. (1971). The genus *Lerista* (Lacertilia : Scincidae) in Western Australia. *J. R. Soc. West. Aust.* **54**: 59–75 [68]. Type data: holotype, WAM R25777, from Salutation Is., 26°32′S, 113°46′E, W.A.

Distribution: NW coastal, W.A. Ecology: lo shrubland, open scrub, woodland, tall shrublan fossorial, nocturnal; (oviparous), arthropod-feede Biological references: Storr, G.M., Smith, L.A. Johnstone, R.E. (1981). *Lizards of Wester Australia. I. Skinks.* Perth : University of Wester Australia Press and Western Australian Museu 200 pp. (taxonomy).

Lerista desertorum (Sternfeld, 1919)

Lygosoma (Rhodona) planiventrale desertoru Sternfeld, R. (1919). Neue Schlangen und Echsen a Zentralaustralien. *Senckenbergiana* **1**: 76–83 [82]. Ty data: lectotype, SMF 15495, from Hermannsbu Mission, upper Finke River, N.T., designation b Mertens, R. (1967). Die herpetologische Sektion d Natur-Museums und Forschungs-Institutes Senckenbe in Frankfurt-a.-M. nebst einem Verzeichnis ihrer Type 1. *Senckenberg. Biol.* **48**: 1–106.

Distribution: W plateau, Lake Eyre basin, S.A W.A., N.T. Ecology: desert, low open shrublan tall shrubland, fossorial, nocturnal; oviparou arthropod-feeder. Biological references: Stor G.M. (1971). The genus *Lerista* (Lacertilia Scincidae) in Western Australia. *J. R. Soc. Wes Aust.* **54**: 59–75 (taxonomy); Storr, G.M., Smit L.A. & Johnstone, R.E. (1981). *Lizards of Wester Australia. I. Skinks.* Perth : University of Wester Australia Press and Western Australian Museu 200 pp. (taxonomy);.

Lerista distinguenda (Werner, 1910)

Ablepharus distinguendus Werner, F. (1910). Reptili (Geckonidae und Scincidae). *in* Michaelsen, W. Hartmeyer, R. (eds.) *Die Fauna Südwest-Australien* Jena : Gustav Fischer 2: 451–493 [490]. Type dat holotype, whereabouts unknown, from Fremantle, W.A.

Distribution: NW coastal, SW coastal, W platea W.A. Ecology: open heath, open scrub, woodlan open forest, fossorial, nocturnal; oviparous arthropod-feeder. Biological references: Stor G.M. (1971). The genus *Lerista* (Lacertilia Scincidae) in Western Australia. *J. R. Soc. Wes Aust.* **54**: 59–75 (taxonomy); Storr, G.M., Smit L.A. & Johnstone, R.E. (1981). *Lizards of Wester Australia. I. Skinks.* Perth : University of Wester Australia Press and Western Australian Museu 200 pp. (taxonomy).

Lerista elegans (Gray, 1845)

Miculia elegans Gray, J.E. (1845). *Catalogue of th Specimens of Lizards in the Collection of the Britis Museum.* London : British Museum xxviii 289 pp. [66 Type data: syntypes, BMNH 1946.8.18.54–56, from W.A

Distribution: NW coastal, SW coastal, W.A Ecology: open heath, woodland, open forest, ta shrubland, fossorial, nocturnal; oviparous arthropod-feeder. Biological references: Storr

G.M. (1971). The genus *Lerista* (Lacertilia : Scincidae) in Western Australia. *J. R. Soc. West. Aust.* **54**: 59–75 (taxonomy); Storr, G.M., Smith, L.A. & Johnstone, R.E. (1981). *Lizards of Western Australia. I. Skinks.* Perth : University of Western Australia Press and Western Australian Museum 200 pp. (taxonomy).

Lerista fragilis (Günther, 1876)

Rhodona fragilis Günther, A. (1876). Descriptions of new species of reptiles from Australia. *J. Mus. Godeffroy* **5**: 45–47 [45]. Type data: syntypes, BMNH 1946.8.18.68–69, ?SMF 15493, from Peak Downs, Qld.

Distribution: NE coastal, Murray-Darling basin, (Bulloo River basin), Lake Eyre basin, (N.S.W.), Qld. Ecology: open scrub, woodland, fossorial, nocturnal; oviparous, arthropod-feeder.

Lerista frosti (Zietz, 1920)

Rhodona tetradactyla Lucas, A.H.S. & Frost, C. (1895). Preliminary notice of a certain species of lizard from central Australia. *Proc. R. Soc. Vict. (ns)* **7**: 264–269 [268] [*non Mocoa tetradactyla* O'Shaugnessy, 1879]. Type data: holotype, NMV D11756, from Tempe Downs, N.T.

Lygosoma frosti Zietz, F.R. (1920). Catalogue of Australian lizards. *Rec. S. Aust. Mus.* **1**: 181–228 [217] [*nom. nov. pro Rhodona tetradactyla* Lucas and Frost, 1895, at that time a junior secondary homonym of *Mocoa tetradactyla* O'Shaugnessy, 1879].

Synonymy that of Cogger, H.G., this work.

Distribution: NW coastal, W plateau, S Gulfs, Murray-Darling basin, Lake Eyre basin, S.A., (Vic.), W.A., N.T. Ecology: desert, hummock grassland, (tussock grassland), low shrubland, open scrub, tall shrubland, fossorial, nocturnal; oviparous, arthropod-feeder. Biological references: Storr, G.M. (1976). Revisionary notes on the *Lerista* (Lacertilia : Scincidae) of Western Australia. *Rec. West. Aust. Mus.* **4**: 241–256 (taxonomy); Storr, G.M. (1971). The genus *Lerista* (Lacertilia : Scincidae) in Western Australia. *J. R. Soc. West. Aust.* **54**: 59–75 (taxonomy); Greer, A.E. (1980). Critical thermal maximum temperatures in Australian scincid lizards: their ecological and evolutionary significance. *Aust. J. Zool.* **28**: 91–102; Storr, G.M., Smith, L.A. & Johnstone, R.E. (1981). *Lizards of Western Australia. I. Skinks.* Perth : University of Western Australia Press and Western Australian Museum 200 pp. (taxonomy).

Lerista gerrardii (Gray, 1864)

Rhodona punctata gerrardii Gray, J.E. (1864). Notice of a new variety of *Rhodona punctata* from the Swan River. *Proc. Zool. Soc. Lond.* **1864**: 296 [296]. Type data: holotype, BMNH 64.10.6.6, from Swan River, W.A.

Distribution: NW coastal, SW coastal, W plateau, W.A. Ecology: low shrubland, open heath, open scrub, woodland, fossorial, nocturnal; oviparous, arthropod-feeder. Biological references: Storr, G.M. (1971). The genus *Lerista* (Lacertilia : Scincidae) in Western Australia. *J. R. Soc. West. Aust.* **54**: 59–75 (taxonomy); Greer, A.E. (1980). Critical thermal maximum temperatures in Australian scincid lizards: their ecological and evolutionary significance. *Aust. J. Zool.* **28**: 91–102; Storr, G.M., Smith, L.A. & Johnstone, R.E. (1981). *Lizards of Western Australia. I. Skinks.* Perth : University of Western Australia Press and Western Australian Museum 200 pp. (taxonomy).

Lerista humphriesi Storr, 1971

Lerista humphriesi Storr, G.M. (1971). The genus *Lerista* (Lacertilia : Scincidae) in Western Australia. *J. R. Soc. West. Aust.* **54**: 59–75 [70]. Type data: holotype, WAM R34048, from Gee Gie Outcamp 21 mi NNW of Murchison House, 27°21'S, 114°09'E, W.A.

Distribution: NW coastal, W.A. Ecology: fossorial, nocturnal; (oviparous), arthropod-feeder. Biological references: Storr, G.M. (1971). The genus *Lerista* (Lacertilia : Scincidae) in Western Australia. *J. R. Soc. West. Aust.* **54**: 59–75 (taxonomy); Storr, G.M., Smith, L.A. & Johnstone, R.E. (1981). *Lizards of Western Australia. I. Skinks.* Perth : University of Western Australia Press and Western Australian Museum 200 pp. (taxonomy).

Lerista ips Storr, 1980

Lerista ips Storr, G.M. (1980). A new *Lerista* and two new *Ctenotus* (Lacertilia : Scincidae) from Western Australia. *Rec. West. Aust. Mus.* **8**: 441–447 [441 pl 1]. Type data: holotype, WAM R63562, from Lake Auld, in 22°07'S, 123°52'E, W.A.

Distribution: W plateau, W.A., N.T. Ecology: desert, fossorial, nocturnal; (oviparous), arthropod-feeder. Biological references: Storr, G.M., Smith, L.A. & Johnstone, R.E. (1981). *Lizards of Western Australia. I. Skinks.* Perth : University of Western Australia Press and Western Australian Museum 200 pp. (taxonomy).

Lerista kalumburu Storr, 1976

Lerista kalumburu Storr, G.M. (1976). Revisionary notes on the *Lerista* (Lacertilia : Scincidae) of Western Australia. *Rec. West. Aust. Mus.* **4**: 241–256 [250]. Type data: holotype, WAM R27915, from Kalumburu, 14°18'S, 126°38'E, W.A.

Distribution: N coastal, W.A. Ecology: woodland, fossorial, nocturnal; (oviparous), arthropod-feeder. Biological references: Storr, G.M., Smith, L.A. & Johnstone, R.E. (1981). *Lizards of Western*

Australia. I. Skinks. Perth : University of Western Australia Press and Western Australian Museum 200 pp. (taxonomy).

Lerista karlschmidti (Marx and Hosmer, 1959)

Rhodona karlschmidti Marx, H. & Hosmer, W. (1959). A new skink from Australia (*Rhodona karlschmidti*, sp. nov.). *Copeia* **1959**: 207–208 [207]. Type data: holotype, FMNH 97957, from Woodstock, about 23 mi S of Townsville, N Qld.

Distribution: NE coastal, N Gulf, N coastal, Qld., N.T. Ecology: tall open shrubland, low woodland, woodland, fossorial, nocturnal; oviparous, arthropod-feeder.

Lerista labialis Storr, 1971

Lerista labialis Storr, G.M. (1971). The genus *Lerista* (Lacertilia : Scincidae) in Western Australia. *J. R. Soc. West. Aust.* **54**: 59–75 [69]. Type data: holotype, WAM R22647, from Poonda, 22°53′S, 119°42′E, W.A.

Distribution: (N Gulf), NW coastal, W plateau, S Gulfs, Murray-Darling basin, Bulloo River basin, Lake Eyre basin, S.A., Vic., N.S.W., Qld., W.A., N.T. Ecology: desert, hummock grassland, low open shrubland, low shrubland, open scrub, tall shrubland, fossorial, nocturnal; oviparous, arthropod-feeder. Biological references: Storr, G.M., Smith, L.A. & Johnstone, R.E. (1981). *Lizards of Western Australia. I. Skinks.* Perth : University of Western Australia Press and Western Australian Museum 200 pp. (taxonomy).

Lerista lineata Bell, 1833

Lerista lineata Bell, T. (1833). Characters of two new genera of reptiles. *Proc. Zool. Soc. Lond.* **1**: 98–99 [99]. Type data: holotype, whereabouts unknown, from Australia.

Distribution: SW coastal, W.A. Ecology: open heath, open forest, fossorial, nocturnal; oviparous, arthropod-feeder. Biological references: Storr, G.M. (1971). The genus *Lerista* (Lacertilia : Scincidae) in Western Australia. *J. R. Soc. West. Aust.* **54**: 59–75 (taxonomy); Jenkins, R.W.G. (1979). The status of endangered Australian reptiles. pp. 169–176 *in, The Status of Endangered Australasian Wildlife.* Adelaide : Royal Zoological Society of South Australia; Storr, G.M., Smith, L.A. & Johnstone, R.E. (1981). *Lizards of Western Australia. I. Skinks.* Perth : University of Western Australia Press and Western Australian Museum 200 pp. (taxonomy).

Lerista lineopunctulata (Duméril and Bibron, 1839)

Rhodona punctata Gray, J.E. (1839). Catalogue of the slender-tongued saurians, with descriptions of many new genera and species. (contd.). *Ann. Mag. Nat. Hist.* **2**: 331–337 [335] [*non Lacerta punctata* Linnaeus, 1758]. Type data: syntypes, BMNH 1946.9.4.73–77, from Australia.

Brachystopus lineopunctulatus Duméril, A.M.C. & Bibron, G. (1839). *Erpétologie Générale* ou Histoire Naturelle Complète des Reptiles. Paris : Roret 5: viii 854 pp. [779] [*ex* Smith ms]. Type data: holotype, MNHP 1246, from Australia (original description cites Cape of Good Hope, southern Africa), see Guibé, J. (1954). *Catalogue des Types de Lézards du Muséum National d'Histoire Naturelle.* Paris : Muséum National d'Histoire Naturelle 119 pp.

Ronia catenulata Gray, J.E. (1841). Description of some new species and four new genera of reptiles from Western Australia, discovered by John Gould, Esq. *Ann. Mag. Nat. Hist.* **7**: 86–91 [87 pl 4 fig 1]. Type data: holotype, BMNH (not found), from W.A.

Soridia miopus Günther, A. (1867). Additions to the knowledge of Australian reptiles and fishes. *Ann. Mag. Nat. Hist. (3)* **20**: 45–68 [49]. Type data: holotype, whereabouts unknown (BMNH records cite 1946.8.15.60 as holotype in error), from Champion Bay, W.A.

Lygosoma (Rhodona) bipes concolor Werner, F. (1910). Reptilia (Geckonidae und Scincidae). *in* Michaelsen, W. & Hartmeyer, R. (eds.) *Die Fauna Südwest-Australiens.* Jena : Gustav Fischer 2: 451–493 [483]. Type data: holotype, ZMB 21464, from Denham, W.A.

Lygosoma (Rhodona) nigriceps Glauert, L. (1962). Herpetological miscellanea. XIII. A new skink from the North-west Cape, Western Australia. *West. Aust. Nat.* **8**: 86–87 [86]. Type data: holotype, WAM R14039, from Vlaming Lighthouse, North West Cape, W.A.

Synonymy that of Cogger, H.G., this work.

Distribution: NW coastal, SW coastal, W.A. Ecology: hummock grassland, open heath, open scrub, woodland, fossorial, nocturnal; oviparous, arthropod-feeder. Biological references: Storr, G.M. (1971). The genus *Lerista* (Lacertilia : Scincidae) in Western Australia. *J. R. Soc. West. Aust.* **54**: 59–75 (taxonomy); Storr, G.M., Smith, L.A. & Johnstone, R.E. (1981). *Lizards of Western Australia. I. Skinks.* Perth : University of Western Australia Press and Western Australian Museum 200 pp. (taxonomy).

Lerista macropisthopus (Werner, 1903)

Lygosoma (Rhodona) macropisthopus Werner, F. (1903). Neue Reptilien und Batrachier aus dem Naturhistorischen Museum in Brussel. *Zool. Anz.* **26**: 246–253 [246]. Type data: holotype, RIB 6481 (type 2016), from Qld.

Distribution: NW coastal, (SW coastal), W plateau, W.A. Ecology: low open shrubland, open heath, open scrub, tall shrubland, fossorial, nocturnal; oviparous, arthropod-feeder. Biological references: Storr, G.M. (1971). The genus *Lerista* (Lacertilia : Scincidae) in Western Australia. *J. R. Soc. West. Aust.* **54**: 59–75 (taxonomy); Storr, G.M., Smith, L.A. & Johnstone, R.E. (1981). *Lizards of Western*

Australia. I. Skinks. Perth : University of Western Australia Press and Western Australian Museum 200 pp. (taxonomy).

Lerista microtis (Gray, 1845)

Mocoa microtis Gray, J.E. (1845). *Catalogue of the Specimens of Lizards in the Collection of the British Museum.* London : British Museum xxviii 289 pp. [83]. Type data: holotype, BMNH 1946.8.18.64, from Swan River, W.A.

Lerista microtis arenicola Storr, G.M. (1971). The genus *Lerista* (Lacertilia : Scincidae) in Western Australia. *J. R. Soc. West. Aust.* **54**: 59–75 [61]. Type data: holotype, WAM R24608, from Eucla, 31°43'S, 128°53'E, W.A.

Synonymy that of Cogger, H.G., this work.

Distribution: SW coastal, W plateau, S Gulfs, S.A., W.A. Ecology: open heath, open scrub, open forest, fossorial, nocturnal; viviparous, arthropod-feeder.

Lerista muelleri (Fischer, 1881)

Phaneropis muelleri Fischer, J.G. (1881). Beschreibung neuer Reptilien. *Arch. Naturgesch.* **47**: 225–238 [236 pl 12 figs 13–15]. Type data: syntypes, ZMB 10074 (not found), SMNS 2057, from W.A., restricted to Nickol Bay, W.A. by Storr, G.M. (1971). The genus *Lerista* (Lacertilia : Scincidae) in Western Australia. *J. R. Soc. West. Aust.* **54**: 59–75.

Ablepharus timidus De Vis, C.W. (1888). A contribution to the herpetology of Queensland. *Proc. Linn. Soc. N.S.W. (2)* **2**: 811–826 [824]. Type data: syntypes, QM J235, QM J13601, from Charleville, Qld.

Ablepharus rhodonoides Lucas, A.H.S. & Frost, C. (1896). Description of a new species of *Ablepharus* from Victoria, with critical notes on two other Australian lizards. *Proc. Linn. Soc. N.S.W.* **21**: 281–283 [281]. Type data: syntypes, AM R4156, other syntype presumed lost, from Mildura, Vic.

Lygosoma (Rhodona) goerlingi Ahl, E. (1935). Beschreibung einer neuen Eidechse aus Westaustralien. *Zool. Anz.* **112**: 204–205 [204] [synonymy with *Lerista muelleri* (Fischer, 1881) in doubt]. Type data: holotype, ZMB 35352 (not found), from Wurarga, Marloo Station, W.A.

Synonymy that of Cogger, H.G., this work.

Distribution: NW coastal, SW coastal, W plateau, S Gulfs, Murray-Darling basin, Bulloo River basin, Lake Eyre basin, S.A., Vic., N.S.W., Qld., W.A., N.T. Ecology: desert, hummock grassland, tussock grassland, low open shrubland, low shrubland, open scrub, tall shrubland, fossorial, nocturnal; oviparous, arthropod-feeder. Biological references: Storr, G.M. (1971). The genus *Lerista* (Lacertilia : Scincidae) in Western Australia. *J. R. Soc. West. Aust.* **54**: 59–75 (taxonomy); Storr, G.M., Smith, L.A. & Johnstone, R.E. (1981). *Lizards of Western Australia. I. Skinks.* Perth : University of Western Australia Press and Western Australian Museum 200 pp. (taxonomy).

Lerista neander Storr, 1971

Lerista neander Storr, G.M. (1971). The genus *Lerista* (Lacertilia : Scincidae) in Western Australia. *J. R. Soc. West. Aust.* **54**: 59–75 [65]. Type data: holotype, WAM R23988, from Mt. Newman, 23°17'S, 119°32'E, W.A.

Distribution: NW coastal, (W plateau), W.A. Ecology: fossorial, nocturnal; (oviparous), arthropod-feeder. Biological references: Greer, A.E. (1980). Critical thermal maximum temperatures in Australian scincid lizards: their ecological and evolutionary significance. *Aust. J. Zool.* **28**: 91–102; Storr, G.M., Smith, L.A. & Johnstone, R.E. (1981). *Lizards of Western Australia. I. Skinks.* Perth : University of Western Australia Press and Western Australian Museum 200 pp. (taxonomy).

Lerista nichollsi (Loveridge, 1933)

Rhodona nichollsi Loveridge, A. (1933). New scincid lizards of the genera *Sphenomorphus*, *Rhodona* and *Lygosoma* from Australia. *Occ. Pap. Bost. Soc. Nat. Hist.* **8**: 95–100 [97]. Type data: holotype, MCZ 33252, from Dalgaranger Station, 50 mi NE of Yalgoo, W.A.

Distribution: NW coastal, W.A. Ecology: low open shrubland, open heath, tall shrubland, fossorial, nocturnal; oviparous, arthropod-feeder. Biological references: Storr, G.M. (1971). The genus *Lerista* (Lacertilia : Scincidae) in Western Australia. *J. R. Soc. West. Aust.* **54**: 59–75 (taxonomy); Storr, G.M., Smith, L.A. & Johnstone, R.E. (1981). *Lizards of Western Australia. I. Skinks.* Perth : University of Western Australia Press and Western Australian Museum 200 pp. (taxonomy).

Lerista orientalis (De Vis, 1889)

Miculia orientalis De Vis, C.W. (1889). Characters of a naked-eyed skink apparently new. *Proc. R. Soc. Qd.* **5**: 160–161 [160] [1888 on title page]. Type data: holotype, from Barkly Tablelands (*ex* Gregory Downs), Qld., QM presumed lost *vide* Covacevich, J. (1971). Amphibian and reptile type specimens in the Queensland Museum. *Mem. Qd. Mus.* **16**: 49–68.

Distribution: NE coastal, N Gulf, N coastal, Qld., N.T. Ecology: low open woodland, low woodland, woodland, fossorial, nocturnal; (oviparous), arthropod-feeder. Biological references: Greer, A.E. (1980). Critical thermal maximum temperatures in Australian scincid lizards: their ecological and evolutionary significance. *Aust. J. Zool.* **28**: 91–102.

Lerista petersoni Storr, 1976

Lerista connivens petersoni Storr, G.M. (1976). Revisionary notes on the *Lerista* (Lacertilia : Scincidae) of Western Australia. *Rec. West. Aust. Mus.* **4**: 241–256 [253]. Type data: holotype, WAM R46197, from Yinnietharra, 24°39'S, 116°09'E, W.A.

Distribution: NW coastal, W.A. Ecology: fossorial, nocturnal; (oviparous), arthropod-feeder. Biological references: Storr, G.M., Smith, L.A. & Johnstone, R.E. (1981). *Lizards of Western Australia. I. Skinks.* Perth : University of Western Australia Press and Western Australian Museum 200 pp. (taxonomy).

Lerista picturata (Fry, 1914)

Lygosoma (Rhodona) picturatum Fry, D.B. (1914). On a collection of reptiles and batrachians from Western Australia. *Rec. West. Aust. Mus. and Art Gall.* **1**: 174–210 [186 pl 27 fig 3 text fig 5]. Type data: holotype, AM R3101, from Boulder, W.A.

Lerista picturata baynesi Storr, G.M. (1971). The genus *Lerista* (Lacertilia : Scincidae) in Western Australia. *J. R. Soc. West. Aust.* **54**: 59–75 [66]. Type data: holotype, WAM R24609, from Eucla, 31°43′S, 128°53′E, W.A.

Synonymy that of Cogger, H.G., this work.

Distribution: W plateau, S.A., W.A. Ecology: low shrubland, open scrub, tall shrubland, fossorial, nocturnal; oviparous, arthropod-feeder. Biological references: Storr, G.M. (1971). The genus *Lerista* (Lacertilia : Scincidae) in Western Australia. *J. R. Soc. West. Aust.* **54**: 59–75 (taxonomy); Storr, G.M., Smith, L.A. & Johnstone, R.E. (1981). *Lizards of Western Australia. I. Skinks.* Perth : University of Western Australia Press and Western Australian Museum 200 pp. (taxonomy).

Lerista planiventralis (Lucas and Frost, 1902)

Rhodona planiventralis Lucas, A.H.S. & Frost, C. (1902). Descriptions of some new lizards from Western Australia. *Proc. R. Soc. Vict.* **15**: 76–79 [78]. Type data: holotype, NMV D11763, from W.A.

Lerista planiventralis decora Storr, G.M. (1978). Taxonomic notes on the reptiles of the Shark Bay region, Western Australia. *Rec. West. Aust. Mus.* **6**: 303–318 [315]. Type data: holotype, WAM R54707, from Denham, Shark Bay, 25°55′S, 113°22′E, W.A.

Synonymy that of Cogger, H.G., this work.

Distribution: NW coastal, SW coastal, W.A. Ecology: hummock grassland, open heath, open scrub, tall shrubland, fossorial, nocturnal; oviparous, arthropod-feeder. Biological references: Warburg, M.R. (1966). On the water economy of several Australian geckos, agamids, and skinks. *Copeia* **1966**: 230–235; Storr, G.M. (1971). The genus *Lerista* (Lacertilia : Scincidae) in Western Australia. *J. R. Soc. West. Aust.* **54**: 59–75 (taxonomy); Storr, G.M., Smith, L.A. & Johnstone, R.E. (1981). *Lizards of Western Australia. I. Skinks.* Perth : University of Western Australia Press and Western Australian Museum 200 pp. (taxonomy).

Lerista praepedita (Boulenger, 1887)

Soridia lineata Gray, J.E. (1839). Catalogue of the slender-tongued saurians, with descriptions of many new genera and species. (contd.). *Ann. Mag. Nat. Hist.* **2**: 331–337 [336] [*non Lerista lineata* Bell, 1833]. Type data: syntypes, BMNH 1946.8.18.70–76, BMNH 1946.8.18.79, MCZ (*ex* BMNH 1946.8.18.77–78), from Australasia.

Pholeophilus capensis Smith, A. (1849). *Illustrations of the Zoology of South Africa.* Reptilia. London : Smith, Elder & Co. [15] [*nom. oblit. vide* Storr, G.M. (1971). The genus *Lerista* (Lacertilia : Scincidae) in Western Australia. *J. R. Soc. West. Aust.* **54**: 59–75]. Type data: syntypes, BMNH 1946.8.18.62–63, from Little Namaqualand, South Africa (in error).

Lygosoma praepeditum Boulenger, G.A. (1887). *Catalogue of the Lizards in the British Museum (Natural History).* 3. London : British Museum xii 575 pp. pls 40 [337] [*nom. nov. pro Soridia lineata* Gray, 1839].

Synonymy that of Cogger, H.G. this work.

Distribution: NW coastal, SW coastal, W.A. Ecology: hummock grassland, open heath, open scrub, tall shrubland, fossorial, nocturnal; oviparous, arthropod-feeder. Biological references: Storr, G.M. (1971). The genus *Lerista* (Lacertilia : Scincidae) in Western Australia. *J. R. Soc. West. Aust.* **54**: 59–75 (taxonomy); Storr, G.M., Smith, L.A. & Johnstone, R.E. (1981). *Lizards of Western Australia. I. Skinks.* Perth : University of Western Australia Press and Western Australian Museum 200 pp. (taxonomy).

Lerista punctatovittata (Günther, 1867)

Rhodona punctatovittata Günther, A. (1867). Additions to the knowledge of Australian reptiles and fishes. *Ann. Mag. Nat. Hist. (3)* **20**: 45–68 [47]. Type data: holotype, BMNH 1946.8.18.80, from Qld.

Rhodona officeri McCoy, F. (1881). The Victorian *Rhodona. Prod. Zool. Vict.* **6**: 7–9 [7 pl 51]. Type data: holotype, NMV (not found), from 2 mi from Swan Hill, Vic.

Synonymy that of Cogger, H.G., this work.

Distribution: (NE coastal), S Gulfs, Murray-Darling basin, Bulloo River basin, Lake Eyre basin, S.A., Vic., N.S.W., Qld. Ecology: low open shrubland, low shrubland, open scrub, low open woodland, woodland, fossorial, nocturnal; oviparous, arthropod-feeder.

Lerista separanda Storr, 1976

Lerista separanda Storr, G.M. (1976). Revisionary notes on the *Lerista* (Lacertilia : Scincidae) of Western Australia. *Rec. West. Aust. Mus.* **4**: 241–256 [246]. Type data: holotype, WAM R40268, from Coulomb Point, 17°21′S, 122°09′E, W.A.

Distribution: N coastal, W.A. Ecology: fossorial, nocturnal; (oviparous), arthropod-feeder, type "collected under bushes of *Crotalaria cunninghami* growing at foot of coastal dunes". Biological references: Storr, G.M., Smith, L.A. & Johnstone, R.E. (1981). *Lizards of Western Australia. I. Skinks.* Perth : University of Western Australia Press and Western Australian Museum 200 pp. (taxonomy).

Lerista stylis (Mitchell, 1955)

Rhodona stylis Mitchell, F.J. (1955). Preliminary account of the Reptilia and Amphibia collected by the National Geographic Society—Commonwealth Government—Smithsonian Institution Expedition to Arnhem Land (April to November, 1948). *Rec. S. Aust. Mus.* **11**: 373–408 [400 fig 6]. Type data: holotype, SAMA R3094, from Yirrkala, N.T.

Distribution: N Gulf, (N coastal), Qld., N.T. Ecology: woodland, fossorial, nocturnal; oviparous, arthropod-feeder.

Lerista terdigitata (Parker, 1926)

Lygosoma (Rhodona) terdigitatum Parker, H.W. (1926). A new lizard from South Australia. *Ann. Mag. Nat. Hist. (9)* **18**: 203–205 [203]. Type data: holotype, BMNH 1946.8.15.61, from Flinders Is., Investigator Group, S.A.

Distribution: (SW coastal), W plateau, S Gulfs, S.A., W.A. Ecology: low shrubland, open scrub, tall shrubland, fossorial, nocturnal; oviparous, arthropod-feeder. Biological references: Storr, G.M. (1971). The genus *Lerista* (Lacertilia : Scincidae) in Western Australia. *J. R. Soc. West. Aust.* **54**: 59–75 (taxonomy); Storr, G.M., Smith, L.A. & Johnstone, R.E. (1981). *Lizards of Western Australia. I. Skinks.* Perth : University of Western Australia Press and Western Australian Museum 200 pp. (taxonomy); Greer, A.E. (1980). Critical thermal maximum temperatures in Australian scincid lizards: their ecological and evolutionary significance. *Aust. J. Zool.* **28**: 91–102.

Lerista walkeri (Boulenger, 1891)

Lygosoma walkeri Boulenger, G.A. (1891). Description of a new scincoid lizard from north Western Australia. *Ann. Mag. Nat. Hist. (6)* **8**: 405 [405]. Type data: syntypes, BMNH 1946.8.15.59, BMNH 1946.8.15.62, from Roebuck Bay and Condillac Is., W.A.

Distribution: N coastal, W.A. Ecology: low open woodland, woodland, fossorial, nocturnal; (oviparous), arthropod-feeder. Biological references: Storr, G.M. (1971). The genus *Lerista* (Lacertilia : Scincidae) in Western Australia. *J. R. Soc. West. Aust.* **54**: 59–75 (taxonomy); Storr, G.M., Smith, L.A. & Johnstone, R.E. (1981). *Lizards of Western Australia. I. Skinks.* Perth : University of Western Australia Press and Western Australian Museum 200 pp. (taxonomy).

Lerista wilkinsi (Parker, 1926)

Lygosoma (Rhodona) wilkinsi Parker, H.W. (1926). New reptiles and a new frog from Queensland. *Ann. Mag. Nat. Hist. (9)* **17**: 665–670 [667]. Type data: holotype, BMNH 1946.8.18.67, from Torrens Creek, 1560 ft, 21°25′S, 145°14′E, N Qld.

Distribution: NE coastal, (N Gulf), Lake Eyre basin, Qld. Ecology: woodland, fossorial, nocturnal; (oviparous), arthropod-feeder.

Lerista xanthura Storr, 1976

Lerista xanthura Storr, G.M. (1976). Revisionary notes on the *Lerista* (Lacertilia : Scincidae) of Western Australia. *Rec. West. Aust. Mus.* **4**: 241–256 [247]. Type data: holotype, WAM R40174, from 7 km SW of the Pollock Hills, 22°54′S, 127°30′E, W.A.

Distribution: W plateau, Murray-Darling basin, (Bulloo River basin), (Lake Eyre basin), S.A., N.S.W., W.A., N.T. Ecology: desert, low shrubland, fossorial, nocturnal; (oviparous), arthropod-feeder. Biological references: Storr, G.M., Smith, L.A. & Johnstone, R.E. (1981). *Lizards of Western Australia. I. Skinks.* Perth : University of Western Australia Press and Western Australian Museum 200 pp. (taxonomy).

Lygosoma Hardwicke and Gray, 1827

Lygosoma Hardwicke, T. & Gray, J.E. (1827). A synopsis of the species of saurian reptiles, collected in India by Major-General Hardwicke. *Zool. J.* **3**: 213–229 [228]. Type species *Lacerta serpens* Bloch, 1776 [=*Lygosoma quadrupes* (Linnaeus, 1766)] by monotypy.

Podophis Wiegmann, A.F.A. (1834). *Herpetologia Mexicana,* seu Descriptio Amphibiorum Novae Hispaniae. Berlin : Lüderitz [11]. Type species *Anguis quadrupes* Linnaeus, 1766 by monotypy.

Riopa Gray, J.E. (1839). Catalogue of the slender-tongued saurians, with descriptions of many new genera and species. (contd.). *Ann. Mag. Nat. Hist.* **2**: 331–337 [332]. Type species *Lygosoma punctata* Gray, 1845 (=*Scincus punctatus* Gmelin, 1799) by subsequent designation, see Smith, M.A. (1937). A review of the genus *Lygosoma* and its allies. *Rec. Ind. Mus.* **39**: 213–234.

Chiamela Gray, J.E. (1839). Catalogue of the slender-tongued saurians, with descriptions of many new genera and species. (contd.). *Ann. Mag. Nat. Hist.* **2**: 331–337 [332]. Type species *Chiamela lineata* Gray, 1839 [=*Lygosoma lineatum* (Gray, 1839)] by subsequent designation, see Smith, M.A. (1937). A review of the genus *Lygosoma* and its allies. *Rec. Ind. Mus.* **39**: 213–234.

Hagria Gray, J.E. (1839). Catalogue of the slender-tongued saurians, with descriptions of many new genera and species. (contd.). *Ann. Mag. Nat. Hist.* **2**: 331–337 [333]. Type species *Hagria vosmaeri* Gray, 1839 [=*Lygosoma vosmaeri* (Gray, 1839)] by monotypy.

Campsodactylus Duméril, A.M.C. & Bibron, G. (1839). *Erpétologie Générale* ou Histoire Naturelle Complète des Reptiles. Paris : Roret 5: viii 854 pp. [761] [*non*

Campsodactylus Duméril, 1837, which is a *nom. nud.*]. Type species *Campsodactylus lamarrei* Duméril and Bibron, 1839 [=*Lygosoma vosmaeri* (Gray, 1839)] by monotypy.

Sphenosoma Fitzinger, L.J. (1843). *Systema Reptilium.* Vienna : Braümüller u. Seidel vi 106 pp. [23]. Type species *Eumeces punctatus* Wiegmann, 1834 (=*Lacerta punctata* Linnaeus, 1766) by original designation.

Mochlus Günther, A. (1864). Report on a collection of reptiles and fish made by Dr Kirk in the Zambesi and Nyssa regions. *Proc. Zool. Soc. Lond.* **1864**: 303–314 [308]. Type species *Mochlus punctulatus* Günther, 1864 [=*Lygosoma sundevalli* (Smith, 1849)] by monotypy.

Sepacontias Günther, A. (1880). Description of new species of reptiles from eastern Africa. *Ann. Mag. Nat. Hist. (5)* **6**: 234–238 [235]. Type species *Sepacontias modestus* Günther, 1880 by monotypy.

Lepidothyris Cope, E.D. (1892). On degenerate types of scapular and pelvic arches in the Lacertilia. *J. Morph.* **7**: 227–244 [233]. Type species *Tiligua fernandi* Burton, 1836 by subsequent designation, see Mittleman, M.B. (1952). A generic synopsis of the lizards of the subfamily Lygosominae. *Smithson. Misc. Coll.* **117**(4069): 1–35.

Paralygosoma Chabanaud, P. (1917). Énumeration des Reptiles non encore étudiés de l'Afrique occidentale, appartenant aux collections du Muséum, avec la description des espèces nouvelles. *Bull. Mus. Natl. Hist. Nat. Paris* **23**: 83–105 [103] [described as a subgenus of *Lygosoma* Hardwicke and Gray, 1827]. Type species *Lygosoma (Paralygosoma) monneti* Chabanaud, 1917 [=*Lygosoma guineensis* (Peters, 1879)] by monotypy.

Squamacilia Mittleman, M.B. (1952). A generic synopsis of the lizards of the subfamily Lygosominae. *Smithson. Misc. Coll.* **117**(4069): 1–35 [9]. Type species *Eumeces isodactylus* Günther, 1864 by original designation.

Synonymy that of Greer, A.E. (1977). The systematics and evolutionary relationships of the scincid lizard genus *Lygosoma. J. Nat. Hist.* **11**: 515–540 [518–519].

This group is also found in south-central and southeastern Asia and sub-Saharan Africa, see Greer, A.E. (1977). The systematics and evolutionary relationships of the scincid lizard genus *Lygosoma. J. Nat. Hist.* **11**: 515–540.

Lygosoma bowringii (Günther, 1864)

Eumeces bowringii Günther, A. (1864). *The Reptiles of British India.* London : Royal Society xxvii 444 pp. [91]. Type data: holotype, BMNH 1946.8.18.82, from Hong Kong.

Euprepes (Riopa) punctatolineatus Peters, W. (1871). Mitteilung über neue Eidechsen (*Egernia krefftii, Euprepes (Riopa) punctatolineatus*). *Mber. K. Preuss. Akad. Wiss. Berl.* **1871**: 30–32 [30] [*errore pro Euprepes (Riopa) punctatostriatus* Peters, 1871].

Euprepes (Riopa) punctatostriatus Peters, W. (1871). Mitteilung über neue Eidechsen (*Egernia krefftii, Euprepes (Riopa) punctatolineatus*). *Mber. K. Preuss. Akad. Wiss. Berl.* **1871**: 30–32 [31]. Type data: holotype, ZMB 7013, from Singapore, type locality erroneously given as Philippine Ils. by Smith, M.A. (1935). *The Fauna of British India; including Ceylon and Burma.* Reptilia and Amphibia. Vol. II. Sauria. London : Taylor & Francis xiv 440 pp.

Lygosoma comotti Boulenger, G.A. (1887). An account of the scincoid lizards collected in Burma, for the Genoa Civic Museum, by Messrs G.B. Comotti and L. Sea. *Ann. Mus. Civ. Stor. Giacomo Doria (2)* **4**: 618–624 [622]. Type data: holotype, MCG 6637, from Minhla, Burma.

Lygosoma whiteheadi Mocquard F. (1890) Diagnoses d'espèces nouvelles de Reptiles et de Batraciens des îles Bornéo et Palawan. *Le Naturaliste* **12**: 144–145 [144]. Type data: holotype, MNHP 89-316, from Kina-Balu, Borneo.

Synonymy that of Cogger, H.G., this work; Smith, M.A. (1935). *The Fauna of British India; including Ceylon and Burma.* Reptilia and Amphibia. Vol. II. Sauria. London : Taylor & Francis xiv 440 pp. [315].

Distribution: Christmas Is.; recent introduction to Christmas Is., widely distributed in Asia and Indo-Malayan Archipelago. Ecology: disturbed habitats, including grassy clearings in secondary growth and coconut plantations.

Menetia Gray, 1845

Menetia Gray, J.E. (1845). *Catalogue of the Specimens of Lizards in the Collection of the British Museum.* London : British Museum xxviii 289 pp. [62,65]. Type species *Menetia greyii* Gray, 1845 by monotypy.

Menetia alanae Rankin, 1979

Menetia alanae Rankin, P.R. (1979). A taxonomic revision of the genus *Menetia* (Lacertilia : Scincidae) in the Northern Territory. *Rec. Aust. Mus.* **32**: 491–499 [492 figs 1–2]. Type data: holotype, AM R52064, from immediately SE of Mt. Carr, Adelaide River township, 13°15′S, 131°06′E, N.T.

Distribution: N coastal, N.T. Ecology: litter, woodland, open forest, terrestrial, diurnal; oviparous, arthropod-feeder.

Menetia amaura Storr, 1978

Menetia amaura Storr, G.M. (1978). Taxonomic notes on the reptiles of the Shark Bay region, Western Australia. *Rec. West. Aust. Mus.* **6**: 303–318 [317]. Type data: holotype, WAM R54724, from False Entrance Well, Carrarang Station, 26°23′S, 113°19′E, W.A.

Distribution: NW coastal, W.A. Ecology: low shrubland, terrestrial, diurnal; oviparous, arthropod-feeder. Biological references: Storr, G.M., Smith, L.A. & Johnstone, R.E. (1981). *Lizards of Western Australia. I. Skinks.* Perth : University of Western Australia Press and Western Australian Museum 200 pp. (taxonomy).

Menetia greyii Gray, 1845

*[**Ablepharus lineoocellatus** (part.)]* Duméril, A.M.C. & Bibron, G. (1839). *Erpétologie Générale* ou Histoire Naturelle Complète des Reptiles. Paris : Roret 5: viii 854

pp. [817] [for lectotype see *Morethia lineoocellata* (Duméril and Bibron, 1839)]. Type data: paralectotypes, MNHP 3092 (2 specimens), from Australia.

Menetia greyii Gray, J.E. (1845). *Catalogue of the Specimens of Lizards in the Collection of the British Museum.* London : British Museum xxviii 289 pp. [66]. Type data: lectotype, BMNH 1946.8.16.9, from W.A., designation by Storr, G.M. (1976). The genus *Menetia* (Lacertilia : Scincidae) in Western Australia. *Rec. West. Aust. Mus.* **4**: 189–200.

Synonymy that of Cogger, H.G., this work.

Distribution: NE coastal, N Gulf, N coastal, NW coastal, SW coastal, W plateau, S Gulfs, Murray-Darling basin, Bulloo River basin, Lake Eyre basin, S.A., Vic., N.S.W., Qld., W.A., N.T., A.C.T. Ecology: litter, desert, hummock grassland, tussock grassland, low open shrubland, low shrubland, open heath, tall open shrubland, open scrub, low open woodland, low woodland, woodland, open forest, terrestrial, diurnal; oviparous, arthropod-feeder. Biological references: Storr, G.M. (1976). The genus *Menetia* (Lacertilia : Scincidae) in Western Australia. *Rec. West. Aust. Mus.* **4**: 189–200 (taxonomy); Ingram, G.J. (1977). Three species of small lizards— two of them new. *Vict. Nat.* **94**: 184–187 (taxonomy); Smyth, M. & Smith, M.J. (1974). Aspects of the natural history of three Australian skinks, *Morethia boulengeri, Menetia greyii* and *Lerista bougainvillii. J. Herpet.* **8**: 329–336; Rankin, P.R. (1979). A taxonomic revision of the genus *Menetia* (Lacertilia : Scincidae) in the Northern Territory. *Rec. Aust. Mus.* **32**: 491–499 (taxonomy); Storr, G.M., Smith, L.A. & Johnstone, R.E. (1981). *Lizards of Western Australia. I. Skinks.* Perth : University of Western Australia Press and Western Australian Museum 200 pp. (taxonomy).

Menetia maini Storr, 1976

Menetia maini Storr, G.M. (1976). The genus *Menetia* (Lacertilia : Scincidae) in Western Australia. *Rec. West. Aust. Mus.* **4**: 189–200 [198 fig 1]. Type data: holotype, WAM R18464, from 23 km SSE of Derby, 17°29′S, 123°43′E, W.A.

Menetia zynja Ingram, G.J. (1977). Three species of small lizards— two of them new. *Vict. Nat.* **94**: 184–187 [186]. Type data: holotype, QM J24454, from Mt. Unbunmaroo, 90 km NW of Boulia, 22°32′S, 140°18′E, W Qld.

Synonymy that of Cogger, H.G., this work.

Distribution: N Gulf, N coastal, W plateau, W.A., N.T., Qld. Ecology: hummock grassland, low open woodland, low woodland, woodland, terrestrial, diurnal; oviparous, arthropod-feeder. Biological references: Rankin, P.R. (1979). A taxonomic revision of the genus *Menetia* (Lacertilia : Scincidae) in the Northern Territory. *Rec. Aust. Mus.* **32**: 491–499 (taxonomy); Storr, G.M., Smith, L.A. & Johnstone, R.E. (1981). *Lizards of Western Australia. I. Skinks.* Perth : University of Western Australia Press and Western Australian Museum 200 pp. (taxonomy).

Menetia surda Storr, 1976

Menetia surda Storr, G.M. (1976). The genus *Menetia* (Lacertilia : Scincidae) in Western Australia. *Rec. West. Aust. Mus.* **4**: 189–200 [196 fig 1]. Type data: holotype, WAM R13249, from Budjan Creek, Corunna Downs, 21°42′S, 119°50′E, W.A.

Distribution: NW coastal, W.A. Ecology: hummock grassland, low open shrubland, low open woodland, tall shrubland, terrestrial, diurnal; oviparous, arthropod-feeder. Biological references: Storr, G.M., Smith, L.A. & Johnstone, R.E. (1981). *Lizards of Western Australia. I. Skinks.* Perth : University of Western Australia Press and Western Australian Museum 200 pp. (taxonomy).

Menetia timlowi Ingram, 1977

Menetia timlowi Ingram, G.J. (1977). Three species of small lizards— two of them new. *Vict. Nat.* **94**: 184–187 [185 figs 1–2]. Type data: holotype, QM J24940, from Barmount, 80 km NW of Marlborough, 22°32′S, 149°06′E, mideastern Qld.

Distribution: NE coastal, Murray-Darling basin, N Gulf, Qld. Ecology: open scrub, woodland, terrestrial, diurnal, oviparous, arthropod-feeder.

Morethia Gray, 1845

Morethia Gray, J.E. (1845). *Catalogue of the Specimens of Lizards in the Collection of the British Museum.* London : British Museum xxviii 289 pp. [62,65]. Type species *Morethia anomalus* Gray, 1845 by monotypy.

Morethia adelaidensis Peters, 1874

A[blepharus] (M[orethia]) anomalus (adelaidensis) Peters, W. (1874). Über einige neue Reptilien. *Mber. K. Preuss. Akad. Wiss. Berl.* **1874**: 368–377 [376] [1875 on title page]. Type data: lectotype, ZMB 4733, from Adelaide, S.A., designated by Smyth, M. (1972). The genus *Morethia* (Lacertilia : Scincidae) in South Australia. *Rec. S. Aust. Mus.* **16**: 1–14; BMNH 64.10.27.9 also designated lectotype by Storr, G.M. (1972). The genus *Morethia* (Lacertilia : Scincidae) in Western Australia. *J. R. Soc. West. Aust.* **55**: 73–79, but Smyth's designation confirmed by Rawlinson, P.A. (1976). The endemic Australian lizard genus *Morethia* (Scincidae : Lygosominae) in southern Australia. *Mem. Natl. Mus. Vict.* **37**: 27–41.

Distribution: W plateau, S Gulfs, Murray-Darling basin, Bulloo River basin, Lake Eyre basin, S.A., Vic., N.S.W., Qld., W.A., N.T. Ecology: low open shrubland, low shrubland, open scrub, low open woodland, terrestrial, diurnal; oviparous, arthropod-feeder. Biological references: Rawlinson, P.A. (1976). The endemic Australian lizard genus *Morethia* (Scincidae : Lygosominae) in southern

Australia. *Mem. Natl. Mus. Vict.* **37**: 27–41 (taxonomy); Greer, A.E. (1980). A new species of *Morethia* (Lacertilia : Scincidae) from northern Australia, with comments on the biology and relationships of the genus. *Rec. Aust. Mus.* **33**: 89–122 (taxonomy); Storr, G.M. (1972). The genus *Morethia* (Lacertilia : Scincidae) in Western Australia. *J. R. Soc. West. Aust.* **55**: 73–79 (taxonomy); Smyth, M. (1972). The genus *Morethia* (Lacertilia : Scincidae) in South Australia. *Rec. S. Aust. Mus.* **16**: 1–14 (taxonomy).

Morethia boulengeri (Ogilby, 1890)

Ablepharus boulengeri Ogilby, J.D. (1890). Re-description of an *Ablepharus* from Australia. *Rec. Aust. Mus.* **1**: 10 [10]. Type data: lectotype, AM R690, from Brawlin, N.S.W., designation by Smyth, M. (1972). The genus *Morethia* (Lacertilia : Scincidae) in South Australia. *Rec. S. Aust. Mus.* **16**: 1–14.

Distribution: SE coastal, NE coastal, (N Gulf), W plateau, S Gulfs, Murray-Darling basin, Bulloo River basin, Lake Eyre basin, S.A., Vic., N.S.W., Qld., W.A., N.T., A.C.T. Ecology: desert, low open shrubland, low shrubland, open scrub, low open woodland, low woodland, woodland, open forest, terrestrial, diurnal; oviparous, arthropod-feeder. Biological references: Rawlinson, P.A. (1976). The endemic Australian lizard genus *Morethia* (Scincidae : Lygosominae) in southern Australia. *Mem. Natl. Mus. Vict.* **37**: 27–41 (taxonomy); Greer, A.E. (1980). A new species of *Morethia* (Lacertilia : Scincidae) from northern Australia, with comments on the biology and relationships of the genus. *Rec. Aust. Mus.* **33**: 89–122 (taxonomy); Storr, G.M. (1972). The genus *Morethia* (Lacertilia : Scincidae) in Western Australia. *J. R. Soc. West. Aust.* **55**: 73–79 (taxonomy); Smyth, M. (1974). Changes in the fat stores of the skinks *Morethia boulengeri* and *Hemiergis peronii* (Lacertilia). *Aust. J. Zool.* **22**: 135–145; Smyth, M. & Smith, M.J. (1974). Aspects of the natural history of three Australian skinks, *Morethia boulengeri, Menetia greyii* and *Lerista bougainvillii. J. Herpet.* **8**: 329–336; Smyth, M. (1972). The genus *Morethia* (Lacertilia : Scincidae) in South Australia. *Rec. S. Aust. Mus.* **16**: 1–14 (taxonomy); Crome, B. (1981). The diet of some ground-layer lizards in three woodlands of the New England Tableland of Australia. *Herpetofauna* **13**: 4–11.

Morethia butleri (Storr, 1963)

Ablepharus butleri Storr, G.M. (1963). *Ablepharus butleri.* A new scincid lizard from Australia. *West. Aust. Nat.* **9**: 46–47 [46]. Type data: holotype, WAM R20615, from 4 mi E of Leonora, 28°52'S, 121°23'E, W.A.

Distribution: NW coastal, SW coastal, W plateau, S Gulfs, S.A., W.A. Ecology: low open shrubland, open scrub, tall shrubland, terrestrial, diurnal; oviparous, arthropod-feeder. Biological references: Rawlinson, P.A. (1976). The endemic Australian lizard genus *Morethia* (Scincidae : Lygosominae) in southern Australia. *Mem. Natl. Mus. Vict.* **37**: 27–41 (taxonomy); Greer, A.E. (1980). A new species of *Morethia* (Lacertilia : Scincidae) from northern Australia, with comments on the biology and relationships of the genus. *Rec. Aust. Mus.* **33**: 89–122 (taxonomy); Storr, G.M. (1972). The genus *Morethia* (Lacertilia : Scincidae) in Western Australia. *J. R. Soc. West. Aust.* **55**: 73–79 (taxonomy); Smyth, M. (1972). The genus *Morethia* (Lacertilia : Scincidae) in South Australia. *Rec. S. Aust. Mus.* **16**: 1–14 (taxonomy).

Morethia lineoocellata (Duméril and Bibron, 1839)

***Ablepharus lineoocellatus* (part.)** Duméril, A.M.C. & Bibron, G. (1839). *Erpétologie Générale* ou Histoire Naturelle Complète des Reptiles. Paris : Roret 5: viii 854 pp. [817]. Type data: lectotype, MNHP 3092 (pentadactyl syntype with snout-vent length 42 mm, tail 51 mm), from Australia, designation by Rawlinson, P.A. (1976). The endemic Australian lizard genus *Morethia* (Scincidae : Lygosominae) in southern Australia. *Mem. Natl. Mus. Vict.* **37**: 27–41. For the other 2 specimens under this number, see synonymy of *Menetia greyii* Gray, 1845.

***Morethia anomalus* (part.)** Gray, J.E. (1845). *Catalogue of the Specimens of Lizards in the Collection of the British Museum.* London : British Museum xxviii 289 pp. [65]. Type data: lectotype, BMNH 1946.8.15.75, from W.A., designation by Storr, G.M. (1972). The genus *Morethia* (Lacertilia : Scincidae) in Western Australia. *J. R. Soc. West. Aust.* **55**: 73–79 [for the paralectotype, BMNH 1946.8.15.74, see under *Morethia obscura* Storr, 1972].

Synonymy that of Cogger, H.G., this work.

Distribution: NW coastal, SW coastal, W.A. Ecology: open heath, woodland, open forest, tall shrubland, terrestrial, diurnal; oviparous, arthropod-feeder. Biological references: Rawlinson, P.A. (1976). The endemic Australian lizard genus *Morethia* (Scincidae : Lygosominae) in southern Australia. *Mem. Natl. Mus. Vict.* **37**: 27–41 (taxonomy); Greer, A.E. (1980). A new species of *Morethia* (Lacertilia : Scincidae) from northern Australia, with comments on the biology and relationships of the genus. *Rec. Aust. Mus.* **33**: 89–122 (taxonomy); Storr, G.M. (1972). The genus *Morethia* (Lacertilia : Scincidae) in Western Australia. *J. R. Soc. West. Aust.* **55**: 73–79 (taxonomy); Smyth, M. (1972). The genus

Morethia (Lacertilia : Scincidae) in South Australia. *Rec. S. Aust. Mus.* **16**: 1–14 (taxonomy).

Morethia obscura Storr, 1972

[Morethia anomalus* (part.)*] Gray, J.E. (1845). *Catalogue of the Specimens of Lizards in the Collection of the British Museum.* London : British Museum xxviii 289 pp. [65] [for the lectotype, BMNH 1946.8.15.75, see under *Morethia lineoocellata* (Duméril and Bibron, 1839)]. Type data: paralectotype, BMNH 1946.8.15.74, from W.A.

Morethia obscura Storr, G.M. (1972). The genus *Morethia* (Lacertilia : Scincidae) in Western Australia. *J. R. Soc. West. Aust.* **55**: 73–79 [76]. Type data: holotype, WAM R16916, from 6 mi E of Kalamunda, 31°58'S, 116°08'E, W.A.

Synonymy that of Cogger, H.G., this work.

Distribution: NW coastal, SW coastal, W plateau, S Gulfs, Murray-Darling basin, S.A., Vic., N.S.W., W.A. Ecology: open heath, open scrub, low open woodland, (woodland), terrestrial, diurnal; oviparous, arthropod-feeder. Biological references: Rawlinson, P.A. (1976). The endemic Australian lizard genus *Morethia* (Scincidae : Lygosominae) in southern Australia. *Mem. Natl. Mus. Vict.* **37**: 27–41 (taxonomy); Greer, A.E. (1980). A new species of *Morethia* (Lacertilia : Scincidae) from northern Australia, with comments on the biology and relationships of the genus. *Rec. Aust. Mus.* **33**: 89–122 (taxonomy).

Morethia ruficauda (Lucas and Frost, 1895)

Ablepharus lineoocellatus ruficaudus Lucas, A.H.S. & Frost, C. (1895). Preliminary notice of a certain species of lizard from central Australia. *Proc. R. Soc. Vict. (ns)* **7**: 264–269 [269]. Type data: lectotype, from either Goyder River or Bagot's Lagoon, Central Australia, N.T., NMV D11757, designation by Coventry, A.J. (1970). Reptile and amphibian type specimens housed in the National Museum of Victoria. *Mem. Natl. Mus. Vict.* **31**: 115–124.

Morethia taeniopleura exquisita Storr, G.M. (1972). The genus *Morethia* (Lacertilia : Scincidae) in Western Australia. *J. R. Soc. West. Aust.* **55**: 73–79 [75]. Type data: holotype, WAM R37709, from Tambrey, 21°37'S, 117°36'E, W.A.

Synonymy that of Greer, A.E. (1980). A new species of *Morethia* (Lacertilia : Scincidae) from northern Australia, with comments on the biology and relationships of the genus. *Rec. Aust. Mus.* **33**: 89–122. Cogger, H.G., this work.

Distribution: N Gulf, N coastal, NW coastal, W plateau, Lake Eyre basin, S.A., W.A., N.T. Ecology: desert, hummock grassland, low open shrubland, low open woodland, low woodland, woodland, open forest, tall shrubland, terrestrial, arboreal, diurnal; oviparous, saxicoline, arthropod-feeder. Biological references: Greer, A.E. (1980). A new species of *Morethia* (Lacertilia : Scincidae) from northern Australia, with comments on the biology and relationships of the genus. *Rec. Aust. Mus.* **33**: 89–122 (taxonomy).

Morethia storri Greer, 1980

Morethia storri Greer, A.E. (1980). A new species of *Morethia* (Lacertilia : Scincidae) from northern Australia, with comments on the biology and relationships of the genus. *Rec. Aust. Mus.* **33**: 89–122 [89 figs 1–3]. Type data: holotype, NTM R1815, from 4.5 km S of Noonamah, in 12°40'S, 131°04'E, N.T.

Distribution: N Gulf, N coastal, W.A., N.T. Ecology: tall open shrubland, woodland, open forest, terrestrial, (arboreal), diurnal; oviparous, arthropod-feeder.

Morethia taeniopleura (Peters, 1874)

Ablepharus (Morethia) taeniopleura Peters, W. (1874). Über einige neue Reptilien. *Mber. K. Preuss. Akad. Wiss. Berl.* **1874**: 368–377 [375] [1875 on title page]. Type data: holotype, ZMB 8051, from Port Clinton (as Port Bowen), Qld.

Distribution: NE coastal, N Gulf, (N.S.W.), Qld. Ecology: open scrub, woodland, terrestrial, diurnal; oviparous, saxicoline, arthropod-feeder. Biological references: Storr, G.M. (1972). The genus *Morethia* (Lacertilia : Scincidae) in Western Australia. *J. R. Soc. West. Aust.* **55**: 73–79 (taxonomy); Greer, A.E. (1980). A new species of *Morethia* (Lacertilia : Scincidae) from northern Australia, with comments on the biology and relationships of the genus. *Rec. Aust. Mus.* **33**: 89–122 (taxonomy); Smyth, M. (1972). The genus *Morethia* (Lacertilia : Scincidae) in South Australia. *Rec. S. Aust. Mus.* **16**: 1–14 (taxonomy).

Notoscincus Fuhn, 1969

Notoscincus Fuhn, I.E. (1969). The "polyphyletic" origin of the genus *Ablepharus* (Reptilia : Scincidae). A case of parallel evolution. *Sond. Aus. Z. f. Zool. Syst. u. Evol. Forsch.* **7**: 67–76 [72]. Type species *Ablepharus wotjulum* Glauert, 1959 by subsequent designation, see Greer, A.E. (1974). The generic relationships of the scincid lizard genus *Leiolopisma* and its relatives. *Aust. J. Zool. Suppl. Ser.* **31**: 1–67.

Notoscincus butleri Storr, 1979

Notoscincus butleri Storr, G.M. (1979). Five new lizards from Western Australia. *Rec. West. Aust. Mus.* **8**: 134–142 [140 pl 4]. Type data: holotype, WAM R61518, from 30 km SSW of Dampier, 20°55'S, 116°39'E, W.A.

Distribution: NW coastal, W.A. Ecology: litter, low open woodland, terrestrial, diurnal; oviparous, arthropod-feeder. Biological references: Storr, G.M., Smith, L.A. & Johnstone, R.E. (1981).

Lizards of Western Australia. I. Skinks. Perth : University of Western Australia Press and Western Australian Museum 200 pp. (taxonomy).

Notoscincus ornatus (Broom, 1896)

Ablepharus ornatus Broom, R. (1896). On two new species of *Ablepharus* from north Queensland. *Ann. Mag. Nat. Hist. (6)* **18**: 342–343 [343]. Type data: holotype, BMNH 1946.8.18.53, from Muldiva, Qld.

Ablepharus wotjulum Glauert, L. (1959). Herpetological miscellanea. IX. *Ablepharus wotjulum*, a new skink from West Kimberley. *West. Aust. Nat.* **6**: 192–193 [192]. Type data: holotype, WAM R11799, from Wotjulum Mission, opposite Yampi Sound, West Kimberley, W.A.

Synonymy that of Cogger, H.G., this work.

Distribution: N Gulf, N coastal, NW coastal, W plateau, Lake Eyre basin, Qld., W.A., N.T. Ecology: litter, desert, hummock grassland, low open shrubland, low open woodland, low woodland, woodland, tall shrubland, terrestrial, diurnal; oviparous, arthropod-feeder. Biological references: Storr, G.M. (1974). The genus *Notoscincus* (Lacertilia : Scincidae) in Western Australia and Northern Territory. *Rec. West. Aust. Mus.* **3**: 111–114 (taxonomy); Greer, A.E. (1974). The generic relationships of the scincid lizard genus *Leiolopisma* and its relatives. *Aust. J. Zool. Suppl. Ser.* **31**: 1–67 (taxonomy); Storr, G.M., Smith, L.A. & Johnstone, R.E. (1981). *Lizards of Western Australia. I. Skinks.* Perth : University of Western Australia Press and Western Australian Museum 200 pp. (taxonomy).

Proablepharus Fuhn, 1969

Proablepharus Fuhn, I.E. (1969). The "polyphyletic" origin of the genus *Ablepharus* (Reptilia : Scincidae). A case of parallel evolution. *Sond. Aus. Z. f. Zool. Syst. u. Evol. Forsch.* **7**: 67–76 [73]. Type species *Ablepharus reginae* Glauert, 1960 by subsequent designation, see Greer, A.E. (1974). The generic relationships of the scincid lizard genus *Leiolopisma* and its relatives. *Aust. J. Zool. Suppl. Ser.* **31**: 1–67.

Proablepharus kinghorni (Copland, 1947)

Ablepharus kinghorni Copland, S.J. (1947). Taxonomic notes on the genus *Ablepharus* (Sauria : Scincidae). I. A new species from the Darling River. *Proc. Linn. Soc. N.S.W.* **71**: 282–286 [282 figs 2–3 pl 18]. Type data: holotype, AM R6458, from Darling River, between Bourke ("30.5 S, 145.58 E") and Wilcannia ("31.28 S, 143.13 E"), N.S.W.

Distribution: (NE coastal), N Gulf, W plateau, Lake Eyre basin, Murray-Darling basin, Bulloo River basin, S.A., N.S.W., Qld., N.T. Ecology: litter, tussock grassland, low open shrubland, low open woodland, terrestrial, diurnal; oviparous, arthropod-feeder.

Proablepharus reginae (Glauert, 1960)

Ablepharus reginae Glauert, L. (1960). Herpetological miscellanea. XII. The family Scincidae in Western Australia. Pt. 3. The genus *Ablepharus. West. Aust. Nat.* **7**: 115–122 [119]. Type data: holotype, WAM R3300, from Queen Victoria Spring, W.A.

Distribution: N coastal, NW coastal, W plateau, Lake Eyre basin, S.A., W.A., N.T. Ecology: litter, desert, low open shrubland, low woodland, tall shrubland, terrestrial, diurnal; oviparous, arthropod-feeder. Biological references: Storr, G.M. (1975). The genus *Proablepharus* (Scincidae : Lacertilia) in Western Australia. *Rec. West. Aust. Mus.* **3**: 335–338 (taxonomy).

Proablepharus tenuis (Broom, 1896)

Ablepharus tenuis Broom, R. (1896). On two new species of *Ablepharus* from north Queensland. *Ann. Mag. Nat. Hist. (6)* **18**: 342–343 [342]. Type data: holotype, BMNH 96.11.21.1, from Muldiva, Qld.

Ablepharus broomensis Lönnberg, E. & Andersson, G.L. (1913). Results of Dr. E. Mjöberg's Swedish Scientific Expeditions to Australia 1910–1913. III. Reptiles. *K. Svenska Vetensk.-Akad. Handl.* **52**(3): 1–17 [11]. Type data: syntypes, NHRM 3098, NHRM 3101, from Broome, W.A.

Ablepharus davisi Copland, S.J. (1952). Taxonomic notes on the genus *Ablepharus* (Sauria : Scincidae). III. A new species from north-west Australia. *Proc. Linn. Soc. N.S.W.* **77**: 121–125 [121 figs 2–3 pl 5]. Type data: holotype, AM R18684, from Harding Ranges, 8 mi W to 8 mi NW from Munja Station, Walcott Inlet, NW Australia, N.T.

Synonymy that of Cogger, H.G., this work.

Distribution: (NE coastal), N Gulf, N coastal, Qld., W.A., N.T. Ecology: litter, low open woodland, low woodland, woodland, terrestrial, diurnal; oviparous, arthropod-feeder. Biological references: Cogger, H.G. & Lindner, D.A. (1974). Frogs and reptiles in fauna survey of the Port Essington District, Cobourg Peninsula, Northern Territory of Australia. Frith, H.J. & Calaby, J.H. (eds.) *CSIRO Div. Wildl. Res. Tech. Pap.* **28**: 63–107 (taxonomy); Storr, G.M. (1975). The genus *Proablepharus* (Scincidae : Lacertilia) in Western Australia. *Rec. West. Aust. Mus.* **3**: 335–338 (taxonomy).

Pseudemoia Fuhn, 1967

Pseudemoia Fuhn, I.E. (1967). *Pseudemoia*, eine neue monotypische Gattung aus Südostaustralien (*Ablepharus/Emoa/spenceri* Lucas and Frost, 1894). *Zool. Anz.* **179**: 243–247 [244]. Type species *Lygosoma (Emoa) spenceri* Lucas and Frost, 1894 by original designation.

Pseudemoia palfreymani Rawlinson, 1974

Pseudemoia palfreymani Rawlinson, P.A. (1974). Revision of the endemic southeastern Australian lizard genus *Pseudemoia* (Scincidae : Lygosominae). *Mem. Natl. Mus. Vict.* **35**: 87–96 [92 fig 1 pl 5 fig 1]. Type data: holotype, NMV D8868, from Pedra Branca (as Pedra Blanca) Is., 43°52'S, 146°59'E, off southern coast of Tas.

Distribution: Tas. Ecology: terrestrial, diurnal; (viviparous), saxicoline, arthropod-feeder.

Pseudemoia spenceri (Lucas and Frost, 1894)

Lygosoma (Emoa) spenceri Lucas, A.H.S. & Frost, C. (1894). The lizards indigenous to Victoria. *Proc. R. Soc. Vict. (ns)* **6**: 24–92 [81 pl 2 figs 1 1a]. Type data: lectotype, NMV D1824, from Brandy Creek, Vic. (original description cites Brandy Creek and Dandenong Ranges, Vic.), designation by Coventry, A.J. (1970). Reptile and amphibian type specimens housed in the National Museum of Victoria. *Mem. Natl. Mus. Vict.* **31**: 115–124.

Lygosoma (Liolepisma) weekesae Kinghorn, J.R. (1929). A new species of *Lygosoma* from New South Wales. *Proc. Linn. Soc. N.S.W.* **54**: 32–33 [32 fig]. Type data: holotype, AM R9745, from Mt. Kosciusko, 5,500 ft, N.S.W.

Synonymy that of Cogger, H.G., this work.

Distribution: SE coastal, (Murray-Darling basin), Vic., N.S.W., A.C.T. Ecology: alpine, woodland, open forest, tall forest, terrestrial, diurnal; viviparous, arthropod-feeder. Biological references: Fuhn, I.E. (1967). *Pseudemoia*, eine neue monotypische Gattung aus Südostaustralien (*Ablepharus/Emoa/spenceri* Lucas and Frost, 1894). *Zool. Anz.* **179**: 243–247; Rawlinson, P.A. (1974). Revision of the endemic southeastern Australian lizard genus *Pseudemoia* (Scincidae : Lygosominae). *Mem. Natl. Mus. Vict.* **35**: 87–96; Greer, A.E. (1974). The generic relationships of the scincid lizard genus *Leiolopisma* and its relatives. *Aust. J. Zool. Suppl. Ser.* **31**: 1–67.

Saiphos Gray, 1831

Saiphos Gray, J.E. (1831). A synopsis of the species of the class Reptilia, pp. 1–110 *in* Griffith, E. *The Animal Kingdom Arranged in Conformity with its Organization by the Baron Cuvier.* London : Whittaker, Treacher & Co. 9: 481+110 pp. [72]. Type species *Seps equalis* Gray, 1825 by monotypy.

Peromeles Wiegmann, A.F.A. (1834). *Herpetologia Mexicana,* seu Descriptio Amphibiorum Novae Hispaniae. Berlin : Lüderitz [11]. Type species *Seps equalis* Gray, 1825 by monotypy.

Siaphos Gray, J.E. (1839). Catalogue of the slender-tongued saurians, with descriptions of many new genera and species. (contd.). *Ann. Mag. Nat. Hist.* **2**: 331–337 [333] [emend. *pro Saiphos* Gray, 1831].

Synonymy that of Cogger, H.G., this work; Smith, M.A. (1937). A review of the genus *Lygosoma* and its allies. *Rec. Ind. Mus.* **39**: 213–234.

Saiphos equalis (Gray, 1825)

Seps equalis Gray, J.E. (1825). A synopsis of the genera of reptiles and Amphibia, with a description of some new species. *Annls. Philos. (2)* **10**: 193–217 [202]. Type data: holotype, BMNH 1946.8.16.88, from unknown locality.

Anguis lacertina Gray, J.E. (1831). A synopsis of the species of the class Reptilia, pp. 1–110 *in* Griffith, E. *The Animal Kingdom Arranged in Conformity with its Organization by the Baron Cuvier.* London : Whittaker, Treacher & Co. 9: 481+110 pp. [72] [*ex* E.W.Gray ms; *nom. nud.*, introduced in synonymy].

Synonymy that of Cogger, H.G., this work.

Distribution: SE coastal, NE coastal, N.S.W., Qld. Ecology: open forest, tall forest, fossorial, nocturnal; oviparous, arthropod-feeder. Biological references: Bustard, H.R. (1964). Reproduction in the Australian rain forest skinks, *Siaphos equalis* and *Sphenomorphus tryoni*. *Copeia* **1964**: 715–716 (reproduction).

Sphenomorphus Fitzinger, 1843

Eulamprus Fitzinger, L.J. (1843). *Systema Reptilium.* Vienna : Braümüller u. Seidel vi 106 pp. [22]. Type species *Lygosoma quoyii* Duméril and Bibron, 1839 by original designation.

Sphenomorphus Fitzinger, L.J. (1843). *Systema Reptilium.* Vienna : Braümüller u. Seidel vi 106 pp. [23] [*Sphenomorphus* given priority over *Eulamprus* on basis of common usage]. Type species *Lygosoma melanopogon* Duméril and Bibron, 1839 [=*Sphenomorphus melanopogon* (Duméril and Bibron, 1839)] by original designation.

Hinulia Gray, J.E. (1845). *Catalogue of the Specimens of Lizards in the Collection of the British Museum.* London : British Museum xxviii 289 pp. [70,74]. Type species *Lygosoma quoyii* Duméril and Bibron, 1839 by subsequent designation, see Mittleman, M.B. (1952). A generic synopsis of the lizards of the subfamily Lygosominae. *Smithson. Misc. Coll.* **117**(4069): 1–35.

Elania Gray, J.E. (1845). *Catalogue of the Specimens of Lizards in the Collection of the British Museum.* London : British Museum xxviii 289 pp. [70,80] [*non Elania* Gray, 1840]. Type species *Scincus mulleri* Schlegel, 1837 [=*Sphenomorphus mulleri* (Schlegel, 1837)] by monotypy.

Lissonota Blyth, E. (1853). Notes and descriptions of various reptiles, new or little known. *J. Asiat. Soc. Beng.* **22**: 639–655 [653] [*non Lissonota* Gravenhorst, 1829]. Type species *Lissonota maculata* Blyth, 1853 [=*Sphenomorphus maculatus* (Blyth, 1853)] by monotypy.

Calyptotis De Vis, C.W. (1886). On a lizard and three species of *Salarias* etc. *Proc. R. Soc. Qd.* **2**: 56–60 [57] [1885 on title page]. Type species *Calyptotis flaviventer* De Vis, 1886 by monotypy.

Ictiscincus Smith, M.A. (1937). A review of the genus *Lygosoma* and its allies. *Rec. Ind. Mus.* **39**: 213–234 [222] [*nom. substit. pro Elania* Gray, 1845]. Type species that of *Elania* Gray, 1845.

Synonymy that of Cogger, H.G., this work.

This group is also found in Asia, Indo-Australian Archipelago, New Guinea and islands of W Pacific.

Sphenomorphus amplus Covacevich and McDonald, 1980

Sphenomorphus amplus Covacevich, J. & McDonald, K.R. (1980). Two new species of skinks from mid-eastern Queensland rainforest. *Mem. Qd. Mus.* **20**: 95–101 [97 pl 16]. Type data: holotype, QM J26054, from Finch Hatton Creek, Eungella National Park, mid-E Qld.

Distribution: NE coastal, Qld. Ecology: closed forest, terrestrial, diurnal; arthropod-feeder.

Sphenomorphus brongersmai Storr, 1972

Sphenomorphus brongersmai Storr, G.M. (1972). Revisionary notes on the *Sphenomorphus isolepis* complex (Lacertilia : Scincidae). *Zool. Meded.* **47**: 1–5 [1]. Type data: holotype, WAM R34707, from Kalumburu, 14°18′S, 126°38′E, W.A.

Distribution: N coastal, W.A. Ecology: litter, woodland, terrestrial, nocturnal; arthropod-feeder. Biological references: Storr, G.M., Smith, L.A. & Johnstone, R.E. (1981). *Lizards of Western Australia. I. Skinks.* Perth : University of Western Australia Press and Western Australian Museum 200 pp. (taxonomy).

Sphenomorphus crassicaudus (Duméril and Duméril, 1851)

L[ygosoma] crassicaudum Duméril, A.M.C. & Duméril, A. (1851). *Catalogue Méthodique de la Collection des Reptiles.* Paris : Mus. Hist. Nat. iv 224 pp. [172] [pl 4 fig 1 in Jacquinot, H. & Guichenot, M.A. (1853). Reptiles et Poissons, *in* Hombron & Jacquinot, H. *in* Dumont d'Urville, M.J. (ed.) *Voyage au Pole Sud et dans l'Océanie sur les corvettes l'*Astrolabe *et la* Zélée, *executé pendant les années 1837–40.* Atlas Sauriens. Zoologie 3:1–28 Paris : Gide et Baudry]. Type data: holotype, MNHP 2979 (Oceania), see Guibé, J. (1954). *Catalogue des Types de Lézards du Muséum National d'Histoire Naturelle.* Paris : Muséum National d'Histoire Naturelle 119 pp., from Oceania (original description cites Australia and Oceania).

Sphenomorphus crassicaudus arnhemicus Storr, G.M. (1967). The genus *Sphenomorphus* (Lacertilia : Scincidae) in Western Australia and the Northern Territory. *J. R. Soc. West. Aust.* **50**: 10–20 [18]. Type data: holotype, WAM R13513, from Yirrkala, 12°15′S, 136°52′E, N.T.

Sphenomorphus crassicaudus darwiniensis Storr, G.M. (1967). The genus *Sphenomorphus* (Lacertilia : Scincidae) in Western Australia and the Northern Territory. *J. R. Soc. West. Aust.* **50**: 10–20 [19]. Type data: holotype, WAM R23624, from Howard Springs (15 air-miles E of Darwin), 12°28′S, 131°03′E, N.T.

Synonymy that of Cogger, H.G., this work.

Distribution: NE coastal, N Gulf, N coastal, Qld., W.A., N.T. Ecology: tall open shrubland, woodland, fossorial, nocturnal; arthropod-feeder. Biological references: Cogger, H.G. & Lindner, D.A. (1974). Frogs and reptiles in fauna survey of the Port Essington District, Cobourg Peninsula, Northern Territory of Australia. Frith, H.J. & Calaby, J.H. (eds.) *CSIRO Div. Wildl. Res. Tech. Pap.* **28**: 63–107 (taxonomy).

Sphenomorphus douglasi Storr, 1967

Sphenomorphus isolepis douglasi Storr, G.M. (1967). The genus *Sphenomorphus* (Lacertilia : Scincidae) in Western Australia and the Northern Territory. *J. R. Soc. West. Aust.* **50**: 10–20 [16]. Type data: holotype, WAM R23446, from Darwin, 12°25′S, 130°49′E, N.T.

Distribution: N coastal, N.T. Ecology: woodland, terrestrial, nocturnal; arthropod-feeder. Biological references: Cogger, H.G. & Lindner, D.A. (1974). Frogs and reptiles in fauna survey of the Port Essington District, Cobourg Peninsula, Northern Territory of Australia. Frith, H.J. & Calaby, J.H. (eds.) *CSIRO Div. Wildl. Res. Tech. Pap.* **28**: 63–107 (taxonomy); Storr, G.M. (1972). Revisionary notes on the *Sphenomorphus isolepis* complex (Lacertilia : Scincidae). *Zool. Meded.* **47**: 1–5 (taxonomy).

Sphenomorphus fuscicaudis Greer, 1979

Sphenomorphus fuscicaudis Greer, A.E. (1979). A new *Sphenomorphus* (Lacertilia : Scincidae) from the rainforests of northeastern Queensland. *.Rec. Aust. Mus.* **32**: 373–383 [373 fig 1]. Type data: holotype, QM J25218, from Mt. Finnigan (3,700 ft), Mt. Finnigan National Park, NE Qld.

Distribution: NE coastal, Qld. Ecology: closed forest, terrestrial, diurnal; oviparous, arthropod-feeder.

Sphenomorphus gracilipes (Steindachner, 1870)

Lygosoma australis Gray, J.E. (1839). Catalogue of the slender-tongued saurians, with descriptions of many new genera and species. (contd.). *Ann. Mag. Nat. Hist.* **2**: 331–337 [332] [*non Tiliqua australis* Gray, 1838; *Tiliqua australis* Gray, 1838 and *Lygosoma australis* Gray, 1839, were both placed in the genus *Lygosoma* by Boulenger, G.A. (1887). *Catalogue of the Lizards in the British Museum (Natural History).* 3. London : British Museum xii 575 pp. pls 1–40, who erroneously treated *Tiliqua australis* as the junior secondary homonym. While it can be argued that *Lygosoma australis* has never been formally rejected as a junior homonym and therefore is available and has priority over *Hinulia gracilipes* Steindachner, 1870, under Article 59 of the International Code of Zoological Nomenclature, we have arbitrarily treated *Lygosoma australis* Gray, 1839 as a junior secondary homonym of *Tiliqua australis* Gray, 1838, rejected prior to 1961 (see Loveridge, A. (1934). Australian reptiles in the Museum of Comparative Zoology, Cambridge, Massachusetts. *Bull. Mus. Comp.*

Zool. **77**: 243–383)]. Type data: neotype, WAM R24980, from Albany, W.A., designation by Storr, G.M. (1967). The genus *Sphenomorphus* (Lacertilia : Scincidae) in Western Australia and the Northern Territory. *J. R. Soc. West. Aust.* **50**: 10–20. Original type locality "Australia".

Hinulia gracilipes Steindachner, F. (1870). Herpetologische Notizen (II). Reptilien gesammelt Während einer Reise in Sengambien. *Sber. Akad. Wiss. Wien* **62**: 326–349 [342 pl 5]. Type data: syntypes (4), NHMW 10141 (2 specimens), ?NHMW 10140, from Australia, perhaps Rockhampton or Cape York, Qld.

Synonymy that of Cogger, H.G., this work.

Distribution: SW coastal, W.A. Ecology: woodland, open forest, terrestrial, fossorial, nocturnal; arthropod-feeder. Biological references: Storr, G.M. (1967). The genus *Sphenomorphus* (Lacertilia : Scincidae) in Western Australia and the Northern Territory. *J. R. Soc. West. Aust.* **50**: 10–20 (taxonomy, as *Sphenomorphus australis*).

Sphenomorphus isolepis (Boulenger, 1887)

Lygosoma isolepis Boulenger, G.A. (1887). *Catalogue of the Lizards in the British Museum (Natural History).* 3. London : British Museum xii 575 pp. pls 40 [234 pl 15 fig 1]. Type data: syntypes, BMNH 1946.8.3.45–46 (Swan River), BMNH 1946.8.17.14 (Nickol Bay), other syntypes not located, from Nickol Bay and Swan River, W.A., the latter in error, see Storr, G.M. (1967). The genus *Sphenomorphus* (Lacertilia : Scincidae) in Western Australia and the Northern Territory. *J. R. Soc. West. Aust.* **50**: 10–20.

Lygosoma (Hinulia) isolepis foresti Kinghorn, J.R. (1932). Herpetological notes. 4. *Rec. Aust. Mus.* **18**: 335–363 [358]. Type data: holotype, AM R10001, from Forrest (as Forest) River, East Kimberley, W.A.

Synonymy that of Cogger, H.G., this work.

Distribution: N Gulf, N coastal, NW coastal, W plateau, W.A., N.T. Ecology: woodland, open forest, terrestrial, nocturnal; arthropod-feeder. Biological references: Storr, G.M. (1972). Revisionary notes on the *Sphenomorphus isolepis* complex (Lacertilia : Scincidae). *Zool. Meded.* **47**: 1–5 (taxonomy).

Sphenomorphus kosciuskoi (Kinghorn, 1932)

Lygosoma (Hinulia) quoyi kosciuskoi Kinghorn, J.R. (1932). Herpetological notes. 4. *Rec. Aust. Mus.* **18**: 335–363 [359]. Type data: holotype, AM R4654, from Mt. Kosciusko, 5,000 ft, N.S.W.

Distribution: SE coastal, Murray-Darling basin, Vic., N.S.W., A.C.T. Ecology: alpine, open forest, tall forest, terrestrial, amphibious, diurnal; viviparous, arthropod-feeder. Biological references: Done, B.S. & Heatwole, H. (1977). Effects of hormones on the aggressive behaviour and social organisation of the scincid lizard, *Sphenomorphus kosciuskoi*. *Z. Tierpsychol.* **44**: 1–12; Done, B.S. & Heatwole, H. (1977). Social behavior of some Australian skinks. *Copeia* **1977**: 419–430; Spellerberg, I.F. (1972). Thermal ecology of allopatric lizards (*Sphenomorphus*) in southeast Australia. I. The environment and lizard critical temperatures. *Oecologia* **9**: 371–383; Spellerberg, I.F. (1972). Thermal ecology of allopatric lizards (*Sphenomorphus*) in southeast Australia. II. Physiological aspects of thermoregulation. *Oecologia* **9**: 385–398; Spellerberg, I.F. (1972). Thermal ecology of allopatric lizards (*Sphenomorphus*) in southeast Australia III. Behavioural aspects of thermo-regulation. *Oecologia* **11**: 1–16; Rawlinson, P.A. (1969). East Gippsland Symposium. 7. The reptiles of East Gippsland. *Proc. R. Soc. Vict.* **82**: 113–128 (taxonomy).

Sphenomorphus luteilateralis Covacevich and McDonald, 1980

Sphenomorphus luteilateralis Covacevich, J. & McDonald, K.R. (1980). Two new species of skinks from mid-eastern Queensland rainforest. *Mem. Qd. Mus.* **20**: 95–101 [96 pl 1a]. Type data: holotype, QM J31685, from Eungella National Park, in 21°03'S, 148°35'E, Qld.

Distribution: NE coastal, Qld. Ecology: closed forest, terrestrial, diurnal; viviparous, arthropod-feeder.

Sphenomorphus mjobergi (Lönnberg and Andersson, 1915)

Lygosoma mjobergi Lönnberg, E. & Andersson, G.L. (1915). Results of Dr. E. Mjöberg's Swedish Scientific Expeditions to Australia 1910–1913. VII. Reptiles collected in northern Queensland. *K. Svenska Vetensk.-Akad. Handl.* **52**(7): 1–9 [6]. Type data: syntypes, NHRM 3217–8, from Malanda and Cedar Creek, N Qld.

Lygosoma darlingtoni Loveridge, A. (1933). New scincid lizards of the genera *Sphenomorphus*, *Rhodona* and *Lygosoma* from Australia. *Occ. Pap. Bost. Soc. Nat. Hist.* **8**: 95–100 [98]. Type data: holotype, QM J5806, from Millaa Millaa, Qld.

Synonymy that of Cogger, H.G., this work.

Distribution: NE coastal, Qld. Ecology: closed forest, fossorial; arthropod-feeder. Biological references: Greer, A.E. (1979). A new *Sphenomorphus* (Lacertilia : Scincidae) from the rainforests of northeastern Queensland. *Rec. Aust. Mus.* **32**: 373–383 (taxonomy).

Sphenomorphus murrayi (Boulenger, 1887)

Lygosoma murrayi Boulenger, G.A. (1887). *Catalogue of the Lizards in the British Museum (Natural History).* 3. London : British Museum xii 575 pp. pls 40 [232 pl 13 fig 1]. Type data: holotype, BMNH 1946.8.21.32, from Qld.

Lygosoma tamburinense Lönnberg, E. & Andersson, G.L. (1915). Results of Dr. E. Mjöberg's Swedish Scientific Expeditions to Australia 1910–1913. VII.

Reptiles collected in northern Queensland. *K. Svenska Vetensk.-Akad. Handl.* **52**(7): 1–9 [5]. Type data: holotype, NHRM 3212, from Mt. Tambourine, Qld.

Lygosoma (Hinulia) tryoni Longman, H.A. (1918). Notes on some Queensland and Papuan reptiles. *Mem. Qd. Mus.* **6**: 37–44 [38]. Type data: holotype, QM J3023, from Macpherson Ranges, 3000 ft, S Qld.

Lygosoma (Hinulia) tenuis intermedius Kinghorn, J.R. (1932). Herpetological notes. 4. *Rec. Aust. Mus.* **18**: 335–363 [358]. Type data: holotype, AM R6485, from Richmond River, N.S.W.

Synonymy that of Cogger, H.G., this work.

Distribution: SE coastal, NE coastal, N.S.W., Qld. Ecology: tall forest, closed forest, terrestrial, noctidiurnal; viviparous, arthropod-feeder. Biological references: Bustard, H.R. (1964). Reproduction in the Australian rain forest skinks, *Siaphos equalis* and *Sphenomorphus tryoni*. *Copeia* **1964**: 715–716 (as *Sphenomorphus tryoni*); Greer, A.E. (1980). Critical thermal maximum temperatures in Australian scincid lizards: their ecological and evolutionary significance. *Aust. J. Zool.* **28**: 91–102.

Sphenomorphus nigricaudis (Macleay, 1877)

Hinulia atrocostata Macleay, W. (1877). The lizards of the *Chevert* Expedition. *Proc. Linn. Soc. N.S.W.* **2**: 60–69 [62] [*non Scincus atrocostata* Lesson, 1830; 1878 on title page]. Type data: holotype, AM R31850, from Mawatta, Binaturi River (as Katow), Papua New Guinea.

Mocoa nigricaudis Macleay, W. (1877). The lizards of the *Chevert* Expedition. *Proc. Linn. Soc. N.S.W.* **2**: 60–69 [63] [1878 on title page]. Type data: lectotype, AM R31840, from Darnley Is., Torres Strait, Qld., designation by Copland, S.J. (1946). Catalogue of reptiles in the Macleay Museum I. *Sphenomorphus pardalis* (Macleay) and *Sphenomorphus nigricaudis nigricaudis* (Macleay). *Proc. Linn. Soc. N.S.W.* **70**: 291–311.

Lygosoma (Hinulia) elegantulum Peters, W. & Doria, G. (1878). Catalogo dei Rettili e dei Batraci raccolti da O. Beccari, L.M. d'Albertis e A.A. Bruijn nella sotto-regione Austro-Malese. *Ann. Mus. Civ. Stor. Nat. Genova* **13**: 323–450 [344]. Type data: syntypes, MCG C.E.27864, remaining syntypes not found, from Somerset, Cape York Peninsula, Qld.

Synonymy that of Cogger, H.G., this work.

Distribution: NE coastal, N Gulf, Qld. Ecology: litter, woodland, terrestrial, nocturnal; oviparous(part), viviparous(part), arthropod-feeder. Biological references: Copland, S.J. (1946). Catalogue of reptiles in the Macleay Museum I. *Sphenomorphus pardalis* (Macleay) and *Sphenomorphus nigricaudis nigricaudis* (Macleay). *Proc. Linn. Soc. N.S.W.* **70**: 291–311 (taxonomy); Greer, A.E. (1979). A new *Sphenomorphus* (Lacertilia : Scincidae) from the rainforests of northeastern Queensland. *Rec. Aust. Mus.* **32**: 373–383 (taxonomy).

Sphenomorphus pardalis (Macleay, 1877)

Hinulia pardalis Macleay, W. (1877). The lizards of the *Chevert* Expedition. *Proc. Linn. Soc. N.S.W.* **2**: 60–69 [63] [1878 on title page]. Type data: holotype, AM R31837, from Barrow Is., Qld.

Lygosoma atromaculatum Garman, S. (1901). Some reptiles and batrachians from Australasia. *Bull. Mus. Comp. Zool.* **39**: 1–14 [8]. Type data: syntypes, MCZ 6475 (2 specimens) Great Barrier Reef, Qld., MCZ 6478 (3 specimens) Qld. (Cooktown) *vide* Loveridge, A. (1934). Australian reptiles in the Museum of Comparative Zoology, Cambridge, Massachusetts. *Bull. Mus. Comp. Zool.* **77**: 243–383, FMNH 73374 from Qld., one MCZ syntype not found.

Sphenomorphus pardalis erro Copland, S.J. (1946). Catalogue of reptiles in the Macleay Museum I. *Sphenomorphus pardalis* (Macleay) and *Sphenomorphus nigricaudis nigricaudis* (Macleay). *Proc. Linn. Soc. N.S.W.* **70**: 291–311 [298 pl 11 fig 2]. Type data: holotype, AM R6352, from unknown locality.

Synonymy that of Cogger, H.G., this work.

Distribution: NE coastal, N Gulf, Qld. Ecology: litter, woodland, terrestrial, nocturnal; arthropod-feeder, cryptozoic. Biological references: Greer, A.E. & Parker, F. (1974). The *fasciatus* species group of *Sphenomorphus* (Lacertilia : Scincidae): notes on eight previously described species and descriptions of three new species. *Papua New Guinea Sci. Soc. Proc.* **25**: 31–61 (taxonomy); Rankin, P.R. (1978). Notes on the biology of the skink *Sphenomorphus pardalis* (Macleay) including a captive breeding record. *Herpetofauna* **10**: 4–7 (natural history).

Sphenomorphus pumilus (Boulenger, 1887)

Lygosoma ornatum Macleay, W. (1877). The lizards of the *Chevert* Expedition. *Proc. Linn. Soc. N.S.W.* **2**: 60–69 [64] [*non Tiliqua ornata* Gray, 1843; 1878 on title page]. Type data: holotype, AM R31848, from Endeavour River, Qld.

Lygosoma pumilum Boulenger, G.A. (1887). *Catalogue of the Lizards in the British Museum (Natural History).* 3. London : British Museum xii 575 pp. pls 40 [325 pl 26 fig 3]. Type data: syntypes, BMNH 1946.8.3.34, BMNH 1946.8.15.63, from Cape York, Qld.

Synonymy that of Cogger, H.G., this work.

Distribution: NE coastal, N Gulf, Qld. Ecology: woodland, fossorial, nocturnal; arthropod-feeder.

Sphenomorphus punctulatus (Peters, 1871)

Lygosoma (Lygosoma) punctulatum Peters, W. (1871). Über einige Arten der herpetologischen Sammlung des Berliner Zoologischen Museums. *Mber. K. Preuss. Akad. Wiss. Berl.* **1871**: 644–652 [646] [1872 on title page]. Type data: holotype, ZMB 7295, from Port Clinton (as Port Bowen), Qld.

Lygosoma heterodactylum Günther, A. (1876). Descriptions of new species of reptiles from Australia. *J.*

Mus. Godeffroy **5**: 45–47 [45]. Type data: holotype, BMNH 73.3.4.16, from Peak Downs, Qld.

Synonymy that of Cogger, H.G., this work.

Distribution: NE coastal, Qld. Ecology: open scrub, woodland, fossorial, nocturnal; arthropod-feeder. Biological references: Greer, A.E. (1980). Critical thermal maximum temperatures in Australian scincid lizards: their ecological and evolutionary significance. *Aust. J. Zool.* **28**: 91–102.

Sphenomorphus quoyii (Duméril and Bibron, 1839)

Scincus vittatus Quoy, J.R.C. & Gaimard, J.P. (1824). Zoologie *in* Freycinet, L.C.D. *Voyage Autour du Monde Exécute sur l'*Uranie *et la* Physicienne *pendant les années 1817–1820.* Paris iv 712 pp. [178 pl 42 fig 1] [*non Scincus vittatus* Olivier, 1804]. Type data: syntypes, MNHP 7112–3, from Neutral Bay, Port Jackson, N.S.W.

***Lygosoma quoyii* (part.)** Duméril, A.M.C. & Bibron, G. (1839). *Erpétologie Générale* ou Histoire Naturelle Complète des Reptiles. Paris : Roret 5: viii 854 pp. [728] [Var. A; extralimital syntype (Var. B) is specimen of *Leiolopisma reevesi* (Gray, 1839), *vide* Guibé, J. (1954). *Catalogue des Types de Lézards du Muséum National d'Histoire Naturelle.* Paris : Muséum National d'Histoire Naturelle 119 pp.]. Type data: syntypes, MNHP 2976 (Port Macquarie), MNHP 2977 (Australia), MNHP 7112–3 (Neutral Bay), from Port Macquarie, N.S.W., Neutral Bay, N.S.W. and Australia (var. A) and China (var. B).

Hinulia gastrosticta Günther, A. (1875). Reptiles. A list of saurians of Australia and New Zealand. pp. 9–19 *Zoology of the Voyage of H.M.S.* Erebus *and* Terror. Vol. 2 London : E.W. Janson [11]. Type data: syntypes, BMNH 1946.8.15.34–35 (Qld.), BMNH 1946.8.15.36 (Australia), BMNH 1946.8.4.99 (Kangaroo Is.), from Qld., Kangaroo Is., S.A. and Australia.

Synonymy that of Cogger, H.G., this work.

Distribution: SE coastal, NE coastal, S Gulfs, Murray-Darling basin, S.A., Vic., N.S.W., Qld. Ecology: low open woodland, woodland, open forest, tall forest, lentic freshwater, lotic freshwater, amphibious, terrestrial, diurnal; viviparous, arthropod-feeder. Biological references: Rawlinson, P.A. (1969). East Gippsland Symposium. 7. The reptiles of East Gippsland. *Proc. R. Soc. Vict.* **82**: 113–128 (taxonomy); Veron, J.E.N. (1969). The reproductive cycle of the water skink, *Sphenomorphus quoyii. J. Herpet.* **3**: 55–63; Weekes, H.C. (1927). Placentation and other phenomena in the scincid lizard *Lygosoma* (*Hinulia*) *quoyi. Proc. Linn. Soc. N.S.W.* **52**: 449–554 (as *Lygosoma (Hinulia) quoyii*); King, D. (1964). The osteology of the water skink, *Lygosoma* (*Sphenomorphus*) *quoyii. Aust. J. Zool.* **12**: 201–216; Spellerberg, I.F. (1972). Thermal ecology of allopatric lizards (*Sphenomorphus*) in southeast Australia III. Behavioural aspects of thermo-regulation. *Oecologia* **11**: 1–16; Spellerberg, I.F. (1972). Thermal ecology of allopatric lizards (*Sphenomorphus*) in southeast Australia. I. The environment and lizard critical temperatures. *Oecologia* **9**: 371–383; Spellerberg, I.F. (1972). Thermal ecology of allopatric lizards (*Sphenomorphus*) in southeast Australia. II. Physiological aspects of thermoregulation. *Oecologia* **9**: 385–398; Veron, J. & Heatwole, H. (1970). Temperature relations of the water skink, *Sphenomorphus quoyi. J. Herpet.* **4**: 141–153.

Sphenomorphus scutirostrus (Peters, 1873)

Lygosoma scutirostrum Peters, W. (1873). Mitteilung über neue Saurier (*Spoeriodactylus, Anolis, Phrynosoma, Tropidolepisma, Lygosoma, Ophioscincus*) aus Centralamerica, Mexico und Australien. *Mber. K. Preuss. Akad. Berl.* **1873**: 738–747 [743] [1874 on title page]. Type data: holotype, ZMB 7823, from Port Clinton (as Port Bowen), Qld.

Cophoscincus obscurus O'Shaughnessy, A.W.E. (1874). Description of a new species of skink. *Ann. Mag. Nat. Hist. (4)* **14**: 35 [35]. Type data: holotype, BMNH 1946.8.16.97, from Qld.

Calyptotis flaviventer De Vis, C.W. (1886). On a lizard and three species of *Salarias* etc. *Proc. R. Soc. Qd.* **2**: 56–60 [57] [1885 on title page]. Type data: none designated, QM (not found), from Breakfast Creek, near Brisbane, Qld.

Synonymy that of Cogger, H.G., this work.

Distribution: SE coastal, NE coastal, N.S.W., Qld. Ecology: woodland, open forest, tall forest, closed forest, fossorial, nocturnal; arthropod-feeder. Biological references: Longman, H.A. (1916). Snakes and lizards from Queensland and the Northern Territory. *Mem. Qd. Mus.* **5**: 46–51 (taxonomy, as *Calyptotis flaviventer*); Czechura, G.V. (1976). Additional notes on the Conondale Range herpetofauna. *Herpetofauna* **8**: 2–4 (taxonomy).

Sphenomorphus tenuis (Gray, 1831)

Tiliqua tenuis Gray, J.E. (1831). A synopsis of the species of the class Reptilia, pp. 1–110 *in* Griffith, E. *The Animal Kingdom Arranged in Conformity with its Organization by the Baron Cuvier.* London : Whittaker, Treacher & Co. 9: 481+110 pp. [71]. Type data: holotype, BMNH 1946.8.17.15, from Australia.

Lygosoma erucata Duméril, A.M.C. & Bibron, G. (1839). *Erpétologie Générale* ou Histoire Naturelle Complète des Reptiles. Paris : Roret 5: viii 854 pp. [726] [*ex* Péron ms]. Type data: holotype, MNHP 7035, from Australia.

Lygosoma brachysoma Lönnberg, E. & Andersson, G.L. (1915). Results of Dr. E. Mjöberg's Swedish Scientific Expeditions to Australia 1910–1913. VII. Reptiles collected in northern Queensland. *K. Svenska*

Vetensk.-Akad. Handl. **52**(7): 1–9 [5]. Type data: holotype, NHRM 3211, from Atherton, Qld.

Synonymy that of Cogger, H.G., this work.

Distribution: SE coastal, NE coastal, N.S.W., Qld. Ecology: woodland, open forest, tall forest, (closed forest), terrestrial, arboreal, diurnal; viviparous, arthropod-feeder. Biological references: Rankin, P.R. (1973). The barred sided skink *Sphenomorphus tenuis tenuis* (Gray) in the Sydney region. *Herpetofauna* **6**: 8–14; Greer, A.E. (1980). Critical thermal maximum temperatures in Australian scincid lizards: their ecological and evolutionary significance. *Aust. J. Zool.* **28**: 91–102; De Lissa, G. (1981). Notes on the skink *Sphenomorphus tenuis. Herpetofauna* **13**: 33 (natural history).

Sphenomorphus tigrinus (De Vis, 1888)

Hinulia tigrina De Vis, C.W. (1888). A contribution to the herpetology of Queensland. *Proc. Linn. Soc. N.S.W. (2)* **2**: 811–826 [817]. Type data: holotype, QM J245, from Innisfail (as Geraldton), Qld.

Distribution: NE coastal, Qld. Ecology: closed forest, terrestrial, diurnal; viviparous, arthropod-feeder. Biological references: Covacevich, J., Ingram, G.J. & Czechura, G.V. (1982). Rare frogs and reptiles of Cape York Peninsula, Australia. *Biol. Conserv.* **22**: 283–294 (distribution).

Sphenomorphus tympanum (Lönnberg and Andersson, 1913)

Lygosoma tympanum Lönnberg, E. & Andersson, G.L. (1913). Results of Dr. E. Mjöberg's Swedish Scientific Expeditions to Australia 1910–1913. III. Reptiles. *K. Svenska Vetensk.-Akad. Handl.* **52**(3): 1–17 [9]. Type data: holotype, NHRM 3094, "... said to have been collected in the neighbourhood of Melbourne ... ", Vic.

Distribution: SE coastal, S Gulfs, Murray-Darling basin, S.A., Vic., N.S.W., A.C.T. Ecology: low open woodland, woodland, open forest, tall forest, closed forest, amphibious, terrestrial, diurnal; viviparous, arthropod-feeder. Biological references: Rawlinson, P.A. (1969). East Gippsland Symposium. 7. The reptiles of East Gippsland. *Proc. R. Soc. Vict.* **82**: 113–128 (taxonomy); Spellerberg, I.F. (1972). Thermal ecology of allopatric lizards (*Sphenomorphus*) in southeast Australia III. Behavioural aspects of thermo-regulation. *Oecologia* **11**: 1–16; Spellerberg, I.F. (1972). Thermal ecology of allopatric lizards (*Sphenomorphus*) in southeast Australia. I. The environment and lizard critical temperatures. *Oecologia* **9**: 371–383; Spellerberg, I.F. (1972). Thermal ecology of allopatric lizards (*Sphenomorphus*) in southeast Australia. II. Physiological aspects of thermoregulation. *Oecologia* **9**: 385–398.

Tiliqua Gray, 1825

Tiliqua Gray, J.E. (1825). A synopsis of the genera of reptiles and Amphibia, with a description of some new species. *Annls. Philos. (2)* **10**: 193–217 [201]. Type species *Lacerta scincoides* White, 1790, subsequent designation by Cogger, H.G., this work.

Cyclodus Wagler, J.G. (1828). *Descriptiones et Icones Amphibiorum.* Part I. Plates I–XIII Stuttgartiae, Tubingae, Monachii : J.G. Cottae [pl 6]. Type species *Cyclodus flavigularis* Wagler, 1828 [= *Tiliqua gigas* (Schneider, 1801)], by monotypy.

Rachites Cocteau, J.T. (1837). Tabulae synopticae Scincoideorum (rapport sur un ouvrage manuscrit de M. le Docteur Cocteau, par M. Duméril). *C. R. Hebd. Séanc. Acad. Sci. Paris* **4**: 14–17 [16] [*nom. nud.*].

Tiligua Cocteau, J.T. (1837). Tabulae synopticae Scincoideorum (rapport sur un ouvrage manuscrit de M. le Docteur Cocteau, par M. Duméril). *C. R. Hebd. Séanc. Acad. Sci. Paris* **4**: 14–17 [16] [*errore pro Tiliqua* Gray, 1825].

Cyclodomorphus Fitzinger, L.J. (1843). *Systema Reptilium.* Vienna : Braümüller u. Seidel vi 106 pp. [23]. Type species *Cyclodus casuarinae* Duméril and Bibron, 1839 by original designation.

Omolepida Gray, J.E. (1845). *Catalogue of the Specimens of Lizards in the Collection of the British Museum.* London : British Museum xxviii 289 pp. [71,87]. Type species *Cyclodus casuarinae* Duméril and Bibron, 1839 by monotypy.

Homolepides Agassiz, L. (1846). *Nomenclatoris Zoologici Index Universalis.* Soloduri : Jent et Gassman 393 pp. [185] [*pro* Omolépides (vernacular) Cocteau, J.T. (1837). Tabulae synopticae Scincoideorum (rapport sur un ouvrage manuscrit de M. le Docteur Cocteau, par M. Duméril). *C. R. Hebd. Séanc. Acad. Sci. Paris* **4**: 14–17 (16)]. Type species, none designated.

Hemisphaeriodon Peters, W. (1867). Herpetologische Notizen. *Mber. K. Preuss. Akad. Wiss. Berl.* **1867**: 13–37 [24]. Type species *Hinulia gerrardii* Gray, 1845 by monotypy.

Homolepida Lütken, C. (1863). Nogle nye krybdyr og padder. *Vidensk. Meddr. Dansk Naturh. Foren.* **1862**: 292–331 [294] [*pro Omolepida* Gray, 1845].

Synonymy that of Cogger, H.G., this work.

This group is also found in E Indonesia and New Guinea.

Tiliqua adelaidensis (Peters, 1863)

Cyclodus adelaidensis Peters, W. (1863). Übersicht der von Hrn. Richard Schomburgk an das zoologische Museum eingesandten Amphibien, aus Buchsfelde bei Adelaide in Südaustralien. *Mber. K. Preuss. Akad. Wiss. Berl.* **1863**: 228–236 [232] [1864 on title page]. Type data: syntypes, ZMB 4710 (2 specimens), from Loos, 4.5 km W Gawler (as Buchsfelde), S.A., see Praite, R. & Tolley, J.C. (1970). *Place Names of South Australia.* Adelaide : Rigby 203 pp.

Distribution: S Gulfs, S.A. Ecology: terrestrial, diurnal; viviparous. Biological references: Mitchell, F.J. (1950). The scincid genera *Egernia* and *Tiliqua* (Lacertilia). *Rec. S. Aust. Mus.* **9**: 275–308 (taxonomy).

Tiliqua branchialis (Günther, 1867)

Hinulia branchialis Günther, A. (1867). Additions to the knowledge of Australian reptiles and fishes. *Ann. Mag. Nat. Hist. (3)* **20**: 45–68 [47]. Type data: syntypes, BMNH 1946.8.19.47–49, from Champion Bay, W.A.

Lygosoma melanops Stirling, E.C. & Zietz, A. (1893). Vertebrata; Reptilia (collected by the Elder Exploring Expedition). *Trans. R. Soc. S. Aust.* **16**: 159–176 [173 pl 6 figs 3 3a]. Type data: syntypes, SAMA R2732, SAMA R8139, from between the Everard and Barrow Ranges, central Australia.

Lygosoma gastrostigma Boulenger, G.A. (1898). Third report on additions to the collection of lizards in the British Museum. *Proc. Zool. Soc. Lond.* **1898**: 912–923 [918, 922 pl 57 fig 2]. Type data: holotype, BMNH 1946.8.9.87, from Sherlock River, Nickol Bay, W.A.

Lygosoma (Homolepida) branchiale elongatum Werner, F. (1910). Reptilia (Geckonidae und Scincidae). *in* Michaelsen, W. & Hartmeyer, R. (eds.) *Die Fauna Südwest-Austaliens.* Jena : Gustav Fischer 2: 451–493 [479]. Type data: holotype, whereabouts unknown, from Boorabbin, W.A.

Lygosoma (Homolepida) woodjonesii Procter, J.B. (1923). The flora and fauna of Nuyts Archipelago and the Investigator group. 5. The lizards. *Trans. R. Soc. S. Aust.* **47**: 79–81 [80]. Type data: syntypes, BMNH 1946.8.17.97–99 (3 adults and 2 embryos), from St. Francis Is., S.A.

Synonymy that of Cogger, H.G., this work.

Distribution: (N Gulf), N coastal, NW coastal, SW coastal, W plateau, S Gulfs, Lake Eyre basin, Murray-Darling basin, N.S.W., S.A., Qld., W.A., N.T. Ecology: desert, hummock grassland, tussock grassland, low open shrubland, low shrubland, open scrub, low woodland, tall shrubland, terrestrial, nocturnal, crepuscular; viviparous, arthropod-feeder. Biological references: Mitchell, F.J. (1950). The scincid genera *Egernia* and *Tiliqua* (Lacertilia). *Rec. S. Aust. Mus.* **9**: 275–308 (taxonomy); Warburg, M.R. (1965). Studies on the environmental physiology of some Australian lizards from arid and semiarid habitats. *Aust. J. Zool.* **13**: 563–575 (as *Tiliqua melanops*); Storr, G.M. (1976). The genus *Omolepida* (Lacertilia, Scincidae) in Western Australia. *Rec. West. Aust. Mus.* **4**: 163–170 (taxonomy, as *Omolepida branchialis*).

Tiliqua casuarinae (Duméril and Bibron, 1839)

Cyclodus casuarinae Duméril, A.M.C. & Bibron, G. (1839). *Erpétologie Générale* ou Histoire Naturelle Complète des Reptiles. Paris : Roret 5. viii 854 pp. [749]. Type data: holotype, MNHP 7131, from Bruny Is., Tas. (original description cites Australia).

Cyclodus (Homolepida) nigricans Peters, W. (1874). Über neue Amphibien. *Mber. K. Preuss. Akad. Wiss. Berl.* **1874**: 616–624 [621] [1875 on title page]. Type data: holotype, ZMB 8193, from Australia.

Lygosoma (Lygosoma) mulleri Peters, W. (1878). Über zwei Scincoiden aus Australien und einer neuen *Amphisbaena* von West Africa. *Sber. Ges. Naturf. Freunde Berl.* **1878**: 191–193 [191] [*non Scincus mulleri* Schlegel, 1837]. Type data: syntypes, two specimens, ?ZMB 9373, from "Süd-Australien".

Hemisphaeriodon tasmanicum Frost, C. & Lucas, A.H.S. (1894). A new skink lizard from Tasmania. *Proc. Linn. Soc. N.S.W. (2)* **8**: 227–228 [227]. Type data: holotype, whereabouts unknown, from Lake St. Clair, Tas.

Lygosoma (Homolepida) petersi Sternfeld, R. (1919). Neue Schlangen und Echsen aus Zentralaustralien. *Senckenbergiana* **1**: 76–83 [81] [*nom. nov. pro Lygosoma mulleri* Peters, 1878; but redescription based on specimens of *Tiliqua branchialis* from Hermannsburg Mission, Upper Finke River, N.T.].

Synonymy that of Cogger, H.G., this work.

Distribution: SE coastal, Murray-Darling basin, Tas., Vic., N.S.W., A.C.T. Ecology: (woodland), open forest, tall forest, terrestrial, noctidiurnal; viviparous, arthropod-feeder. Biological references: Mitchell, F.J. (1950). The scincid genera *Egernia* and *Tiliqua* (Lacertilia). *Rec. S. Aust. Mus.* **9**: 275–308 (taxonomy); Mebs, D. (1974). Haltungs-erfahrungen mit *Tiliqua casuarinae* (Sauria, Scincidae). *Salamandra* **10**: 104–106; Timms, B.V. (1977). Notes on the she-oak skink. *Hunter Nat. Hist.* **May 1977**: 99–102 (natural history); Rankin, P.R. (1973). Lizard mimicking a snake - juvenile *Tiliqua casuarinae. Herpetofauna* **5**: 13–14; Shea, G.M. (ed.) (1982). Observations on some members of the genus *Tiliqua. Herpetofauna* **13**:18–20 (natural history).

Tiliqua gerrardii (Gray, 1845)

Hinulia gerrardii Gray, J.E. (1845). *Catalogue of the Specimens of Lizards in the Collection of the British Museum.* London : British Museum xxviii 289 pp. [75]. Type data: syntypes, BMNH xv. 5a–b, from Australia?

Hinulia picta Macleay, W. (1885). On some Reptilia lately received from the Herbert River District, Queensland. *Proc. Linn. Soc. N.S.W.* **10**: 64–68 [65] [1886 on title page]. Type data: holotype, MMUS (not found), from vicinity of the Herbert River, Qld.

Tiliqua longicauda De Vis, C.W. (1888). A contribution to the herpetology of Queensland. *Proc. Linn. Soc. N.S.W. (2)* **2**: 811–826 [816]. Type data: syntypes, QM J1186–7, from Rockhampton and Johnstone River, Qld.

Synonymy that of Cogger, H.G., this work.

Distribution: SE coastal, NE coastal, N.S.W., Qld. Ecology: (woodland), open forest, tall forest, closed forest, terrestrial, noctidiurnal; viviparous, arthropod-feeder. Biological references: Mitchell, F.J. (1950). The scincid genera *Egernia* and *Tiliqua* (Lacertilia). *Rec. S. Aust. Mus.* **9**: 275–308 (taxonomy); Wilhoft, D.C. (1960). Observations on adults and juveniles of *Hemisphaeriodon gerrardii* in captivity. *N. Qd. Nat.* **28**: 3–4 (natural history); Longley, G. (1938). Notes on a pink tongued skink. *Proc. R. Zool. Soc. N.S.W.* **1937–38**: 19–21 (natural history); Matz, G. (1973). Biologie et reproduction de *Tiliqua gerrardii* (Gray) (Scincidae). *Bull. Soc. Zool. Fr.* **98**: 590; Field, R. (1980). The pink-tongued skink (*Tiliqua gerrardii*) in captivity. *Herpetofauna* **11**: 6–10 (natural history); Shea, G.M. (ed.) (1982). Observations on some members of the genus *Tiliqua. Herpetofauna* **13**:18–20 (natural history).

Tiliqua maxima (Storr, 1976)

Omolepida maxima Storr, G.M. (1976). The genus *Omolepida* (Lacertilia, Scincidae) in Western Australia. *Rec. West. Aust. Mus.* **4**: 163–170 [169]. Type data: holotype, WAM R27760, from Kalumburu, 14°18′S, 126°38′E, W.A.

Distribution: N coastal, W.A. Ecology: woodland, terrestrial; viviparous. Biological references: Storr, G.M., Smith, L.A. & Johnstone, R.E. (1981). *Lizards of Western Australia. I. Skinks.* Perth : University of Western Australia Press and Western Australian Museum 200 pp. (taxonomy).

Tiliqua multifasciata Sternfeld, 1919

Tiliqua occipitalis multifasciata Sternfeld, R. (1919). Neue Schlangen und Echsen aus Zentralaustralien. *Senckenbergiana* **1**: 76–83 [79]. Type data: lectotype, SMF 14037, from Hermannsburg Mission, Upper Finke River, N.T., designation by Mertens, R. (1967). Die herpetologische Sektion des Natur-Museums und Forschungs-Institutes Senckenberg in Frankfurt-a.-M. nebst einem Verzeichnis ihrer Typen. 1. *Senckenberg. Biol.* **48**: 1–106.

Tiliqua occipitalis nossiteri Glauert, L. (1923). Contributions to the fauna of Western Australia. 3. Annotated list of lizards from Wallal. *J. R. Soc. West. Aust.* **9**: 57–60 [59]. Type data: syntypes, from Wallal, W.A., WAM R1013, remaining three syntypes not cited by number in original description.

Tiliqua occipitalis auriculare Kinghorn, J.R. (1931). Herpetological notes. 2. *Rec. Aust. Mus.* **18**: 85–91 [88]. Type data: holotype, WAM R10080, from Broome, W.A.

Synonymy that of Cogger, H.G., this work.

Distribution: N Gulf, N coastal, NW coastal, W plateau, Lake Eyre basin, S.A., Qld., W.A., N.T. Ecology: desert, hummock grassland, tussock grassland, low open shrubland, low shrubland, low open woodland, low woodland, woodland, terrestrial, diurnal; viviparous, omnivore. Biological references: Mitchell, F.J. (1950). The scincid genera *Egernia* and *Tiliqua* (Lacertilia). *Rec. S. Aust. Mus.* **9**: 275–308 (taxonomy); Pianka, E.R. (1972). Zoogeography and speciation of Australian desert lizards: an ecological perspective. *Copeia* **1972**: 127–144 (ecology); Christian, T. (1977). Notes on centralian bluetongues (*Tiliqua multifasciata*). *Vict. Herp. Soc. Newsletter* **1**: 8–9 (natural history); Barnett, B. (1977). Additional notes on new-born centralian bluetongues, *Tiliqua multifasciata. Vict. Herp. Soc. Newsletter* **1**: 10 ; Shea, G. & Peterson, M. (1981). Observations on sympatry of the social lizards *Tiliqua multifasciata* Sternfeld and *T. occipitalis* (Peters). *Aust. J. Herp.* **1**: 27–28; Storr, G.M., Smith, L.A. & Johnstone, R.E. (1981). *Lizards of Western Australia. I. Skinks.* Perth : University of Western Australia Press and Western Australian Museum 200 pp. (taxonomy).

Tiliqua nigrolutea (Quoy and Gaimard, 1824)

Scincus nigroluteus Quoy, J.R.C. & Gaimard, J.P. (1824). Zoologie *in* Freycinet, L.C.D. *Voyage Autour du Monde Exécute sur l'*Uranie *et la* Physicienne *pendant les années 1817–1820.* Paris iv 712 pp. [176 pl 41] [as only 2 specimens are cited in the original description, the status of one of the labelled types, probably the juvenile, is doubtful]. Type data: syntypes, MNHP 7134 (adult), MNHP 3024 (adult), MNHP 3023 (juvenile), from Blue Mts., N.S.W. (MNHP catalogue cites New Holland).

Scinc[us] erucotis Gray, J.E. (1838). Catalogue of the slender-tongued saurians, with descriptions of many new genera and species. (contd.) *Ann. Mag. Nat. Hist.* **2**: 287–293 [290] [*nom. nud.*; Péron ms name introduced in synonymy by Gray; 1839 on title page].

Synonymy that of Cogger, H.G., this work.

Distribution: SE coastal, Murray-Darling basin, S.A., Tas., Vic., N.S.W., A.C.T. Ecology: alpine, woodland, open forest, tall forest, terrestrial, diurnal; viviparous, omnivore. Biological references: Mitchell, F.J. (1950). The scincid genera *Egernia* and *Tiliqua* (Lacertilia). *Rec. S. Aust. Mus.* **9**: 275–308 (taxonomy); Stettler, P.H. (1953). *Tiliqua nigrolutea* Gray. *Lacerta* **11**: 63–66 (natural history); Longley, G. (1944). Notes on a hybrid blue tongue lizard. *Proc. R. Zool. Soc. N.S.W.* **1943–44**: 22–23; Baehr, M. (1976). Beiträge zur Verbreitung und Ökologie tasmanischer Reptilien. *Stuttgart Beitr. Naturk. (A)* **292**: 1–24 (natural history); Spellerberg, I.F. (1972). Temperature tolerances of southeast Australian reptiles examined in relation to reptile thermoregulation behaviour and distribution. *Oecologia* **9**: 23–46; Bissinger, B.E. & Simon, C.A. (1979). Comparison of tongue extrusions in representatives of six families of lizards. *J. Herpet.* **13**: 133–139.

Tiliqua occipitalis (Peters, 1863)

Cyclodus occipitalis Peters, W. (1863). Übersicht der von Hrn. Richard Schomburgk an das zoologische Museum eingesandten Amphibien, aus Buchsfelde bei Adelaide in Südaustralien. *Mber. K. Preuss. Akad. Wiss. Berl.* **1863**: 228–236 [231] [1864 on title page]. Type data: holotype, ZMB 4709, from Loos, 4.5 km W Gawler (as Buchsfelde), S.A., see Praite, R. & Tolley, J.C. (1970). *Place Names of South Australia.* Adelaide : Rigby 203 pp.

Cyclodus fasciatus Lütken, C. (1863). Nogle nye krybdyr og padder. *Vidensk. Meddr. Dansk Naturh. Foren.* **1862**: 292–331 [292 pl 1 fig 1] [1862 on title page]. Type data: holotype, UZM R47563, from Australia.

Synonymy that of Cogger, H.G., this work.

Distribution: NW coastal, SW coastal, W plateau, S Gulfs, Murray-Darling basin, Lake Eyre basin, S.A., Vic., N.S.W., Qld., W.A., N.T. Ecology: desert, low open shrubland, low shrubland, open heath, open scrub, (woodland), tall shrubland, terrestrial, diurnal; viviparous, omnivore. Biological references: Mitchell, F.J. (1950). The scincid genera *Egernia* and *Tiliqua* (Lacertilia). *Rec. S. Aust. Mus.* **9**: 275–308 (taxonomy); Pianka, E.R. (1972). Zoogeography and speciation of Australian desert lizards: an ecological perspective. *Copeia* **1972**: 127–144 (ecology); Shea, G. & Peterson, M. (1981). Observations on sympatry of the social lizards *Tiliqua multifasciata* Sternfeld and *T. occipitalis* (Peters). *Aust. J. Herp.* **1**: 27–28; Storr, G.M., Smith, L.A. & Johnstone, R.E. (1981). *Lizards of Western Australia. I. Skinks.* Perth : University of Western Australia Press and Western Australian Museum 200 pp. (taxonomy).

Tiliqua scincoides (White, 1790)

Lacerta scincoides White, J. (1790). *Journal of a Voyage to New South Wales,* with sixty five plates of non-descript animals, birds, lizards, serpents, curious cones of trees and other natural productions. Appendix. London : Debrett 299 pp. [242 pl 30 fig 1] [*ex* Shaw ms]. Type data: syntypes, BMNH 1946.8.21.94, other syntype presumed lost, from Australia.

Scincus crotaphomelas Lacépède, B.G.E. (1804). Mémoire sur plusieurs animaux de la Nouvelle-Hollande dont la description n'a pas encore été publiée. *Ann. Mus. Natl. Hist. Nat. Paris* **4**: 184–211 [209]. Type data: none designated, whereabouts unknown, from Australia. Maury (*in litt.*) cites 3 MNHP specimens bearing this name, from Port Jackson, in the original lisst of specimens collected by Péron and Lesueur during Baudin's voyages (1800–1804).

Scincus tuberculatus Merrem, B. (1820). *Versuch eines Systems Amphibien. Tentamen Systematis Amphibiorum.* Marburg : Krieger xv 199 pp. [73] [this name is here regarded as applying to a newly-described species. However Merrem's original description is ambiguous in that the name *tuberculatus* could be interpreted as a substitute name for *Lacerta scincoides* White, 1790]. Type data: none designated, whereabouts unknown, from Australia.

Tiliqua whitii Gray, J.E. (1831). A synopsis of the species of the class Reptilia, pp. 1–110 *in* Griffith, E. *The Animal Kingdom Arranged in Conformity with its Organization by the Baron Cuvier.* London : Whittaker, Treacher & Co. 9: 481+110 pp. [67] [*non Scincus whitii* Lacépède, 1804; ?*nom. substit. pro Lacerta scincoides* White, 1790].

***Cyclodus boddaertii* (part.)** Duméril, A.M.C. & Bibron, G. (1839). *Erpétologie Générale* ou Histoire Naturelle Complète des Reptiles. Paris : Roret 5: viii 854 pp. [752] [for extralimital syntype (MNHP 3025, Java) see Guibé, J. (1954). *Catalogue des Types de Lézards du Mulnon Scincus whitii* Lacépède, 1804; ?*nom. substit. pro Lacerta scincoides* White, 1790].

***Cyclodus boddaertii* (part.)** Duméril, A.M.C. & Bibron, G. (1839). *Erpétologie Générale* ou Histoire Naturelle Complète des Reptiles. Paris : Roret 5: viii 854 pp. [752] [for extralimital syntype (MNHP 3025, Java) see Guibé, J. (1954). *Catalogue des Types de Lézards du Muséum National d'Histoire Naturelle.* Paris : Muséum National d'Histoire Naturelle 119 pp.]. Type data: syntypes, MNHP (Australian syntype(s) not found), from Australia and Java.

Tiliqua scincoides intermedia Mitchell, F.J. (1955). Preliminary account of the Reptilia and Amphibia collected by the National Geographic Society—Commonwealth Government—Smithsonian Institution Expedition to Arnhem Land (April to November, 1948). *Rec. S. Aust. Mus.* **11**: 373–408 [393]. Type data: holotype, SAMA R3095, from Yirrkala (as Yirrkalla), N.T.

Synonymy that of Cogger, H.G., this work.

Distribution: SE coastal, NE coastal, N Gulf, N coastal, W plateau, S Gulfs, Murray-Darling basin, (Bulloo River basin), Lake Eyre basin, S.A., Vic., N.S.W., Qld., W.A., N.T., A.C.T. Ecology: low open shrubland, low shrubland, open heath, tall open shrubland, open scrub, low open woodland, low woodland, woodland, open forest, terrestrial, diurnal; viviparous, omnivore. Biological references: Cabanac, M., Hammel, T. & Hardy, J.D. (1967). *Tiliqua scincoides*: temperature-sensitive units in lizard brain. *Science* **158**: 1050–1051; Flynn, T.T. (1923). On the occurrence of a true allanto-placenta of the conjoint type in an Australian lizard. *Rec. Aust. Mus.* **14**: 72–77; Mitchell, F.J. (1950). The scincid genera *Egernia* and *Tiliqua* (Lacertilia). *Rec. S. Aust. Mus.* **9**: 275–308 (taxonomy); Goddard, C.K. (1969). Mechanics of the urinary bladder of the lizard *Tiliqua scincoides scincoides* (Shaw). *Aust. J. Exp. Biol. Med. Sci.* **47**: 701–711; Hammel, H.T. Caldwell, F.J. & Abrams, R.M. (1967). Regulation of body temperature in the blue-tongued lizards. *Science* **156**: 1260–1262; Bartholomew, G.A., Tucker, V.A. & Lee, A.K. (1965). Oxygen consumption, thermal conductance, and heart rate in the Australian skink *Tiliqua scincoides. Copeia* **1965**: 169–173; Cabanac, H.P. & Hammel, H.T. (1971). Peripheral sensitivity and temperature regulation in *Tiliqua scincoides. Int. J. Biometeorol.* **15**: 239–243; Firth, B.T., Webb, G.J.W. & Johnson, C.R. (1972). Effect of time of day and photoperiod on heart rate in the scincid lizard, *Tiliqua scincoides. Comp. Biochem. Physiol.* **43A**: 805–808; Myhre, K. & Hammel, H.T. (1969). Behavioral regulation of internal temperature in the lizard *Tiliqua scincoides. Am. J. Physiol.* **217**: 1490–1495; Georges, A. (1979). Head-body temperature differences in the Australian blue-tongued lizard, *Tiliqua scincoides* during radiant heating *J. Thermal Biol.* **4**: 213–217; Bissinger, B.E. & Simon, C.A. (1979). Comparison of tongue extrusions in representatives of six families of lizards. *J. Herpet.* **13**: 133–139; Drane, C.R. & Webb, G.J.W. (1980). Functional morphology of the dermal vascular system of the Australian lizard *Tiliqua scincoides. Herpetologica* **36**: 60–66; Shea, G. (1981). Notes on the reproductive biology of the eastern blue-tongue skink, *Tiliqua scincoides* (Shaw). *Herpetofauna* **12**: 16–23; Storr, G.M., Smith, L.A. & Johnstone, R.E. (1981). *Lizards of Western Australia. I. Skinks.* Perth : University of Western Australia Press and Western Australian Museum 200 pp. (taxonomy).

Trachydosaurus Gray, 1825

Trachydosaurus Gray, J.E. (1825). A synopsis of the genera of reptiles and Amphibia, with a description of some new species. *Annls. Philos. (2)* **10**: 193–217 [201]. Type species *Trachydosaurus rugosus* Gray, 1825 by monotypy.

Trachysaurus Gray, J.E. (1826). Reptilia. Appendix *in* King, P.P. *Narrative of a survey of the Intertropical and Western Coasts of Australia performed between the years 1818 and 1822.* London : John Murray 2: 424–434 [430] [1827 on title page; emend. *pro Trachydosaurus* Gray, 1825].

Brachydactylus Smith, A. (1834). A saurian reptile from Swan River. *South African Quart. J.* **2**: 143–144 [144]. Type species *Brachydactylus typicus* Smith, 1834 by monotypy.

Tachydosaurus Gray, J.E. (1838). Catalogue of the slender-tongued saurians, with descriptions of many new genera and species. (contd.). *Ann. Mag. Nat. Hist.* **2**: 287–293 [288] [1839 on title page; *pro Trachydosaurus* Gray, 1825].

Silubolepis Duméril, A.M.C. & Bibron, G. (1839). *Erpétologie Générale* ou Histoire Naturelle Complète des Reptiles. Paris : Roret 5: viii 854 pp. [754] [*nom. nud.*; Cocteau ms name introduced in synonymy].

Synonymy that of Cogger, H.G., this work.

Trachydosaurus rugosus Gray, 1825

Scincus tropisurus Péron, F. (1807). *Voyage de Découvertes aux Terres Australes, Fait par Ordre du Gouvernement, sur les Corvettes le* Géographe, *le* Naturaliste, *et la goülette le* Casuarina, *pendant les années 1800, 1801, 1802, 1803 et 1804. Historique.* Vol. 1 Paris : Arthus Bertrand [118] [*nom. oblit.*?, *vide* Douglas, A.M. & Ride, W.D.L. (1962). Reptiles. *in* Ride, W.D.L., *et al.* The results of an Expedition to Bernier and Dorre Islands, Shark Bay, Western Australia in July, 1959. *Western Australia Fisheries Dept. Fauna Bull.* **2**: 113–119]. Type data: none designated, whereabouts unknown, from Bernier Is., W.A.

Trachydosaurus rugosus Gray, J.E. (1825). A synopsis of the genera of reptiles and Amphibia, with a description of some new species. *Annls. Philos. (2)* **10**: 193–217 [201]. Type data: holotype, BMNH 1946.8.5.1, no locality specified in original description; (King George's Sound, W.A. *vide* Gray, J.E. (1826). Reptilia. Appendix *in* King, P.P. *Narrative of a survey of the Intertropical and Western Coasts of Australia performed between the years 1818 and 1822.* London : John Murray 2: 424–434).

Scincus peronii Wagler, J.G. (1830). *Natürliches System der Amphibien, mit vorangehender Classification der Säugethiere und Vögel.* München : Cotta'schen vi 354 pp. [163] [*nom. nud.*; Duméril ms name introduced in synonymy].

Trachysaurus peronii Wagler, J. (1833). *Descriptiones et Icones Amphibiorum.* Part III. Plates XXV–XXXVI Stuttgartiae, Tubingae, Monachii : J.G. Cottae [pl 36 and accompanying text] [*nom. nov. pro Trachydosaurus rugosus* Gray, 1825 (as *Trachysaurus rugosus* Gray, 1827)].

Brachydactylus typicus Smith, A. (1834). A saurian reptile from Swan River. *South African Quart. J.* **2**: 143–144 [144 fig]. Type data: holotype, BMNH stuffed, unnumbered, from Swan River, W.A.

Scincus pachyurus Gray, J.E. (1838). Catalogue of the slender-tongued saurians, with descriptions of many new genera and species. (contd.) *Ann. Mag. Nat. Hist.* **2**: 287–293 [288] [1839 on title page; *nom. nud.*; Péron ms name introduced in synonymy of *T. rugosus* but Péron's specimen was from Parramatta and is more likely to have been *Tiliqua scincoides*].

Trachydosaurus asper Gray, J.E. (1845). *Catalogue of the Specimens of Lizards in the Collection of the British Museum.* London : British Museum xxviii 289 pp. [103]. Type data: syntypes, BMNH xv.81.a–b, BMNH 42.6.29.58–59, from Adelaide, S.A.

Trachydosaurus rugosus konowi Mertens, R. (1958). Neue Eidechsen aus Australien. *Senckenberg. Biol.* **39**: 51–56 [52]. Type data: holotype, WAM 12721, from Rottnest Is., W.A.

Synonymy that of Cogger, H.G., this work.

Distribution: NW coastal, SW coastal, W plateau, S Gulfs, Murray-Darling basin, Bulloo River basin, Lake Eyre basin, S.A., Vic., N.S.W., Qld., W.A., A.C.T. Ecology: desert, tussock grassland, low open shrubland, low shrubland, open heath, open scrub, low open woodland, low woodland, woodland, open forest, terrestrial, diurnal; viviparous, omnivore. Biological references: Satrawaha, R. & Bull, C.M. (1981). The area occupied by an omnivorous lizard, *Trachydosaurus rugosus. Aust. Wildl. Res.* **8**: 435–442; Mitchell, F.J. (1950). The scincid genera *Egernia* and *Tiliqua* (Lacertilia). *Rec. S. Aust. Mus.* **9**: 275–308 (taxonomy); Simpson, D.A. (1949). The epiphyseal complex in *Trachysaurus rugosus. Trans. R. Soc. S. Aust.* **73**: 1–5; Kirby, S. & Burnstock, G. (1969).

Pharmacological studies of the cardiovascular system in the anaesthetized sleepy lizard (*Tiliqua rugosa*) and toad (*Bufo marinus*). *Comp. Biochem. Physiol.* **28**: 321–331; Bentley, P.J. (1959). Studies on the water and electrolyte metabolism of the lizard *Trachysaurus rugosus* (Gray). *J. Physiol.* **145**: 37–47; Saint Girons, H., Lemire, M. & Bradshaw, S.D. (1977). Structure de la glande nasale externe de *Tiliqua rugosa* (Reptilia, Scincidae) et rapports avec sa fonction. *Zoomorph.* **88**: 277–288; Bourne, A.R. & Seamark, R.F. (1978). Seasonal variation in steroid biosynthesis by the testis of the lizard *Tiliqua rugosa. Comp. Biochem. Physiol.* **59B**: 363–367; Braysher, M.L. (1971). The structure and function of the nasal salt gland from the sleepy lizard *Trachydosaurus* (formerly *Tiliqua*) *rugosus*: family Scincidae. *Physiol. Zool.* **44**: 129–136; Burnstock, G. & Wood, M. (1967). Innervation of the urinary bladder of the sleepy lizard (*Trachysaurus rugosus*). II. Physiology and pharmacology. *Comp. Biochem. Physiol.* **20**: 675–690; Firth, B.T., Kennaway, D.J. & Rozenbilds, M.A.M. (1979). Plasma melatonin in the scincid lizard, *Trachydosaurus rugosus*: diel rhythm, seasonality and the effect of constant darkness. *Gen. Comp. Endocrinol.* **37**: 493–500; Bourne, A.R. (1981). Progesterone-like activity in the plasma of the viviparous skink *Trachydosaurus rugosus* (stump-tailed lizard). pp. 14–16 *in* Banks, C.B. & Martin, A.A. (eds.) *Proceedings of the Melbourne Herpetological Symposium.* Melbourne : Zoological Board of Victoria; Bull, C.M. & Satrawaha, R. (1981). Dispersal and social organisation in *Trachydosaurus rugosus.* p. 24 *in* Banks, C.B. & Martin, A.A. (eds.) Proceedings of the Melbourne Herpetological Symposium. Melbourne : Zoological Board of Victoria; Saint Girons, H. & Bradshaw, S.D. (1981). Donées histo-physiologiques sur l'hypophyse de *Tiliqua rugosa* (Reptilia : Scincidae) et notamment la signification fonctionelle des cellules gamma. *Arch. Anat. Microsc. Morph. Exp.* **70**: 129–140.

Tropidophorus Duméril and Bibron, 1839

Tropidophorus Duméril, A.M.C. & Bibron, G. (1839). *Erpétologie Générale* ou Histoire Naturelle Complète des Reptiles. Paris : Roret 5: viii 854 pp. [554]. Type species *Tropidophorus cocincinensis* Duméril and Bibron, 1839 by monotypy.

Norbea Gray, J.E. (1845). *Catalogue of the Specimens of Lizards in the Collection of the British Museum.* London : British Museum xxviii 289 pp. [101]. Type species *Norbea brookei* Gray, 1845 by monotypy.

Aspris Blyth, E. (1853). Notes and descriptions of various reptiles, new or little known. *J. Asiat. Soc. Beng.* **22**: 639–655 [650]. Type species *Aspris berdmorei* Blyth, 1853 by monotypy.

Amphixestus Peters, W. (1871). Über neue Reptilien aus Ostafrica und Sarawak (Borneo), vorzüglich aus der Sammlung des Hrn. Marquis J. Doria zu Genua. *Mber. K. Preuss. Akad. Wiss. Berl.* **1871**: 566–581 [573] [1872 on title page]. Type species *Amphixestus beccarii* Peters, 1871 by monotypy.

Enoplosaurus Sauvage, H.E. (1879). Enoplosaurus. *Bull. Soc. Philom. Paris (7)* **3**: 210–211 [211]. Type species *Enoplosaurus insignis* Sauvage, 1879 by monotypy.

Synonymy that of Cogger, H.G., this work.

This group is also found in SE Asia.

Tropidophorus queenslandiae De Vis, 1890

Tropidophorus queenslandiae De Vis, C.W. (1890). Descriptions of two lizards of genera new to Australian herpetology. *Proc. Linn. Soc. N.S.W. (2)* **4**: 1034–1036 [1034] [1889 on title page]. Type data: syntypes, QM J233, QM J252, ?QM J19744–19751, from Bellenden Ker and Herberton, Qld.

Distribution: NE coastal, Qld. Ecology: closed forest, rotting wood, terrestrial, nocturnal, crepuscular; viviparous, arthropod-feeder. Biological references: Greer, A.E. (1979). A phylogenetic subdivision of Australian skinks. *Rec. Aust. Mus.* **32**: 339–371 (taxonomy); Greer, A.E. (1980). Critical thermal maximum temperatures in Australian scincid lizards: their ecological and evolutionary significance. *Aust. J. Zool.* **28**: 91–102; Naylor, L.M. (1980). The maintenance of a group of prickly forest skinks (*Tropidophorus queenslandiae* de Vis), in captivity. *Thylacinus, J. Aust. Soc. Zookeepers* **5**: 5–6 (natural history).

TYPHLOPIDAE

INTRODUCTION

A family of worm-like burrowing snakes containing about 150 species in two genera; one genus and 29 species are currently recognised from Australia.

Characterised in Australia by: fewer than 40 subequal scales around the middle of the body; eyes vestigial, reduced to pigmented spots below the head scales; tail very short, terminating in a blunt spine; dentititon reduced and lacking venom glands.

References

Cogger, H.G. (1979). *Reptiles and Amphibians of Australia.* Sydney : A.H. & A.W. Reed 608 pp.

Hahn, D.E. (1980). List der rezenten Amphibien und Reptilien; Anomalepididae, Leptotyphlopidae, Typhlopidae. *Das Tierreich* **101**: i–xii 1–93

Robb, J. (1966). The generic status of the Australasian typhlopids (Reptilia : Squamata). *Ann. Mag. Nat. Hist. (13)* **9**: 675–679

Waite, E.R. (1918). Review of the Australian blind snakes. *Rec. S. Aust. Mus.* **1**: 1–34

Worrell, E. (1970). *Reptiles of Australia.* 2nd edn. Sydney : Angus & Robertson xv 169 pp.

Ramphotyphlops Fitzinger, 1843

[Typhlina* (part.)*] Wagler, J.G. (1830). *Natürliches System der Amphibien, mit vorangehender Classification der Säugethiere und Vögel.* München : Cotta'schen vi 354 pp. [196]. Type species *Acontias lineatus* Reinwardt, *nom. nud.* and *Typhlops sentemstriatus* [*sic*] Schneider, 1801 [=*Leptotyphlops septemstriatus* (Schneider, 1801)].

Ramphotyphlops Fitzinger, L.J. (1843). *Systema Reptilium.* Vienna : Braümüller u. Seidel vi 106 pp. [24] [name conserved in 1982 by the International Commission on Zoological Nomenclature under Opinion 1207, wherein the generic name *Typhlina* Wagler, 1830 was suppressed for the purposes of the Law of Priority]. Type species *Typhlops multilineatus* Schlegel, 1839 by original designation.

Pseudotyphlops Fitzinger, L.J. (1843). *Systema Reptilium.* Vienna : Braümüller u. Seidel vi 106 pp. [24] [*non Pseudotyphlops* Schlegel, 1837]. Type species *Typhlops polygrammicus* Schlegel, 1839 by original designation.

Pilidion Duméril, A.M.C. & Bibron, G. (1844). *Erpétologie Générale* ou Histoire Naturelle Complète des Reptiles. Paris : Roret 6: xii 609 pp. [257] [*nom. substit. pro Typhlina* Wagler, 1830].

Typhlinalis Gray, J.E. (1845). *Catalogue of the Specimens of Lizards in the Collection of the British Museum.* London : British Museum xxviii 289 pp. [134] [*nom. substit. pro Typhlina* Wagler, 1830].

Synonymy that of Cogger, H.G., this work; Waite, E.R. (1918). Review of the Australian blind snakes. *Rec. S. Aust. Mus.* **1**: 1–34; Robb, J. (1966). The structure and possible function of the cloacal pouches of male Australian typhlopids. *Aust. J. Zool.* **14**: 27–30; McDowell, S.B. (1974). A catalogue of the snakes of New Guinea and the Solomons, with special reference to those in the Bernice P. Bishop Museum. Part I. Scolecophidia. *J. Herpet.* **8**: 1–57; Storr, G.M. (1981). The genus *Ramphotyphlops* (Serpentes : Typhlopidae) in Western Australia. *Rec. West. Aust. Mus.* **9**: 235–271.

This group is also found in Indo-Malayan Archipelago, Philippine Ils., New Guinea and islands of western Pacific; if the parthenogenetic species *braminus* is correctly assigned to this genus, distribution is extended to Mexico, the Indian subcontinent, Africa and Madagascar and many islands of the central Pacific.

Ramphotyphlops affinis (Boulenger, 1889)

[Typhlops (Onychocephalus) unguirostris* (part.)*] Peters, W. (1867). Über Felderthiere und Amphibien. *Mber. K. Preuss. Akad. Wiss. Berl.* **1867**: 703–713 [708] [1868 on title page]. Type data: paralectotype, from Rockhampton, Qld., smaller of 2 original syntypes ZMB

5884 (*vide* McDowell, S.B. (1974). A catalogue of the snakes of New Guinea and the Solomons, with special reference to those in the Bernice P. Bishop Museum. Part I. Scolecophidia. *J. Herpet.* **8**: 1–57). For the lectotype, see *Ramphotyphlops unguirostris* (Peters, 1867).

Typhlops affinis Boulenger, G.A. (1889). Description of new Typhlopidae in the British Museum. *Ann. Mag. Nat. Hist. (6)* **4**: 360–363 [363]. Type data: holotype, BMNH 1946.1.10.56, from Qld.

Typhlops kenti Boulenger, G.A. (1914). Descriptions of new species of snakes in the collection of the British Museum. *Ann. Mag. Nat. Hist. (8)* **14**: 482–485 [482]. Type data: holotype, BMNH 1946.1.11.86, from "Northern Queensland".

Typhlops cumingii mansuetus Barbour, T. (1921). Reptiles and amphibians from the British Solomon Islands. *Proc. New Engl. Zool. Club* **7**: 91–112 [108 pl 6]. Type data: holotype, MCZ 14270, from Keri Keri, San Cristobal, Solomon Ils.

Synonymy that of Cogger, H.G., this work.

Distribution: (SE coastal), NE coastal, N Gulf, Murray-Darling basin, (N coastal), N.S.W., Qld., (N.T.). Ecology: fossorial, nocturnal, soil; oviparous, feeds primarily on ant and termite larvae and pupae. Biological references: McDowell, S.B. (1974). A catalogue of the snakes of New Guinea and the Solomons, with special reference to those in the Bernice P. Bishop Museum. Part I. Scolecophidia. *J. Herpet.* **8**: 1–57 (taxonomy); Waite, E.R. (1918). Review of the Australian blind snakes. *Rec. S. Aust. Mus.* **1**: 1–34 (taxonomy).

Ramphotyphlops australis (Gray, 1845)

Anilios australis Gray, J.E. (1845). *Catalogue of the Specimens of Lizards in the Collection of the British Museum.* London : British Museum xxviii 289 pp. [135]. Type data: lectotype, BMNH 1946.1.10.61; for the paralectotype see under *Ramphotyphlops endoterus* (Waite, 1918), designation by Cogger, H.G., this work, from W.A.

Onychocephalus (Ophthalmidion) bicolor Peters, W. (1857). Abhandlung über die Typhlopinen oder blodsichtigen Schlangen und über neue dahin gehörige Arten. *Mber. K. Preuss. Akad. Wiss. Berl.* **1857**: 508–509 [509] [*nom. nud.; ex* Ph. Schmidt ms].

Onychocephalus bicolor Peters, W. (1860). Eine neue zu der Gattung *Onycocephalus* gehörige Wurmschlange, *Onychocephalus macrurus*, und verglich sie mit den bisher von dieser Gattung bekannt gewordenen Arten. *Mber. K. Preuss. Akad. Wiss. Berl.* **1860**: 77–83 [81] [*ex* Ph. Schmidt ms]. Type data: syntypes (probable), ZMB 4721–22, from Adelaide, S.A., original description cites Melbourne, Vic.

Typhlops preissi Jan, G. (1864). *Iconographie Générale des Ophidiens.* Milan : G. Jan 100 pp. [15]. Type data: holotype, RMNH 3689, from "South-western Australia".

Synonymy that of Cogger, H.G., this work.

Distribution: NW coastal, SW coastal, W plateau, S Gulfs, Murray-Darling basin, Bulloo River basin, Lake Eyre basin, W.A., S.A., Vic., N.S.W., (N.T.). Ecology: fossorial, nocturnal, soil; oviparous, feeds primarily on ant and termite larvae and pupae. Biological references: Storr, G.M. (1981). The genus *Ramphotyphlops* (Serpentes : Typhlopidae) in Western Australia. *Rec. West. Aust. Mus.* **9**: 235–271 (taxonomy); Waite, E.R. (1918). Review of the Australian blind snakes. *Rec. S. Aust. Mus.* **1**: 1–34 (taxonomy).

Ramphotyphlops batillus (Waite, 1894)

Typhlops batillus Waite, E.R. (1894). Notes on Australian Typhlopidae. *Proc. Linn. Soc. N.S.W. (2)* **9**: 9–14 [9 pl 1 fig 1–3] [1895 on title page; type locality suspect; status uncertain]. Type data: holotype F, AM R42756, from Wagga Wagga, N.S.W.

Biological references: Waite, E.R. (1918). Review of the Australian blind snakes. *Rec. S. Aust. Mus.* **1**: 1–34; McDowell, S.B. (1974). A catalogue of the snakes of New Guinea and the Solomons, with special reference to those in the Bernice P. Bishop Museum. Part I. Scolecophidia. *J. Herpet.* **8**: 1–57.

Ramphotyphlops bituberculatus (Peters, 1863)

Onychocephalus bituberculatus Peters, W. (1863). Übersicht der von Hrn. Richard Schomburgk an das zoologische Museum eingesandten Amphibien, aus Buchsfelde bei Adelaide in Südaustralien. *Mber. K. Preuss. Akad. Wiss. Berl.* **1863**: 228–236 [233] [1864 on title page]. Type data: syntypes, ZMB 4723, ZMB 4724 (2 specimens), from Loos, 4.5 km W Gawler (as Buchsfelde), S.A., see Praite, R. & Tolley, J.C. (1970). *Place Names of South Australia.* Adelaide : Rigby 203 pp.

Distribution: W plateau, Lake Eyre basin, Bulloo River basin, Murray-Darling basin, S Gulfs, NE coastal, W.A., N.T., S.A., Qld., N.S.W., Vic. Ecology: fossorial, nocturnal, soil; oviparous, feeds primarily on ant and termite larvae and pupae. Biological references: Waite, E.R. (1918). Review of the Australian blind snakes. *Rec. S. Aust. Mus.* **1**: 1–34 (taxonomy); Storr, G.M. (1981). The genus *Ramphotyphlops* (Serpentes : Typhlopidae) in Western Australia. *Rec. West. Aust. Mus.* **9**: 235–271 (taxonomy).

Ramphotyphlops braminus (Daudin, 1803)

Eryx braminus Daudin, F.M. (1803). *Histoire Naturelle, générale et particulière des Reptiles.* Paris : Dufart Tom. 7 436 pp. [279]. Type data: description based on plate 43 in Russell, P. (1796). *An Account of Indian Serpents, Collected on the Coast of Coromandel.* London : Shakespeare Press 91 pp. and description pp. 48–49, Vol. 1, from Vizagapatam, India.

Typhlops bramineus Meyer, A.B. (1887). Verzeichniss der von mir in den Jahren 1870–1873 in ostindischen Archipel gesammelten Reptilien und Batrachier. *Abh.*

Ber. K. Zool. Anthro.-Ethno. Mus. Dresden **1886/87**: 1–16 [8] [*errore pro Eryx braminus* Daudin, 1803].

Synonymy that of McDowell, S.B. (1974). A catalogue of the snakes of New Guinea and the Solomons, with special reference to those in the Bernice P. Bishop Museum. Part I. Scolecophidia. *J. Herpet.* **8**: 1–57.

Distribution: N coastal, N.T., Cocos (Keeling) Is., Christmas Is. Ecology: fossorial, nocturnal, soil; oviparous, parthenogenetic, feeds primarily on ant and termite larvae and pupae. Biological references: McDowell, S.B. (1974). A catalogue of the snakes of New Guinea and the Solomons, with special reference to those in the Bernice P. Bishop Museum. Part I. Scolecophidia. *J. Herpet.* **8**: 1–57 (taxonomy, parthenogenesis); Storr, G.M. (1968). First Australian record of the Asian blind-snake *Typhlops braminus. J. Herpet.* **1**: 98; Swanson, S. (1981). *Typhlina bramina*: an arboreal blind snake? *N. Territory Nat.* **4**: 13 (arboreal habit).

Ramphotyphlops broomi (Boulenger, 1898)

Typhlops broomi Boulenger, G.A. (1898). Description of two new snakes from Queensland. *Ann. Mag. Nat. Hist. (7)* **2**: 414 [414]. Type data: holotype, BMNH 1946.1.10.64, from Muldiva, Qld.

Distribution: N coastal, N Gulf, NE coastal, SE coastal, Lake Eyre basin, Murray-Darling basin, N.T., Qld., N.S.W., Vic., S.A. Ecology: fossorial, nocturnal, soil; oviparous, feeds primarily on ant and termite larvae and pupae. Biological references: Waite, E.R. (1918). Review of the Australian blind snakes. *Rec. S. Aust. Mus.* **1**: 1–34 (taxonomy); Gillam, M.W. (1979). Notes on the status of the "blind" snake *Typhlina tovelli* (Loveridge). *Herpetofauna* **11**: 2–4 (taxonomy).

Ramphotyphlops diversus (Waite, 1894)

Typhlops diversus Waite, E.R. (1894). Notes on Australian Typhlopidae. *Proc. Linn. Soc. N.S.W. (2)* **9**: 9–14 [10 pl 1 figs 4–6] [1895 on title page]. Type data: holotype, QM J2943, from Mowen, Central Railway, Qld.

Typhlops ammodytes Montague, P.D. (1914). A report on the fauna of the Monte Bello Islands. *Proc. Zool. Soc. Lond.* **1914**: 625–652 [642 pl 1 figs 8–10]. Type data: holotype, BMNH 1946.1.11.23, from Hermite Is., Monte Bello Ils., W.A.

Synonymy that of Cogger, H.G., this work.

Distribution: NW coastal, W plateau, N coastal, N Gulf, Lake Eyre basin, NE coastal, (Bulloo River basin), (Murray-Darling basin), W.A., N.T., Qld. Ecology: fossorial, nocturnal, soil; oviparous, feeds primarily on ant and termite larvae and pupae. Biological references: Waite, E.R. (1918). Review of the Australian blind snakes. *Rec. S. Aust. Mus.* **1**: 1–34 (taxonomy); Storr, G.M. (1981). The genus *Ramphotyphlops* (Serpentes : Typhlopidae) in Western Australia. *Rec. West. Aust. Mus.* **9**: 235–271 (taxonomy).

Ramphotyphlops endoterus (Waite, 1918)

[Anilios australis* (part.)*] Gray, J.E. (1845). *Catalogue of the Specimens of Lizards in the Collection of the British Museum.* London : British Museum xxviii 289 pp. [135]. Type data: paralectotype, BMNH 1946.1.11.14 (Australia), for the lectotype (BMNH 1946.1.10.61) see under *Ramphotyphlops australis* (Gray, 1845), designation by Cogger, H.G., this work, from Australia.

Typhlops endoterus Waite, E.R. (1918). Review of the Australian blind snakes. *Rec. S. Aust. Mus.* **1**: 1–34 [32 fig 24]. Type data: holotype, SAMA R88, from Hermannsburg, N.T.

Typhlops leonhardii Sternfeld, R. (1919). Neue Schlangen und Echsen aus Zentralaustralien. *Senckenberg. Biol.* **1**: 76–83 [77]. Type data: lectotype, SMF 16666, designation by Mertens, R. (1967). Die herpetologische Sektion des Natur-Museums und Forschungs-Institutes Senckenberg in Frankfurt-a.-M. nebst einem Verzeichnis ihrer Typen. 1. *Senckenberg. Biol.* **48**: 1–106, from Hermannsburg Mission, Upper Finke River, N.T.

Synonymy that of Cogger, H.G., this work.

Distribution: W plateau, S Gulfs, Lake Eyre basin, (Bulloo River basin), (Murray-Darling basin), W.A., N.T., Qld., S.A., N.S.W. Ecology: fossorial, nocturnal, soil; oviparous, feeds primarily on ant and termite larvae and pupae. Biological references: Storr, G.M. (1981). The genus *Ramphotyphlops* (Serpentes : Typhlopidae) in Western Australia. *Rec. West. Aust. Mus.* **9**: 235–271 (taxonomy).

Ramphotyphlops exocoeti (Boulenger, 1887)

Typhlops exocoeti Boulenger, G.A. (1887). Reptiles *in* Günther, A. (ed.). Report on a zoological collection made by the officers of H.M.S. *Flying Fish* at Christmas Island, Indian Ocean. *Proc. Zool. Soc. Lond.* **1887**: 516–517 [517]. Type data: syntypes, BMNH 1946.1.11.78–79, from Christmas Is., Indian Ocean.

Distribution: Christmas Is. Ecology: closed forest, fossorial, nocturnal.

Ramphotyphlops grypus (Waite, 1918)

Typhlops grypus Waite, E.R. (1918). Review of the Australian blind snakes. *Rec. S. Aust. Mus.* **1**: 1–34 [17 fig 7]. Type data: holotype, NMV D12351, from unknown locality.

Typhlops nigroterminatus Parker, H.W. (1931). A new species of blind snake from N.W. Australia. *Ann. Mag. Nat. Hist. (10)* **8**: 604–605 [605]. Type data: holotype, BMNH 1946.1.11.58, from Roebuck Bay, N W.A.

Synonymy that of Cogger, H.G., this work.

Distribution: NW coastal, W plateau, N coastal, N Gulf, Lake Eyre basin, W.A., N.T., Qld. Ecology: fossorial, nocturnal, soil; oviparous, feeds primarily on ant and termite larvae and pupae. Biological references: Storr, G.M. (1981). The genus *Ramphotyphlops* (Serpentes : Typhlopidae) in Western Australia. *Rec. West. Aust. Mus.* **9**: 235–271 (taxonomy).

Ramphotyphlops guentheri (Peters, 1865)

Typhlops (Onychocephalus) guntheri Peters, W. (1865). Einen ferneren Nachtrag zu seiner Abhandlung über *Typhlopina. Mber. K. Preuss. Akad. Wiss. Berl.* **1865**: 259–263 [259 fig 1] [1866 on title page; as *Güntheri*, automatically corrected to guentheri by Boulenger, G.A. (1893). *Catalogue of snakes in the British Museum (Natural History).* 2. London : British Museum xi 382 pp.]. Type data: holotype, BMNH 1946.1.10.88, from "North Australia".

Typhlops nigricauda Boulenger, G.A. (1895). Descriptions of a new snake and a new frog from north Australia. *Proc. Zool. Soc. Lond.* **1895**: 867 [867 pl 49 fig 1]. Type data: syntypes, BMNH 1946.1.11.13, UZMO K2011, from Daly River, N.T.

Synonymy that of Cogger, H.G., this work.

Distribution: N coastal, (N Gulf), W.A., N.T. Ecology: fossorial, nocturnal, soil; oviparous, feeds primarily on ant and termite larvae and pupae. Biological references: Waite, E.R. (1918). Review of the Australian blind snakes. *Rec. S. Aust. Mus.* **1**: 1–34 (taxonomy); Storr, G.M. (1981). The genus *Ramphotyphlops* (Serpentes : Typhlopidae) in Western Australia. *Rec. West. Aust. Mus.* **9**: 235–271 (taxonomy).

Ramphotyphlops hamatus Storr, 1981

Ramphotyphlops hamatus Storr, G.M. (1981). The genus *Ramphotyphlops* (Serpentes : Typhlopidae) in Western Australia. *Rec. West. Aust. Mus.* **9**: 235–271 [251 figs 15–16]. Type data: holotype, WAM R69572, from Marandoo, 22°38'S, 118°06'E, W.A.

Distribution: NW coastal, W plateau, W.A. Ecology: fossorial, nocturnal, soil; oviparous, feeds primarily on ant and termite larvae and pupae.

Ramphotyphlops kimberleyensis Storr, 1981

Ramphotyphlops kimberleyensis Storr, G.M. (1981). The genus *Ramphotyphlops* (Serpentes : Typhlopidae) in Western Australia. *Rec. West. Aust. Mus.* **9**: 235–271 [254 fig 18]. Type data: holotype, WAM R41456, from Bigge Is., 14°32'S, 125°08'E, W.A.

Distribution: N coastal, W.A. Ecology: fossorial, nocturnal, soil; oviparous, feeds primarily on ant and termite larvae and pupae.

Ramphotyphlops leptosoma Robb, 1972

Ramphotyphlops leptosoma Robb, J. (1972). A new species of the genus *Rhamphotyphlops* (Serpentes : Typhlopidae) from Western Australia. *J. R. Soc. West. Aust.* **55**: 39–40 [39]. Type data: holotype, WAM R29623, from "The Loop", lower Murchison River, W.A.

Distribution: NW coastal, W.A. Ecology: fossorial, nocturnal, soil; oviparous, feeds primarily on ant and termite larvae and pupae. Biological references: Storr, G.M. (1981). The genus *Ramphotyphlops* (Serpentes : Typhlopidae) in Western Australia. *Rec. West. Aust. Mus.* **9**: 235–271 (taxonomy).

Ramphotyphlops leucoproctus (Boulenger, 1889)

Typhlops leucoproctus Boulenger, G.A. (1889). Description of new Typhlopidae in the British Museum. *Ann. Mag. Nat. Hist. (6)* **4**: 360–363 [361]. Type data: syntypes, BMNH 1946.1.11.84 (Fly River); BMNH, 1946.1.11.97 (Qld.); BMNH 1946.1.10.58–60 (Murray Is.), from Fly River, Papua New Guinea; Qld. and Murray Is., Torres Strait, Qld.

Distribution: NE coastal, (N Gulf), Qld. Ecology: fossorial, nocturnal, soil; oviparous, feeds primarily on ant and termite larvae and pupae. Biological references: McDowell, S.B. (1974). A catalogue of the snakes of New Guinea and the Solomons, with special reference to those in the Bernice P. Bishop Museum. Part I. Scolecophidia. *J. Herpet.* **8**: 1–57 (taxonomy).

Ramphotyphlops ligatus (Peters, 1879)

Typhlops ligatus Peters, W. (1879). Mitteilung über neue Amphibien des Kgl. zoologischen Museums (*Euprepes, Acontias, Typhlops, Zamenis, Spilotes, Oedipus*). *Mber. K. Preuss. Akad. Wiss. Berl.* **1879**: 773–779 [775 fig 3] [1880 on title page]. Type data: holotype, ZMB 6231, from Port Mackay, Qld.

Typhlops curtus Ogilby, J.D. (1892). On some undescribed reptiles and fishes from Australia. *Rec. Aust. Mus.* **2**: 23–26 [23]. Type data: holotype, AM R1132, from Walsh River, Gulf of Carpenteria, Qld.

Synonymy that of Cogger, H.G., this work.

Distribution: N coastal, NE coastal, Murray-Darling basin, Bulloo River basin, Lake Eyre basin, (N Gulf), (W plateau), W.A., N.T., Qld., N.S.W., Vic., S.A. Ecology: fossorial, nocturnal, soil; oviparous, feeds primarily on ant and termite larvae and pupae. Biological references: Waite, E.R. (1918). Review of the Australian blind snakes. *Rec. S. Aust. Mus.* **1**: 1–34 (taxonomy); Storr, G.M. (1981). The genus *Ramphotyphlops* (Serpentes : Typhlopidae) in Western Australia. *Rec. West. Aust. Mus.* **9**: 235–271 (taxonomy); Wells, R. (1979). New reptile records for the Northern Territory. *N. Territory Nat.* **1**: 3–4 (distribution).

Ramphotyphlops margaretae Storr, 1981

Ramphotyphlops margaretae Storr, G.M. (1981). The genus *Ramphotyphlops* (Serpentes : Typhlopidae) in

Western Australia. *Rec. West. Aust. Mus.* **9**: 235–271 [259 fig 23]. Type data: holotype, WAM R15710, from Lake Throssell, 27°25′S, 124°18′E, W.A.

Distribution: W plateau, W.A. Ecology: fossorial, nocturnal, soil; oviparous, feeds primarily on ant and termite larvae and pupae.

Ramphotyphlops micromma Storr, 1981

Ramphotyphlops micromma Storr, G.M. (1981). The genus *Ramphotyphlops* (Serpentes : Typhlopidae) in Western Australia. *Rec. West. Aust. Mus.* **9**: 235–271 [260 fig 24]. Type data: holotype, WAM R1341, from Leopold Downs, 17°52′S, 125°26′E, W.A.

Distribution: N coastal, W.A. Ecology: fossorial, nocturnal, soil; oviparous, feeds primarily on ant and termite larvae and pupae.

Ramphotyphlops minimus (Kinghorn, 1929)

Typhlops minimus Kinghorn, J.R. (1929). Two new snakes from Australia. *Rec. Aust. Mus.* **17**: 190–193 [190 fig 1]. Type data: holotype, AM R9692, from Groote Eylandt, Gulf of Carpentaria, N.T.

Distribution: N Gulf, N.T.; Groote Eylandt. Ecology: fossorial, nocturnal, soil; oviparous, feeds primarily on ant and termite larvae and pupae.

Ramphotyphlops nigrescens (Gray, 1845)

Anilios nigrescens Gray, J.E. (1845). *Catalogue of the Specimens of Lizards in the Collection of the British Museum.* London : British Museum xxviii 289 pp. [135]. Type data: lectotype, from Parramatta, N.S.W., BMNH 1946.1.11.17, designation by Cogger, H.G., this work.

Typhlops ruppelli Jan, G. (1864). *Iconographie Générale des Ophidiens.* Milan : G. Jan 100 pp. [14]. Type data: lectotype, SMF 16617, designation by Mertens, R. (1967). Die herpetologische Sektion des Natur-Museums und Forschungs-Institutes Senckenberg in Frankfurt-a.-M. nebst einem Verzeichnis ihrer Typen. 1. *Senckenberg. Biol.* **48**: 1–106, from Sydney, N.S.W.

Typhlops reginae Boulenger, G.A. (1889). Description of new Typhlopidae in the British Museum. *Ann. Mag. Nat. Hist. (6)* **4**: 360–363 [362]. Type data: syntypes, BMNH 1946.1.11.81–83, from Qld.

Synonymy that of Cogger, H.G., this work.

Distribution: NE coastal, SE coastal, Murray-Darling basin, Qld., N.S.W., A.C.T., Vic. Ecology: fossorial, nocturnal, soil, open forest, woodland, closed forest; oviparous, feeds primarily on ant and termite larvae and pupae. Biological references: Waite, E.R. (1918). Review of the Australian blind snakes. *Rec. S. Aust. Mus.* **1**: 1–34 (taxonomy); Shine, R. (1980). Reproduction, feeding and growth in the Australian burrowing snake *Vermicella annulata. J. Herpet.* **14**: 71–77 (as prey).

Ramphotyphlops pinguis (Waite, 1897)

Typhlops pinguis Waite, E.R. (1897). Notes on Australian Typhlopidae. *Trans. R. Soc. S. Aust.* **21**: 25–27 pl 3 [25 pl 3]. Type data: holotype, SAMA R803, from S.A.

Distribution: SW coastal, Murray-Darling basin, W.A., S.A., Vic. Ecology: fossorial, nocturnal, soil; oviparous, feeds primarily on ant and termite larvae and pupae. Biological references: Waite, E.R. (1918). Review of the Australian blind snakes. *Rec. S. Aust. Mus.* **1**: 1–34 (taxonomy); Storr, G.M. (1981). The genus *Ramphotyphlops* (Serpentes : Typhlopidae) in Western Australia. *Rec. West. Aust. Mus.* **9**: 235–271 (taxonomy).

Ramphotyphlops polygrammicus (Schlegel, 1839)

Typhlops polygrammicus Schlegel, H. (1839). *Abbildungen neuer oder unvollständig bekannter Amphibien, nach der Natur der dem Leben entworfen.* Düsseldorf : Arnz xiv 141 pp. (1837–1844) [40 (1839), pl 32 (1838)]. Type data: holotype, RMNH 3712, from Timor, Indonesia.

Typhlops torresianus Boulenger, G.A. (1889). Description of new Typhlopidae in the British Museum. *Ann. Mag. Nat. Hist. (6)* **4**: 360–363 [362]. Type data: syntypes, BMNH 1946.1.11.48–49, from Murray Is., Torres Strait, Qld.

Typhlops soensis de Jong, J.K. (1930). Notes on some reptiles from the Dutch-East-Indies. *Treubia* **12**: 115–119 [117]. Type data: holotype, MZB, from Soë, South Timor, alt. 88m, Indonesia.

Synonymy that of Cogger, H.G., this work.

Distribution: N Gulf, NE coastal, Qld. Ecology: fossorial, nocturnal, soil; oviparous, feeds primarily on ant and termite larvae and pupae. Biological references: Waite, E.R. (1918). Review of the Australian blind snakes. *Rec. S. Aust. Mus.* **1**: 1–34 (taxonomy); McDowell, S.B. (1974). A catalogue of the snakes of New Guinea and the Solomons, with special reference to those in the Bernice P. Bishop Museum. Part I. Scolecophidia. *J. Herpet.* **8**: 1–57 (taxonomy); Rieppel, O. (1979). The braincase of *Typhlops* and *Leptotyphlops* (Reptilia : Serpentes). *Zool. J. Linn. Soc.* **65**: 161–176 (morphology).

Ramphotyphlops proximus (Waite, 1893)

Typhlops proximus Waite, E.R. (1893). Notes on Australian Typhlopidae. *Rec. Aust. Mus.* **2**: 57–62 pl 15 [60 pl 15 fig 3–4]. Type data: holotype, AM 6411, from N.S.W.

Distribution: NE coastal, SE coastal, Murray-Darling basin, Qld., N.S.W., Vic., S.A. Ecology: fossorial, nocturnal, soil; oviparous, feeds primarily on ant and termite larvae and pupae.

Ramphotyphlops troglodytes Storr, 1981

Ramphotyphlops troglodytes Storr, G.M. (1981). The genus *Ramphotyphlops* (Serpentes : Typhlopidae) in Western Australia. *Rec. West. Aust. Mus.* **9**: 235–271 [263 fig 27]. Type data: holotype, WAM R51043, from Tunnel Cave, Napier Range, 17°37′S, 125°14′E, W.A.

Distribution: N coastal, W.A. Ecology: fossorial, nocturnal, soil; oviparous, feeds primarily on ant and termite larvae and pupae.

Ramphotyphlops tovelli (Loveridge, 1945)

Typhlops tovelli Loveridge, A. (1945). A new blind snake (*Typhlops tovelli*) from Darwin, Australia. *Proc. Biol. Soc. Wash.* **58**: 111–112 [111]. Type data: holotype, MCZ 48844, from Koonowarra Sports Ground, about 5 mi S Darwin, N.T.

Distribution: N coastal, N.T. Ecology: fossorial, nocturnal, soil; oviparous, feeds primarily on ant and termite larvae and pupae. Biological references: Gillam, M.W. (1979). Notes on the status of the "blind" snake *Typhlina tovelli* (Loveridge). *Herpetofauna* **11**: 2–4 (taxonomy).

Ramphotyphlops unguirostris (Peters, 1867)

***Typhlops (Onychocephalus) unguirostris* (part.)** Peters, W. (1867). Über Flederthiere und Amphibien. *Mber K. Preuss. Akad. Wiss. Berl.* **1867**: 703–713 [708 fig 3] [1868 on title page]. Type data: lectotype, from Rockhampton, Qld., ZMB 5884 (Rockhampton), larger of 2 original syntypes; for the paralectotype, see under *Ramphotyphlops affinis* (Boulenger, 1867), designation by McDowell, S.B. (1974). A catalogue of the snakes of New Guinea and the Solomons, with special reference to those in the Bernice P. Bishop Museum. Part I. Scolecophidia. *J. Herpet.* **8**: 1–57.

Typhlops curvirostris Peters, W. (1879). Mitteilung über neue Amphibien des Kgl. zoologischen Museums (*Euprepes*, *Acontias*, *Typhlops*, *Zamenis*, *Spilotes*, *Oedipus*). *Mber. K. Preuss. Akad. Wiss. Berl.* **1879**: 773–779 [776 fig 5] [1880 on title page]. Type data: holotype, ZMB 7525, from Port Clinton (as Port Bowen), Qld.

Synonymy that of Cogger, H.G., this work.

Distribution: N coastal, N Gulf, NE coastal, Lake Eyre basin, Bulloo River basin, Murray-Darling basin, S Gulfs, W.A., N.T., Qld., N.S.W., S.A., Vic. Ecology: fossorial, nocturnal, soil; oviparous, feeds primarily on ant and termite larvae and pupae. Biological references: Waite, E.R. (1918). Review of the Australian blind snakes. *Rec. S. Aust. Mus.* **1**: 1–34 (taxonomy).

Ramphotyphlops waitii (Boulenger, 1895)

Typhlops waitii Boulenger, G.A. (1895). A new *Typhlops* previously confounded with *Typhlops unguirostris*. *Proc. Linn. Soc. N.S.W. (2)* **9**: 718–719 [718]. Type data: holotype, BMNH 1946.1.11.74, from "N.W. Australia".

Distribution: NW coastal, SW coastal, W plateau, W.A. Ecology: fossorial, nocturnal, soil; oviparous, feeds primarily on ant and termite larvae and pupae. Biological references: Storr, G.M. (1981). The genus *Ramphotyphlops* (Serpentes : Typhlopidae) in Western Australia. *Rec. West. Aust. Mus.* **9**: 235–271 (taxonomy).

Ramphotyphlops wiedii (Peters, 1867)

Typhlops wiedii Peters, W. (1867). Herpetologische Notizen. *Mber. K. Preuss. Akad. Wiss. Berl.* **1867**: 13–37 [24] [1868 on title page]. Type data: holotype, ZMB 5845, from Brisbane, Qld.

Distribution: N coastal, N Gulf, NE coastal, SE coastal, Murray-Darling basin, N.T., Qld., N.S.W., (Vic.), (S.A.). Ecology: fossorial, nocturnal, soil; oviparous, feeds primarily on ant and termite larvae and pupae. Biological references: Waite, E.R. (1918). Review of the Australian blind snakes. *Rec. S. Aust. Mus.* **1**: 1–34 (taxonomy); Hoser, R.T. (1980). Further records of aggregations of various species of Australian snakes. *Herpetofauna* **12**: 16–22 (aggregation).

Ramphotyphlops yampiensis Storr, 1981

Ramphotyphlops yampiensis Storr, G.M. (1981). The genus *Ramphotyphlops* (Serpentes : Typhlopidae) in Western Australia. *Rec. West. Aust. Mus.* **9**: 235–271 [268 fig 32]. Type data: holotype, WAM R26839, from Koolan Is., 16°08′S, 123°45′E, W.A.

Distribution: N coastal, W.A. Ecology: fossorial, nocturnal, soil; oviparous, feeds primarily on ant and termite larvae and pupae.

Ramphotyphlops yirrikalae (Kinghorn, 1942)

Typhlops yirrikalae Kinghorn, J.R. (1942). Herpetological notes. 4. *Rec. Aust. Mus.* **21**: 118–121 [118 fig 1]. Type data: holotype, AM R12381, from Yirrikala Mission Station, near Caledon Bay, Arnhem Land, N.T.

Distribution: N coastal, N Gulf, N.T. Ecology: fossorial, nocturnal, soil; oviparous, feeds primarily on ant and termite larvae and pupae.

BOIDAE

INTRODUCTION

A family of moderate to very large terrestrial and arboreal constricting snakes which occurs throughout the tropical and temperate regions of the world. Of the 23 genera and 85 species currently recognised, the 4 genera and 13 species in Australia belong to the subfamily Pythoninae (pythons).

Characterised in Australia by: a pair of cloacal spurs; enlarged ventral scales, much larger than adjacent body scales; more than 30 mid-body scale rows; usually some labial scales with pits; a series of loreal scales; with solid teeth and lacking venom glands.

Within this family a number of combinations are introduced which require explanation. McDowell (1975: 30) correctly points out that except for the west African genus *Calabaria* and the Australian genus *Aspidites* the remaining genera within the Pythoninae "are only weakly defined". He went on to state that "... a good case could be made for referring them all to *Python* Daudin." We believe that such an action would obscure relationships within this morphologically conservative subfamily. In his taxonomic arrangement, McDowell emphasised a suite of morphological characters including the presence/absence of a prehensile tail and of apical pits in the body scales. Using these characters, he recognised the *Liasis* group (scale pits present, tail not prehensile) and the *Python* group (containing *Morelia sensu stricto* and *Chondropython*: no scale pits and strongly prehensile tail).

McDowell's conclusions regarding relationships within the Australo-Papuan boids were supported by Schwaner & Dessauer (1981) using immuno-diffusion techniques. They also showed, however, that the immunological distances between McDowell's New Guinean species of *"Python"* and *Python regius* from Africa, were much greater than those between *Liasis* and any New Guinean members of *Python*. As the type-species of *Python* is *Python tigris* Daudin, 1803 (=*Python molurus* Linnaeus, 1758 of eastern Asia), the relationship between the Australasian snakes assigned to *"Python"* by McDowell and the type species of *Python (tigris)* remains an open question. Although we have adopted McDowell's findings on relationships within the Australo-Papuan pythonine snakes, we believe that to assign any of them to the genus *Python* would be inconsistent with the continued recognition of a separate genus *Liasis* (*sensu* McDowell). We have resurrected, therefore, the genus *Morelia* Gray, 1842 from the synonymy of the genus *Python*, to which it was recently relegated by McDowell (1975: 30), to accommodate those prehensile-tailed Australasian pythons which McDowell transferred to *Python*.

As indicated above, the other major Australo-Papuan genus recognised by McDowell is *Liasis* of Gray, 1842. This long-standing genus presents some complex taxonomic and nomenclatural problems; the latter are difficult to resolve without recourse to a ruling by the International Commission on Zoological Nomenclature. The taxonomic problems have been examined at length by McDowell (1975) who concluded that *"Liasis" amethistinus* should be removed to the genus *Python*, while the monotypic genus *Bothrochilus* (containing the species *boa* from the Bismarck Archipelago) should be relegated to the synonymy of *Liasis*. While formally proposing this taxonomic arrangement McDowell indicated, however, that it included the following reasoning: if *Bothrochilus* warranted recognition as a monotypic genus then there would be an even stronger case for hiving off *Liasis childreni* from other *"Liasis"* into its own monotypic genus. Rather than take this new step, McDowell retained *Liasis* in its traditional sense (excluding *amethistinus* and *boeleni*) and expanded it by the inclusion of *Bothrochilus boa*. Taxonomists may ultimately choose McDowell's alternative schema of treating *boa* and *childreni* in different and separate genera (which we believe more accurately reflects the phylogeny of this group based on morphological criteria) but we have followed McDowell's taxonomic conclusions in the following pages.

The nomenclatural consequences of his actions, however, are more complex than he envisaged. McDowell regarded *Liasis childreni* as the type species of *Liasis* on the following basis. Gray (1842: 44–45), in erecting the genus *Liasis*, did not designate a type species but included three nominal species - *childreni, amethystinus* and *olivacea*. Subsequently, Gray (1849: 91–93) erected three subgenera in *Liasis*: *Simalia* (containing *amethystinus* and *macklotii*), *Liasis* (containing only *childrenii*) and *Lisalia* (containing only *olivacea*). McDowell regarded this as the subsequent designation of *childreni* as type-species of *Liasis*. Unfortunately, as has been pointed out by Dr L.D. Brongersma (*in litt.*), there is an earlier type-species designation for *Liasis*. Desmarest (1846: 337) designated "*Boa amethistinus* Schneid., Daud." as type-species of Gray's *Liasis*, so *Liasis* should become a synonym of that genus (in McDowell's taxonomic arrangement, the genus *Python*) to which *amethistina* is assigned. A further minor complication is that, as pointed out by McDowell, Gray's concept of *amethystinus* as included in the original description of *Liasis*, was apparently based not on *amethistina* of Schneider but on *mackloti* of Duméril and Bibron. This is indicated by Gray's citation of *amethystinus* in the Leyden (Leiden) Museum, and his subsequent referral (1849: 92) of the Leyden specimen(s) to *macklotii*. Had Gray nominated *amethystinus* as type-species of *Liasis*, his designation might have been regarded as based on a misidentification and therefore subject to resolution by the International Commission on Zoological Nomenclature. There is no ambiguity, however, in Desmarest's designation of the nominal species *amethistina* of Schneider as type-species of *Liasis* and although it might be argued that Schneider's *amethistina* was not among the originally-included species in *Liasis* (and therefore not eligible as type-species under Article 69(a) of the International Code of Zoological Nomenclature), we believe such an argument could not be sustained, and we regard *Boa amethistina* Schneider, 1801 as the validly designated type species of *Liasis* Gray, 1842.

The genus *Liasis*, consequently, has been relegated to the synonymy of *Morelia*, which now contains the Australian species *spilota, oenpelliensis, carinata, bredli* and *amethistina*. The genus *Bothrochilus* becomes the next available name for those Australian species previously included in *Liasis, viz fuscus, olivaceus, childreni, albertisii* and *perthensis*. It might be argued that *Liasis* rather than *Morelia* should be chosen as the senior synonym (they were published simultaneously by Gray, 1842 on consecutive pages: *Morelia* on page 43, *Liasis* on page 44). The use of *Liasis* in this context, however, would be the antithesis of long-standing usage and would lead to unnecessary confusion.

References

Cogger, H.G. (1979). *Reptiles and Amphibians of Australia.* Sydney : A.H. & A.W. Reed 608 pp.

Gray, J.E. (1849). *Catalogue of the specimens of snakes in the collection of the British Museum.* London : British Museum xv 125 pp.

McDowell, S.B. (1975). A catalogue of the snakes of New Guinea and the Solomons, with special reference to those in the Bernice P. Bishop Museum. Part II. Anilioidea and Pythoninae. *J. Herpet.* **9**: 1–79

Schwaner, T.D. & Dessaur, H.G. (1981). Immunodiffusion evidence for the relationships of papuan boids. *J. Herpet.* **15**: 250–253.

Stimson, A.F. (1969). Liste der rezenten Amphibien und Reptilien. Boidae. *Das Tierreich* **89**: 1–49

Worrell, E. (1951). Classification of Australian Boidae. *Proc. R. Zool. Soc. N.S.W.* **1949–50**: 20–25

Worrell, E. (1970). *Reptiles of Australia.* 2nd edn. Sydney : Angus & Robertson xv 169 pp.

Aspidites Peters, 1876

Aspidiotes Krefft, G. (1864). Description of *Aspidiotes melanocephalus*, a new snake from Port Denison, N.E. Australia. *Proc. Zool. Soc. Lond.* **1864**: 20–22 [20] [*non Aspidiotes* Bouché, 1844 and *non Aspidiotes* Schoenherr, 1847]. Type species *Aspidiotes melanocephalus* Krefft, 1864 by monotypy.

Aspidiotus Peters, W. (1876). Berichtgung zu s.533; *Aspidiotus/Aspidites. Mber. K. Preuss. Akad. Wiss. Berl.* **1876**: 914 [*lapsus pro Aspidiotes* Krefft, 1864; *non Aspidiotus* Bouché, 1833].

Aspidites Peters, W. (1876). Berichtigung zu s.533; *Aspidiotus/Aspidites. Mber. K. Preuss. Akad. Wiss. Berl.* **1876**: 914 [914] [*nom. nov. pro Aspidiotus* [*sic*] Krefft, 1864].

Synonymy that of Cogger, H.G., this work.

Aspidites melanocephalus (Krefft, 1864)

Aspidiotes melanocephalus Krefft, G. (1864). Description of *Aspidiotes melanocephalus*, a new snake from Port Denison, N.E. Australia. *Proc. Zool. Soc. Lond.* **1864**: 20–22 [20]. Type data: holotype, BMNH 1946.1.8.2, from Bowen (as Port Denison), Qld.

Distribution: N coastal, N Gulf, (W plateau), Lake Eyre basin, NE coastal, W.A., N.T., Qld. Ecology: terrestrial, nocturnal, predator, closed forest, open forest, tall shrubland, woodland; oviparous, lacertiphagous, ophiophagous, constrictor. Biological references: Smith, L.A. (1981). A revision of the python genera *Aspidites* and *Python* (Serpentes : Boidae) in Western Australia. *Rec. West. Aust. Mus.* **9**: 211–226 (taxonomy); Stull, O.G. (1935). A checklist of the family Boidae. *Proc. Bost. Soc. Nat. Hist.* **40**: 387–408 (taxonomy).

Aspidites ramsayi (Macleay, 1882)

Aspidiotes ramsayi Macleay, W. (1882). Description of two new species of snakes. *Proc. Linn. Soc. N.S.W.* **6**: 811–813 [813]. Type data: holotype, MMUS, from Fort Bourke, N.S.W., presumed lost.

Aspidites collaris Longman, H.A. (1913). Herpetological notes. I. Systematic. Including the description of one new species. Part II. Ethological. *Mem. Qd. Mus.* **2**: 39–45 [40]. Type data: holotype, QM J944, from Avondale Station, via Cunnamulla, Qld.

Synonymy that of Cogger, H.G., this work.

Distribution: NW coastal, W plateau, Lake Eyre basin, Bulloo River basin, Murray-Darling basin, W.A., N.T., S.A., Qld., N.S.W. Ecology: terrestrial, nocturnal, predator, tall shrubland, desert; oviparous, lacertiphagous, ophiophagous, constrictor. Biological references: Smith, L.A. (1981). A revision of the python genera *Aspidites* and *Python* (Serpentes : Boidae) in Western Australia. *Rec. West. Aust. Mus.* **9**: 211–226 (taxonomy); Waite, E.R. (1894). Redescription of *Aspidites ramsayi* Macleay. *Proc. Linn. Soc. N.S.W. (2)* **9**: 715–717 (taxonomy); Fyfe, G. & Harvey, C. (1981). Some observations on the woma (*Aspidites ramsayi*) in captivity. *Herpetofauna* **18**: 23–25 (natural history).

Bothrochilus Fitzinger, 1843

Liasis Gray, J.E. (1840). *Synopsis of the Contents of the British Museum.* 42nd edn. London : British Museum pp. 40–42 [42] [*nom. nud.*].

Nardoa Gray, J.E. (1842). Synopsis of the species of prehensile-tailed snakes, or family Boidae. *in* Gray, J.E. (ed.) *Zoological Miscellany.* London : Treuttel, Würtz & Co. pp. 41–46 [45] [*non Nardoa* Gray, 1840]. Type species *Nardoa schlegeli* Gray, 1842 [=*Bothrochilus boa* (Schlegel, 1837)] by subsequent designation, see Stimson, A.F. (1969). Liste der rezenten Amphibien und Reptilien. Boidae. *Das Tierreich* **89**: 1–49.

Bothrochilus Fitzinger, L.J. (1843). *Systema Reptilium.* Vienna : Braümüller u. Seidel vi 106 pp. [24]. Type species *Tortrix boa* Schlegel, 1837 by original designation.

***Simalia* (part.)** Gray, J.E. (1849). *Catalogue of the Specimens of Snakes in the Collection of the British Museum.* London : British Museum xv 125 pp. [91]. Type species, none designated, *Liasis macklotii* (*sic*) one of two included species, see also synonymy of *Morelia*.

Lisalia Gray, J.E. (1849). *Catalogue of the Specimens of Snakes in the Collection of the British Museum.* London : British Museum xv 125 pp. [92] [described as subgenus of *Liasis*]. Type species *Liasis olivacea* Gray, 1842 by monotypy.

Leiopython Hubrecht, A.A.W. (1879). On a new genus and species of Pythonidae from Salawatti. *Notes Leyden Mus.* **1**: 14–15 [14]. Type species *Leiopython gracilis* Hubrecht, 1879 by monotypy.

Nardoana Berg, C. (1901). Herpetological notes. *Comm. Mus. Nacl. Buenos Aires* **1**: 289–291 [289] [*nom. substit. pro Nardoa* Gray, 1842].

Nordoana Berg, C. (1901). Herpetological notes. *Comm. Mus. Nacl. Buenos Aires* **1**: 289–291 [289] [*lapsus pro*

Nardoana Berg, 1901].

Synonymy that of Cogger, H.G., this work.

This group is also found in Timor and New Guinea, including the Bismarck Archipelago, see McDowell, S.B. (1975). A catalogue of the snakes of New Guinea and the Solomons, with special reference to those in the Bernice P. Bishop Museum. Part II. Anilioidea and Pythoninae. *J. Herpet.* **9**: 1–79.

Bothrochilus albertisii (Peters and Doria, 1878)

Liasis albertisii Peters, W. & Doria, G. (1878). Catalogo dei Rettili e dei Batraci raccolti da O. Beccari, L.M. d'Albertis e A.A. Bruijn nella sotto-regione Austro-Malese. *Ann. Mus. Civ. Stor. Nat. Genova* **13**: 323–450 [401 pl 3 fig 2]. Type data: syntypes, MCG C.E. 29989 (near Andai), MCG C.E. 29990 (Kapaor), from near Andai, and Kapaor, Irian Jaya, Indonesia.

Leiopython gracilis Hubrecht, A.A.W. (1879). On a new genus and species of Pythonidae from Salawatti. *Notes Leyden Mus.* **1**: 14–15 [15]. Type data: holotype, RMNH 4796, from Salawati, Irian Jaya, Indonesia.

Synonymy that of Cogger, H.G., this work.

Distribution: Qld.; northern-most Torres Strait islands. Ecology: terrestrial, nocturnal, predator, open forest; oviparous, general carnivore, constrictor. Biological references: McDowell, S.B. (1975). A catalogue of the snakes of New Guinea and the Solomons, with special reference to those in the Bernice P. Bishop Museum. Part II. Anilioidea and Pythoninae. *J. Herpet.* **9**: 1–79 (taxonomy); Schwaner, T.D. & Dessauer, H.C. (1981). Immunodiffusion evidence for the relationships of Papuan boids. *J. Herpet.* **15**: 250–253 (biochemical taxonomy); Ross, R. & Larman, R. (1977). Captive breeding in two species of python, *Liasis albertisii* and *L. mackloti*. *Int. Zoo Yearb.* **17**: 133–136 (reproduction); Brongersma, L.D. (1956). Notes on New Guinea reptiles and amphibians IV. *Proc. K. Ned. Akad. Wet.* **59C**: 447–453 (taxonomy).

Bothrochilus childreni (Gray, 1842)

Liasis childreni Gray, J.E. (1842). Synopsis of the species of prehensile-tailed snakes, or family Boidae. *in* Gray, J.E. (ed.) *Zoological Miscellany*. London : Treuttel, Würtz & Co. pp. 41–46 [44]. Type data: holotype, BMNH 1946.1.16.78, from unknown locality.

Nardoa gilbertii Gray, J.E. (1842). Synopsis of the species of prehensile-tailed snakes, or family Boidae. *in* Gray, J.E. (ed.) *Zoological Miscellany*. London : Treuttel, Würtz & Co. pp. 41–46 [45]. Type data: holotype, BMNH 1946.1.16.69, from Port Essington, N.T.

Liasis maculosus Peters, W. (1873). Über eine neue Schildkrötenart, *Cinosternon effeldtii* und einige andere neue oder weniger bekannte Amphibien. *Mber. K. Preuss. Akad. Wiss. Berl.* **1873**: 603–618 [608] [1874 on title page]. Type data: syntypes, ZMB 5948 (Port Mackay), ZMB 5860 (Rockhampton), ZMB 7513 (Port, Clinton), from Port Mackay, Rockhampton and Port Clinton (as Port Bowen), Qld.

Synonymy that of Cogger, H.G., this work.

Distribution: SW coastal, NW coastal, W plateau, N coastal, N Gulf, NE coastal, Lake Eyre basin, Bulloo River basin, Murray-Darling basin, (S Gulfs), W.A., N.T., S.A., Qld., N.S.W. Ecology: nocturnal, predator, desert, tall shrubland, closed forest, tall forest, open forest, woodland, tall open shrubland; oviparous, general carnivore, constrictor, semi-arboreal. Biological references: Dunn, R.W. (1979). Breeding Children's pythons, *Liasis childreni*, at Melbourne Zoo. *Int. Zoo Yearb.* **19**: 89–90 (reproduction); Barnett, B. (1980). Captive breeding and a novel egg incubation technique of the Children's python (*Liasis childreni*). *Herpetofauna* **11**: 15–18 (reproduction); Stull, O.G. (1935). A checklist of the family Boidae. *Proc. Bost. Soc. Nat. Hist.* **40**: 387–408 (taxonomy); Worrell, E. (1951). Classification of Australian Boidae. *Proc. R. Zool. Soc. N.S.W.* **1949–50**: 20–25 (taxonomy); Sheargold, T. (1979). Notes on the reproduction of Children's pythons (*Liasis childreni* Gray, 1842). *Herpetofauna* **10**: 2–4.

Bothrochilus fuscus (Peters, 1873)

Liasis fuscus Peters, W. (1873). Über eine neue Schildkrötenart, *Cinosternon effeldtii* und einige andere neue oder weniger bekannte Amphibien. *Mber. K. Preuss. Akad. Wiss. Berl.* **1873**: 603–618 [607] [1874 on title page]. Type data: holotype, ZMB 7840 (presumed lost), from Port Clinton (as Port Bowen), Qld.

Liasis cornwallisius Günther, A. (1879). Notice of a collection of reptiles from islands of Torres Straits. *Ann. Mag. Nat. Hist. (5)* **3**: 84–87 [85]. Type data: holotype, BMNH 1946.1.10.41, from Dauan (as Cornwallis) Is., Torres Strait, Qld.

Nardoa crassa Macleay, W. (1885). On some Reptilia lately received from the Herbert River District, Queensland. *Proc. Linn. Soc. N.S.W.* **10**: 64–68 [66] [1886 on title page]. Type data: syntypes, AM B5940, AM R31886, third syntype presumed lost, from "vicinity of Herbert River", Qld.

Synonymy that of Cogger, H.G., this work.

Distribution: N coastal, N Gulf, NE coastal, N.T., W.A., Qld. Ecology: terrestrial, nocturnal, predator, closed forest, open forest, woodland, low open woodland, swamp; oviparous, general carnivore, constrictor, semi-aquatic. Biological references: McDowell, S.B. (1975). A catalogue of the snakes of New Guinea and the Solomons, with special reference to those in the Bernice P. Bishop Museum. Part II. Anilioidea and Pythoninae. *J. Herpet.* **9**: 1–79 (taxonomy, as *L. mackloti*); Smith, L.A. (1981). A revision of the *Liasis olivaceus* species-group (Serpentes : Boidae) in

Western Australia. *Rec. West. Aust. Mus.* **9**: 227–233 (taxonomy, as *L. mackloti*); Ross, R. & Larman, R. (1977). Captive breeding in two species of python, *Liasis albertisii* and *L. mackloti*. *Int. Zoo Yearb.* **17**: 133–136 (reproduction, as *L. mackloti*); Churchill, G. (1971). Notes on water pythons (*Liasis fuscus*). *Herpetofauna* **3**: 8 (natural history).

Bothrochilus olivaceus (Gray, 1842)

Liasis olivacea Gray, J.E. (1842). Synopsis of the species of prehensile-tailed snakes, or family Boidae. *in* Gray, J.E. (ed.) *Zoological Miscellany*. London : Treuttel, Würtz & Co. pp. 41–46 [45]. Type data: holotype, BMNH 1946.1.1.56, from Port Essington, N.T.

Liasis papuanus Peters, W. & Doria, G. (1878). Catalogo dei Rettili e dei Batraci raccolti da O. Beccari, L.M. d'Albertis e A.A. Bruijn nella sotto-regione Austro-Malese. *Ann. Mus. Civ. Stor. Nat. Genova* **13**: 323–450 [400]. Type data: holotype, MCG C.E. 29988, from Romoi, near Soron, Irian Jaya, Indonesia.

Liasis tornieri Werner, F. (1897). Über einige noch unbeschriebene Reptilian und Batrachier. *Zool. Anz.* **20**: 261–267 [261]. Type data: holotype, NHMW 15078, from Bogadjim, 5°25'S, 145°45'E (as Stephansort), Papua New Guinea.

Liasis olivaceus barroni Smith, L.A. (1981). A revision of the *Liasis olivaceus* species-group (Serpentes : Boidae) in Western Australia. *Rec. West. Aust. Mus.* **9**: 227–233 [231]. Type data: holotype, WAM R55384, from Tambrey, W.A. in 21°35'S, 117°34'E.

Synonymy that of Cogger, H.G., this work.

Distribution: NW coastal, N coastal, N Gulf, Lake Eyre basin, W.A., N.T., Qld. Ecology: terrestrial, arboreal, nocturnal, predator, open forest, woodland, low open shrubland; oviparous, general carnivore, constrictor, often saxicoline. Biological references: Smith, L.A. (1981). A revision of the *Liasis olivaceus* species-group (Serpentes : Boidae) in Western Australia. *Rec. West. Aust. Mus.* **9**: 227–233 (taxonomy); Worrell, E. (1951). Classification of Australian Boidae. *Proc. R. Zool. Soc. N.S.W.* **1949–50**: 20–25 (taxonomy); McDowell, S.B. (1975). A catalogue of the snakes of New Guinea and the Solomons, with special reference to those in the Bernice P. Bishop Museum. Part II. Anilioidea and Pythoninae. *J. Herpet.* **9**: 1–79 (taxonomy).

Bothrochilus perthensis (Stull, 1932)

Liasis childreni perthensis Stull, O.G. (1932). Five new subspecies of the family Boidae. *Occ. Pap. Bost. Soc. Nat. Hist.* **8**: 25–30 [26]. Type data: holotype, MCZ 24426, from Perth, W.A.

Distribution: SW coastal, (NW coastal), (W plateau), W.A., (N.T.). Ecology: terrestrial, nocturnal, predator; general carnivore, constrictor. Biological references: Mitchell, F.J. (1965). Australian geckos assigned to the genus *Gehyra* Gray. (Reptilia : Gekkonidae). *Senckenberg. Biol.* **46**: 287–319(natural history); Hoser, R.T. (1981). Australian pythons (Part 2). The smaller *Liasis*. *Herptile* **6**: 13–19.

Chondropython Meyer, 1874

Chondropython Meyer, A.B. (1874). Eine Mitteilung von Hrn. Dr. Adolf Bernhard Meyer über die von ihm auf Neu-Guinea den Inseln Jobi, Mysore und Mafoor im Jahr 1873 gesammelten Amphibien. *Mber. K. Preuss. Akad. Wiss. Berl.* **1874**: 128–140 [134] [1875 on title page]. Type species *Chondropython azureus* Meyer, 1874 by monotypy.

This group is also found in New Guinea.

Chondropython viridis (Schlegel, 1872)

Python viridis Schlegel, H. (1872). *De Dierentuin van het Koninklijk Zoölogische Genootschap "Natura Artis Magistra" te Amsterdam zoölogisch geschetst door Prof H. Schlegel, met historische herinneringen van P.H. Witkampf.* Amsterdam : Gebr, Van Es 5 pts. [pt. 4, 54]. Type data: syntypes, RMNH 4672 (2 specimens), from Aru (as Aroe) Ils., Indonesia.

Chondropython azureus Meyer, A.B. (1874). Eine Mitteilung von Hrn. Dr. Adolf Bernhard Meyer über die von ihm auf Neu-Guinea den Inseln Jobi, Mysore und Mafoor im Jahr 1873 gesammelten Amphibien. *Mber. K. Preuss. Akad. Wiss. Berl.* **1874**: 128–140 [134] [1875 on title page]. Type data: syntypes (probable), from Kordo, Mysore, Schouten Ils., Irian Jaya, ZMB 8832 (labelled holotype); MTKD 638–9 (destroyed), *vide* Obst, F.J. (1977). Die herpetologische sammlung des staatlichen museums für Tierkunde Dresden und ihre Typusexemplare. *Zool. Abh. Mus. Tierk. Dresden* **34**: 171–186 who cites MTKD 638–9 as syntypes.

Chondropython pulcher Sauvage, H.E. (1878). Essai sur la faune herpétologique de la Nouvelle-Guinée, suivi de la description de quelques espèces nouvelles ou peu connues. *Bull. Sci. Soc. Philomath. Paris (7)* **2**: 25–44 [37]. Type data: syntypes, MNHP 5087–89, MNHP 5168–70, from Mansinam, Irian Jaya, Indonesia.

Synonymy that of Cogger, H.G., this work.

Distribution: NE coastal, Qld. Ecology: arboreal, nocturnal, predator, closed forest; oviparous, general carnivore, constrictor. Biological references: McDowell, S.B. (1975). A catalogue of the snakes of New Guinea and the Solomons, with special reference to those in the Bernice P. Bishop Museum. Part II. Anilioidea and Pythoninae. *J. Herpet.* **9**: 1–79 (taxonomy); Groombridge, B.C. (1979). Variations in morphology of the superficial palate of henophidian snakes and some possible systematic implications. *J. Nat. Hist.* **13**: 447–475 (morphology); Murphy, J.B., Carpenter, C.C. & Gillingham, J.C. (1978). Caudal luring in the green tree python, *Chondropython viridis* (Reptilia, Serpentes, Boidae). *J. Herpet.* **12**: 117–119 (behaviour).

Morelia Gray, 1842

Morelia Gray, J.E. (1842). Synopsis of the species of prehensile-tailed snakes, or family Boidae. *in* Gray, J.E. (ed.) *Zoological Miscellany.* London : Treuttel, Würtz & Co. pp. 41–46 [43]. Type species *Morelia variegata* Gray, 1842 by subsequent designation, see McDowell, S.B. (1975). A catalogue of the snakes of New Guinea and the Solomons, with special reference to those in the Bernice P. Bishop Museum. Part II. Anilioidea and Pythoninae. *J. Herpet.* **9**: 1–79.

Liasis Gray, J.E. (1842). Synopsis of the species of prehensile-tailed snakes, or family Boidae. *in* Gray, J.E. (ed.) *Zoological Miscellany.* London : Treuttel, Würtz & Co. pp. 41–46 [44]. Type species *Boa amethistina* Schneider, 1801 by subsequent designation, see Desmarest, E. (1846). Reptiles. *in* d'Orbigny, C. (ed.) *Dictionnaire Universal D'Histoire Naturelle.* Paris : Houssiaux 7: 337.

***Simalia* (part.)** Gray, J.E. (1849). *Catalogue of the Specimens of Snakes in the Collection of the British Museum.* London : British Museum xv 125 pp. [91]. Type species, none designated, *Liasis amethystinus* (*sic*) one of two included species, see also synonymy of *Bothrochilus.*

Aspidopython Meyer, A.B. (1874). Eine Mitteilung von Hrn. Dr. Adolf Bernhard Meyer über die von ihm auf Neu-Guinea den Inseln Jobi, Mysore und Mafoor im Jahr 1873 gesammelten Amphibien. *Mber. K. Preuss. Akad. Wiss. Berl.* **1874**: 128–140 [135] [1875 on title page]. Type species *Aspidopython jakati* Meyer, 1874 by monotypy.

Hypaspistes Ogilby, J.D. (1891). Descriptions of three new Papuan snakes. *Rec. Aust. Mus.* **1**: 192–194 [192] *non Hypaspistes* Waterhouse, 1886]. Type species *Hypaspistes dipsadides* Ogilby, 1891 by monotypy.

Synonymy that of Cogger, H.G., this work.

This group is also found in the Moluccas, Ceram, Ambon, Kei and Aru Ils., Indonesia; the island of New Guinea including the Bismarck Archipelago; and southern Philippine Ils., see McDowell, S.B. (1975). A catalogue of the snakes of New Guinea and the Solomons, with special reference to those in the Bernice P. Bishop Museum. Part II. Anilioidea and Pythoninae. *J. Herpet.* **9**: 1–79.

Morelia amethistina (Schneider, 1801)

Boa amethistina Schneider, J.G. (1801). *Historiae Amphibiorum naturalis et literariae.* fasc. sec. Jena : Frommani vi 374 pp. [254]. Type data: holotype, ZMB 485 (presumed lost), from unknown locality.

Aspidopython jakati Meyer, A.B. (1874). Eine Mitteilung von Hrn. Dr. Adolf Bernhard Meyer über die von ihm auf Neu-Guinea den Inseln Jobi, Mysore und Mafoor im Jahr 1873 gesammelten Amphibien. *Mber. K. Preuss. Akad. Wiss. Berl.* **1874**: 128–140 [135]. Type data: holotype (presumed lost), from Jakati, Irian Jaya, Indonesia, see Stimson, A.F. (1969). Liste der rezenten Amphibien und Reptilien. Boidae. *Das Tierreich* **89**: –49.

Liasis duceboracensis Günther, A. (1879). Notice of a collection of reptiles from islands of Torres Straits. *Ann. Mag. Nat. Hist. (5)* **3**: 84–87 [86]. Type data: syntypes, BMNH 1946.1.10.34–35, from Duke of York Is., Papua New Guinea.

Hypaspistes dipsadides Ogilby, J.D. (1891). Descriptions of three new Papuan snakes. *Rec. Aust. Mus.* **1**: 192–194 [192]. Type data: holotype, AM R1087, from New Guinea (Geographical Society Expedition).

Liasis clarki Barbour, T. (1914). On some Australian reptiles. *Proc. Biol. Soc. Wash.* **27**: 201–206 [202]. Type data: holotype, MCZ 9600, from Mer Is., Murray Ils., Torres Strait, Qld.

Liasis amethistinus kinghorni Stull, O.G. (1933). Two new subspecies of the family Boidae. *Occ. Pap. Mus. Zool. Univ. Mich.* **12**(267): 1–4 [3]. Type data: holotype, MCZ 35022, from Lake Barrine, Qld.

Synonymy that of Cogger, H.G., this work.

Distribution: N Gulf, NE coastal, Qld. Ecology: terrestrial, arboreal, nocturnal, predator, closed forest, open forest, mangrove; oviparous, general carnivore, constrictor. Biological references: McDowell, S.B. (1975). A catalogue of the snakes of New Guinea and the Solomons, with special reference to those in the Bernice P. Bishop Museum. Part II. Anilioidea and Pythoninae. *J. Herpet.* **9**: 1–79 (taxonomy); Schwaner, T.D. & Dessauer, H.C. (1981). Immunodiffusion evidence for the relationships of Papuan boids. *J. Herpet.* **15**: 250–253 (biochemical taxonomy); Mengden, G.A. & Stock, A.D. (1980). Chromosomal evolution in Serpentes; a comparison of G and C chromosome banding patterns of some colubrid and boid genera. *Chromosoma (Berl.)* **79**: 53–64 (karyology); Webb, G.J.W. & Heatwole, H. (1971). Patterns of heat distribution within the bodies of some Australian pythons. *Aust. Zool.* **1971**: 209–220 (physiology).

Morelia bredli (Gow, 1981)

Python bredli Gow, G.F. (1981). A new species of python from central Australia. *Aust. J. Herp.* **1**: 29–34 [29]. Type data: holotype, NTM R4535, from Pitchie Ritchie Park, Alice Springs, 23°42'S, 133°51'E, N.T.

Distribution: Lake Eyre basin, W plateau, N.T. Ecology: terrestrial, arboreal, nocturnal, predator, desert; oviparous, general carnivore, constrictor.

Morelia carinata (Smith, 1981)

Python carinatus Smith, L.A. (1981). A revision of the python genera *Aspidites* and *Python* (Serpentes : Boidae) in Western Australia. *Rec. West. Aust. Mus.* **9**: 211–226 [220]. Type data: holotype, WAM R45352, from Mitchell River Falls, W.A. in 14°50'S, 125°42'E.

Distribution: N coastal, W.A.

Morelia oenpelliensis (Gow, 1977)

Python oenpelliensis Gow, G.F. (1977). A new species of python from Arnhem Land. *Aust. Zool.* **19**: 133–139 [133 figs 1–3]. Type data: holotype, NTM R0840, from 6.5 km SW Oenpelli, 12°21'S, 133°01'E, N.T.

Distribution: N coastal, N.T. Ecology: terrestrial, arboreal, nocturnal, predator, open forest; oviparous, general carnivore, constrictor, saxicoline. Biological references: Swanson, S. (1979). Some rock-dwelling reptiles of the Arnhem Land escarpment. *N. Territory Nat.* **1**: 14–18 (natural history); Jenkins, R.W.G. (1979). The status of endangered Australian reptiles. pp. 169–176 *in, The Status of Endangered Australasian Wildlife.* Adelaide : Royal Zoological Society of South Australia (conservation).

Morelia spilota (Lacépède, 1804)

[Coluber arges] Linnaeus, C. (1758). *Systema Naturae.* Classis III. Amphibia. Holmiae : Laurentii Salvii 824 pp. 10th edn. 1: 194–238 [227] [this name has been widely but erroneously applied to the species represented by *Coluber spilotus* Lacépède, 1804. However, the snake shown in the illustration in Seba, A. (1735). *Locupletissimi rerum naturalium thesauri accurata descriptio et iconibus artificiosissimis expressio, per universam physices historiam.* Amsterdam vol. 4 [pl 103 fig 1], on which Linnaeus' *arges* was based, purportedly came from "Africa". It is not a specimen of any Australian species, and was described 35 years prior to the discovery of eastern Australia].

Coluber spilotus Lacépède, B.G.E. (1804). Mémoire sur plusieurs animaux de la Nouvelle-Hollande dont la description n'a pas encore été publiée. *Ann. Mus. Natl. Hist. Nat. Paris* **4**: 184–211 [209]. Type data: holotype, MNHP 3272, from Australia.

Python punctatus Merrem, B. (1820). *Versuch eines Systems Amphibien. Tentamen Systematis Amphibiorum.* Marburg : Krieger xv 199 pp. [90]. Type data: presumed lost, from Australia.

Python peronii Wagler, J.G. (1828). *Descriptiones et Icones Amphibiorum.* Part I. Plates I–XIII Stuttgartiae, Tubingae, Monachii : J.G. Cottae [pl 1 and description] [*ex* Cuvier ms]. Type data: syntypes, MNHP, RMNH, from Australia ("Dusky-Bay"), presumed lost, *vide* Stimson, A.F. (1969). Liste der rezenten Amphibien und Reptilien. Boidae. *Das Tierreich* **89**: 1–49.

Python spilotes Gray, J.E. (1841). A catalogue of the species of reptiles and Amphibia hitherto described as inhabiting Australia, with a description of some new species from Western Australia. Appendix *in* Grey, G. *Journals of Two Expeditions, of Discovery in north-west and western Australia, during the years 1837, 38 and 39, under the authority of Her Majesty's Government.* London : T. & W. Boone 2: 422–449 [432] [invalid emend. *pro Coluber spilotus* Lacépède, 1804].

Morelia variegata Gray, J.E. (1842). Synopsis of the species of prehensile-tailed snakes, or family Boidae. *in* Gray, J.E. (ed.) *Zoological Miscellany.* London : Treuttel, Würtz & Co. pp. 41–46 [43]. Type data: holotype, BMNH iv.1.2a, from Port Essington, N.T.

Morelia argus fasciolata Jan, G. (1863). *Elenco Sistematico Degli Ofidi Descritti e Disegnati per l'Iconografia Generale.* Milan : A. Lombardi 143 pp. [25] [*nom. nud.*].

Morelia argus fasciolata Jan, G. (1864). *Iconographie Générale des Ophidiens.* Milan : G. Jan 100 pp. [95 and Atlas livr. 7: pl 5 fig c]. Type data: figure of specimen in Milan, from Australia.

Python spilotes macrospila Werner, F. (1909). Über neue oder seltene Reptilien des Musée Royal d'Histoire Naturelle de Belgique in Brussell. *Zool. Jb. Abt. Syst.* **28**: 263–288 [274]. Type data: holotype, RIB General No. 4414 (Type No. 2496), from unknown locality.

Python spilotus imbricatus Smith, L.A. (1981). A revision of the python genera *Aspidites* and *Python* (Serpentes : Boidae) in Western Australia. *Rec. West. Aust. Mus.* **9**: 211–226 [222 fig 7]. Type data: holotype, WAM R54340, from Jurien Bay, 30°18'S, 115°02'E, W.A.

Synonymy that of Cogger, H.G., this work.

Distribution: N coastal, W plateau, N Gulf, NE coastal, Lake Eyre basin, Bulloo River basin, Murray-Darling basin, SE coastal, S Gulfs, SW coastal, W.A., N.T., Qld., N.S.W., Vic., S.A. Ecology: terrestrial, arboreal, nocturnal, predator, closed forest, tall forest, open forest, woodland, low woodland, low open woodland, tall shrubland, open scrub, tall open shrubland, open heath, low open shrubland; oviparous, general carnivore, constrictor. Biological references: McDowell, S.B. (1975). A catalogue of the snakes of New Guinea and the Solomons, with special reference to those in the Bernice P. Bishop Museum. Part II. Anilioidea and Pythoninae. *J. Herpet.* **9**: 1–79 (taxonomy); Smith, L.A. (1981). A revision of the *Liasis olivaceus* species-group (Serpentes : Boidae) in Western Australia. *Rec. West. Aust. Mus.* **9**: 227–233 (taxonomy); Worrell, E. (1951). Classification of Australian Boidae. *Proc. R. Zool. Soc. N.S.W.* **1949–50**: 20–25 (taxonomy); Schwaner, T.D. & Dessauer, H.C. (1981). Immunodiffusion evidence for the relationships of Papuan boids. *J. Herpet.* **15**: 250–253 (biochemical taxonomy); Hay, M. (1971). Notes on breeding and growth rate of *Morelia spilotes spilotes*. *Herpetofauna* **3**: 10 (reproduction); Covacevich, J. & Limpus, C. (1973). Two large winter aggregations of three species of tree-climbing snakes in south-eastern Queensland. *Herpetofauna* **6**: 16–21 (aggregation); Mackay, R. (1950). Notes on the diamond snake and carpet snake. *Proc. R. Zool. Soc. N.S.W.* **1948–49**: 35–36(natural history).

ACROCHORDIDAE

INTRODUCTION

A family (commonly called File Snakes) of three aquatic species within a single genus; two species occur in Australia. This group is found from India through south-east Asia and the Indo-Malayan Archipelago to New Guinea and Australia.

Characterised in Australia by: skin with small finely keeled and pointed scales the texture of a coarse file or rasp; scales more or less uniform in size around the body, without enlarged ventral scales; more than 80 scales around middle of body; a series of loreal scales; tail prehensile; teeth solid and lacking venom glands.

Acrochordus Hornstedt, 1787

Acrochordus Hornstedt, C.F. (1787). Beskrifning pa en Ny Orm fran Java. *K. Vet. Acad. nya Handl. (2)* **8**: 306–308 [307]. Type species *Acrochordus javanicus* Hornstedt, 1787 by monotypy.

Chersydrus Cuvier, G.L.C.F.D. (1817). *Le Règne Animal, distribué d'après son organisation.* Paris : Déterville Vol. 2 [75]. Type species *Acrochordus fasciatus* Shaw, 1802 (=*Hydrus granulatus* Schneider, 1799) by original designation.

Acrocordus Gray, J.E. (1825). A synopsis of the genera of reptiles and amphibia, with a description of some new species. *Annls. Philos. (2)* **10**: 193–217 [207] [*err. typ.*; incorrect subsequent spelling].

Chersydreas Gray, J.E. (1825). A synopsis of the genera of reptiles and amphibia, with a description of some new species. *Annls. Philos. (2)* **10**: 193–217 [207] [*err. typ.*; incorrect subsequent spelling].

Verrucator Schlegel, H. (1837). (*ex* Merrem ms). *Essai sur la Physionomie des Serpens.* 2. Partie descriptive. La Haye : Kips & Stockum [428] [*nom. nud.* introduced in synonymy].

Potamophis Schmidt, P. (1852). Beiträge zur ferneren Kenntniss der Meerschlangen. *Abh. Geb. Naturw., Hamburg* **2**: 69–86 [75]. [*non Potamophis* Cantor, 1836 and *non Potamophis* Fitzinger, 1843; *nom. substit. pro Acrochordus* Hornstedt, 1787].

Synonymy that of McDowell, S.B. (1979). A catalogue of the snakes of New Guinea and the Solomons, with special reference to those in the Bernice P. Bishop Museum. Part III. Boinae and Acrochordoidea (Reptilia : Serpentes). *J. Herpet.* **13**: 1–92.

This group is also found in India, SE Asia, Indo-Malaysian Archipelago and New Guinea. Prior to 1979, the name *Acrochordus javanicus* Hornstedt, 1787 was applied to the Australian species *Acrochordus arafurae* McDowell, 1979.

Acrochordus arafurae McDowell, 1979

[Acrochordus javanicus] Hornstedt, C.F. (1787). Beskrifning pa en Ny Orm fran Java. *K. Vet. nya Handl. (2)* **8**: 306–308 [307] [Australian records of *Acrochordus javanicus* apply to *Acrochordus arafurae* McDowell, 1979].

Acrochordus arafurae McDowell, S.B. (1979). A catalogue of the snakes of New Guinea and the Solomons, with special reference to those in the Bernice P. Bishop Museum. Part III. Boinae and Acrochordoidea (Reptilia : Serpentes). *J. Herpet.* **13**: 1–92 [75 fig 27]. Type data: holotype, AMNH 59887, from Lake Daviumbo, Western Province, Papua New Guinea.

Distribution: N Gulf, N coastal, (W.A.), N.T., Qld.; Western Australian records unconfirmed. Ecology: aquatic, nocturnal, nectonic, lentic freshwater, lotic freshwater, predator; semi-gregarious, perennially active, viviparous, piscivorous. Biological references: Magnusson, W.A. (1979). Production of an embryo by an *Acrochordus javanicus* isolated for seven years. *Copeia* **1979**: 744–745 (*arafurae* as *javanicus*); Longman, H.A. (1916). Snakes and lizards from Queensland and the Northern Territory. *Mem. Qd. Mus.* **5**: 46–51 (as *Acrochordus javanicus*; natural history).

Acrochordus granulatus (Schneider, 1799)

Hydrus granulatus Schneider, J.G. (1799). *Historiae Amphibiorum Naturalis et Literariae.* fasc. prim. Jena : Frommani xvi 266 pp [243]. [Type material not located by McDowell, S.B. (1979). A catalogue of the snakes of New Guinea and the Solomons, with special reference to those in the Bernice P. Bishop Museum. Part III. Boinae and Acrochordoidea (Reptilia : Serpentes). *J. Herpet.* **13**: 1–92]. Type data: status unknown, Lampi Collection, presumed lost.

Acrochordus fasciatus Shaw, G. (1802). *General Zoology*, or Systematic Natural History. Vol. 3 Pt. 2 Amphibia. London : Kearsley [576 pl 130]. Type data: syntypes, BMNH; presumed lost, from unknown locality

(=India), see Gray, J.E. (1849). *Catalogue of the Specimens of Snakes in the Collection of the British Museum.* London : British Museum xv 125 pp.

Chersydrus annulatus Gray, J.E. (1849). *Catalogue of the Specimens of Snakes in the Collection of the British Museum.* London : British Museum xv 125 pp. [61]. Type data: syntypes, BMNH 1946.1.1.43–4 & BMNH 1946.1.15.68 (Madras), BMNH 1946.1.1.40, (Singapore), BMNH 1946.1.15.65 (Sumatra), from Madras, India; Singapore; Sumatra, Indonesia.

Synonymy that of Cogger, H.G., this work.

Distribution: NE coast, Gulf of Carpentaria, N coast, NW shelf, Qld., N.T., W.A. Ecology: noctidiurnal, aquatic, nectonic, predator; viviparous, piscivorous, primarily mangrove or estuarine littoral, but sometimes lotic freshwater, non-gregarious, perennially active. Biological references: Heatwole, H. & Seymour, R. (1975). Pulmonary and cutaneous oxygen uptake in sea snakes and a file snake. *Comp. Biochem. Physiol.* **51A**: 339–405 (physiology); Heatwole, H. (1975). Voluntary submergence times of marine snakes. *Marine Biol.* **32**: 205–213; Voris, H.K. & Glodek, G.S. (1980). Habitat, diet and reproduction of the file snake, *Acrochordus granulatus*, in the Straits of Malacca. *J. Herpet.* **14**: 108–111 (natural history); Heatwole, H. (1977). Heart rate during breathing and apnea in marine snakes (Reptilia, Serpentes). *J. Herpet.* **11**: 67–76 (physiology); Heatwole, H. (1976). *Reptile Ecology.* Brisbane : Queensland University Press pp. 87–89 (physiology); Heatwole, H. (1981). Temperature relations of some sea snakes. *The Snake* **13**: 53–57 (thermal ecology); Lemen, C.A. & Voris, H.K. (1981). A comparison of reproductive strategies among marine snakes. *J. Anim. Ecol.* **50**: 89–101 (reproduction); Feder, M.E. (1980). Blood oxygen stores in the file snake, *Acrochordus granulatus*, and in other marine snakes. *Physiol. Zool.* **53**: 394–401 (physiology); Dunson, W.A. & Minton, S.A. (1978). Diversity, distribution, and ecology of Philippine marine snakes (Reptilia, Serpentes). *J. Herpet.* **12**: 281–286.

COLUBRIDAE

INTRODUCTION

The most speciose group of snakes in the world, containing approximately 1,600 species in 300 genera, of which only ten species in eight genera occur in Australia. The family contains species with solid teeth and lacking venom glands and species with grooved rear maxillary teeth and venom glands. The family is represented on all continents; in Australia two subfamilies are represented: the Colubrinae and Homalopsinae.

Characterised in Australia by: more or less cylindrical tail; a single row of enlarged ventral scales; fewer than 30 mid-body scale rows; loreal shield(s) present or, if absent, 23 or more mid-body scale rows and a divided anal scale.

The taxonomic limits of this family are subject to widely differing interpretations, with little agreement among specialists. The family Colubridae is used in this volume in its traditional sense to exclude those related colubroid groups treated as full families: the Acrochordidae, Elapidae, Hydrophiidae and Laticaudidae. The relationships within and between these groups are the subject of considerable current research. The very detailed classification proposed by Smith, Smith and Sawin (1977) without supporting data remains to be critically evaluated.

References

Cogger, H.G. (1979). *Reptiles and Amphibians of Australia.* Sydney : A.H. & A.W. Reed 608 pp.

Smith, H.M., Smith, R.B. & Sawin, H.L. (1977). A summary of snake classification (Reptilia, Serpentes). *J. Herpet.* **11**: 115–121

Worrell, E. (1970). *Reptiles of Australia.* 2nd edn. Sydney : Angus & Robertson xv 169 pp.

Boiga Fitzinger, 1826

Boiga Fitzinger L.J. (1826). *Neue Classification der Reptilien nach ihren natürlichen Verwandtschaften.* Vienna: Heubner 66 pp. [60]. Type species *Hurria irregularis* (Merrem, 1802) by subsequent designation, Wall, F. (1924). A hand-list of the snakes of the Indian Empire. Part III, *J. Bombay Nat. Hist. Soc.* **29**: 864–878.

Pappophis Macleay, W. (1877). The ophidians of the *Chevert* Expedition. *Proc. Linn. Soc. N.S.W.* **2**: 33–41 [39] [1878 on title page]. Type species *Pappophis laticeps* Macleay, 1877 by subsequent designation, Cogger, H.G., this work.

Synonymy that of Cogger, H.G., this work.

This group is also found in Africa, Asia, Indo-Malayan Archipelago, Philippine Ils., New Guinea and islands of the western Pacific.

Boiga irregularis (Merrem, 1802)

Coluber irregularis Merrem, B. *in* Bechstein, J.M. (1802). *Herrn de Lacépède's Naturgeschichte der Amphibien oder der eyerlegen den vierfüssigen Thiere und der Schlangen. Eine Fortsetzung von Buffon's Naturgeschichte aus dem Französischen übersetzt und mit Anmerkungen und Zusätzen versehen.* 4 Weimar : Industrie Comptoir [239 pl 37 fig 1]. Type data: presumed lost, from unknown locality.

Hurria pseudoboiga Daudin, F.M. (1803). *Histoire Naturelle, générale et particulière des Reptiles.* Paris : Dufart Tom. 5 365 pp. [277 pl 59 figs 8–9, pl 66 figs 1, 3] [*nom. nov. pro Coluber irregularis* Merrem, 1802]. Type data: that of *Coluber irregularis* Merrem, 1802.

Dendrophis (Ahetula) fusca Gray, J.E. (1842). Description of some hitherto unrecorded species of Australian reptiles and batrachians. *in* Gray, J.E. (ed.)

Zoological Miscellany. London : Treuttel, Würtz & Co. pp. 51–57 [54]. Type data: holotype, BMNH 1946.1.1.28, from Port Essington, N.T.

Pappophis laticeps Macleay, W. (1877). The ophidians of the *Chevert* Expedition. *Proc. Linn. Soc. N.S.W.* **2**: 33–41 [39] [1878 on title page]. Type data: syntypes, AM R31888–91, from Hall Sound, Papua New Guinea.

Pappophis flavigastra Macleay, W. (1877). The ophidians of the *Chevert* Expedition. *Proc. Linn. Soc. N.S.W.* **2**: 33–41 [40] [1878 on title page]. Type data: holotype, AM R31887, from Mawatta, Binaturi River (as Katow), Papua New Guinea.

Dipsas boydii Macleay, W. (1884). Census of Australian snakes with descriptions of two new species. *Proc. Linn. Soc. N.S.W.* **9**: 548–568 [548]. Type data: holotype, AM R31892, from "collection from J.A. Boyd of Ripple Creek, Ingham, in northern Qld."

Dipsas ornata Macleay, W. (1888). Notes on some ophidians from King's Sound, north-west Australia. *Proc. Linn. Soc. N.S.W. (2)* **3**: 416–418 [416]. Type data: holotype, AM R31893, from "vicinity of King's Sound", W.A.

Synonymy that of Cogger, H.G., this work.

Distribution: N coastal, N Gulf, NE coastal, SE coastal, W.A., N.T., Qld., N.S.W. Ecology: terrestrial, arboreal, nocturnal, predator, caves, closed forest, tall forest, open forest, woodland, low open woodland, tall open shrubland; oviparous, general carnivore. Biological references: Covacevich, J. & Limpus, C. (1973). Two large winter aggregations of three species of tree-climbing snakes in south-eastern Queensland. *Herpetofauna* **6**: 16–21; Irvine, W. (1954). Notes on the brown tree snake. *Reptilia (Aust. Rept. Club)* **1**: 5 (natural history); Johnson, C.R. (1975). Thermoregulation in the Papuan-New Guinean boid and colubrid snakes, *Candoia carinata, Candoia asper* and *Boiga irregularis. Zool. J. Linn. Soc.* **56**: 283–290.

Cerberus Cuvier, 1829

Cerberus Cuvier, G.L.C.F.D. (1829). *Le Règne Animal.* 2nd edn. Paris : Déterville Vol. 2 xv 406 pp. [81]. Type species *Coluber cerberus* Daudin, 1803 (= *Hydrus rhynchops* Schneider, 1799) by tautonymy.

This group is also found in India, south-east Asia, Indo-Malayan Archipelago, Philippine Ils., and New Guinea.

Cerberus rynchops (Schneider, 1799)

Hydrus rynchops Schneider, J.G. (1799). *Historiae Amphibiorum naturalis et literariae.* fasc. prim. Jena : Frommani xvi 266 pp. [246]. Type data: not found, from unknown locality.

Homalopsis australis Gray, J.E. (1842). Monographic synopsis of the water snakes, or the family Hydridae. *in* Gray, J.E. (ed.) *Zoological Miscellany.* London : Treuttel, Würtz & Co. pp. 59–68 [65]. Type data: holotype, BMNH 1946.1.2.40, from Port Essington, N.T.

Synonymy that of Cogger, H.G., this work; for extralimital synonyms, see Ko Ko Gyi (1970). A revision of colubrid snakes of the sub-family Homalopsinae. *Univ. Kans. Publs. Mus. Nat. Hist.* **20**: 44–223.

Distribution: N coastal, N Gulf, W.A., N.T., Qld. Ecology: amphibious, noctidiurnal, predator, mangrove, estuarine; feeds on fish and small crustaceans. Biological references: Ko Ko Gyi (1970). A revision of colubrid snakes of the sub-family Homalopsinae. *Univ. Kans. Publs. Mus. Nat. Hist.* **20**: 44–223 (taxonomy); Gorman, G.C., Licht, P. & McCollum, F. (1981). Annual reproductive patterns in three species of marine snakes from the Central Philippines. *J. Herpet.* **15**: 335–354 (reproduction); Feder, M.E. (1980). Blood oxygen stores in the file snake *Acrochordus granulatus* and in other marine snakes. *Physiol. Zool.* **53**: 394–401 (physiology); Bergman, R.A.M. (1955). L'anatomie de *Cerberus rhynchops. Arch. Neerl. Zool.* **11**: 113–126 (anatomy); Kopstein, F. (1931). Herpetologische Notizen, 4. *Fordonia leucobalia* Schlegel and *Cerberus rhynchops* Schneider. *Treubia* **13**: 1–4 (natural history).

Dendrelaphis Boulenger, 1890

Dendrelaphis Boulenger, G.A. (1890). *The Fauna of British India; including Ceylon and Burma.* Reptilia and Batrachia. London : Taylor & Francis xviii 541 pp. [339]. Type species *Ahaetula caudolineata* Gray, 1834 by original designation.

Tachyophis Mertens, R. (1934). Die Schlangengattung *Dendrelaphis* Boulenger in systematischer und zoogeographischer Beziehung, I. *Arch. Naturgesch. Abt B, ns* **3**: 187–204 [197] [described as subgenus or section of *Dendrelaphis* Boulenger, 1890; *non Tachyophis* Rochebrune, 1884]. Type species *Coluber pictus* Gmelin, 1789 [= *Dendrelaphis boiga* (Lacépède, 1789)], by original designation. Compiled from secondary source: Savage, J.M. (1952). Two centuries of confusion : the history of the snake name *Ahaetulla. Bull. Chicago Acad. Sci.* **9**: 203–216.

This group is also found in India, south-east Asia, Indo-Malayan Archipelago, Philippine Ils., New Guinea and Solomon Ils.

Dendrelaphis calligastra (Günther, 1867)

*[**Dendrophis lineolata** (**part.**)]* Duméril, A.M.C., Bibron, G. & Duméril, A. (1854). *Erpétologie Générale* ou Histoire Naturelle Complète des Reptiles. Paris : Roret 7(1): xvi 780 pp. [200] [This name is usually mistakenly attributed to Hombron and Jacquinot *in* Dumont d'Urville, M.J. (ed.) (1842-1853). *Voyage au Pole Sud et dans l'Océanie sur les corvettes l'*Astrolabe *et la* Zélée, *executé pendant les années 1837–40.* Atlas Sauriens (1842, pl 2), who published only a figure without application of binomial nomenclature. It was subsequently described by Jacquinot, H. & Guichenot, M.A. (1853). Reptiles et Poissons, *in* Dumont d'Urville, M.J. (ed.) *Voyage au Pole Sud et dans l'Océanie sur les corvettes l'*Astrolabe *et la* Zélée, *executé pendant les années*

1837–40. Zoologie 3: 1–28. Paris : Gide et Baudry, p 20, who attributed the name to C. Duméril (in ms). The subsequent year Duméril, A.M.C., Bibron, G. & Duméril, A. (1854). *Erpétologie Générale* ou Histoire Naturelle Complète des Reptiles. Paris : Roret 7(1): xvi 780 pp., also described this taxon as a new "unpublished species". The series on which Duméril, Bibron and Duméril based their description included material "probably from Australia"; this description is included in the above synonymy pending resolution of the authorship and type fixation of *Dendrophis lineolata*.] Type data: MNHP syntypes, status unknown, from Papua New Guinea; Waigeo Ils., Irian Jaya, Indonesia; "probably Australia".

Dendrophis calligastra Günther, A. (1867). Additions to the knowledge of Australian reptiles and fishes. *Ann. Mag. Nat. Hist. (3)* **20**: 45–68 [53]. Type data: holotype, BMNH 67.5.6.71, from Cape York, Qld.

Dendrophis salomonis Günther, A. (1872). Seventh account of new species of snakes in the collection of the British Museum. *Ann. Mag. Nat. Hist. (4)* **9**: 13–37 [25]. Type data: syntypes, BMNH, from Solomon Ils.

Dendrophis aruensis Doria, G. (1874). Enumerazione dei rettili raccolti dal Dott. O. Beccari in Amboina alle Isole Aru ed alle Isole Kei durante gli anni. *Ann. Mus. Civ. Stor. Nat. Giacomo Doria* **6**: 325–357 [349 pl 12 fig g]. Type data: syntypes, MCG C.E. 30594, from Wokan, Aru Ils., Indonesia.

Dendrophis katowensis Macleay, W. (1877). The ophidians of the *Chevert* Expedition. *Proc. Linn. Soc. N.S.W.* **2**: 33–41 [37] [1878 on title page]. Type data: syntypes, AM R31916–17, from Mawatta, Binaturi River (as Katow), Papua New Guinea.

Dendrophis darnleyensis Macleay, W. (1877). The ophidians of the *Chevert* Expedition. *Proc. Linn. Soc. N.S.W.* **2**: 33–41 [38] [1878 on title page]. Type data: syntypes, AM R31914–15, from Darnley Is., Torres Strait, Qld.

Dendrophis culligaster Boulenger, G.A. (1894). *Catalogue of snakes in the British Museum (Natural History).* 2. London : British Museum xi 382 pp. pls 1–20 [80] [emend. *pro Dendrophis calligastra* Günther, 1867].

Dendrelaphis papuensis Boulenger, G.A. (1895). Descriptions of two new reptiles obtained by Mr A.S. Meek in the Trobriand Islands, British New Guinea. *Ann. Mag. Nat. Hist. (6)* **16**: 408–409 [409]. Type data: syntypes, BMNH, from Trobriand Ils., Papua New Guinea.

Dendrelaphis schlenckeri Ogilby, J.D. (1898). Ophidia and Pisces *in* T. Steel (ed.) Contributions to a knowledge of the fauna of British New Guinea. *Proc. Linn. Soc. N.S.W.* **23**: 361–363 [361 fig]. Type data: holotype, AM R2380, from Fife Bay, in 10°35'S, 150°00'E, Papua New Guinea.

Dendrophis calligastra keiensis Mertens, R. (1926). Über die Rassen einiger indo-australischer Reptilien. *Senckenbergiana* **8**: 272–279 [277]. Type data: holotype, SMF 18662 F, from Dulah, Kei Ils., Indonesia.

Dendrophis calligastra distinguendus Meise, W. & Hennig, W. (1932). Die Schlangengattung *Dendrophis. Zool. Anz.* **99**: 273–279 [279]. Type data: holotype, MTKD D2107, from Muliama, Süd-Mecklenburg, Papua New Guinea; destroyed, see Obst, F.J. (1977). Die herpetologische Sammlung des Staatlichen Museums für Tierkunde Dresden und ihre Typusexemplare. *Zool. Abh. Mus. Tierk. Dresden* **34**: 171–186.

Synonymy that of Cogger, H.G., this work; however, the status of many of the synonyms cited above is uncertain, and some have been treated as distinct species by various authors, see references below.

Distribution: N Gulf, NE coastal, Qld.; also occurs in New Guinea, Solomon Ils. and Indonesian islands of the Arafura Sea. Ecology: arboreal, diurnal, predator, closed forest, woodland, mangrove; oviparous, general carnivore. Biological references: Mertens, R. (1926). Über die Rassen einiger indo-australischer Reptilien. *Senckenberg. Biol.* **8**: 272–279(taxonomy); Roux, J. (1919). Note sur quelques reptiles provenant de la Nouvelle-Guinée. *Rev. Suisse Zool.* **27**: 347–351 (natural history); Meise, W. & Hennig, W. (1932). Die Schlangengattung *Dendrophis. Zool. Anz.* **99**: 273–297; Loveridge, A. (1948). New Guinean reptiles and amphibians in the Museum of Comparative Zoology and United States National Museum. *Bull. Mus. Comp. Zool. Harv.* **101**: 303–430; Whitaker, R., Whitaker, Z. & Mills, D. (1982). Reptiles of Papua New Guinea. *Wildl. in Papua New Guinea* **82/2**: 1–53.

Dendrelaphis punctulata (Gray, 1826)

Leptophis punctulatus Gray, J.E. (1826). Reptilia. Appendix *in* King, P.P. *Narrative of a survey of the intertropical and western coasts of Australia performed between the years 1818 and 1822.* London : John Murray 2: 424–434 [432]. Type data: holotype, BMNH 1946.1.23.34, from Careening Bay, 15°06'S, 125°00'E, W.A., see King, P.P. (1826). *Narrative of a survey of the intertropical and western coasts of Australia.* London : John Murray 243 pp. Appendix.

Elaps lewisii Gray, J.E. (1841). A catalogue of the species of reptiles and Amphibia hitherto described as inhabiting Australia, with a description of some new species from Western Australia. Appendix E *in* Grey, G. *Journals of Two Expeditions of Discovery in North-west and Western Australia, during the years 1837, 38, and 39, under the authority of Her Majesty's Government.* London : T. & W. Boone 2: 422–449 [444]. Type data: presumed lost, from unknown locality.

Dendrophis (Ahetula) olivacea Gray, J.E. (1842). Description of some hitherto unrecorded species of Australian reptiles and batrachians. *in* Gray, J.E. (ed.) *Zoological Miscellany.* London : Treuttel, Würtz & Co. pp. 51–57 [54]. Type data: syntypes (probable), BMNH 1946.1.23.31, BMNH 1946.1.23.41, ?BMNH 1964.1514 from Port Essington, N.T. and N coast of Australia.

Dendrophis prasinus Girard, C. (1858). Descriptions of some new reptiles, collected by the U.S. Exploring Expedition, under the command of Capt. C. Wilkes, U.S.N. Third Part. Including the species of ophidians, exotic to North America. *Proc. Acad. Nat. Sci. Philad.* **1857**. 181–182 [181]. Type data: not found, from Australia.

Dendrophis gracilis Macleay, W. (1875). Notes on a new species of *Dendrophis* from Cleveland Bay. *Proc. Linn. Soc. N.S.W.* **1**: 15–16 [15] [1877 on title page]. Type data: syntypes, AM R31908–31910, from Cleveland Bay, Townsville, Qld.

Dendrophis breviceps Macleay, W. (1877). The ophidians of the *Chevert* Expedition. *Proc. Linn. Soc. N.S.W.* **2**: 33–41 [37] [1878 on title page]. Type data: syntypes, AM R31911–13, from Mawatta, Binaturi River (as Katow), Papua New Guinea.

Dendrophis olivacea Macleay, W. (1878). Notes on a collection of snakes from Port Darwin. *Proc. Linn. Soc. N.S.W.* **2**: 219–222 [220] [*non Dendrophis (Ahetula) olivacea* Gray, 1842]. Type data: syntypes, presumed lost, from Port Darwin, N.T.

Dendrophis bilorealis Macleay, W. (1884). Notes on some reptiles from the Herbert River, Queensland. *Proc. Linn. Soc. N.S.W.* **8**: 432–436 [435] [1883 on title page]. Type data: holotype, AM R5942, from Ripple Creek, Herbert River, Qld.

Synonymy that of Cogger, H.G., this work.

Distribution: N coastal, N Gulf, NE coastal, SE coastal, Murray-Darling basin, W.A., N.T., Qld., N.S.W.; also occurs on New Guinea. Ecology: arboreal, diurnal, predator, closed forest, tall forest, open forest, woodland, tall open shrubland, low open woodland, mangrove; oviparous, general carnivore. Biological references: Covacevich, J. & Limpus, C. (1973). Two large winter aggregations of three species of tree-climbing snakes in south-eastern Queensland. *Herpetofauna* **6**: 16–21 (aggregation); Cogger, H.G. & Lindner, D.A. (1974). Frogs and reptiles in fauna survey of the Port Essington District, Cobourg Peninsula, Northern Territory of Australia. H.J. Frith & J.H. Calaby (eds.) *CSIRO Div. Wildl. Res. Tech. Pap.* **28**: 63–107 (natural history); Swan, G. (1975). Notes on the incubation and hatching of eggs of the green tree snake (*Dendrelaphis punctulatus*). *Herpetofauna* **7**: 18–20 (reproduction).

Enhydris Sonnini and Latreille, 1802

Enhydris Sonnini de Manoncourt, C.S. & Latreille, P.A. (1802). *Histoire Naturelle des Reptiles*, avec Figures Déssinnées d'après Nature. *in* Buffon, G.L.L. de *Histoire Naturelle de Buffon, classée d'après le système de Linné par R.R. Castel.* Nouvelle édition. (Suites). Paris : Déterville Vol. 4 [200]. Publication date established from Anon (1903). *Catalogue of the Books, Manuscripts, Maps and Drawings in the British Museum (Natural History). A —D.* London : British Museum 1: viii 500 pp. Type species *Enhydris caerulea* Sonnini and Latreille, 1802 (= *Hydris enhydris* Schneider, 1799) by subsequent designation, see Ko Ko Gyi (1970). A revision of colubrid snakes of the sub-family Homalopsinae. *Univ. Kans. Publs. Mus. Nat. Hist.* **20**: 44–223.

Pseudoferania Ogilby, J.D. (1890). Description of a new snake from the Herbert River District. *Proc. Linn. Soc. N.S.W. (2)* **5**: 51–52 [51]. Type species *Pseudoferania macleayi* Ogilby, 1890 by monotypy.

Synonymy that of Cogger, H.G., this work; for extra-limital synonyms see Ko Ko Gyi (1970). A revision of colubrid snakes of the sub-family Homalopsinae. *Univ. Kans. Publs. Mus. Nat. Hist.* **20**: 44–223.

This group is also found in the Indian subcontinent, China, Formosa, south-east Asia, Indo-Malayan Archipelago and New Guinea. The Asian species *Enhydris punctata* has been removed from the Australian fauna as early Australian literature records have not been substantiated.

Enhydris polylepis (Fischer, 1886)

Hypsirhina polylepis Fischer, J.G. (1886). Herpetologische Notizen. *Abh. Geb. Naturw. Hamb.* **9**: 51–67 [62]. Type data: holotype, from Fly River, Papua New Guinea, MTKD D437 (destroyed), *vide* Obst, F.J. (1977). Die herpetologische Sammlung des Staatlichen Museums für Tierkunde Dresden und ihre Typusexemplare. *Zool. Abh. Mus. Tierk. Dresden* **34**: 171–186.

Pseudoferania macleayi Ogilby, J.D. (1890). Description of a new snake from the Herbert River District. *Proc. Linn. Soc. N.S.W. (2)* **5**: 51–52 [51]. Type data: holotype, AM R460, from Ripple Creek, Herbert River, Qld.

Distribution: N coastal, N Gulf, NE coastal, N.T., Qld. Ecology: amphibious, noctidiurnal, predator, lentic freshwater, swamp; piscivore. Biological references: Griffiths, K. (1981). Macleay's water snake *Enhydris polylepis. Herpetofauna* **12**: 31–32 (natural history); Ko Ko Gyi (1970). A revision of colubrid snakes of the sub-family Homalopsinae. *Univ. Kans. Publs. Mus. Nat. Hist.* **20**: 44–223 (taxonomy); Thomson, D.F. (1935). Preliminary notes on a collection of snakes from Cape York Peninsula. *Proc. Zool. Soc. Lond.* **1935**: 723–731 (natural history).

Fordonia Gray, 1842

Fordonia Gray, J.E. (1842). Monographic synopsis of the water snakes, or the family Hydridae. *in* Gray, J.E. (ed.) *Zoological Miscellany.* London : Treuttel, Würtz & Co. pp. 59–68 [67]. Type species *Homalopsis leucobalia* Schlegel, 1837 by monotypy.

Hydropsis Fitzinger, L.J. (1843). *Systema Reptilium.* Vienna : Braümüller u. Seidel vi 106 pp. [25]. Type species *Homalops[is] leucobalia* Schlegel, 1837 by original designation.

Hemiodontus Duméril, A.M.C. (1853). Prodrome général de la classification des serpents. *Mém. Acad. Sci. Inst. Fr.* **23**: 399–536 [494]. Type species *Homalopsis leucobalia* Schlegel, 1837 by monotypy.

Synonymy that of Cogger, H.G., this work.

This group is also found in south-east Asia, Indo-Malayan Archipelago, Philippine Ils. and New Guinea.

Fordonia leucobalia (Schlegel, 1837)

Homalopsis leucobalia Schlegel, H. (1837). *Essai sur la Physionomie des Serpens.* Partie Générale et Partie Descriptive. 2 Parts La Haye : Kips & Stockum [345 pl 13]. Type data: holotype, RMNH 1161, from Timor, Indonesia.

Fordonia unicolor Gray, J.E. (1849). *Catalogue of the Specimens of Snakes in the Collection of the British Museum.* London : British Museum xv 125 pp. [77]. Type data: syntypes, BMNH iii.22.2a–b, from Borneo.

Hemiodontus chalybaeus Jan, G. (1863). *Elenco Sistematico Degli Ofidi Descritti e Disegnati per l'Iconografia Generale.* Milan : A. Lombardi 143 pp. [79]. Type data: holotype, MSNM, from Singapore.

Fordonia bicolor Theobald, W. (1868). Catalogue of the reptiles of British Birma, embracing the Provinces of Pegu, Martaban, and Tenasserim; with descriptions of new or little-known species. *J. Linn. Soc. Zool.* **10**: 4–67 [56]. Type data: holotype, not found, from near Rangoon, Burma.

Fordonia papuensis Macleay, W. (1877). The ophidians of the *Chevert* Expedition. *Proc. Linn. Soc. N.S.W.* **2**: 33–41 [35]. Type data: syntypes, AM R31903–5, from Mawatta, Binaturi River (as Katow), Papua New Guinea.

Fordonia variabilis Macleay, W. (1878). Notes on a collection of snakes from Port Darwin. *Proc. Linn. Soc. N.S.W.* **2**: 219–222 [219]. Type data: syntypes, 3 specimens, not found, from Port Darwin, N.T.

Synonymy that of Cogger, H.G., this work.

Distribution: NW Shelf, N coast, Gulf of Carpentaria, NE coast, W.A., N.T., Qld. Ecology: amphibious, mangrove, estuarine. Biological references: Kopstein, F. (1931). Herpetologische Notizen, 4. *Fordonia leucobalia* Schlegel and *Cerberus rhynchops* Schneider. *Treubia* **13**: 1–4 (natural history).

Myron Gray, 1849

Myron Gray, J.E. (1849). *Catalogue of the Specimens of Snakes in the Collection of the British Museum.* London : British Museum xv 125 pp. [70]. Type species *Myron richardsonii* Gray, 1849 by subsequent designation, see Boulenger, G.A. (1896). *Catalogue of Snakes in the British Museum (Natural History).* 3. London : British Museum xiv 727 pp. pls 1–25.

Neospades De Vis, C.W. (1889). List of birds, lizards, and snakes collected at Cambridge Gulf, and descriptions of two new vertebrates in Mr Saville-Kent's collection. *Proc. R. Soc. Qd.* **6**: 236–239 [238]. [*non Neospades* Blackburn, 1887]. Type species *Neospades kentii* De Vis, 1889 by monotypy.

Synonymy that of Cogger, H.G., this work.

This group is also found in New Guinea.

Myron richardsonii Gray, 1849

Myron richardsonii Gray, J.E. (1849). *Catalogue of the Specimens of Snakes in the Collection of the British Museum.* London : British Museum xv 125 pp. [70]. Type data: holotype, BMNH 1946.1.2.43, from "N.-West Australia".

Neospades kentii De Vis, C.W. (1889). List of birds, lizards, and snakes collected at Cambridge Gulf, and descriptions of two new vertebrates in Mr Saville-Kent's collection. *Proc. R. Soc. Qd.* **6**: 236–239 [238 pl 14] [1890 on title page]. Type data: holotype, QM J681, from Cambridge Gulf, W.A.

Synonymy that of Cogger, H.G., this work.

Distribution: NW Shelf, N coast, Gulf of Carpentaria, W.A., N.T. Ecology: amphibious, predator, mangrove, estuarine. Biological references: Ko Ko Gyi (1970). A revision of colubrid snakes of the sub-family Homalopsinae. *Univ. Kans. Publs. Mus. Nat. Hist.* **20**: 44–223 (taxonomy).

Stegonotus Duméril, Bibron and Duméril, 1854

Stegonotus Duméril, A.M.C., Bibron, G. & Duméril, A. (1854). *Erpétologie Générale* ou Histoire Naturelle Complète des Reptiles. Paris : Roret 7(1): xvi 780 pp. [680]. Type species *Stegonotus mulleri* Duméril, Bibron and Duméril, 1854 by monotypy.

Zamenophis Günther, A. (1872). Seventh account of new species of snakes in the collection of the British Museum. *Ann. Mag. Nat. Hist. (4)* **9**: 13–37 [21]. Type species *Zamenophis australis* Günther, 1872 by monotypy.

Synonymy that of Cogger, H.G., this work.

This group is also found in Philippine Ils., Moluccas and New Guinea.

Stegonotus cucullatus (Duméril, Bibron and Duméril, 1854)

Lycodon cucullatum Duméril, A.M.C., Bibron, G. & Duméril, A. (1854). *Erpétologie Générale* ou Histoire Naturelle Complète des Reptiles. Paris : Roret 7(1): xvi 780 pp. [376]. Type data: holotype, MNHP 3412, from Dore, Japen Is., Indonesia (as Hâvre Dorey, Nouvelle Guinée).

Zamenophis australis Günther, A. (1872). Seventh account of new species of snakes in the collection of the British Museum. *Ann. Mag. Nat. Hist. (4)* **9**: 13–37 [21]. Type data: holotype, BMNH 1946.1.14.93, from Cape York, Qld.

Lycodon keyensis Doria, G. (1874). Enumerazione dei rettili raccolti dal Dott. O. Beccari in Amboina alle Isole Aru ed alle Isole Kei durante gli anni. *Ann. Mus. Civ. Stor. Nat. Giacomo Doria* **6**: 325–357 [351 pl 12 fig h]. Type data: holotype, MCG C.E. 7521, from Kei Ils., Indonesia.

Lycodon magnus Meyer, A.B. (1874). Eine Mitteilung von Hrn. Dr. Adolf Bernhard Meyer über die von ihm auf Neu-Guinea den Inseln Jobi, Mysore und Mafoor im Jahr 1873 gesammelten Amphibien. *Mber. K. Preuss. Akad. Wiss. Berl.* **1874**: 128–140 [136] [1875 on title page]. Type data: syntypes, not found, from Biak Is. (as Mysore), Irian Jaya, Indonesia; New Guinea.

Lycodon darnleyensis Macleay, W. (1877). The ophidians of the *Chevert* Expedition. *Proc. Linn. Soc. N.S.W.* **2**: 33–41 [38] [1878 on title page]. Type data: holotype, AM R31894, from Darnley Is., Torres Strait, Qld.

Herbertophis plumbeus Macleay, W. (1884). Notes on some reptiles from the Herbert River, Queensland. *Proc. Linn. Soc. N.S.W.* **8**: 432–436 [434] [1883 on title page]. Type data: holotype, AM R31895, from Ripple Creek, Herbert River, Qld.

Stegonotus reticulatus Boulenger, G.A. (1895). On a collection of reptiles and batrachians from Ferguson Island, D'Entrecasteaux group British New Guinea. *Ann. Mag. Nat. Hist. (6)* **16**: 28–32 [31]. Type data: syntypes, BMNH 1946.1.14.87–88, from Fergusson Is., Papua New Guinea.

Synonymy that of Cogger, H.G., this work.

Distribution: N coastal, N Gulf, NE coastal,(W.A.), N.T., Qld. Ecology: terrestrial, (arboreal), nocturnal, predator, closed forest, tall forest, open forest, woodland, low open woodland; oviparous, general carnivore. Biological references: McDowell, S.B. (1972). The species of *Stegonotus* (Serpentes : Colubridae) in Papua New Guinea. *Zool. Meded.* **47**: 6–26 (taxonomy); Cogger, H.G. & Lindner, D.A. (1974). Frogs and reptiles in fauna survey of the Port Essington District, Cobourg Peninsula, Northern Territory of Australia. Frith, H.J. & Calaby, J.H. (eds.) *CSIRO Div. Wildl. Res. Tech. Pap.* **28**: 63–107 (natural history, as *S. modestus*).

Stegonotus parvus (Meyer, 1874)

Lycodon parvus Meyer, A.B. (1874). Eine Mitteilung von Hrn. Dr. Adolf Bernhard Meyer über die von ihm auf Neu-Guinea den Inseln Jobi, Mysore und Mafoor im Jahr 1873 gesammelten Amphibien. *Mber. K. Preuss. Akad. Wiss. Berl.* **1874**: 128–140 [137] [1875 on title page]. Type data: syntypes, not found, from Japen Is. (as Jobi), Irian Jaya, Indonesia; New Guinea.

Distribution: NE coastal, Qld.; Murray Is., Torres Strait. Ecology: terrestrial, (arboreal), nocturnal, predator, closed forest, woodland; oviparous, general carnivore. Biological references: McDowell, S.B. (1972). The species of *Stegonotus* (Serpentes : Colubridae) in Papua New Guinea. *Zool. Meded.* **47**: 6–26 (taxonomy).

Styporhynchus Peters, 1863

Styporhynchus Peters, W. (1863). Mittheilung über eine neue Schlangengattung, *Styporhynchus*, und verschiedene andere Amphibien des zoologischen Museums. *Mber. K. Preuss. Akad. Wiss. Berl.* **1863**: 399–413 [399]. Type species *Styporhynchus truncatus* Peters, 1863 by monotypy.

Katophis Macleay, W. (1877). The ophidians of the *Chevert* Expedition. *Proc. Linn. Soc. N.S.W.* **2**: 33–41 [36] [1878 on title page]. Type species *Katophis plumbea* Macleay, 1877 by monotypy.

Synonymy that of Cogger, H.G., this work; Underwood, G. (1982). Parallel evolution in the context of character analysis. *Zool. J. Linn. Soc.* **74**: 245–266 [248].

This group is also found in eastern Indonesia and Papua New Guinea.

Styporhynchus mairii (Gray, 1841)

Tropidonotus mairii Gray, J.E. (1841). A catalogue of the species of reptiles and Amphibia hitherto described as inhabiting Australia, with a description of some new species from Western Australia. Appendix E *in* Grey, G. *Journals of Two Expeditions of Discovery in North-west and Western Australia, during the years 1837, 38, and 39, under the authority of Her Majesty's Government.* London : T. & W. Boone 2: 422–449 [442]. Type data: presumed lost, from Australia.

Tropidonotus australis Gray, J.E. (1842). Description of some hitherto unrecorded species of Australian reptiles and batrachians. *in* Gray, J.E. (ed.) *Zoological Miscellany.* London : Treuttel, Würtz & Co. pp. 51–57 [54]. Type data: syntypes, BMNH 1946.1.13.52, BMNH 1946.1.12.88–90, from Port Essington, N.T.

Tropidonotus semicinctus **(?part.)** Duméril, A.M.C., Bibron, G. & Duméril, A. (1854). *Erpétologie Générale* ou Histoire Naturelle Complète des Reptiles. Paris : Roret 7(1): xvi 780 pp. [603]. Type data: syntype, MNHP 7419, from Australia; other syntypes not found, original description cites New Guinea and Australia.

Katophis plumbea Macleay, W. (1877). The ophidians of the *Chevert* Expedition. *Proc. Linn. Soc. N.S.W.* **2**: 33–41 [36]. Type data: syntypes, AM R31898–R31901, from Mawatta, Binaturi River (as Katow), Papua New Guinea.

Tropidonotus angusticeps Macleay, W. (1884). Notes on some reptiles from the Herbert River, Queensland. *Proc. Linn. Soc. N.S.W.* **8**: 432–436 [433] [1883 on title page]. Type data: holotype, AM R31902, from Ripple Creek, Herbert River, Qld.

Tropidonotus ater Macleay, W. (1885). On some Reptilia lately received from the Herbert River District, Queensland. *Proc. Linn. Soc. N.S.W.* **10**: 64–68 [67] [1886 on title page]. Type data: holotype, AM B5943, from "vicinity of Herbert River", Qld.

Synonymy that of Cogger, H.G., this work.

Distribution: N coastal, N Gulf, NE coastal, SE coastal, W.A., N.T., Qld., N.S.W. Ecology: amphibious, nocturnal, predator, closed forest, tall forest, open forest, woodland; oviparous, primarily anuriphagous. Biological references: Lyon, B. (1973). Observations on the common keelback snake, *Natrix mairii*, in Brisbane, south-eastern Queensland. *Herpetofauna* **6**: 2–5 (natural history); Cogger, H.G. & Lindner, D.A. (1974). Frogs and reptiles. *in* Fauna survey of the Port Essington District, Cobourg Peninsula, Northern Territory of Australia. Frith, H.J. & Calaby, J.H. (eds.) *CSIRO Div. Wildl. Res. Tech. Pap.* **28**: 63–107 (natural history, as *Amphiesma mairii*); Covacevich, J. (1974). An unusual aggregation of snakes following major flooding in the Ipswich-Brisbane area, south-eastern Queensland. *Herpetofauna* **7**(1): 21–24 (as *Amphiesma mairii*); Covacevich, J. & Archer, M. (1975). The distribution of the cane toad, *Bufo marinus*, in Australia and its effects on indigenous vertebrates. *Mem. Qd. Mus.* **17**: 305–310 (as predator, as *Amphiesma mairii*); Worrell, E. (1946). The northern river snake (*Natrix mairii*). *Proc. R. Zool. Soc. N.S.W.* **1945–46**: 32 (natural history); Brongersma, L.D. (1948). A new subspecies of *Natrix mairii* (Gray) from Dutch New Guinea. *Proc. K. Ned. Akad. Wet.* **51**: 372–381 (taxonomy).

ELAPIDAE

INTRODUCTION

This family of front-fanged venomous snakes contains some 200 species in about 50 genera. In Australia, 25 genera and 75 species are currently recognized. As treated here (see comments under Colubridae), the family occurs from Africa, through Asia and Indo-Malaysia to Australia and New Guinea, the Solomon Islands and Fiji, with one New World group (the coral snakes). The family contains some of the world's deadliest snakes, including the mambas, cobras and kraits.

Characterised in Australia by: a pair of functionally hollow enlarged fangs at the front of the maxilla; no loreal scales; scales in 23 or fewer rows at mid-body.

The taxonomic and phylogenetic limits of this "family" are in a state of flux. Despite a number of recent refinements to the traditional classification, the results of current research in a number of disciplines suggest only that the Australian elapids, sea snakes (Hydrophiidae) and sea kraits (Laticaudidae) are very closely related. Within the Australian elapids, the generic classification remains extremely labile. Cogger (1979) broadly adopted the taxonomic conclusions of McDowell (1967), but in applying available names introduced a number of new combinations which are here explained or justified for the first time. Cogger's nomenclature has been widely adopted (with some notable exceptions), but it must be emphasised that the generic taxonomy of the group is in great need of revision.

Several genera warrant special mention. The close relationship between the genera *Furina* and *Glyphodon* has long been recognised. Storr (1981) formally synonymised the two. We have somewhat arbitrarily retained their separate identity pending a more rigorous examination of their relationships. We also have rejected on morphological grounds Storr's transfer of *Cacophis warro* De Vis, 1884 to *Furina*; the relationships of *Cacophis warro* remain obscure but appear to lie with *Simoselaps bertholdi*. Recognition of the genus *Parademansia* to accommodate *Diemenia microlepidota* McCoy, 1879 is contrary to the conclusions of Covacevich *et al.* (1980) that this species is congeneric with the taipan, *Oxyuranus scutellatus*. This is largely a matter of interpretation of the available data (which were clearly documented by Covacevich *et al.*), but we believe that the differences between the two taxa are such as to warrant generic separation.

The genus *Simoselaps*, as treated, is clearly polyphyletic and awaits effective revision. Intensive work is currently being undertaken on a variety of fronts—morphological, ecological, karyological and biochemical—to establish relationships within Australian elapids and between them and their relatives elsewhere. The classification adopted below should be seen, therefore, as an interim one.

References

Cogger, H.G. (1979). *Reptiles and Amphibians of Australia.* Sydney : A.H. & A.W. Reed 608 pp.

Covacevich, J., McDowell, S.B., Tanner, C. & Mengden, G.A. (1981). The relationship of the taipan, *Oxyuranus scutellatus*, and the small-scaled snake, *Oxyuranus microlepidotus* (Serpentes : Elapidae). pp. 160–168 *in* Banks, C.B. & Martin, A.A. (eds.) *Proceedings of the Melbourne Herpetological Symposium, 1980.* Melbourne :

Zoological Board of Victoria

Klemmer, K. (1963). Liste der rezenten Giftschlangen; Elapidae, Hydrophiidae, Viperidae und Crotalidae. Marburg : N.G. Elwert Universitäts- U. Verlagsbuchhand pp. 253–464

McDowell, S.B. (1967). *Aspidomorphus*, a genus of New Guinea snakes of the family Elapidae, with notes on related genera. *J. Zool.* **151**: 497–543

Storr, G.M. (1981). The genus *Furina* (Serpentes : Elapidae) in Western Australia. *Rec. West. Aust. Mus.* **9**: 119–123

Worrell, E. (1970). *Reptiles of Australia.* 2nd edn. Sydney : Angus & Robertson xv 169 pp.

Acanthophis Daudin, 1803

Acanthophis Daudin, F.M. (1803). *Histoire Naturelle, générale et particulière des Reptiles.* Paris : Dufart Tom. 5 365 pp. [287]. Type species *Acanthophis cerastinus* Daudin, 1803 by monotypy.

Ophryas Merrem, B. (1820). *Versuch eines Systems Amphibien. Tentamen Systematis Amphibiorum.* Marburg : Krieger xv 199 pp. [146]. Type species *Ophryas acantophis* Merrem, 1820 by monotypy.

Acanthopis Gray, J.E. (1825). A synopsis of the genera of reptiles and Amphibia, with a description of some new species. *Annls. Philos. (2)* **10**: 193–217 [205] [emend. *pro Acanthophis* Daudin, 1803].

Acantophis Berthold, A.A. (1827). Anmerkungen und Zusatzen. *in* Latreille, P.A. *Natürliche Familien des Thierreichs.* Weimar : Landes-Industrie-Comptoirs 604 pp. [100] [*errore pro Acanthophis* Daudin, 1803].

Ophrias Cuvier, G.L.C.F.D. (1829). *Le Règne Animal.* 2nd edn. Paris : Déterville Vol. 2 xv 406 pp. [95] [*non Ophrias* Rafinesque, 1815 (*nom. nud.*); emend. *pro Ophryas* Merrem, 1820].

Synonymy that of Cogger, H.G., this work; Storr, G.M. (1981). The genus *Acanthophis* (Serpentes : Elapidae) in Western Australia. *Rec. West. Aust. Mus.* **9**: 203–210.

This group is also found in eastern Indonesia and Papua New Guinea.

Acanthophis antarcticus (Shaw and Nodder, 1802)

Boa palpebrosa Shaw, G. (1802). *General Zoology*, or Systematic Natural History. Vol. 3 Pt. 2 Amphibia. London : Kearsley vii pp. 313–615 [362] [*nom. oblit.*; as vol.3 of Shaw's *Gen. Zool.* was acknowledged in *Phil. Trans.* on Feb. 1802, the date of publication must have been prior to Feb. of that year]. Type data: presumed lost, from unknown locality.

Boa antarctica Shaw, G. & Nodder, F.P. (1802). Boa antarctica. *Naturalist's Miscellany* 13 pl 535 [pl 535] [note that date of publication (1794) cited by Boulenger, G.A. (1896). *Catalogue of Snakes in the British Museum (Natural History).* 3. London : British Museum xiv 727 pp. pls 25 [355] and subsequent authors (Worrell, E. (1963). *Reptiles of Australia;* Crocodiles, Turtles, Tortoises, Lizards, Snakes. Sydney : Angus & Robertson xv 207 pp. [190]; Klemmer, K. (1963). Liste der rezenten Giftschlangen; Elapidae, Hydrophiidae, Viperidae und Crotalidae. Marburg : N.G. Elwert Universitäts- U. Verlagsbuchhand pp. 253–464 [272]) is in error, *vide* Sherborn C.D. (1895). On the dates of Shaw and Nodder's "Naturalist's Miscellany". *Ann. Mag. Nat. Hist. (6)* **15**: 375–376; according to Sherborn's formula, the date of publication could not have been prior to May, 1802]. Type data: presumed lost, from "Australasia".

Acanthophis cerastinus Daudin, F.M. (1803). *Histoire Naturelle, générale et particulière des Reptiles.* Paris : Dufart Tom. 5 365 pp. [289]. Type data: presumed lost, from unknown locality.

Acanthophis brownii Leach, W.E. (1814). *The Zoological Miscellany*; Being Descriptions of New or Interesting Animals. London : E. Nodder & Son Vol. 1. (illustr. by R.P. Nodder) 144 pp. [12 pl 3]. Type data: holotype, presumed lost, from Sydney (as Port Jackson), N.S.W.

Boa ambigua Leach, W.E. (1814). *The Zoological Miscellany*; Being Descriptions of New or Interesting Animals. London : E. Nodder & Son Vol. 1. (illustr. by R.P. Nodder) 144 pp. [12] [*nom. nud.*, name introduced in synonymy; *ex* Brown ms].

Ophryas acantophis Merrem, B. (1820). *Versuch eines Systems Amphibien. Tentamen Systematis Amphibiorum.* Marburg : Krieger xv 199 pp. [147]. Type data: holotype, from Australia, presumed lost.

Vipera sorda Salvado, R. (1851). *Memorie Stariche dell'Australia.* Rome : S. Congreg. de Propaganda Fide xii 389 pp. [48]. Type data: status unknown, from Australia.

Boa aculeata Boulenger, G.A. (1896). *Catalogue of Snakes in the British Museum (Natural History).* 3. London : British Museum xiv 727 pp. pls 25 [356] [*nom. nud.*; apparently a ms name of Shaw introduced in synonymy].

Synonymy that of Cogger, H.G., this work.

Distribution: SW coastal, W plateau, S Gulfs, SE coastal, Murray-Darling basin, Bulloo River basin, Lake Eyre basin, NE coastal, N Gulf, W.A., S.A., N.S.W., Qld., N.T., A.C.T., (Vic.). Ecology: terrestrial, nocturnal, predator, tall forest, open forest, woodland, low open woodland, low shrubland, low open shrubland, tussock grassland; viviparous, general carnivore. Biological references:

Shine, R. (1980). Ecology of the Australian death adder *Acanthophis antarcticus* (Elapidae): evidence for convergence with the Viperidae. *Herpetologica* **36**: 281–289 (ecology); Storr, G.M. (1981). The genus *Acanthophis* (Serpentes : Elapidae) in Western Australia. *Rec. West. Aust. Mus.* **9**: 203–210 (taxonomy); Lillywhite, H.B. (1980). Behavioral thermoregulation in Australian elapid snakes. *Copeia* **1980**: 452–458; Kim, H.S. & Tamiya, N. (1981). Isolation, properties and amino acid sequence of a long-chain neurotoxin, Acanthophis antarcticus B, from the venom of an Australian snake (the common death adder, *Acanthophis antarcticus*). *Biochem. J.* **193**: 899–906 (venom); Hudson, P. (1979). On the birth and breeding of death adders in captivity. *Herpetofauna* **11**: 11–13 (reproduction); Carpenter, C.C., Murphy, J.B. & Carpenter, G.C. (1978). Tail luring in the death adder, *Acanthophis antarcticus* (Reptilia, Serpentes, Elapidae). *J. Herpet.* **12**: 574–577; Mackay, W.J. (1889). The osteology and myology of the death adder (*Acanthophis antarctica*). *Proc. Linn. Soc. N.S.W. (2)* **4**: 896–986; Campbell, C.H. (1966). The death adder. Effect of the bite and its treatment. *Med. J. Aust.* **11**: 922–925; Sheumack, D.D., Howden, M.E.H. & Spence, I. (1979). Isolation and partial characterisation of a lethal neurotoxin from the venom of the Australian death adder (*Acanthophis antarcticus*). *Toxicon* **17**: 609.

Acanthophis praelongus Ramsay, 1877

Acanthophis praelongus Ramsay, E.P. (1877). Description of a supposed new species of *Acanthophis* from North Australia. *Proc. Linn. Soc. N.S.W.* **2**: 72–74 [72]. Type data: holotype, AM R451, from Cape York Peninsula (probably near Somerset), Qld.

Acanthophis antarcticus rugosus Loveridge, A. (1948). New Guinean reptiles and amphibians in the Museum of Comparative Zoology and United States National Museum. *Bull. Mus. Comp. Zool.* **101**: 303–430 [392] Type data: holotype, MCZ 22812 M, from Merauke, Irian Jaya, Indonesia.

Synonymy that of Cogger, H.G., this work.

Distribution: N coastal, N Gulf, NE coastal, W.A., N.T., Qld. Ecology: terrestrial, nocturnal, predator, open forest, woodland, low open woodland; viviparous, general carnivore. Biological references: Storr, G.M. (1981). The genus *Acanthophis* (Serpentes : Elapidae) in Western Australia. *Rec. West. Aust. Mus.* **9**: 203–210 (taxonomy); Campbell, C.H. (1966). The death adder. Effect of the bite and its treatment. *Med. J. Aust.* **11**: 922–925 (bite and treatment as *A. antarcticus*).

Acanthophis pyrrhus Boulenger, 1898

Acanthophis pyrrhus Boulenger, G.A. (1898). Description of a new death adder (*Acanthophis*) from central Australia. *Ann. Mag. Nat. Hist. (7)* **2**: 75 [75]. Type data: holotype, BMNH 1946.1.18.62, from Station Point, N.T.

Distribution: NW coastal, W plateau, N coastal, Lake Eyre basin, W.A., N.T., S.A., Qld. Ecology: terrestrial, nocturnal, predator, desert, tall shrubland, low open shrubland, low woodland; viviparous, general carnivore. Biological references: Storr, G.M. (1981). The genus *Acanthophis* (Serpentes : Elapidae) in Western Australia. *Rec. West. Aust. Mus.* **9**: 203–210 (taxonomy); Gow, G.F. (1981). Notes on the desert death adder (*Acanthophis pyrrhus* Boulenger, 1898), with the first reproductive record. *N. Territory Nat.* **4**: 21–22 (natural history).

Austrelaps Worrell, 1963

Austrelaps Worrell, E. (1963). A new elapine generic name. *Aust. Reptile Park Rec.* **1**: 1–7 [3]. Type species *Hoplocephalus superbus* Günther, 1858 by original designation.

Austrelaps superbus (Günther, 1858)

***Hoplocephalus superbus* (part.)** Günther, A. (1858). *Catalogue of Colubrine Snakes in the Collection of the British Museum.* London : British Museum xvi 281 pp. [217]. Type data: syntypes, BMNH 1946.1.19.6, BMNH 1946.1.20.34–36, BMNH 1946.1.20.40, BMNH 1946.1.20.42 (Australia); BMNH 1946.1.19.5, BMNH 1946.1.20.37–38 (Tasmania); for other syntypes, see *Elapognathus minor* (Günther, 1863); type localities originally Australia and Tas. but restricted to Tas. by Günther, A. (1863). Third account of new species of snakes in the collection of the British Museum. *Ann. Mag. Nat. Hist. (3)* **12**: 348–365 in describing *Hoplocephalus minor* from part of the original syntype series of *H. superbus*.

Alecto labialis Jan, G. (1859). Plan d'une *Iconographie descriptive des Ophidiens*, et description sommaire de nouvelles espèces de Serpents. *Rev. Mag. Zool. (2)* **11–12**: 122–130; Jan, G. (1859). Prodrome d'une *Iconographie descriptive des Ophidiens* et Description sommaire de nouvelles espèces de Serpents venimeux. Reprinted from: *Rev. Mag. Zool. (2)* **10** and *(2)* **11–12**, 1858 and 1859. Paris : Bouchard-Huzard 32 pp. [128; pl C]. Type data: holotype, ZMG (presumed lost), from Australia.

Hoplocephalus ramsayi Krefft, G. (1864). Descriptions of three new species of Australian snakes. *Proc. Zool. Soc. Lond.* **1864**: 180–182 [180]. Type data: holotype, not found, from "neighbourhood of Braidwood", N.S.W.

Alecto schmidti Jan, G. & Sordelli, F. (1873). *Iconographie Générale des Ophidiens.* Atlas 1860–1881 Milan : Jan & Sordelli 50 livr. [pl 1 fig 4]. Type data: holotype, ZMH 478, from Australia.

Hoplocephalus bransbyi Macleay, W. (1878). On a new species of *Hoplocephalus*, from Sutton Forest. *Proc. Linn. Soc. N.S.W.* **3**: 52–54 [52]. Type data: holotype, AM R31922, from Sutton Forest, N.S.W.

Hoplocephalus brankysi Goldman, J., Hill, L. & Stanbury, P.J. (1969). Type specimens in the Macleay

Museum, University of Sydney. II. Amphibians and Reptiles. *Proc. Linn. Soc. N.S.W.* **93**: 427–438 [436] [*lapsus pro Hoplocephalus bransbyi* Macleay, 1878].

Synonymy that of Cogger, H.G., this work.

Distribution: S Gulfs, SE coastal, Murray-Darling basin, S.A., Vic., N.S.W., Tas., A.C.T. Ecology: terrestrial, noctidiurnal, predator, alpine, open forest, woodland; viviparous, general carnivore, but primarily anuriphagous. Biological references: Shine, R. & Allen, S. (1980). Ritual combat in the Australian copperhead, *Austrelaps superbus* (Serpentes : Elapidae). *Vict. Nat.* **97**: 188–190; Shine, R. (1979). Activity patterns in Australian elapid snakes (Squamata : Serpentes : Elapidae). *Herpetologica* **35**: 1–11 (behaviour); Rawlinson, P.A. (1974). Biogeography and ecology of the reptiles of Tasmania and the Bass Strait area. Chapt. 11 pp. 291–338 *in* Williams, W.D. (ed.) *Biogeography and Ecology in Tasmania.* The Hague : Junk (biogeography); Baehr, M. (1976). Beiträge zur Verbreitung und Ökologie tasmanischer Reptilien. *Stuttgart Beitr. Naturk. (A)* **292**: 1–24 (natural history); Rawlinson, P.A. (1969). East Gippsland Symposium. 7. The reptiles of East Gippsland. *Proc. R. Soc. Vict.* **82**: 113–128; Kellaway, C.H. (1929). A preliminary note on the venom of the Australian copperhead (*Denisonia superba*): its toxic effects in the common laboratory animals. *Med. J. Aust.* **1929**: 56–63.

Cacophis Günther, 1863

Cacophis Günther, A. (1863). Third account of new species of snakes in the collection of the British Museum. *Ann. Mag. Nat. Hist. (3)* **12**: 348–365 [361]. Type species *Cacophis krefftii* Günther, 1863 by monotypy.

Petrodymon Krefft, G. (1866). On snakes observed in the neighbourhood of Sydney. *Trans. Philos. Soc. N.S.W.* **1862–65**: 34–60 [55]. Type species *Diemansia cucullata* Günther, 1862 by monotypy.

Synonymy that of Cogger, H.G., this work.

Cacophis harriettae Krefft, 1869

Cacophis harriettae Krefft, G. (1869). *The Snakes of Australia*; an Illustrated and Descriptive Catalogue of All the Known Species. Sydney : Govt. Printer xxv 100 pp. [76 pl 11 fig 13]. Type data: holotype, AM 6676, from Warro, Port Curtis, Qld.

[Glyphodon sp.] Worrell, E. (1963). *Reptiles of Australia;* Crocodiles, Turtles, Tortoises, Lizards, Snakes. Sydney : Angus & Robertson xv 207 pp. [125 pl 56] [this is the form referred to by McDowell as *Cacophis harriettae flavicollis* and probably represents a species distinct from *Cacophis harriettae* Krefft, 1869].

Cacophis h[arriettae] flavicollis McDowell, S.B. (1967). *Aspidomorphus*, a genus of New Guinea snakes of the family Elapidae, with notes on related genera. *J. Zool.* **151**: 497–543 [536] [*ex* Hosmer ms; *nom. nud.*].

Synonymy that of Cogger, H.G., this work.

Distribution: NE coastal, Qld. Ecology: terrestrial, nocturnal, predator, tall forest, closed forest, woodland; oviparous, lacertiphagous. Biological references: Shine, R. (1980). Comparative ecology of three Australian snake species of the genus *Cacophis* (Serpentes : Elapidae). *Copeia* **1980**: 831–838 (ecology); Kellaway, C.H. (1934). The venoms of some of the small and rare Australian venomous snakes. *Med. J. Aust.* **1934**: 74–78.

Cacophis krefftii Günther, 1863

Cacophis krefftii Günther, A. (1863). Third account of new species of snakes in the collection of the British Museum. *Ann. Mag. Nat. Hist. (3)* **12**: 348–365 [361]. Type data: holotype, BMNH 1946.1.17.60, labelled "North of the Clarence River, N.S.W. but original description cites "probably from Port Macquarie, N.S.W."

Cacophis fordei Krefft, G. (1869). *The Snakes of Australia*; an Illustrated and Descriptive Catalogue of All the Known Species. Sydney : Govt. Printer xxv 100 pp. [75 pl 12 fig 8]. Type data: holotype, AM 6672, from Pine Mountain, near Ipswich, Qld.

Synonymy that of Cogger, H.G., this work.

Distribution: SE coastal, NE coastal, N.S.W., Qld. Ecology: terrestrial, nocturnal, predator, closed forest, tall forest; oviparous, lacertiphagous. Biological references: Shine, R. (1980). Comparative ecology of three Australian snake species of the genus *Cacophis* (Serpentes : Elapidae). *Copeia* **1980**: 831–838 (ecology); Wells, R. (1980). Notes on Krefft's dwarf snake (*Cacophis krefftii*). *Herpetofauna* **11**: 18–19 (natural history).

Cacophis squamulosus (Duméril, Bibron and Duméril, 1854)

Pseudelaps squamulosus Duméril, A.M.C. (1853). Prodrome général de la classification des serpents. *Mém. Acad. Sci. Inst. Fr.* **23**: 399–536 [517] [*nom. nud.*].

Pseudelaps squamulosus Duméril, A.M.C., Bibron, G. & Duméril, A. (1854). *Erpétologie Générale* ou Histoire Naturelle Complète des Reptiles. Paris : Roret 7(2): xii pp. 781–1536 [1235]. Type data: holotype, MNHP 7667, from unknown locality (as Tasmania, *errore*).

Diemansia cucullata Günther, A. (1862). On new species of snakes in the collection of the British Museum. *Ann. Mag. Nat. Hist. (3)* **9**: 52–59, 124–132 [129]. Type data: holotype, BMNH 1946.1.17.58, from "neighbourhood of Sydney", N.S.W.

Pseudoelaps atropolios Jan, G. & Sordelli, F. (1873). *Iconographie Générale des Ophidiens.* Atlas 1860–1881 Milan : Jan & Sordelli 50 livr. [pl 5 fig 1]. Type data. holotype, ZMH 531 (specimen from G. Krefft, Sydney). from Australia (by implication).

Synonymy that of Cogger, H.G., this work.

Distribution: SE coastal, NE coastal, N.S.W., Qld. Ecology: terrestrial, nocturnal, predator, closed forest, tall forest, open forest; oviparous, primarily lacertiphagous. Biological references: Shine, R. (1980). Comparative ecology of three Australian snake species of the genus *Cacophis* (Serpentes : Elapidae). *Copeia* **1980**: 831–838 (ecology); Wells, R. (1980). Notes on *Cacophis squamulosus*. *Herpetofauna* **11**: 26 (natural history).

Cryptophis Worrell, 1961

Cryptophis Worrell, E. (1961). Herpetological name changes. *West. Aust. Nat.* **8**: 18–27 [26]. Type species *Hoplocephalus pallidiceps* Günther, 1858 by original designation.

Cryptophis nigrescens (Günther, 1862)

Hoplocephalus nigrescens Günther, A. (1862). On new species of snakes in the collection of the British Museum. *Ann. Mag. Nat. Hist. (3)* **9**: 52–59, 124–132 [131 pl 9, fig 12]. Type data: holotype, BMNH 1946.1.20.58, from "environs of Sydney", N.S.W.

Alecto permixta Jan, G. (1863). *Elenco Sistematico Degli Ofidi Descritti e Disegnati per l'Iconografia Generale.* Milan : A. Lombardi 143 pp. [117] [*nom. nud.*]. Type data: status unknown, ZMH, from Sydney, N.S.W.

Alecto melanotus Jan, G. (1863). *Elenco Sistematico Degli Ofidi Descritti e Disegnati per l'Iconografia Generale.* Milan : A. Lombardi 143 pp. [117] [*nom. nud.*]. Type data: status unknown, SMNS, from Sydney, N.S.W.

Alecto permixta Jan, G. & Sordelli, F. (1873). *Iconographie Générale des Ophidiens.* Atlas 1860–1881 Milan : Jan & Sordelli 50 livr. [pl 1 fig 2]. Type data: lectotype, MNHP 949 (two syntypes with this number; the lectotype is the specimen with the catalogue number attached to it); both specimens received from G. Krefft of the Australian Museum in 1864 and the one which was illustrated is the designated lectotype, second paralectotype, ZMH 776, designation by Cogger, H.G., this work, from Sydney, N.S.W.

Hoplocephalus assimilis Macleay, W. (1885). On some Reptilia lately received from the Herbert River District, Queensland. *Proc. Linn. Soc. N.S.W.* **10**: 64–68 [68] [1886 on title page]. Type data: syntypes, AM B5944, AM R31923–5, from "vicinity of Herbert River", Qld.

Synonymy that of Cogger, H.G., this work.

Distribution: NE coastal, SE coastal, (Murray-Darling basin), Qld., N.S.W., A.C.T., Vic. Ecology: terrestrial, nocturnal, predator, closed forest, tall forest, open forest; sometimes diurnal in closed forest, viviparous, primarily lacertiphagous. Biological references: Pollitt, C.C. (1981). Studies on the venom and blood of the eastern small-eyed snake *Cryptophis nigrescens* (Günther). pp. 44–54 *in* Banks, C.B. & Martin, A.A. (eds.) *Proceedings of the Melbourne Herpetological Symposium, 1980.* Melbourne : Zoological Board of Victoria (venom, blood); Bridge, G. (1979). The small-eyed snake (*Cryptophis nigrescens*). *Vict. Herp. Soc. Newsletter* **13**: 17–18 (natural history); Hoser, R.T. (1980). Further records of aggregations of various species of Australian snakes. *Herpetofauna* **12**: 16–22 (aggregation).

Cryptophis pallidiceps (Günther, 1858)

Hoplocephalus pallidiceps Günther, A. (1858). *Catalogue of Colubrine Snakes in the Collection of the British Museum.* London : British Museum xvi 281 pp. [214]. Type data: syntypes, BMNH 1946.1.18.88 (NE Australia); BMNH 1946.1.20.64–65 (Port Essington), from Port Essington, N.T. and NE Australia (the latter is erroneous); original BMNH records show that the correct locality is NW Australia.

Distribution: N coastal, (N Gulf), W.A., N.T. Ecology: terrestrial, nocturnal, predator, open forest, woodland, tall open shrubland; viviparous, primarily lacertiphagous. Biological references: Smith, L.A. (1978). The elapid snakes *Denisonia pallidiceps* and *Denisonia suta* in the Kimberleys of Western Australia. *West. Aust. Nat.* **14**: 75–76 (distribution); Cogger, H.G. & Lindner, D.A. (1974). Frogs and reptiles in fauna survey of the Port Essington District, Cobourg Peninsula, Northern Territory of Australia. Frith, H.J. & Calaby, J.H. (eds.) *CSIRO Div. Wildl. Res. Tech. Pap.* **28**: 63–107 (natural history).

Demansia Günther, 1858

Demansia Gray, J.E. (1842). Description of some hitherto unrecorded species of Australian reptiles and batrachians. *in* Gray, J.E. (ed.) *Zoological Miscellany.* London : Treuttel, Würtz & Co. pp. 51–57 [54] [*nom. nud.*, introduced in synonymy].

Demansia Günther, A. (1858). *Catalogue of Colubrine Snakes in the Collection of the British Museum.* London : British Museum xvi 281 pp. [211]. Type species *Lycodon reticulatus* Gray, 1842 by subsequent designation, see Klemmer, K. (1963). Liste der rezenten Giftschlangen; Elapidae, Hydrophiidae, Viperidae and Crotalidae. Marburg : N.G. Elwert Universitäts - U. Verlagsbuchhand pp. 256–464.

Diemansia Günther, A. (1858). *Catalogue of Colubrine Snakes in the Collection of the British Museum.* London : British Museum xvi 281 pp. [254] [probably valid emend. *pro Demansia* Günther, 1858].

Diemennia Günther, A. (1863). Notes on *Diemenia superciliosa. Proc. Zool. Soc. Lond.* **1863**: 17–18 [17] [invalid emend. *pro Diemenia* Günther ms by editor of *Proc. Zool. Soc. Lond.*, see description of *Diemenia* Günther, 1863 (350)].

Diemenia Günther, A. (1863). Third account of new species of snakes in the collection of the British Museum. *Ann. Mag. Nat. Hist. (3)* **12**: 348–365 [350] [*non Diemenia* Spinola, 1850; emend. *pro Demansia* Günther, 1858].

Elapocephalus Macleay, W. (1878). Notes on a collection of snakes from Port Darwin. *Proc. Linn. Soc. N.S.W.* **2**: 219–222 [221] [*non Elapocephalus* Günther,

1858]. Type species *Elapocephalus ornaticeps* Macleay, 1878 by monotypy.

Elapocranium Macleay, W. (1878). On a new species of *Hoplocephalus*, from Sutton Forest. *Proc. Linn. Soc. N.S.W.* **3**: 52–54 [54] [*nom. nov. pro Elapocephalus* Macleay, 1878].

Elapidocephalus Macleay, W. (1884). Census of Australian snakes with descriptions of two new species. *Proc. Linn. Soc. N.S.W.* **9**: 548–568 [560] [*errore pro Elapocephalus* Macleay, 1878].

Synonymy that of Cogger, H.G., this work.

This group is also found in New Guinea.

Demansia atra (Macleay, 1884)

Diemenia atra Macleay, W. (1884). Census of Australian snakes with descriptions of two new species. *Proc. Linn. Soc. N.S.W.* **9**: 548–568 [549] [1885 on title page]. Type data: syntypes, AM B5941, AM R31920, from unspecified locality, but types collected by J.A. Boyd of Ripple Creek, Ingham, Qld.

Hoplocephalus vestigiatus De Vis, C.W. (1884). Descriptions of a new snake with a synopsis of the genus *Hoplocephalus*. *Proc. R. Soc. Qd.* **1**: 138–140 [138]. Type data: holotype, from "uncertain" locality, QM J206 *vide* Covacevich, J. (1971). Amphibian and reptile type specimens in the Queensland Museum. *Mem. Qd. Mus.* **16**: 49–68.

Diemenia maculiceps Boettger, O. (1898). *Katalog der Reptilien-Sammlung im Museum der Senckenbergischen Naturforschenden Gesellschaft in Frankfurt-am-Main.* II. Teil (Schlangen). Frankfurt-am-Main : Knauer ix 160 pp. [116]. Type data: holotype, SMF 20498, from Burnett River, Qld.

Synonymy that of Cogger, H.G., this work.

Distribution: N coastal, N Gulf, NE coastal, W.A., N.T., Qld. Ecology: terrestrial, diurnal, predator, closed forest, tall forest, open forest, woodland, low open woodland; oviparous, lacertiphagous. Biological references: Storr, G.M. (1978). Whip snakes (*Demansia* : Elapidae) of Western Australia. *Rec. West. Aust. Mus.* **6**: 287–301 (taxonomy); Cogger, H.G. & Lindner, D.A. (1974). Frogs and reptiles in fauna survey of the Port Essington District, Cobourg Peninsula, Northern Territory of Australia. Frith, H.J. & Calaby, J.H. (eds.) *CSIRO Div. Wildl. Res. Tech. Pap.* **28**: 63–107 (natural history); Shine, R. (1980). Ecology of eastern Australian whipsnakes of the genus *Demansia. J. Herpet.* **14**: 381–389 (ecology); Kellaway, C.H. (1934). The venoms of some of the small and rare Australian venomous snakes. *Med. J. Aust.* **1934**: 74–78.

Demansia olivacea (Gray, 1842)

Lycodon olivaceus Gray, J.E. (1842). Description of some hitherto unrecorded species of Australian reptiles and batrachians. *in* Gray, J.E. (ed.) *Zoological Miscellany.* London : Treuttel, Würtz & Co. pp. 51–57 [54]. Type data: syntypes, from Port Essington, N.T., presumed lost; 2 syntypes listed by Boulenger, G.A. (1896). *Catalogue of Snakes in the British Museum (Natural History).* 3. London : British Museum xiv 727 pp. pls 25 cannot be Gray's specimens, as locality and other collection data do not correspond *vide* Cogger, H.G. & Lindner, D.A. (1974). Frogs and reptiles in fauna survey of the Port Essington District, Cobourg Peninsula, Northern Territory of Australia. Frith, H.J. & Calaby, J.H. (eds.) *CSIRO Div. Wildl. Res. Tech. Pap.* **28**: 63–107.

Elapocephalus ornaticeps Macleay, W. (1878). Notes on a collection of snakes from Port Darwin. *Proc. Linn. Soc. N.S.W.* **2**: 219–222 [221]. Type data: holotype, AM R31918, from Port Darwin, N.T.

Diemenia angusticeps Macleay, W. (1888). Notes on some ophidians from King's Sound, north-west Australia. *Proc. Linn. Soc. N.S.W. (2)* **3**: 416–418 [417]. Type data: holotype, AM R31921, from "vicinity of King's Sound", W.A.

Demansia olivacea rufescens Storr, G.M. (1978). Whip snakes (*Demansia* : Elapidae) of Western Australia. *Rec. West. Aust. Mus.* **6**: 287–301 [292]. Type data: holotype, WAM R52747, from Marandoo minesite, near Mt. Bruce, 22°40'S, 118°10'E, W.A.

Demansia olivacea calodera Storr, G.M. (1978). Whip snakes (*Demansia* : Elapidae) of Western Australia. *Rec. West. Aust. Mus.* **6**: 287–301 [294 fig 2]. Type data: holotype, WAM R54992, from Tamala, 26°42'S, 113°42'E, W.A.

Synonymy that of Cogger, H.G., this work.

Distribution: NW coastal, W plateau, N coastal, N Gulf, W.A., N.T., Qld. Ecology: terrestrial, diurnal, predator, open forest, woodland, low open woodland; oviparous, lacertiphagous. Biological references: Storr, G.M. (1978). Whip snakes (*Demansia* : Elapidae) of Western Australia. *Rec. West. Aust. Mus.* **6**: 287–301 (taxonomy); Cogger, H.G. & Lindner, D.A. (1974). Frogs and reptiles in fauna survey of the Port Essington District, Cobourg Peninsula, Northern Territory of Australia. Frith, H.J. & Calaby, J.H. (eds.) *CSIRO Div. Wildl. Res. Tech. Pap.* **28**: 63–107 (natural history); Shine, R. (1980). Ecology of eastern Australian whipsnakes of the genus *Demansia. J. Herpet.* **14**: 381–389 (ecology).

Demansia papuensis (Macleay, 1877)

Diemenia papuensis Macleay, W. (1877). The ophidians of the *Chevert* Expedition. *Proc. Linn. Soc. N.S.W.* **2**: 33–41 [40]. Type data: holotype, AM R31919, from Papua New Guinea (probably Hall Sound).

Demansia papuensis melaena Storr, G.M. (1978). Whip snakes (*Demansia* : Elapidae) of Western Australia. *Rec. West. Aust. Mus.* **6**: 287–301 [300]. Type data: holotype, WAM R47590, from Katherine, 14°28'S, 132°16'E, N.T.

Synonymy that of Cogger, H.G., this work.

Distribution: N coastal, (N Gulf), NE coastal, W.A., N.T., Qld. Ecology: terrestrial, diurnal, predator; oviparous. Biological references: Storr, G.M. (1978). Whip snakes (*Demansia* : Elapidae) of Western Australia. *Rec. West. Aust. Mus.* **6**: 287–301 (taxonomy).

Demansia psammophis (Schlegel, 1837)

E[laps] psammophis Schlegel, H. (1837). *Essai sur la Physionomie des Serpens.* Partie Générale et Partie Descriptive. 2 Parts La Haye : Kips & Stockum [455]. Type data: syntypes, MNHP 7663A; MNHP 7666, from Australia.

Lycodon reticulatus Gray, J.E. (1842). Description of some hitherto unrecorded species of Australian reptiles and batrachians. *in* Gray, J.E. (ed.) *Zoological Miscellany.* London : Treuttel, Würtz & Co. pp. 51–57 [54]. Type data: lectotype (as "hololectotype"), BMNH 1946.1.19.79, from Australia. designation by Storr, G.M. (1978). Whip snakes (*Demansia* : Elapidae) of Western Australia. *Rec. West. Aust. Mus.* **6**: 287–301.

Pseudelaps psammophidius Duméril, A.M.C., Bibron, G. & Duméril, A. (1854). *Erpétologie Générale* ou Histoire Naturelle Complète des Reptiles. Paris : Roret 7(2): xii pp. 781–1536 [1234] [*nom. nov. pro Elaps psammophis* Schlegel, 1837].

Demansia reticulata cupreiceps Storr, G.M. (1978). Whip snakes (*Demansia* : Elapidae) of Western Australia. *Rec. West. Aust. Mus.* **6**: 287–301 [295]. Type data: holotype, WAM R34555, from Callagiddy, 25°02'S, 114°01'E, W.A.

Synonymy that of Cogger, H.G., this work.

Distribution: N coastal, NW coastal, SW coastal, W plateau, S Gulfs, Murray-Darling basin, Bulloo River basin, Lake Eyre basin, SE coastal, NE coastal, N Gulf, W.A., N.T., S.A., Vic., N.S.W., Qld., A.C.T. Ecology: terrestrial, diurnal, predator, closed forest, tall forest, open forest, woodland, low woodland, low open woodland, tall shrubland, open scrub, open heath, low shrubland; oviparous, lacertiphagous. Biological references: Covacevich, J. & Limpus, C. (1972). Observations on community egg-laying by the yellow-faced whip snake, *Demansia psammophis* (Schlegel, 1837) (Squamata : Elapidae). *Herpetologica* **28**: 208–210; Storr, G.M. (1978). Whip snakes (*Demansia* : Elapidae) of Western Australia. *Rec. West. Aust. Mus.* **6**: 287–301 (taxonomy); Shine, R. (1980). Ecology of eastern Australian whipsnakes of the genus *Demansia. J. Herpet.* **14**: 381–389 (ecology); Kellaway, C.H. (1934). The venoms of some of the small and rare Australian venomous snakes. *Med. J. Aust.* **1934**: 74–78.

Demansia simplex Storr, 1978

Demansia simplex Storr, G.M. (1978). Whip snakes (*Demansia* : Elapidae) of Western Australia. *Rec. West. Aust. Mus.* **6**: 287–301 [290]. Type data: holotype, WAM R13841, from Kalumburu, 14°17'S, 126°40'E, W.A.

Distribution: N coastal, W.A. Ecology: terrestrial, diurnal, predator; oviparous.

Demansia torquata (Günther, 1862)

Diemansia torquata Günther, A. (1862). On new species of snakes in the collection of the British Museum. *Ann. Mag. Nat. Hist. (3)* **9**: 52–59, 124–132 [130 pl 9 fig 10]. Type data: holotype, BMNH 1946.1.17.41, from Percy Ils., Qld.

Distribution: N Gulf, NE coastal, Lake Eyre basin, Murray-Darling basin, Bulloo River basin, N.T., Qld., N.S.W., (S.A.). Ecology: terrestrial, diurnal, predator, closed forest, tall forest, open forest, low woodland, low open woodland, open scrub, tall shrubland, desert; oviparous, lacertiphagous. Biological references: Shine, R. (1980). Ecology of eastern Australian whipsnakes of the genus *Demansia. J. Herpet.* **14**: 381–389 (ecology).

Denisonia Krefft, 1869

Denisonia Krefft, G. (1869). *The Snakes of Australia*; an Illustrated and Descriptive Catalogue of All the Known Species. Sydney : Govt. Printer xxv 100 pp. [82]. Type species *Denisonia ornata* Krefft, 1869 by monotypy.

Denisonia devisi Waite and Longman, 1920

Hoplocephalus ornatus De Vis, C.W. (1884). On a new species of *Hoplocephalus. Proc. R. Soc. Qd.* **1**: 100 [100 pl 15] [*non Denisonia ornata* Krefft, 1869]. Type data: holotype, QM J199, from near Surat, Qld.

Denisonia maculata devisi Waite, E.R. & Longman, H.A. (1920). Description of little-known Australian snakes. *Rec. S. Aust. Mus.* **1**: 173–180 [178] [*nom. nov. pro Hoplocephalus ornatus* De Vis, 1884].

Synonymy that of Cogger, H.G., this work.

Distribution: (NE coastal), Murray-Darling basin, Bulloo River basin, Lake Eyre basin, Qld., N.S.W. Ecology: terrestrial, nocturnal, predator, woodland, low woodland, low open woodland; viviparous, primarily anuriphagous. Biological references: Mackay, R.D. (1956). The status of *Denisonia maculata* Steindachner and related species. (Ophidia : Elapidae). *Proc. R. Zool. Soc. N.S.W.* **1954–55**: 29–33 (taxonomy).

Denisonia fasciata Rosén, 1905

Denisonia fasciata Rosén, N. (1905). List of snakes in the Zoological Museums of Lund and Malmö, with descriptions of a new species and a new genus. *Ann. Mag. Nat. Hist. (7)* **15**: 168–181 [179]. Type data: syntypes, ZMLU Cat.Typ.No.129, from "West Australia".

Distribution: NW coastal, W plateau, (SW coastal), W.A. Ecology: terrestrial, nocturnal, predator, low woodland, low shrubland, tall shrubland, desert; viviparous, lacertiphagous. Biological references:

Smith, L.A. (1980). Taxonomy of *Denisonia punctata* and *Denisonia fasciata* (Serpentes : Elapidae). *Rec. West. Aust. Mus.* **8**: 327–333 (taxonomy).

Denisonia maculata (Steindachner, 1867)

Hoplocephalus maculatus Steindachner, F. (1867). Reptilien. *in, Reise der österreichischen Fregatte* Novara *um die Erde in den Jahren 1857, 1858, 1859.* Zoologie 1(3): 1–98 Vienna : State Printer [81 pl 3 figs 3–5] [1869 on title page]. Type data: syntypes, NHMW unnumbered (2 specimens), from N.S.W., subsequent restriction to Rockhampton, Qld. proposed by Mackay, R.D. (1956). The status of *Denisonia maculata* Steindachner and related species. (Ophidia : Elapidae). *Proc. R. Zool. Soc. N.S.W.* **1954–55**: 29–33.

Denisonia ornata Krefft, G. (1869). *The Snakes of Australia*; an Illustrated and Descriptive Catalogue of All the Known Species. Sydney : Govt. Printer xxv 100 pp. [82 pl 11 fig 4]. Type data: holotype, AM 6697, from near Rockhampton, Qld.

Hoplocephalus muelleri Fischer, J.G. (1885). Herpetologische Bemerkungen. *Jb. Hamb. Wiss. Anst.* **2**: 82–121 [109]. Type data: holotype, SMNS 2377 (=10821), from Qld.

Synonymy that of Cogger, H.G., this work.

Distribution: NE coastal, Qld. Ecology: terrestrial, nocturnal, predator, open forest, woodland; viviparous, anuriphagous. Biological references: Mackay, R.D. (1956). The status of *Denisonia maculata* Steindachner and related species. (Ophidia : Elapidae). *Proc. R. Zool. Soc. N.S.W.* **1954–55**: 29–33 (taxonomy).

Denisonia punctata Boulenger, 1896

Denisonia punctata Boulenger, G.A. (1896). *Catalogue of Snakes in the British Museum (Natural History).* 3. London : British Museum xiv 727 pp. pls 25 [341 pl 18 fig 4]. Type data: holotype, BMNH 1946.1.20.76, from Port Walcott, W.A.

Distribution: NW coastal, W plateau, N coastal, N Gulf, Lake Eyre basin, W.A., N.T., Qld. Ecology: terrestrial, nocturnal, predator, open forest, woodland, low shrubland, tall shrubland, desert, tussock grassland; viviparous, lacertiphagous. Biological references: Smith, L.A. (1980). Taxonomy of *Denisonia punctata* and *Denisonia fasciata* (Serpentes : Elapidae). *Rec. West. Aust. Mus.* **8**: 327–333 (taxonomy).

Drysdalia Worrell, 1961

Drysdalia Worrell, E. (1961). Herpetological name changes. *West. Aust. Nat.* **8**: 18–27 [25]. Type species *Hoplocephalus coronoides* Günther, 1858 by original designation.

Drysdalia coronata (Schlegel, 1837)

Elaps coronatus Schlegel, H. (1837). *Essai sur la Physionomie des Serpens.* Partie Générale et Partie Descriptive. 2 Parts La Haye : Kips & Stockum [454]. Type data: lectotype, MNHP 321, from King George's Sound, W.A. (original description cites Australia) designation by Coventry, A.J. & Rawlinson, P.A. (1980). Taxonomic revision of the elapid snake genus *Drysdalia* Worrell, 1961. *Mem. Natl. Mus. Vict.* **41**: 65–78.

Trimesurus olivaceus Gray, J.E. (1841). A catalogue of the species of reptiles and Amphibia hitherto described as inhabiting Australia, with a description of some new species from Western Australia. Appendix E *in* Grey, G. *Journals of Two Expeditions of Discovery in North-west and Western Australia, during the years 1837, 38, and 39, under the authority of Her Majesty's Government.* London : T. & W. Boone 2: 422–449 [443]. Type data: presumed lost, from Australia.

Elaps melanocephalus Gray, J.E. & Neill, J. (1845). Catalogue of reptiles and fish, found at King George's Sound. Appendix *in* Eyre, E.J. *Journals of Expeditions of Discovery into Central Australia and Overland from Adelaide to King George's Sound in the Years 1840–1.* London : T. & W. Boone 1: 412–418 [416]. Type data: figure prepared by Neill and deposited BMNH, from King George's Sound, W.A.

Synonymy that of Cogger, H.G., this work.

Distribution: SW coastal, W plateau, (NW coastal), W.A. Ecology: terrestrial, nocturnal, predator, open forest, low open woodland, open scrub; viviparous, lacertiphagous, ophiophagous. Biological references: Shine, R. (1981). Venomous snakes in cold climates: ecology of the Australian genus *Drysdalia* (Serpentes : Elapidae). *Copeia* **1981**: 14–25; Coventry, A.J. & Rawlinson, P.A. (1980). Taxonomic revision of the elapid snake genus *Drysdalia* Worrell, 1961. *Mem. Natl. Mus. Vict.* **41**: 65–78 (taxonomy); Fleay, D. (1952). The crowned or coronated snake. *Vict. Nat.* **68**: 146–150 (as *Denisonia coronata*).

Drysdalia coronoides (Günther, 1858)

Hoplocephalus coronoides Günther, A. (1858). *Catalogue of Colubrine Snakes in the Collection of the British Museum.* London : British Museum xvi 281 pp. [215]. Type data: lectotype, BMNH 1946.1.17.39, from Tas. (original description cites Tas. and Australia), designation by Coventry, A.J. & Rawlinson, P.A. (1980). Taxonomic revision of the elapid snake genus *Drysdalia* Worrell, 1961. *Mem. Natl. Mus. Vict.* **41**: 65–78.

Alecto labialis Jan, G. & Sordelli, F. (1873). *Iconographie Générale des Ophidiens.* Atlas 1860–1881 Milan : Jan & Sordelli 50 livr. [pl 1 fig 1]. Type data: holotype, MNHP (status unknown), from Australia.

Denisonia nigra De Vis, C.W. (1905). A new genus of lizards. *Ann. Qd. Mus.* **6**: 46–52 pl 15 [50]. Type data: holotype, QM J196, from Tas.

Synonymy that of Cogger, H.G., this work.

Distribution: SE coastal, Murray-Darling basin, S.A., Vic., A.C.T., N.S.W., Tas. Ecology: terrestrial, nocturnal, predator, alpine, tall forest, open forest, woodland; viviparous, lacertiphagous.

Biological references: Coventry, A.J. & Rawlinson, P.A. (1980). Taxonomic revision of the elapid snake genus *Drysdalia* Worrell, 1961. *Mem. Natl. Mus. Vict.* **41**: 65–78 (taxonomy); Shine, R. (1981). Venomous snakes in cold climates: ecology of the Australian genus *Drysdalia* (Serpentes : Elapidae). *Copeia* **1981**: 14–25 (ecology); Kellaway, C.H. (1934). The venoms of some of the small and rare Australian venomous snakes. *Med. J. Aust.* **1934**: 74–78.

Drysdalia mastersii (Krefft, 1866)

Hoplocephalus mastersii Krefft, G. (1866). Descriptions of three species of snakes of the genus *Hoplocephalus. Proc. Zool. Soc. Lond.* **1866**: 370–371 [370]. Type data: syntypes (7), from Flinders Ranges (as Flinder's Range), S.A., BMNH 1946.1.17.42 is the only syntype located.

Distribution: W plateau, S Gulfs, Murray-Darling basin, W.A., S.A., Vic., (N.S.W.). Ecology: terrestrial, nocturnal, predator, open scrub; viviparous, lacertiphagous. Biological references: Coventry, A.J. & Rawlinson, P.A. (1980). Taxonomic revision of the elapid snake genus *Drysdalia* Worrell, 1961. *Mem. Natl. Mus. Vict.* **41**: 65–78 (taxonomy); Shine, R. (1981). Venomous snakes in cold climates: ecology of the Australian genus *Drysdalia* (Serpentes : Elapidae). *Copeia* **1981**: 14–25 (ecology).

Drysdalia rhodogaster (Jan and Sordelli, 1873)

Alecto rhodogaster Jan, G. (1863). *Elenco Sistematico Degli Ofidi Descritti e Disegnati per l'Iconografia Generale.* Milan : A. Lombardi 143 pp. [117] [?*nom. nud.*]. Type data: holotype, from Australia, ZMH 477 (=R00747) *vide* Coventry, A.J. & Rawlinson, P.A. (1980). Taxonomic revision of the elapid snake genus *Drysdalia* Worrell, 1961. *Mem. Natl. Mus. Vict.* **41**: 65–78.

Alecto rhodogaster Jan, G. & Sordelli, F. (1873). *Iconographie Générale des Ophidiens.* Atlas 1860–1881 Milan : Jan & Sordelli 50 livr. [pl 2 fig 2]. Type data: holotype, from Australia, ZMH 477 (=R00747) *vide* Coventry, A.J. & Rawlinson, P.A. (1980). Taxonomic revision of the elapid snake genus *Drysdalia* Worrell, 1961. *Mem. Natl. Mus. Vict.* **41**: 65–78.

Hoplocephalus collaris Macleay, W. (1887). Description of a new species of *Hoplocephalus. Proc. Linn. Soc. N.S.W. (2)* **1**: 1111–1112 [1111]. Type data: holotype, MMUS (presumed lost), from near Bega, N.S.W.

Pseudelaps minutus Fry, D.B. (1915). Herpetological notes. *Proc. R. Soc. Qd.* **27**: 60–95 [93 fig 7]. Type data: holotype, AM R3971, from Colo Vale, N.S.W.

Synonymy that of Cogger, H.G., this work.

Distribution: SE coastal, N.S.W. Ecology: terrestrial, nocturnal, predator, open forest, tall forest, woodland; viviparous, lacertiphagous. Biological references: Coventry, A.J. & Rawlinson, P.A. (1980). Taxonomic revision of the elapid snake genus *Drysdalia* Worrell, 1961. *Mem. Natl. Mus. Vict.* **41**: 65–78 (taxonomy); Shine, R. (1981). Venomous snakes in cold climates: ecology of the Australian genus *Drysdalia* (Serpentes : Elapidae). *Copeia* **1981**: 14–25 (ecology).

Echiopsis Fitzinger, 1843

Echiopsis Fitzinger, L.J. (1843). *Systema Reptilium.* Vienna : Braümüller u. Seidel vi 106 pp. [28]. Type species *Naja curta* Schlegel, 1837 by monotypy.

Brachyaspis Boulenger, G.A. (1896). *Catalogue of Snakes in the British Museum (Natural History).* 3. London : British Museum xiv 727 pp. pls 25 [353] [*non Brachyaspis* Salter, 1866]. Type species *Naja curta* Schlegel, 1837 by original designation.

Synonymy that of Cogger, H.G., this work.

Echiopsis atriceps (Storr, 1980)

Brachyaspis atriceps Storr, G.M. (1980). A new *Brachyaspis* (Serpentes : Elapidae) from Western Australia. *Rec. West. Aust. Mus.* **8**: 397–399 [397 fig 1]. Type data: holotype, WAM R67330, from Lake Cronin, 32°23'S, 119°45'E, W.A.

Distribution: SW coastal, (W plateau), W.A. Ecology: terrestrial, nocturnal, predator, woodland.

Echiopsis curta (Schlegel, 1837)

Naja curta Schlegel, H. (1837). *Essai sur la Physionomie des Serpens.* Partie Générale et Partie Descriptive. 2 Parts La Haye : Kips & Stockum [486]. Type data: holotype, MNHP 953, from vicinity of King George's Sound, W.A.

Hoplocephalus temporalis Günther, A. (1862). On new species of snakes in the collection of the British Museum. *Ann. Mag. Nat. Hist. (3)* **9**: 52–59, 124–132 [130 pl 9 fig 11]. Type data: syntypes, from S.A., BMNH 1946.1.17.87–89; 2 syntypes cited in original description not found.

Synonymy that of Cogger, H.G., this work.

Distribution: SW coastal, W plateau, (S Gulfs), Murray-Darling basin, W.A., S.A., Vic., N.S.W. Ecology: terrestrial, nocturnal, predator, open scrub, low open woodland, low shrubland, hummock grassland; viviparous, general carnivore. Biological references: Storr, G.M. (1980). A new *Brachyaspis* (Serpentes : Elapidae) from Western Australia. *Rec. West. Aust. Mus.* **8**: 397–399 (taxonomy).

Elapognathus Boulenger, 1896

Elapognathus Boulenger, G.A. (1896). *Catalogue of Snakes in the British Museum (Natural History).* 3. London : British Museum xiv 727 pp. pls 25 [356]. Type species *Hoplocephalus minor* Günther, 1863 by monotypy.

Elapognathus minor (Günther, 1863)

*[**Hoplocephalus superbus*** **(part.)***]* Günther, A. (1858). *Catalogue of Colubrine Snakes in the Collection of the British Museum.* London : British Museum xvi 281 pp.

[217]. Type data: syntypes, from "S. Australia" and Swan River, W.A. (original description cites Australia and Tas. as type localities), BMNH 1946.1.19.14–15 ("S.Australia"); BMNH 1946.1.18.64 F (Swan River); BMNH 1946.1.20.71 M (Swan River), ?ZMB 4928 (not found) [for other syntypes see *Austrelaps superbus* (Günther, 1858)].

Hoplocephalus minor Günther, A. (1863). Third account of new species of snakes in the collection of the British Museum. *Ann. Mag. Nat. Hist. (3)* **12**: 348–365 [362]. Type data: syntypes, BMNH 1946.1.18.64 F (Swan River); BMNH 1946.1.19.14–15 ("S.Australia"); BMNH 1946.1.20.71 M (Swan River); BMNH 1964.962 (the skull from MCZ 76635), ZMB 4928 (not found), MCZ 76635("S. Australia", skull removed), from "S. Australia" and Swan River, W.A. (both as "S. W. Australia").

Synonymy that of Cogger, H.G., this work.

Distribution: SW coastal, W.A. Ecology: terrestrial, nocturnal, predator. Biological references: Glauert, L. (1960). *A Handbook of Snakes of Western Australia.* Perth : Western Australian Naturalists' Club 2nd edn. 62 pp. (description).

Furina Duméril, 1853

Brachysoma Fitzinger, L.J. (1843). *Systema Reptilium.* Vienna : Braümüller u. Seidel vi 106 pp. [25] [*non Brachysoma* Brandt, 1835]. Type species *Calamaria diadema* Schlegel, 1837 by monotypy.

Furina Duméril, A.M.C. (1853). Prodrome général de la classification des serpents. *Mém. Acad. Sci. Inst. Fr.* **23**: 399–536 [517]. Type species *Calamaria diadema* Schlegel, 1837 because it is the only clearly available included species and hence type species by monotypy.

[Glyphodon* (part.)*] Günther, A. (1858). *Catalogue of Colubrine Snakes in the Collection of the British Museum.* London : British Museum xvi 281 pp. [210]. Type species *Glyphodon tristis* Günther, 1858 by subsequent designation, see Boulenger, G.A. (1896). *Catalogue of Snakes in the British Museum (Natural History).* 3. London : British Museum xiv 727 pp. pls 25.; originally included species *Elaps ornatus* Gray, 1842 and *Glyphodon tristis* Günther, 1858.

Lunelaps Worrell, E. (1961). Herpetological name changes. *West. Aust. Nat.* **8**: 18–27 [22]. Type species *Pseudelaps christieanus* Fry, 1915 by monotypy.

Synonymy that of Cogger, H.G., this work.

Furina diadema (Schlegel, 1837)

C[alamaria] diadema Schlegel, H. (1837). *Essai sur la Physionomie des Serpens.* Partie Générale et Partie Descriptive. 2 Parts La Haye : Kips & Stockum [32]. Type data: syntypes, MNHP 3941, MNHP 7668, from Australia, but restricted to Port Jackson, N.S.W. by Duméril, A.M.C., Bibron, G. & Duméril, A. (1854). *Erpétologie Générale* ou Histoire Naturelle Complète des Reptiles. Paris : Roret 7(2): xii pp. 781–1536.

Rabdion occipitale Girard, C. (1858). Descriptions of some new reptiles, collected by the U.S. Exploring Expedition, under the command of Capt. C. Wilkes, U.S.N. Third Part. Including the species of ophidians, exotic to North America. *Proc. Acad. Nat. Sci. Philad.* **1857**: 181–182 [181]. Type data: not located, from Australia.

Cacophis blackmanii Krefft, G. (1869). *The Snakes of Australia*; an Illustrated and Descriptive Catalogue of All the Known Species. Sydney : Govt. Printer xxv 100 pp. [77 pl 12 fig 9]. Type data: holotype, AM 6674, from Pine Mt., near Ipswich, Qld.

Denisonia bancrofti De Vis, C.W. (1911). Description of snakes apparently new. *Ann. Qd. Mus.* **10**: 22–25 [23]. Type data: ?syntypes, from Stannary Hills, Qld., QM J195, QM J12881 *vide* Covacevich, J. (1971). Amphibian and reptile type specimens in the Queensland Museum. *Mem. Qd. Mus.* **16**: 49–68 [62].

Synonymy that of Cogger, H.G., this work.

Distribution: SE coastal, NE coastal, Murray-Darling basin, S Gulfs, Lake Eyre basin, Qld., N.S.W., A.C.T., S.A. Ecology: terrestrial, nocturnal, predator, tall forest, open forest, woodland, open scrub, tall shrubland, tall open shrubland; oviparous, lacertiphagous, cryptozoic. Biological references: Shine, R. (1981). Ecology of Australian elapid snakes of the genera *Furina* and *Glyphodon. J. Herpet.* **15**: 219–224; Lorking, W. (1954). The red-naped snake. *Reptilia (Aust. Rept. Club)* **1**: 10 (natural history).

Furina ornata (Gray, 1842)

Elaps ornatus Gray, J.E. (1842). Description of some hitherto unrecorded species of Australian reptiles and batrachians. *in* Gray, J.E. (ed.) *Zoological Miscellany.* London : Treuttel, Würtz & Co. pp. 51–57 [55]. Type data: holotype, BMNH 1946.1.21.76, from "Australia, (Swan River?)".

Brachysoma simile Macleay, W. (1878). Notes on a collection of snakes from Port Darwin. *Proc. Linn. Soc. N.S.W.* **2**: 219–222 [221]. Type data: syntypes, AM R31928–31, from Port Darwin, N.T.

Pseudelaps christieanus Fry, D.B. (1915). Herpetological notes. *Proc. R. Soc. Qd.* **27**: 60–95 [91 fig 6]. Type data: holotype, AM R4989, from Port Darwin, N.T.

Synonymy that of Cogger, H.G., this work.

Distribution: NW coastal, W plateau, N coastal, N Gulf, Lake Eyre basin, NE coastal, W.A., S.A., N.T., Qld. Ecology: terrestrial, nocturnal, predator, closed forest, tall forest, woodland, low open woodland, low shrubland, desert; oviparous, lacertiphagous, cryptozoic. Biological references: Storr, G.M. (1981). The genus *Furina* (Serpentes : Elapidae) in Western Australia. *Rec. West. Aust. Mus.* **9**: 119–123 (taxonomy); Cogger, H.G. & Lindner, D.A. (1974). Frogs and reptiles in fauna survey of the Port Essington District, Cobourg Peninsula, Northern Territory of Australia. Frith, H.J. & Calaby, J.H. (eds.) *CSIRO Div. Wildl. Res. Tech. Pap.* **28**: 63–107 (taxonomy as *F. diadema*);

Worrell, E. (1945). The orange-naped whipsnake. *Proc. R. Zool. Soc. N.S.W.* **1944–45**: 32–33 (natural history as *F. christieanus*).

Glyphodon Günther, 1858

***Glyphodon* (part.)** Günther, A. (1858). *Catalogue of Colubrine Snakes in the Collection of the British Museum.* London : British Museum xvi 281 pp. [210]. Type species *Glyphodon tristis* Günther, 1858 by subsequent designation, see Boulenger, G.A. (1896). *Catalogue of Snakes in the British Museum (Natural History).* 3. London : British Museum xiv 727 pp. pls 25.

Mainophis Macleay, W. (1877). The ophidians of the *Chevert* Expedition. *Proc. Linn. Soc. N.S.W.* **2**: 33–41 [36] [1878 on title page]. Type species *Mainophis robusta* Macleay, 1877 by monotypy.

Synonymy that of Cogger, H.G., this work.

This group is also found in New Guinea, including the Aru Ils.

Glyphodon barnardi Kinghorn, 1939

Glyphodon barnardi Kinghorn, J.R. (1939). Two Queensland snakes. *Rec. Aust. Mus.* **20**: 257–260 [258 figs 1–2]. Type data: holotype, CAS 77798, from 15 mi S Duaringa, Qld.

Distribution: NE coastal, N Gulf, (Lake Eyre basin), Qld. Ecology: terrestrial, nocturnal, predator, tall forest, open forest, woodland; oviparous, cryptozoic. Biological references: Shine, R. (1981). Ecology of Australian elapid snakes of the genera *Furina* and *Glyphodon. J. Herpet.* **15**: 219–224.

Glyphodon dunmalli Worrell, 1955

Glyphodon dunmalli Worrell, E. (1955). A new elapine snake from Queensland. *Proc. R. Zool. Soc. N.S.W.* **1953/54**: 41–43 figs 5 [41]. Type data: holotype, AM R14809, from Glenmorgan, Qld.

Distribution: NE coastal, (Murray-Darling basin), Qld. Ecology: terrestrial, nocturnal, predator, woodland; oviparous, cryptozoic. Biological references: Shine, R. (1981). Ecology of Australian elapid snakes of the genera *Furina* and *Glyphodon. J. Herpet.* **15**: 219–224.

Glyphodon tristis Günther, 1858

Glyphodon tristis Günther, A. (1858). *Catalogue of Colubrine Snakes in the Collection of the British Museum.* London : British Museum xvi 281 pp. [211]. Type data: holotype, BMNH 1946.1.18.99, from NE coast of Australia.

Mainophis robusta Macleay, W. (1877). The ophidians of the *Chevert* Expedition. *Proc. Linn. Soc. N.S.W.* **2**: 33–41 [36] [1878 on title page]. Type data: syntypes, AM R31896–7, from Mawatta, Binaturi River (as Katow), Papua New Guinea.

Denisonia fenestrata De Vis, C.W. (1905). A new genus of lizards. *Ann. Qd. Mus.* **6**: 46–52 pl 15 [50]. Type data: syntypes, QM J200, from Qld.; other syntype presumed lost.

Synonymy that of Cogger, H.G., this work.

Distribution: NE coastal, N Gulf, Qld. Ecology: terrestrial, nocturnal, predator, closed forest, tall forest, open forest; oviparous, cryptozoic. Biological references: Shine, R. (1981). Ecology of Australian elapid snakes of the genera *Furina* and *Glyphodon. J. Herpet.* **15**: 219–224 (ecology).

Hemiaspis Fitzinger, 1860

Hemiaspis Fitzinger, L. (1860). Ausbeute der österreichischen Naturforscher an Säugethiere und Reptilien während der Weltumsegelung Sr. Majestat Fregatte *Novara. Sber. Akad. Wiss. Wien* **42**: 384–416 [410]. Type species *Alecto signata* Jan, 1859 by monotypy.

Drepanodontis Worrell, E. (1961). A new generic name for a nominal species of *Denisonia. Proc. R. Zool. Soc. N.S.W.* **1958–59**: 54–55 [54]. Type species *Hoplocephalus damelii* Günther, 1876 by original designation.

Synonymy that of Cogger, H.G., this work.

Hemiaspis damelii (Günther, 1876)

Hoplocephalus damelii Günther, A. (1876). Descriptions of new species of reptiles from Australia. *J. Mus. Godeffroy* **5**: 45–47 [46]. Type data: syntypes, from Rockhampton and Peak Downs, Qld., BMNH 1946.1.18.89–90 (Peak Downs), ZMH 824 (Qld.); other syntypes not located.

Hoplocephalus suboccipitalis Ogilby, J.D. (1892). On some undescribed reptiles and fishes from Australia. *Rec. Aust. Mus.* **2**: 23–26 [23]. Type data: holotype, AM R1127, from Moree, N.S.W.

Denisonia daemelii Boulenger, G.A. (1896). *Catalogue of Snakes in the British Museum (Natural History).* 3. London : British Museum xiv 727 pp. pls 25 [339] [emend. *pro Hoplocephalus damelii* Günther, 1876].

Synonymy that of Cogger, H.G., this work.

Distribution: NE coastal, Murray-Darling basin, Qld., N.S.W. Ecology: terrestrial, nocturnal, predator, open forest, woodland; viviparous, anuriphagous, lacertiphagous.

Hemiaspis signata (Jan, 1859)

Alecto signata Jan, G. (1859). Plan d'une *Iconographie descriptive des Ophidiens*, et description sommaire de nouvelles espèces de Serpents. *Rev. Mag. Zool. (2)* **11–12**: 122–130; Jan, G. (1859). Prodrome d'une *Iconographie descriptive des Ophidiens* et Description sommaire de nouvelles espèces de Serpents venimeux. Reprinted from: *Rev. Mag. Zool. (2)* **10** and *(2)* **11–12**, 1858 and 1859. Paris : Bouchard-Huzard 32 pp. [128; pl C]. Type data: holotype, presumed lost, from Australia.

Denisonia vagrans Garman, S. (1901). Some reptiles and batrachians from Australasia. *Bull. Mus. Comp. Zool.* **39**:

1-14 [13] [as variety of *Denisonia signata*]. Type data: holotype, MCZ 6490, from Dunk Is., Qld.

Synonymy that of Cogger, H.G., this work.

Distribution: SE coastal, NE coastal, N.S.W., Qld. Ecology: terrestrial, noctidiurnal, predator, closed forest, tall forest, open forest, swamp; viviparous, anuriphagous, lacertiphagous. Biological references: Shine, R. (1979). Activity patterns in Australian elapid snakes (Squamata : Serpentes : Elapidae). *Herpetologica* **35**: 1-11 (ecology); Rankin, P.R. (1972). Notes on the swamp snake (*Drepanodontis signata*) in captivity. *Herpetofauna* **5**: 15-17 (natural history); Ormsby, A.I. (1950). The marsh snake. *Proc. R. Zool. Soc. N.S.W.* **1948**: 29-31 (natural history).

Hoplocephalus Wagler, 1830

Hoplocephalus Wagler, J.G. (1830). *Natürliches System der Amphibien, mit vorangehender Classification der Säugethiere und Vögel.* München : Cotta'schen vi 354 pp. [172]. Type species *Naja bungaroides* Schlegel, 1837 by subsequent designation, see Klemmer, K. (1963). Liste der rezenten Giftschlangen; Elapidae, Hydrophiidae, Viperidae und Crotalidae. Marburg : N.G. Elwert Universitäts- U. Verlagsbuchhand pp. 253-464; Klemmer cites *Naja bungaroides* Boie, 1828 as type species, but Mees, G.F. (1957). Over het belang van Temminck's *Discours Préliminaire* voor de zoologische nomenclatuur. *Zool. Meded.* **35**: 205-227 [214-215] has shown that name is a *nom. nud.*; consequently Schlegel's name is no longer a junior homonym and becomes the type species.

Elapocormus Fitzinger, L.J. (1843). *Systema Reptilium.* Vienna : Braümüller u. Seidel vi 106 pp. [28]. Type species *Naja bungaroides* Schlegel, 1837 by original designation.

Synonymy that of Cogger, H.G., this work.

Hoplocephalus bitorquatus (Jan, 1859)

A[lecto] bitorquata Jan, G. (1859). Plan d'une *Iconographie descriptive des Ophidiens*, et description sommaire de nouvelles espèces de Serpents. *Rev. Mag. Zool. (2)* **11-12**: 122-130; Jan, G. (1859). Prodrome d'une *Iconographie descriptive des Ophidiens* et Description sommaire de nouvelles espèces de Serpents venimeux. Reprinted from: *Rev. Mag. Zool. (2)* **10** and *(2)* **11-12**, 1858 and 1859. Paris : Bouchard-Huzard 32 pp. [128; pl B]. Type data: holotype, SMF 20533, from Australia (data with holotype erroneously cites Sydney, N.S.W. as type locality, coll. W. Kirchner, 1848).

Hoplocephalus sulcans De Vis, C.W. (1884). Descriptions of a new snake with a synopsis of the genus *Hoplocephalus. Proc. R. Soc. Qd.* **1**: 138-140 [138]. Type data: syntypes, presumed lost, from Mitchell district, Qld.

Hoplocephalus waitii Ogilby, J.D. (1894). Description of a new Australian snake. *Proc. Linn. Soc. N.S.W.* **9**: 261-262 [261]. Type data: holotype, AM 6590, from unknown locality.

Denisonia angulata De Vis, C.W. (1905). A new genus of lizards. *Ann. Qd. Mus.* **6**: 46-52 pl 15 [51]. Type data: holotype, QM J194, from Qld.

Denisonia revelata De Vis, C.W. (1911). Description of snakes apparently new. *Ann. Qd. Mus.* **10**: 22-25 [22]. Type data: holotype, QM J2957, from Stannary Hills, Qld.

Synonymy that of Cogger, H.G., this work.

Distribution: NE coastal, SE coastal, Murray-Darling basin, N.S.W., Qld. Ecology: terrestrial, nocturnal, predator, open forest, woodland; viviparous, anuriphagous, lacertiphagous, semi-arboreal. Biological references: Covacevich, J. (1968). *The Snakes of Brisbane.* Brisbane : Queensland Museum 27 pp. (natural history).

Hoplocephalus bungaroides (Schlegel, 1837)

Naja bungaroides Schlegel, H. (1837). *Essai sur la Physionomie des Serpens.* Partie Générale et Partie Descriptive. 2 Parts La Haye : Kips & Stockum [477] [*non Naja bungaroides* Boie, 1828]. Type data: syntypes, MNHP 7678 (Port Jackson), MNHP 7679 (Indes?), from Port Jackson, N.S.W. and "Indes?", see Guibé, J. & Roux-Estève, R. (1972). Les types de Schlegel (Ophidiens) présents dans les collections du Muséum National d'Histoire Naturelle de Paris. *Zool. Meded.* **47**: 129-134; however Duméril, A.M.C., Bibron, G. & Duméril, A. (1854). *Erpétologie Générale* ou Histoire Naturelle Complète des Reptiles. Paris: Roret 7(2): xii pp. 781-1536 cite "Nouvelle Hollande" as the locality for MNHP 7679.

Alecto variegata Duméril, A.M.C., Bibron, G. & Duméril, A. (1854). *Erpétologie Générale* ou Histoire Naturelle Complète des Reptiles. Paris : Roret 7(2): xii pp. 781-1536 [1254 pl 76 fig 1]. Type data: holotype, MNHP 7679, from Australia.

Synonymy that of Cogger, H.G., this work.

Distribution: SE coastal, N.S.W. Ecology: terrestrial, nocturnal, predator, open forest, woodland; viviparous, anuriphagous, lacertiphagous. Biological references: Anstis, M. (1973). Notes on *Hoplocephalus bungaroides* (broad headed snake) in captivity. *R. Zool. Soc. N.S.W. Bull. Herpet.* **1**: 5-7; Adams, D. (1973). Broad-headed snake (*Hoplocephalus bungaroides*). *Herpetofauna* **5**: 19-22 (natural history); Wells, R. (1981). Remarks on the prey preferences of *Hoplocephalus bungaroides. Herpetofauna* **12**: 25-28; Jenkins, R.W.G. (1979). The status of endangered Australian reptiles. pp. 169-176 *in, The Status of Endangered Australasian Wildlife.* Adelaide : Royal Zoological Society of South Australia; Ormsby, A.I. (1947). Notes on the broad-headed snake (*Hoplocephalus bungaroides*). *Proc. R. Zool. Soc. N.S.W.* **1946-47**: 19-21 (natural history); Hosmer, W. (1952). The broadheaded snake *Hoplocephalus bungaroides* (Boie) also known as the fierce snake, and night snake. *N. Qd. Nat.* **20**: 14–16; Kellaway, C.H. (1934). The venoms of the broad-headed snake (*Hoplocephalus bungaroides*) and of the yellow-banded snake (*Hoplocephalus stephensi*). *Med. J. Aust.* **1934**: 249–255; Hayes, D. (1973). Observation on distribution of the broad-headed snake *Hoplocephalus bungaroides* (Boie). *Herpetofauna* **6**: 27–28.

Hoplocephalus stephensii Krefft, 1869

Hoplocephalus stephensii Krefft, G. (1869). *The Snakes of Australia*; an Illustrated and Descriptive Catalogue of All the Known Species. Sydney : Govt. Printer xxv 100 pp. [58 pl 6 fig 7]. Type data: holotype (probable), AM R37361, from Port Macquarie, on the Hastings River, N.S.W., *vide* Cogger, H.G. (1979). Type specimens of reptiles, and amphibians in The Australian Museum. *Rec. Aust. Mus.* **32**:, 163–210; ZMB 8547 labelled "typus" but probably in error.

Distribution: SE coastal, NE coastal, Qld., N.S.W. Ecology: terrestrial, (arboreal), nocturnal, predator, closed forest, tall forest, open forest; viviparous, general carnivore.

Neelaps Günther, 1863

Neelaps Günther, A. (1863). On new species of snakes in the collection of the British Museum. *Ann. Mag. Nat. Hist. (3)* **11**: 20–25 [24]. Type species *Furina calonotus* [*sic*] Duméril, Bibron and Duméril, 1854 by monotypy.

Narophis Worrell, E. (1961). Herpetological name changes. *West. Aust. Nat.* **8**: 18–27 [27]. Type species *Furina bimaculata* Duméril, Bibron and Duméril, 1854 by monotypy.

Synonymy that of Cogger, H.G., this work.

Neelaps bimaculatus (Duméril, Bibron and Duméril, 1854)

Furina bimaculata Duméril, A.M.C., Bibron, G. & Duméril, A. (1854). *Erpétologie Générale* ou Histoire Naturelle Complète des Reptiles. Paris : Roret 7(2): xii pp. 781–1536 [1240]. Type data: holotype, MNHP 3942, from Tas. (*errore*); data with holotype cites "Australia" as locality.

Distribution: NW coastal, SW coastal, W plateau, W.A., S.A. Ecology: terrestrial, burrowing, nocturnal, predator, woodland, open forest, open scrub, low shrubland, tall shrubland, desert; oviparous, lacertiphagous, cryptozoic. Biological references: Storr, G.M. (1978). Taxonomic notes on the reptiles of the Shark Bay region, Western Australia. *Rec. West. Aust. Mus.* **6**: 303–318 (taxonomy); Storr, G.M. (1968). The genus *Vermicella* (Serpentes : Elapidae) in Western Australia and the Northern Territory. *J. R. Soc. West. Aust.* **50**: 80–92 (taxonomy).

Neelaps calonotus (Duméril, Bibron and Duméril, 1854)

Furina calonotos Duméril, A.M.C. (1853). Prodrome général de la classification des serpents. *Mém. Acad. Sci. Inst. Fr.* **23**: 399–536 [517] [?*nom. nud.*]. Type data: syntypes, MNHP 3943 (2 specimens), from Australia.

Furina calonotos Duméril, A.M.C., Bibron, G. & Duméril, A. (1854). *Erpétologie Générale* ou Histoire Naturelle Complète des Reptiles. Paris : Roret 7(2): xii pp. 781–1536 [1241 pl 75 fig b]. Type data: syntypes, MNHP 3943 (2 specimens), from Tas. (*errore*).

Neelaps calonotus Günther, A. (1863). On new species of snakes in the collection of the British Museum. *Ann. Mag. Nat. Hist. (3)* **11**: 20–25 [24] [valid emend. *pro Furina calonotos* Duméril, Bibron and Duméril, 1854].

Neelaps caledonicus Hoffman, C.K. (1890). Reptilien. III. Schlangen und Entwicklungsgeschichte der Reptilien. *in* Bronn, H.G. *Klassen und Ordnungen Thier-Reichs, wissenshaftlich dargestellt in Wort und Bild.* Leipzig : C.F. Winter'sche Verlagshandlung **6**(3): 1301–2089 [1788]. [*lapsus pro Neelaps calonotus* Duméril, Bibron and Duméril, 1854 (cited as "*Neelaps caledonicus* Günther"); see Trouessart, E. (1898). Sur la non-existence des serpents venimeux terrestres à la Nouvelle-Calédonie. *Bull. Soc. Zool. Fr.* **23**: 186–187].

Neelaps neocaledonicus Palacky, J. (1898). La distribution des Ophidiens sur le globe. *Mém. Soc. Zool. Fr.* **11**: 88–125 [121]. [*lapsus pro Neelaps caledonicus* Trouessart "(douteux d'après Vaillant)" (=*Neelaps caledonicus* Hoffman, 1890)].

Synonymy that of Cogger, H.G., this work.

Distribution: SW coastal, W.A. Ecology: terrestrial, burrowing, nocturnal, predator, open forest, woodland, open heath; oviparous, lacertiphagous, cryptozoic. Biological references: Storr, G.M. (1968). The genus *Vermicella* (Serpentes : Elapidae) in Western Australia and the Northern Territory. *J. R. Soc. West. Aust.* **50**: 80–92 (taxonomy); Glauert, L. (1960). *A Handbook of Snakes of Western Australia.* Perth : Western Australian Naturalists' Club 2nd edn. 62 pp. (description).

Notechis Boulenger, 1896

Oplocephalus Gray, J.E. (1831). A synopsis of the species of the class Reptilia, pp. 1–110 *in* Griffith, E. *The Animal Kingdom Arranged in Conformity with its Organization by the Baron Cuvier.* London : Whittaker, Treacher & Co. 9: 481+110 pp. [Synopsis, 84] [*ex* Cuvier vernacular; *nom. oblit.*, availability in doubt (see under *Notechis scutatus* (Peters, 1861)]. Type species *Oplocephalus cuvieri* Gray, 1830 by monotypy.

Notechis Boulenger, G.A. (1896). *Catalogue of Snakes in the British Museum (Natural History).* 3. London : British Museum xiv 727 pp. pls 25 [351]. Type species *Naja (Hamadryas) scutata* Peters, 1861 by monotypy.

Synonymy that of Cogger, H.G., this work.

Notechis ater (Krefft, 1866)

Hoplocephalus ater Krefft, G. (1866). Descriptions of three species of snakes of the genus *Hoplocephalus*. *Proc. Zool. Soc. Lond.* **1866**: 370–371 [370]. Type data: holotype, AM 6577, from Flinders Range, S.A. (original register entry cites Port Lincoln, S.A. as the type locality).

Notechis scutatus niger Kinghorn, J.R. (1921). Studies in Australian reptiles. 2. *Rec. Aust. Mus.* **13**: 143–154 [145 pl 26 figs 6–8]. Type data: holotype, AM R7124, from Deep Creek, 20 mi from Kingscote, Kangaroo Is., S.A.

Notechis ater humphreysi Worrell, E. (1963). Two new subspecies of the elapine genus *Notechis* from Bass Strait. *Aust. Rept. Park Rec.* **2**: 2–11 [5 fig 2a–c]. Type data: holotype, AM R17955, from New Year Is. near King Is., Bass Strait, Tas.

Notechis ater serventyi Worrell, E. (1963). Two new subspecies of the elapine genus *Notechis* from Bass Strait. *Aust. Rept. Park Rec.* **2**: 2–11 [3 figs 1a–c]. Type data: holotype, AM R17953, from Chappell Is., Furneaux Group, Bass Strait, Tas.

Synonymy that of Cogger, H.G., this work.

Distribution: SW coastal, W plateau, S Gulfs, W.A., S.A., Tas. Ecology: terrestrial, diurnal, predator, open forest, woodland, open heath, tussock grassland, swamp; occasionally nocturnal, viviparous, general carnivore. Biological references: Rawlinson, P.A. (1974). Biogeography and ecology of the reptiles of Tasmania and the Bass Strait area. Chapt. 11 pp. 291–338 *in* Williams, W.D. (ed.) *Biogeography and Ecology in Tasmania.* The Hague : Junk (biogeography); Baehr, M. (1976). Beiträge zur Verbreitung und Ökologie tasmanischer Reptilien. *Stuttgart Beitr. Naturk. (A)* **292**: 1–24 (natural history).

Notechis scutatus (Peters, 1861)

Oploceph[alus] cuvieri Gray, J.E. (1831). A synopsis of the species of the class Reptilia. pp. 1–110 *in* Griffith, E. *The Animal Kingdom Arranged in Conformity with its Organization by the Baron Cuvier.* London : Whittaker, Treacher & Co. 9: 481+110 pp. [synopsis, 84] [*nom. oblit.*; Sherborn, C.D. (1924). *Index Animalium.* London : British Museum 2: 945–1771 regards *Oplocephalus cuvieri* as a *nom. nud.* apparently without justification]. Type data: syntypes (probable), MNHP, presumed lost, from unspecified locality.

Naja (Hamadryas) scutata Peters, W. (1861). Eine zweite übersicht (Vergl. Monatsbericht 1859 p. 269) der von Hrn. F. Jagor auf Malacca, Java, Borneo und den Philippinen gesammelten und dem Kgl. zoologischen Museum übersandten Schlangen. *Mber. K. Preuss. Akad. Wiss. Berl.* **1861**: 683–691 [690] [1862 on title page]. Type data: holotype, ZMB 2815 (presumed lost), from Java (*errore*).

A[lecto] fasciolata Jan, G. (1863). *Elenco Sistematico Degli Ofidi Descritti e Disegnati per l'Iconografia Generale.* Milan : A. Lombardi 143 pp. [117] [?*nom. nud.*]. Type data: syntypes (probable), MSNM, not located, from Australia.

Hoplocephalus fuscus Steindachner, F. (1867). Reptilien. *in Reise der österreichischen Fregatte* Novara *um die Erde in den Jahren 1857, 1858, 1859.* Zoologie 1(3): 1–98 Vienna : State Printer [82 pl 3 figs 11–12]. Type data: syntypes, not found, from Australia.

Alecto fasciolata Jan, G. & Sordelli, F. (1873). *Iconographie Générale des Ophidiens.* Atlas 1860–1881 Milan : Jan & Sordelli 50 livr. [pl 6 fig 4]. Type data: syntypes (probable), MSNM specimen figured, not found, from Australia.

Notechis scutatus occidentalis Glauert, L. (1948). A western tiger snake, *Notechis scutatus occidentalis*, subsp. nov. *West. Aust. Nat.* **1**: 139–141 [139]. Type data: holotype, WAM R9430, from unspecified locality, but collector from Bassendean, W.A.

Synonymy that of Cogger, H.G., this work.

Distribution: SE coastal, Murray-Darling basin, NE coastal, S.A., Vic., N.S.W., A.C.T., Qld. Ecology: terrestrial, diurnal, predator, closed forest, tall forest, open forest, woodland, open heath, tussock grassland, swamp; occasionally nocturnal, viviparous, general carnivore. Biological references: Shine, R. (1979). Activity patterns in Australian elapid snakes (Squamata : Serpentes : Elapidae). *Herpetologica* **35**: 1–11 (behaviour); Lillywhite, H.B. (1980). Behavioral thermoregulation in Australian elapid snakes. *Copeia* **1980**: 452–458; Webb, G.A. (1981). A note on climbing ability in tiger snakes (*Notechis scutatus*) and predation on arboreal nesting birds. *Vict. Nat.* **98**: 159–160 (climbing and food); Shine, R. (1977). Reproduction in Australian elapid snakes. I. Testicular cycles and mating seasons. *Aust. J. Zool.* **25**: 647–653 (reproduction); Shine, R. (1977). Reproduction in Australian elapid snakes. II. Female reproductive cycles. *Aust. J. Zool.* **25**: 655–666 (reproduction); Shine, R. (1977). Habitats, diets, and sympatry in snakes: a study from Australia. *Can. J. Zool.* **55**: 1118–1128 (ecology); Shine, R. (1978). Growth rates and sexual maturation in six species of Australian elapid snakes. *Herpetologica* **34**: 73–79 (growth); Campbell, C.H. (1977). The tiger snake: a review of the toxicology of the venom and the effect of the bite. *Herpetofauna* **9**: 7–17.

Oxyuranus Kinghorn, 1923

Oxyuranus Kinghorn, J.R. (1923). A new genus of elapine snake from north Australia. *Rec. Aust. Mus.* **14**: 42–45 [42]. Type species *Oxyuranus maclennani* Kinghorn, 1923 by monotypy.

This group is also found in New Guinea.

Oxyuranus scutellatus (Peters, 1867)

Pseudechis scutellatus Peters, W. (1867). Über Flederthiere und Amphibien. *Mber. K. Preuss. Akad. Wiss. Berl.* **1867**: 703–713 [710] [1868 on title page]. Type data: holotype, ZMB 5883, from Rockhampton, Qld.

Pseudechis wilesmithii De Vis, C.W. (1911). Description of snakes apparently new. *Ann. Qd. Mus.* **10**: 22–25 [24]. Type data: holotype, QM J201, from Walsh River, Qld.

Oxyuranus maclennani Kinghorn, J.R. (1923). A new genus of elapine snake from north Australia. *Rec. Aust. Mus.* **14**: 42–45 [42]. Type data: holotype, AM R7901, from Coen, Cape York Peninsula, Qld.

Oxyuranus scutellatus canni Slater, K.R. (1956). On the New Guinea taipan. *Mem. Natl. Mus. Vict.* **20**: 1–5 [2 pl]. Type data: holotype, NMV D8614, from Napa Napa, Port Moresby, Papua New Guinea.

Synonymy that of Cogger, H.G., this work.

Distribution: N coastal, N Gulf, NE coastal, SE coastal, W.A., N.T., N.S.W., Qld. Ecology: terrestrial, diurnal, predator, closed forest, tall forest, open forest, woodland, low open woodland; occasionally nocturnal, oviparous, general carnivore. Biological references: Fohlman, J. (1979). Comparison of two highly toxic Australian snake venoms: the taipan (*Oxyuranus s. scutellatus*) and the fierce snake (*Parademansia microlepidotus*). *Toxicon* **17**: 170–172; Covacevich, J., McDowell, S.B., Tanner, C. & Mengden, G.A. (1981). The relationship of the taipan, *Oxyuranus scutellatus*, and the small-scaled snake, *Oxyuranus microlepidotus* (Serpentes : Elapidae). pp. 160–168 *in* Banks, C.B. & Martin, A.A. (eds.) *Proceedings of the Melbourne Herpetological Symposium, 1980.* Melbourne : Zoological Board of Victoria (taxonomy); Gow, G.F. (1980). History of the taipan *Oxyuranus scutellatus*—with two new distribution records. *N. Territory Nat.* **1**: 15–19 (natural history); Slater, K.R. (1956). On the New Guinea taipan. *Mem. Natl. Mus. Vict.* **20**: 1–5 (natural history); Mackay, R. (1952). The taipan. *Proc. R. Zool. Soc. N.S.W.* **1950–51**: 24–25 (natural history); Thomson, D.F. (1933). Notes on the Australian snakes of the genera *Pseudechis* and *Oxyuranus. Proc. Zool. Soc. Lond.* **1933**: 855–860 (natural history); Johnson, C.R. (1975). Head-body thermal control, thermal preferenda, and voluntary maxima in the taipan, *Oxyuranus scutellatus* (Serpentes : Elapidae). *Zool. J. Linn. Soc.* **56**: 1–12.

Parademansia Kinghorn, 1955

Parademansia Kinghorn, J.R. (1955). Herpetological notes. No.5. *Rec. Aust. Mus.* **23**: 283–286 [285]. Type species *Pseudechis microlepidotus* McCoy (= *Diemenia microlepidota* McCoy, 1879) by original designation.

Parademansia microlepidota (McCoy, 1879)

Diemenia microlepidota McCoy, F. (1879). Small-scaled brown snake. *Prod. Zool. Vict.* **3**: 12–13 [12 pl 23 figs 2–3]. Type data: lectotype, NMV D12354, from junction of Murray and Darling Rivers, Vic., designation by Coventry, A.J. (1970). Reptile and amphibian type specimens housed in the National Museum of Victoria. *Mem. Natl. Mus. Vict.* **31**: 115–124

Diemenia ferox Macleay, W. (1882). Description of two new species of snakes. *Proc. Linn. Soc. N.S.W.* **6**: 811–813 [812]. Type data: holotype, MMUS presumed lost, from Fort Bourke, N.S.W.

Synonymy that of Cogger, H.G., this work.

Distribution: Lake Eyre basin, Bulloo River basin, Murray-Darling basin, N.T., Qld., N.S.W., S.A. Ecology: terrestrial, diurnal, predator, low shrubland, tall shrubland, tussock grassland; oviparous, general carnivore. Biological references: Covacevich, J., McDowell, S.B., Tanner, C. & Mengden, G.A. (1981). The relationship of the taipan, *Oxyuranus scutellatus*, and the small-scaled snake, *Oxyuranus microlepidotus* (Serpentes : Elapidae). pp. 160–168 *in* Banks, C.B. & Martin, A.A. (eds.) *Proceedings of the Melbourne Herpetological Symposium, 1980.* Melbourne : Zoological Board of Victoria (taxonomy); Mirtschin, P. (1981). South Australian records of the inland taipan (*Oxyuranus microlepidotus* McCoy, 1879). *Herpetofauna* **13**: 20–23 (distribution); Fohlman, J. (1979). Comparison of two highly toxic Australian snake venoms: the taipan (*Oxyuranus s. scutellatus*) and the fierce snake (*Parademansia microlepidotus*). *Toxicon* **17**: 170–172 (venom); Broad, A.J., Sutherland, S.K., Tanner, C. & Covacevich, J. (1979). Electrophoretic, enzyme, and preliminary toxicity studies of the venom of the small-scaled snake, *Parademansia microlepidota* (Serpentes : Elapidae), with additional data on its distribution. *Mem. Qd. Mus.* **19**: 319–329 (venom, distribution).

Pseudechis Wagler, 1830

Pseudechis Wagler, J.G. (1830). *Natürliches System der Amphibien, mit vorangehender Classification der Säugethiere und Vögel.* München : Cotta'schen vi 354 pp. [171]. Type species *Coluber porphyricus* [*sic*] Shaw, 1794 by monotypy.

Alecto Wagler, J.G. (1830). *Natürliches System der Amphibien, mit vorangehender Classification der Säugethiere und Vögel.* München : Cotta'schen vi 354 pp. [173]. Type species *Trimeresurus leptocephalus* Lacépède, 1804 by monotypy.

Pseudechys Stirling, E.C. & Zietz, A. (1893). Vertebrata; Reptilia (collected by the Elder Exploring Expedition). *Trans. R. Soc. S. Aust.* **16**: 159–176 [175] [*lapsus pro Pseudechis* Wagler, 1830].

Synonymy that of Cogger, H.G., this work; Mackay, R.D. (1955). A revision of the genus *Pseudechis. Proc. R. Zool. Soc. N.S.W.* **1953–54**: 15–23.

This group is also found in New Guinea.

Pseudechis australis (Gray, 1842)

Naja australis Gray, J.E. (1842). Description of some hitherto unrecorded species of Australian reptiles and batrachians. *in* Gray, J.E. (ed.) *Zoological Miscellany.* London : Treuttel, Würtz & Co. pp. 51–57 [55]. Type data: holotype, BMNH 1946.1.20.39, from Port Essington, N.T.

Pseudechis darwiniensis Macleay, W. (1878). Notes on a collection of snakes from Port Darwin. *Proc. Linn. Soc. N.S.W.* **2**: 219–222 [220]. Type data: holotype, AM R31927, from Port Darwin, N.T.

***Pseudechis cupreus* (part.)** Boulenger, G.A. (1896). *Catalogue of Snakes in the British Museum (Natural History).* 3. London : British Museum xiv 727 pp. pls 25 [329]. Type data: none; description based on literature records (see also under *Pseudonaja textilis* Duméril, Bibron and Duméril, 1854), from "Murray River", N.S.W./Vic.

Pseudechis denisonioides Werner, F. (1909). Reptilia exkl. Geckonidae und Scincidae. *in* Michaelsen, W. & Hartmeyer, R. (eds.) *Die Fauna Südwest-Australiens.* Jena : Gustav Fischer 2: 251–278 [258]. Type data: holotype not found, from Eradu, W.A.

Pseudechis platycephalus Thomson, D.F. (1933). Notes on the Australian snakes of the genera *Pseudechis* and *Oxyuranus. Proc. Zool. Soc. Lond.* **1933**: 855–860 [859 pl 3 figs 1–2]. Type data: holotype, NMV D12355, from East Alligator River, Arnhem Land, N.T.

Denisonia brunnea Mitchell, F.J. (1951). The South Australian reptile fauna. Part 1. Ophidia. *Rec. S. Aust. Mus.* **9**: 545–557 [551 fig 2]. Type data: holotype, SAMA R3151, from Mount Wedge, near Elliston on the W coast of Eyre Peninsula, S.A.

Synonymy that of Cogger, H.G., this work.

Distribution: NW coastal, SW coastal, W plateau, N coastal, N Gulf, NE coastal, Lake Eyre basin, S Gulfs, Murray-Darling basin, Bulloo River basin, W.A., N.T., Qld., S.A., N.S.W. Ecology: terrestrial, noctidiurnal, predator, tall forest, open forest, woodland, low woodland, low open woodland, open scrub, tall shrubland, low shrubland, tussock grassland, hummock grassland, desert; oviparous, general carnivore. Biological references: Leonardi, T.M., Howden, M.E.H. & Spence, I. (1979). A lethal myotoxin isolated from the venom of the Australian king brown snake (*Pseudechis australis*). *Toxicon* **17**: 549 (venom); Mackay, R.D. (1955). A revision of the genus *Pseudechis. Proc. R. Zool. Soc. N.S.W.* **1953–54**: 15–23 (taxonomy); Fitzgerald, M. & Pollitt, C. (1981). Oviparity and captive breeding in the mulga or king brown snake *Pseudechis australis* (Serpentes : Elapidae). *Aust. J. Herp.* **1**: 57–60.

Pseudechis colletti Boulenger, 1902

Pseudechis colletti Boulenger, G.A. (1902). Description of a new snake of the genus *Pseudechis* from Queensland. *Ann. Mag. Nat. Hist. (7)* **10**: 494–495 [494]. Type data: holotype, BMNH 1946.1.20.69, from Qld., "locality unknown".

Distribution: Lake Eyre basin, Murray-Darling basin, (Bulloo River basin), Qld. Ecology: terrestrial, diurnal, predator, low open shrubland, low open woodland; oviparous, general carnivore. Biological references: Mackay, R.D. (1955). A revision of the genus *Pseudechis. Proc. R. Zool. Soc. N.S.W.* **1953–54**: 15–23 (taxonomy); Anon. (1979). Eggs under incubation. *Vict. Herp. Soc. Newsletter* **16**: 22 (reproduction).

Pseudechis guttatus De Vis, 1905

Pseudechis guttata De Vis, C.W. (1905). A new genus of lizards. *Ann. Qd. Mus.* **6**: 46–52 pl 15 [49]. Type data: holotype, QM J189, from Cecil Plains, Qld.

Pseudechis mortonensis De Vis, C.W. (1911). Description of snakes apparently new. *Ann. Qd. Mus.* **10**: 22–25 [24]. Type data: holotype (probable), QM J207a, from "Brisbane Suburbs", Qld., *vide* Covacevich, J. (1971). Amphibian and reptile type specimens in the Queensland Museum. *Mem. Qd. Mus.* **16**: 49–68 [64].

Synonymy that of Cogger, H.G., this work.

Distribution: SE coastal, NE coastal, Murray-Darling basin, Qld., N.S.W. Ecology: terrestrial, diurnal, predator, tall forest, open forest, woodland, tussock grassland; oviparous, general carnivore. Biological references: Mackay, R.D. (1955). A revision of the genus *Pseudechis. Proc. R. Zool. Soc. N.S.W.* **1953–54**: 15–23 (taxonomy); Charles, N., Whitaker, P. & Shine, R. (1980). Oviparity and captive breeding in the spotted black snake, *Pseudechis guttatus* (Serpentes : Elapidae). *Aust. Zool.* **20**: 361–364 (reproduction).

Pseudechis porphyriacus (Shaw, 1794)

Coluber porphyriacus Shaw, G. (1794). *The Zoology of New Holland.* Illustrated by J. Sowerby. London 33 pp. [27 pl 10]. Type data: syntypes (probable), presumed lost, from unspecified locality.

Trimeresurus leptocephalus Lacépède, B.G.E. (1804). Mémoire sur plusieurs animaux de la Nouvelle-Hollande dont la description n'a pas encore été publiée. *Ann. Mus. Natl. Hist. Nat. Paris* **4**: 184–211 [209 pl 56 fig 1]. Type data: lectotype (only syntype located), RMNH unnumbered, from Australia, designation by Cogger, H.G., this work.

Acanthophis tortor Lesson, R.P. (1829–1831). Description de quelques reptiles nouveaux ou peu connus. *in* Duperrey, L.I. (ed.) (1826–1838) *Voyage autour du monde, executé par ordre du Roi, sur la corvette de Sa Majesté,* La Coquille, *pendant les années 1822, 1823, 1824 et 1825.* Zoologie 2(1): 34–65 and Atlas (Reptiles pl 1–7) Paris : Arthus Bertrand [livraison 11 pl 6 (1829); 55 (1831)]. [Sherborn, C.D. & Woodward, B.B. (1901). Notes on the dates of publication of the natural history portions of some French voyages. Part 1. *Ann. Mag. Nat. Hist. (7)* **7**: 388–392 established publication date of livraison; Anon.

(1829). [Review of] *Voyage autour du monde*. Livraison 11. *Bull. Sci. Nat. Géol.* **17**: 267 identifies livraison in which plate appears; Sherborn, C.D. & Woodward, B.B. (1906). Notes on the date of publication of the natural history portions of some French voyages. "*Voyage autour du monde ... sur ... la* Coquille *pendant ... 1822–25 ...* par L.J. Duperry," etc. A correction. *Ann. Mag. Nat. Hist. (7)* **7**: 335–336 established date of publication of text]. Type data: syntypes (probable), not found, from Australasia (Port Jackson, N.S.W. and New South Wales).

Naja porphyrica Schlegel, H. (1837). *Essai sur la Physionomie des Serpens.* Partie Générale et Partie Descriptive. 2 Parts La Haye : Kips & Stockum [479] [*errore pro Coluber porphyriacus* Shaw, 1794].

Pseudechis porphyraicus McCoy, F. (1867). On the recent zoology and palaeontology of Victoria. *Ann. Mag. Nat. Hist. (3)* **20**: 175–202 [182] [*errore pro Coluber porphyriacus* Shaw, 1794].

Synonymy that of Cogger, H.G., this work.

Distribution: NE coastal, SE coastal, Murray-Darling basin, S Gulfs, Qld., N.S.W., A.C.T., Vic., S.A. Ecology: terrestrial, diurnal, predator, closed forest, tall forest, open forest, tussock grassland, low open woodland, swamp; viviparous, primarily anuriphagous. Biological references: Lillywhite, H.B. (1980). Behavioral thermoregulation in Australian elapid snakes. *Copeia* **1980**: 452–458 (thermoregulation); Heatwole, H. & Johnson, C.R. (1979). Thermoregulation in the red-bellied black snake, *Pseudechis porphyriacus* (Serpentes : Elapidae). *Zool. J. Linn. Soc.* **65**: 83–102 (thermoregulation); Shine, R. (1979). Activity patterns in Australian elapid snakes (Squamata : Serpentes : Elapidae). *Herpetologica* **35**: 1–11 (behaviour); Craig, J. (1978). The birth of red-bellied black snakes - *Pseudechis porphyriacus. S. Aust. Herpet. Gp. Newsletter* **1978**: 7 (reproduction); Shine, R., Grigg, G.C., Shine, T.G. & Harlow, P. (1981). Mating and male combat in Australian blacksnakes, *Pseudechis porphyriacus. J. Herpet.* **15**: 101–107 (behaviour); Mackay, R.D. (1955). A revision of the genus *Pseudechis. Proc. R. Zool. Soc. N.S.W.* **1953–54**: 15–23 (taxonomy); Shine, R. (1977). Reproduction in Australian elapid snakes. I. Testicular cycles and mating seasons. *Aust. J. Zool.* **25**: 647–653 (reproduction); Shine, R. (1977). Reproduction in Australian elapid snakes. II. Female reproductive cycles. *Aust. J. Zool.* **25**: 655–666 (reproduction); Shine, R. (1977). Habitats, diets, and sympatry in snakes: a study from Australia. *Can. J. Zool.* **55**: 1118–1128 (ecology), Shine, R. (1978). Growth rates and sexual maturation in six species of Australian elapid snakes. *Herpetologica* **34**: 73–79 (growth).

Pseudonaja Günther, 1858

Pseudonaja Günther, A. (1858). *Catalogue of Colubrine Snakes in the Collection of the British Museum.* London : British Museum xvi 281 pp. [227]. Type species *Pseudonaja nuchalis* Günther, 1858 by monotypy.

Euprepiosoma Fitzinger, L. (1860). Ausbeute der österreichischen Naturforscher an Säugethiere und Reptilien während der Weltumsegelung Sr. Majestat Fregatte *Novara. Sber. Akad. Wiss. Wien* **42**: 384–416 [410]. Type species *Furina textilis* Duméril, Bibron and Duméril, 1854 by monotypy.

Synonymy that of Cogger, H.G., this work.

This group is also found in New Guinea.

Pseudonaja affinis Günther, 1872

Pseudonaja affinis Günther, A. (1872). Seventh account of new species of snakes in the collection of the British Museum. *Ann. Mag. Nat. Hist. (4)* **9**: 13–37 [35 pl 4 fig C]. Type data: holotype, BMNH 1946.1.19.77, from Australia.

Demansia nuchalis tanneri Worrell, E. (1961). A new insular brown snake. *Proc. R. Zool. Soc. N.S.W.* **1958–59**: 56–58 [56]. Type data: holotype, NMV D9819, from Boxer Is., Recherche Archipelago, W.A.

Synonymy that of Cogger, H.G., this work.

Distribution: SW coastal, W plateau, S Gulfs, W.A., S.A. Ecology: terrestrial, diurnal, predator, open forest, woodland, open heath, low shrubland, tall shrubland; oviparous, general carnivore. Biological references: Storr, G.M. (1979). *Dangerous Snakes of Western Australia.* Perth : Western Australian Museum Press 3rd edn. 24 pp. (natural history).

Pseudonaja guttata (Parker, 1926)

Demansia guttata Parker, H.W. (1926). New reptiles and a new frog from Queensland. *Ann. Mag. Nat. Hist. (9)* **17**: 665–670 [668]. Type data: holotype, BMNH 1946.1.20.67, from Winton, 22°19'S, 143°04'E, Qld.

Distribution: W plateau, N Gulf, Lake Eyre basin, (Murray-Darling basin), N.T., Qld, S.A. Ecology: terrestrial, diurnal, predator, tussock grassland, low shrubland; oviparous. Biological references: Gillam, M.W. (1979). The genus *Pseudonaja* (Serpentes : Elapidae) in the Northern Territory. *Territory Parks Wildl. Commn. Res. Bull.* **1**: 1–28 (taxonomy).

Pseudonaja ingrami (Boulenger, 1908)

Diemenia ingrami Boulenger, G.A. (1908). Description of a new elapine snake from Australia. *Ann. Mag. Nat. Hist. (8)* **1**: 333–334 [334]. Type data: holotype, BMNH 1946.1.20.32, from Alexandria, N.T.

Distribution: W plateau, N Gulf, Lake Eyre basin, N.T., Qld. Ecology: terrestrial, diurnal, predator, tussock grassland; oviparous, general carnivore. Biological references: Gillam, M.W. (1979). The genus *Pseudonaja* (Serpentes : Elapidae) in the Northern Territory. *Territory Parks Wildl. Commn. Res. Bull.* **1**: 1–28 (taxonomy).

Pseudonaja modesta (Günther, 1872)

Cacophis modesta Günther, A. (1872). Seventh account of new species of snakes in the collection of the British Museum. *Ann. Mag. Nat. Hist. (4)* **9**: 13–37 [35 pl 3 fig C]. Type data: syntypes, BMNH 1946.1.17.46 (W.A.), BMNH 1946.1.18.42 (Perth), BMNH 1946.1.18.44, ("North-western Australia"), from W.A., NW Australia and Perth, W.A.

Furina ramsayi Macleay, W. (1885). On a new snake from the Barrier Ranges. *Proc. Linn. Soc. N.S.W.* **10**: 61–62 [61] [1886 on title page]. Type data: syntypes, AM B5945, AM B5947–8, from Milparinka, Barrier Ranges, N.S.W. (original register entry and type labels cite Silverton, N.S.W.).

Synonymy that of Cogger, H.G., this work.

Distribution: SW coastal, NW coastal, N coastal, W plateau, Lake Eyre basin, Bulloo River basin, Murray-Darling basin, (S Gulfs), N Gulf, W.A., N.T., S.A., Qld., N.S.W. Ecology: terrestrial, diurnal, predator, desert, low open shrubland, low shrubland, tall shrubland, open scrub, woodland; oviparous. Biological references: Gillam, M.W. (1979). The genus *Pseudonaja* (Serpentes : Elapidae) in the Northern Territory. *Territory Parks Wildl. Commn. Res. Bull.* **1**: 1–28 (taxonomy).

Pseudonaja nuchalis Günther, 1858

Pseudonaja nuchalis Günther, A. (1858). *Catalogue of Colubrine Snakes in the Collection of the British Museum.* London : British Museum xvi 281 pp. [227]. Type data: syntypes, BMNH 1946.1.20.33 and BMNH 1946.1.20.57 ("North West Australia"); BMNH 1946.1.20.41 (Port Essington), from NW Australia and Port Essington, N.T.

Diemenia aspidorhyncha McCoy, F. (1879). The shield-fronted brown snake. *Prod. Zool. Vict.* **3**: 13–14 [13 pl 23 fig 4]. Type data: holotype, NMV D12352, from unstated locality; Vic. by implication, restricted to junction of Murray and Darling Rivers by Coventry, A.J. (1970). Reptile and amphibian type specimens housed in the National Museum of Victoria. *Mem. Natl. Mus. Vict.* **31**: 115–124.

Brachysoma sutherlandi De Vis, C.W. (1884). Descriptions of a new snake with a synopsis of the genus *Hoplocephalus. Proc. R. Soc. Qd.* **1**: 138–140 [139]. Type data: holotype, QM J190, from Carl Creek, Norman River, Qld.

Pseudelaps bancrofti De Vis, C.W. (1911). Description of snakes apparently new. *Ann. Qd. Mus.* **10**: 22–25 [25]. Type data: holotype, QM J187, from Stannary Hills, Qld.

Diemenia carinata Longman, H.A. (1915). Reptiles from Queensland and the Northern Territory. *Mem. Qd. Mus.* **3**: 30–34 [31 pl 14]. Type data: holotype, QM J1508, from Cane Grass Station, via Charleville, Qld.

Demansia acutirostris Mitchell, F.J. (1951). The South Australian reptile fauna. Part 1. Ophidia. *Rec. S. Aust. Mus.* **9**: 545–557 [547 fig 1]. Type data: holotype, SAMA R3133, from island in Lake Eyre, 28°26'S, 137°24'E, S.A.

Synonymy that of Cogger, H.G., this work.

Distribution: SW coastal, NW coastal, N coastal, W plateau, N Gulf, Lake Eyre basin, Bulloo River basin, Murray-Darling basin, S Gulfs, W.A., N.T., S.A., N.S.W., Qld., (Vic.). Ecology: terrestrial, noctidiurnal, predator, open forest, woodland, low woodland, low open woodland, open scrub, tall shrubland, tall open shrubland, low shrubland, desert; oviparous, general carnivore. Biological references: Gillam, M.W. (1979). The genus *Pseudonaja* (Serpentes : Elapidae) in the Northern Territory. *Territory Parks Wildl. Commn. Res. Bull.* **1**: 1–28 (taxonomy); Lillywhite, H.B. (1980). Behavioral thermoregulation in Australian elapid snakes. *Copeia* **1980**: 452–458 (thermoregulation); Banks, C. (1981). Notes on seasonal colour change in a western brown snake. *Herpetofauna* **13**: 29–30.

Pseudonaja textilis (Duméril, Bibron and Duméril, 1854)

Furina textilis Duméril, A.M.C., Bibron, G. & Duméril, A. (1854). *Erpétologie Générale* ou Histoire Naturelle Complète des Reptiles. Paris : Roret 7(2): xii pp. 781–1536 [1242]. Type data: holotype, MNHP 3944 (not found), from Australia.

Pseudoelaps superciliosus Fischer, J.G. (1856). Neue Schlangen des Hamburgischen Naturhistorischen Museums. *Abh. Geb. Natur. Hamb.* **3**: 79–116 [107 pl 2 fig 3]. Type data: holotype, ZMH 362, from "New Holland (Sydney)" in original description; type labelled Sydney, N.S.W.

Demansia annulata Günther, A. (1858). *Catalogue of Colubrine Snakes in the Collection of the British Museum.* London : British Museum xvi 281 pp. [213]. Type data: holotype, BMNH 1946.1.17.54, from Australia.

P[seudoelaps] kubingii Jan, G. (1859). Plan d'une *Iconographie descriptive des Ophidiens*, et description sommaire de nouvelles espèces de Serpents. *Rev. Mag. Zool. (2)* **11–12**: 122–130; Jan, G. (1859). Prodrome d'une *Iconographie descriptive des Ophidiens* et Description sommaire de nouvelles espèces de Serpents venimeux. Reprinted from: *Rev. Mag. Zool. (2)* **10** and *(2)* **11–12**, 1858 and 1859. Paris : Bouchard-Huzard 32 pp. [127; pl C]. Type data: holotype, Pesth (Budapest), presumed lost or destroyed, from N.S.W.

P[seudoelaps] sordellii Jan, G. (1859). Plan d'une *Iconographie descriptive des Ophidiens*, et description sommaire de nouvelles espèces de Serpents. *Rev. Mag. Zool. (2)* **11–12**: 122–130; Jan, G. (1859). Prodrome d'une *Iconographie descriptive des Ophidiens* et Description sommaire de nouvelles espèces de Serpents venimeux. Reprinted from: *Rev. Mag. Zool. (2)* **10** and

(2) **11–12**, 1858 and 1859. Paris : Bouchard-Huzard 32 pp. [127; pl C]. Type data: holotype, MSNM not found, from Australia.

P[seudoelaps] kubinyi Jan, G. (1863). *Elenco Sistematico Degli Ofidi Descritti e Disegnati per l'Iconografia Generale.* Milan : A. Lombardi 143 pp. [116] [emend. or *lapsus pro Pseudoelaps kubingii* Jan, 1859].

Cacophis guntheri Steindachner, F. (1867). Reptilien. *in, Reise der österreichischen Fregatte* Novara *um die Erde in den Jahren 1857, 1858, 1859.* Zoologie 1(3): 1–98 Vienna : State Printer [91 fig] [1869 on title page]. Type data: holotype, not found, from Australia.

Diemenia superciliaris McCoy, F. (1867). On the recent zoology and palaeontology of Victoria. *Ann. Mag. Nat. Hist. (3)* **20**: 175–202 [182] [*lapsus pro Pseudoelaps superciliosus* Fischer, 1856].

Pseudoelaps beckeri Jan, G. & Sordelli, F. (1873). *Iconographie Générale des Ophidiens.* Atlas 1860–1881 Milan : Jan & Sordelli 50 livr. [pl 4 fig 2]. Type data: holotype, ZMH not found, from Sydney, N.S.W.

Furina bicucullata McCoy, F. (1879). The two-hooded furina snake. *Prod. Zool. Vict.* **4**: 13–15 [13 pl 32]. Type data: lectotype, NMV D1832, from Longwood, Vic.; original description cites Longwood, Vic. and Benalla, Vic., designation by Coventry, A.J. (1970). Reptile and amphibian type specimens housed in the National Museum of Victoria. *Mem. Natl. Mus. Vict.* **31**: 115–124.

Furina cucullata Boulenger, G.A. (1896). *Catalogue of Snakes in the British Museum (Natural History).* 3. London : British Museum xiv 727 pp. pls 25 [325] [*lapsus pro Furina bicucullata* McCoy, 1879].

***Pseudechis cupreus* (part.)** Boulenger, G.A. (1896). *Catalogue of Snakes in the British Museum (Natural History).* 3. London : British Museum xiv 727 pp. pls 25 [329]. Type data: no types; description based on literature records; see also under *Pseudechis australis* (Gray, 1842), from Murray River (N.S.W./Vic.).

Demansia textilis inframacula Waite, E. R. (1925). Field notes on some Australian reptiles and batrachians. *Rec. S. Aust. Mus.* **3**: 17–32 [26 fig 10–11]. Type data: syntypes, SAMA, not found, from N end of Coffin's Bay Peninsula, on W coast of Eyre Peninsula, S.A.; not included in Houston, T.F. (1976). Vertebrate type-specimens in the South Australian Museum. III. Reptiles. *Rec. S. Aust. Mus.* **17**: 181–187.

Synonymy that of Cogger, H.G., this work.

Distribution: W plateau, Lake Eyre basin, N Gulf, NE coastal, SE coastal, Murray-Darling basin, S Gulfs, N.T., Qld., N.S.W., A.C.T., Vic., S.A. Ecology: terrestrial, diurnal, predator, tall forest, open forest, woodland, low woodland, low open woodland, open scrub, open heath, low open shrubland, tussock grassland; oviparous, general carnivore. Biological references: Wells, R. (1980). Eggs and young of *Pseudonaja textilis textilis. Herpetofauna* **12**: 30–32 (reproduction); Gillam, M.W. (1979). The genus *Pseudonaja* (Serpentes : Elapidae) in the Northern Territory. *Territory Parks Wildl. Commn. Res. Bull.* **1**: 1–28 (taxonomy); Shine, R. (1979). Activity patterns in Australian elapid snakes (Squamata : Serpentes : Elapidae). *Herpetologica* **35**: 1–11 (behaviour); Coulter, A.R., Broad, A.J. & Sutherland, S.K. (1979). Isolation and properties of a high molecular weight neurotoxin from the eastern brown snake (*Pseudonaja textilis*). p. 160 *in* Chubb, I.W. & Greffen, L.B. (eds.), *Neurotoxins : Fundamental and Clinical Advances.* Adelaide : Adelaide Univ. Union Press (venom).

Rhinoplocephalus Müller, 1885

Rhinoplocephalus Müller, F. (1885). Vierter Nachtrag zum Katalog der herpetologischen Sammlung der Basel Museums. *Verh. Naturf. Ges. Basel* **7**: 668–717 pl 9–11 [690]. Type species *Rhinoplocephalus bicolor* Müller, 1885 by monotypy.

Rhinhoplocephalus Boulenger, G.A. (1896). *Catalogue of Snakes in the British Museum (Natural History).* 3. London : British Museum xiv 727 pp. pls 25 [352]. [emend. *pro Rhinoplocephalus* Müller, 1885]. Type species that of *Rhinoplocephalus* Müller, 1885.

Synonymy that of Cogger, H.G., this work.

Rhinoplocephalus bicolor Müller, 1885

Rhinoplocephalus bicolor Müller, F. (1885). Vierter Nachtrag zum Katalog der herpetologischen Sammlung der Basel Museums. *Verh. Naturf. Ges. Basel* **7**: 668–717 pl 9–11 [690 pl 9 figs f–i]. Type data: holotype, NHMB 2190, from "Southern Australia".

Distribution: SW coastal, W.A. Ecology: terrestrial, predator. Biological references: Glauert, L. (1960). *A Handbook of Snakes of Western Australia.* Perth : Western Australian Naturalists' Club 2nd edn. 62 pp. (description).

Simoselaps Jan, 1859

Rhynchoelaps Jan, G. (1858). Plan d'une *Iconographie descriptive des Ophidiens,* et Description sommaire de nouvelles espèces de Serpents. *Rev. Mag. Zool. (2)* **10**: 514–527 [518]. [described as subgenus of *Elaps* Schneider, 1801; both genus and species *nom. nud.*]. Type species *E[laps] bertholdi* Jan, 1858 by monotypy.

Simoselaps Jan, G. (1859). Plan d'une *Iconographie descriptive des Ophidiens*, et description sommaire de nouvelles espèces de Serpents. *Rev. Mag. Zool. (2)* **11–12**: 122–130 [123]. Type species *Elaps bertholdi* Jan, 1859 by monotypy.

Brachyurophis Günther, A. (1863). On new species of snakes in the collection of the British Museum. *Ann. Mag. Nat. Hist. (3)* **11**: 20–25 [21]. Type species *Brachyurophis semifasciata* Günther, 1863 by monotypy.

Rhinelaps Günther, A. (1872). Seventh account of new species of snakes in the collection of the British Museum. *Ann. Mag. Nat. Hist. (4)* **9**: 13–37 [33]. Type species *Rhinelaps fasciolatus* Günther, 1872 by monotypy.

Rhynchelaps Boulenger, G.A. (1896). *Catalogue of the Snakes in the British Museum (Natural History).* 3. London : British Museum xiv 727 pp. [361] [emend. *pro Rhynchoelaps* Jan, 1858].

Hornea Lucas, A.H.S. & Frost, C. (1896). Reptilia. *Rep. Horn Expedition to Central Australia.* 2: 112–151 [150]. Type species *Hornea pulchella* Lucas and Frost, 1896 by monotypy.

Melwardia Worrell, E. (1960). A new elapine snake from Western Australia. *West. Aust. Nat.* **7**: 132–134 [132]. Type species *Melwardia minima* Worrell, 1960 by original designation.

Synonymy that of Cogger, H.G., this work.

Simoselaps anomalus (Sternfeld, 1919)

Rhynchelaps anomalus Sternfeld, R. (1919). Neue Schlangen und Echsen aus Zentralaustralien. *Senckenberg. Biol.* **1**: 76–83 [77]. Type data: lectotype, SMF 20557, from Hermannsburg Mission, upper Finke River, N.T., designation by Mertens, R. (1967). Die herpetologische Sektion des Natur-Museums und Forschungs-Institutes Senckenberg in Frankfurt-a.-M. nebst einem Verzeichnis ihrer Typen. 1. *Senckenberg. Biol.* **48**: 1–106.

Distribution: NW coastal, W plateau, N coastal, W.A., N.T., (S.A.). Ecology: fossorial, noctidiurnal, predator, low shrubland, desert; oviparous, lacertiphagous. Biological references: Storr, G.M. (1968). The genus *Vermicella* (Serpentes : Elapidae) in Western Australia and the Northern Territory. *J. R. Soc. West. Aust.* **50**: 80–92 (taxonomy); Storr, G.M. (1979). Revisionary notes on the genus *Vermicella* (Serpentes : Elapidae). *Rec. West. Aust. Mus.* **8**: 75–79 (taxonomy).

Simoselaps approximans (Glauert, 1954)

Rhynchoelaps approximans Glauert, L. (1954). Herpetological miscellanea. III. A new burrowing snake from north-western Australia (*Rhynchoelaps approximans*, sp. nov.). *West. Aust. Nat.* **4**: 85 [85 fig]. Type data: holotype, WAM R10768, from Muccan Station, 23°30'S, 120°00'E, W.A.

Distribution: NW coastal, W plateau, W.A. Ecology: fossorial, nocturnal, predator, low shrubland, desert; oviparous. Biological references: Storr, G.M. (1979). Revisionary notes on the genus *Vermicella* (Serpentes : Elapidae). *Rec. West. Aust. Mus.* **8**: 75–79 (taxonomy); Storr, G.M. (1968). The genus *Vermicella* (Serpentes : Elapidae) in Western Australia and the Northern Territory. *J. R. Soc. West. Aust.* **50**: 80–92 (taxonomy).

Simoselaps australis (Krefft, 1864)

Simotes australis Krefft, G. (1864). Descriptions of three new species of Australian snakes. *Proc. Zool. Soc. Lond.* **1864**: 180–182 [180 fig]. Type data: syntypes, 2 specimens, not found, from Port Curtis, Qld. and Clarence River, N.S.W.

Distribution: NE coastal, SE coastal, Murray-Darling basin, (S Gulfs), (Lake Eyre basin), Qld., N.S.W., Vic., S.A. Ecology: fossorial, nocturnal, predator, tall forest, open forest, woodland, low woodland, low open woodland, open scrub; oviparous, lacertiphagous. Biological references: Mackay, R. (1949). The Australian coral snake. *Proc. R. Zool. Soc. N.S.W.* **1947–48**: 36–37 (natural history); Thomson, D.F. (1934). A new snake from north Queensland. *Proc. Zool. Soc. Lond.* **1934**: 529–531 (taxonomy).

Simoselaps bertholdi (Jan, 1859)

E[laps] bertholdi Jan, G. (1858). Plan d'une *Iconographie descriptive des Ophidiens,* et description sommaire de nouvelles espèces de Serpents. *Rev. Mag. Zool. (2)* **10**: 514–527 [518] [*nom. nud.*].

Elaps bertholdi Jan, G. (1859). Plan d'une *Iconographie descriptive des Ophidiens*, et description sommaire de nouvelles espèces de Serpents. *Rev. Mag. Zool. (2)* **11–12**: 122–130 [123]. Type data: holotype, ZMG 191a, from Australia (by implication).

Elaps mattozoi Ferreira, J.B. (1891). Sur quelques espèces du genre *Elaps* déposés au Muséum de Lisbonne. *Jorn. Sci. Math. Phys. Nat. (2)* **2**: 89–95 [93]. Type data: holotype, MB (presumed lost), from Vic.

Furina robusta De Vis, C.W. (1905). A new genus of lizards. *Ann. Qd. Mus.* **6**: 46–52 pl 15 [51]. Type data: holotype, QM J205, from Coolgardie, W.A.

Synonymy that of Cogger, H.G., this work.

Distribution: SW coastal, W plateau, Lake Eyre basin, S Gulfs, W.A., N.T., S.A. Ecology: fossorial, noctidiurnal, predator, desert, woodland, open heath, open scrub, tall shrubland; oviparous, lacertiphagous. Biological references: Ogilby, J.D. (1890). Description of *Vermicella bertholdi. Rec. Aust. Mus.* **1**: 80–81; Storr, G.M. (1979). Revisionary notes on the genus *Vermicella* (Serpentes : Elapidae). *Rec. West. Aust. Mus.* **8**: 75–79 (taxonomy); Storr, G.M. (1978). Taxonomic notes on the reptiles of the Shark Bay region, Western Australia. *Rec. West. Aust. Mus.* **6**: 303–318 (taxonomy); Storr, G.M. (1968). The genus *Vermicella* (Serpentes : Elapidae) in Western Australia and the Northern Territory. *J. R. Soc. West. Aust.* **50**: 80–92 (taxonomy); Waite, E.R. (1929). *The Reptiles and Amphibians of South Australia.* Adelaide : Harrison Weir 270 pp. (natural history); Thomson, D.F. (1934). A new snake from north Queensland. *Proc. Zool. Soc. Lond.* **1934**: 529–531 (taxonomy).

Simoselaps fasciolatus (Günther, 1872)

Rhinelaps fasciolatus Günther, A. (1872). Seventh account of new species of snakes in the collection of the British Museum. *Ann. Mag. Nat. Hist. (4)* **9**: 13–37 [34 pl 5 fig B]. Type data: holotype, BMNH 1946.1.19.12, from Perth, W.A.

Vermicella fasciata Stirling, E.C. & Zietz, A. (1893). Vertebrata; Reptilia (collected by the Elder Exploring Expedition). *Trans. R. Soc. S. Aust.* **16**: 159–176 [175 pl 6]. Type data: holotype, SAMA R2935, from near the

Barrow Range, W.A., *vide* Houston, T.F. (1976). Vertebrate type-specimens in the South Australian Museum. III. Reptiles. *Rec. S. Aust. Mus.* **17**: 181–187.

Hornea pulchella Lucas, A.H.S. & Frost, C. (1896). Reptilia. *Rep. Horn Expedition to Central Australia.* 2: 112–151 [150 pl 12 fig 6]. Type data: holotype, NMV D11226, from Charlotte Waters, N.T.

Synonymy that of Cogger, H.G., this work.

Distribution: SW coastal, NW coastal, W plateau, Lake Eyre basin, (Bulloo River basin), (Murray-Darling basin), W.A., N.T., S.A., Qld., N.S.W. Ecology: fossorial, nocturnal, predator, desert, woodland, open heath, open scrub, tall shrubland; oviparous, lacertiphagous. Biological references: Storr, G.M. (1968). The genus *Vermicella* (Serpentes : Elapidae) in Western Australia and the Northern Territory. *J. R. Soc. West. Aust.* **50**: 80–92 (taxonomy); Thomson, D.F. (1934). A new snake from north Queensland. *Proc. Zool. Soc. Lond.* **1934**: 529–531 (taxonomy).

Simoselaps incinctus (Storr, 1968)

Vermicella semifasciata incincta Storr, G.M. (1968). The genus *Vermicella* (Serpentes : Elapidae) in Western Australia and the Northern Territory. *J. R. Soc. West. Aust.* **50**: 80–92 [89] [1967 on title page]. Type data: holotype, WAM R26844, from near Alice Springs, N.T.

Distribution: (W plateau), Lake Eyre basin, N Gulf, N.T., Qld. Ecology: fossorial, nocturnal, predator, tall shrubland, desert; oviparous.

Simoselaps littoralis (Storr, 1968)

Vermicella bertholdi littoralis Storr, G.M. (1968). The genus *Vermicella* (Serpentes : Elapidae) in Western Australia and the Northern Territory. *J. R. Soc. West. Aust.* **50**: 80–92 [84] [1967 on title page]. Type data: holotype, WAM R12770, from 7 mi S Geraldton, 28°52'S, 114°38'E, W.A.

Distribution: NW coastal, W.A. Ecology: fossorial, noctidiurnal, predator, open heath, low shrubland; oviparous, lacertiphagous. Biological references: Storr, G.M. (1979). Revisionary notes on the genus *Vermicella* (Serpentes : Elapidae). *Rec. West. Aust. Mus.* **8**: 75–79 (taxonomy); Storr, G.M. (1978). Taxonomic notes on the reptiles of the Shark Bay region, Western Australia. *Rec. West. Aust. Mus.* **6**: 303–318 (taxonomy).

Simoselaps minimus (Worrell, 1960)

Melwardia minima Worrell, E. (1960). A new elapine snake from Western Australia. *West. Aust. Nat.* **7**: 132–134 [132 figs 1–2]. Type data: holotype, AM R16494, from Broome, W.A.

Distribution: N coastal, W.A. Ecology: fossorial, noctidiurnal, predator. Biological references: Storr, G.M. (1979). Revisionary notes on the genus *Vermicella* (Serpentes : Elapidae). *Rec. West. Aust. Mus.* **8**: 75–79 (taxonomy).

Simoselaps semifasciatus (Günther, 1863)

Brachyurophis semifasciata Günther, A. (1863). On new species of snakes in the collection of the British Museum. *Ann. Mag. Nat. Hist. (3)* **11**: 20–25 [21 pl 3 fig B]. Type data: syntypes, BMNH 1946.1.19.10–11, from Baranquilla, New Granada (*errore*); corrected to "West Australia(?)", see Boulenger, G.A. (1896). Descriptions of a new snake and a new frog from north Australia. *Proc. Zool. Soc. Lond.* **1895**: 867.

Pseudoelaps rhinostomus Jan, G. & Sordelli, F. (1873). *Iconographie Générale des Ophidiens.* Atlas 1860–1881 Milan : Jan & Sordelli 50 livr. [pl 6 fig 1]. Type data: holotype, originally in RMNH, not found, from "Southwestern Australia".

Rhynchoelaps campbelli Kinghorn, J.R. (1929). Two new snakes from Australia. *Rec. Aust. Mus.* **17**: 190–193 [191 fig 2]. Type data: holotype, AM R9387, from Almaden, Qld.

Rhynchoelaps roperi Kinghorn, J.R. (1931). Herpetological notes. 3. *Rec. Aust. Mus.* **18**: 267–269 [267 fig 1]. Type data: holotype, AM R9930, from Roper River, N.T.

Rhynchoelaps smithii Kinghorn, J.R. (1931). Herpetological notes. 3. *Rec. Aust. Mus.* **18**: 267–269 [267] [*errore pro Rhynchoelaps roperi* Kinghorn, 1931].

Rhynchelaps woodjonesii Thomson, D.F. (1934). A new snake from north Queensland. *Proc. Zool. Soc. Lond.* **1934**: 529–531 [529 pl 1]. Type data: holotype, NMV (D.F. Thomson Coll. No. 69), from Lower Archer River, Gulf of Carpentaria, Qld.

Synonymy that of Cogger, H.G., this work.

Distribution: SW coastal, NW coastal, N coastal, W plateau, N Gulf, NE coastal, (S Gulfs), (Lake Eyre basin), W.A., N.T., S.A., Qld. Ecology: fossorial, nocturnal, predator, desert, tall shrubland, low shrubland, open scrub, low open woodland, low woodland, woodland, open forest, tall forest; oviparous, feeds only on reptile eggs. Biological references: Storr, G.M. (1968). The genus *Vermicella* (Serpentes : Elapidae) in Western Australia and the Northern Territory. *J. R. Soc. West. Aust.* **50**: 80–92 (taxonomy); Thomson, D.F. (1934). A new snake from north Queensland. *Proc. Zool. Soc. Lond.* **1934**: 529–531 (taxonomy).

Simoselaps warro (De Vis, 1884)

Cacophis warro De Vis, C.W. (1884). Descriptions of a new snake with a synopsis of the genus *Hoplocephalus. Proc. R. Soc. Qd.* **1**: 138–140 [139]. Type data: holotype, QM J188, from Warro Station, Port Curtis, Qld.

Pseudelaps albiceps Boulenger, G.A. (1898). Description of two new snakes from Queensland. *Ann. Mag. Nat. Hist. (7)* **2**: 414 [414]. Type data: holotype, BMNH 1946.1.17.50, from Port Douglas, Qld.

Denisonia rostralis De Vis, C.W. (1911). Description of snakes apparently new. *Ann. Qd. Mus.* **10**: 22–25 [23]. Type data: holotype, presumed lost, from Stannary Hills, Qld.

Rhynchelaps fuscicollis Lönnberg, E. & Andersson, G.L. (1915). Results of Dr. E. Mjöberg's Swedish Scientific Expeditions to Australia 1910–1913. VII. Reptiles collected in northern Queensland. *K. Svenska Vetensk.-Akad. Handl.* **52**(7): 1–9 [9]. Type data: holotype, NHRM 2454, from Cairns, Qld.

Synonymy that of Cogger, H.G., this work.

Distribution: N Gulf, NE coastal, Qld. Ecology: fossorial, burrowing, nocturnal, predator, open forest, woodland; oviparous. Biological references: Storr, G.M. (1979). Revisionary notes on the genus *Vermicella* (Serpentes : Elapidae). *Rec. West. Aust. Mus.* **8**: 75–79 (taxonomy).

Suta Worrell, 1961

Suta Worrell, E. (1961). Herpetological name changes. *West. Aust. Nat.* **8**: 18–27 [25]. Type species *Hoplocephalus sutus* Peters, 1863 by tautonymy.

Suta suta (Peters, 1863)

Hoplocephalus sutus Peters, W. (1863). Übersicht der von Hrn. Richard Schomburgk an das zoologische Museum eingesandten Amphibien, aus Buchsfelde bei Adelaide in Südaustralien. *Mber. K. Preuss. Akad. Wiss. Berl.* **1863**: 228–236 [234] [1864 on title page]. Type data: holotype, ZMB 4715, from Loos, 4.5 km W Gawler (as Buchsfelde), S.A., see Praite, R. & Tolley, J.C. (1970). *Place Names of South Australia.* Adelaide : Rigby 203 pp.

Hoplocephalus frenatus Peters, W. (1870). Eine Mitteilung über neue Amphibien (*Hemidactylus, Urosaura, Tropidolepisma* ...) des Königlich Zoologischen Museum. *Mber. K. Preuss. Akad. Wiss. Berl.* **1870**: 641–652 [646] [1871 on title page]. Type data: holotype, ZMB 6922, from Lake Elphinstone, Qld.

Hoplocephalus carpentariae Macleay, W. (1887). On a new *Hoplocephalus* from the Gulf of Carpentaria. *Proc. Linn. Soc. N.S.W. (2)* **2**: 403–404 [403] [1888 on title page. Type status of AM R31926 uncertain; while Goldman, J., Hill, L. & Stanbury, P.J. (1969). Type specimens in the Macleay Museum, University of Sydney. II. Amphibians and Reptiles. *Proc. Linn. Soc. N.S.W.* **93**: 427–438 cite this specimen as holotype, it differs in several major characters from the original description. We have tentatively accepted its type status here, and have discussed the nomenclatural consequences of such acceptance under the genus *Unechis*, of which Macleay's *carpentariae* is the nominal type species]. Type data: holotype, AM R31926, from near Normanton, Qld.

Hoplocephalus frontalis Ogilby, J.D. (1890). Description of a new snake belonging to the genus *Hoplocephalus. Proc. Linn. Soc. N.S.W. (2)* **4**: 1027–1030 [1027] [1889 on title page]. Type data: holotype, AM R655, from Narrabri, N.S.W.

Hoplocephalus stirlingi Lucas, A.H.S. & Frost, C. (1896). Reptilia. *Rep. Horn Expedition to Central Australia.* 2: 112–151 [149 pl 12 fig 5]. Type data: lectotype (only extant syntype), NMV D11758, from Charlotte Waters, N.T., (other type localities in original description were Oodnadatta, S.A., Alice Springs, N.T., and Hermannsburg, N.T.), designation by Coventry, A.J. (1970). Reptile and amphibian type specimens housed in the National Museum of Victoria. *Mem. Natl. Mus. Vict.* **31**: 115–124.

Denisonia frontalis propinqua De Vis, C.W. (1905). A new genus of lizards. *Ann. Qd. Mus.* **6**: 46–52 pl 15 [51]. Type data: holotype, QM J198, from Qld.

Denisonia forresti Boulenger, G.A. (1906). Description of a new lizard and a new snake from Australia. *Ann. Mag. Nat. Hist. (7)* **18**: 440–441 [440]. Type data: syntypes, BMNH 1946.1.20.72–3, from Alexandria, N.T.

Synonymy that of Cogger, H.G., this work.

Distribution: W plateau, N coastal, S Gulfs, N Gulf, NE coastal, Bulloo River basin, Lake Eyre basin, Murray-Darling basin, W.A., N.T., S.A., N.S.W., Vic., Qld. Ecology: terrestrial, nocturnal, predator, desert, open scrub, tall shrubland, low open woodland, low woodland, woodland, open forest; viviparous, general carnivore. Biological references: Smith, L.A. (1978). The elapid snakes *Denisonia pallidiceps* and *Denisonia suta* in the Kimberleys of Western Australia. *West. Aust. Nat.* **14**: 75–76 (distribution); Coventry, A.J. (1971). Identification of the black-headed snakes (*Denisonia*) within Victoria. *Vict. Nat.* **88**: 304–306 (general, in Vic.).

Tropidechis Günther, 1863

Tropidechis Günther, A. (1863). Third account of new species of snakes in the collection of the British Museum. *Ann. Mag. Nat. Hist. (3)* **12**: 348–365 [363]. Type species *Hoplocephalus carinatus* Krefft, 1863 by original designation.

Tropidechis carinatus (Krefft, 1863)

Hoplocephalus carinatus Krefft, G. (1863). Description of a new species of *Hoplocephalus* with keeled scales. *Proc. Zool. Soc. Lond.* **1863**: 86 [86 fig]. Type data: holotype, not found, from near Grafton, N.S.W.; Boulenger, G.A. (1896). *Catalogue of Snakes in the British Museum (Natural History).* 3. London : British Museum xiv 727 pp. pls 25 [351] was in error in citing BMNH 1946.1.18.49 as one of the types.

Distribution: SE coastal, NE coastal, N.S.W., Qld. Ecology: terrestrial, noctidiurnal, predator, closed forest, tall forest, open forest; viviparous, general carnivore. Biological references: Trinca, J.C., Graydon, J.J., Covacevich, J. & Limpus, C. (1971). The rough-scaled snake (*Tropidechis carinatus*) a dangerously venomous Australian snake. *Med. J. Aust.* **1971**: 801–809 (natural history); Beard, D.J. (1979). Rough-scaled snake, *Tropidechis carinatus. Herpetofauna* **10**: 26–28 (natural history).

Unechis Worrell, 1961

Unechis Worrell, E. (1961). Herpetological name changes. *West. Aust. Nat.* **8**: 18–27 [25]. Type species *Unechis carpentariae* Worrell, 1961 [*non Hoplocephalus carpentariae* Macleay, 1887; see *Suta suta* (Peters,

1863)] by monotypy. [Worrell's genus *Unechis* was based on a misidentification of the type species, *Hoplocephalus carpentariae* Macleay, 1887; Worrell regarded as his type species the taxon represented by *Denisonia carpentariae auctorum*, a species which is here regarded as synonymous with *Denisonia boschmai* Brongersma and Knaap-van Meeuwen, 1961. Under Article 70 of the International Code of Zoological Nomenclature such a case must be referred to the Commission for resolution. Consequently we have maintained current usage pending such a submission].

Parasuta Worrell, E. (1961). Herpetological name changes. *West. Aust. Nat.* **8**: 18–27 [26]. Type species *Elaps gouldii* Gray, 1841 by original designation.

Synonymy that of Cogger, H.G., this work.

This group is also found in New Guinea.

Unechis boschmai (Brongersma and Knaap-van Meeuwen, 1961)

Denisonia boschmai Brongersma, L.D. & Knaap-van Meeuwen, M.S. (1961). On a new species of *Denisonia* (Reptilia, Serpentes) from New Guinea. *Zool. Meded.* **39**: 550–554 [553]. Type data: holotype, RMNH 10874 M, from Merauke, Irian Jaya, Indonesia.

[Unechis carpentariae] Worrell, E. (1961). Herpetological name changes. *West. Aust. Nat.* **8**: 18–27 [25] [*non Unechis carpentariae* Macleay, 1887; see *Suta suta*].

Synonymy that of Parker, F. (1972). Snakes of the elapid genus *Suta* in New Guinea. *Papua New Guinea Sci. Soc. Proc.* **23**: 13–17.

Distribution: N Gulf, Lake Eyre basin, NE coastal, Qld. Ecology: terrestrial, noctidiurnal, predator, open forest, woodland; cryptozoic.

Unechis dwyeri (Worrell, 1956)

Denisonia dwyeri Worrell, E. (1956). A new snake from Queensland. *Aust. Zool.* **12**: 202–205 [202 fig 1] [taxonomic status uncertain]. Type data: holotype, AM R14808, from Glenmorgan, Qld.

Biological references: Coventry, A.J. (1971). Identification of the black-headed snakes (*Denisonia)* within Victoria. *Vict. Nat.* **88**: 304–306 (general, in Vic.).

Unechis flagellum (McCoy, 1878)

H[oplocephalus] flagellum McCoy, F. (1867). On the recent zoology and palaeontology of Victoria. *Ann. Mag. Nat. Hist. (3)* **20**: 175–202 [182] [*nom. nud.*].

Hoplocephalus flagellum McCoy, F. (1878). The little whip snake. *Prod. Zool. Vict.* **2**: 7–9 pl 11 fig 1 [7 pl 11 fig 1]. Type data: lectotype, NMV D4259, from Boroondara, Vic. (original type localities were Boroondara, Prahran, South Brighton, Caulfield and Mount Martha, Vic.), designation by Coventry, A.J. (1970). Reptile and amphibian type specimens housed in the National Museum of Victoria. *Mem. Natl. Mus. Vict.* **31**: 115–124.

Synonymy that of Cogger, H.G., this work.

Distribution: SE coastal, Murray-Darling basin, S.A., Vic., N.S.W., A.C.T. Ecology: terrestrial, nocturnal, predator, open forest, woodland; viviparous, cryptozoic. Biological references: Lillywhite, H.B. (1980). Behavioral thermoregulation in Australian elapid snakes. *Copeia* **1980**: 452–458 (thermoregulation); Fyfe, G. (1980). Breeding of the little whipsnakes (*Unechis flagellum*). *Vict. Herp. Soc. Newsletter* **20**: 2 (reproduction); Coventry, A.J. (1971). Identification of the black-headed snakes (*Denisonia)* within Victoria. *Vict. Nat.* **88**: 304–306 (general, in Vic.); Kellaway, C.H. (1934). The venoms of some of the small and rare Australian venomous snakes. *Med. J. Aust.* **1934**: 74–78

Unechis gouldii (Gray, 1841)

Elaps gouldii Gray, J.E. (1841). Description of some new species and four new genera of reptiles from Western Australia, discovered by John Gould, Esq. *Ann. Mag. Nat. Hist. (7)* **42**: 86–91 [91]. Type data: holotype, BMNH 1946.1.21.73, from W.A.

Distribution: NW coastal, SW coastal, W plateau, W.A. Ecology: terrestrial, nocturnal, predator; viviparous, cryptozoic. Biological references: Storr, G.M. (1981). The *Denisonia gouldii* species-group (Serpentes : Elapidae) in Western Australia. *Rec. West. Aust. Mus.* **8**: 501–515 (taxonomy).

Unechis monachus (Storr, 1964)

Denisonia monachus Storr, G.M. (1964). *Denisonia monachus*, a new elapid snake from Western Australia. *West. Aust. Nat.* **9**: 89–90 [89]. Type data: holotype, WAM R20606, from Kalgoorlie, 30°43'S, 121°27'E, W.A.

Distribution: NW coastal, W plateau, Lake Eyre basin, Murray-Darling basin, (SW coastal), W.A., N.T., S.A., N.S.W. Ecology: terrestrial, nocturnal, predator; viviparous, cryptozoic. Biological references: Storr, G.M. (1981). The *Denisonia gouldii* species-group (Serpentes : Elapidae) in Western Australia. *Rec. West. Aust. Mus.* **8**: 501–515 (taxonomy).

Unechis nigriceps (Günther, 1863)

Hoplocephalus nigriceps Günther, A. (1863). Third account of new species of snakes in the collection of the British Museum. *Ann. Mag. Nat. Hist. (3)* **12**: 348–365 [362]. Type data: holotype, BMNH 1946.1.20.70, "probably from Australia".

Denisonia nigrostriata brevicauda Mitchell, F.J. (1951). The South Australian reptile fauna. Part 1. Ophidia. *Rec. S. Aust. Mus.* **9**: 545–557 [550]. Type data: holotype, SAMA R3137, from Fowler's Bay, S.A.

Synonymy that of Storr, G.M. (1981). The *Denisonia gouldii* species-group (Serpentes : Elapidae) in Western Australia. *Rec. West. Aust. Mus.* **8**: 501–515.

Distribution: SW coastal, W plateau, S Gulfs, Murray-Darling basin, W.A., S.A., Vic., N.S.W. Ecology: terrestrial, nocturnal, predator; viviparous, cryptozoic. Biological references: Storr, G.M. (1981). The *Denisonia gouldii* species-group (Serpentes : Elapidae) in Western Australia. *Rec. West. Aust. Mus.* **8**: 501–515 (taxonomy); Coventry, A.J. (1971). Identification of the black-headed snakes (*Denisonia*) within Victoria. *Vict. Nat.* **88**: 304–306 (general, in Vic., as *Denisonia brevicauda*).

Unechis nigrostriatus (Krefft, 1864)

Alecto dorsalis Jan, G. (1863). *Elenco Sistematico Degli Ofidi Descritti e Disegnati per l'Iconografia Generale.* Milan : A. Lombardi 143 pp. [117] [?*nom. nud.*]. Type data: holotype, ZMH 479, from Australia.

Hoplocephalus nigrostriatus Krefft, G. (1864). Descriptions of three new species of Australian snakes. *Proc. Zool. Soc. Lond.* **1864**: 180–182 [181]. Type data: holotype, presumed lost, from neighbourhood of Rockhampton, Qld.

Alecto dorsalis Jan, G. & Sordelli, F. (1873). *Iconographie Générale des Ophidiens.* Atlas 1860–1881 Milan : Jan & Sordelli 50 livr. [pl 2 fig 1]. Type data: holotype, ZMH 479, from Australia.

Synonymy that of Cogger, H.G., this work.

Distribution: N Gulf, NE coastal, Qld. Ecology: terrestrial, nocturnal, predator, closed forest, tall forest, open forest, woodland; cryptozoic.

Unechis spectabilis (Krefft, 1869)

Hoplocephalus spectabilis Krefft, G. (1869). *The Snakes of Australia*; an Illustrated and Descriptive Catalogue of All the Known Species. Sydney : Govt. Printer xxv 100 pp. [61 pl 12 fig 4]. Type data: holotype, not found, from Port Lincoln, S.A.

Denisonia spectabilis nullarbor Storr, G.M. (1981). The *Denisonia gouldii* species-group (Serpentes : Elapidae) in Western Australia. *Rec. West. Aust. Mus.* **8**: 501–515 [512 fig 6]. Type data: holotype, WAM R66626, from 4 km S of (new) Eucla, 31°43'S, 128°52'E, W.A.

Synonymy that of Cogger, H.G., this work.

Distribution: W plateau, S Gulfs, Murray-Darling basin, SE coastal, W.A., S.A., Vic., N.S.W., A.C.T., Qld. Ecology: terrestrial, nocturnal, predator, low shrubland, open heath, tall shrubland, open scrub, low open woodland; viviparous. Biological references: Storr, G.M. (1981). The *Denisonia gouldii* species-group (Serpentes : Elapidae) in Western Australia. *Rec. West. Aust. Mus.* **8**: 501–515 (taxonomy); Shine, R. (1979). Activity patterns in Australian elapid snakes (Squamata : Serpentes : Elapidae). *Herpetologica* **35**: 1–11 (behaviour, as *Unechis gouldii*).

Vermicella Günther, 1858

Vermicalla Günther, A. (1858). *Catalogue of Colubrine Snakes in the Collection of the British Museum.* London : British Museum xvi 281 pp. [210] [*lapsus pro Vermicella* Günther, 1858].

Vermicella Günther, A. (1858). *Catalogue of Colubrine Snakes in the Collection of the British Museum.* London : British Museum xvi 281 pp. [236]. Type species *Calamaria annulata* Gray, 1841 by monotypy.

Homaloselaps Jan, G. (1858). Plan d'une *Iconographie descriptive des Ophidiens,* et Description sommaire de nouvelles espèces de Serpents. *Rev. Mag. Zool. (2)* **10**: 514–527 [518] [described as subgenus of *Elaps* Schneider, 1801]. Type species *Elaps occipitalis* Duméril, Bibron and Duméril, 1854 by monotypy.

Synonymy that of Cogger, H.G., this work; Storr, G.M. (1968). The genus *Vermicella* (Serpentes : Elapidae) in Western Australia and the Northern Territory. *J. R. Soc. West. Aust.* **50**: 80–92.

Vermicella annulata (Gray, 1841)

Calamaria annulata Gray, J.E. (1841). A catalogue of the species of reptiles and Amphibia hitherto described as inhabiting Australia, with a description of some new species from Western Australia. Appendix E *in* Grey, G. *Journals of Two Expeditions of Discovery in North-west and Western Australia, during the years 1837, 38, and 39, under the authority of Her Majesty's Government.* London : T. & W. Boone 2: 422–449 [443]. Type data: holotype, BMNH 1946.1.19.21, from Australia.

Elaps occipitalis Duméril, A.M.C., Bibron, G. & Duméril, A. (1854). *Erpétologie Générale* ou Histoire Naturelle Complète des Reptiles. Paris : Roret 7(2): xii pp. 781–1536 [1220]. Type data: holotype, MNHP 3936, from Rio de Janeiro (*errore*) and Australia.

Vermicella lunulata Krefft, G. (1869). *The Snakes of Australia*; an Illustrated and Descriptive Catalogue of All the Known Species. Sydney : Govt. Printer xxv 100 pp. [79 pl 12 fig 14]. Type data: holotype, AM 6684, from upper Burdekin River district, Qld.

Rhynchelaps latizonatus De Vis, C.W. (1905). A new genus of lizards. *Ann. Qd. Mus.* **6**: 46–52 pl 15 [49]. Type data: holotype, QM J192, from Qld., Herberton, *vide* Covacevich, J. (1971). Amphibian and reptile type specimens in the Queensland Museum. *Mem. Qd. Mus.* **16**: 49–68 [64].

Vermicella annulata snelli Storr, G.M. (1968). The genus *Vermicella* (Serpentes : Elapidae) in Western Australia and the Northern Territory. *J. R. Soc. West. Aust.* **50**: 80–92 [82] [1967 on title page]. Type data: holotype, WAM R19203, from Mundiwindi, 23°51'S, 120°09'E, W.A.

Synonymy that of Cogger, H.G., this work.

Distribution: NW coastal, W plateau, N Gulf, Lake Eyre basin, Bulloo River basin, Murray-Darling basin, SE coastal, NE coastal, S Gulfs, W.A., N.T., S.A., Vic., N.S.W., A.C.T., Qld. Ecology: burrowing, nocturnal, predator, desert, tall shrubland, low shrubland, open heath, open scrub, low open woodland, low woodland, woodland, open forest, tall forest; oviparous, feeds primarily on blind snakes (Typhlopidae). Biological references: Shine, R. (1980). Reproduction, feeding and growth in the Australian burrowing snake *Vermicella annulata*. *J. Herpet.* **14**: 71–77 (reproduction and ecology); McDowell, S.B. (1969). *Toxicocalamus*, a New Guinea genus of snakes of the family Elapidae. *J. Zool. Lond.* **159**: 443–511 (taxonomy); Storr, G.M. (1968). The genus *Vermicella* (Serpentes : Elapidae) in Western Australia and the Northern Territory. *J. R. Soc. West. Aust.* **50**: 80–92 (taxonomy); Bustard, H.R. (1969). Defensive display behaviour in the bandy-bandy *Vermicella annulata* (Serpentes : Elapidae). *Herpetologica* **25**: 319–320 (behaviour); Kellaway, C.H. (1934). The venoms of some of the small and rare Australian venomous snakes. *Med. J. Aust.* **1934**: 74–78.

Vermicella multifasciata (Longman, 1915)

Furina multifasciata Longman, H.A. (1915). Reptiles from Queensland and the Northern Territory. *Mem. Qd. Mus.* **3**: 30–34 [30]. Type data: holotype, QM J2019, from Port Darwin, N.T.

Distribution: N coastal, W.A., N.T. Ecology: burrowing, nocturnal, predator, woodland; oviparous. Biological references: Waite, E.R. & Longman, H.A. (1920). Description of little-known Australian snakes. *Rec. S. Aust. Mus.* **1**: 173–180 (description); Storr, G.M. (1968). The genus *Vermicella* (Serpentes : Elapidae) in Western Australia and the Northern Territory. *J. R. Soc. West. Aust.* **50**: 80–92 (taxonomy).

HYDROPHIIDAE

INTRODUCTION

This family contains the viviparous sea snakes, a group of marine, front-fanged venomous snakes containing about 55 species in 18 genera. Of these, 29 species in 12 genera occur in Australian waters. Sea snakes, which are closely allied to the elapids, are largely confined to the tropical waters of Asia, Indo-Malaysia, Australasia and the western Pacific. One pelagic species (*Pelamis platurus*) ranges across the Indian Ocean to east Africa and across the Pacific to the west coast of the Americas. The majority of species are highly venomous.

Characterised in Australia by: a pair of enlarged fangs at the front of the maxilla; tail laterally compressed and paddle-shaped; nasal scales in contact; nostrils valvular.

References

Burger, W.L. & Natsuno, T. (1974). A new genus for the Arafura smooth seasnake and redefinitions of other seasnake genera. *The Snake* **6**: 61–75

Cogger, H.G. (1975). The sea snakes of Australia and New Guinea. pp. 50–139 *in* Dunson, W. (ed.) *The Biology of Sea Snakes.* Baltimore : University Park Press

Cogger, H.G. (1979). *Reptiles and Amphibians of Australia.* Sydney : A.H. & A.W. Reed 608 pp.

Smith, M.A. (1926). *Monograph on the Sea Snakes. (Hydrophiidae).* London : British Museum xvii 130 pp.

Voris, H.K. (1977). A phylogeny of the sea snakes (Hydrophiidae). *Fieldiana Zool.* **70**: 79–166

Acalyptophis Boulenger, 1896

Acalyptus Duméril, A.M.C. (1853). Prodrome général de la classification des serpents. *Mém. Acad. Sci. Inst. Fr.* **23**: 399–536 [522] [*non Acalyptus* Schonheer, 1836]. Type species *Acalyptus peronii* Duméril, 1853 by monotypy.

Acalyptophis Boulenger, G.A. (1896). *Catalogue of Snakes in the British Museum (Natural History).* 3. London : British Museum xiv 727 pp. pls 25 [269] [*nom. substit. pro Acalyptus* Duméril, 1853].

Pseudodistira Kinghorn, J.R. (1926). A new genus of sea snake and other reptiles collected by H.M.A.S. *Geranium* during a cruise to north Australia. *Proc. Zool. Soc. Lond.* **1926**: 71–74 [71]. Type species *Pseudodistira horrida* Kinghorn, 1926 by monotypy.

Synonymy that of Cogger, H.G., this work; Smith, M.A. (1926). *Monograph on the Sea Snakes. (Hydrophiidae).* London : British Museum xvii 130 pp.

This group is also found in waters of southern New Guinea and South China Sea.

Acalyptophis peronii (Duméril, 1853)

Acalyptus peronii Duméril, A.M.C. (1853). Prodrome général de la classification des serpents. *Mém. Acad. Sci. Inst. Fr.* **23**: 399–536 [522]. Type data: holotype, MNHP 7711, from ?New Holland.

Acalyptus superciliosus vel peroni Duméril, A.M.C., Bibron, G. & Duméril, A. (1854). *Erpétologie Générale* ou Histoire Naturelle Complète des Reptiles. Paris : Roret 7(2): xii pp. 781–1536 [1340]. Type data: holotype, MNHP 7711, from Australia.

Pseudodistira horrida Kinghorn, J.R. (1926). A new genus of sea snake and other reptiles collected by H.M.A.S. *Geranium* during a cruise to north Australia. *Proc. Zool. Soc. Lond.* **1926**: 71–74 [71 fig 1 pl 1]. Type data: holotype, AM R8640, from McCulloch Reef, in 17°20'S, 146°30'E, Great Barrier Reef, Qld.

Synonymy that of Cogger, H.G., this work; Smith, M.A. (1926). *Monograph on the Sea Snakes. (Hydrophiidae).* London : British Museum xvii 130 pp.

Distribution: Indian Ocean, NW Shelf, N coast, Arafura Sea, Gulf of Carpentaria, Great Barrier Reef, NE coast, (W coast), (SE coast), W.A., N.T., Qld., (N.S.W.); Timor Sea. Ecology: aquatic, noctidiurnal, nektonic, predator, inshore, continental shelf; perennially active, viviparous, piscivore, coral reefs. Biological references: Smith, M.A. (1926). *Monograph on the Sea Snakes. (Hydrophiidae).* London : British Museum xvii 130 pp. (taxonomy); Cogger, H.G. (1975). The sea snakes of Australia and New Guinea. pp. 59–139 *in* Dunson, W. (ed.) *The Biology of Sea Snakes.* Baltimore : University Park Press (taxonomy); Lemen, C.A. & Voris, H.K. (1981). A comparison of reproductive strategies among marine snakes. *J. Anim. Ecol.* **50**: 89–101 (reproduction); Smith, L.A. (1974). The sea snakes of Western Australia (Serpentes : Elapidae : Hydrophiinae) with a description of a new subspecies. *Rec. West. Aust. Mus.* **3**: 93–110 (taxonomy); Seymour, R.S. & Webster, M.E.D. (1975). Gas transport and blood acid-base balance in diving sea snakes. *J. Exp. Zool.* **191**: 169–182 (physiology); Redfield, J.A., Holmes, J.C. & Holmes, R.D. (1978). Sea snakes of the eastern Gulf of Carpentaria. *Aust. J. Mar. Freshwat. Res.* **29**: 325–334 (distribution and abundance).

Aipysurus Lacépède 1804

Aipysurus Lacépède, B.G.E. (1804). Mémoire sur plusieurs animaux de la Nouvelle-Hollande dont la description n'a pas encore été publiée. *Ann. Mus. Natl. Hist. Nat. Paris* **4**: 184–211 [210]. Type species *Aipysurus laevis* Lacépède, 1804 by monotypy.

Stephanohydra Tschudi, J.J. (1837). Neues Genus von Wasserschlangen. *Arch. Naturgesch.* **3**: 331–335 pl 8 [335]. Type species *Stephanohydra fusca* Tschudi, 1837 by monotypy.

Aspisurus Gray, J.E. (1841). A catalogue of the species of reptiles and Amphibia hitherto described as inhabiting Australia, with a description of some new species from Western Australia. Appendix E *in* Grey, G. *Journals of Two Expeditions of Discovery in North-west and Western Australia, during the years 1837, 38, and 39, under the authority of Her Majesty's Government.* London : T. & W. Boone 2: 422–449 [433] [*non Aspisurus* Lacépède, 1802; *lapsus pro Aipysurus* Lacépède, 1804].

Hypotropis Gray, J.E. (1846). A new genus of sea-snake from Port Essington. *Ann. Mag. Nat. Hist.* **18**: 284 [284]. Type species *Hypotropis jukesii* Gray, 1846 by monotypy.

Aepyurus Agassiz, L. (1846). *Nomenclatoris Zoologici Index Universalis.* Soloduri : Jent et Gassman 393 pp. [9] [emend. *pro Aipysurus* Lacépède, 1804].

Tomogaster Gray, J.E. (1849). *Catalogue of the Specimens of Snakes in the Collection of the British Museum.* London : British Museum xv 125 pp. [59] [*ex* Bibron ms]. Type species *Tomogaster eydouxii* Gray, 1849 by monotypy.

Pelagophis Peters, W. & Doria, G. (1878). Catalogo dei Rettili e dei Batraci raccolti da O. Beccari, L.M. d'Albertis e A.A. Bruijn nella sotto-regione Austro-Malese. *Ann. Mus. Civ. Stor. Nat. Genova* **13**: 323–450 [413 pl 5]. Type species *Pelagophis lubricus* Peters and Doria, 1878 by monotypy.

Synonymy that of Cogger, H.G., this work; Smith, M.A. (1926). *Monograph on the Sea Snakes. (Hydrophiidae).* London : British Museum xvii 130 pp. [13].

This group is also found in waters of South China Sea, Indonesia, Papua New Guinea and islands of western Pacific.

Aipysurus apraefrontalis Smith, 1926

Aipysurus apraefrontalis Smith, M.A. (1926). *Monograph on the Sea Snakes. (Hydrophiidae).* London : British Museum xvii 130 pp. [24 fig 13]. Type data: holotype, BMNH 1946.1.1.94, from Ashmore Reefs, Australia.

Distribution: Ashmore Is., NW Shelf, W.A.; Timor Sea. Ecology: aquatic, diurnal, nektonic, predator, inshore; perennially active, viviparous, piscivore, coral reefs. Biological references: Cogger, H.G. (1975). The sea snakes of Australia and New Guinea. pp. 59–139 *in* Dunson, W. (ed.) *The Biology of Sea Snakes.* Baltimore : University Park Press (taxonomy); Smith, L.A. (1974). The sea snakes of Western Australia (Serpentes : Elapidae : Hydrophiinae) with a description of a new subspecies. *Rec. West. Aust. Mus.* **3**: 93–110 (taxonomy).

Aipysurus duboisii Bavay, 1869

Aipysurus duboisii Bavay, A. (1869). Catalogue des reptiles de la Nouvelle-Calédonie et description d'espèces nouvelles. *Mém. Soc. Linn. Normandie* **15**: 1–37 [33]. Type data: holotype, Marine Museum Brest *vide* Smith, M.A. (1926). *Monograph on the Sea Snakes. (Hydrophiidae).* London : British Museum xvii 130 pp., from Lifu (as Lifou), Loyalty Ils., New Caledonia.

Aipysurus australis Sauvage, H.E. (1877). Sur quelques ophidiens d'espèces nouvelles ou peu connues de la collection du Muséum. *Bull. Soc. Philomath. Paris (7)* **1**: 107–115 [114]. Type data: holotype, MNHP (not found), from Australia.

Pelagophis lubricus Peters, W. & Doria, G. (1878). Catalogo dei Rettili e dei Batraci raccolti da O. Beccari, L.M. d'Albertis e A.A. Bruijn nella sotto-regione Austro-Malese. *Ann. Mus. Civ. Stor. Nat. Genova* **13**: 323–450 [414 pl 5] [erroneously referred to *Aipysurus laevis* by Capocaccia, L. (1961). Catalogo dei tipi di Rettili del Museo Civico di Storia Naturale di Genova. *Ann. Mus. Civ. Stor. Nat. Giacomo Doria* **72**: 86–111]. Type data: holotype, MCG C.E.30765, from Yule Is., Papua New Guinea.

Synonymy that of Cogger, H.G., this work; Smith, M.A. (1926). *Monograph on the Sea Snakes. (Hydrophiidae).* London : British Museum xvii 130 pp.

Distribution: NW Shelf, N coast, Arafura Sea, Gulf of Carpentaria, Great Barrier Reef, NE coast, Coral Sea, W.A., N.T., Qld., Ashmore Is., Cartier Is.; Timor Sea. Ecology: aquatic, diurnal, nektonic, predator, inshore, continental shelf; perennially active, viviparous, piscivore, coral reefs. Biological references: Smith, M.A. (1926). *Monograph on the Sea Snakes. (Hydrophiidae).* London : British Museum xvii 130 pp. (taxonomy); Cogger, H.G. (1975). The sea snakes of Australia and New Guinea. pp. 59–139 *in* Dunson, W. (ed.) *The Biology of Sea Snakes.* Baltimore : University Park Press (taxonomy); Smith, L.A. (1974). The sea snakes of Western Australia (Serpentes : Elapidae : Hydrophiinae) with a description of a new subspecies. *Rec. West. Aust. Mus.* **3**: 93–110 (taxonomy); Heatwole, H. (1981). Temperature relations of some sea snakes. *The Snake* **13**: 53–57 (thermoregulation); Redfield, J.A., Holmes, J.C. & Holmes, R.D. (1978). Sea snakes of the eastern Gulf of Carpentaria. *Aust. J. Mar. Freshwat. Res.* **29**: 325–334 (distribution and abundance).

Aipysurus eydouxii (Gray, 1849)

Tomogaster eydouxii Gray, J.E. (1849). *Catalogue of the Specimens of Snakes in the Collection of the British Museum.* London : British Museum xv 125 pp. [59]. Type data: holotype, BMNH 1946.1.6.86, from Indian Ocean.

Thalassophis anguillaeformis Schmidt, P. (1852). Beiträge zur ferneren Kenntniss der Meerschlangen. *Abh. Geb. Naturw. Hamb.* **2**: 69–86 [76 pl 1]. Type data: holotype, ZMH 386, from Java, Indonesia *vide* Smith, M.A. (1926). *Monograph on the Sea Snakes. (Hydrophiidae).* London : British Museum xvii 130 pp.

Thalassophis muraeneformis Schmidt, P. (1852). Beiträge zur ferneren Kenntniss der Meerschlangen. *Abh. Geb. Naturw. Hamb.* **2**: 69–86 [77]. Type data: holotype, ZMH 385, from Samarang, Java, Indonesia *vide* Smith, M.A. (1926). *Monograph on the Sea Snakes. (Hydrophiidae).* London : British Museum xvii 130 pp.

Aipysurus margaritophorus Bleeker, P. (1858). Eene Zeeslang (*Aipysurus margaritophorus*). *Natuurk. Tijdschr. Ned.-Indie. (4)* **16**: 49–50 [49]. Type data: holotype, BMNH 1946.1.7.1, from Java, Indonesia.

Synonymy that of Cogger, H.G., this work; Smith, M.A. (1926). *Monograph on the Sea Snakes. (Hydrophiidae).* London : British Museum xvii 130 pp.

Distribution: NW Shelf, N coast, Arafura Sea, Gulf of Carpentaria, Great Barrier Reef, NE coast, Coral Sea, W.A., N.T., Qld., Ashmore Is., Cartier Is.; Timor Sea. Ecology: aquatic, diurnal, nektonic, predator, inshore, continental shelf; perennially active, viviparous, piscivore, coral reefs. Biological references: Smith, M.A. (1926). *Monograph on the Sea Snakes. (Hydrophiidae).* London : British Museum xvii 130 pp. (taxonomy); Lemen, C.A. & Voris, H.K. (1981). A comparison of reproductive strategies among marine snakes. *J. Anim. Ecol.* **50**: 89–101 (reproduction); Cogger, H.G. (1975). The sea snakes of Australia and New Guinea. pp. 59–139 *in* Dunson, W. (ed.) *The Biology of Sea Snakes.* Baltimore : University Park Press (taxonomy); Smith, L.A. (1974). The sea snakes of Western Australia (Serpentes : Elapidae : Hydrophiinae) with a description of a new subspecies. *Rec. West. Aust. Mus.* **3**: 93–110 (taxonomy); Redfield, J.A., Holmes, J.C. & Holmes, R.D. (1978). Sea snakes of the eastern Gulf of Carpentaria. *Aust. J. Mar. Freshwat. Res.* **29**: 325–334 (distribution and abundance).

Aipysurus foliosquama Smith, 1926

Aipysurus foliosquama Smith, M.A. (1926). *Monograph on the Sea Snakes. (Hydrophiidae).* London : British Museum xvii 130 pp. [22 figs 11–12 pl 2 fig 3]. Type data: holotype, BMNH 1946.1.1.96, from Ashmore Reefs, Australia.

Distribution: Ashmore Is.; Timor Sea. Ecology: aquatic, diurnal, nektonic, predator, inshore; perennially active, viviparous, piscivore, coral reefs. Biological references: Cogger, H.G. (1975). The sea snakes of Australia and New Guinea. pp. 59–139 *in* Dunson, W. (ed.) *The Biology of Sea Snakes.* Baltimore : University Park Press (taxonomy); Heatwole, H., Minton, S.A., Taylor, R. & Taylor, V. (1978). Underwater observations on sea snake behaviour. *Rec. Aust. Mus.* **31**: 737–761 (behaviour); Heatwole, H. (1975). Predation on sea snakes. pp. 233–249 *in* Dunson, W. (ed.) *The Biology of Sea Snakes.* Baltimore : University Park Press (as prey).

Aipysurus fuscus (Tschudi, 1837)

Stephanohydra fusca Tschudi, J.J. (1837). Neues Genus von Wasserschlangen. *Arch. Naturgesch.* **3**: 331–335 [335 pl 8]. Type data: holotype, ZMB 2824, from Asia (from Celebes, *vide* Smith, M.A. (1926). *Monograph on the Sea Snakes. (Hydrophiidae).* London : British Museum xvii 130 pp., who also erroneously referred to an Ashmore Reef specimen as the type).

Distribution: Ashmore Is.; Timor Sea. Ecology: aquatic, diurnal, nektonic, predator, inshore; perennially active, viviparous, piscivore, coral reefs. Biological references: Smith, M.A. (1926). *Monograph on the Sea Snakes. (Hydrophiidae).* London : British Museum xvii 130 pp. (taxonomy); Cogger, H.G. (1975). The sea snakes of Australia and New Guinea. pp. 59–139 *in* Dunson, W. (ed.) *The Biology of Sea Snakes.* Baltimore : University Park Press (taxonomy); Heatwole, H. (1975). Predation on sea snakes. pp. 233–249 *in* Dunson, W. (ed.) *The Biology of Sea Snakes.* Baltimore : University Park Press (as prey); Heatwole, H., Minton, S.A., Taylor, R. & Taylor, V. (1978). Underwater observations on sea snake behaviour. *Rec. Aust. Mus.* **31**: 737–761 (behaviour).

Aipysurus laevis Lacépède, 1804

Aipysurus laevis Lacépède, B.G.E. (1804). Mémoire sur plusieurs animaux de la Nouvelle-Hollande dont la description n'a pas encore été publiée. *Ann. Mus. Natl. Hist. Nat. Paris* **4**: 184–211 [210 pl 56 fig 3]. Type data: neotype, WAM R22384, designation by Smith, L.A. (1974). The sea snakes of Western Australia (Serpentes : Elapidae : Hydrophiinae) with a description of a new subspecies. *Rec. West. Aust. Mus.* **3**: 93–110 [99], from near Locker Is., off Onslow, Western Australia, in 21°44'S, 114°46'E.

Hypotropis jukesii Gray, J.E. (1846). A new genus of sea-snake from Port Essington. *Ann. Mag. Nat. Hist.* **18**: 284 [284]. Type data: holotype, BMNH 1946.1.9.53, from near Darnley Is., Torres Strait (as near Darnley Islands, Port Essington), Qld.

Aipysurus fuliginosus Duméril, A.M.C., Bibron, G. & Duméril, A. (1854). *Erpétologie Générale* ou Histoire Naturelle Complète des Reptiles. Paris : Roret 7(2): xii pp. 781–1536 [1327 pl 77 fig 1]. Type data: holotype, MNHP 639, from New Caledonia.

Aipysurus laevis pooleorum Smith, L.A. (1974). The sea snakes of Western Australia (Serpentes : Elapidae : Hydrophiinae) with a description of a new subspecies. *Rec. West. Aust. Mus.* **3**: 93–110 [97]. Type data: holotype, WAM R21366, from Shark Bay, W.A.

Synonymy that of Cogger, H.G., this work; Smith, M.A. (1926). *Monograph on the Sea Snakes. (Hydrophiidae).* London : British Museum xvii 130 pp.

Distribution: NW Shelf, N coast, Arafura Sea, Gulf of Carpentaria, Great Barrier Reef, NE coast, Coral Sea, W.A., N.T., Qld., (N.S.W.), Ashmore Is., Cartier Is.; Timor Sea. Ecology: aquatic, diurnal, nektonic, predator, estuarine, inshore, continental shelf; perennially active, viviparous, piscivore, coral reefs. Biological references: Smith, M.A. (1926). *Monograph on the Sea Snakes. (Hydrophiidae).* London : British Museum xvii 130 pp. (taxonomy); Cogger, H.G. (1975). The sea snakes of Australia and New Guinea. pp. 59–139 *in* Dunson, W. (ed.) *The Biology of Sea Snakes.* Baltimore : University Park Press (taxonomy); Smith, L.A. (1974). The sea snakes of Western Australia (Serpentes : Elapidae : Hydrophiinae) with a description of a new subspecies. *Rec. West. Aust. Mus.* **3**: 93–110 (taxonomy); Heatwole, H. (1981). Temperature relations of some sea snakes. *The Snake* **13**: 53–57 (thermoregulation); Seymour, R.S. & Webster, M.E.D. (1975). Gas transport and blood acid-base balance in diving sea snakes. *J. Exp. Zool.* **191**: 169–182 (physiology); Tamiya, N. & Maeda, N. (1979). Neurotoxins of Australian sea snakes. pp. 95–101 *in* Chubb, I.W. & Greffen, L.B. (eds.) *Neurotoxins : Fundamental and Clinical Advances.* Adelaide : Adelaide Univ. Union Press (venom); Heatwole, H., Heatwole, E. & Johnson, C.R. (1974). Shark predation on sea snakes. *Copeia* **1974**: 780–781 (as prey); Redfield, J.A., Holmes, J.C. & Holmes, R.D. (1978). Sea snakes of the eastern Gulf of Carpentaria. *Aust. J. Mar. Freshwat. Res.* **29**: 325–334 (distribution and abundance); Maeda, N. & Tamiya, N. (1976). Isolation, properties and amino acid sequences of three neurotoxins from the venom of a sea snake, *Aipysurus laevis. Biochem. J.* **153**: 79–87.

Aipysurus tenuis Lönnberg and Andersson, 1913

Aipysurus tenuis Lönnberg, E. & Andersson, G.L. (1913). Results of Dr. E. Mjöberg's Swedish Scientific Expeditions to Australia 1910–1913. III. Reptiles. *K. Svenska Vetensk.-Akad. Handl.* **52**(3): 1–17 [13]. Type data: syntypes, NHRM 2400 (3 specimens), from Cape Jaubert, near Broome, W.A.

Distribution: NW Shelf, W.A. Biological references: Smith, M.A. (1926). *Monograph on the Sea Snakes. (Hydrophiidae).* London : British Museum xvii 130 pp. (taxonomy); Cogger, H.G. (1975). The sea snakes of Australia and New Guinea. pp. 59–139 *in* Dunson, W. (ed.) *The Biology of Sea Snakes.* Baltimore : University Park Press (taxonomy); Smith, L.A. (1974). The sea snakes of Western Australia (Serpentes : Elapidae : Hydrophiinae) with a description of a new subspecies. *Rec. West. Aust. Mus.* **3**: 93–110 (taxonomy).

Astrotia Fischer, 1856

Astrotia Fischer, J.G. (1856). Die Familie der Seeschlangen. *Abh. Geb. Naturw. Hamburg* **3**: 1–78 [38]. Type species *Hydrophis schizopholis* Schmidt, 1846 by monotypy.

This group is also found in waters of India, Sri Lanka, Indo-Malayan Archipelago and New Guinea.

Astrotia stokesii (Gray, 1846)

*[**Hydrus major** (**part.**)]* Shaw, G. (1802). *General Zoology*, or Systematic Natural History. Vol. 3 Pt. 2 Amphibia. London : Kearsley vii pp. 313–615 [558] [original syntype series comprised two species, see *Disteira major*]. Type data: paralectotype, BMNH iii.9.1.a, from Indian Ocean (as Indian Seas), see Smith, M.A. (1926). *Monograph on the Sea Snakes. (Hydrophiidae).* London : British Museum xvii 130 pp.

Hydrus stokesii Gray, J.E. (1846). Descriptions of some new Australian reptiles. *in* Stokes, J.L. (ed.) *Discoveries in Australia*; With an Account of the Coasts and Rivers Explored and Surveyed During the Voyage of H.M.S. *Beagle* in the years 1837-38-39-40-41-42-43. London : T. & W. Boone 1: 498–504 [502 pl 3]. Type data: holotype, BMNH 1946.1.17.12, from Australian Seas.

Hydrophis schizopholis Schmidt, P. (1846). Beschreibung zweier neuen Reptilien, aus dem Naturhistorischen Museum zu Hamburg. *Abh. Geb. Naturw. Hamb.* **1**: 164–172 [166 pl 15 figs 1–7]. Type data: holotype, ZMH 387, *vide* Smith, M.A. (1926). *Monograph on the Sea Snakes. (Hydrophiidae).* London : British Museum xvii 130 pp., from China Sea.

Hydrus annulatus Gray, J.E. (1849). *Catalogue of the Specimens of Snakes in the Collection of the British Museum.* London : British Museum xv 125 pp. [59]. Type data: holotype, BMNH 1946.1.19.74, from Singapore.

Hydrophis guentheri Theobald, W. (1868). Catalogue of reptiles in the Museum of the Asiatic Society of Bengal; second sub-order, Serpentes Colubrini Venenati. *J. Asiat. Soc. Beng.* **37**: 67–74 [69] [as *Güntheri* in original description]. Type data: holotype from mouth of Hooghly River, India (as "probably from the Sandheads") presumed lost, *vide* Smith, M.A. (1926). *Monograph on the Sea Snakes. (Hydrophiidae).* London : British Museum xvii 130 pp.

Hydrophis granosa Anderson, J. (1871). On some Indian reptiles. *Proc. Zool. Soc. Lond.* **1871**: 149–211 [190]. Type data: holotype, IM, from mouth of Hooghly River, India (as Sand Heads).

Hydrophis guttata Murray, J.A. (1887). Description of three new species of *Hydrophis* from the Bombay harbour and the Mekran coast. *J. Bombay Nat. Hist. Soc.* **2**: 32–35 [33]. Type data: holotype, BMNH 1946.1.19.73, from Makran coast, Iran and Pakistan (as Mekran Coast).

Synonymy that of Cogger, H.G., this work; Smith, M.A. (1926). *Monograph on the Sea Snakes. (Hydrophiidae).* London : British Museum xvii 130 pp.

Distribution: NW Shelf, N coast, Arafura Sea, Gulf of Carpentaria, Great Barrier Reef, NE coast, Coral Sea, SE coast, W.A., N.T., Qld., N.S.W.; Timor Sea. Ecology: aquatic, noctidiurnal, nektonic, predator, estuarine, inshore, continental shelf; perennially active, viviparous, piscivore. Biological references: Smith, M.A. (1926). *Monograph on the Sea Snakes. (Hydrophiidae).* London : British Museum xvii 130 pp. (taxonomy); Cogger, H.G. (1975). The sea snakes of Australia and New Guinea. pp. 59–139 *in* Dunson, W. (ed.) *The Biology of Sea Snakes.* Baltimore : University Park Press (taxonomy); McDowell, S.B. (1972). The genera of sea-snakes of the *Hydrophis* group (Serpentes : Elapidae). *Trans. Zool. Soc. Lond.* **32**: 189–247 (taxonomy); Tamiya, N. & Maeda, N. (1979). Neurotoxins of Australian sea snakes. pp. 95–101 *in* Chubb, I.W. & Greffen, L.B. (eds.) *Neurotoxins : Fundamental and Clinical Advances.* Adelaide : Adelaide Univ. Union Press (venom); Dunson, W.A. & Minton, S.A. (1978). Diversity, distribution, and ecology of Philippine marine snakes (Reptilia, Serpentes). *J. Herpet.* **12**: 281–286; Maeda, N. & Tamiya, N. (1978). Three neurotoxins from the venom of a sea snake *Astrotia stokesii*, including two long-chain neurotoxic proteins with amidated C-termini. *Biochem. J.* **175**: 507–517; Smith, L.A. (1974). The sea-snakes of Western Australia (Serpentes : Elapidae, Hydrophiinae) with a description of a new subspecies. *Rec. West. Aust. Mus.* **3**: 93–110 (taxonomy); Redfield, J.A., Holmes, J.C. & Holmes, R.D. (1978). Sea snakes of the eastern Gulf of Carpentaria. *Aust. J. Mar. Freshwat. Res.* **29**: 325–334 (distribution and abundance).

Disteira Lacépède, 1804

Disteira Lacépède, B.G.E. (1804). Mémoire sur plusieurs animaux de la Nouvelle-Hollande dont la description n'a pas encore été publiée. *Ann. Mus. Natl. Hist. Nat. Paris* **4**: 184–211 [210]. Type species *Disteira doliata* Lacépède, 1804 by monotypy.

Distira Boulenger, G.A. (1896). *Catalogue of Snakes in the British Museum (Natural History).* 3. London : British Museum xiv 727 pp. pls 25 [285] [invalid emend. *pro Disteira* Lacépède, 1804].

Melanomystax Wall, F. (1921). *Ophidia Taprobanica* or the Snakes of Ceylon. Colombo : Cottle xxii 581 pp. [381]. Type species *Hydrophis nigrocinctus* Daudin, 1803 by monotypy.

Synonymy that of Cogger, H.G., this work; Smith, M.A. (1926). *Monograph on the Sea Snakes. (Hydrophiidae).* London : British Museum xvii 130 pp.

This group is also found in waters of India and Burma, Indo-Malayan Archipelago and New Guinea.

Disteira kingii (Boulenger, 1896)

Hydrophis kingii Boulenger, G.A. (1896). *Catalogue of Snakes in the British Museum (Natural History).* 3. London : British Museum xiv 727 pp. pls 25 [276]. Type data: holotype, BMNH 1946.1.10.10, from N. Australia.

Distribution: NW Shelf, N coast, Arafura Sea, Gulf of Carpentaria, Great Barrier Reef, NE coast, Coral Sea, W.A., N.T., Qld., (N.S.W.); Timor Sea. Ecology: aquatic, nektonic, predator, estuarine, inshore, continental shelf; perennially active, viviparous, piscivore. Biological references: Smith, M.A. (1926). *Monograph on the Sea Snakes. (Hydrophiidae).* London : British Museum xvii 130 pp. (taxonomy as *Hydrophis kingii*); Cogger, H.G. (1975). The sea snakes of Australia and New Guinea. pp. 59–139 *in* Dunson, W. (ed.) *The Biology of Sea Snakes.* Baltimore : University Park Press (taxonomy); McDowell, S.B. (1972). The genera of sea-snakes of the *Hydrophis* group (Serpentes : Elapidae). *Trans. Zool. Soc. Lond.* **32**: 189–247 (taxonomy); Smith, L.A. (1974). The sea snakes of Western Australia (Serpentes : Elapidae : Hydrophiinae) with a description of a new subspecies. *Rec. West. Aust. Mus.* **3**: 93–110 (taxonomy, as *Hydrophis kingii*); Redfield, J.A., Holmes, J.C. & Holmes, R.D. (1978). Sea snakes of the eastern Gulf of Carpentaria. *Aust. J. Mar. Freshwat. Res.* **29**: 325–334 (distribution and abundance).

Disteira major (Shaw, 1802)

***Hydrus major* (part.)** Shaw, G. (1802). *General Zoology*, or Systematic Natural History. Vol. 3 Pt. 2

Amphibia. London : Kearsley vii pp. 313–615 [558 pl 124]. Type data: lectotype [for the identity of paralectotype see *Astrotia stokesii*], BMNH 1946.1.9.24, from Indian Ocean (as Indian Seas), see Smith, M.A. (1926). *Monograph on the Sea Snakes. (Hydrophiidae).* London : British Museum xvii 130 pp., designation by Cogger, H.G., this work.

Disteira doliata Lacépède, B.G.E. (1804). Mémoire sur plusieurs animaux de la Nouvelle-Hollande dont la description n'a pas encore été publiée. *Ann. Mus. Natl. Hist. Nat. Paris* **4**: 184–211 [199, 210 pl 57 fig 2]. Type data: presumed lost, from Australia.

Pelamis shavii Merrem, B. (1820). *Versuch eines Systems Amphibien. Tentamen Systematis Amphibiorum.* Marburg : Krieger xv 199 pp. [139] [*nom. substit. pro Hydrus major* Shaw, 1802].

Hydrophis mentalis Gray, J.E. (1842). Monographic synopsis of the water snakes, or the family Hydridae. *in* Gray, J.E. (ed.) *Zoological Miscellany.* London : Treuttel, Würtz & Co. pp. 59–68 [62]. Type data: holotype, BMNH 1946.1.9.24, from Indian Ocean.

Disteira dumerilii Jan, G. (1859). Plan d'une *Iconographie descriptive des Ophidiens*, et Description sommaire de nouvelles espèces de Serpents: les Serpents de mer. *Rev. Mag. Zool. (2)* **11**: 148–157 [149]. Type data: holotype, MNHP 7705, from Australia.

Hydrophis lacepedei Jan, G. (1859). Prodrome d'une *Iconographie descriptive des Ophidiens* et Description sommaire de nouvelles espèces de Serpents venimeux. Reprinted from: *Rev. Mag. Zool. (2)* **10** and *(2)* **11–12**, 1858 and 1859. Paris : Bouchard-Huzard 32 pp. [pl D] [?*lapsus pro Disteira dumerilii* Jan, 1859 *vide* Smith, M.A. (1926). *Monograph on the Sea Snakes. (Hydrophiidae).* London : British Museum xvii 130 pp.].

Distira nasalis De Vis, C.W. (1905). A new genus of lizards. *Ann. Qd. Mus.* **6**: 46–52 pl 15 [48]. Type data: holotype, QM J203, from Qld. coast.

Synonymy that of Cogger, H.G., this work; Smith, M.A. (1926). *Monograph on the Sea Snakes. (Hydrophiidae).* London : British Museum xvii 130 pp.

Distribution: NW Shelf, (W. Coast), N coast, Arafura Sea, Gulf of Carpentaria, Great Barrier Reef, NE coast, (SE coast), Coral Sea, W.A., N.T., Qld., (N.S.W.); Timor Sea. Ecology: aquatic, nektonic, predator, estuarine, inshore, continental shelf; perennially active, viviparous, piscivore. Biological references: Smith, M.A. (1926). *Monograph on the Sea Snakes. (Hydrophiidae).* London : British Museum xvii 130 pp. (taxonomy); Cogger, H.G. (1975). The sea snakes of Australia and New Guinea. pp. 59–139 *in* Dunson, W. (ed.) *The Biology of Sea Snakes.* Baltimore : University Park Press (taxonomy); McDowell, S.B. (1972). The genera of sea-snakes of the *Hydrophis* group (Serpentes : Elapidae). *Trans. Zool. Soc. Lond.* **32**: 189–247 (taxonomy); Okonogi, T., Hattori, Z. Fukami, M., Inoue, T., Sato, M., Ajiki, Y. & Kogure, Y. (1980). A miscellany report of sea snakes. *The Snake* **12**: 19–24 (natural history); Smith, L.A. (1974). The sea-snakes of Western Australia (Serpentes : Elapidae, Hydrophiinae) with a description of a new subspecies. *Rec. West. Aust. Mus.* **3**: 93–110 (taxonomy, as *Hydrophis major*); Redfield, J.A., Holmes, J.C. & Holmes, R.D. (1978). Sea snakes of the eastern Gulf of Carpentaria. *Aust. J. Mar. Freshwat. Res.* **29**: 325–334 (distribution and abundance).

Emydocephalus Krefft, 1869

Emydocephalus Krefft, G. (1869). *The Snakes of Australia*; an Illustrated and Descriptive Catalogue of All the Known Species. Sydney : Govt. Printer xxv 100 pp. [92]. Type species *Emydocephalus annulatus* Krefft, 1869 by subsequent designation, Cogger, H.G., this work.

This group is also found in waters of Ryukyu Archipelago, New Guinea and islands of western Pacific.

Emydocephalus annulatus Krefft, 1869

Emydocephalus annulatus Krefft, G. (1869). *The Snakes of Australia*; an Illustrated and Descriptive Catalogue of All the Known Species. Sydney : Govt. Printer xxv 100 pp. [8 Sept. 1869] [92]. Type data: syntypes, AM R454, AM R6633, from "probably the Australian Seas."

Emydocephalus tuberculatus Krefft, G. (1869). *The Snakes of Australia*; an Illustrated and Descriptive Catalogue of All the Known Species. Sydney : Govt. Printer xxv 100 pp. [8 Sept. 1869] [93]. Type data: holotype, AM R455, from "probably the Australian Seas."

Aipysurus chelonicephalus Bavay, A. (1869). Catalogue des reptiles de la Nouvelle-Calédonie et description d'espèces nouvelles. *Mém. Soc. Linn. Normandie* **15**: 1–37 [34]. Type data: holotype, Marine Museum Brest *vide* Smith, M.A. (1926). *Monograph on the Sea Snakes. (Hydrophiidae).* London : British Museum xvii 130 pp., from Lifu (as Lifou), Loyalty Ils., New Caledonia.

Synonymy that of Cogger, H.G., this work; Smith, M.A. (1926). *Monograph on the Sea Snakes. (Hydrophiidae).* London : British Museum xvii 130 pp.

Distribution: NW Shelf, N coast, Gulf of Carpentaria, Arafura Sea, Great Barrier Reef, NE coast, Coral Sea, W.A., N.T., Qld., Ashmore Is., Cartier Is.; Timor Sea. Ecology: aquatic, diurnal, nektonic, inshore; perennially active, viviparous, coral reefs, feed exclusively on fish eggs. Biological references: Smith, M.A. (1926). *Monograph on the Sea Snakes. (Hydrophiidae).* London : British Museum xvii 130 pp. (taxonomy); Cogger, H.G. (1975). The sea snakes of Australia and New Guinea. pp. 59–139 *in* Dunson, W. (ed.) *The Biology of Sea Snakes.* Baltimore : University Park Press (taxonomy); McDowell, S.B. (1969). Notes on the Australian sea-snake *Ephalophis greyi* M. Smith (Serpentes : Elapidae : Hydrophiinae) and the origin and classification of sea-snakes. *J. Linn. Soc. Zool.* **48**: 333–349 (taxonomy); Heatwole, H. (1981). Temperature relations of some sea snakes. *The Snake* **13**: 53–57 (thermoregulation);

Seymour, R.S. & Webster, M.E.D. (1975). Gas transport and blood acid-base balance in diving sea snakes. *J. Exp. Zool.* **191**: 169–182 (diving physiology); Voris, H.K. (1966). Fish eggs as the apparent sole food item for a genus of sea snake, *Emydocephalus* (Krefft). *Ecology* **47**: 152–154 (diet); Feder, M.E. (1980). Blood oxygen stores in the file snake *Acrochordus granulatus* and in other marine snakes. *Physiol. Zool.* **53**: 394–401 (physiology).

Enhydrina Gray, 1849

Enhydrina Gray, J.E. (1849). *Catalogue of the Specimens of Snakes in the Collection of the British Museum.* London : British Museum xv 125 pp. [47]. Type species *Hydrus valakadyn* Boie, 1827 by subsequent designation, see, Klemmer, K. (1963). Liste der rezenten Giftschlangen; Elapidae, Hydrophiidae, Viperidae und Crotalidae. Marburg : N.G. Elwert Universitats- U. Verlagsbuchhand pp. 253–464.

This group is also found in waters of Persian Gulf, Indian subcontinent, south-east Asia, Indo-Malayan Archipelago, Philippine Ils. and New Guinea.

Enhydrina schistosa (Daudin, 1803)

Hydrophis schistosus Daudin, F.M. (1803). *Histoire Naturelle, générale et particulière des Reptiles.* Paris : Dufart Tom. 7 436 pp. [386] [description based on pl 10 in Russell, P. (1801). *A Continuation of an Account of Indian Serpents.* London : Shakespeare Press 15 pp., in which no type locality was specified]. Type data: holotype, BMNH 1946.1.10.7 (*ex* BMNH 1921.7.28.1), *vide* Smith, M.A. (1926). *Monograph on the Sea Snakes. (Hydrophiidae).* London : British Museum xvii 130 pp., from Tranquebar, India.

Hydrus valakadyn Boie, F. (1827). Bemerkungen über Merrem's Versuch eines Systems der Amphibien. *Isis Von Oken, Jena* **20**: 508–566 [554]. Type data: holotype, pl 11 in Russell, P. (1801). *A Continuation of an Account of Indian Serpents.* London : Shakespeare Press 15 pp., from Tranquebar, India.

Disteira russelii Fitzinger, L.J. (1827). Über die Hydren oder Wasserschlangen. *Isis Von Oken, Jena* **20**: 731–741 [733]. Type data: holotype (probable), pl 11 in Russell, P. (1801). *A Continuation of an Account of Indian Serpents.* London : Shakespeare Press 15 pp., from Asia, Indian Ocean.

Hydrophis bengalensis Gray, J.E. (1842). Monographic synopsis of the water snakes, or the family Hydridae. *in* Gray, J.E. (ed.) *Zoological Miscellany.* London : Treuttel, Würtz & Co. pp. 59–68 [62]. Type data: holotype, BMNH 1946.1.1.73, from Bengal, India.

Hydrophis subfasciata Gray, J.E. (1842). Monographic synopsis of the water snakes, or the family Hydridae. *in* Gray, J.E. (ed.) *Zoological Miscellany.* London : Treuttel, Würtz & Co. pp. 59–68 [62]. Type data: holotype, BMNH 1946.1.1.90, from Bengal, India.

Thalassophis werneri Schmidt, P. (1852). Beiträge zur ferneren Kenntniss der Meerschlangen. *Abh. Geb. Naturw. Hamb.* **2**: 69–86 [84 pl 6]. Type data: holotype, ZMH 392, from Samarang, Java, Indonesia.

Hydrophis schistotus Jan, G. (1859). Prodrome d'une *Iconographie descriptive des Ophidiens* et Description sommaire de nouvelles espèces de Serpents venimeux. Reprinted from: *Rev. Mag. Zool. (2)* **10** and *(2)* **11–12**, 1858 and 1859. Paris : Bouchard-Huzard 32 pp. [23] [*errore pro Hydrophis schistosus* Daudin, 1803].

Synonymy that of Cogger, H.G., this work; Smith, M.A. (1926). *Monograph on the Sea Snakes. (Hydrophiidae).* London : British Museum xvii 130 pp.

Distribution: N coast, Arafura Sea, Great Barrier Reef, Gulf of Carpentaria, NE coast, N.T., Qld. Ecology: aquatic, nektonic, predator, inshore, continental shelf; perennially active, viviparous. Biological references: Smith, M.A. (1926). *Monograph on the Sea Snakes. (Hydrophiidae).* London : British Museum xvii 130 pp. (taxonomy); Cogger, H.G. (1975). The sea snakes of Australia and New Guinea. pp. 59–139 *in* Dunson, W. (ed.) *The Biology of Sea Snakes.* Baltimore : University Park Press (taxonomy); McDowell, S.B. (1972). The genera of sea-snakes of the *Hydrophis* group (Serpentes : Elapidae). *Trans. Zool. Soc. Lond.* **32**: 189–247 (taxonomy); Voris, H.K. & Moffett, M.W. (1981). Size and proportion relationship between the beaked sea snake and its prey. *Biotropica* **13**: 15–19 (ecology); Lemen, C.A. & Voris, H.K. (1981). A comparison of reproductive strategies among marine snakes. *J. Anim. Ecol.* **50**: 89–101 (reproduction); Bergman, R.A.M. (1955). L'anatomie de *Enhydrina schistosa. Arch. Neerl. Zool.* **11**: 127–142; Karlsson, E., Eaker, D., Fryklund, L. & Kadin, S. (1972). Chromatographic separation of *Enhydrina schistosa* venom and the characterization of two principal neurotoxins. *Biochem.* **11**: 4628–4633; Kasturirangen, L.R. (1951). Placentation in the sea snake, *Enhydrina schistosa* (Daudin). *Proc. Indian Acad. Sci.* **348**: 1–32; Thireau, M. (1968). Analyse descriptivé et biométrique de la colonne vertébrale du serpent marin *Enhydrina schistosa* Daudin (Hydrophiinés). Caractères vertebraux des hydrophiidés. *Bull. Mus. Natl. Hist. Nat.* **39**: 1044–1056; Voris, H.K., Voris, H.H. & Liat, L.B. (1978). The food and feeding behaviour of the marine snake, *Enhydrina schistosa* (Hydrophiidae). *Copeia* **1978**: 134–146; Voris, H.K. & Jayne, B.C. (1979). Growth, reproduction and population structure of a marine snake, *Enhydrina schistosa* (Hydrophiidae). *Copeia* **1979**: 307–318; Redfield, J.A., Holmes, J.C. & Holmes, R.D. (1978). Sea snakes of the eastern Gulf of Carpentaria. *Aust. J. Mar. Freshwat. Res.* **29**: 325–334 (distribution and abundance).

Ephalophis Smith, 1931

Ephalophis Smith, M.A. (1931). Description of a new genus of sea snake from the coast of Australia, with a note on the structure providing for complete closure of the mouth in aquatic snakes. *Proc. Zool. Soc. Lond.* **1931**: 397–398 [397]. Type species *Ephalophis greyi* Smith, 1931 by monotypy.

Ephalophis greyi Smith, 1931

Ephalophis greyi Smith, M.A. (1931). Description of a new genus of sea snake from the coast of Australia, with a note on the structure providing for complete closure of the mouth in aquatic snakes. *Proc. Zool. Soc. Lond.* **1931**: 397–398 [397 pl 1]. Type data: holotype, BMNH 1946.1.6.92, from Cape Boileau, W.A.

Distribution: NW Shelf, (N coast), W.A., (N.T.). Ecology: amphibious, predator, sub-littoral, estuarine, inshore; perennially active, viviparous. Biological references: Smith, M.A. (1926). *Monograph on the Sea Snakes. (Hydrophiidae).* London : British Museum xvii 130 pp. (taxonomy); McDowell, S.B. (1969). Notes on the Australian sea-snake *Ephalophis greyi* M. Smith (Serpentes : Elapidae : Hydrophiinae) and the origin and classification of sea-snakes. *J. Linn. Soc. Zool.* **48**: 333–349 (taxonomy); Cogger, H.G. (1975). The sea snakes of Australia and New Guinea. pp. 59–139 *in* Dunson, W. (ed.) *The Biology of Sea Snakes.* Baltimore : University Park Press (taxonomy); McDowell, S.B. (1974). Additional notes on the rare and primitive sea snake, *Ephalophis greyi. Trans. Zool. Soc. Lond.* **32**: 195–247 (taxonomy); Smith, L.A. (1974). The sea-snakes of Western Australia (Serpentes : Elapidae, Hydrophiinae) with a description of a new subspecies. *Rec. West. Aust. Mus.* **3**: 93–110 (taxonomy).

Hydrelaps Boulenger, 1896

Hydrelaps Boulenger, G.A. (1896). *Catalogue of Snakes in the British Museum (Natural History).* 3. London : British Museum xiv 727 pp. pls 25 [270]. Type species *Hydrelaps darwiniensis* Boulenger, 1896 by monotypy.

This group is also found in southern New Guinea west of Torres Strait.

Hydrelaps darwiniensis Boulenger, 1896

Hydrelaps darwiniensis Boulenger, G.A. (1896). *Catalogue of Snakes in the British Museum (Natural History).* 3. London : British Museum xiv 727 pp. pls 25 [270 pl 12 fig 1]. Type data: syntypes, BMNH 1946.1.1.91–92, from Port Darwin, N.T.

Distribution: NW Shelf, N coast, Gulf of Carpentaria, W.A., N.T., Qld. Ecology: amphibious, predator, sub-littoral, estuarine, inshore; perennially active, viviparous. Biological references: Smith, M.A. (1926). *Monograph on the Sea Snakes. (Hydrophiidae).* London : British Museum xvii 130 pp. (taxonomy); McDowell, S.B. (1972). The genera of sea-snakes of the *Hydrophis* group (Serpentes : Elapidae). *Trans. Zool. Soc. Lond.* **32**: 189–247 (taxonomy); Cogger, H.G. (1975). The sea snakes of Australia and New Guinea. pp. 59–139 *in* Dunson, W. (ed.) *The Biology of Sea Snakes.* Baltimore : University Park Press (taxonomy); Brongersma, L.D. (1956). Notes on New Guinean reptiles and amphibians. V. *Proc. Ned. Akad. Wet.* **59C**: 599–610 (taxonomy); Smith, L.A. (1974). The sea-snakes of Western Australia (Serpentes : Elapidae, Hydrophiinae) with a description of a new subspecies. *Rec. West. Aust. Mus.* **3**: 93–110 (taxonomy).

Hydrophis Sonnini and Latreille, 1802

Hydrophis Sonnini de Manoncourt, C.S. & Latreille, P.A. (1802). *Histoire Naturelle des Reptiles*, avec Figures Déssinnées d'après Nature *in* Buffon, G.L.L. de *Histoire Naturelle de Buffon, classée d'après le systême de Linné par R.R. Castel. Nouvelle édition. (Suites).* Paris : Déterville Vol. 4 [193]. Publication date established from Anon (1903). *Catalogue of the Books, Manuscripts, Maps and Drawings in the British Museum (Natural History). A—D.* London : British Museum 1: viii 500 pp. Type species *Hydrophis laticauda* Sonnini and Latreille, 1802 [=*Hydrophis fasciatus* (Schneider, 1799)] by subsequent designation, Smith, M.A. (1926). *Monograph on the Sea Snakes. (Hydrophiidae).* London : British Museum xvii 130 pp.

Leioselasma Lacépède, B.G.E. (1804). Mémoire sur plusieurs animaux de la Nouvelle-Hollande dont la description n'a pas encore été publiée. *Ann. Mus. Natl. Hist. Nat. Paris* **4**: 184–211 [198, 210]. Type species *Leioselasma striata* Lacépède, 1804 [*species inquirenda*] by monotypy.

Microcephalophis Lesson, R.-P. (1834). Reptiles, pp. 289–336 *in* Bélanger, C. *Voyage aux Indes-Orientales, par le Nord de l'Europe pendant les Années 1825, 1826, 1827, 1828 et 1829.* Paris : Arthus Bertrand xxxix 535 pp. [320]. Type species *Microcephalophis gracilis* Lesson, 1834 by monotypy. [?*non Hydrus gracilis* Shaw, 1802; Smith, M.A. (1926). *Monograph on the Sea Snakes. (Hydrophiidae).* London : British Museum xvii 130 pp. and McDowell, S.B. (1972). The genera of sea-snakes of the *Hydrophis* group (Serpentes : Elapidae). *Trans. Zool. Soc. Lond.* **32**: 189–247, both regard *Microcephalophis gracilis* Lesson, 1834 as a new combination for *Hydrus gracilis* Shaw, 1802. Lesson, however, makes no reference to Shaw's species, and includes only several species figured by Russell, P. (1801). *A Continuation of an Account of Indian Serpents.* London : Shakespeare Press 15 pp., [pls 6–10] which were not conspecific with *H. gracilis* of Shaw, the latter being represented by pl 13 in Russell].

Liopala Gray, J.E. (1842). Monographic synopsis of the water snakes, or the family Hydridae. *in* Gray, J.E. (ed.) *Zoological Miscellany.* London : Treuttel, Würtz & Co. pp. 59–68 [60]. Type species *Hydrus gracilis* Shaw, 1802 by subsequent designation, Gray, J.E. (1849). *Catalogue* of the Specimens of Snakes in the Collection of the British Museum. London : British Museum xv 125 pp.

Liopola Gray, J.E. (1842). Monographic synopsis of the water snakes, or the family Hydridae. *in* Gray, J.E. (ed.) *Zoological Miscellany.* London : Treuttel, Würtz & Co. pp. 59–68 [60] [*errore pro Liopala* Gray, 1842].

Asturia Gray, J.E. (1842). Description of some hitherto unrecorded species of Australian reptiles and batrachians. *in* Gray, J.E. (ed.) *Zoological Miscellany.* London : Treuttel, Würtz & Co. pp. 51–57 [55] [*nom. nud.*; date of publication, Apr. 1842].

Aturia Gray, J.E. (1842). Monographic synopsis of the water snakes, or the family Hydridae. *in* Gray, J.E. (ed.) *Zoological Miscellany.* London : Treuttel, Würtz & Co. pp. 59–68 [61] [date of publication, May 1842]. Type species *Aturia ornata* Gray, 1842 by subsequent designation, see Smith, M.A. (1926). *Monograph on the Sea Snakes. (Hydrophiidae).* London : British Museum xvii 130 pp.

Chitulia Gray, J.E. (1849). *Catalogue of the Specimens of Snakes in the Collection of the British Museum.* London : British Museum xv 125 pp. [36, 56]. Type species *Chitulia inornata* Gray, 1849 by subsequent designation, see Smith, M.A. (1926). *Monograph on the Sea Snakes. (Hydrophiidae).* London : British Museum xvii 130 pp.

Porrecticollis Wall, F. (1921). *Ophidia Taprobanica* or the Snakes of Ceylon. Colombo : Cottle xxii 581 pp. [335]. Type species *Hydrophis obscurus* Daudin, 1803 by monotypy.

Dolichodira Wall, F. (1921). *Ophidia Taprobanica* or the Snakes of Ceylon. Colombo : Cottle xxii 581 pp. [399]. Type species *Hydrophis diadema* Günther, 1864 (=*Hydrophis torquatus* Günther, 1864) by monotypy.

Polyodontognathus Wall, F. (1921). *Ophidia Taprobanica* or the Snakes of Ceylon. Colombo : Cottle xxii 581 pp. [374]. Type species *Hydrus caerulescens* Shaw, 1802 by monotypy.

Polypholophis Wall, F. (1921). *Ophidia Taprobanica* or the Snakes of Ceylon. Colombo : Cottle xxii 581 pp. [380]. Type species *Hydrophis neglectus* Wall, 1906 (=*Hydrophis stricticollis* Günther, 1864) by monotypy.

Micromastophis Wall, F. (1921). *Ophidia Taprobanica* or the Snakes of Ceylon. Colombo : Cottle xxii 581 pp., [*errata* insert; *nom. substit. pro Aturia* Wall, F. (1921). *Ophidia Taprobanica* or the Snakes of Ceylon. Colombo : Cottle xxii 581 pp. [344]; *non Aturia* Gray, 1842]. Type species *Hydrus fasciatus* Schneider, 1799 [=*Hydrophis fasciatus* (Schneider, 1799)] by monotypy.

Synonymy that of Cogger, H.G., this work; Smith, M.A. (1926). *Monograph on the Sea Snakes. (Hydrophiidae).* London : British Museum xvii 130 pp.

This group is also found in waters of Persian Gulf, Indian subcontinent, south-east Asia, Indo-Malayan Archipelago, Philippine Ils. New Guinea and islands of western Pacific.

Hydrophis atriceps Günther, 1864

Hydrophis atriceps Günther, A. (1864). *The Reptiles of British India.* London : Royal Society xxvii 444 pp. [371 pl 25 fig 1]. Type data: syntypes, BMNH 1946.1.2.62, BMNH 63.9.29.5, from Thailand.

Hydrophis alcocki Wall, F. (1906). A descriptive list of the sea-snakes in the Indian Museum. *Mem. Asiat. Soc. Beng.* **1**: 277–299 [288 pl 15 fig 3]. Type data: holotype, IM 14470, *vide* Smith, M.A. (1926). *Monograph on the Sea Snakes. (Hydrophiidae).* London : British Museum xvii 130 pp., from Puri, Orissa, India.

Disteira cincinnatii Van Denburgh, J. & Thompson, J.C. (1908). Description of a new species of sea snake from the Philippine Islands, with a note on the palatine teeth in the Proteroglypha. *Proc. Calif. Acad. Sci.* **3**: 41–48 [42 pl 1]. Type data: holotype, CAS 15016, from 1 mi. NE of Cavite, Manila Bay, Philippines.

Synonymy that of Cogger, H.G., this work; Smith, M.A. (1926). *Monograph on the Sea Snakes. (Hydrophiidae).* London : British Museum xvii 130 pp.

Distribution: (NW Shelf), N coast, Arafura Sea, (Gulf of Carpentaria), W.A., N.T., (Qld.); Timor Sea. Ecology: aquatic, nektonic, predator, inshore, continental shelf; perennially active, viviparous, piscivore. Biological references: Smith, M.A. (1926). *Monograph on the Sea Snakes. (Hydrophiidae).* London : British Museum xvii 130 pp. (taxonomy); McDowell, S.B. (1972). The genera of sea-snakes of the *Hydrophis* group (Serpentes : Elapidae). *Trans. Zool. Soc. Lond.* **32**: 189–247 (taxonomy); Cogger, H.G. (1975). The sea snakes of Australia and New Guinea. pp. 59–139 *in* Dunson, W. (ed.) *The Biology of Sea Snakes.* Baltimore : University Park Press (taxonomy); Bergman, R.A.M. (1962). The anatomy of *Hydrophis fasciatus atriceps. Biol. Jaarb.* **30**: 389–416.

Hydrophis belcheri (Gray, 1849)

Aturia belcheri Gray, J.E. (1849). *Catalogue of the Specimens of Snakes in the Collection of the British Museum.* London : British Museum xv 125 pp. [46]. Type data: holotype, BMNH 1946.1.1.97, from New Guinea.

Hydrophis pachycercos Fischer, J.G. (1856). Die Familie der Seeschlangen. *Abh. Geb. Naturw. Hamburg* **3**: 1–78 [44 pl 2]. Type data: holotype, ZMH 390, *vide* Smith, M.A. (1926). *Monograph on the Sea Snakes. (Hydrophiidae).* London : British Museum xvii 130 pp., from Indian Seas.

Hydrophis pachycerios Jan, G. (1859). Prodrome d'une *Iconographie descriptive des Ophidiens* et Description sommaire de nouvelles espèces de Serpents venimeux. Reprinted from: *Rev. Mag. Zool. (2)* **10** and *(2)* **11–12**, 1858 and 1859. Paris : Bouchard-Huzard 32 pp. [24] [*lapsus pro Hydrophis pachycercos* Fischer, 1856].

Synonymy that of Cogger, H.G., this work; Smith, M.A. (1926). *Monograph on the Sea Snakes. (Hydrophiidae).* London : British Museum xvii 130 pp.

Distribution: N coast, Arafura Sea, Gulf of Carpentaria, Coral Sea, N.T., Qld. Ecology: aquatic, nektonic, inshore, continental shelf; viviparous, piscivore. Biological references: Smith, M.A. (1926). *Monograph on the Sea Snakes. (Hydrophiidae).* London : British Museum xvii 130 pp. (taxonomy); McDowell, S.B. (1972). The genera of sea-snakes of the *Hydrophis* group (Serpentes : Elapidae). *Trans. Zool. Soc. Lond.* **32**: 189–247 (taxonomy); Cogger, H.G. (1975). The sea snakes of Australia and New Guinea. pp. 59–139 *in* Dunson, W. (ed.) *The Biology of Sea Snakes.* Baltimore : University Park Press (taxonomy).

Hydrophis caerulescens (Shaw, 1802)

Hydrus caerulescens Shaw, G. (1802). *General Zoology,* or Systematic Natural History. Vol. 3 Pt. 2 Amphibia. London : Kearsley vii pp. 313–615 [561] [label on holotype cites type locality as "Indian Ocean : Vizagapatam"]. Type data: holotype, BMNH 1946.1.3.90 M, from Indian Ocean (as East-Indian), see Smith, M.A. (1926). *Monograph on the Sea Snakes. (Hydrophiidae).* London : British Museum xvii 130 pp.

Polyodontes annulatus Lesson, R.-P. (1834). Reptiles, pp. 289–336 *in* Bélanger, C. *Voyage aux Indes-Orientales, par le Nord de l'Europe pendant les Années 1825, 1826, 1827, 1828 et 1829.* Paris : Arthus Bertrand xxxix 535 pp. [325 pl 4] [synonymy doubtful, *vide* Smith, M.A. (1926). *Monograph on the Sea Snakes. (Hydrophiidae).* London : British Museum xvii 130 pp.]. Type data: type(s) presumed lost, from Malabar coast, India and Gulf of Martaban, Burma.

Hydrophis hybrida Schlegel, H. (1844). *Abbildungen neuer oder unvollständig bekannter Amphibien, nach der Natur der dem Leben entworfen.* Düsseldorf : Arnz xiv 141 pp. [115 pl 37]. Type data: holotype, RMNH 1492 M, *vide* Smith, M.A. (1926). *Monograph on the Sea Snakes. (Hydrophiidae).* London : British Museum xvii 130 pp., from Moluccas, Indonesia.

H[ydrophis] protervus Jan, G. (1859). Plan d'une *Iconographie descriptive des Ophidiens,* et Description sommaire de nouvelles espèces de Serpents: les Serpents de mer. *Rev. Mag. Zool. (2)* **11**; 148–157; Jan, G. (1859). Prodrome d'une *Iconographie descriptive des Ophidiens* et Description sommaire de nouvelles espèces de Serpents venimeux. Reprinted from: *Rev. Mag. Zool. (2)* **10** and *(2)* **11–12**, 1858 and 1859. Paris : Bouchard-Huzard 32 pp. [148; pl D]. Type data: holotype (probable), MNHP *vide* Smith, M.A. (1926). *Monograph on the Sea Snakes. (Hydrophiidae).* London : British Museum xvii 130 pp., from "Mer de la Chine?".

H[ydrophis] frontalis Jan, G. (1863). *Elenco Sistematico Degli Ofidi Descritti e Disegnati per l'Iconografia Generale.* Milan : A. Lombardi 143 pp.; Jan, G. & Sordelli, F. (1881). *Iconographie Générale des Ophidiens.* Atlas 1860–1881 Milan : Jan & Sordelli 50 livr. [110; livr. 39 pl 5 fig 2]. Type data: holotype, RMNH 1477 M, from Indian Ocean.

H[ydrophis] polydonta Jan, G. (1863). *Elenco Sistematico Degli Ofidi Descritti e Disegnati per l'Iconografia Generale.* Milan : A. Lombardi 143 pp.; Jan, G. & Sordelli, F. (1881). *Iconographie Générale des Ophidiens.* Atlas 1860–1881 Milan : Jan & Sordelli 50 livr. [110; livr. 41 pl 1]. Type data: holotype, MSNM *vide* Smith, M.A. (1926). *Monograph on the Sea Snakes. (Hydrophiidae).* London : British Museum xvii 130 pp., from unknown locality.

Hydrophis caerulescens thai Smith, M.A. (1920). On sea snakes from the coasts of the Malay Peninsula, Siam and Cochin-China. *J. Fed. Malay States Mus.* **10**(1): 1–63 [18]. Type data: holotype, BMNH 1921.2.11.83 M, from Hua Hin, Gulf of Siam, Thailand.

Synonymy that of Cogger, H.G., this work; Smith, M.A. (1926). *Monograph on the Sea Snakes. (Hydrophiidae).* London : British Museum xvii 130 pp.

Distribution: Gulf of Carpentaria, Qld. Ecology: aquatic, nektonic, predator, inshore, continental shelf; perennially active, viviparous, piscivore. Biological references: Smith, M.A. (1926). *Monograph on the Sea Snakes. (Hydrophiidae)* London : British Museum xvii 130 pp. (taxonomy); McDowell, S.B. (1972). The genera of sea-snakes of the *Hydrophis* group (Serpentes : Elapidae). *Trans. Zool. Soc. Lond.* **32**: 189–247 (taxonomy); Cogger, H.G. (1975). The sea snakes of Australia and New Guinea. pp. 59–139 *in* Dunson, W. (ed.) *The Biology of Sea Snakes.* Baltimore : University Park Press (taxonomy); Redfield, J.A., Holmes, J.C. & Holmes, R.D. (1978). Sea snakes of the eastern Gulf of Carpentaria. *Aust. J. Mar. Freshwat. Res.* **29**: 325–334 (distribution and abundance); Bal, D.V. & Nawathe, K.V. (1949). The circulatory system of *Hydrophis caerulescens* (Shaw, 1802). *J. Univ. Bombay (ns)* **17**: 1–14.

Hydrophis elegans (Gray, 1842)

Aturia elegans Gray, J.E. (1842). Monographic synopsis of the water snakes, or the family Hydridae. *in* Gray, J.E. (ed.) *Zoological Miscellany.* London : Treuttel, Würtz & Co. pp. 59–68 [61]. Type data: holotype, BMNH 1946.1.3.89, from Port Essington, N.T.

Distira grandis Boulenger, G.A. (1896). *Catalogue of Snakes in the British Museum (Natural History).* 3. London : British Museum xiv 727 pp. pls 25 [293]. Type data: syntypes, BMNH 1946.1.17.10–11 (Malay Archipelago), BMNH 70.11.30.71 (Qld.), from Malay Archipelago and Qld.

Synonymy that of Cogger, H.G., this work; Smith, M.A. (1926). *Monograph on the Sea Snakes. (Hydrophiidae).* London : British Museum xvii 130 pp.

Distribution: W coast, NW Shelf, N coast, Arafura Sea, Gulf of Carpentaria, Great Barrier Reef, NE coast, Coral Sea, SE coast, W.A., N.T., Qld., N.S.W.; Timor Sea. Ecology: aquatic, noctidiurnal, nektonic, predator, inshore, continental shelf; perennially active, viviparous, piscivore, coral reefs. Biological references: Smith, M.A. (1926). *Monograph on the Sea Snakes. (Hydrophiidae).* London : British Museum xvii 130 pp. (taxonomy); McDowell, S.B. (1972). The genera of sea-snakes of the *Hydrophis* group (Serpentes : Elapidae). *Trans. Zool. Soc. Lond.* **32**: 189–247 (taxonomy); Cogger, H.G. (1975). The sea snakes of Australia and New Guinea. pp. 59–139 *in* Dunson, W. (ed.) *The Biology of Sea Snakes.* Baltimore : University Park Press (taxonomy); Heatwole, H. (1981). Temperature relations of some sea snakes. *The Snake* **13**: 53–57 (thermoregulation); Seymour, R.S. & Webster, M.E.D. (1975). Gas transport and blood acid-base balance in diving sea snakes. *J. Exp. Zool.* **191**: 169–182 (diving physiology); Redfield, J.A., Holmes, J.C. & Holmes, R.D. (1978). Sea snakes of the eastern Gulf of Carpentaria. *Aust. J. Mar. Freshwat. Res.* **29**: 325–334 (distribution and abundance).

Hydrophis gracilis (Shaw, 1802)

Hydrus gracilis Shaw, G. (1802). *General Zoology*, or Systematic Natural History. Vol. 3 Pt. 2 Amphibia. London : Kearsley vii pp. 313–615 [560]. Type data: holotype, BMNH 1946.1.17.37, from unknown locality.

Hydrus kadellnagam Boie, F. (1827). Bemerkungen über Merrem's Versuch eines Systems der Amphibien. *Isis Von Oken, Jena* **20**: 508–566 [554] [as *Hydrus Kadell-Nagam*]. Type data: holotype, pl 13 in Russell, P. (1801). *A Continuation of an Account of Indian Serpents.* London : Shakespeare Press 15 pp., from unknown locality.

Thalassophis microcephala Schmidt, P. (1852). Beiträge zur ferneren Kenntniss der Meerschlangen. *Abh. Geb. Naturw. Hamb.* **2**: 69–86 [78 pl 2]. Type data: syntypes, ZMH 396, ZMH 400, *vide* Smith, M.A. (1926). *Monograph on the Sea Snakes. (Hydrophiidae).* London : British Museum xvii 130 pp., from coast of Java, Indonesia.

Hydrophis leprogaster Duméril, A.M.C. & Bibron, G. (1856). *in* Fischer, J.G. Die Familie der Seeschlangen. *Abh. Geb. Naturw., Hamburg* **3**: 1–78 [53] [*nom. nud.* introduced in synonymy].

Hydrophis guntherii Murray, J.A. (1884). *The Vertebrate Zoology of the Sind.* London xvi 424 pp. [396]. Type data: holotype, BMNH 1946.1.17.38, from Karachi, Pakistan.

Hydrophis rostralis Smith, M.A. (1917). Preliminary diagnoses of four new sea snakes. *J. Nat. Hist. Soc. Siam* **2**: 340–342 [340]. Type data: holotype, "The Federated Malay States Museum, Kuala Lumpur 1102" *vide,* Smith, M.A. (1926). *Monograph on the Sea Snakes. (Hydrophiidae).* London : British Museum xvii 130 pp.

Synonymy that of Cogger this work; Smith, M.A. (1926). *Monograph on the Sea Snakes. (Hydrophiidae).* London : British Museum xvii 130 pp.

Distribution: Gulf of Carpentaria, Great Barrier Reef, Coral Sea, Qld. Ecology: aquatic, nektonic, inshore, continental shelf; perennially active, viviparous, piscivore. Biological references: Smith, M.A. (1926). *Monograph on the Sea Snakes. (Hydrophiidae).* London : British Museum xvii 130 pp. (taxonomy); McDowell, S.B. (1972). The genera of sea-snakes of the *Hydrophis* group (Serpentes : Elapidae). *Trans. Zool. Soc. Lond.* **32**: 189–247 (taxonomy); Cogger, H.G. (1975). The sea snakes of Australia and New Guinea. pp. 59–139 *in* Dunson, W. (ed.) *The Biology of Sea Snakes.* Baltimore : University Park Press (taxonomy).

Hydrophis inornatus (Gray, 1849)

Chitulia inornata Gray, J.E. (1849). *Catalogue of the Specimens of Snakes in the Collection of the British Museum.* London : British Museum xv 125 pp. [56]. Type data: holotype, BMNH 1946.1.1.27, from Indian Ocean.

Chitulia fasciata Gray, J.E. (1849). *Catalogue of the Specimens of Snakes in the Collection of the British Museum.* London : British Museum xv 125 pp. [56]. Type data: holotype, BMNH 1946.1.1.59, from Indian Ocean.

Thalassophis schlegelii Schmidt, P. (1852). Beiträge zur ferneren Kenntniss der Meerschlangen. *Abh. Geb. Naturw. Hamb.* **2**: 69–86 [83 pl 5]. Type data: syntypes, ZMH 393–395, *vide* Smith, M.A. (1926). *Monograph on the Sea Snakes. (Hydrophiidae).* London : British Museum xvii 130 pp., from Java, Indonesia.

Hydrophis manillae Owen, R. (1859). *Descriptive Catalogue of the Specimens of Natural History in Spirit Contained in the Museum of the Royal College of Surgeons. Vertebrata.* London : Royal College of Surgeons p. 77 [*nom. nud.*]. Type data: holotype, Museum Royal College of Surgeons no. 2164, from Manila, Philippines.

Hydrophis longiceps Günther, A. (1864). *The Reptiles of British India.* London : Royal Society xxvii 444 pp. [375 pl 25 fig 0] [*nom. substit. pro Chitulia fasciata* Gray, 1849].

Distira cyanosoma Wall, F. (1913). Some new snakes from the oriental region. *J. Bombay Nat. Hist. Soc.* **22**: 514–516 [516]. Type data: holotype, BMNH 1946.1.1.58, from Manila Bay, Philippines.

Synonymy that of Cogger, H.G., this work; Smith, M.A. (1926). *Monograph on the Sea Snakes. (Hydrophiidae).* London : British Museum xvii 130 pp.

Distribution: (NW Shelf), N coast, (Gulf of Carpentaria), (Arafura Sea), (W.A.), N.T.; (Timor Sea). Ecology: aquatic, noctidiurnal, nektonic, inshore, continental shelf; perennially active, viviparous, piscivore. Biological references: Smith, M.A. (1926). *Monograph on the Sea Snakes. (Hydrophiidae).* London : British Museum xvii 130 pp. (taxonomy); McDowell, S.B. (1972). The genera of sea-snakes of the *Hydrophis* group (Serpentes : Elapidae). *Trans. Zool. Soc. Lond.* **32**: 189–247 (taxonomy); Cogger, H.G. (1975). The sea snakes of Australia and New Guinea. pp. 59–139 *in* Dunson, W. (ed.) *The Biology of Sea Snakes.* Baltimore : University Park Press (taxonomy).

Hydrophis melanocephalus Gray, 1849

***Hydrophis sublaevis melanocephala* (part.)** Gray, J.E. (1849). *Catalogue of the Specimens of Snakes in the Collection of the British Museum.* London : British Museum xv 125 pp. [53]. Type data: lectotype, BMNH 1946.1.9.22 from Indian Ocean (type series from Indian Ocean and Madras, India), designation by Cogger, H.G., this work.

Disteira orientalis Stejneger, L. (1901). Diagnoses of eight new batrachians and reptiles from the Riu Kiu Archipelago, Japan. *Proc. Biol. Soc. Wash.* **14**: 189–191 [191]. Type data: holotype, TIU 29 *vide* Smith, M.A. (1926). *Monograph on the Sea Snakes. (Hydrophiidae).* London : British Museum xvii 130 pp., from Okinawa Shima, Japan.

Synonymy that of Cogger, H.G., this work; Smith, M.A. (1926). *Monograph on the Sea Snakes. (Hydrophiidae).* London : British Museum xvii 130 pp.

Distribution: (NW Shelf), (N coast), (Arafura Sea), (Coral Sea), (W.A.), (N.T.), (Qld.); Timor Sea. Ecology: aquatic, noctidiurnal, nektonic, inshore, continental shelf; perennially active, viviparous, piscivore, coral reefs. Biological references: Smith, M.A. (1926). *Monograph on the Sea Snakes. (Hydrophiidae).* London : British Museum xvii 130 pp. (taxonomy); McDowell, S.B. (1972). The genera of sea-snakes of the *Hydrophis* group (Serpentes : Elapidae). *Trans. Zool. Soc. Lond.* **32**: 189–247 (taxonomy); Cogger, H.G. (1975). The sea snakes of Australia and New Guinea. pp. 59–139 *in* Dunson, W. (ed.) *The Biology of Sea Snakes.* Baltimore : University Park Press (taxonomy); Seymour, R.S. & Webster, M.E.D. (1975). Gas transport and blood acid-base balance in diving sea snakes. *J. Exp. Zool.* **191**: 169–182 (diving physiology).

Hydrophis melanosoma Günther, 1864

Hydrophis melanosoma Günther, A. (1864). *The Reptiles of British India.* London : Royal Society xxvii 444 pp. [367 pl 25 fig E]. Type data: holotype, BMNH 1946.1.10.6, from unknown locality.

Hydrophis floweri Boulenger, G.A. (1898). Description of a new sea-snake from Borneo. *Proc. Zool. Soc. Lond.* **1898**: 106–107 [106 pl 9]. Type data: syntypes, BMNH 1946.1.9.48–51, from Brunei Bay, Borneo.

Synonymy that of Cogger, H.G., this work; Smith, M.A. (1926). *Monograph on the Sea Snakes. (Hydrophiidae).* London : British Museum xvii 130 pp.

Distribution: (NW Shelf), (N coast), (Arafura Sea), (Gulf of Carpentaria), Great Barrier Reef, Coral Sea, (W.A.), (N.T.), Qld. Ecology: aquatic, nektonic, inshore, continental shelf; perennially active, viviparous, piscivore. Biological references: Smith, M.A. (1926). *Monograph on the Sea Snakes. (Hydrophiidae).* London : British Museum xvii 130 pp. (taxonomy); McDowell, S.B. (1972). The genera of sea-snakes of the *Hydrophis* group (Serpentes : Elapidae). *Trans. Zool. Soc. Lond.* **32**: 189–247 (taxonomy); Cogger, H.G. (1975). The sea snakes of Australia and New Guinea. pp. 59–139 *in* Dunson, W. (ed.) *The Biology of Sea Snakes.* Baltimore : University Park Press (taxonomy).

Hydrophis obscurus Daudin, 1803

Hydrophis obscura Daudin, F.M. (1803). *Histoire Naturelle, générale et particulière des Reptiles.* Paris : Dufart Tom. 7 436 pp. [375]. Type data: holotype, pl 8 in Russell, P. (1801). *A Continuation of an Account of Indian Serpents.* London : Shakespeare Press 15 pp., from unknown locality.

Hydrophis cloris Daudin, F.M. (1803). *Histoire Naturelle, générale et particulière des Reptiles.* Paris : Dufart Tom. 7 436 pp. [377]. Type data: holotype, pl 7 in Russell, P. (1801). *A Continuation of an Account of Indian Serpents.* London : Shakespeare Press 15 pp., from unknown locality.

Hydrophis subcinctus Gray, J.E. (1842). Monographic synopsis of the water snakes, or the family Hydridae. *in* Gray, J.E. (ed.) *Zoological Miscellany.* London : Treuttel, Würtz & Co. pp. 59–68 [63]. Type data: holotype, presumed lost, from Indian Ocean.

Hydrophis latifasciata Günther, A. (1864). *The Reptiles of British India.* London : Royal Society xxvii 444 pp. [372 pl 25 fig T]. Type data: holotype F, BMNH 56.5.6.104, from coast of Mergui, Burma.

Hydrophis coronata Günther, A. (1864). *The Reptiles of British India.* London : Royal Society xxvii 444 pp. [372 pl 25 fig M]. Type data: syntypes, BMNH iii.3.6.a–b, *vide* Smith, M.A. (1926). *Monograph on the Sea Snakes. (Hydrophiidae).* London : British Museum xvii 130 pp., from coast of Bengal, India.

Synonymy that of Cogger, H.G., this work; Smith, M.A. (1926). *Monograph on the Sea Snakes. (Hydrophiidae).* London : British Museum xvii 130 pp.

Distribution: (Gulf of Carpentaria), Great Barrier Reef, Coral Sea, Qld. Ecology: aquatic, nektonic, inshore, continental shelf; perennially active, viviparous, piscivore. Biological references: Smith, M.A. (1926). *Monograph on the Sea Snakes. (Hydrophiidae).* London : British Museum xvii 130 pp. (taxonomy); McDowell, S.B. (1972). The genera of sea-snakes of the *Hydrophis* group (Serpentes : Elapidae). *Trans. Zool. Soc. Lond.* **32**: 189–247 (taxonomy); Cogger, H.G. (1975). The sea snakes of Australia and New Guinea. pp. 59–139 *in* Dunson, W. (ed.) *The Biology of Sea Snakes.* Baltimore : University Park Press (taxonomy).

Hydrophis ornatus (Gray, 1842)

Aturia ornata Gray, J.E. (1842). Monographic synopsis of the water snakes, or the family Hydridae. *in* Gray, J.E. (ed.) *Zoological Miscellany.* London : Treuttel, Würtz & Co. pp. 59–68 [61]. Type data: holotype, BMNH 1946.1.23.72, from unknown locality.

Hydrophis ocellata Gray, J.E. (1849). *Catalogue of the Specimens of Snakes in the Collection of the British Museum.* London : British Museum xv 125 pp. [53]. Type data: holotype, BMNH 1946.1.3.91, from Australia.

Hydrophis laevis Lütken, C. (1863). Nogle nye krybdyr og padder. *Vidensk. Meddr. Dansk Naturh. Foren.* **1862**: 292–331 [309 pl 1, pl 2 fig 6]. Type data: syntypes (probable), whereabouts unknown, from "easterly part of the globe".

Hydrophis ellioti Günther, A. (1864). *The Reptiles of British India.* London : Royal Society xxvii 444 pp. [377 pl 25 fig N]. Type data: syntypes, BMNH 1946.1.4.85 (Madras), BMNH 1946.1.6.85 (Ceylon), BMNH 1946.1.6.84, 4th syntype not located, from Madras, India; Sri Lanka; Thailand and unspecified.

Hydrophis godeffroyi Peters, W. (1872). Über den *Hydrus fasciatus* Schneider und andere Seeschlangen. *Mber. K. Preuss. Akad. Wiss. Berl.* **1872**: 848–861 [856 pl 1 fig 3]. Type data: syntypes, ZMB 7593 (2), ZMB 28399, *vide* Smith, M.A. (1926). *Monograph on the Sea Snakes. (Hydrophiidae).* London : British Museum xvii 130 pp., from Gilbert Ils. (as Kings-Mills Ils.), Kiribati.

Distira andamanica Annandale, N. (1905). Additions to the collection of oriental snakes in the Indian Museum. Part 2. Specimens from the Andamans and Nicobars. *J. Asiat. Soc. Beng. (ns)* **1**: 173–176 [174]. Type data: holotype, IM 15238, *vide* Smith, M.A. (1926). *Monograph on the Sea Snakes. (Hydrophiidae).* London : British Museum xvii 130 pp., from Andaman Ils., India.

Distira mjobergi Lönnberg, E. & Andersson, G.L. (1913). Results of Dr. E. Mjöberg's Swedish Scientific Expeditions to Australia 1910–1913. III. Reptiles. *K. Svenska Vetensk.-Akad. Handl.* **52**(3): 1–17 [13]. Type data: holotype, NHRM 2402, from Cape Jaubert, W.A.

Hydrophis lamberti Smith, M.A. (1917). Preliminary diagnoses of four new sea snakes. *J. Nat. Hist. Soc. Siam* **2**: 340–342 [340]. Type data: holotype, BMNH 1946.1.9.20, from mouth of the Mekong (as Meklong) River, Gulf of Siam, Vietnam.

Synonymy that of Cogger, H.G., this work; Smith, M.A. (1926). *Monograph on the Sea Snakes. (Hydrophiidae).* London : British Museum xvii 130 pp.

Distribution: W coast, NW Shelf, N coast, Arafura Sea, Gulf of Carpentaria, Great Barrier Reef, NE coast, Coral Sea, SE coast, W.A., N.T., Qld., N.S.W. Ecology: aquatic, noctidiurnal, nektonic, inshore, continental shelf; perennially active, viviparous, piscivore. Biological references: Smith, M.A. (1926). *Monograph on the Sea Snakes. (Hydrophiidae).* London : British Museum xvii 130 pp. (taxonomy); McDowell, S.B. (1972). The genera of sea-snakes of the *Hydrophis* group (Serpentes : Elapidae). *Trans. Zool. Soc. Lond.* **32**: 189–247 (taxonomy); Cogger, H.G. (1975). The sea snakes of Australia and New Guinea. pp. 59–139 *in* Dunson, W. (ed.) *The Biology of Sea Snakes.* Baltimore : University Park Press (taxonomy); Heatwole, H. (1981). Temperature relations of some sea snakes. *The Snake* **13**: 53–57 (thermoregulation); Lemen, C.A. & Voris, H.K. (1981). A comparison of reproductive strategies among marine snakes. *J. Anim. Ecol.* **50**: 89–101 (reproduction); Tamiya, N. & Maeda, N. (1979). Neurotoxins of Australian sea snakes. pp. 95–101 *in* Chubb, I.W. & Greffen, L.B. (eds.) *Neurotoxins : Fundamental and Clinical Advances.* Adelaide : Adelaide Univ. Union Press (venom); Mittleman, M.B. (1947). Geographic variation in the sea snake *Hydrophis ornatus* (Gray). *Proc. Biol. Soc. Wash.* **60**: 1–8; Smith, L.A. (1974). The sea-snakes of Western Australia (Serpentes : Elapidae, Hydrophiinae) with a description of a new subspecies. *Rec. West. Aust. Mus.* **3**: 93–110 (taxonomy, as *H. ocellatus*); Redfield, J.A., Holmes, J.C. & Holmes, R.D. (1978). Sea snakes of the eastern Gulf of Carpentaria. *Aust. J. Mar. Freshwat. Res.* **29**: 325–334 (distribution and abundance).

Hydrophis pacificus Boulenger, 1896

Hydrophis pacificus Boulenger, G.A. (1896). *Catalogue of Snakes in the British Museum (Natural History).* 3. London : British Museum xiv 727 pp. pls 25 [278]. Type data: holotype, BMNH 1946.1.10.14, from New Britain, Papua New Guinea.

Distira macfarlani Boulenger, G.A. (1896). *Catalogue of Snakes in the British Museum (Natural History).* 3. London : British Museum xiv 727 pp. pls 25 [294 pl 17 fig 1]. Type data: syntypes, BMNH 1946.1.3.92–93, from Murray Is., Torres Strait, Qld.

Synonymy that of Cogger, H.G., this work.

Distribution: N coast, Gulf of Carpentaria, Great Barrier Reef, Coral Sea, N.T., Qld. Ecology: aquatic, nektonic, inshore, continental shelf; perennially active, viviparous, piscivore. Biological references: Smith, M.A. (1926). *Monograph on the Sea Snakes. (Hydrophiidae).* London : British Museum xvii 130 pp. (taxonomy); McDowell, S.B. (1972). The genera of sea-snakes of the *Hydrophis* group (Serpentes : Elapidae). *Trans. Zool. Soc. Lond.* **32**: 189–247 (taxonomy); Cogger, H.G. (1975). The sea snakes of Australia and New Guinea. pp. 59–139 *in* Dunson, W. (ed.) *The Biology of Sea Snakes.* Baltimore : University Park Press (taxonomy).

Lapemis Gray, 1835

Lapemis Gray, J.E. (1835). *Illustrations of Indian Zoology; Chiefly Selected from the Collection of Major-General Hardwicke.* London 2(19–20): pl 87 [pl 87 fig 2]. Type species *Lapemis hardwickii* Gray, 1835 by monotypy.

Praescutata Wall, F. (1921). *Ophidia Taprobanica* or the Snakes of Ceylon. Colombo : Cottle xxii 581 pp. [390]. Type species *Thalassophis viperina* Schmidt, 1852 [= *Lapemis viperinus* (Schmidt, 1852)] by monotypy.

Thalassophina Smith, M.A. (1926). *Monograph on the Sea Snakes. (Hydrophiidae).* London : British Museum xvii 130 pp. [33]. Type species *Thalassophis viperina* Schmidt, 1852 [=*Lapemis viperinus* (Schmidt, 1852)] by monotypy.

Kolpophis Smith, M.A. (1926). *Monograph on the Sea Snakes. (Hydrophiidae).* London : British Museum xvii 130 pp. [106]. Type species *Distira annandalei* Laidlaw, 1901 [=*Lapemis annandalei* (Laidlaw, 1901)] by monotypy.

Synonymy that of McDowell, S.B. (1972). The genera of sea-snakes of the *Hydrophis* group (Serpentes : Elapidae). *Trans. Zool. Soc. Lond.* **32**: 189–247.

This group is also found in waters of Persian Gulf, Indian subcontinent, south-east Asia, Indo-Malayan Archipelago, Philippine Ils. and New Guinea.

Lapemis hardwickii Gray, 1835

Lapemis hardwickii Gray, J.E. (1835). *Illustrations of Indian Zoology; Chiefly Selected from the Collection of Major-General Hardwicke.* London 2(19–20): pl 87 [pl 87]. Type data: holotype, BMNH 1946.1.18.39, from India.

Hydrophis pelamidoides Schlegel, H. (1837). *Essai sur la Physionomie des Serpens.* Partie Générale et Partie Descriptive. 2 Parts La Haye : Kips & Stockum [512 pl 18 figs 16–17]. Type data: not designated, from unknown locality.

Lapemis loreatus Gray, J.E. (1843). Description of two new species of reptiles from the collection made during the voyages of H.M.S. *Sulphur. Ann. Mag. Nat. Hist.* **11**: 46 [46]. Type data: holotype, BMNH 1946.1.17.48, from ? Borneo, see Smith, M.A. (1926). *Monograph on the Sea Snakes. (Hydrophiidae).* London : British Museum xvii 130 pp.

Hydrophis (Pelamis) pelamidoides annulata Fischer, J.G. (1856). Die Familie der Seeschlangen. *Abh. Geb. Naturw., Hamburg* **3**: 1–78 [67 pl 3]. Type data: syntypes, ZMH 419–420, *vide* Smith, M.A. (1926). *Monograph on the Sea Snakes. (Hydrophiidae).* London : British Museum xvii 130 pp., from Java, Indonesia.

H[ydrophis] problematicus Jan, G. (1859). Plan d'une *Iconographie descriptive des Ophidiens*, et Description sommaire de nouvelles espèces de Serpents: les Serpents de mer. *Rev. Mag. Zool. (2)* **11**: 148–157; Jan, G. (1859). Prodrome d'une *Iconographie descriptive des Ophidiens* et Description sommaire de nouvelles espèces de Serpents venimeux. Reprinted from: *Rev. Mag. Zool. (2)* **10** and *(2)* **11–12**, 1858 and 1859. Paris : Bouchard-Huzard 32 pp. [150; pl D]. Type data: syntypes, MNHP (not found), MSNM (not found), from Manila, Philippines.

H[ydrophis] abbreviatus Jan, G. (1863). *Elenco Sistematico Degli Ofidi Descritti e Disegnati per l'Iconografia Generale.* Milan : A. Lombardi 143 pp. [109]. Type data: holotype, MCZ 20647, from Manila, Philippines.

H[ydrophis] brevis Jan, G. (1863). *Elenco Sistematico Degli Ofidi Descritti e Disegnati per l'Iconografia Generale.* Milan : A. Lombardi 143 pp. [109]. Type data: holotype, MSNM (not found), from Manila, Philippines.

Hydrophis pelamoides Hilgendorf, F. (1876). Die japanischen Schlangen. *Mitt. Dt. Ges. Nat.-U. Volkerk. Ostasiens* **1**: 29–34 [31] [?*lapsus pro Hydrophis pelamidoides* Schlegel, 1837].

Synonymy that of Cogger, H.G., this work; Smith, M.A. (1926). *Monograph on the Sea Snakes. (Hydrophiidae).* London : British Museum xvii 130 pp.

Distribution: NW Shelf, N coast, Gulf of Carpentaria, Arafura Sea, Great Barrier Reef, NE coast, Coral Sea, W.A., N.T., Qld.; Timor Sea. Ecology: aquatic, noctidiurnal, nektonic, inshore, continental shelf; perennially active, viviparous, piscivore. Biological references: Smith, M.A. (1926). *Monograph on the Sea Snakes. (Hydrophiidae).* London : British Museum xvii 130 pp. (taxonomy); McDowell, S.B. (1972). The genera of sea-snakes of the *Hydrophis* group (Serpentes : Elapidae). *Trans. Zool. Soc. Lond.* **32**: 189–247 (taxonomy); Cogger, H.G. (1975). The sea snakes of Australia and New Guinea. pp. 59–139 *in* Dunson, W. (ed.) *The Biology of Sea Snakes.* Baltimore : University Park Press (taxonomy); Heatwole, H. (1981). Temperature relations of some sea snakes. *The Snake* **13**: 53–57 (thermoregulation); Lemen, C.A. & Voris, H.K. (1981). A comparison of reproductive strategies among marine snakes. *J. Anim. Ecol.* **50**: 89–101 (reproduction); Seymour, R.S. & Webster, M.E.D. (1975). Gas transport and blood acid-base balance in diving sea snakes. *J. Exp. Zool.* **191**: 169–182 (diving physiology); Bergman, R.A.M. (1949). The anatomy of *Lapemis hardwickii* Gray. *Proc. K. Ned. Akad. Wet. Amsterdam* **58**:882–898; Redfield, J.A., Holmes, J.C. & Holmes, R.D. (1978). Sea snakes of the eastern Gulf of Carpentaria. *Aust. J. Mar. Freshwat. Res.* **29**: 325–334 (distribution and abundance).

Parahydrophis Burger and Natsuno, 1974

Parahydrophis Burger, W.L. & Natsuno, T. (1974). A new genus for the Arafura smooth sea snake and redefinitions of other sea snake genera. *The Snake* **6**: 61–75 [65]. Type species *Distira mertoni* Roux, 1910 by original designation.

This group is also found in Aru Ils. and southern New Guinea.

Parahydrophis mertoni (Roux, 1910)

Distira mertoni Roux, J. (1910). Reptilien und Amphibien der Aru- und Kei-Inseln. *Abh. Senckenb. Naturf. Ges.* **33**: 210–247 [222 pl 13 fig 4]. Type data: holotype, NHMB, from Sungei Waskei, Wokam, Aru Is., Indonesia.

Distribution: N coast, Gulf of Carpentaria, N.T., Qld. Ecology: amphibious, predator, sub-littoral, inshore estuarine; perennially active, viviparous. Biological references: Smith, M.A. (1926).

Monograph on the Sea Snakes. (Hydrophiidae). London : British Museum xvii 130 pp. (taxonomy); McDowell, S.B. (1969). Notes on the Australian sea-snake *Ephalophis greyi* M. Smith (Serpentes : Elapidae : Hydrophiinae) and the origin and classification of sea-snakes. *J. Linn. Soc. Zool.* **48**: 333–349 (taxonomy); Cogger, H.G. (1975). The sea snakes of Australia and New Guinea. pp. 59–139 *in* Dunson, W. (ed.) *The Biology of Sea Snakes.* Baltimore : University Park Press (taxonomy).

Pelamis Daudin, 1803

Pelamis Daudin, F.M. (1803). *Histoire Naturelle, générale et particulière des Reptiles.* Paris : Dufart Tom. 7 436 pp. [357]. Type species *Pelamis bicolor* Schneider, 1799 by subsequent designation, see Gray, J.E. (1825). A synopsis of the genera of reptiles and Amphibia, with a description of some new species. *Annls. Philos. (2)* **10**: 193–217.

Pelamydrus Stejneger, L. (1910). The batrachians and reptiles of Formosa. *Proc. U.S. Natl. Mus.* **38**: 91–114 [111]. Type species *Anguis platura* Linnaeus, 1766 by original designation.

Synonymy that of Cogger, H.G., this work; Smith, M.A. (1926). *Monograph on the Sea Snakes. (Hydrophiidae).* London : British Museum xvii 130 pp.

This group is also found in tropical and subtropical waters of Pacific and Indian Oceans.

Pelamis platurus (Linnaeus, 1766)

Anguis platura Linnaeus, C. (1766). *Systema Naturae.* Amphibia. 12th edn. Holmiae : Laurenti Salvii 1: 347–393 [391]. Type data: presumed lost, from unknown locality.

Hydrus bicolor Schneider, J.G. (1799). *Historiae Amphibiorum naturalis et literariae.* fasc. prim. Jena : Frommani xvi 266 pp. [242]. Type data: not designated, from Tahitian seas.

Pelamis schneideri Rafinesque, C.S. (1817). Dissertation on water snakes, sea snakes and sea serpents. *Am. Month. Mag.* **1**: 431–436 [432] [*nom. substit. pro Hydrus bicolor* Schneider, 1799].

H[ydrophis] pelamis Schlegel, H. (1837). *Essai sur la Physionomie des Serpens.* Partie Générale et Partie Descriptive. 2 Parts La Haye : Kips & Stockum [508 pl 18 figs 13–15]. Type data: not designated, from unknown locality.

Pelamis ornata Gray, J.E. (1842). Monographic synopsis of the water snakes, or the family Hydridae. *in* Gray, J.E. (ed.) *Zoological Miscellany.* London : Treuttel, Würtz & Co. pp. 59–68 [60]. Type data: holotype, BMNH 1946.1.17.51, from Borneo (as India), see Smith, M.A. (1926). *Monograph on the Sea Snakes. (Hydrophiidae).* London : British Museum xvii 130 pp.

Pelamis variegata Duméril, A.M.C., Bibron, G. & Duméril, A. (1854). *Erpétologie Générale* ou Histoire Naturelle Complète des Reptiles. Paris : Roret 7(2): xii pp. 781–1536 [1337]. Type data: holotype, MNHP 3976, from Macassar, Celebes Is., Indonesia.

Pelamis bicolor sinuata Duméril, A.M.C., Bibron, G. & Duméril, A. (1854). *Erpétologie Générale* ou Histoire Naturelle Complète des Reptiles. Paris : Roret 7(2): xii pp. 781–1536 [1338]. Type data: syntypes (probable), MNHP 8683 (5 specimens), from unspecified locality.

Hydrophis (Pelamis) bicolor alternans Fischer, J.G. (1856). Die Familie der Seeschlangen. *Abh. Geb. Naturw., Hamburg* **3**: 1–78 [63] [*nom. substit. pro Pelamis variegata* Duméril, Bibron and Duméril, 1854].

Hydrophis bicolor maculata Jan, G. (1863). *Elenco Sistematico Degli Ofidi Descritti e Disegnati per l'Iconografia Generale.* Milan : A. Lombardi 143 pp. [109] [?*nom. nud.*]. Type data: those of *Hydrophis bicolor maculata* Jan and Sordelli, 1872.

Hydrophis bicolor maculata Jan, G. & Sordelli, F. (1872). *Iconographie Générale des Ophidiens.* Atlas 1860–1881 Milan : Jan & Sordelli 50 livr. [livr. 40 pl 3 fig 1]. Type data: syntypes (possible), MNHP 7710, MNHP 3975, from Indian Ocean.

Synonymy that of Cogger, H.G., this work; Smith, M.A. (1926). *Monograph on the Sea Snakes. (Hydrophiidae).* London : British Museum xvii 130 pp..

Distribution: Indian Ocean, W coast, NW Shelf, N coast, Gulf of Carpentaria, Arafura Sea, Great Barrier Reef, NE coast, Coral Sea, SE coast, Tasman Sea, W.A., N.T., Qld., N.S.W.; Timor Sea. Ecology: aquatic, diurnal, nektonic, inshore, continental shelf, oceanic; perennially active, viviparous, piscivore. Biological references: Smith, M.A. (1926). *Monograph on the Sea Snakes. (Hydrophiidae).* London : British Museum xvii 130 pp. (taxonomy); Cogger, H.G. (1975). The sea snakes of Australia and New Guinea. pp. 59–139 *in* Dunson, W. (ed.) *The Biology of Sea Snakes.* Baltimore : University Park Press (taxonomy); Kropach, C. (1975). The yellow-bellied sea snake, *Pelamis*, in the eastern Pacific. pp. 185–213 *in* Dunson, W. (ed.) *The Biology of Sea Snakes.* Baltimore : University Park Press (biology), Heatwole, H. (1981). Temperature relations of some sea snakes. *The Snake* **13**: 53–57 (thermoregulation); Dunson, W. (1968). Salt gland secretion in the pelagic sea snake *Pelamis. Am. J. Physiol.* **215**: 1512–1517 (physiology); Bolanos, R., Flores, A., Taylor, R.T. & Cerdas, L. (1974). Color patterns and venom characteristics in *Pelamis platurus. Copeia* **1974**: 909–911; Dunson, W.A. & Ehlert, G.W. (1971). Effects of temperature, salinity, and surface water flow on distribution of the sea snake *Pelamis. Limnol. Oceanogr.* **16**: 845–853; Graham, J.B. (1974). Aquatic respiration in the sea snake *Pelamis platurus. Resp. Physiol.* **21**: 1–7; Graham, J.B., Gee, J.H. & Robinson, F.S. (1975). Hydrostatic and gas exchange functions of the lung of the sea snake, *Pelamis platurus. Comp. Biochem. Physiol.* **50A**: 477–482; Hecht, M.K.,

Kropach, C. & Hecht, B.M. (1974). Distribution of the yellow-bellied sea snake *Pelamis platurus*, and its significance in relation to the fossil record. *Herpetologica* **30**: 387–395; Graham, J.B. (1974). Body temperatures of the sea snake *Pelamis platurus. Copeia* **1974**: 531–533.

LATICAUDIDAE

INTRODUCTION

This family ranges from south-east Asia to the central Pacific and comprises a single genus with four species, two of which occur in Australian waters. Generally known as sea kraits, the members of this family are front-fanged venomous snakes; they are oviparous and while primarily marine, are often found in rocky or coral crevices along shores or in mangrove swamps.

Characterised in Australia by: a pair of enlarged fangs at the front of the maxilla; tail laterally compressed, paddle-shaped; nasal scales separated by internasals; valvular nostrils.

References

Cogger, H.G. (1975). The sea snakes of Australia and New Guinea. pp. 59–139 *in* Dunson, W. (ed.) *The Biology of Sea Snakes.* Baltimore : University Park Press

Cogger, H.G. (1979). *Reptiles and Amphibians of Australia.* Sydney : A.H. & A.W. Reed 608 pp.

Smith, M.A. (1926). *Monograph on the Sea Snakes. (Hydrophiidae).* London : British Museum xvii 130 pp.

Laticauda Laurenti, 1768

Laticauda Laurenti, J.N. (1768). *Synopsin Reptilium.* Vienna : J. Thomae 214 pp. [109]. Type species *Laticauda scutatus* Laurenti, 1768 by subsequent designation, Smith, M.A. (1926). *Monograph on the Sea Snakes. (Hydrophiidae).* London : British Museum xvii 130 pp.

Platurus Sonnini de Manoncourt, C.S. & Latreille, P.A. (1802). *Histoire Naturelle des Reptiles,* avec Figures Déssinnées d'après Nature. *in* Buffon, G.L.L. de *Histoire Naturelle de Buffon, classée d'après le système de Linné par R.R. Castel.* Nouvelle édition. (Suites). Paris : Déterville Vol. 4 [183]. Type species *Platurus fasciatus* Sonnini and Latreille, 1802 *vide* Smith, M.A. (1926). *Monograph on the Sea Snakes. (Hydrophiidae).* London : British Museum xvii 130 pp.

Platyurus Agassiz, L. (1846). *Nomenclatoris Zoologici Index Universalis.* Soloduri : Jent et Gassman 393 pp. [297] [emend. *pro Platurus* Sonnini and Latreille, 1802].

Synonymy that of Cogger, H.G., this work; Smith, M.A. (1926). *Monograph on the Sea Snakes. (Hydrophiidae).* London : British Museum xvii 130 pp.

This group is also found in India, south-east Asia, Ryukyu Archipelago, Indo-Malayan Archipelago, Philippine Ils., New Guinea and islands of western and central Pacific.

Laticauda colubrina (Schneider, 1799)

***Coluber laticaudatus* (part.)** Linnaeus, C. (1758). *Systema Naturae.* Classis III. Amphibia. Holmiae : Laurentii Salvii 824 pp. 10th edn. 1: 194–238 [222]. Type data: syntype, NHRM (one of two extant syntypes), from "Indiis", *vide* Andersson, L.G. (1899). Catalogue of Linnean type-specimens of snakes in the Royal Museum in Stockholm. *K. Svenska Vet.-Akad. Handl. afd. 4* **24**: 1–35 [18].

Hydrus colubrinus Schneider, J.G. (1799). *Historiae Amphibiorum naturalis et literariae.* fasc. prim. Jena : Frommani xvi 266 pp. [238]. Type data: holotype, ZMB 9078, from unknown locality *vide* Smith, M.A. (1926). *Monograph on the Sea Snakes. (Hydrophiidae).* London : British Museum xvii 130 pp.

Coluber platycaudatus Oken, L. (1836). *Allgemeine Naturgeschichte.* Amphibien oder Lurche. Stuttgart : Hoffmann'sche Verlags-Buchhandlung 6: 419–683 [566] [*nom. substit. pro Hydrus colubrinus* Schneider, 1799].

Platurus frontalis De Vis, C.W. (1905). A new genus of lizards. *Ann. Qd. Mus.* **6**: 46–52 pl 15 [48]. Type data: holotype, QM J202, from New Guinea.

Synonymy that of Cogger, H.G., this work; Smith, M.A. (1926). *Monograph on the Sea Snakes. (Hydrophiidae).* London : British Museum xvii 130 pp.

Distribution: Coral Sea, Great Barrier Reef, NE coast, SE coast, Qld., N.S.W. Ecology: amphibious, noctidiurnal, nektonic, littoral, supralittoral, sub-littoral, estuarine, inshore; perennially active, oviparous, piscivore, coral reefs, no resident or breeding populations in Australia. Biological references: Gorman, G.C., Licht, P. & McCollum, F. (1981). Annual reproductive patterns in three species of marine snakes from the Central Philippines. *J. Herpet.* **15**: 335–354 (reproduction); Smith, M.A. (1926). *Monograph on the Sea Snakes. (Hydrophiidae).* London : British Museum xvii 130 pp. (taxonomy); Cogger, H.G. (1975). The sea snakes of Australia and New Guinea. pp. 59–139 *in* Dunson, W. (ed.) *The Biology of Sea Snakes.* Baltimore : University Park Press (taxonomy); Saint Girons, H. (1964). Notes sur l'écologie et la structure des populations des Laticaudinae (Serpents, Hydrophidae) en Nouvelle Calédonie. *La Terre et La Vie* **1964**: 185–214 (ecology).

Laticauda laticaudata (Linnaeus, 1758)

***Coluber laticaudatus* (part.)** Linnaeus, C. (1758). *Systema Naturae*. Classis III. Amphibia. Holmiae : Laurentii Salvii 824 pp. 10th edn. 1: 194–238 [222]. Type data: syntype, NHRM (one of two extant syntypes), from "Indiis", *vide* Andersson, L.G. (1899). Catalogue of Linnean type-specimens of snakes in the Royal Museum in Stockholm. *K. Svenska Vet.-Akad. Handl. afd. 4* **24**: 1–35 [18].

Laticauda scutata Laurenti, J.N. (1768). *Synopsin Reptilium.* Vienna : J. Thomae 214 pp. [109]. Type data: holotype, pl 16 in Linnaeus, C. (1754). *Museum S.R.M. Adolphis Friderici Regis ... in quo Animalia Rariora Imprimis, et Exotica ... Describuntur et Determinantur, ... cum Iconibus.* Holmiae : Typ. Reg. xxx 96 pp., from "Indiis".

Platurus laurenti Rafinesque, C.S. (1817). Dissertation on water snakes, sea snakes and sea serpents. *Am. Month. Mag.* **1**: 431–436 [432] [*non Platurus laurenti* Daudin, 1803; Smith, M.A. (1926). *Monograph on the Sea Snakes. (Hydrophiidae).* London : British Museum xvii 130 pp. regards *P. laurenti* as a substitute name, but there is no indication of this in the original description, nor does Smith indicate the name being substituted]. Type data: type(s) not located, from unknown locality.

Platurus fischeri Jan, G. (1859). Plan d'une *Iconographie descriptive des Ophidiens*, et Description sommaire de nouvelles espèces de Serpents: les Serpents de mer. *Rev. Mag. Zool. (2)* **11**: 148–157 [149]. Type data: syntypes, MSNM (not found), MNHP (not found), from Indian Ocean.

P[laturus] affinis Anderson, J. (1871). On some Indian reptiles. *Proc. Zool. Soc. Lond.* **1871**: 149–211 [190]. Type data: holotype, IM 8289, from Tolly's Nullah (as Fallah's Nullah), a tidal stream, Calcutta, India, see Smith, M.A. (1926). *Monograph on the Sea Snakes. (Hydrophiidae).* London : British Museum xvii 130 pp.

Platurus muelleri Boulenger, G.A. (1896). *Catalogue of Snakes in the British Museum (Natural History).* 3. London : British Museum xiv 727 pp. pls 25 [309] [*ex* Müller ms]. Type data: type(s) not located, from South Pacific.

Synonymy that of Cogger, H.G., this work; Smith, M.A. (1926). *Monograph on the Sea Snakes. (Hydrophiidae).* London : British Museum xvii 130 pp.

Distribution: Coral Sea, Great Barrier Reef, NE coast, SE coast, Qld., N.S.W. Ecology: amphibious, nektonic, noctidiurnal, littoral, supralittoral, sub-littoral, estuarine, inshore; perennially active, oviparous, piscivore, coral reefs, no resident or breeding populations in Australia. Biological references: Smith, M.A. (1926). *Monograph on the Sea Snakes. (Hydrophiidae).* London : British Museum xvii 130 pp. (taxonomy); Cogger, H.G. (1975). The sea snakes of Australia and New Guinea. pp. 59–139 *in* Dunson, W. (ed.) *The Biology of Sea Snakes.* Baltimore : University Park Press (taxonomy); Tamiya, N., Arai, H. & Sato, S. (1967). Studies on sea snake venoms: crystallization of erabutoxins a and b from *Laticauda semifasciata* venom and of laticotoxin a from *Laticauda laticaudata* venom. pp. 249–258 *in* Russell, F.E. & Saunders, P.R. (eds.) *Animal Toxins.* Oxford : Pergamon Press (venom); Klemmer, K. (1967). Observations on the sea snake, *Laticauda laticaudata* in captivity. *Int. Zoo Yearb.* **7**: 229–231.

APPENDIX I

SPECIES INQUIRENDA

The following names, purportedly based on Australian type specimens, cannot at this time be assigned to any known taxa within the Australian herpetofauna and consequently are relegated to the status of *species inquirenda*. In most cases the original type specimens could not be located.

Myobatrachidae

Bufo leucogaster

Péron, F. (1807). *Voyage de Découvertes aux Terres Australes, Fait par Ordre du Gouvernement, sur les Corvettes le* Géographe, *le* Naturaliste, *et la goülette le* Casuarina, *pendant les années 1800, 1801, 1802, 1803 et 1804. Historique.* Vol. 1 Paris : Arthus Bertrand [406].
Type data: none designated, from Parramatta, N.S.W. Maury (*in litt.*) cites 11 MNHP specimens bearing this name, from Port Jackson, in the original list of specimens collected by Péron and Lesueur during Baudin's voyages (1800–1804).

Bufo proteus

Péron, F. (1807). *Voyage de Découvertes aux Terres Australes, Fait par Ordre du Gouvernement, sur les Corvettes le* Géographe, *le* Naturaliste, *et la goülette le* Casuarina, *pendant les années 1800, 1801, 1802, 1803 et 1804. Historique.* Vol. 1 Paris : Arthus Bertrand [406].
Type data: none designated, from Parramatta, N.S.W.

Rana pollicifera

Péron, F. (1807). *Voyage de Découvertes aux Terres Australes, Fait par Ordre du Gouvernement, sur les Corvettes le* Géographe, *le* Naturaliste, *et la goülette le* Casuarina, *pendant les années 1800, 1801, 1802, 1803 et 1804. Historique.* Vol. 1 Paris : Arthus Bertrand [406].
Type data: none designated, from Parramatta, N.S.W.

Rana pustulosa

Péron, F. (1807). *Voyage de Découvertes aux Terres Australes, Fait par Ordre du Gouvernement, sur les Corvettes le* Géographe, *le* Naturaliste, *et la goülette le* Casuarina, *pendant les années 1800, 1801, 1802, 1803 et 1804. Historique.* Vol. 1 Paris : Arthus Bertrand [406].
Type data: none designated, from Parramatta, N.S.W. Maury (*in litt.*) cites 5 MNHP specimens bearing this name, from Port Jackson, in the original list of specimens collected by Péron and Lesueur during Baudin's voyages (1800–1804).

Hylidae

Hyla citripoda

Péron, F. (1807). *Voyage de Découvertes aux Terres Australes, Fait par Ordre du Gouvernement, sur les Corvettes le* Géographe, *le* Naturaliste, *et la goülette le* Casuarina, *pendant les années 1800, 1801, 1802, 1803 et 1804. Historique.* Vol. 1 Paris : Arthus Bertrand [407].
Type data: none designated, from Parramatta, N.S.W.

Hyla cyanea

Péron, F. (1807). *Voyage de Découvertes aux Terres Australes, Fait par Ordre du Gouvernement, sur les Corvettes le* Géographe, *le* Naturaliste, *et la goülette le* Casuarina, *pendant les années 1800, 1801, 1802, 1803 et 1804. Historique.* Vol. 1 Paris : Arthus Bertrand [406]. [*non Hyla cyanea* Daudin, 1803; ?=*Litoria aurea* (Lesson, 1829)].
Type data: none designated, from Parramatta, N.S.W. Maury (*in litt.*) cites 3 MNHP specimens bearing this name, from Port Jackson, in the original list of specimens collected by Péron and Lesueur during Baudin's voyages (1800–1804).

Hyla dayi

Günther, A. (1897). Description of new species of lizards and of a tree frog from north-eastern Queensland. *Novit. Zool.* **4**: 403–406 [406] [for discussion of the status of this species see Tyler, M.J. (1968). An additional Australian species of the hylid frog genus *Nyctimystes* with notes on *Hyla dayi* Günther. *Ark. Zool. (2)* **20**: 501–504].
Type data: holotype, from Bartle Frere Mts., Qld., presumed lost.

Hyla ianopoda

Péron, F. (1807). *Voyage de Découvertes aux Terres Australes, Fait par Ordre du Gouvernement, sur les Corvettes le* Géographe, *le* Naturaliste, *et la goülette le* Casuarina, *pendant les années 1800, 1801, 1802, 1803 et 1804. Historique.* Vol. 1 Paris : Arthus Bertrand [407].
Type data: none designated, from Parramatta, N.S.W.

Hyla nebulosa

Péron, F. (1807). *Voyage de Découvertes aux Terres Australes, Fait par Ordre du Gouvernement, sur les Corvettes le* Géographe, *le* Naturaliste, *et la goülette le* Casuarina, *pendant les années 1800, 1801, 1802, 1803 et 1804. Historique.* Vol. 1 Paris : Arthus Bertrand [407].
Type data: none designated, from Parramatta, N.S.W. Maury (*in litt.*) cites a single MNHP specimen bearing this name in the original list of specimens collected by Péron and Lesueur during Baudin's voyages (1800–1804).

Hyla ocellata

Péron, F. (1807). *Voyage de Découvertes aux Terres Australes, Fait par Ordre du Gouvernement, sur les Corvettes le* Géographe, *le* Naturaliste, *et la goülette le* Casuarina, *pendant les années 1800, 1801, 1802, 1803 et 1804. Historique.* Vol. 1 Paris : Arthus Bertrand [407].
Type data: none designated, from Parramatta, N.S.W. Maury (*in litt.*) cites 2 MNHP specimens bearing this name, from Port Jackson, in the original list of specimens collected by Péron and Lesueur during Baudin's voyages (1800–1804).

Hyla rubeola

Péron, F. (1807). *Voyage de Découvertes aux Terres Australes, Fait par Ordre du Gouvernement, sur les Corvettes le* Géographe, *le* Naturaliste, *et la goülette le* Casuarina, *pendant les années 1800, 1801, 1802, 1803 et 1804. Historique.* Vol. 1 Paris : Arthus Bertrand [407].
Type data: none designated, from Parramatta, N.S.W. Maury (*in litt.*) cites a single MNHP specimen bearing this name, from Port Jackson, in the original list of specimens collected by Péron and Lesueur during Baudin's voyages (1800–1804).

Microhylidae?

Pachybatrachus petersii

Keferstein, W. (1868). Über die Batrachier Australiens. *Arch. Naturgesch.* **34**: 253–290 [274 pl 6 figs 8–10] [described in monotypic genus within family Engystomatidae; family allocation uncertain].
Type data: holotype, ZFMK (*ex* ZMG), from "Neu-Sud-Wales".

Gekkonidae

Gecko dorreensis

Péron, F. (1807). *Voyage de Découvertes aux Terres Australes, Fait par Ordre du Gouvernement, sur les Corvettes le* Géographe, *le* Naturaliste, *et la goülette le* Casuarina, *pendant les années 1800, 1801, 1802, 1803 et 1804. Historique.* Vol. 1 Paris : Arthus Bertrand [118].
Type data: none designated, from Bernier Is., W.A.

Gecko trachyloemus

Peters, W. (1872). Über neue oder weniger bekannte Saurier. *Mber. K. Preuss. Akad. Wiss. Berl.* **1872**: 774–776 [774].
Type data: holotype, ?ZMB, from N Australia.

Gymnodactylus asper

Boulenger, G.A. (1913). Description of new lizards in the collection of the British Museum. *Ann. Mag. Nat. Hist. (8)* **12**: 563–566 [563] [?= *Underwoodisaurus millii* (Bory de Saint-Vincent, 1825)].
Type data: holotype, BMNH, from Milparinka (as Milparinha), N.S.W.

Peropus pusillus

Cope, E.D. (1869). Observations on reptiles of the Old World. Part 2. *Proc. Acad. Nat. Sci. Philad.* **1868**: 316–323 [319] [treated as a valid species of *Lepidodactylus* by Boulenger, G.A. (1885). *Catalogue of the Lizards in the British Museum (Natural History).* 1. London : British Museum xii 436 pp. pls 32 [167] and by Wermuth, H. (1965). Liste der rezenten Amphibien und Reptilien; Gekkonidae, Pygopodidae, Xantusiidae. *Das Tierreich* **80**: i–xxii 1–246 [100]; Loveridge, A. (1934). Australian reptiles in the Museum of Comparative Zoology, Cambridge, Massachusetts. *Bull. Mus. Comp. Zool.* **77**: 243–383 [303] rejected type status of MCZ 5725, which had previously been regarded as Cope's holotype and assigned to *Diplodactylus spinigerus*].
Type data: holotype, from SW Australia, presumed lost.

Agamidae

Agama grandoculis

Lacépède, B.G.E. (1804). Mémoire sur plusieurs animaux de la Nouvelle-Hollande dont la description n'a pas encore été publiée. *Ann. Mus. Natl. Hist. Nat. Paris* **4**: 184–211 [191, 202] [probably a member of the *Amphibolurus barbatus* species-group].
Type data: not located, presumed lost, from Australia.

Varanidae

Tupinambis endrachtensis

Péron, F. (1807). *Voyage de Découvertes aux Terres Australes, Fait par Ordre du Gouvernement, sur les Corvettes le* Géographe, *le* Naturaliste, *et la goülette le* Casuarina, *pendant les années 1800, 1801, 1802, 1803 et 1804. Historique.* Vol. 1 Paris : Arthus Bertrand [118] [almost certainly refers to *Varanus gouldii* (Gray, 1838); *nom. oblit.* and/or *nom. nud.*].
Type data: none designated, from Bernier Is., W.A.

Scincidae

Ablepharus cupreus

Gray, J.E. (1839). Catalogue of the slender-tongued saurians, with descriptions of many new genera and species. (contd.). *Ann. Mag. Nat. Hist.* **2**: 331–337 [335] [status discussed by Cogger, H.G. (1971). The reptiles of Lord Howe Island. *Proc. Linn. Soc. N.S.W.* **96**: 23–38].
Type data: holotype, BMNH, from unknown locality.

Eumeces serratus

Fischer, J.G. (1886). Herpetologische Notizen. *Abh. Geb. Naturw. Hamb.* **9**: 51–67 [56 pl 2 fig 3] [while the original description suggests that this taxon belongs to the genus *Emoia*, a number of characters preclude it from being assigned to either of the species of *Emoia* known from the type locality].
Type data: holotype, MTKD D510, from Murray Is., Torres Strait, Qld. (destroyed, see Obst, F.J. (1977). Die herpetologische Sammlung des Staatlichen Museums für Tierkunde Dresden und ihre Typusexemplare. *Zool. Abh. Mus. Tierk. Dresden* **34**: 171–186).

Heteropus lateralis

De Vis, C.W. (1885). A conspect of the genus, *Heteropus*. *Proc. R. Soc. Qd.* **1**: 166–173 [168] [*non Scincus lateralis* Say, 1828 and *non Lygosoma laterale* Günther, 1867; ? = *Carlia vivax* (De Vis, 1884); assigned to synonymy of *Carlia pectoralis* (De Vis, 1885) by Mitchell, F.J. (1953). A brief revision of the four-fingered members of the genus *Leiolopisma* (Lacertilia). *Rec. S. Aust. Mus.* **11**: 75–90].
Type data: holotype (probable), QM J234, from Pine River, SE Qld. (as Moreton Bay district), see Covacevich, J. (1971). Amphibian and reptile type specimens in the Queensland Museum. *Mem. Qd. Mus.* **.16**: 49–68 [56].

Hinulia domina

De Vis, C.W. (1888). A contribution to the herpetology of Queensland. *Proc. Linn. Soc. N.S.W. (2)* **2**: 811–826 [818].
Type data: holotype, from Qld., presumed lost, *vide* Covacevich, J. (1971). Amphibian and reptile type specimens in the Queensland Museum. *Mem. Qd. Mus.* **16**: 49–68.

Hinulia elegans

Gray, J.E. (1838). Catalogue of the slender-tongued saurians, with descriptions of many new genera and species. (contd.). *Ann. Mag. Nat. Hist.* **2**: 287–293 [288].
Type data: none designated, from unknown locality.

Lygosoma devisii

Boulenger, G.A. (1890). First report on additions to the lizard collection in the British Museum (Natural History). *Proc. Zool. Soc. Lond.* **1890**: 77–86 [79] [*nom. nov. pro Heteropus lateralis* De Vis, 1885, at that time a junior secondary homonym within the genus *Lygosoma*].
Type data: that of *Heteropus lateralis* De Vis, 1885.

Tetradactylus guttulatus

Macleay, W. (1885). On some Reptilia lately received from the Herbert River District, Queensland. *Proc. Linn. Soc. N.S.W.* **10**: 64–68 [66].
Type data: holotype, presumed lost, from "vicinity of Herbert River", Qld.

Tiliqua stoddartii

Gray, J.E. (1838). Catalogue of the slender-tongued saurians, with descriptions of many new genera and species. (contd.). *Ann. Mag. Nat. Hist.* **2**: 287–293 [291].
Type data: none designated, from Australia.

Tiliqua vachelli

Gray, J.E. (1838). Catalogue of the slender-tongued saurians, with descriptions of many new genera and species. (contd.). *Ann. Mag. Nat. Hist.* **2**: 287–293 [291].
Type data: none designated, from Australia.

Elapidae

A[lecto] wiebeli

Jan, G. (1863). *Elenco Sistematico Degli Ofidi Descritti e Disegnati per l'Iconografia Generale.* Milan : A. Lombardi 143 pp. [117].
Type data: ?holotype, ZMH 480, from Australia.

Boa laevis

Lacépède, B.G.E. (1804). Mémoire sur plusieurs animaux de la Nouvelle-Hollande dont la description n'a pas encore été publiée. *Ann. Mus. Natl. Hist. Nat. Paris* **4**: 184–211 [209].
Type data: not located, presumed lost, from Australia.

F[urina] bilineata

Jan, G. (1859). Plan d'une *Iconographie descriptive des Ophidiens*, et description sommaire de nouvelles espèces de Serpents. *Rev. Mag. Zool. (2)* **11–12**: 122–130 [123, 125]; Jan, G. (1859). Prodrome d'une *Iconographie descriptive des Ophidiens* et Description sommaire de nouvelles espèces de Serpents venimeux. Reprinted from: *Rev. Mag. Zool. (2)* **10** and *(2)* **11–12**, 1858 and 1859. Paris : Bouchard-Huzard 32 pp. [pl 6].
Type data: holotype, MSNM (not found), from ?Australia.

Leptaspis inornata

Fitzinger, L. (1860). Ausbeute der österreichischen Naturforscher an Säugethiere und Reptilien während der Weltumsegelung Sr. Majestat Fregatte *Novara. Sber. Akad. Wiss. Wien* **42**: 384–416 [410] [*nom. nud.*].
Type data: none designated, from Australia.

Tortrix australis

Gray, J.E. (1841). A catalogue of the species of reptiles and Amphibia hitherto described as inhabiting Australia, with a description of some new species from Western Australia. Appendix E *in* Grey, G. *Journals of Two Expeditions of Discovery in North-west and Western Australia, during the years 1837, 38, and 39, under the authority of Her Majesty's Government.* London : T. & W. Boone 2: 422–449 [444].
Type data: none designated, from unspecified locality (Australia by implication).

Hydrophiidae

Leioselasma striata

Lacépède, B.G.E. (1804). Mémoire sur plusieurs animaux de la Nouvelle-Hollande dont la description n'a pas encore été publiée. *Ann. Mus. Natl. Hist. Nat. Paris* **4**: 184–211 [(198, 210] [McDowell, S.B. (1972). The genera of sea-snakes of the *Hydrophis* group (Serpentes : Elapidae). *Trans. Zool. Soc. Lond.* **32**: 189–247 [215] presents an unconvincing case for regarding this species as a synonym of *Hydrophis cyanocinctus* Daudin, 1803, a species not known from Australian waters; McDowell also suggested, but did not formally propose, that the specimen represented by pl 18 figs 4–5 in Schlegel, H. (1837). *Essai sur la Physionomie des Serpens.* Partie Générale et Partie Descriptive. 2 Parts La Haye : Kips & Stockum be made the neotype of *Leioselasma striata* Lacépède, 1804].

Type data: holotype, from Australia, MNHP presumed lost, *vide* Boie, F. (1827). Bemerkungen über Merrem's Versuch eines Systems der Amphibien. *Isis Von Oken, Jena* **20**: 508–566 [553].

NAMES NO LONGER APPLICABLE TO AUSTRALIAN TAXA

The following names, purportedly based on Australian type specimens, have been shown to belong to taxa which do not occur as wild populations in Australia. In most cases the specified type locality is clearly in error.

Iguanidae

Oreodeira gracilipes

Girard, C. (1858). *United States Exploring Expedition during the years 1838, 1839, 1840, 1841, 1842.* Herpetology. Philadelphia : J.B. Lippincott 20: xvi 496 pp. [420] [=*Strobilurus torquatus* Wiegmann, 1834 *vide* Kluge (*in litt.*); originally described as a member of the sub-family Acrodontes].
Type data: holotype, ANSP 6307, from N.S.W.

Scincidae

Lygosoma (Hinulia) compressicaudum

Werner, F. (1897). Über einige neue oder seltene Reptilien und Frösche der zoologischen Sammlung des Staates in München. *Sber. Akad. Wiss. München* **27**: 203–220 [210] [not assignable to any Australian species, type locality almost certainly erroneous; identification and type data from P.A. Rawlinson, *in litt.*].
Type data: holotype (probable), ZSM 80/1947 (labelled "lectotypus"), from Australia.

Lygosoma infrapunctatum

Boulenger, G.A. (1887). *Catalogue of the Lizards in the British Museum (Natural History).* 3. London : British Museum xii 575 pp. pls 40 [274 pl 21 fig 1] [an endemic New Zealand lizard, *vide* McCann, C. (1955). The lizards of New Zealand: Gekkonidae and Scincidae. *Dominion Mus. Bull.* **17**: i–viii 1–127 (122)].
Type data: holotype, BMNH 1946.8.16.43, from Australia.

Panaspis aeneus

Cope, E.D. (1869). Observations on reptiles of the Old World. Part 2. *Proc. Acad. Nat. Sci. Philad.* **1868**: 316–323 [317] [=*Panaspis cabindae* (Bocage, 1866) *vide* Perret, J.-L. (1975). La différenciation dans le genre *Panaspis* Cope (Reptilia, Scincidae). *Bull. Soc. Neuchâteloise Sci. Nat.* **98**: 5–16].
Type data: syntypes (2), MCZ 5784, from "probably Swan River, Australia".

Anguidae

Macrogongylus brauni

Werner, F. (1901). Neue Reptilien des Königsberger Zoologischen Museums. *Zool. Anz.* **24**: 297–301 [299] [originally described as a scincid lizard; Loveridge, A. (1934). Australian reptiles in the Museum of Comparative Zoology, Cambridge, Massachusetts. *Bull. Mus. Comp. Zool.* **77**: 243–383 [344] has referred it to the synonymy of the anguid lizard, *Celestus occiduus* Shaw, 1802].
Type data: holotype, Königsberg Zoologischen Museum, from Australia (in error).

Typhlopidae

Typhlops labialis

Waite, E.R. (1918). Review of the Australian blind snakes. *Rec. S. Aust. Mus.* **1**: 1–34 [30] [=*Typhlops diardi* Schlegel, 1839, *vide* McDowell, S.B. (1974). A catalogue of the snakes of New Guinea and the Solomons, with special reference to those in the Bernice P. Bishop Museum. Part I. Scolecophidia. *J. Herpet.* **8**: 1–57 (11)].
Type data: holotype, WAM R630, from "Western Australia".

Dasypeltidae

Trophidechis [*sic*] *dunensis*

De Vis, C.W. (1911). Description of snakes apparently new. *Ann. Qd. Mus.* **10**: 22–25 [21] [=*Dasypeltis scabra* (Linnaeus, 1758), *vide* Cogger, H.G. (1966). The status of the "elapid" snake *Tropidechis dunensis* De Vis. *Copeia* **1966**: 893–894; originally described as a member of the family Elapidae].
Type data: holotype, QM J191, from "Darro, Darling Downs", Qld.

APPENDIX II

MUSEUM ACRONYMS

AM	The Australian Museum, Sydney, N.S.W., Australia
AMNH	American Museum of Natural History, New York, N.Y., U.S.A.
ANSP	Academy of Natural Sciences, Philadelphia, Pa., U.S.A.
ANWC	Australian National Wildlife Collection, C.S.I.R.O. Division Wildlife Research, Canberra, A.C.T., Australia
BMNH	British Museum (Natural History), London SW7 5BD, England
CAM	Central Australian Museum, Alice Springs, N.T., Australia
CAS	California Academy of Sciences, San Francisco, Calif., U.S.A.
COM	Colombo Museum, Colombo, Sri Lanka
FMNH	Field Museum of Natural History, Chicago, Ill., U.S.A.
GZM	Giessener Zoologischen Museum, Federal Republic of Germany
IM	Indian Museum, Calcutta, India
LACM	Los Angeles County Museum of Natural History, Los Angeles, Calif., U.S.A.
MB	Museu Bocage, Faculdade de Ciencias, Universidade Lisboa, Lisbon, Portugal
MCG	Museo Civico di Storia Naturale di Genova "Giacomo Doria", Genoa, Italy
MCZ	Museum of Comparative Zoology, Harvard University, Cambridge, Mass., U.S.A.
MDUG	Museo Alfred Dugès, Universidad Guanajuato, Guanajuato, Mexico
MMUS	Macleay Museum, University of Sydney, Sydney, N.S.W., Australia
MNHN	Museo Nacional de Historia Natural, Santiago, Chile
MNHP	Muséum National d'Histoire Naturelle, Paris, France
MSNM	Museo Civico di Storia Naturale di Milano, Milan, Italy
MTKD	Museum für Tierkunde, Dresden, German Democratic Republic (formerly Königlich Zoologischen Museum)
MZB	Museum Zoologicum Bogoriense, Bogor, Indonesia
MZUP	Museo Zoologia, Universita di Padova, Padua, Italy
NHMB	Naturhistorisches Museum, Basel, Switzerland
NHMW	Naturhistorisches Museum, Vienna, Austria
NHRM	Naturhistoriska Riksmuseet, Stockholm, Sweden
NMV	National Museum of Victoria, Melbourne, Vic., Australia
NTM	Northern Territory Art Gallery and Museum, Darwin, N.T., Australia
QM	Queensland Museum, Brisbane, Qld., Australia
QVM	Queen Victoria Museum, Launceston, Tas., Australia
RIB	Institut Royal des Sciences Naturelles de Belgique, Brussels, Belgium
RMNH	Rijksmuseum van Natuurlijke Historie, Leiden, Holland
SAMA	South Australian Museum, Adelaide, S.A., Australia
SMF	Natur-Museum Senckenberg, Frankfurt-am-Main, Federal Republic of Germany
SMNS	Staatliches Museum für Naturkunde, Stuttgart, Federal Republic of Germany
TIU	Science College Museum, Tokyo Imperial University, Tokyo, Japan
USNM	National Museum of Natural History, Washington, D.C., U.S.A.
UZM	Zoologisk Museum, Universitets Kobenhaven, Copenhagen, Denmark
UZMO	Zoolgisk Museum, Universitetets Oslo, Oslo, Norway
WAM	Western Australian Museum, Perth, W.A., Australia
WNS	Wiesbaden Naturwissenschaftliche Sammlung des Städt, Wiesbaden, Federal Republic of Germany

ZFMK	Zoologisches Forschungsinstitut und Museum Alexander Koenig, Bonn, Federal Republic of Germany
ZIUU	Zoologisk Institut, Universitetets Uppsala, Uppsala, Sweden
ZMA	Zoologisch Museum, Universiteit van Amsterdam, Amsterdam, Holland
ZMB	Zoologisches Museum, Universität Humboldt, Berlin, German Democratic Republic
ZMG	Zoologisches Museum, Universität Göttingen, Göttingen, Federal Republic of Germany
ZMH	Zoologisches Museum für Hamburg, Hamburg, Federal Republic of Germany
ZMLU	Zoologisk Museum, Universitets Lund, Lund, Sweden
ZSM	Zoologisches Sammlung des Bayerischen Staates, Munich, Federal Republic of Germany

APPENDIX III

ABBREVIATIONS AND SYMBOLS

For definitions of nomenclatural terms which appear throughout the *Catalogue*, the reader should refer to the text and glossary of the 1961 edn. of the International Code of Zoological Nomenclature.

A.C.T	Australian Capital Territory
alt.	altitude
Bd.	Band
E	east, eastern
ed./eds.	editor/editors
edn.	edition
emend.	emendation
err. typ.	*error typographicus*
F	female
fasc.	fascicule
fig/figs	figure/figures
ft	feet
Is./Ils.	Island/Islands
km	kilometre
livr.	livraison
m	metre
M	male
mi	mile
ms	manuscript
Mt./Mts.	Mount, Mountain/Mountains
N	north, northern
Nat.	Natural
Natl.	National
no.	number
nom. dub.	*nomen dubium*
nom. nov.	*nomen novum*
nom. nud.	*nomen nudum*
nom. oblit.	*nomen oblitum*
nom. substit.	substitute name
ns	new series
N.S.W.	New South Wales
N.T.	Northern Territory
pl/pls	plate/plates
pro	for
pt/pts	part/parts
Qd./Qld.	Queensland
S	south, southern
S.A.	South Australia
ser.	series
sp.	species
Tas./Tasm.	Tasmania
var./Var.	variety
Verl.	Verlag
Vic./Vict.	Victoria
vol./Vol.	volume
W	west, western
W.A.	Western Australia
[name]	square brackets enclosing a valid or available name indicate a qualification of the use of that name in the context in which it appears.

TAXONOMIC INDEX

R81/1546 Cat. No. 83 1536 3